ZEOLITE MOLECULAR SIEVES

ZEOLITE MOLECULAR SIEVES

STRUCTURE, CHEMISTRY, AND USE

DONALD W. BRECK

*Senior Research Fellow
Union Carbide Corporation
Tarrytown, New York*

ROBERT E. KRIEGER PUBLISHING COMPANY
MALABAR, FLORIDA

Original Edition 1974
Reprint 1984 w/corrections & revisions

Printed and Published by
ROBERT E. KRIEGER PUBLISHING COMPANY, INC.
KRIEGER DRIVE
MALABAR, FL 32950

Copyright © 1974 by John Wiley & Sons, Inc.
Reprint by arrangement

Reproduction or translation of any part of this work beyond that permitted by Sections 107 or 108 of the 1976 United States Copyright Act without the permission of the copyright owner is unlawful. Requests for permission or further information should be addressed to the Permissions Department

Library of Congress Cataloging in Publication Data

Breck, Donald W.
    Zeolite molecular sieves.

    Reprint. Originally published: New York: Wiley, 1973, c1974.
    Includes index.
    1. Molecular sieves. 2. Zeolites. I. Title.
[TP159.M6B7 1984]   660.2′8423   83-26809
ISBN 0-89874-648-5

Printed in the United States of America

10 9 8 7 6 5 4 3

PREFACE

> Long winded writers I abhor,
> And glib, prolific chatters:
> Give me the ones who tear and gnaw
> their hair and pens to tatters:
> who find their writing such a chore
> they write what only matters.
>   – *From "Grooks" by Piet Hein*

Several years ago, I undertook the preparation of the manuscript for this book confident that zeolite science had attained a reasonable level of maturity and that the time had come for publication of an appropriate monograph. Today, with file boxes filled with obsolete manuscripts, I am much wiser. Each chapter has been subjected to several revisions and the manuscript size has increased to the point of becoming unwieldy. However, I believe that most of the important developments published and available by the end of 1972 are included. As author, I have had to use my judgment, hopefully objectively, as to the subject matter.

In order to be comprehensive, much of the detailed information on structure, synthesis, ion exchange, and adsorption has been summarized in tables. Appropriate illustrations and photographs are used. Chapter 2 includes a collection of stereophotographs of all of the basic zeolite framework structures. Most readers will find a stereoviewer desirable. A considerable amount of previously unpublished information obtained by my associates at Union Carbide Corporation during the last 25 years is included in the book.

The reader will quickly observe by a perusal of the Table of Contents that the important subject of zeolite catalysis has been deliberately omitted. This was a decision we had to make due to size and time limitations. We hope to prepare a second volume which will be concerned with zeolite catalysis and certain other areas of interest to the user of zeolites.

In the preparation of this volume, I have relied very heavily on friends and colleagues. I especially want to thank Professor J. V. Smith for his comments and review of Chapters 1, 2, and 3. He also supplied many of the models illustrated in Chapter 2. I wish to thank Ms. E. M. Flanigen and Mr. R. L. Mays for their critical review of the manuscript.

I especially want to thank my editor, Ms. N. L. Marcus, for her perserverance. The work would have been impossible without her help. I want to thank Dr. W. H. Flank for proofreading the entire manuscript and for his constructive comments. Finally, I wish to thank Union Carbide Corporation for their support and encouragement.

                                       D. W. Breck
                                       July, 1973

PREFACE TO THE REPRINT EDITION

Due to popular demand, Dr. Breck's reference work on ZEOLITE MOLECULAR SIEVES has been returned to print.

As an aid to the user, Union Carbide has generously given of their time to review the original book, making minor revisions and corrections. A few revisions have been included covering information published since 1974. We wish to express our thanks to Dr. W. H. Frank and Mr. H. Garcia for their efforts in this regard.

Publisher

TABLE OF CONTENTS

Preface
Chapter One   INTRODUCTION   1
    A.   *Molecular sieve technology*   1
    B.   *The scope of the book*   2
    C.   *Types of adsorbents*   3
    D.   *Types of molecular sieves*   4
    E.   *Early observations*   10
    F.   *Classification and nomenclature*   19
Chapter Two   STRUCTURE OF ZEOLITES   29
    A.   *Introduction*   29
    B.   *Review of silicate structures*   29
    C.   *Classification of zeolite structures*   45
    D.   *Theoretical zeolite structures*   47
    E.   *Internal channel structure in zeolites*   59
    F.   *Framework density*   62
    G.   *Aperture sizes in dehydrated zeolites*   64
    H.   *Zeolites of group 1*   67
    I.   *Zeolites of group 2*   77
    J.   *Zeolites of group 3*   83
    K.   *Zeolites of group 4*   92
    L.   *Zeolites of group 5*   117
    M.   *Zeolites of group 6*   122
    N.   *Zeolites of group 7*   128
    O.   *Other zeolites*   132
    P.   *Tables of zeolite structural data*   133
Chapter Three   MINERAL ZEOLITES   186
    A.   *Types of occurrences*   187
    B.   *Igneous zeolites*   189
    C.   *Sedimentary zeolites*   192
    D.   *Origin of zeolite minerals*   200
    E.   *Physical properties*   205
    F.   *Glossary of terms*   207
    G.   *Tables of mineral zeolite data*   209

Chapter Four	**THE SYNTHETIC ZEOLITES**	**245**
A.	Methods of zeolite synthesis	245
B.	Review of early work	251
C.	Formation of zeolites from aluminosilicate gels of alkali metals	257
D.	Zeolites of the alkaline earths	294
E.	The alkylammonium or nitrogeneous zeolites	304
F.	The synthesis of zeolites from clays	313
G.	Synthetic zeolites with other framework atoms	320
H.	Synthesis of salt-containing zeolites	331
I.	Kinetics and mechanism of zeolite crystallization	333
J.	Crystal growth	344
K.	Summary list of synthetic zeolites	346
L.	X-ray powder data for synthetic zeolites	347
Chapter Five	**PHYSICAL PROPERTIES OF CRYSTALLINE ZEOLITES**	**379**
A.	Physical properties of zeolite crystals	379
B.	Uniformity of composition in synthetic zeolites	389
C.	Optical properties	390
D.	Dielectric properties	392
E.	Electrical conductivity	397
F.	Thermochemistry	410
G.	Zeolitic water	412
H.	Structure of zeolites by infrared spectroscopy	415
I.	Pore volume in dehydrated zeolites	425
Chapter Six	**CHEMICAL PROPERTIES AND REACTIONS OF ZEOLITES**	**441**
A.	Dehydration of zeolites	442
B.	Cation hydrolysis and structural hydroxyl groups	460
C.	Transformation reactions	483
D.	Reactions in solution	502
E.	Defect structures–stabilization–superstable zeolites	507
F.	Radiation effects on zeolites	523
Chapter Seven	**ION EXCHANGE REACTIONS IN ZEOLITES**	**529**
A.	Ion exchange theory	530
B.	Ion exchange equilibria in aqueous solution	536
C.	Thermodynamics of ion exchange processes	565
D.	Hydrogen exchange in zeolites	567
E.	Ion exchange kinetics and ion diffusion	571
F.	Cation sieve effects in zeolites	579
G.	Cation exchange in nonaqueous solvents	580

H.	Salt occlusion or imbibition	*585*
I.	Ion exchange in fused salts	*587*
J.	Zeolite ion exchange applications	*588*

Chapter Eight   ADSORPTION BY DEHYDRATED ZEOLITE CRYSTALS   **593**

A.	Equilibrium adsorption of gases and vapors on dehydrated zeolites	*596*
B.	The application of isotherm equations to zeolite adsorption	*628*
C.	The molecular sieve effect	*633*
D.	Heat of adsorption	*645*
E.	Character of the adsorbed phase in zeolites	*660*
F.	Specificity of adsorption selectivity effects	*664*
G.	Adsorption kinetics and diffusion	*671*
H.	Adsorption equilibria for binary mixtures on zeolites	*689*
I.	Adsorption separation of mixtures	*699*

Chapter Nine   MANUFACTURE AND PROPERTIES OF COMMERCIAL MOLECULAR SIEVE ADSORBENTS   **725**

A.	Manufacturing processes	*725*
B.	Pelletization of synthetic zeolite powders	*742*
C.	Properties of commercial molecular sieves	*746*

Appendix   **756**

Index   **761**

ZEOLITE MOLECULAR SIEVES

Chapter One

INTRODUCTION

Rarely in our technological society does the discovery of a new class of inorganic materials result in such a wide scientific interest and kaleidoscopic development of applications as has happened with the zeolite molecular sieves. From the first industrial research efforts in 1948 at Union Carbide Corporation until the end of 1972, over 7,000 papers and 2,000 United States patents have been published dealing with zeolite science and technology. The extent of international scientific and commercial interest in the zeolite molecular sieves was evidenced by several large conferences. In the USSR, the Second All-Union Conference on Zeolites sponsored by the USSR Academy of Sciences was held in Leningrad in 1964; a total of 81 papers were presented covering the science and application of zeolites. An international conference on molecular sieves organized by R. M. Barrer and sponsored by the Society of Chemical Industry in Great Britain was held in London in April 1967; 31 papers were presented to the 200 scientists from over 18 countries. The Second International Conference, chaired by E. M. Flanigen and L. B. Sand, was held in the United States in September 1970; the published proceedings included 77 papers (1).

The Third International Conference held in Switzerland in September 1973 was organized under the chairmanship of W. M. Meier.

A. MOLECULAR SIEVE TECHNOLOGY

The properties and uses of zeolites are being explored in many scientific disciplines: modern inorganic and organic chemistry, physical chemistry, colloid chemistry, biochemistry, mineralogy, geology, surface chemistry, oceanography, crystallography, catalysis, and in all types of chemical engineering process technology. The wide variety of applications includes separation and recovery of normal paraffin hydrocarbons, catalysis of hydrocarbon reactions, drying of refrigerants, separation of air components, carrying catalysts in the curing of plastics and rubber, recovering radioactive ions from radioactive waste solutions, removing carbon dioxide and sulfur compounds from natural gas, cryopumping, sampling air at high altitudes, solubilizing enzymes, separating hydrogen isotopes, and removal of atmospheric pollutants such as sulfur dioxide.

Cracking catalysts containing crystalline zeolite molecular sieves were first used in 1962 (2,3) and at present close to 95% of the installed capacity, which is in excess of 4 million barrels of oil per day, employs zeolite catalysts (4). Annual savings of greater than $250 million in operating expenses and several hundred million more in capital investment have been reported.

The first experimental observations of the adsorption of gases on zeolites and their behavior as molecular sieves were conducted on zeolite minerals. The first definitive experiments on the separation of mixtures using the dehydrated zeolite mineral chabazite as a molecular sieve were performed by Barrer in 1945 (5). He classified zeolites into three groups based upon their ability to adsorb or exclude molecular species of different sizes. The classification defined the approximate intrachannel dimensions.

In 1948, the first industrial research efforts by Milton and his associates at Union Carbide Corporation resulted in the synthesis and the manufacture of synthetic zeolite molecular sieves which had never existed as minerals (6). This controlled synthesis was a key research achievement.

B. THE SCOPE OF THE BOOK

Although review articles and conference proceedings cover various aspects of the science and application of zeolite molecular sieves, a researcher or engineer is still at a loss to readily obtain specific information about zeolite molecular sieves and their use. It is this void that we hope to fill with this book.

There are two main purposes in preparing this book: It is an introduction to the subject of zeolite molecular sieves for the newcomer to the field, and a reference for additional information and background.

Details on structure, properties, characterization, synthesis, chemistry (including ion exchange, adsorption) and commercial materials are included.

Each chapter is followed by a bibliography of the pertinent published literature including patents. There are many examples in zeolite science where an issued patent is either a primary reference or the only source of essential technical information.

The industrial application of zeolite molecular sieves is a separate subject in itself. Engineering procedures and the design of processes are not covered in detail in this book.

C. TYPES OF ADSORBENTS

Commercial adsorbents which exhibit ultraporosity and which are generally used for the separation of gas and vapor mixtures include the activated carbons, activated clays, inorganic gels such as silica gel and activated alumina, and the crystalline aluminosilicate zeolites.

Activated carbons, activated alumina, and silica gel do not possess an ordered crystal structure and consequently the pores are nonuniform. The distribution of the pore diameters within the adsorbent particles may be narrow (20 to 50 A) or it may range widely (20 to several thousand A) as is the case for some activated carbons. Hence, all molecular species, with the possible exception of high molecular weight polymeric materials, may enter the pores. Zeolite molecular sieves have pores of uniform size (3 A to 10 A) which are uniquely determined by the unit structure of the crystal. These pores will completely exclude molecules which are larger than their diameter.

The pore size distribution for a zeolite molecular sieve, a typical silica gel, and activated carbon are illustrated schematically in Fig. 1.1.

The term "molecular sieve" was originated by J. W. McBain to define porous solid materials which exhibit the property of acting as sieves on a molecular scale (7).

At present, the most important molecular sieve effects are shown by the dehydrated crystalline zeolites. These materials all have a high internal surface area available for adsorption due to the channels or pores which uniformly penetrate the entire volume of the solid. The external surface of the adsorbent particles contributes only a small amount of the total available surface area.

4 INTRODUCTION

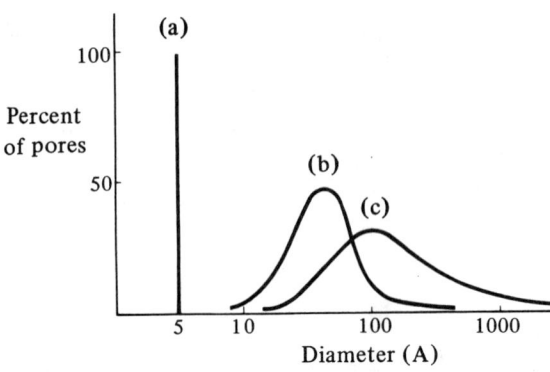

Figure 1.1 Distribution of pore sizes in microporous adsorbents. (a) Dehydrated zeolite, e.g., type A; (b) typical silica gel; (c) activated carbon.

external surface of the adsorbent particles contributes only a small amount of the total available surface area.

The zeolites selectively adsorb or reject different molecules. Molecular sieve action may be total or partial. If total, the diffusion of one species into the solid may be wholly prevented while the diffusion of a second species occurs; if partial, the components of a binary mixture diffuse into the solid at different rates depending upon the conditions.

In some cases, the *activated* diffusion of the species in the solid is of particular interest. As the size of the diffusing molecule approaches the size of the pores in the zeolite, the interaction energy between the species and the aperture increases in importance. If the aperture is sufficiently small relative to the size of the diffusing species, the repulsive interaction becomes dominant and the diffusing species needs a specific activation energy to pass through the aperture (see Chapter 8).

Although zeolite molecular sieves are today being used in diverse applications such as catalysis and ion exchange, the term "molecular sieve" is still retained although it does not completely imply this wide sphere of application. Stereospecific catalytic and ion-sieve behavior are well known.

D. TYPES OF MOLECULAR SIEVES

Crystalline Zeolites

Zeolites are crystalline, hydrated aluminosilicates of group I and group

II elements,* in particular, sodium, potassium, magnesium, calcium, strontium, and barium. Structurally the zeolites are "framework" aluminosilicates which are based on an infinitely extending three-dimensional network of AlO_4 and SiO_4 tetrahedra linked to each other by sharing all of the oxygens.

Zeolites may be represented by the empirical formula

$$M_{2/n}O \cdot Al_2O_3 \cdot xSiO_2 \cdot yH_2O$$

In this oxide formula, x is generally equal to or greater than 2 since AlO_4 tetrahedra are joined only to SiO_4 tetrahedra, n is the cation valence. The framework contains channels and interconnected voids which are occupied by the cation and water molecules. The cations are quite mobile and may usually be exchanged, to varying degrees, by other cations. Intracrystalline "zeolitic" water in many zeolites is removed continuously and reversibly. In many other zeolites, mineral and synthetic, cation exchange or dehydration may produce structural changes in the framework. Ammonium and alkyammonium cations may be incorporated in synthetic zeolites, e.g., NH_4, CH_3NH_3, $(CH_3)_2NH_2$, $(CH_3)_3NH$, and $(CH_3)_4N$. In some synthetic zeolites, aluminum cations may be substituted by gallium ions and silicon ions by germanium or phosphorus ions. The latter necessitates a modification of the structural formula.

The structural formula of a zeolite is best expressed for the crystallographic unit cell as: $M_{x/n}[(AlO_2)_x (SiO_2)_y] \cdot wH_2O$ where M is the cation of valence n, w is the number of water molecules and the ratio y/x usually has values of 1 - 5 depending upon the structure.†
The sum $(x + y)$ is the total number of tetrahedra in the unit cell. The portion with [ ] represents the framework composition.

Although there are 34 species of zeolite minerals and about 100 types of synthetic zeolites, only a few have practical significance at the present time. Many of the zeolites, after dehydration, are permeated by very small channel systems which are not interpenetrating and which may contain serious diffusion blocks. In other cases dehydration irreversibly disturbs the framework structure and the positions of metal cations, so that the structure partially collapses and dehydration is not completely reversible. To be used as a molecular sieve, the structure of the zeolite after complete dehydration must remain intact.

*As formed in nature or synthesized. Higher polyvalent ions, e.g., rare earths, are readily introduced by cation exchange.
† Occlusion of AlO_2^- species within the zeolite structure may lower the ratio below 1.

6 INTRODUCTION

Molecular sieve action has been reported for other solids, crystalline and noncrystalline. These include coal, special active carbons, porous glass, microporous beryllium oxide powders, and layer silicates modified by exchange with organic cations.

Coal

As a result of adsorption studies, several coals were found to have pore diameters approaching molecular sizes (8). A comparison of the relative adsorption of *n*-butane and isobutane indicated that certain coals had pores with openings between 4.9 and 5.6 A in diameter at 0° C and about 4 A at 145° C. The adsorption capacity, however, was low and the rate of adsorption very slow.

A typical anthracite has a pore volume of 0.07 cc/g and 0.10 cc/cc coal. An anthracite coal which was activated by heating in carbon dioxide exhibited selective adsorption behavior toward *n*-butane (9). Monolayer volumes in the raw devolatized and activated anthracites are shown in Table 1.1. The activated anthracite has a capacity of about 130 cc of *n*-butane/g and 50 cc of isobutane/g. A selectivity for *n*-butane similar to that observed in crystalline zeolites is indicated.

Table 1.1 Monolayer Volumes for Hydrocarbons on Raw and Activated Anthracite (9)

Material	V_m (cc/g)(STP) Raw	V_m (cc/g)(STP) Activated
n-C_4H_{10}	12.6	134
iso-C_4H_{10}	8.6	54.5
neo-C_5H_{12}	2.3	15.5

Oxides

The controlled thermal decomposition of beryllium hydroxide *in vacuo* produces BeO consisting of porous aggregates of 30 A crystallites (10). Only 20% of the internal pore volume is available to adsorption by carbon tetrachloride and 50% by nitrogen. It was concluded that there are pores which are less than 6 A and others less than 3 A. The micropore size depends upon whether the hydroxide is decomposed *in vacuo* or in the presence of water vapor. The range of pore sizes is 6 - 20 A if the decomposition occurs in water vapor where the pore size increases at relatively low temperatures (500°C) due to crystal growth.

Glasses

Leaching alkali silicate glasses with acids results in porous adsorbents which are reported to exhibit molecular sieve-type adsorption toward small molecules (11). In general, the glasses consist of three components: an alkali such as sodium or potassium, another oxide such as B_2O_3 or Al_2O_3, and silica. The porosity of the resulting solid depends upon the conditions of leaching and the composition. Results show that the adsorption pore volume is much lower than that in the zeolites and generally in the range of 0.01 to 0.04 cc/g as measured by water adsorption. Porous glass with uniform pore sizes of 3 - 10 Å has been reported. Small variations in the composition of the glass significantly change the behavior toward water adsorption. Porous glass molecular sieves have been reported to be more stable in acid media than crystalline zeolites. However, the fact that the porosity is produced by acid leaching of the glass and that the pore size distribution is controlled by the degree of acid leaching contradicts the argument of acid stability. It is known, moreover, that some of the zeolites are quite stable toward acid conditions. (See Chap. 6).

Recently, silica gels with pore radii of about 4 Å have been reported; silica gel commonly has pore size distributions which range from 10 to several hundred Å (12). The silica was prepared by the evaporation of a water solution of normal silicic acid at 0°C. The adsorptive properties varied with the pH of the preparation and mean pore radii were 7 - 9 Å (Table 1.2). Type I adsorption isotherms were observed and larger molecules such as benzene and CCl_4 exhibited the lowest adsorption. These silica gels, however, were found to be unstable when treated with water; the pore system collapsed.

Table 1.2 Adsorptive Properties of Silica Samples (12)

Sample	pH	$SA(m^2/g)$	$V(cc/g)$	$r(Å)$
1	0.97	742	0.345	9.3
2	0.94	429	0.184	8.6
3	1.87	255	0.090	7.1
4	1.67	322	0.122	7.6

SA = specific surface area, V = pore volume, r = mean pore radius.

Carbon

Present methods for the preparation of carbon adsorbents with small, narrow pore size distributions are largely empirical (13, 14, 15). In general the method has involved the controlled pyrolysis of polymers such as polyvinychloride $(C_2H_2Cl_2)_n$ and Saran (a copolymer of 80 - 90% polyvinylidene chloride with various amounts of polyvinyl chloride and plasticizer). The pyrolysis results in the loss of HCl as indicated by Eq. 1.1.

$$(C_2H_2Cl_2)_n \rightarrow 2n \text{ HCl} + 2n \text{ C} \qquad (1.1)$$

This leaves essentially pure carbon which is 25 wt% of the starting polymer (Fig. 1.2). At present little is known about the relationship between the original polymer structure in terms of the size and shape of the polymer units and the resulting pore size and shape of the microporous carbon. Originally it was thought that the resulting carbon might have a uniform pore size that was related to the initial regular polymer structure. Most of the studies have been confined to the Saran type of polymers since thermoplastic polymers do not yield good cokes upon carbonization. It is believed that the mechanism of the release of HCl during carbonization involves several steps. More than half of the HCl is lost at temperatures below 200° C; further liberation requires higher temperatures with the process being complete at 600° C. The first step is the loss of one HCl molecule per pair of carbon atoms which leads to the formation of the carbonized chain. Crosslinking then occurs followed by the elimination of another HCl.

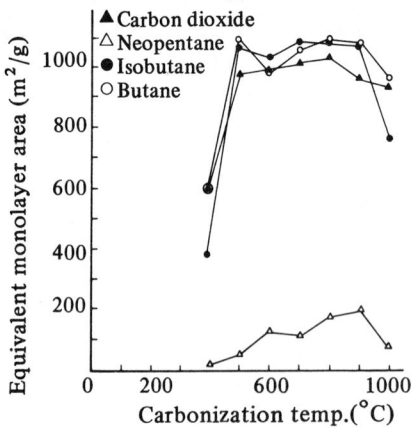

Figure 1.2 Surface area of Saran carbons heated to different carbonization temperatures (15).

The molecular sieve properties of the resulting carbons were determined by measuring the adsorption of nitrogen, carbon dioxide, n-butane, isobutane, and neopentane. The minimum cross-sectional diameter of these molecules (see Chapter 6) suggests their use as probes for determining the pore size. Typical results (Fig. 1.2) for a Saran-428 carbon show that except for neopentane, adsorption reached a maximum with carbonization temperatures in the range of 500 - 800°C. The very high surface areas, $\sim$ 1000 m^2/g, for butane and isobutane as opposed to the low surface area for neopentane are noteworthy. The ability of this carbon to separate isobutane and neopentane is apparent. In addition the adsorption of benzene and cyclohexane on a Saran-derived carbon was found to be about ten times as great as the adsorption of neopentane. This indicates that the pores in Saran-derived carbons are slit-like in shape; straight-chain hydrocarbons and flat molecules are preferentially adsorbed but the adsorption of a large spherical molecule such as neopentane is slow. Fig. 1.3 compares the separation ability of two Saran-derived carbon molecular sieves with zeolite 5A and a typical activated carbon. The separation of n-heptane from a mixture with isooctane (2, 2, 4-trimethylpentane) by a carbon molecular sieve and zeolite 5A are shown. The carbon sieves are concluded to have slit-shaped pore constrictions ranging in size between 4.5 and 5.7 A in thickness which connect cavities of about 12 A in thickness. The adsorption capacities may be comparable to those of the zeolites. At present, commercial products based upon

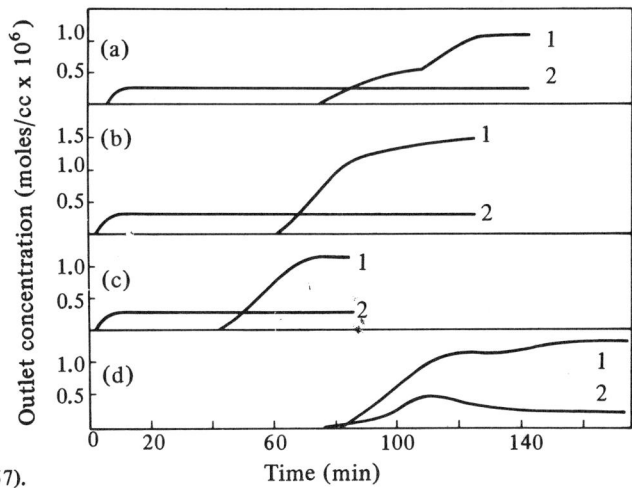

Figure 1.3 Dynamic adsorption of n-heptane (1) and 2,2,4-trimethylpentane (2). (a) and (b) molecular sieve carbons; (c) molecular sieve zeolite type 5A; (d) commercial activated carbon (57).

10 INTRODUCTION

carbons derived from polymers are not available.

Alkali Graphite Intercalation Compounds

Intercalation compounds of graphite and alkali metals adsorb small gas molecules (16). The compounds of composition $C_{24}M$ (where M = K, Rb, Cs) adsorb the gases H_2, D_2, N_2, CH_4, Ar to different degrees depending on the size of the intercalated atom. Adsorption isotherms of type I are exhibited by $C_{24}Rb$ for D_2 at 113°K. The compound $C_{24}K$ has an adsorption capacity for H_2 of 144 cc/g but does not adsorb N_2 at −196°C. The mole ratio is 2.1 $H_2/C_{24}K$.

As the size of the alkali atom increases, the size of the interlayer pores increases. Thus $C_{24}Cs$ adsorbs both H_2 and N_2 as well as CH_4. In the latter case, a type III isotherm was observed. This is associated with swelling of the interlayer compound in the c-direction perpendicular to the carbon atom layers.

Selective adsorption of D_2 from a D_2-argon mixture by $C_{24}K$ was observed at 77°K. D_2 was nearly completely adsorbed while argon adsorption was negligible. Separation of a D_2-N_2 mixture was incomplete and substantial adsorption of N_2 occurred.

Table 1.3 Adsorptive Capacity of Alkali Graphites (63°K to 196°K)(16)

Compound	\multicolumn{4}{c}{Adsorbate *(molecules adsorbed gas/ atoms of alkali)*}
	H_2, D_2	N_2	Ar	CH_4
$C_{24}K$	2.1	none	none	none
$C_{24}Rb$	2.05	1.0	1.2	0.9
$C_{24}Cs$	2.00	1.3	1.4	1.2

He and Ne were not adsorbed.

E. EARLY OBSERVATIONS

Zeolites were first recognized by Cronstedt (17) as a new group of minerals consisting of hydrated aluminosilicates of the alkali and alkaline earths with his discovery of stilbite in 1756. Because the mineral exhibited intumescence when heated in a blowpipe flame, he called the mineral a zeolite which comes from two Greek words meaning "to boil" and "a stone." Studies of zeolite minerals have taken place over the last two centuries. The most recently discovered zeolite mineral is garronite (1962) (18).

Ion Exchange by Soil Minerals

The misuse of the term "zeolite" to include all inorganic ion-exchange materials originated in the earliest studies of the phenomenon of ion exchange in soils. In 1845, H. S. Thompson (19) conducted experiments which showed that certain soils have the power of "decomposing" and retaining ammonium salts. When a solution of ammonium sulfate was filtered through the soil, the filtrate contained calcium sulfate and the ammonium salt was retained in the soil. Later J. T. Way (20) showed that the hydrated silicates in the soil produced this phenomenon. He was able to show that only the ammonium or the potassium was exchanged for the calcium in the soil. Subsequently Way prepared an artificial base exchanger consisting of a sodium aluminosilicate. Several years later H. Eichhorn published a paper on the action of dilute salt solutions on silicates (21), showing that the base exchange principle discovered by Way is reversible. He studied the quantitative behavior of the zeolite minerals, chabazite and natrolite, in contact with dilute salt solutions and found that sodium and calcium could reversibly replace each other in the zeolite.

The Permutites

The accepted term for synthetic aluminosilicates which are crystallographically amorphous and are prepared for their ion exchange properties is "permutite." This term should always be employed when referring to synthetic amorphous aluminosilicate cation exchange materials. A complete summary of the literature covering the preparation of permutite materials and their application up to 1930 is given by Shreve (22).

The chemical composition of most permutites is represented by an empirical formula in terms of the oxides: $Na_2O \cdot Al_2O_3 \cdot xSiO_2 \cdot yH_2O$ in which x often has a value of 5 - 6.* The chemical composition must be shown in terms of the oxide ratios since they are noncrystalline, gel-like solids. A structural formula, as is used for the crystalline zeolites, is not possible.

The first usable synthetic cation exchangers were made by a fusion method (23). However, this method had several inherent disadvan-

---

* Examples are Culligan zeolite: $0.9\ Na_2O \cdot Al_2O_3 \cdot 5.5\ SiO_2 \cdot 19\ H_2O$ and Super Nalcolite: $0.7\ Na_2O \cdot Al_2O_3 \cdot 6.1\ SiO_2 \cdot 15\ H_2O$.

tages and frequently nonhydrated silicates were formed. The product, if treated with water at a temperature of about 70°C, showed a tendency to become slimy. This method was generally abandoned in favor of a "wet method."

A second method claimed the production of a "zeolitic" water softener by the addition of a dilute sodium aluminate solution to an alkali silicate solution. The precipitate that was formed was washed, filtered, and dried before use (24).

In a third method, permutites are synthesized by mixing dilute solutions of sodium aluminate and sodium silicate (25). The precipitated aluminosilicate was heated, filtered, washed, and ignited. In a modification of this method very dilute solutions of sodium aluminate and sodium silicate were mixed and heated to the boiling point. A flocculent precipitate that formed was removed and filter pressed. The commercial water softener Doucil® is prepared by the precipitate method from dilute solutions of sodium aluminate and sodium silicate at temperatures below 20°C. The gelatinous precipitate containing 6 - 16% alumina is dried slowly to a 60% moisture content, washed, redried for a period of one day, and crushed to the desired size (26).

In a fourth method, similar types of reagents are used in concentrated rather than dilute solutions so that a flocculent precipitate is not formed; rather, the whole mass is converted into a homogeneous gel with a relatively low water content. (This is similar to the gel "phase" that is prepared as an initial step in the synthesis of crystalline zeolites from aluminosilicate gels). (See Chapter 4.) The gel was pressed to remove excess liquid, dried, and the cake placed in water. **This leached the alkali and salts, but also caused diminution of the cake into small granules.** A procedure that gave control of the Si/Al ratio in these artificial permutites was claimed by Bruce (27). He notes that previous attempts to prepare permutites by the addition of aluminum sulfate mixtures to sodium silicate and sodium aluminate solutions failed to produce materials of uniform composition and durability. Synthetic permutites prepared from sodium silicate and aluminum sulfate give products with a high Si/Al ratio resulting in unsatisfactory water softening properties. The cation exchange capacity is directly related to the aluminum content of the gel. In general, the permutites prepared by the addition of a sodium silicate to a sodium aluminate solution have a sodium to alumina ratio of about 1, with varying amounts of silica.

During the first part of the twentieth century considerable effort was expended on the preparation of inorganic aluminosilicates, both artificial and natural, for use in softening water by the ion exchange process. The use of naturally occurring glauconite (green sand) in water treatment was common. Glauconites are ferrous aluminosilicates which contain exchangeable potassium on the crystal surfaces.

Because of the industrial incentive, extensive investigation was made on the occurrence and availablity of glauconite. However, glauconite is not a zeolite but a member of the mica group (28).

Misconceptions have appeared in the literature concerning the relationship between clays and zeolites. Some clay minerals, such as bentonite, have well-known cation exchange properties. A number of patents have been issued on processes and modifications for the slurrying, pelletizing, heating, drying, dehydration, and rehydrating of clays. The product has been erroneously labeled "zeolite." Other similar products have been produced by crushing and extracting with acids of set hydraulic cement (29), or prepared from a mixture of sand, cement, and powdered iron oxide (30).

This brief summary of the extensive literature on the permutite "zeolites" covered only the highlights of early research efforts. The Shreve bibliography is recommended as a detailed source (22).

Adsorption by Zeolites

The first experimental observations of the adsorption of gases on zeolites and their behavior as molecular sieves were conducted on the zeolite minerals.

Various terms have been used to describe the penetration by, and containment of, molecules in dehydrated, crystalline zeolites. These include occlusion, imbibition, intercalation, persorption, sorption, and adsorption. We shall use adsorption because it best describes the interaction between a molecule and a surface, whether it is the external surface of a solid or the internal convoluted surface of a dehydrated, microporous zeolite crystal.

The study of the adsorption of gases and vapors on solids dates from Gideon's tests on fleece in 1100 B.C.(31) and the early experiments of Scheele and Abbé F. Fontana (7). The reversible dehydration-hydration phenomena of zeolite minerals attracted many early investigators. Much of the early work on the adsorption of gases was performed on the stable zeolites which retained their structure upon dehydration.

In 1840 A. Damour observed that crystals of zeolites could be reversibly dehydrated without any apparent change in their transparency or morphology (32). The idea that the structures of dehydrated zeolites consist of open spongy-frameworks is due to G. Friedel (33) who observed that various liquids such as alcohol, benzene, chloroform, carbon disulfide, and mercury were occluded by zeolites. He found that the refractive indices of the zeolite changed during this adsorption. The adsorption of gases by dehydrated zeolites was studied by F. Grandjean in 1909 (34,35). He observed that chabazite adsorbs ammonia, air, hydrogen, carbon disulfide, hydrogen sulfide, iodine, and bromine. When heated to higher temperatures, even mercury vapor was adsorbed.

Dehydrated chabazite crystals were observed to rapidly adsorb the vapors of water, methyl alcohol, ethyl alcohol, and formic acid; however, when exposed to acetone, ether, or benzene, essentially no adsorption occurred (36). McBain deduced from these observations that the pore openings in the chabazite crystals must be less than 5 A in diameter. To describe this phenomenon of selective adsorption, or "persorption," as he termed it, McBain originated the term "molecular sieve" (7). He distinguished between zeolite-type adsorption and a solid solution; the interstices or ultrapores within the crystal lattice could contain any molecules which are small enough to permeate the crystal. The adsorbed phase dispersed through the internal voids of the crystal without displacing any atoms which make up the permanent crystal structure.

The quantity of gas or liquid (x) adsorbed by a solid depends upon pressure, temperature, the nature of the gas, and the nature of the solid. When exposed to a gas or liquid, the intracrystalline voids and channels of a dehydrated zeolite fill with the molecular species concerned and when the filling is complete, no more adsorption occurs. This results in an isotherm contour of the Langmuir type, or type I (see Chap. 8), as illustrated in Fig. 1.4 for the adsorption of nitrogen on dehydrated chabazite.

It is evident that at a temperature of 89.2°K, the adsorption voids

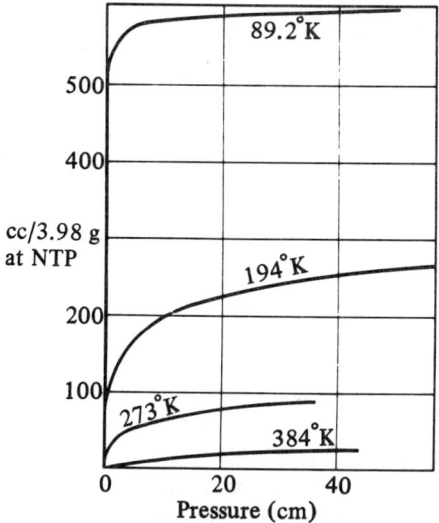

Figure 1.4 Adsorption of nitrogen on chabazite at different temperatures. These isotherms are of the BET type I. (37)

fill at a very low pressure, below 100 torr. Because of the inherent characteristics of the crystal structure of the zeolite, the adsorption of a guest molecule depends partially upon its polarity and polarizability. The shape of the isotherm is determined by the molecule-zeolite interaction energies. A polar molecule such as water is adsorbed very strongly; the isotherm is rectangular in shape.

Early investigators tested zeolite minerals which showed the greatest stability toward dehydration, in particular chabazite and mordenite. In addition to extensive studies of the adsorption of permanent gases, Barrer and Ibbitson determined the adsorption of hydrocarbons on chabazite (38). They found that hydrocarbons such as propane, *n*-butane, *n*-pentane, and *n*-heptane were adsorbed quite rapidly at temperatures above 100°C. However, branched-chain hydrocarbons, such as isobutane and isooctane, were totally excluded. The adsorption equilibria were reversible. Following his initial studies, Barrer classified the zeolites into three groups based upon their ability to adsorb or exclude certain molecular species (Table 1.3).

This classification defined the approximate interstitial channel dimensions. Molecular sieve separations were classified into two types: (1) total molecular sieve action and (2) partial molecular sieve action. In the case of total molecular sieve action it was found that monosubstituted methanes could be separated from monosubstituted ethanes using dehydrated mordenite. Partial molecular sieve action was based

16 INTRODUCTION

Table 1.3 Three Categories of Molecular Sieve Zeolites (39)

Class 1
  Chabazite
  Gmelinite
  Synthetic zeolite
    $(BaAl_2Si_4O_{12}, n\,H_2O)$
	Do not occlude iso-paraffins or aromatics.
	Occlude *n*-paraffins slowly.
	Occlude CH_4, C_2H_6, and molecules of smaller cross section very rapidly.
	Diameter of narrowest cross section of interstitial channel between 4.89 and 5.58 A.

Class 2 Mordenite (Na-rich)	Does not occlude *n*-paraffins, isoparaffins, or aromatics. Occludes CH_4 and C_2H_6 slowly. Occludes N_2, O_2, and molecules of smaller cross section rapidly. Diameter of narrowest cross section of interstitial channel between 4.0 and 4.89 A.
Class 3 Ca- and Ba- rich mordenites[a]	Do not occlude hydrocarbons, including CH_4, C_2H_6. Occlude Ar, N_2, and molecules of smaller cross section. Diameter of narrowest cross section of interstitial channel between 3.84 and 4.0 A.

[a] Prepared hydrothermally by cation exchange.

on the differences in adsorption rates as evidenced in column 2 of Table 1.4. Partial or complete separations were observed in mixtures such as ethane-propane, and ethanol-*n*-heptane. The rate of adsorption decreased as the number of carbon atoms in the hydrocarbon increased; diffusion of the molecule occurred in a stretched-out configuration and the ability of chabazite to adsorb the paraffin hydrocarbon depends on the shape and cross-sectional diameter of the molecules rather than on their molecular volume. For example, isobutane with an estimated cross-sectional diameter of 5.6 A and a molar volume of 96 cc was not adsorbed whereas *n*-heptane with a cross-sectional diameter of 4.9 A and a molar volume of 145 cc was adsorbed. The separation of normal hydrocarbons from mixtures with various branched-chain hydrocarbons and cyclic hydrocarbons in both the

Table 1.4 Molecules Occluded or Excluded by Three Classes of Molecular Sieves (5, 39)

Typical Molecules Rapidly Occluded at Room Temp. or Below.	Typical Molecules Moderately, Rapidly or Slowly Occluded at Room Temp. or Above.	Typical Molecules Not Appreciably Occluded at Room Temp. or Above.
	Class 1 Minerals	
He, Ne, Ar	C_3H_8 and simple higher	Aromatic hydrocarbons, cyclo-
H_2, N_2, O_2	n-paraffins	and isoparaffins.
CO, CO_2	CH_3CH_2OH	Derivatives of these hydro-
COS, CS_2	$CH_3CH_2NH_2$	carbons.
H_2O	$CH_3CH_2F, CH_3CH_2Cl,$	Heterocyclic compounds (e.g.,
HCl, HBr	CH_3CH_2Br	thiophene, pyrrole, pyridine).
NO	I_2, HI	$CHCl_3, CCl_4, CHCl:CCl_2,$
NH_3	CH_2Br_2	$CH_3CHCl_2, CHCl_2CCl_3,$
CH_3NH_2	CH_3I	CCl_3CCl_3 and analogous
CH_3OH	CH_3CH_2CN	bromo- and iodo-compounds.
CH_3CN	CH_3CH_2SH	Secondary straight-chain alco-
HCN	$HCO_2CH_3, HCO_2CH_2CH_3$	hols, thiols, nitriles, and
Cl_2	CH_3COCH_3	halides
CH_3Cl, CH_3Br, CH_3F	$CH_3CO_2CH_3$	Primary amines with NH_2
CH_2Cl_2, CH_2F_2	$(CH_3)_2NH, (CH_3CH_2)_2NH$	group attached to a second-
CH_4, C_2H_6		ary carbon atom. Tertiary
C_2H_2		amines.
CH_2O, H_2S, CH_3SH		Branched-chain ethers, thio-
		ethers, and secondary amines.
	Class 2 Minerals	
He, Ne, Ar	CH_4, C_2H_6	All classes of molecules in
H_2, O_2, N_2	CH_3OH	cols. 2 and 3, section 1.
CO	CH_3NH_2	
NH_3	CH_3CN	
H_2O	CH_3Cl, CH_3F	
	HCN	
	Cl_2	
	Class 3 Minerals	
He, Ne	Ar	All classes of molecules in
H_2, O_2, N_2	HCl	cols. 2 and 3, section 1.
H_2O	NH_3	

liquid and vapor phase was achieved with some limitations. Although the *n*-paraffins could be completely desorbed from chabazite by evacuation at elevated temperatures, the rate of recovery was limited in the case of higher molecular weight hydrocarbons. Some organic materials such as pyridine poisoned the adsorbent; in an isobutylene mixture, polymerization occurred. Factors such as pressure, temperature, particle size, and the conditions of dehydration and desorption were found to influence the rate of adsorption in zeolites. The adsorption of molecules was interpreted as a diffusion process and a diffusion equation was successfully employed. In terms of its molecular sieve character, gmelinite behaved similar to chabazite. In contrast, however, the adsorption properties of dehydrated mordenite were found to be quite different. Gases such as O_2, N_2, and Ar were rapidly adsorbed. Methane and ethane were slowly adsorbed and hydrocarbons such as n-propane, n-butane, and isoparaffins were excluded.

Variations in the chemical composition of the zeolite were found to affect adsorption (see Chapter 8). Calcium exchange for sodium in mordenite reduces adsorption for methane and ethane to an insignificant quantity. In the case of chabazite, it was observed that samples from different localities with different compositions showed a considerable difference in their adsorption character (40). It was assumed that this was due to a higher content of alkali ions in contrast to the normally calcium-rich type of chabazite. The calcium-rich form and an original mineral which contained principally calcium were nearly identical in their adsorption isotherms for nitrogen; sodium exchange for calcium considerably decreased nitrogen adsorption. The potassium-rich form adsorbed almost no nitrogen. This effect was further investigated in a study of the adsorption of hydrogen, carbon dioxide, and propylene.

Not all specimens of analcime were found to be adsorbents (41). Of several specimens only one was observed to adsorb gases such as argon, nitrogen, methane, and ethane. Chemical analysis indicated that the adsorbing specimen contained calcium in contrast to the other specimens which contained only sodium as the cation.

The so-called lamellar and fibrous zeolites, as typified by heulandite, stilbite, and natrolite, did not behave as molecular sieve adsorbents after complete dehydration.

Various theories were advanced to explain the adsorption behavior of zeolites. Barrer and Ibbitson proposed that the adsorbed molecules form a regularly disposed, frequently mobile, interstitial component of the dehydrated zeolite crystals—an institial solid solution. The structurally stable three-dimensional zeolites were able to adsorb a quantity of gas equivalent to the water of hydration on a volume basis.

F. CLASSIFICATION AND NOMENCLATURE

Mineral Zeolites

The first analysis of the crystal structure of a zeolite, analcime, was reported by Taylor in 1930 (42). In the same year, Pauling proposed structures for the framework of natrolite, as well as for the related framework structures of sodalite, scapolite, and cancrinite (43). Additional investigations of the structures of fibrous zeolites were reported in 1933 (44). The first investigation of the structure of chabazite (45) was shown to be incorrect by the later investigation of Dent and Smith (46). From 1933 until the publication of the structure of the new synthetic zeolite A in 1956 by Reed and Breck of Union Carbide Corporation, essentially no new zeolite structures were reported (47). The interpretation of the behavior of zeolites upon dehydration, rehydration and adsorption of other molecular species was based upon these early structure results.

During this period, three types of zeolitic frameworks were considered to exist: (1) three-dimensional framework structures with uniform bonding, (2) lamellar-type structures with aluminosilicate sheets weakly bonded to one another, and (3) fiber-like structures with weakly crosslinked aluminosilicate chain units. This classification, introduced by Bragg, has been used until recent years (48,49). It should be reiterated that all zeolites are framework aluminosilicates and all of the zeolite structures are three-dimensional. In some zeolites, the bond density in certain crystallographic directions is not uniform. Some of these have been referred to as lamellar and/or fibrous. The terms are used to indicate the crystal habit or cleavage, i.e., the lamellar zeolites possess fibrous habits. No zeolite is, in a true sense, a layer structure or a one-dimensional structure.

The framework structure of the cubic zeolite mineral faujasite (50) and detailed structure of the cubic synthetic zeolite A (47) were reported in 1956. A list of zeolites according to the early classification is shown in Table 1.5.

Several of these zeolites were not properly classified; in some instances no direct relationship between habit and detailed structure existed. For example, the mineral erionite is fibrous in habit but has a robust, three-dimensional structure. Recent structural analyses of

20 INTRODUCTION

many zeolites have led to systems for the identification and classification of zeolites which are based upon structural and physicochemical characteristics rather than external appearance (see Chapter 2).

Similar problems have existed with other large mineral groups such as the clay minerals; the structures of clay minerals were not resolved in detail until the last two decades. An understanding of the feldspar minerals, which is so important to an understanding of the formation and geological history of rocks and an understanding of the chemistry of the earth's crust, is fairly recent. Many unsolved problems

Table 1.5 Early Classification of Zeolite Minerals

Name	Composition Ideal Formula (51)	Interstitial Volume Available to H_2O (cc/g Dehydrated Zeolite)
A. Three-dimensional framework structures		
Analcime	$NaAlSi_2O_6 \cdot H_2O$	0.089
Chabazite	$(Ca, Na_2)Al_2Si_4O_{12} \cdot 6 H_2O$	0.29
Gmelinite	$(Ca, Na_2)Al_2Si_4O_{12} \cdot 6 H_2O$	0.27
Harmotome	$(Ba, K_2)Al_2Si_5O_{14} \cdot 5 H_2O$	0.18
Levynite	$CaAl_2Si_3O_{10} \cdot 5 H_2O$	0.23
Mordenite	$(Ca, Na_2, K_2)Al_2Si_{10}O_{24} \cdot 6.7 H_2O$	0.15
B. Lamellar structures—weakly bonded aluminosilicate sheets		
Epistilbite	$CaAl_2Si_6O_{16} \cdot 5 H_2O$	0.19
Heulandite	$CaAl_2Si_6O_{16} \cdot 5 H_2O$	0.21
Stilbite	$(Ca,Na_2)Al_2Si_6O_{16} \cdot 6 H_2O$	0.22
C. Fibrous structures—weakly crosslinked chains		
Edingtonite	$BaAl_2Si_3O_{10} \cdot 3 H_2O$	0.15
Mesolite	$Na_2Ca_2Al_6Si_9O_{30} \cdot 8 H_2O$	0.15
Natrolite	$Na_2Al_2Si_3O_{10} \cdot 2 H_2O$	0.11
Scolecite	$CaAl_2Si_3O_{10} \cdot 3 H_2O$	0.16
Thomsonite	$(Ca, Na_2)Al_2Si_2O_8 \cdot 2.5 H_2O$	0.16
D. Unclassified		
Brewsterite	$(Sr, Ca, Ba)Al_2Si_6O_{16} \cdot 5 H_2O$	0.17
Faujasite	$(Ca, Na_2)Al_2Si_5O_{14} \cdot 10 H_2O$	0.33
Gismondine	$CaAl_2Si_2O_8 \cdot 4 H_2O$	0.28
Laumontite	$CaAl_2Si_4O_{12} \cdot 4 H_2O$	0.17
Phillipsite	$(Ca, K_2)Al_2Si_4O_{12} \cdot 4\text{-}1/2 H_2O$	0.20

still exist in classifying these mineral groups. These minerals continue to receive attention as more sophisticated experimental techniques are developed. Some of the methods employed in the characterization and identification of zeolite minerals and synthetic zeolites are the same as those used extensively in these other problems. For many years minerals were characterized and identified on the basis of chemical composition, optical properties, certain physicochemical properties, and morphology. Modern methods include x-ray diffraction, which is based on the crystal structure, for identifying fine-grained materials.

The mineralogist, E. S. Dana, employed a classification of the minerals and zeolites which was based upon these factors (51). For example, zeolite minerals were classified as highly acid species, (i.e., high in silica content) followed by metasilicates and orthosilicates.

Winchell, in a discussion of "What is a mineral?" defines a mineral species as a crystalline phase found in inorganic nature (52). He proposes that a mineral should not be defined in terms of a simple formula because the formulas assigned to most minerals give an approximate composition. Minerals cannot be defined in terms of composition alone.

A mineral species may vary in composition and different minerals may have nearly identical chemical compositions. For example, the three different feldspar minerals, adularia, sanidine, and microcline have essentially the same composition, $KAlSi_3O_8$. Twenty-two different phases of the simple oxide, SiO_2, have been defined (53). The feldspar plagioclase includes six subspecies. Similarly, hornblende has 18 subspecies identified by a combination of properties such as chemical composition, optics, and x-ray powder patterns.

Since the advent of x-ray crystallography, many early zeolite names have been discredited and much of the confusion that existed in the naming of zeolite minerals has been removed. The current list of zeolite minerals includes 34 species as defined by their chemical composition, structure, and related physical properties. The basic framework structures of 30 zeolite minerals are known. Synthetic zeolites which are structurally and topologically related to the zeolite minerals faujasite and mordenite are manufactured for industrial use; but zeolite minerals have not yet found extensive application as commercial molecular sieve materials even though a few occur abundantly in sediments. These include the zeolites analcime, chabazite, clinoptilolite, erionite, mordenite, laumontite, phillipsite (Chapter 3).

It appears that many zeolites are metastable toward other more stable phases such as analcime or feldspars, and may transform during long periods of geological time. Although well crystallized, zeolite minerals vary considerably in chemical composition. Conflicting experimental results have been reported on the same mineral by various investigators. These are due, in part, to structural differences caused by the presence of different metal cations.

The synthetic zeolites are better suited for research and industrial applications because of their greater uniformity in composition and purity. This is particularly essential where a high degree of reproducibility is required in an industrial separation process and where minor impurities such as iron, commonly found in minerals, can effect a major difference in applications such as in heterogeneous catalysts.

Synthetic Zeolites

The characterization and identification of complex, synthetic aluminosilicates such as the zeolites, is hindered by the absence of a definitive system of chemical nomenclature. One might use a system similar to the IUPAC rules for naming complex compounds which would be based on the unit cell composition. The mineral analcime, with the unit cell contents of $Na_{16}(AlO_2)_{16}(SiO_2)_{32} \cdot 16\ H_2O$ would be named sodium 16-alumino-32-silicate-16-water under the IUPAC rules. Jadeite, $Na_4 Al_4 Si_8 O_{24}$ would be called sodium 4-alumino-8-silicate. Both minerals have the same anhydrous composition in terms of anhydrous oxide mole ratios, $Na_2O \cdot Al_2O_3 \cdot 4\ SiO_2$. Nepheline, $Na_8(AlO_2)_8(SiO_2)_8$, would be called sodium 8-alumino-8-silicate and synthetic zeolite type A, $Na_{12}(AlO_2)_{12}(SiO_2)_{12} \cdot 27\ H_2O$ would be named sodium 12-alumino-12-silicate-27 water. A nomenclature system of this type is unwieldy and requires detailed knowledge of the unit cell. The latter is not always available.

Various systems of naming the synthetic zeolites, and other synthetic aluminosilicates have been employed by different investigators. These systems include: (1) assigning to the synthetic species the name of the mineral relative, if one exists, based on a similarity in crystal structure as might be evidenced by similar x-ray powder patterns; (2) assigning to the synthetic zeolite species a code or letter designation. Several investigators have assigned different code letters to the same synthetic zeolite species and the same letter to different species. This has led to much confusion in the literature.*

The designation of a synthetic zeolite by an arbitrary code is no less subjective than the historical methods for naming minerals. The characteristic properties which are the basis for designating the synthetic zeolite as a separate species first must be determined carefully. It is unlikely that any proposed system for zeolite nomenclature will meet with complete approval by all concerned. However, the following practices are generally applicable (54).

1. The synthetic zeolite is designated by the letter(s) assigned to the zeolite by the original investigator (law of priority), for example, zeolite A, zeolite K-G, zeolite ZK-5, etc.
2. These letters designate the zeolite as synthesized. Thus zeolite A designates the synthetic zeolite, $Na_{12}[(AlO_2)_{12}(SiO_2)_{12}] \cdot 22\ H_2O$ as prepared in the Na_2O, Al_2O_3, SiO_2, H_2O system. This formula refers to typical compositions. We also have used the terms "type A," "type X," etc.
3. In some cases, various investigators have referred to a synthetic zeolite by the name of a related mineral, e.g., "synthetic analcime," "synthetic mordenite," etc. This approach is inadequate. The terms "mordenite-type," "analcime-type," etc., are preferable for indicating that the synthetic zeolite is related structurally to a mineral. In light of current knowledge of the effect of cation type and location, Si and Al distribution, and Si/Al ratio on the properties of the zeolite, a statement of identity is inaccurate; only similarity is established.
4. Unfortunately some confusion is unavoidable. Thus Na-B refers to a synthetic analcime-type zeolite and Na-D indicates a mordenite type. I have used the letter B to refer to the synthetic zeolite phases which Barrer and others have referred to as "P." Since the use of P for these phases has precedence in the published literature, it is preferred. The letter D has been used to refer to a chabazite-type of synthetic zeolite.

---

* A Zeolite Nomenclature Committee under the IUPAC Commission on Colloid and Surface Science headed by R. M. Barrer is currently at work on the problem of zeolite nomenclature.

In summary, where a letter has been employed to refer to more than one synthetic zeolite, additional letter(s) are necessary; in these examples, the symbol for the predominant alkali metal is used.

5. A problem arises if the synthetic zeolite contains tetrahedral atoms other than Al and Si – P, Ga, Ge. The symbol P is used as a prefix, as P-L, to indicate zeolite L which contains substantial phosphorus substitution in the framework; in this case, typical unit cell contents are: $K_{21} [(AlO_2)_{34}(SiO_2)_{25}(PO_2)_{13}] \cdot 42 H_2O$.

6. It may be necessary to indicate a particular composition in cases where the letter(s) refer to a synthetic zeolite which can vary in framework composition. This is accomplished by giving the Si/Al ratio, unit cell contents, etc.

7. We in Union Carbide Corporation have adopted the use of "N" to refer to a synthetic zeolite which is prepared from systems that contain alkylammonium bases. Thus "N-A" refers to a synthetic tetramethylammonium zeolite that has the type A framework.

8. When a different cationic form of a synthetic zeolite is prepared by ion exchange, it may be so referred to, i.e., calcium exchanged (zeolite) A abbreviated as $Ca^{ex}A$ or CaA. A hyphen between Ca and A, Ca-A, refers to a completely different zeolite. Thus:

$$Ca \text{ exchanged } A = Ca^{ex}A = CaA \neq Ca\text{-}A$$

Of course, these terms refer to an unspecified level of exchange. It is assumed that the major cation component is, in this example, Ca^{2+} (greater than 50% exchange). In most cases, it is necessary to specify the degree of exchange as percent of exchange equivalents, or in terms of the unit cell contents. Thus $Ca_2 Na_8 [(AlO_2)_{12}(SiO_2)_{12}] \cdot x H_2O$ is equal to 33% exchange.

9. Obviously, any alteration of the parent zeolite by ion exchange, dealumination, decationization, etc. must be specified carefully.

The characterization of a previously unknown mineral or synthetic material, such as a zeolite, requires structural, compositional, and physicochemical information to include (1) the basic crystal structure as evidenced by x-ray crystallography, (2) chemical composition, and (3) the chemical and physical properties which are characteristic of zeolites. These include stability, dehydration behavior, cation exchange behavior and adsorption behavior of gases and vapors. These properties reflect important structural differences that may not be adequately indicated by other means.

Once a new material is properly characterized as a zeolite by means of its properties, it can later be identified by means of its x-ray powder data and chemical composition.

Most synthetic zeolites are homocationic or dicationic; they have one or two different cation species when synthesized. The minerals, on the other hand, exhibit considerable compositional variation due to their genesis and subsequent alteration by cation exchange and/or recrystallization. The cation exchange behavior and concurrent compositional differences depend upon the history and the initial composition. Cation exchange and dehydration may cause large changes in the zeolite structures.

A tacit assumption exists that two related zeolites, which often exhibit property differences such as those just described and which have the same type of framework structure but different cation composition, may be interconverted by cation exchange. This is *not* a general rule. For example, magnesium which is present in many zeolite minerals is not readily exchangeable (55). Polyvalent ions may exchange partially, or not at all, with univalent ions in synthetic zeolites (Chapter 7). Univalent ions, such as K^+, in zeolites L or T do not readily exchange out.

Considerable importance must be given to the cation species and concentration in distinguishing a given zeolite because of the resultant property differences. One might argue that the well-known mineral microcline is not to be differentiated from albite (both of these are long recognized as separate mineral species), because by the process of cation exchange one can be converted into the other. However, because of the density of the structure of these materials this process will not occur under mild conditions but requires a high temperature situation (56). The chemical composition of the feldspar microcline, $KAlSi_3O_8$, is, with the exception of the different alkali ion, the same as that of the mineral albite, $NaAlSi_3O_8$.

There are several examples within the group of zeolite minerals where isostructural zeolites (zeolites having the same aluminosilicate framework arrangement) have been clearly identified and classified as separate species by mineralogists for many years. For example, phillipsite and harmotome have essentially the same aluminosilicate framework structure but differ primarily in their cation composition and Si/Al ratio and Si-Al ordering. Natrolite is a sodium zeolite, scolecite a calcium zeolite, and mesolite a sodium-calcium zeolite. Thomsonite and gonnardite, although they have the same structure, differ in chemical composition and Si/Al distribution. Another similar example is analcime and wairakite. Chabazite and herschelite differ in cation compositon and to some degree in Si/Al distribution (Table 1.6)

Thus there is sound scientific precedence for the differentiation of separate species of synthetic zeolites and related materials based on differences in composition, structure, and properties.

Table 1.6 Isostructural Zeolite Minerals[a]

1	Phillipsite $(K, Na)_{10} [(AlO_2)_{10}(SiO_2)_{22}] \cdot 20 H_2O$	Harmotome $Ba_2 [(AlO_2)_4(SiO_2)_{12}] \cdot 12 H_2O$
2	Natrolite $Na_{16} [(AlO_2)_{16}(SiO_2)_{24}] \cdot 16 H_2O$	Scolecite $Ca_8 [(AlO_2)_{16}(SiO_2)_{24}] \cdot 24 H_2O$
3	Thomsonite $Na_4 Ca_8 [(AlO_2)_{20}(SiO_2)_{20}] \cdot 24 H_2O$	Gonnardite $Na_4 Ca_2 [(AlO_2)_8(SiO_2)_{12}] \cdot 14 H_2O$
4	Chabazite $Ca_2 [(AlO_2)_4(SiO_2)_8] \cdot 13 H_2O$	Herschelite $Na_4 [(AlO_2)_4(SiO_2)_8] \cdot 12 H_2O$
5	Analcime $Na_{16} [(AlO_2)_{16}(SiO_2)_{32}] \cdot 16 H_2O$	Wairakite $Ca_8 [(AlO_2)_{16}(SiO_2)_{32}] \cdot 16 H_2O$

[a]Isostructural is defined as having the same topology of the aluminosilicate framework. Other structural features such as Si/Al ratio, Si-Al distribution and ordering, cation species and population, and cation locations may not necessarily be identical.

REFERENCES

1. D. W. Breck, Molecular Sieve Zeolites, *Advan. Chem. Ser., 101, 102,* American Chemical Society, Washington, D. C., 1971, p. 1.
2. S. C. Eastwood, R. D. Drew, and F. D. Hartzell, *Oil Gas J.,* **60**:152(1962).
3. K. M. Elliott and S. C. Eastwood, *Oil Gas J.,* **60**:142(1962).

4. E. J. Demmel, A. V. Perrella, W. A. Stover, and J. P. Shambaugh, *Oil Gas J.*, **64**:178 (1966).
5. R. M. Barrer, *J. Soc. Chem. Ind.*, **64**:130(1945).
6. R. M. Milton, Molecular Sieves, Society of Chemical Industry, London, 1968, p.199.
7. J. W. McBain, The Sorption of Gases and Vapors by Solids, Chap. 5, Rutledge and Sons, London, 1932.
8. R. B. Anderson, W. K. Hall, J. A. Lecky, and K. C. Stein, *J. Phys. Chem.*, **60**:1548 (1956).
9. J. E. Metcalfe, III, M. Kawahata, and P. L. Walker, Jr., *Fuel*, **42**:233(1963).
10. R. F. Horlock and P. J. Anderson, *Trans. Faraday Soc.*, **63**: 717 (1967).
11. L. S. Yastrebova, A. A. Bessonov, and S. S. Khvoshchev, *Materialy Vses. Soveshch. Tseolitam*, **2**: 229, Leningrad, 1965.
12. D. Dollimore and G. R. Heal, *Trans. Faraday Soc.*, **59**:2386 (1963).
13. J. J. Kipling and R. B. Wilson, *Trans. Faraday Soc.*, **56**:562(1960).
14. M.M. Dubinin, O. Kadlec, and A. Zukal, *Nature*, **207**:75(1965).
15. T. G. Lamond, J. E. Metcalfe, III, and P. L. Walker, Jr., *Carbon*, **3**;59(1965).
16. K. Watanabe, T. Kondow, M. Soma, T. Onishi, and K. Tamaru, *Chem. Lett.*, **1972**:477.
17. A. Cronstedt, *Akad. Handl. Stockholm*, **18**:120(1756); *Mineral.*, Stockholm, 102 (1758).
18. G. P. L. Walker, *Mineral. Mag.*, **33**:173(1962).
19. H. S. Thompson, *J. Roy. Agr. Soc. Engl.*, **11**:68(1850).
20. J. T. Way, *J. Roy. Agr. Soc. Engl.*, **11**:313(1850).
21. H. Eichhorn, *Poggendorf Ann. Phys. Chem.*, **105**:126(1858).
22. R. N. Shreve, "Greensand Bibliography to 1930," *U. S. Bur. Mines, Bull*, **328**(1930).
23. R. Gans, *Ger. Pat. 211,118* (1908); *Jahrb. Preuss. Geol. Landenstalt*, **26**:179(1905).
24. B. R. Boehringer and A. E. Gessler, *U. S. Pat. 1,050,204* (1913).
25. P. DeBrunn, *Brit. Pat. 15,090* (1913).
26. Anon., *Chem. Met. Eng.*, **27**:1211(1922).
27. W. M. Bruce, *U. S. Pat. 1,906,202* (1933).
28. R. E. Grimm, Clay Mineralogy, McGraw-Hill Book Co., New York, 1953, p.65.
29. S. J. Kocsor, *Ind. Eng. Chem.*, **7**:259(1915).
30. P. Siedler, *Z. Angew. Chem.*, **22**:1019(1909).
31. C. H. Giles, *J. Chem. Educ.*, **39**:584(1962).
32. A. Damour, *Ann. Mines*, **17**:191(1840).
33. G. Friedel, *Bull. Soc. Fr. Mineral. Cristallogr.*, **19**:14,96(1896).
34. F. Grandjean, *Compt. Rendu*, **149**:866(1909).
35. W. Eitel, Physical Chemistry of Silicates, Univ. of Chicago Press, Chicago, 1954, p. 994.
36. O. Weigel and E. Steinhoff, *Z. Kristallogr.*, **61**:125(1925).
37. R. M. Barrer, *Proc. Roy. Soc.*, **A167**:393(1938).
38. R. M. Barrer and D. A. Ibbitson, *Trans. Faraday Soc.*, **40**:206(1944).
39. R. M. Barrer, *Quart. Rev.*, **3**:293(1949).
40. E. Rabinowitsch and W. C. Wood, *Trans. Faraday Soc.*, **32**:947(1936).
41. R. M. Barrer and D. A. Ibbitson, *Trans. Faraday Soc.*, **40**:195(1944).
42. W. H. Taylor, *Z. Kristallogr.*, **74**:1(1930).
43. L. Pauling, *Proc. Nat. Acad. Sci.*, **16**:453(1930); *Z. Kristallogr.*, **74**:213(1930).
44. W. H. Taylor, C. A. Meek, and W. W. Jackson, *Z. Kristallogr.*, **84**:373(1933).
45. J. Wyart, *Bull. Soc. Fr. Mineral. Cristallogr.*, **56**:81(1933).
46. L. S. Dent and J. V. Smith, *Nature*, **181**:1794(1958).
47. T. B. Reed and D. W. Breck, *J. Amer. Chem. Soc.*, **78**:5972(1956).
48. W. L. Bragg, Atomic Structure of Minerals, Cornell University Press, Ithaca, 1937, p. 251.

REFERENCES

49. R. M. Barrer, *Ann. Rep. Prog. Chem.*, **41**:31(1944).
50. G. Bergerhoff, H. Koyama, and W. Nowacki,*Experientia*, **12**:418(1956).
51. E. S. Dana, System of Mineralogy, 6th Ed., John Wiley & Sons, New York, 1942.
52. A. N. Winchell, *Amer. Mineral.*, **34**:220(1949).
53. R.B. Sosman, The Phases of Silica, Rutgers University Press, New Brunswick,1965.
54. D. W. Breck, Molecular Sieve Zeolites, *Advan. Chem. Ser., 101,* American Chemical Society, Washington, D. C., 1971.
55. P. A. Vaughan, *Acta Crystallogr.*, **21**:983(1966).
56. J. R. O'Neil and H. P. Taylor, Jr., *Amer. Mineral.*, **52**:1414(1967).
57. J. E. Metcalfe, III, Ph.D. Thesis, Pennsylvania State University, 1965.

Chapter Two

STRUCTURE OF ZEOLITES

A. INTRODUCTION

Crystalline zeolite molecular sieves are complex materials, chemically and structurally, comprising the major group of the framework silicates (1). As a result of structural studies during the last two decades. there is extensive information available on the structure of zeolites; it is detailed in some instances and cursory in others. Many of the properties of zeolites can now be interpreted on a structural basis. Thus the zeolites can be classified by framework-structure type.

The major source of structural information is x-ray crystal structure analysis. In addition, important information has been obtained from infrared absorption (ir), nuclear magnetic resonance (nmr), and electron spin resonance (esr). (See Chapter 5.)

Zeolites are subject to unusual types of defects. For the most part, defect structures will be discussed in Chapter 6; this chapter is concerned with the more classical aspects of zeolite structures.

B. REVIEW OF SILICATE STRUCTURES

General

Current concepts of silicate structures are based on the principles governing the structures of complex ionic crystals as developed by Pauling (2). The fundamental unit is a tetrahedral complex consisting of a small cation, such as Si^{4+}, in tetrahedral coordination with four oxygens (Pauling's first rule) (Fig. 2.1). The Al^{3+} ion commonly coordinates tetrahedrally as well as octahedrally with oxygen in silicates. This has a profound effect on aluminosilicate structures and their composition. Other ions in tetrahedral coordination include P^{5+}, Ga^{2+}, and Ge^{4+}. The complexity of silicate structures occurs because of the

30 STRUCTURE OF ZEOLITES

Figure 2.1 The tetrahedron of oxygen coordinated with silicon. In the isolated complex ion the overall charge is 4^-. Occupancy of the tetrahedral site by Al^{3+} would produce an overall charge of 5^-. O–O = $a\sqrt{2}$ = 2.62 Å; Si–O = $a\sqrt{3/2}$ = 1.61 Å; O–Si–O angle = 109° 28'.

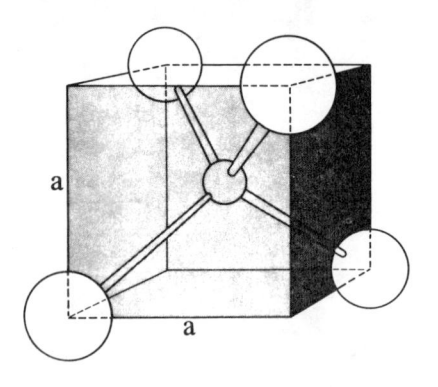

various ways in which the tetrahedral groups may link by the common sharing of oxygen ions to form polynuclear complexes. Considerable variations in chemical composition results from the substitution of cations, such as those listed in Table 2.1, in tetrahedral and octahedral sites as permitted by Pauling's electrovalence rules.

The substitution of aluminum for a silicon produces a deficiency in electrical charge that must be locally neutralized by the presence of an additional positive ion (generally one of the alkali metals, R^+, or alkaline earths, R^{2+}) within the interstices of the structure. Differ-

Table 2.1 Coordination of Cations with Oxygen in Silicate Structures [a]

Ion	Radius (Å)	Radius Ratio	Coordination Number [b]	Bond Strengths
B^{3+}	0.20	0.20	3, 4	1 or 3/4
Be^{2+}	0.31	0.25	4	1/2
Li^+	0.60	0.34	4	1/4
Si^{4+}	0.41	0.37	<u>4</u>,6	1
P^{5+}	0.34	0.34	4,6	5/4 or 5/6
Al^{3+}	0.50	0.41	<u>4</u>,5,<u>6</u>	3/4 or 1/2
Ge^{4+}	0.44	0.43	4,6	1 or 2/3
Ga^{3+}	0.62	0.46	4,6	3/4 or 1/2
Mg^{2+}	0.65	0.47	6	1/3
Na^+	0.95	0.54	<u>6</u>,8	1/6 or 1/8
Ca^{2+}	0.99	0.67	7,<u>8</u>,9	1/4
K^+	1.33	0.75	6,7,8,<u>9</u>,10,12	1/9
NH_4^+	1.50	0.85	<u>9</u>,12	1/9 or 1/12

[a] Ref. 1
[b] An underscore indicates common coordination.

ent types of aluminosilicates result from differences in the way in which the tetrahedra may link in space in one, two, or three dimensions and from the types of other ions that substitute within the interstices.

In some structures, the tetrahedra link to form infinite chains which result in fibrous needlelike crystals. In other structures, the tetrahedra are linked in layers or sheets as in mica minerals. Similar arrangements are found in the clay minerals wherein two types of sheets may exist, one consisting of aluminum, iron, or magnesium ions in a six-fold coordination with oxygens. These layer or sheet structures do not have three-dimensional stability and may expand if the layers are forced apart by water, other molecules, or ions.

If SiO_4 or AlO_4 tetrahedra are linked in three dimensions by a mutual sharing of oxygen alone, a framework structure results.

Silicate Structure Classes

A better understanding of zeolite structural chemistry may be achieved by a brief review of the five main types of silicates and aluminosilicates (1). These include island structures, isolated group structures, chain structures, sheet structures, and framework structures.

Four kinds of models are used to illustrate silicate crystal structures. These are:

1. Solids—Some regular solids such as the tetrahedron and octahedron are used to represent MO_4 and MO_6 groupings (Fig 2.2b); Other solids include the cube, hexagonal prism, and Archimedean semi-regular solids such as the truncated octahedron and cuboctahedron (see Fig. 2.18b). Some of the geometrical characteristics of these solids are given in Table 2.2.
2. Framework models— The MO_4 tetrahedron can be portrayed by four wires corresponding to M—O linkages; the points where they are linked represent the positions of oxygen atoms (Fig. 2.2c). A modification of this was proposed by Meier (3). We will use this method to illustrate the framework structures of aluminosilicates and zeolites.
3. Space filling models—These models are drawn or constructed to represent oxygen atoms in the structure. This approach gives a

32 STRUCTURE OF ZEOLITES

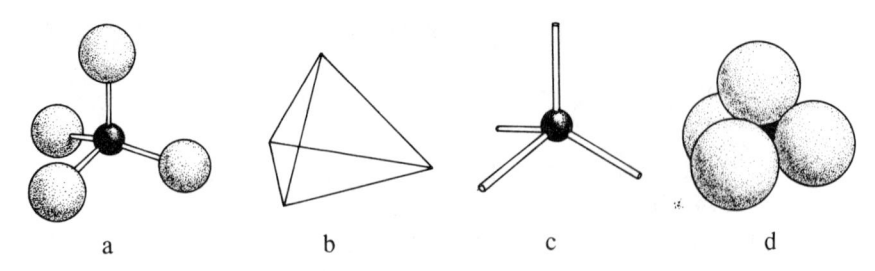

Figure 2.2 The methods of representing the tetrahedral coordination of oxygen ions with aluminum and silicon by means of (b) solid tetrahedron (c) skeletal tetrahedron, (d) a space-filling model based on packed spheres, and (a) ball and stick model.

more realistic view of the true structure, but the models are much more difficult to construct. Figure 2.2d illustrates a space-filling model of oxygen atoms in a tetrahedron.

4. Ball and stick models—These models represent spatial arrangements of atoms in crystals (Fig. 2.2a).

Island Structures

The SiO_4^{4-} tetrahedra exist as discrete ions in silicates such as sodium calcium orthosilicate, Na_2CaSiO_4, and in $Na_2H_2SiO_4 \cdot 8 H_2O$ (4) (Fig. 2.3). The four oxygen atoms around a silicon are not linked to any other silicon atoms. The structures are described as an assembly of SiO_4^{4-} groups with sodium and calcium ions between them in octahedral sites. In $Na_2H_2SiO_2 \cdot 8 H_2O$, two hydrogens are probably attached to each SiO_4^{4-} ion.

Table 2.2 Some Polyhedra in Structures[a]

Polyhedron	Type of Face, n-gon: 3 4 5 6 8	Vertices	Edges	Edge/Cube edge	Figure
Tetrahedron	4 − − − −	4	6	−	2.2
Cube	− 6 − − −	8	12	−	2.18a
Octahedron	8 − − − −	6	12	−	−
Dodecahedron	− − 12 − −	20	30	−	−
Cuboctahedron	8 6 − − −	10	24	0.707	−
Truncated octahedron	− 6 − 8 −	24	36	0.354	2.18b
Truncated cuboctahedron	− 12 − 8 6	48	72	0.261	2.18b

[a] A detailed atlas of polyhedra is given by H. M. Cundy and A. P. Rollett, *Mathematical Models*, 2nd Ed., Oxford University Press, 1960. See also A. F. Wells, *The Third Dimension in Chemistry*, Oxford University Press, 1956.

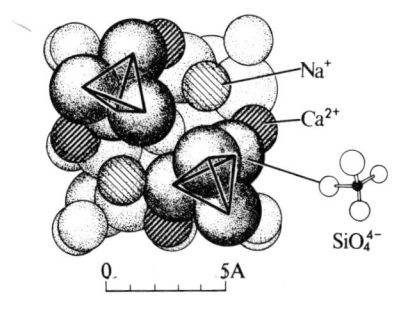

Figure 2.3 A packing drawing of the structure of Na$_2$CaSiO$_4$ (5). The location of Na$^+$ and Ca^{2+} ions between the individual SiO$_4^{4-}$ ions is indicated.

Isolated Group Structures or Complexes

By the mutual sharing of oxygens at the tetrahedron corners, two or more SiO$_4$ groups may link to form polyanionic groups (Fig. 2.4). If two oxygens are shared mutually between adjacent tetrahedra the stability of the structure is considerably lowered due to cation coulombic repulsion. This is the basis for Pauling's third rule. The sharing of two oxygens, edge sharing, brings the cation centers closer together by a factor of 0.58 when compared to that which occurs in corner sharing only. Edge sharing is present in stishovite, a recently discovered high pressure form of SiO$_2$ (7). Figures 2.5 and 2.6 illustrate structures based on the Si$_3$O$_9$ and Si$_6$O$_{16}$ group. A group of five, linked tetrahedra occurs in zunyite (8). In these structures,

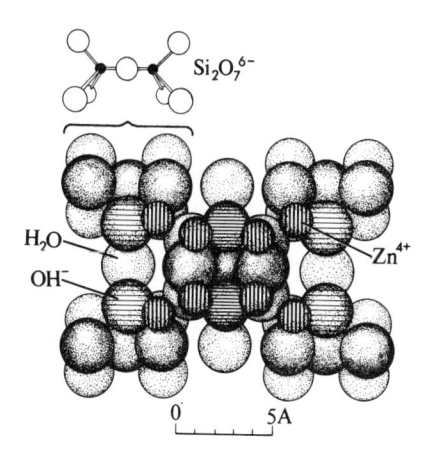

Figure 2.4 Two tetrahedra link to form the (Si$_2$O$_7$)$^{6-}$ ion which is a structural unit in hemimorphite. The zinc atoms occupy sites in tetrahedral coordination with 3 oxygens and a hydroxyl group (6).

34 STRUCTURE OF ZEOLITES

Figure 2.5 In benitoite, BaTiSi$_3$O$_9$, three tetrahedra form a ring structure. The Ba^{2+} and the Ti^{4+} are 6-coordinated with oxygen (9).

Figure 2.6 The structure of beryl, Be$_3$Al$_2$Si$_6$O$_{18}$, is based on Si$_6$O$_{18}^{12}$ units; six SiO$_4$ tetrahedra linked in a hexagonal ring. These ring units are joined laterally and vertically by Be^{2+} and Al^{3+} ions in tetrahedral and octahedral coordination, respectively. Parallel, continuous channels, formed by the 6-rings, pass through the beryl structure and are frequently occupied by water molecules and other ions such as Cs$^+$ (10).

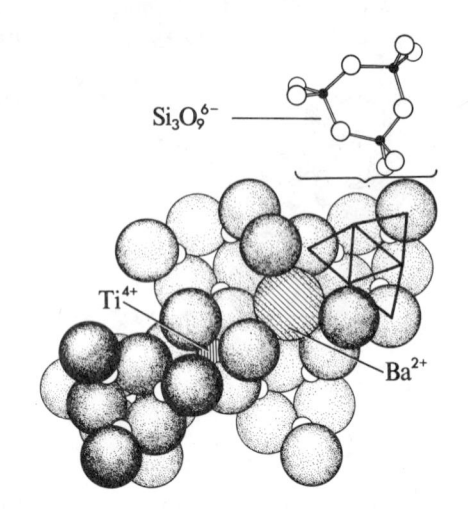

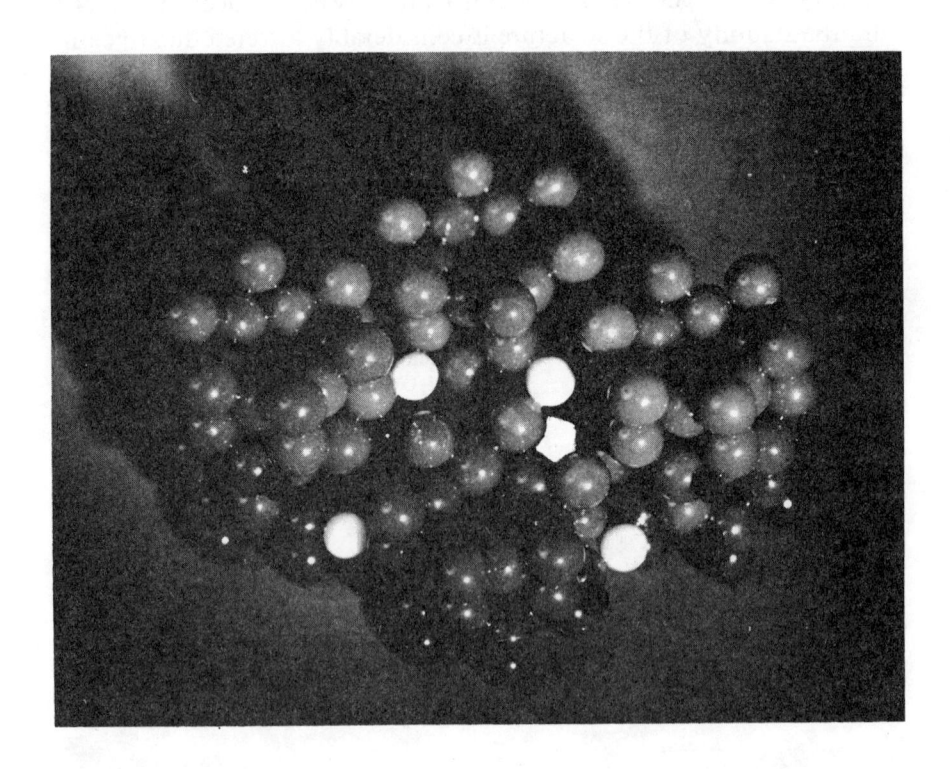

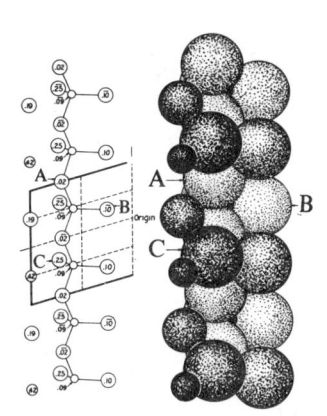

Figure 2.7 Packing drawing of the silicate chain as found in jadeite (5). NaAl(SiO$_3$)$_2$ showing Al^{3+} ions (small spheres) and Na$^+$ ions (intermediate size spheres).

shared oxygen atoms do not lie near metal cations.

Chain Structures

Single chains of (SiO$_3$)$_n$ result when SiO$_4$ tetrahedra are joined at one oxygen. The pyroxenes such as enstatite (MgSiO$_3$) and jadeite [NaAl(SiO$_3$)$_2$] are characterized by this unit (11). The structure of jadeite provides for sixfold coordination of the Al^{3+} ions as shown in Fig. 2.7. Two oxygens of each tetrahedron are shared and two are active. The double chains of the amphibole minerals are represented by the (Si$_4$O$_{11}$)$^{6-}$ unit as found in tremolite. In tremolite, each magnesium ion is surrounded by six oxygens, Ca$_2$Mg$_5$Si$_8$O$_{22}$(OH)$_2$. (See Fig. 2.8.) The unit cell, which consists of two of the formula units, contains four OH groups, each of which is attached to three magnesium ions to give an electrostatic bond strength of one. There are many compositional variants due to isomorphous substitution. Soda tremolite, Na$_2$CaMg$_5$Si$_8$O$_{22}$(OH)$_2$, results from the substitution of two sodium ions for one calcium ion; in hornblende, aluminum ions may replace silicon ions up to Al$_2$Si$_6$O$_{22}$ in the tetrahedral positions (12). Other types of double chain structures have been reviewed (13).

Sheet Structures

By linking three corners of each tetrahedron to neighboring tetrahedra, structure types of a large group of silicates are found as a two-dimensional planar network of tetrahedra linked to form infinite anions of composition (Si$_2$O$_5$)$^{2-}_\infty$. The largest family of layer structures is built of one or two silicon-oxygen sheets, illustrated in Fig.

36 STRUCTURE OF ZEOLITES

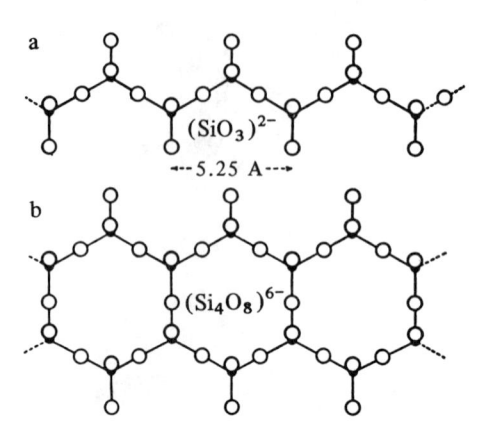

Figure 2.8 The arrangement of tetrahedra in chains as found in (a) jadeite and (b) tremolite (1).

2.9, combined with layers of Mg^{2+} or Al^{3+} ions and OH^- groups. The vertices of all the tetrahedra in the sheets point in the same direction.

In the mineral chrysotile, the Si_2O_5 sheets are joined with a $Mg(OH)_2$ octahedral layer to form the composite structure (Fig. 2.10) of idealized composition $Mg_3Si_2O_5(OH)_4$. Because of a slight misfit between the Mg-OH layer and the Si_2O_5 layer, the composite rolls up and tubular morphology results (14). The clay mineral kaolinite, $Al_2Si_2O_5(OH)_4$, has the same structure with Al^{3+} ions in the octahedral sites (Fig. 2.11).

The other common type of composite layer structure is produced

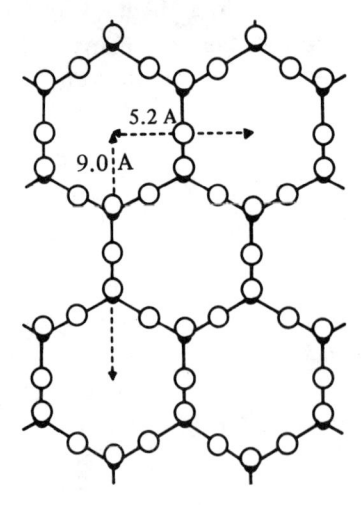

Figure 2.9 The hexagonal arrangement of SiO_4 tetrahedra in an infinitely extending silicon-oxygen sheet of composition Si_2O_5 (1).

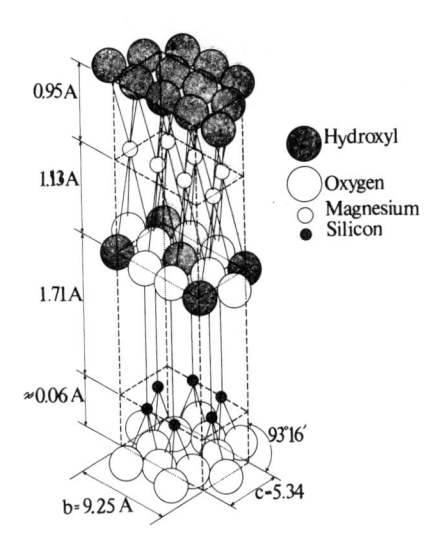

Figure 2.10 An expanded view of the structure of chrysotile (14).

by interposing a sheet of octahedral Al(OH)$_3$ or Mg(OH)$_2$ between two Si$_2$O$_5$ sheets. The magnesium composition is given by talc, Mg$_3$Si$_4$O$_{10}$(OH)$_2$ and the aluminum composition by pyrophyllite, Al$_2$Si$_4$O$_{10}$(OH)$_2$. Charged layers result when Al^{3+} substitutes for Si^{4+} in the tetrahedral sheets.

The clay mineral montmorillonite results when some of the Al^{3+} ions in the octahedral sites in pyrophyllite are replaced by Mg^{2+} to yield a charged layer. Hydrated cations occupy space between the layers. A typical composition is Na$_{0.66}$[Al$_{3.34}$Mg$_{0.66}$Si$_8$O$_{20}$](OH)$_4$ (see Fig. 2.12)

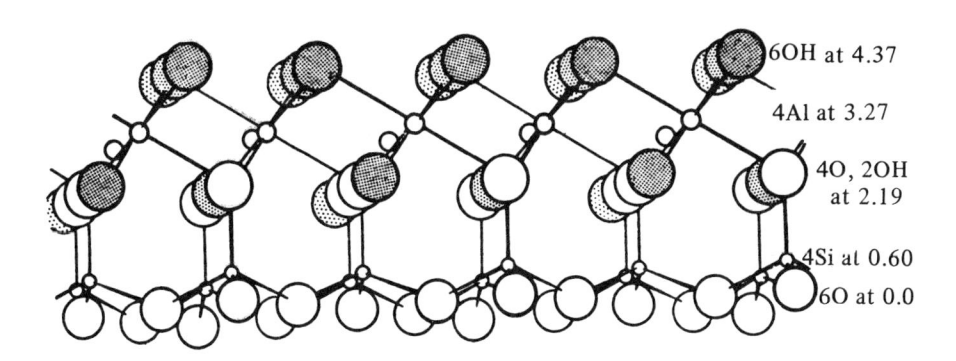

Figure 2.11 A view of the layer structure of kaolinite showing the arrangement of atoms and hydroxyl groups in a single layer (15).

38 STRUCTURE OF ZEOLITES

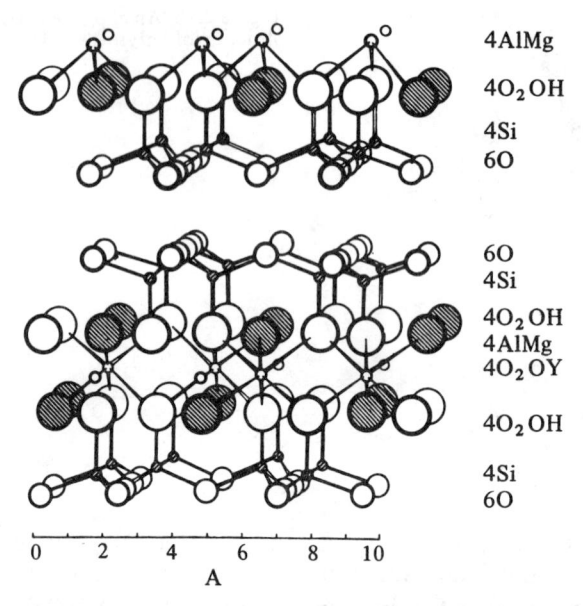

```
4AlMg
4O₂OH
4Si
6O

6O
4Si
4O₂OH
4AlMg
4O₂OY
4O₂OH
4Si
6O
```

Figure 2.12 The structure of montmorillonite. Under ordinary conditions, one molecular water layer gives a spacing of 12.5 A. When dehydrated, it shrinks to 9.6 A (15).

(15). In addition to cations, such as Na⁺, between the layers, H₂O molecules are present. The c-axis spacing varies with the degree of hydration. The effect is reversible. Other organic molecules can also enter between the layers to produce greater expansion depending on their size. The cations between the layers are exchangeable and may be replaced by many different types including large organic cations (16), which can result in a free distance between the layers of as much as 5.6 A. More complex chemical compositions result if Al^{3+} also substitutes for Si^{4+} in the tetrahedral layer as in the mineral saponite:

$$Na_{0.66}(Mg_{5.34}Al_{0.66})(Al_{1.34}Si_{6.66}O_{20})(OH)_4 \cdot n\ H_2O$$
$$\text{interlayer}\quad \text{octahedral}\quad \text{tetrahedral}\quad \text{interlayer}$$

Many other variants are possible through substitution and cation exchange.

Framework Structures

Three-dimensional, continuous framework structures result when all of the oxygen atoms in the tetrahedra are mutually shared between tetrahedral silicons or aluminums.

Silica Structures, SiO_2. There are 17 crystalline varieties of SiO_2. The crystalline polymorphs of SiO_2 vary in the arrangement of the SiO_4 tetrahedra (17). The arrangement of tetrahedra in the structures of hexagonal β-cristobalite is shown in Stereo 2.1. In coesite, a high-pressure form of SiO_2 with a density of 3.01 g/cc, the SiO_4 tetrahedra are very densely packed.

When Al^{3+} is tetrahedrally substituted for Si^{4+} in a three-dimensional framework of tetrahedra, the polyanionic framework has the composition $[(Al_x Si_{1-x})O_2]^{x-}$. Positive ions must be present to balance the electrostatic charge. This can be represented by:

$$(SiO_2)_n \xrightarrow{xNaAl} Na_x(AlO_2)_x(SiO_2)_{n-x}$$

In framework aluminosilicates the ratio $O/(Al + Si)$ is 2. The three main groups of framework structures are feldspars, feldspathoids, and zeolites.

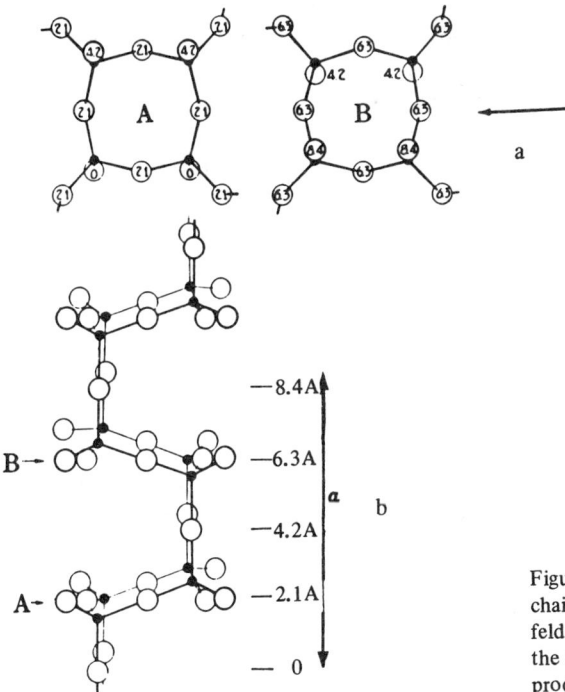

Figure 2.13 The double-crankshaft chain of linked tetrahedra in the feldspar structure. (a) Plain view of the rings A and B which, when joined, produce a chain as shown in (b) (1).

40 STRUCTURE OF ZEOLITES

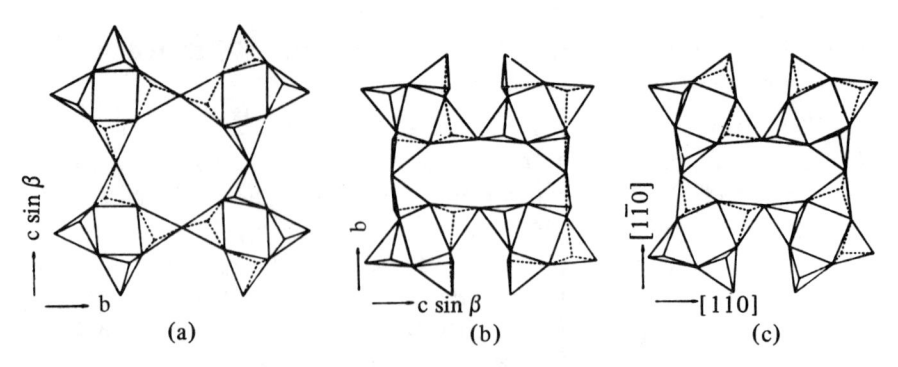

Figure 2.14 Solid tetrahedra projections illustrating (a) harmotome (as viewed down the UUDD chains); (b) the structure of feldspar; and (c) paracelsian (18).

Feldspar Structure. The detailed structures of the feldspars differ, but the basic framework is topologically the same. The feldspar structure consists of crosslinked, "double-crankshaft" chains of (Al, SiO$_4$) tetrahedra. Each chain is composed of parallel rings of four tetrahedra, two of which point one way and two the other way (Fig. 2.13). A shorthand notation for this is UUDD. The sharing of oxygen atoms of opposing tetrahedra produces chains; crosslinking produces the aluminosilicate framing containing parallel 8-rings. Possible framework structures formed from parallel 4-rings have been discussed by Smith and Rinaldk (18). There are two basic ways in which the chains can crosslink: One way results in a flexible framework allowing internal cavities to be large or small; the other type is inflexible and results in small cavities as found in feldspars (19) (Fig. 2.14). Cations which occupy cavities in the framework include Na$^+$, as in albite, NaAlSi$_3$O$_8$, Ca^{2+} in anorthite, CaAl$_2$Si$_2$O$_8$, and K$^+$ in ortho-

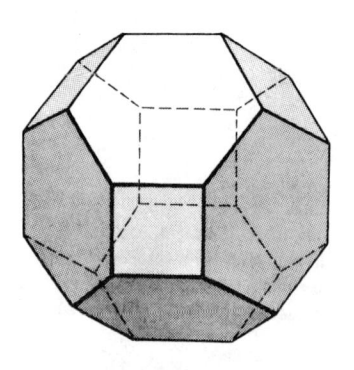

Figure 2.15 The truncated octahedron. Euler's theorem for polyhedra states that $F + V = E + 2$ where F is the number of faces, V the number of vertices, and E the number of edges. There are 36 edges, 24 vertices, 6 square faces, and 8 hexagonal faces (Table 2.1). A tetrahedral atom (Al, Si) is located at each vertex. Oxygen atoms are located between the tetrahedral atoms but not necessarily on the edge. The edges are not meant to portray bonds, but merely the geometry of the polyhedron. In sodalite, this polyhedral arrangement is distorted so that the exact arrangement of oxygen atoms is not regular. In sodalite-type structure, these polyhedra are close-packed to fill space so that each face is shared between two polyhedra. The central cage, the β-cage, contains the anions.

clase, KAlSi$_3$O$_8$. The K$^+$ ions are irregularly coordinated with nine oxygens and the Na$^+$ ions in albite with six. Certain zeolite structures are based on combinations of 4-rings in the flexible framework.

The Feldspathoids. The feldspathoids are more open in structure than the feldspars, and due to the presence of larger cavities, contain additional cations and in some instances anions within the channels. Water is present in some synthetic types; in these cases, the materials may fall within our definition of a zeolite since they may have ion exchange properties and limited adsorption for water water vapor.

Typical feldspathoids include sodalite, nepheline, scapolite, cancrinite, and synthetic types (Table 2.3).

The cubic feldspathoids (sodalite, nosean, ultramarine) all have essentially the same type of aluminosilicate framework which was first determined for ultramarine (20). The framework can be described in terms of either 4-rings or 6-rings (Al$_2$Si$_2$O$_8$ or Al$_3$Si$_3$O$_{12}$). The positions of the tetrahedral atoms are those of the vertices of a regular truncated octahedron (Fig. 2.15 and Table 2.2). Anions (Cl$^-$, SO$_4$$^{2-}$) occupy positions in the center of the "cage" (21). In sodalite, the Cl$^-$ ions are tetrahedrally coordinated to Na$^+$ ions which lie on the 3-fold axes. A synthetic analog of sodalite (see Chapter 4) contains additional Na$^+$ and OH$^-$ ions and water molecules; this is referred to as "hydroxy" or "basic" sodalite. The sodalite framework structure provides for intersecting channels which run parallel to the [111] direction and are restricted by the 6-rings (Stereo 2.2 and Fig. 2.16).

The cubic, tetracalcium trialuminate, 4 CaO · 3Al$_2$O$_3$ is reported

Table 2.3 Feldspathoids

Mineral	Unit Cell Contents	Crystallographic Data
Sodalite	Na$_8$[(AlO$_2$)$_6$(SiO$_2$)$_6$]Cl$_2$	Cubic, a = 8.87 A
Ultramarine	Na$_8$[(AlO$_2$)$_6$(SiO$_2$)$_6$]S$_4$	Cubic, a = 9.16
Nosean	Na$_8$[(AlO$_2$)$_6$(SiO$_2$)$_6$]SO$_4$	Cubic, a = 9.03 A
Nepheline	Na$_6$K$_2$[(AlO$_2$)$_8$(SiO$_2$)$_8$]	Hexagonal, a = 10.0 A, c = 8.4 A
Cancrinite	(Na, Ca, K)$_{6-8}$[(AlO$_2$)$_6$(SiO$_2$)$_6$](CO$_3$, SO$_4$, Cl)$_{1-2}$	Hexagonal, a = 12.6 A, c = 5.18 A
Scapolite (marialite)	Na$_8$[(AlO$_2$)$_6$(SiO$_2$)$_{18}$]Cl$_2$	Tetragonal, a = 12.06 A, c = 7.57 A

42 STRUCTURE OF ZEOLITES

(a)

(b)

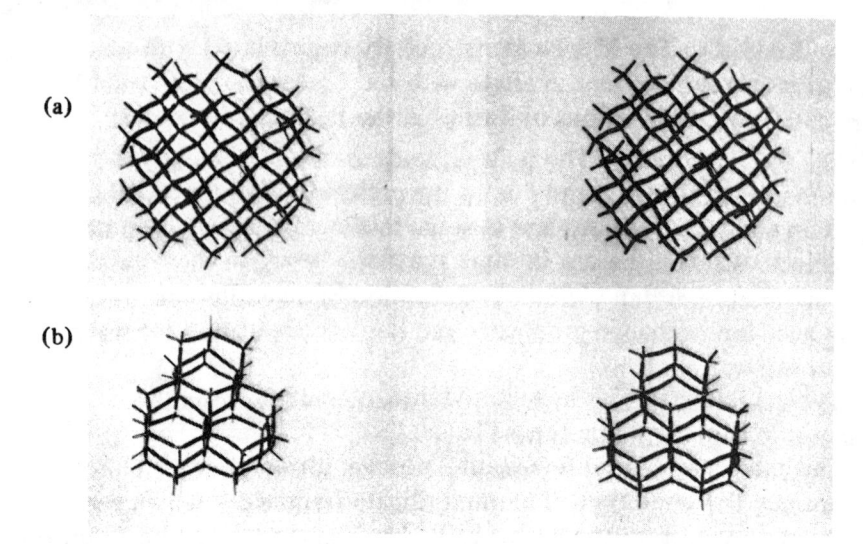

Stereo 2.1 Stereoscopic view of the framework of the structure of (a) β-cristobalite and (b) β-tridymite. Compared to quartz, density 2.655 g/cc, tridymite and cristobalite are relatively open structures with densities of 2.30 and 2.27.

Stereo 2.2 Stereoscopic view of the aluminosilicate framework of the sodalite-type feldspathoids viewed parallel to the cubic axis. The individual truncated octahedra are evident and are arranged in 8-fold coordination so that each 6-ring is shared between two cages.

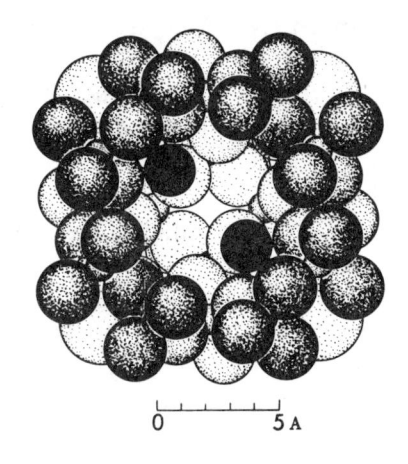

Figure 2.16 A packing drawing of sodalite viewed along the cubic axis. The Na$^+$ ions are black and the Al^{3+} ions are small and dotted, nearly concealed in the tetrahedra centers. The central large anion is omitted from this drawing. The free diameter of the central cavity is 6.2 A. When fully hydrated, the synthetic hydroxysodalite contains about 4H$_2$O and 4 Na$^+$ which are located within the cages on the 3-fold axis (x = 0.140 and 2 A from the plane of 6-rings (5).

to be the aluminate analog of sodalite (22). The structure has the sodalite framework and a unit cell composition of Ca$_8$ [(AlO$_2$)$_{12}$]O$_2$. Single oxygen atoms occupy positions in the center of the truncated octahedral units like chlorine in sodalite. If true, this structure seriously violates the stability rule that aluminum atoms do not share the same oxygen when in tetrahedral coordination (23).

A synthetic type of sodalite containing tetramethyl ammonium (TMA) ions in the center of the cage has been reported. Due to space limitation, one TMA occupies each central site so that the framework is necessarily rich in Si in order to maintain charge balance. Thus the unit cell has the composition (24): [(CH$_3$)$_4$N]$_2$ [(AlO$_2$)$_2$(SiO$_2$)$_{10}$].

The framework structure of cancrinite is based upon another type of cage-like unit consisting of 18 tetrahedra (25,26). Synthetic analogs of cancrinite containing additional OH$^-$ (as NaOH) and water have been prepared and show ion exchange properties (27). Cations are present in both the small and large channels and anions (CO$_3$$^{2-}$, OH$^-$) are found in the large channels (Stereo 2.3).

The sodalite structure may also be considered in terms of parallel 6-rings stacked in a way analogous to closest packed spheres (see Section D). In sodalite, the sequence is ABCABC... whereas the framework of cancrinite consists of parallel 6-rings in the ABAB... sequence. The internal free diameter of the cancrinite cage is about 5 A in cancrinite itself.

The nepheline structure is based on a tridymite-type framework (see Stereo 2.1) and contains 16 tetrahedra per unit cell (28). Synthe-

44 STRUCTURE OF ZEOLITES

tic hydrous types of nepheline of composition $[Na_8(AlO_2)_8(SiO_2)_8]\cdot$ 4-8 H_2O have been reported (27).

Tetrahedra with apices pointing in one direction along the c-axis are occupied by silicon ions; those which point in the opposite direction by aluminum ions. The negative charge is balanced by sodium and potassium ions. Actually the framework is distorted from the ideal, hexagonal, tridymite framework. Channels which run parallel to the c-axis are formed by distorted 6-rings.

The scapolites are generally represented by the minerals marialite, $Na_8[AlO_2)_6(SiO_2)_{18}]Cl_2$, and meionite, $Ca_4[AlO_2)_6(SiO_2)_6]CO_3$ (29). The tetragonal unit cell contains 24 tetrahedra (Fig. 2.17). Synthetic, hydrous scapolites with OH^- ions in the cavities have not been reported in the literature.

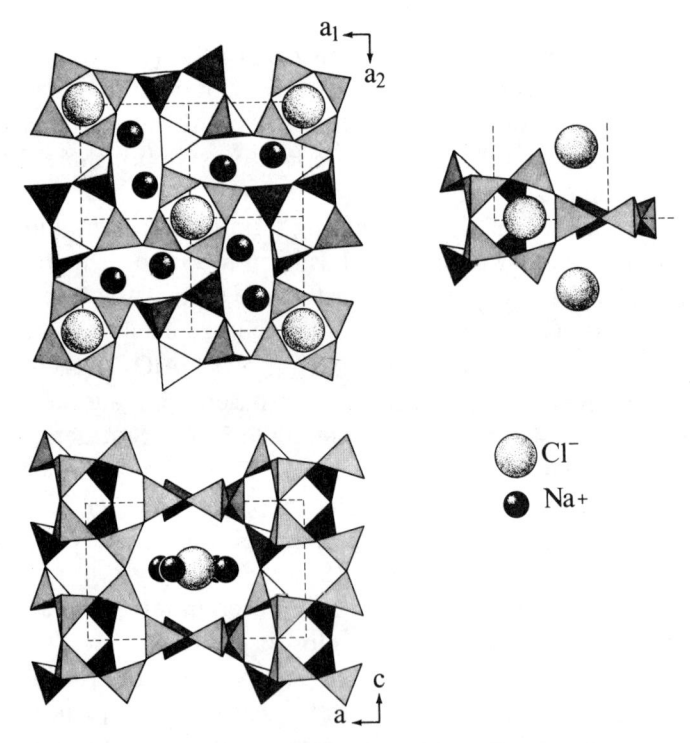

Figure 2.17 The structure of scapolite (marialite) (25). The structure is represented by solid tetrahedra to show the location of tetrahedra in the framework. A spherical opening at the center is surrounded by four elliptical shaped channels which run parallel to the c-axis. The large central opening is occupied by the anion (such as Cl^-) and the vertical channel by the alkali ion or alkaline earth ion.

C. CLASSIFICATION OF ZEOLITE STRUCTURES

The zeolites comprise the largest group of aluminosilicates with framework structures since there are over 35 known, different framework topologies and an infinite number possible.

Early interpretations of the physical and chemical properties of zeolites were based upon fragmentary structural information. As a result of investigations during the last ten years, there is extensive information on the crystal structure of over 35 different zeolites; this is detailed in a few instances and cursory in others.[a] Nearly 100 synthetic types have been reported. Zeolites are classified into groups according to common features of the aluminosilicate framework structures. Important properties now have a structural interpretation. The properties which are structure-related include:

1. High degree of hydration and the behavior of "zeolitic" water.
2. Low density and large void volume when dehydrated.
3. Stability of the crystal structure of many zeolites when dehydrated and when as much as 50 vol% of the dehydrated crystals are void.
4. Cation exchange properties.
5. Uniform molecular-sized channels in the dehydrated crystals.
6. Various physical properties such as electrical conductivity.
7. Adsorption of gases and vapors.
8. Catalytic properties.

In order to understand and relate these properties, new concepts were needed concerning the spatial arrangement of the basic structural components, i.e., the tetrahedra, cations, and H_2O molecules.

Structural classifications of zeolites have been proposed by Smith (30) and Fischer and Meier (31,32) and Breck (33). Earlier classifications were based on morphological properties (see Chap. 1, Section G). The classification used in this book is based on the framework topology of the zeolites for which the structures are known. The classification consists of seven groups; within each group, zeolites have a common subunit of structure which is a specific array of $(Al, Si)O_4$ tetrahedra. In the classification, the Si-Al distribution is neglected. For example, the two simplest units are the ring of four tetrahedra (4-ring) and six tetrahedra (6-ring) as found in many other framework aluminosilicates. These subunits have been called *secondary building units* (SBU) by Meier (32). (The primary units are of course the SiO_4

---

[a] A later compilation of zeolite structures has been published by Meier and Olson. See ref. 222.

46 STRUCTURE OF ZEOLITES

and AlO_4 tetrahedra.) In some cases, the zeolite framework can be considered in terms of polyhedral units, such as the truncated octahedron. Some of the SBU are probably involved in crystal growth processes. The SBU's as proposed by Meier are shown in Fig. 2.18a; these are associated with characteristic configurations of tetrahedra. Some of the polyhedra which are involved in zeolite structures are shown in Fig. 2.18b. These are cagelike units designated by Greek letters: $\alpha, \beta,$

Figure 2.18 (a) The secondary building units (SBU) in zeolite structures according to Meier (32). Only the positions of tetrahedral (T) silicons and aluminums are shown. Oxygen atoms lie near the connecting solid lines, which are not intended to mean bonds. The 4-1 unit is based on the configuration of 5 tetrahedra present in the structures of group 5 (Fig. 2.44). The 5-1 unit is based on the configuration of 5-rings found in group 6 (Fig. 2.47). The 4-4-1 unit is based on the configuration of tetrahedra found in group 7 (Fig. 2.50).

(b) Some polyhedra in zeolite frameworks: α (26-hedron Type I) or truncated cuboctahedron; β (14-hedron Type I) (Table 2.1) or truncated octahedron; δ or double 8-ring; D6R or double 6-ring (hexagonal prism); γ or 18-hedron; and ϵ or the 11-hedron (34).

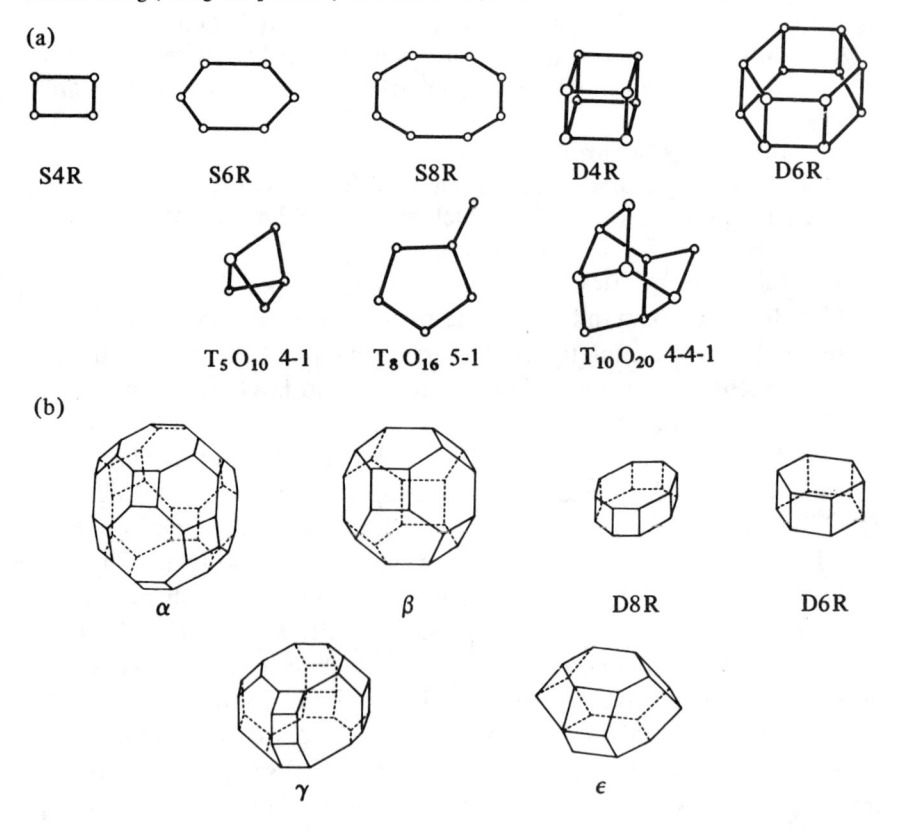

γ, etc. The α-cage refers to the largest unit—the truncated cuboctahedron.

The classification which follows is based on seven groups. Although, in other classifications, each group has been named after a representative member, an arbitrary designation by number is preferable since no single member is more representative than any other. The seven groups are as follows:

Group	Secondary Building Unit (SBU)
1	Single 4-ring, S4R
2	Single 6-ring, S6R
3	Double 4-ring, D4R
4	Double 6-ring, D6R
5	Complex 4-1, T_5O_{10} unit
6	Complex 5-1, T_8O_{16} unit
7	Complex 4-4-1, $T_{10}O_{20}$ unit

The zeolite classification in Table 2.4 lists the typical unit cell contents, the polyhedral cage type, the void space of the dehydrated zeolite based on the amount of water contained in the fully hydrated zeolite, the type of channel system, and the free aperture size of the main channels (see Sections E, F, and G) for each zeolite.

In many instances aperture sizes determined for hydrated zeolites are not consistent with the molecular sieve properties of the dehydrated crystals; this is due to distortion of the framework upon dehydration and the presence of cations in channels. Accurate data on the positions of cations and water molecules are available for some of the hydrated zeolites; positions of cations in some dehydrated zeolites are known. Upon dehydration some of the cations assume different positions in the structure.

Zeolites A, X, Y, and faujasite have frameworks consisting of linked truncated octahedra (β-cages) characteristic of the structure of sodalite. Some of the synthetic zeolites (see Chap. 4) are structurally related to minerals and are included in this classification; however, for most of the synthetic zeolites the structures are not known (Table 2.4b).

D. THEORETICAL ZEOLITE STRUCTURES

The framework structures of the zeolites are composed of assem-

48 STRUCTURE OF ZEOLITES

Table 2.4 Classification of Zeolites

Name	Typical Unit Cell Contents	Type of Polyhedral Cage [a]	Framework Density, g/cc [b]	Void Fraction [c]	Type of Channels [d]	Free Aperture of Main Channels, Å [e]
Group 1 (S4R)						
Analcime	$Na_{16}[(AlO_2)_{16}(SiO_2)_{32}] \cdot 16 H_2O$		1.85	0.18	One	2.6
Harmotome	$Ba_2[(AlO_2)_4(SiO_2)_{12}] \cdot 12 H_2O$		1.59	0.31	Three	4.2 × 4.4
Phillipsite	$(K, Na)_{10}[(AlO_2)_{10}(SiO_2)_{22}] \cdot 20 H_2O$		1.58	0.31	Three	4.2 × 4.4, 2.8 × 4.8
Gismondine	$Ca_4[(AlO_2)_8(SiO_2)_8] \cdot 16 H_2O$		1.52	0.46	Three	3.1 × 4.4
P	$Na_6[(AlO_2)_6(SiO_2)_{10}] \cdot 15 H_2O$		1.57	0.41	Three	3.5
Paulingite	$(K_2,Na,Ca,Ba)_{76}[(AlO_2)_{152}(SiO_2)_{520}] \cdot 700 H_2O$	α, γ, δ, (10-hedron)	1.54	0.49	Three	3.9
Laumontite	$Ca_4[(AlO_2)_8(SiO_2)_{16}] \cdot 16 H_2O$		1.77	0.34	One	4.6 × 6.3
Yugawaralite	$Ca_2[(AlO_2)_4(SiO_2)_{12}] \cdot 8 H_2O$		1.81	0.27	Two	3.6 × 2.8
Group 2 (S6R)						
Erionite [f]	$(Ca,Mg,K_2,Na_2)_{4.5}[(AlO_2)_9(SiO_2)_{27}] \cdot 27 H_2O$	ϵ, 23-hedron	1.51	0.35	Three	3.6 × 5.2
Offretite [f]	$(K_2,Ca)_{2.7}[(AlO_2)_{5.4}(SiO_2)_{12.6}] \cdot 15 H_2O$	ϵ, 14-hedron (II)	1.55	0.40	Three	3.6 × 5.2, ‖a 6.4, ‖c
T	$(Na_{1.2}K_{2.8}[(AlO_2)_4(SiO_2)_{14}] \cdot 14 H_2O$	ϵ, 23, 14-hedron	1.50	0.40	Three	3.6 × 4.8
Levynite [m]	$Ca_3[(AlO_2)_6(SiO_2)_{12}] \cdot 18 H_2O$	Ellipsoidal 17-hedron	1.54	0.40	Two	3.2 × 5.1
Omega [g]	$Na_{6.8}$, $TMA_{1.6}[(AlO_2)_8(SiO_2)_{28}] \cdot 21 H_2O$	14-hedron (II)	1.65	0.38	One	7.5
Sodalite Hydrate	$Na_6[(AlO_2)_6(SiO_2)_6] \cdot 7.5 H_2O$	β	1.72	0.35	Three	2.2
Losod	$Na_{12}[(AlO_2)_{12}(SiO_2)_{12}] \cdot 19 H_2O$	ϵ, 17-hedron	1.58	0.33	Three	2.2
Group 3 (D4R) [n]						
A	$Na_{12}[(AlO_2)_{12}(SiO_2)_{12}] \cdot 27 H_2O$	α, β	1.27	0.47	Three	4.2
N-A	$Na_4TMA_3[(AlO_2)_7(SiO_2)_{17}] \cdot 21 H_2O$	α, β	1.3	0.5	Three	4.2
ZK-4	$Na_8TMA[(AlO_2)_9(SiO_2)_{15}] \cdot 28 H_2O$	α, β	1.3	0.47	Three	4.2

Table 2.4 Classification of Zeolites *(continued)*

Name	Typical Unit Cell Contents	Type of Polyhedral Cage[a]	Framework Density, g/cc[b]	Void Fraction[c]	Type of Channels[d]	Free Aperture of Main Channels, Å[e]
Group 4 (D6R)						
Faujasite	$(Na_2, K_2, Ca, Mg)_{29.5}[(AlO_2)_{59}(SiO_2)_{133}] \cdot 235\ H_2O$	β, 26-hedron(II)	1.27	0.47	Three	7.4
X	$Na_{86}[(AlO_2)_{86}(SiO_2)_{106}] \cdot 264\ H_2O$	β, 26-hedron (II)	1.31	0.50	Three	7.4
Y	$Na_{56}[(AlO_2)_{56}(SiO_2)_{136}] \cdot 250\ H_2O$	β, 26-hedron (II)	1.25 - 1.29	0.48	Three	7.4
Chabazite	$Ca_2[(AlO_2)_4(SiO_2)_8] \cdot 13\ H_2O$	20-hedron	1.45	0.47	Three	3.7 × 4.2
Gmelinite	$Na_8[(AlO_2)_8(SiO_2)_{16}] \cdot 24\ H_2O$	14-hedron (II)	1.46	0.44	Three	3.6 x 3.9, ‖a 7.0, ‖c
ZK-5[O]	$(R,Na_2)_{15}[(AlO_2)_{30}(SiO_2)_{66}] \cdot 98\ H_2O$	α, γ	1.46	0.44	Three	3.9
L[h]	$K_9[(AlO_2)_9(SiO_2)_{27}] \cdot 22\ H_2O$	ε	1.61	0.32	One	7.1
Group 5 (T_5O_{10})[i]						
Natrolite	$Na_{16}[(AlO_2)_{16}(SiO_2)_{24}] \cdot 16\ H_2O$		1.76	0.23	Two	2.6 x 3.9
Scolecite	$Ca_8[(AlO_2)_{16}(SiO_2)_{24}] \cdot 24\ H_2O$		1.75	0.31	Two	2.6 x 3.9
Mesolite	$Na_{16}Ca_{16}[(AlO_2)_{48}(SiO_2)_{72}] \cdot 64\ H_2O$		1.75	0.30	Two	2.6 x 3.9
Thomsonite	$Na_4Ca_8[(AlO_2)_{20}(SiO_2)_{20}] \cdot 24\ H_2O$		1.76	0.32	Two	2.6 x 3.9
Gonnardite	$Na_4Ca_2[(AlO_2)_8(SiO_2)_{12}] \cdot 14\ H_2O$		1.74	0.31	Two	2.6 x 3.9
Edingtonite	$Ba_2[(AlO_2)_4(SiO_2)_6] \cdot 8\ H_2O$		1.68	0.36	Two	3.5 x 3.9
Group 6 (T_8O_{16})[j]						
Mordenite	$Na_8[(AlO_2)_8(SiO_2)_{40}] \cdot 24\ H_2O$		1.70	0.28	Two	6.7 x 7.0, ‖c 2.9 x 5.7, ‖b
Dachiardite	$Na_5[(AlO_2)_5(SiO_2)_{19}] \cdot 12\ H_2O$		1.72	0.32	Two	3.7 x 6.7, ‖b 3.6 x 4.8, ‖c
Ferrierite	$Na_{1.5}Mg_2[(AlO_2)_{5.5}(SiO_2)_{30.5}] \cdot 18\ H_2O$		1.76	0.28	Two	4.3 x 5.5 ‖c 3.4 x 4.8 ‖b
Epistilbite	$Ca_3[(AlO_2)_6(SiO_2)_{18}] \cdot 18\ H_2O$		1.76	0.25	Two	3.2 x 5.3, ‖a 3.7 x 4.4, ‖c
Bikitaite	$Li_2[(AlO_2)_2(SiO_2)_4] \cdot 2\ H_2O$		2.02	0.23	One	3.2 x 4.9

Table 2.4 Classification of Zeolites

Name	Typical Unit Cell Contents	Type of Polyhedral Cage [a]	Framework Density, g/cc [b]	Void Fraction [c]	Type of Channels [d]	Free Aperture of Main Channels, Å [e]
Group 7 ($T_{10}O_{20}$) [k]						
Heulandite	$Ca_4[(AlO_2)_8(SiO_2)_{28}] \cdot 24 H_2O$		1.69	0.39	Two	4.0 x 5.5, ‖a 4.0 x 7.2, ‖c
Clinoptilolite	$Na_6[(AlO_2)_6(SiO_2)_{30}] \cdot 24 H_2O$		1.71	0.34	?	?
Stilbite	$Ca_4[(AlO_2)_8(SiO_2)_{28}] \cdot 28 H_2O$		1.64	0.39	Two	4.1 x 6.2, ‖a 2.7 x 5.7, ‖c
Brewsterite	$(Sr, Ba, Ca)_2[(AlO_2)_4(SiO_2)_{12}] \cdot 10 H_2O$		1.77	0.26	Two	2.7 x 4.1 ‖c 2.3 x 5.0 ‖a

[a] Of the five space-filling solids of Federov, three (cube, hexagonal prism, and truncated octahedron) are found as polyhedral units in zeolite frameworks. The cube is the double 4-ring (D4R) as shown here. The double 6-ring (D6R) is the hexagonal prism or 8-hedron. The α-cage is the Archimedean semiregular, solid, truncated cuboctahedron referred to also as a 26-hedron, type I. The β-cage is the truncated octahedron or 14-hedron, type I. The γ-cage is the 18-hedron and the ε-cage the 11-hedron. Other polyhedral units are as given by Barrer (58).

[b] The framework density is based on the dimensions of the unit cell of the hydrated zeolite and framework contents only. Multiplication by 10 gives the density in units of tetrahedra/1000 Å^3.

[c] The void fraction is determined from the water content of the hydrated zeolite.

[d] Refers to the network of channels which permeate the structure of the hydrated zeolite. Considerable distortion may occur in the group 5 and 7 zeolites upon dehydration.

[e] Based upon the structure of the hydrated zeolite.

[f] Erionite and offretite may also be considered to consist of double 6-rings linked by single 6-rings.

[g] Zeolite Ω may be considered to consist of either strips of 4-rings or 5-rings parallel to c-axis. (See p. 81).

[h] Zeolite L consists of double 6-rings linked by single 12-rings.

[i] The T_5O_{10} refers to the unit of 5 tetrahedra as given by Meier for the 4-1 type of SBU. Fig. 67a (32).

[j] The T_8O_{16} unit refers to the characteristic configuration of tetrahedra shown in Fig. 72 (32).

[k] The $T_{10}O_{20}$ unit is the characteristic configuration of tetrahedra shown in Fig. 80 (32).

[l] Synthetic zeolites classified in group 1 include ZK-19 (35), W (36), and P-W (37) related to phillipsite, and various synthetic phases related to analcite—See Chapter 4.

[m] The synthetic zeolite, ZK-20 (38), is reported to have the levynite-type structure.

[n] Other zeolites with the A-type structure include zeolite α (39), ZK-21 (40) and ZK-22 (40).

[o] R = [1,4-dimethyl- 1,4-diazoniabicyclo (2,2,2) octane]$^{2+}$.

Table 2.4b Some Synthetic Zeolites with Unknown Structures

Name	Typical Composition or Unit Cell Contents	Crystal Data [f]	Density g/cc [a]	Void Volume cc/g	Approximate Pore Size, A [c]	Reference
Li-A	$Li_2Al_2Si_2O_8 \cdot 4\ H_2O$ [b]	Orthorhombic	—	0.15	2.6	36
F	$K_{11}[(AlO_2)_{11}(SiO_2)_{11}] \cdot 16\ H_2O$	Tetragonal $a = 10.4, c = 13.9$	2.28	0.16	2.6+	41
Z	$K_2Al_2Si_2O_8 \cdot 3\ H_2O$ [b]	—	—	0.14	2.6	42
H	$K_{14}[(AlO_2)_{14}(SiO_2)_{14}] \cdot 28\ H_2O$	Hexagonal $a = 13.4, c = 13.2$	2.18	0.22	2.6	43
Li-H	$Li_2Al_2Si_8O_{20} \cdot 5\ H_2O$ [b]	Tetragonal	—	0.16	2.6	36
J	$K_7[(AlO_2)_7(SiO_2)_7 \cdot 4\ H_2O$	Tetragonal $a = 9.45, c = 9.92$	2.22	0.08	2.6	44
E	$(K,Na)_2Al_2Si_2O_8 \cdot 3.3\ H_2O$	—	—	0.21	2.6+	45
M	$K_{14}[(AlO_2)_{14}\ (SiO_2)_{14}] \cdot 12\ H_2O$	Tetragonal $a = 13.1, c = 10.5$	2.37	0.10	2.6	46
Q	$K_{40}[(AlO_2)_{40}(SiO_2)_{44}] \cdot 86\ H_2O$	Tetragonal $a = 13.5, c = 35.2$	2.11	0.23	2.6	47
W	$K_{42}[(AlO_2)_{42}(SiO_2)_{76}] \cdot 107\ H_2O$	Cubic, $a = 20.1$	2.18	0.22	3.6	59
N [d]	$(Na,TMA)_2O \cdot Al_2O_3 \cdot 1.8$-$2.2\ SiO_2 \cdot Y\ H_2O$	Cubic, 37.2	—	0.16	2.6	48
ZSM-2	$Li_2O \cdot Al_2O_3 \cdot 3.3$-$4.0\ SiO_2 \cdot Y\ H_2O$	Tetragonal, $a = 27.4, c = 28.1$	—	0.22	6+	49
ZSM-3	$(Na,Li)_2O \cdot Al_2O_3 \cdot 2.8$-$4.5\ SiO_2 \cdot 9\ H_2O$	Hexagonal, $a = 17.5$	—	0.30	6+	50
ZSM-4	$(TMA,Na)_2O \cdot Al_2O_3 \cdot 6$-$15\ SiO_2 \cdot 5\ H_2O$	Cubic, $a = 22.2$	—	0.14	6+	51
ZSM-5 [e]	$(TPA,Na)_2O \cdot Al_2O_3 \cdot 5$-$100\ SiO_2 \cdot Y\ H_2O$	Tetragonal $a = 23.2, c = 19.9$	—	0.10	6	52
ZSM-10 [g]	$(R,K_2)O \cdot Al_2O_3 \cdot 5$-$7\ SiO_2 \cdot 9\ H_2O$	—	—	0.14	6+	53
BETA [h]	$(TEA,Na)_2O \cdot Al_2O_3 \cdot 5$-$100\ SiO_2 \cdot 4\ H_2O$	Cubic, $a = 12.04$	—	0.20	6+	54
Z-21	$Na_2O \cdot Al_2O_3 \cdot 1.7$-$2.1\ SiO_2 \cdot 9\ H_2O$	Cubic, $a = 36.7$	—	0.14	2.6	55

[a] Density measured for fully hydrated zeolite.
[b] Unit cell constants not determined.
[c] Pore size determined by adsorption. Value of 2.6 A, for example, indicates water and ammonia adsorbed by dehydrated zeolite.
[d] TMA = tetramethylammonium
[e] TPA = tetrapropylammonium; structure reported, see ref. 224 and p. 373.
[f] Cell dimensions from x-ray powder data are uncertain.
[g] R = [1,4-dimethyl-1,4-diazoniabicyclo (2,2,2) octane]$^{2+}$.
[h] TEA = tetraethylammonium.

52 STRUCTURE OF ZEOLITES

Figure 2.19 The 17 simplest ways of linking together UUDD chains. The smallest unit cell of the idealized structure is outlined by the rectangle in each case. The configuration of tetrahedra in the 8-rings which are formed by linking the double crankshaft chains is shown. Of these 17 structures, 4 are associated with known framework aluminosilicates, two of which are harmotome-phillipsite and gismondine (18).

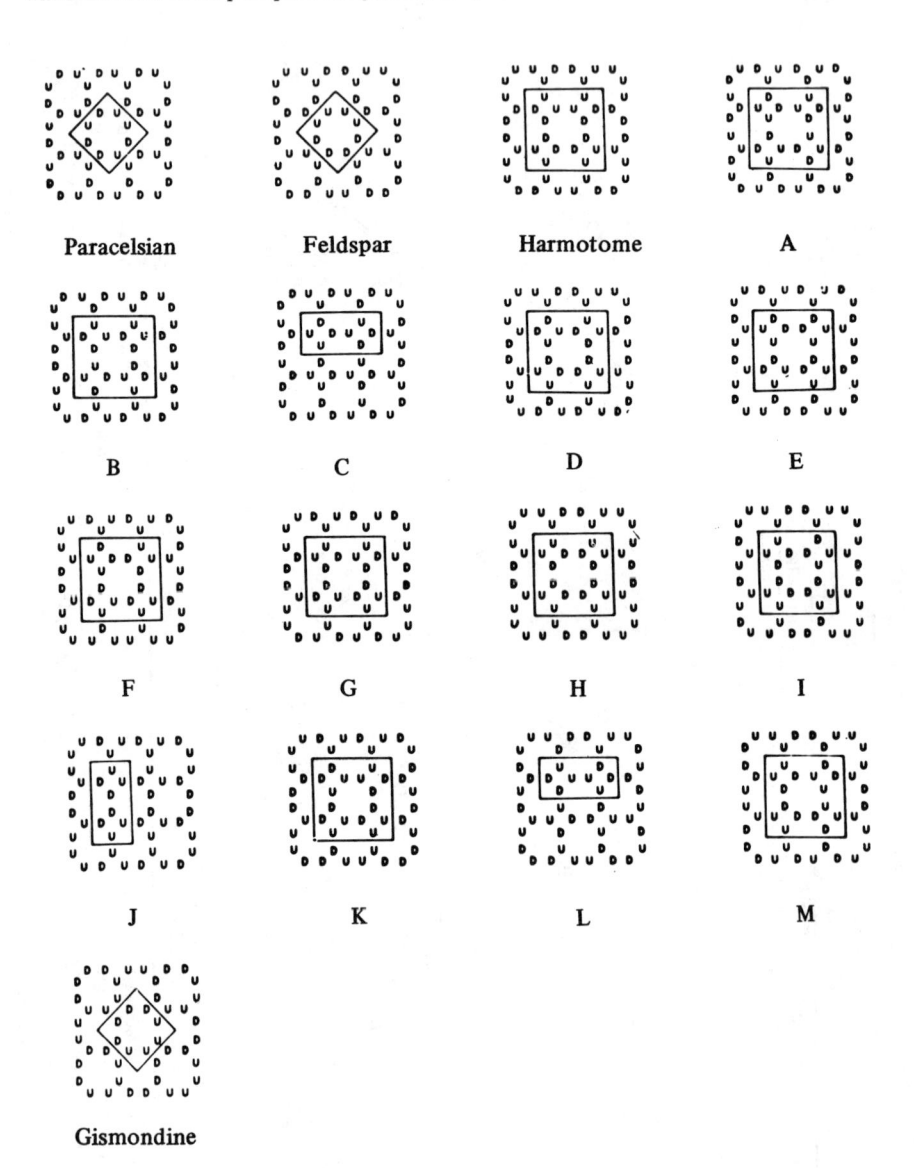

Paracelsian   Feldspar   Harmotome   A

B   C   D   E

F   G   H   I

J   K   L   M

Gismondine

blages of tetrahedra in building units which range from simple 4-rings to large polyhedra. It is evident that other assemblages of these units are possible which are not now represented by known zeolite structures. This section will review briefly some of the simplest possibilities that have been proposed. When structural analyses of all known zeolites are completed, the framework structures may correspond to some of these proposed theoretical structures.

Structures Based on Parallel 4- and 8-Rings

The classification and description of framework structures which may be formed from parallel 4- and 8-rings has been discussed in detail by Smith and Rinaldi (18). A shorthand is used to describe up (U) or down (D) position of the tetrahedra. One group, which is based upon the UUDD configuration of tetrahedra in the 4-ring,

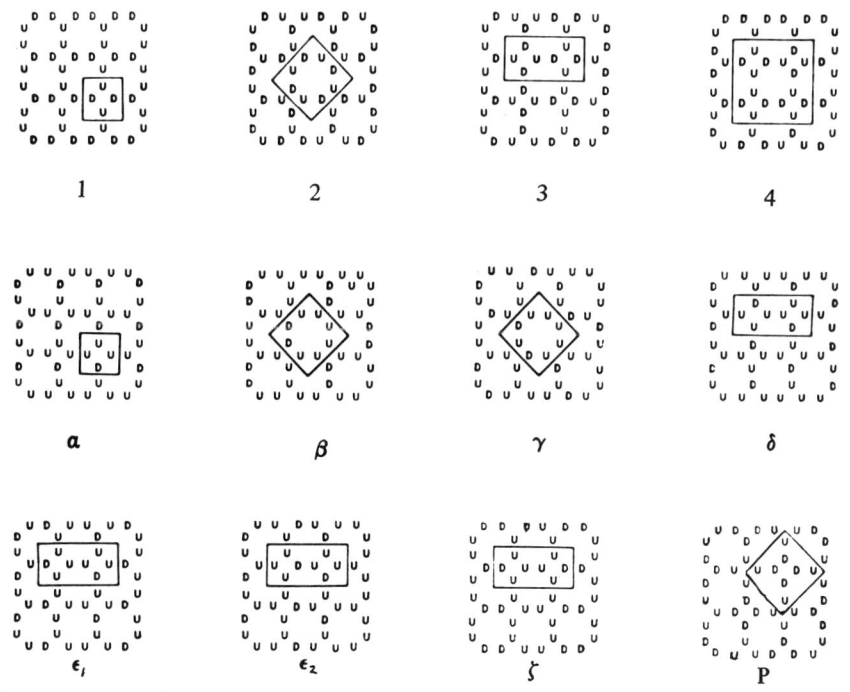

Figure 2.20 The four methods of linking UDUD chains are shown in structures 1, 2, 3, and 4. The next seven structures represent the methods of linking UUUD chains, the last is a structure once proposed for zeolite P as formed from UUUU rings. This sequence, UUUU, does not permit chain formation and the framework structure which results occurs by a sharing of the corner oxygen atoms as discussed (18) (Section H).

54 STRUCTURE OF ZEOLITES

contains 17 simplest ways of linking together the double-crankshaft chains of tetrahedra (Fig. 2.19). Using a UDUD configuration in the 4-rings, four structures can be formed (Fig. 2.20).

Two types of frameworks may be built from UUDD rings based on the linkage between the sheets of tetrahedra (19). Each sheet consists of alternating 4- and 8-rings. One type, called the flexible type can expand or contract by rotation of alternate UUDD rings. In the other type, referred to as inflexible, expansion-contraction is difficult. The type illustrated in Fig. 2.19 is the flexible type. Feldspar has the inflexible type. There are only thirteen ways of linking UUDD chains in the inflexible structure.

The third sequence of UUUD permits chain formation which may be crosslinked, and the possible framework structures which result (with restrictions applied to the volume of the repeat unit) are illustrated in Fig. 2.20. Distortions of these ideal frameworks, which are known to occur in the case of harmotome for example, can alter the actual dimensions of the unit cell by as much as 10%.

Structures Based on Linked Chains of 5-Rings

Six possible methods for linking the 5-ring chains such as those found in mordenite (see Section M) have been proposed and are illustrated in Fig. 2.21 (see Table 2.5) (56). Projections are shown of these structures parallel to the c-axis. The six structures result from linking the 5-ring chains together by square 4-rings. Of the six possible arrangements, two correspond to the structures of mordenite and dachiardite.

Structures Based on the Stacking of 6-Rings in Layers

An infinite number of structures can be built by arranging parallel 6-rings. The simplest structures are based upon linking a 6-ring to another 6-ring vertically superimposed, resulting in a hexagonal prism,

Table 2.5 Possible Group 6 Structures (56)

		Symmetry	a (Å)	b (Å)	γ
1	Mordenite	Cmcm	18.13	20.5	—
2	Unknown	Immm	18.13	20.5	—
3	Unknown	Bmma	18.7	19.6	—
4	Unknown	Bmmm	18.7	10.6	—
5	Dachiardite	B2/m	18.7	10.3	106°45
6	Unknown	B2/m	18.7	11.7	123°

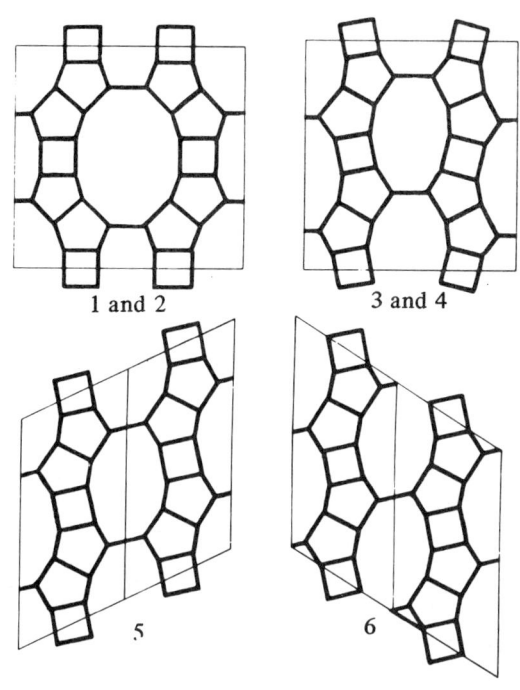

Figure 2.21 The 6 simple structures based on the 5-ring chain. The numbers 1-6 correspond with those in the first column of Table 2.5 (56).

1 and 2

3 and 4

5

6

or by linking to adjacent, horizontal 6-rings through a tilted 4-ring (30,57). The centers of the 6-rings are considered as if they follow the same pattern as close-packed spheres in simple structures. Using the sequences of ABC or AB, and additional sequences, a large number of structures are possible (see Table 2.6). Fig. 2.22 shows two projections parallel to the hexagonal axis of double 6-rings with linked, tilted 4-rings.

Structures Formed by Arrangements of Archimedean Polyhedra

The framework structures of some of the group 4 zeolites can be considered in terms of linked polyhedra, in particular, the truncated octahedron and truncated cuboctahedron. Possible linkages of Archimedean polyhedra have been reviewed in detail (58). Many types of polyhedra are not suitable for framework structures because four edges do not meet at a corner. Table 2.7 lists some of the possible and known structures.

The truncated octahedra may be joined to each other by means of a square face S, cube face S', hexagonal face H, or hexagonal prism

56 STRUCTURE OF ZEOLITES

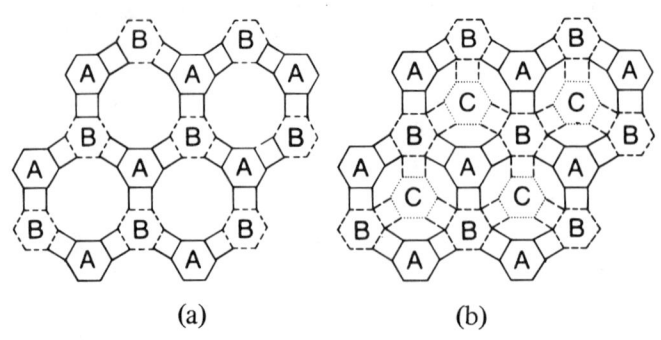

Figure 2.22 The AABB sequence (a) is characteristic of gmelinite and the AABBCC (b) arrangement is characteristic of chabazite (57).

H'. In joining the truncated cuboctahedra, the linkage may be through the octagonal faces as indicated by O for a single, or O' for a double linkage. Certain combinations do not result in frameworks because the structures do not close and small re-entrant angles are left. Structure 6 was once proposed as the framework structure for zeolite L (59). This is produced by linking truncated octahedra in the same manner as the SiO_4 tetrahedra in β-tridymite. In zeolite X (see Fig. 2.38) the orientation of the truncated octahedra across the interposing hexagonal prisms is uniform and the hexagonal faces bordering on the hexagonal prism alternate, resulting in an inverse three-fold axis; facing across the 4-ring of the hexagonal prism are the hexagonal face and 4-ring of the two adjacent truncated octahedra. In the proposed hexagonal structure (see Table 2.7, number 6) this same orientation occurs between truncated octahedra in the individual layers but the linkage in the c-direction involves the hexagonal faces of the adjacent truncated octahedra facing each other across the 4-rings (see Fig. 2.23).

The interpreted crystal structure of zunyite, $Al_{13}Si_5O_{20}F_2(OH,F)_{16}Cl$ is based on a tetrahedral framework of truncated rhombidodecahedra linked through double 4-rings. The framework composition is $XY_5\square_4O_{20}$ with X = Al, Y = Si with AlO_4 tetrahedra separated from Si_5O_{16} groups by vacant $\square O_4$ tetrahedra (Fig. 2.24) (60). Superficially, this unit resembles the "sodalite" unit, truncated octahedron. Further, it is apparent that these units cannot be arranged in more than this one configuration to form an infinite, three-dimensional

Table 2.6 Some Possible Zeolite Structures from 6-Rings (32,57)

Layer Sequence	No. of Layers	c-axis	Name
AB	2	5	Cancrinite
ABC	3	7.5	Sodalite
AAB	3	7.5	Offretite
AABB	4	10.0	Gmelinite
ABAC	4	10.5	LOSOD
AABBCC	6	15.0	Chabazite
AABAAC	6	15.1	Erionite
AABBCCBB	8	20	I[a]
AABCCABBC	9	22.5	Levynite
AABBCCAACCBB	12	30	II[a]
etc.			

[a]Unknown structures I and II are proposed structures (57). These are based upon arrangements of double 6-rings. One of these (I) is a four-layer sequence of AABBCCBB and the other (II) is a six-layer sequence of AABBCCAACCBB. In structure I, the cavities would be oriented in the c-direction of the hexagonal unit cell, stacked one above the other, and joined through the double 6-rings. Each cavity would have nine distorted 8-ring apertures. Six of the apertures are shared by the same type of cavity which are displaced in the c-direction by c/2, and the three central 8-rings are shared with a gmelinite-type cavity. Structure II contains the same large cavity but the chabazite-type cavity is also present. The three known cavity types of gmelinite, chabazite, and the hypothetical structure would all exist in a unit cell with a c-dimension of 30 A.

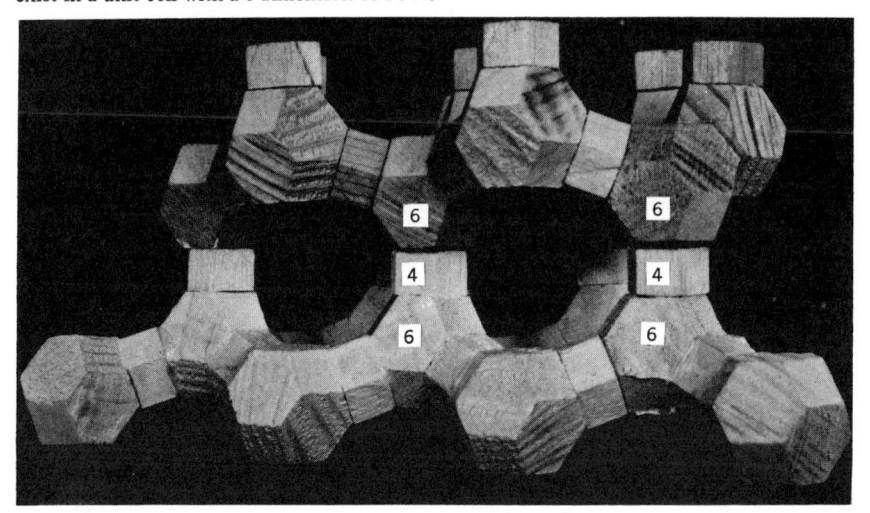

Figure 2.23 The proposed structure 6 based upon D6R and β-cage units. This hexagonal structure is produced by taking alternate layers of the truncated octahedral β-cages as they occur in (111) planes of zeolite X, rotating by 60 degrees and relinking them together.

Structures Based on ε-Cages

Zeolite L (Section K) can be considered in terms of chains of linked 4-rings which parallel the c-axis (61). A series of interesting theoretical structures related to zeolite L has been proposed based upon various combinations of two types of chains. The two types of chains are based on the unconnected oxygen as being either near (N) or far (F) from the reference axis. For example, the feldspathoid cancrinite consists of NF chains and zeolite L, NNF chains. Although a large number of hexagonal zeolite structures are theoretically possible, few are known. One series consists of structures which in projection all appear as in Fig. 2.25. They all have an a-dimension of 22 A. This type of structure provides for an 18-ring surrounding the main c-axis channels and a free aperture of about 15 A. It is also estimated that such a zeolite would have a void space approximating 0.55 cc/cc. (See Chap. 5.)

Table 2.7 Theoretical Structures (58)

No.	Arrangement	Space Group	Estimated Cell Dimensions (A) a	c	Number of Tetrahedra	Known Structure
A. Frameworks from Truncated Octahedra							
1	S(H-H)	Pm3m	8.8	–	12	Sodalite
2	S'(H-H)	Pm3m	12.3	–	24	Zeolite A
3	H(H-S)	Fd3m	17.5	–	96	Sodalite
4	H'(H-S)	Fd3m	24.7	–	192	Zeolite X
5	H(H-S), (H-H)	P6₃/mmc	12.4	20.5	64	–
6	H'(H-S),(H-H)	P6₃/mmc	17.5	28.5	96	–
B. Frameworks from Truncated Cuboctahedra							
7	O(S-S)≡S'(H-H)	Pm3m	12.3	–	24	Zeolite A
8	O'(S-S)≡H(O-S)	Im3m	15.1	–	48	RhO
9	H'(O-S)	Im3m	18.7	–	96	ZK-5
C. Truncated Octahedra and Truncated Cuboctahedra							
10	H'(S-S)	Fm3m	31.1	–	384	

Figure 2.24 The cubododecahedral (truncated rhombidodecahedron) cage and cubic cages in the framework structure of zunyite. ○ = silicon, ⊙ = aluminum, ● = vacancy.

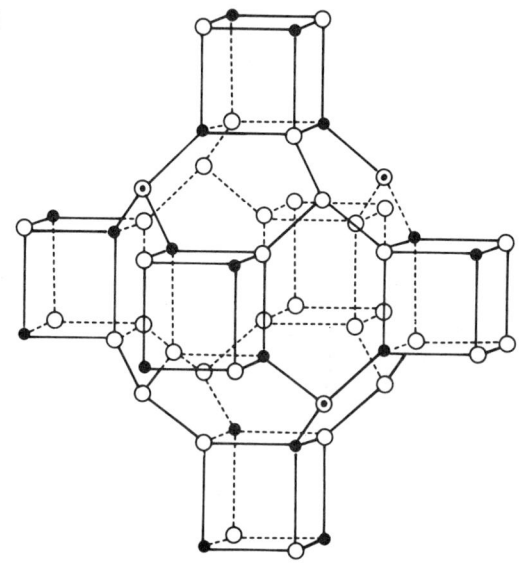

E. INTERNAL CHANNEL STRUCTURE IN ZEOLITES

The nature of the void spaces and the interconnecting channels in dehydrated zeolites is important in determining the physical and chemical properties. Information on the type of channel and the void space in each zeolite is given in the table of structural data for each specific zeolite. Three types of channel systems are identified:

1. A one-dimensional system which does not permit intersection of the channels is illustrated by analcime (Fig. 2.26).
2. Two-dimensional systems, as found in the zeolites of groups 5, 6, and 7 (Fig. 2.27).
3. There are two types of three-dimensional, intersecting channels. In one type, the channels are equidimensional; the free diameter of all the channels is equal, regardless of direction. The second type consists of three-dimensional intersecting channels, but the channels are not equidimensional; the diameter depends upon the crystallographic direction. The first type includes chabazite (Fig. 2.29a), faujasite (Fig. 2.30), erionite (Fig. 2.29c) the synthetic zeolites X, Y (Fig. 2.30), ZK-5 (Fig. 2.28b), and zeolite A (Fig. 2.28a). In all of these zeolites, at least four or

60 STRUCTURE OF ZEOLITES

Figure 2.25 (a) Theoretical structure based on ϵ-cages and NNF chains; (b) projection of the framework of hexagonal structures related to zeolite L along the c-axis. The large 18-ring aperture is shown (61).

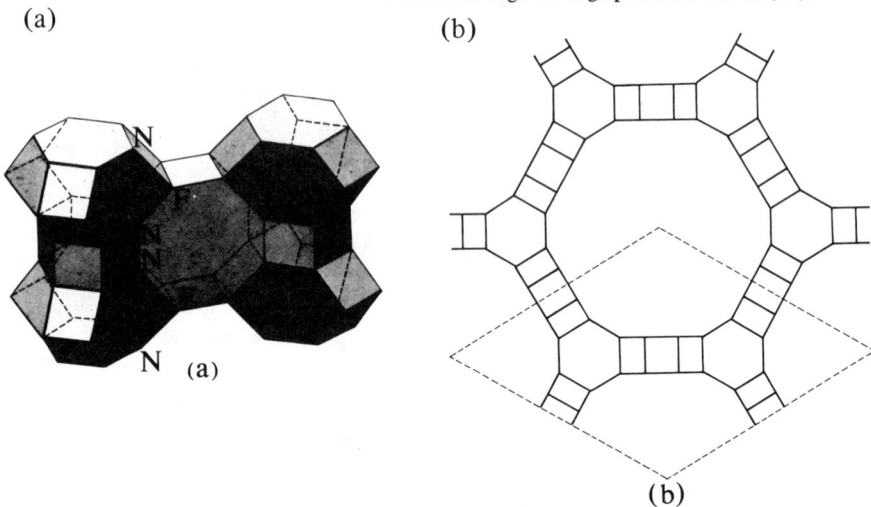

six equidimensional channels are available for access to any single cavity.

In erionite (Fig. 2.29c) molecules must diffuse in a zigzag path, since the main channel in the c-direction does not run continuously. The second type of three-dimensional channel systems contain one or two equidimensional channels that intersect a smaller second or third channel.

In gmelinite and offretite (Fig. 2.29b), the main channel runs parallel to the c-axis intersecting two smaller channels that run parallel

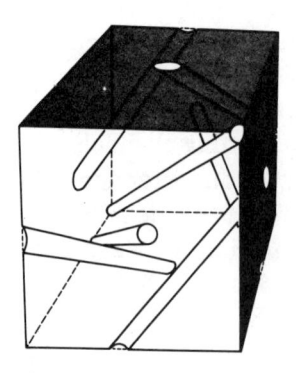

Figure 2.26 Schematic illustration of the nonintersecting channels in analcime. The channels parallel [111] and are formed by 6-rings.

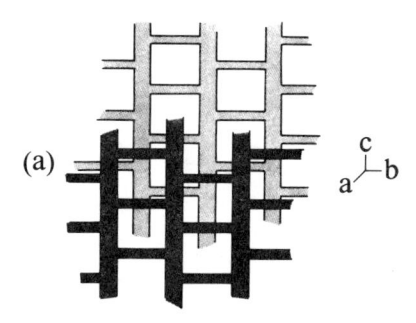

Figure 2.27 (a) Schematic illustration of the two-dimensional channels in mordenite. The main channels parallel c and are linked by small channels parallel to a. (b) Schematic illustration of two-dimensional channels as observed in the group 5 zeolites, such as natrolite.

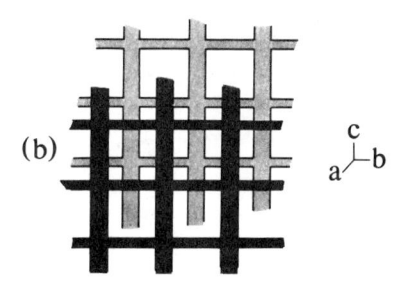

to the a-axis. Ideally, the large c-direction channels would permit very rapid diffusion, but they are subject to blocking by stacking faults in the crystal. Thus the adsorption character of gmelinite is governed mainly by the two-dimensional channel system.

Levynite (Fig. 2.29d) contains a two-dimensional channel network which runs in a plane perpendicular to the c-direction with a spacing of 7.5 A between planes. The only free channel open to molecules to move from the two-dimensional network into one above or below is through a small channel of about 2.7 A free diameter.

The system of channels in faujasite-type zeolites may be described in terms of a stacking of close-packed intersecting tubes as shown in Fig. 2.30. These tubes or channels run parallel to the [110] direction. The dimensions and nature of the cavities in the zeolites and the dimensions of the apertures leading into the cavities are listed in the tables at the end of this chapter. (See Chap. 5.)

62 STRUCTURE OF ZEOLITES

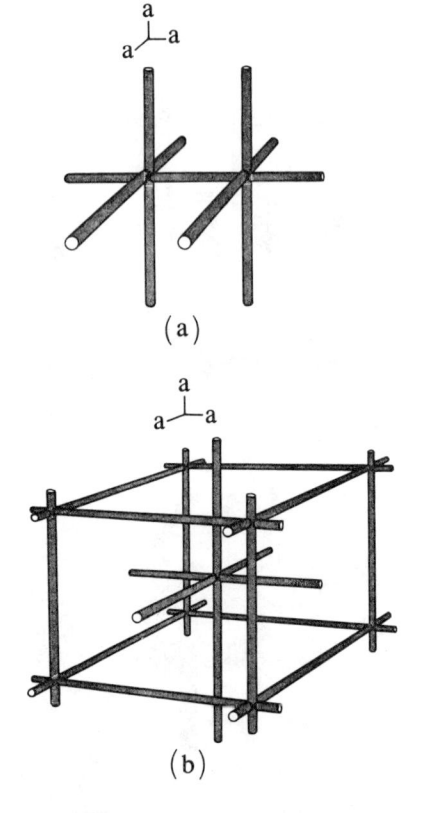

Figure 2.28 (a) Diagram of the three-dimensional system of intracrystalline channels in zeolite A. Channels formed by connected voids parallel to the [100] directions. A second system of interconnected α-cages and β-cages is not shown because it does not have any function except in ion exchange. (b) Diagram of the two independent sets of channels in paulingite. Both sets run parallel to [100], are of the same size but do not intersect.

F. FRAMEWORK DENSITY

The character of the water in hydrated zeolite crystals varies in behavior from what is best described as an isolated liquid to actual involvement in clusters or in bonding between the cations and the framework oxygen molecule. This is dependent to some extent upon the free space or the size of the cavity that is available. This subject is discussed at length in Chapter 5.

The intracrystalline volume which is occupied by water may amount to as much as 50% of the volume of the crystal. When zeolites which are stable and do not undergo appreciable dimensional change upon dehydration are dehydrated, this free space is available for occupancy by other molecular species. The capacity of the zeolite for adsorption is usually related to the free space or void volume

Figure 2.29 Schematic illustration of the main intracrystalline channels in zeolites which have frameworks based on parallel arrays of 6-rings: (a) chabazite, (b) gmelinite, (c) erionite, and (d) levynite.

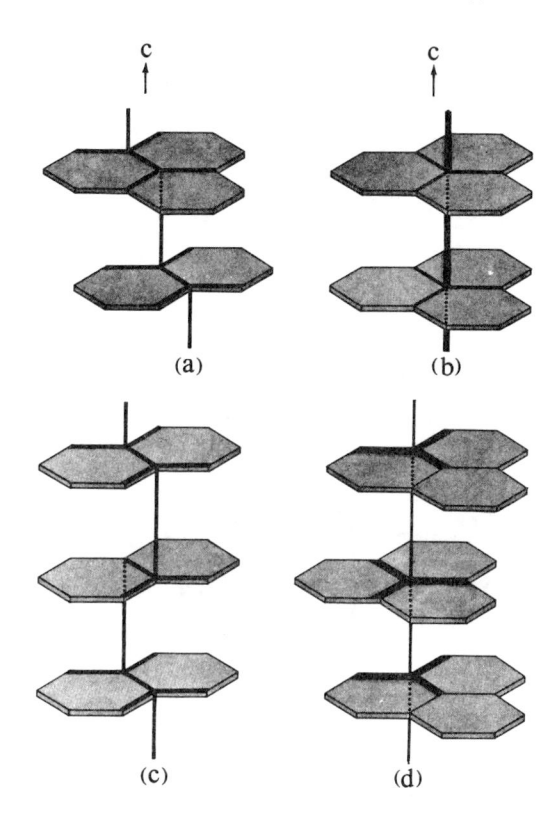

as determined by the quantity of contained water when fully hydrated. Void volumes of the zeolites are listed in the individual zeolite tables at the end of this chapter. The density of the zeolite used in this calculation varies with composition, thus the void volumes are not necessarily exact quantities.

Included in each table is the framework density of the dehydrated zeolite. This is calculated from the measured densities of the hydrated zeolites and the aluminosilicate framework composition of the crystal. The framework density (d_f) in g/cc is given by:

$$d_f = \frac{1.66[M]}{V} = \frac{1.66[59x + 60y]}{V}$$

where M is the formula weight of the framework, x and y are the number of AlO_2 and SiO_2 units per unit cell and V is the unit cell volume in $Å^3$. This quantity expresses the density of the aluminosilicate framework without the water or exchangeable cations and is equal to ten

64 STRUCTURE OF ZEOLITES

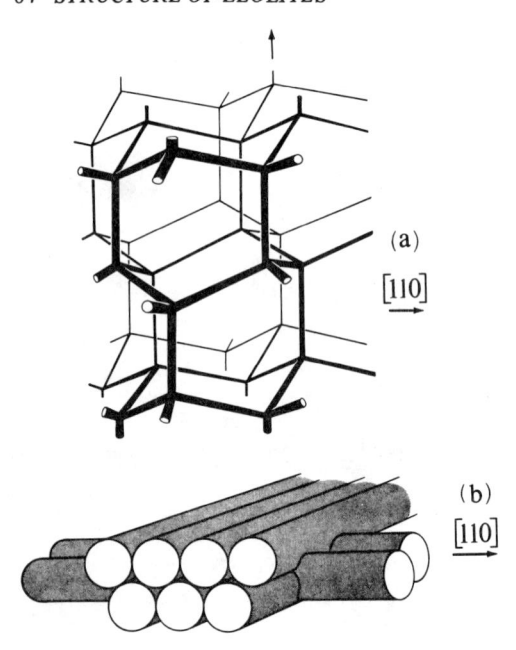

Figure 2.30 Illustration of the channel system in the [110] direction of faujasite and zeolites X and Y. (a) The 4-connected network, (b) the channels as an array of overlapping tubes.

times the number of tetrahedra per 1000 Å^3

G. APERTURE SIZES IN DEHYDRATED ZEOLITES

Consideration of adsorption and ion-exchange mechanisms in zeolite structures involves an understanding of the function of the openings to the channels in dehydrated and hydrated zeolite crystals and the factors involved in the interaction of molecular and ionic species with these channel openings. The apertures are bounded by oxygen atoms of connected tetrahedra and, in the first approximation, the limiting size of the aperture is governed by the size of the ring. In general, these rings involve 6, 8, 10, or 12 oxygen atoms; the maximum opening is formed by regular rings (Table 2.8). The rings are distorted to varying degrees; for example, in the zeolite A structure, the main aperture is formed by an 8-oxygen ring which is quite regular. The free diameter of 4.2 - 4.3 Å is based on the assumption that the oxygen ions are rigid hard spheres and is calculated from the interatomic distance of two opposing oxygens across the ring. The accepted diameter of the oxygen ion in silicate crystals is 2.7 A. This is an oversimplification of the real situation since the electron distribution of the atom does not fall off abruptly at the limit of the ionic radius. For strong bonds, such as Si-O, the hard-sphere approximation is quite good. For weaker bonds, such as Na-O and K-O, this approximation is quite poor. To consider the interaction between molecules and the oxygen atoms in the aperture rings as hard spheres during adsorption is incorrect; other factors are important. For example, the kinetic energy possessed by the diffusing molecule will be important in determining whether or not it can surmount the potential energy barrier created by the aperture into the crystal interstices. The effect of temperature and thermal vibration of the oxygens in the aperture rings is important in determing the aperture size. At room temperature, the root mean square displacement of the atoms can amount to 0.1 to 0.2 A; the effective size then will increase with increasing temperature and decrease with decreasing temperature. This is borne out by the adsorption of critically sized molecules (Chap. 8). In general, a correlation of the size of the diffusing molecule and the size of the aperture, as derived from the hard-sphere model, is quite good.

Another factor that must be considered is the change in the size of the aperture as the process of adsorption occurs. In the case of chabazite, the position of certain ions is determined by the absence of an adsorbed phase; the cations may change position due to coordination with adsorbed molecules and hence the framework will also change. Thus as a molecular species is adsorbed in a structure

Table 2.8 Apertures Formed by Rings of Tetrahedra Found in Zeolite Structures (34)

No. of Tetrahedra in Ring	Maximum Free Dimension (A)
4	1.6
5	1.5
6	2.8
8	4.3
10	6.3
12	8.0
18 (proposed but not observed)	15

66 STRUCTURE OF ZEOLITES

(a)

(b)

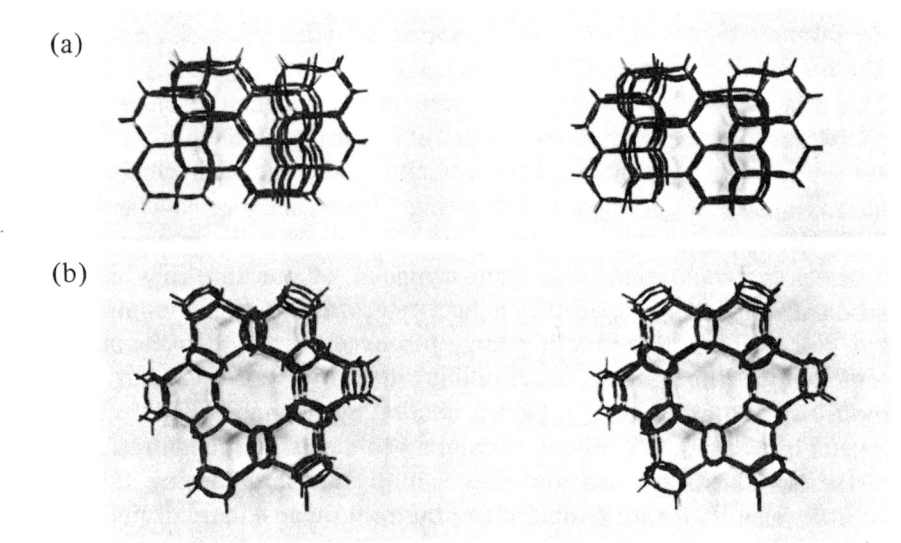

Stereo 2.3 Stereoscopic view of cancrinite structure (a) viewed perpendicular to c-axis along the a-axis, showing cancrinite, ϵ, units and ABAB stacking; (b) viewed along c-axis sharing chains formed by 12-rings.

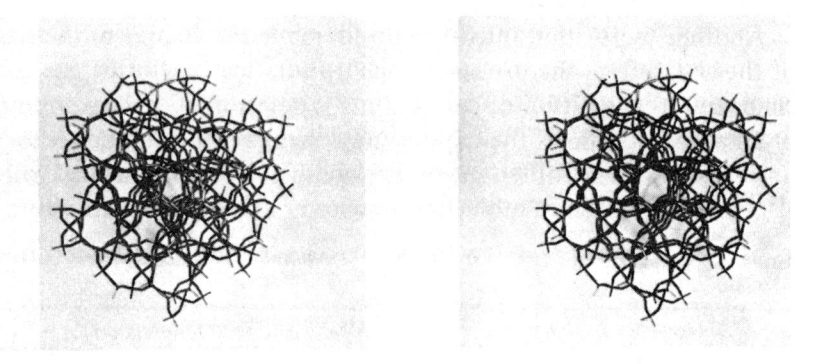

Stereo 2.4 Stereoscopic view of the framework structure of analcime viewed in the direction of the 6-rings which form the nonintersecting channel parallel to [111].

The simplest structural units in the framework are 4-rings and 6-rings which are linked to create additional 8- and 12-rings. The 6-rings are situated parallel to the [111] direction or 3-fold axis and the 4-rings are perpendicular to the 4-fold axes. The 8-rings, highly distorted, are parallel to the (100) plane. The framework encompasses 16 cavities which form continuous channels that run parallel to the 3-fold axes without intersection. These channels are occupied by the water molecules. 24 smaller cavities are situated adjacent to these channels.

such as chabazite, the aperture size and shape changes with movement of the cations as they are influenced by coordination with the incoming molecules. The size of the apertures cannot be absolutely determined from the structure of the hydrated zeolite because of distortions and cation movement upon dehydration. The effect of hydration on cation locations has been demonstrated for zeolites A, X, Y, faujasite, and chabazite. Cation blocking effects in zeolites A, X, Y, chabazite, and mordenite are known in some detail.

Compositional and structural information for each of the zeolites is summarized in Tables 2.18 through 2.67. Each zeolite is discussed in order as it appears in Table 2.4, according to the structural groups. The oxide formulas give typical chemical compositions. The information in each table is arranged in three groups; chemical composition, crystallographic data, and structural properties. For convenience, these tables are grouped at the end of the chapter, in alphabetical order.

The x-ray powder data reference is for the appropriate table as given in Chap. 3 (Minerals) or Chap. 4 (Synthetic Zeolites). Framework density and void volume are discussed in Chap. 5; the kinetic diameter is described in Chap. 8.

H. ZEOLITES OF GROUP 1

Analcime (Table 2.20) Stereo 2.4

Analcime occurs very commonly and extensively; synthetic types have been reported by many investigators over a period of several decades. The crystal structure of analcime was one of the first to be determined (62). The zeolite minerals analcime, wairakite, viseite, and kehoeite have basically the same framework structure. Despite extensive recent attempts at structural refinement, considerable detail in terms of the cations, water positions, and ordering is still unknown. Studies on analcime have not resulted in its application as a molecular sieve adsorbent.

The framework structure of analcime is complex, and illustration in a simple way is difficult. The model of the aluminosilicate framework shown in Stereo 2.4 indicates this complexity. In hydrated analcime, the Na^+ ions are in a distorted coordination octahedron with four framework oxygens and two H$_2$O molecules. The water content of analcime varies linearly with the silica content. As the silica content increases, the cation population decreases and there is a concurrent increase in the number of water molecules. The Na$^+$ ions can be exchanged by alkali metal cations (Li, K), NH$_4^+$, Ag, and divalent ions (Ca, Mg) at high temperatures (225°C) (63). If the ions are too large, they occupy the water positions. If total potassium replacement is achieved, the "zeolite" that results has no intracrystalline water. For example, replacement of the Na$^+$ in analcime by K$^+$ or Cs removes H$_2$O molecules due to occupancy of the water sites by the alkali metal ion; the degree of hydration therefore varies with the degree of ion exchange. A synthetic hydrated type of potassium analcime has been reported. The K$^+$ ions must occupy the normal cation positions (64). A calcium-rich variety of analcime, when dehydrated, was found to occlude molecules such as methane and ethane at ambient temperatures (65). This result is inconsistent with the calculated diameter of the channel system since kinetic diameters of the molecules concerned are considerably larger ($\sim$ 1 A) than the free diameter of the channel. A neutron diffraction study of analcime has shown that the H$_2$O molecules are not hydrogen bonded (66).

Wairakite (Table 3.35)

The mineral wairakite is the calcium-rich form of analcime (67, 68). Synthetic types have been prepared by direct synthesis (69, 70). The detailed structure of wairakite is not known but the aluminosilicate framework is probably the same as that of analcime. It can be concluded, however, that the H$_2$O molecules and Ca^{2+} ions occupy the channels.

Viseite and Kehoeite (Tables 3.34 and 3.22)

Viseite (71) and kehoeite (72) are varieties of analcime with rather disordered structures. These are the only known zeolite minerals which contain ions other than Si^{4+} or Al^{3+} in the tetrahedral framework, in both cases phosphorus. As indicated by the unit cell contents, some tetrahedral atoms are missing, presumably replaced by protons. A close inspection of the x-ray powder data for viseite shows that it might be interpreted as being a mixture of crandallite and analcime.

Crandallite is a calcium-aluminum-phosphate of composition:
$CaAl_3(PO_4)_2(OH)_5 \cdot H_2O$.
Published data for kehoeite seem to conclusively show that this zeolite is a separate species. Although it can be concluded that the basic framework structure is the same as that of analcime, little else is known concerning the location of phosphorus atoms or the nature of the defect tetrahedra.

Harmotome and Phillipsite (Tables 2.38, 2.56) Stereo 2.5

Harmotome and phillipsite contain the 4-ring $(Al, Si)_4O_8$ unit as the smallest structural unit. The individual tetrahedra in a 4-ring can be oriented in several ways. One orientation which is observed in the feldspar structures and in harmotome, phillipsite, and gismondine is shown in Fig. 2.13 and 2.19.

The oxygens which point outwards from the double crankshaft chains are shared to form the framework structure. This crosslinking produces 8-rings normal to the chains. Although an infinite number of crosslinked patterns are possible, it was shown in Section D that with certain restrictions imposed, there are 17 simple structures of the flexible type that can be derived (14, 15). Two of these are zeolite frameworks; one is the framework of harmotome and phillipsite and the other of gismondine. The sequence of tetrahedra around the 8-ring in harmotome is UDDDDDDU; and in gismondine, UUUUDDDD. In the inflexible feldspar structure, the chains are so distorted that the channels which are formed by the 8-rings are actually collapsed about the cations (Na^+, K^+) which bond directly with oxygens on both sides. In zeolites the chains are twisted to provide for wide channels to permit occupancy by cations and water molecules. These framework distortions alter the symmetry so that there does not seem to be much resemblance between ideal and actual structures. Additionally, it is believed that dehydration of the zeolite results in considerable framework distortion as the water molecules are removed from coordination about the cations. The joining oxygens act like hinges for displacive transformation of the framework.

Structure

The framework structures of harmotome and phillipsite are basically the same. The zeolites differ in overall symmetry and chemical composition. The linkage of the 4-rings in harmotome is compared with that in feldspar in Fig. 2.14. The framework structure based upon a model constructed from the skeletal tetrahedra is shown in Stereo 2.5.

Structural studies have been performed on harmotome which contains Ba^{2+} as the primary cation (73). The Ba^{2+} ions, coordinated with six of the framework oxygens and four H_2O molecules, are located in the voids at the channel intersections. The cation locations in the dehydrated zeolite are unknown, but one may presume that Ba^{2+} ions would relocate into the 8-rings. Eight of the 12 H_2O molecules in the unit cell are coordinated with the Ba^{2+} ions. Dehydrated harmotome shows zeolitic adsorption of small polar molecules such as ammonia. Since the kinetic diameter of the ammonia molecule is 2.6 A, this is in agreement with the observed structure.

Phillipsite has the same aluminosilicate framework and two-dimensional channel system as harmotome but a different chemical composition. The structural analysis was based on fully hydrated phillipsite (74). The positions of the Na^+ and K^+ cations and the H_2O molecules are not definitely known. When dehydrated, phillipsite is not very stable and decomposes at temperatures above 250°C. Adsorption studies have shown that it does readsorb water, which is in agreement with the size of the channels. In phillipsite and harmotome, the channels intersect in a plane perpendicular to the c-axis; a three-dimensional channel system is not present. It is probable that the channel system is subject to diffusion blocks, and the dehydrated zeolite is occupied by cations. The synthetic zeolites ZK-19, W, and P-W are reported to have the phillipsite-type structure (35, 36, 37).

Gismondine (Table 2.34) Stereo 2.6

The framework structure of gismondine is also based upon the UUDD configuration of tetrahedra, linked so the array around the 8-ring aperture is UUUUDDDD. In gismondine, the UUDD chains run parallel to both the a- and b-axis; in phillipsite ahd harmotome they are parallel to the a-axis only.

The tetrahedron framework of gismondine is illustrated in Stereo 2.6; the lattice constants (~ 10 A) and the unit cell content of gismondine are the same as for synthetic zeolite P (see Table 2.54), and recently the framework structure of P has been established to be the

ZEOLITES OF GROUP 1 71

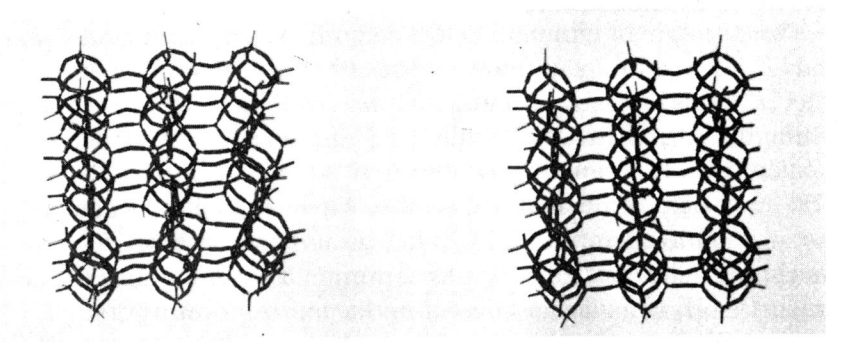

Stereo 2.5 Stereoscopic view of the framework structure of phillipsite and harmotome. The view is parallel to the main 8-ring channel. The double chains are folded in an S-shaped configuration in the direction of the b-axis. The 4- and 8-rings are evident as well as the channels which parallel the a- and b-axis. The wider channel, parallel to a, is bounded by 8-rings with free dimensions of 4.2 x 4.8 A and intersects channels parallel to the b-axis that have free dimension of 2.8 and 4.8 A. The channel intersections provide cavities.

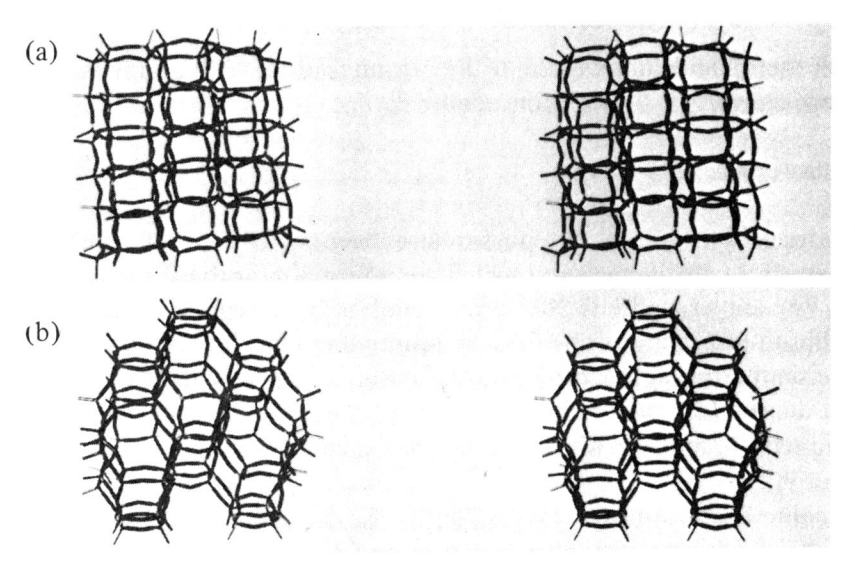

(a)

(b)

Stereo 2.6 Stereoscopic view of gismondine (a) view in c direction parallel to the 4-fold axis and (b) in direction of the main 8-ring channels.

Similar to harmotome and phillipsite, the channels in gismondine run parallel to the a and b directions. The main apertures in the channels are formed by 8-rings which have a free aperture size of 3 to 4 A. In the channels each calcium ion is surrounded by 4 water molecules and coordinated with 2 of the framework oxygens. The water positions are only partially occupied, so that the average coordination number with water is four, to give a total average coordination of six.

same as gismondine (206).

The structure of gismondine was determined for the hydrated mineral (75, 76). Nothing is known of the effect of dehydration upon the Ca^{2+} ions; it can be assumed that they will move from the spaces within the channel to sites within the 8-ring apertures. Complete dehydration of gismondine removes a substantial quantity of water, 21% by weight of the hydrated zeolite. When exposed to water vapor at room temperature, dehydrated gismondine adsorbs 15.6% by weight on a dehydrated basis; this corresponds to about nine of the original 16 H_2O molecules present in the unit cell. Adsorption of other small gas molecules has not been observed; this indicates substantial occupation of the channels by blocking Ca^{2+} ions and distortion of the framework upon dehydration.

A synthetic type of gismondine was prepared containing $(CH_3)_4N^+$ as the cation. The unit cell composition was:(77)

$$[(CH_3)_4N]_4[(AlO_2)_4(SiO_2)_{12}] \cdot 4H_2O$$

After thermal decomposition of the organic cation, sodium ion exchange converts it to the cubic zeolite P.

Zeolite P (Table 2.54)

A series of synthetic zeolite phases have been reported which exhibit rather subtle differences and which are referred to as the P zeolites (78, 79) and as zeolite B (80). These zeolites have been referred to as phillipsite-like, harmotome-like, or gismondine-like. This is due to close similarities in the x-ray powder patterns of the synthetic zeolites with those of the minerals. The two most common types are the cubic zeolite, referred to as zeolite B or P_c, and a tetragonal type, called P_t.

Zeolite P is readily synthesized in the Na_2O-Al_2O_3-SiO_2-H_2O system (Chap. 4). The structure first proposed was based upon a body-centered cubic arrangement of the D4R units of eight tetrahedra (Stereo 2.7). This has been shown to be incorrect and the framework structure has been shown to be the same as that of gismondine (206). There appears to be no basic difference in the framework structures of the cubic P and tetragonal P. Dehydration behavior is similar to that observed in the zeolites of group 5.

Zeolite P is closely related to garronite. The Ca^{2+} exchanged form

Table 2.9 Some Data on Various Cation Exchanged Forms of Zeolites P (B)

Cation Form	Unit Cell Dimensions (A)	H_2O (wt%)	Density (g/cc)	Refractive Index
Li^+ [b]	a = 9.99 c = 9.91	18.7	2.02	1.489
Na^+, cubic[a]	a = 10.05	19.3	2.01 - 2.08	1.40
Na^+, tet.[b]	a = 10.11 c = 9.83	16.9	2.15	1.482
K^+ [b]	a = 9.93 c = 9.67	10.8	2.28	1.504
Ca^{2+} [b]	a = 9.89 c = 10.30	18	2.16	1.505
Ba^{2+} [b]	a = 9.92 c = 10.36	14.6	2.57	1.531
Mg^{2+}, cubic[a]	a = 10.03	14	–	
Zn^{2+}, cubic[a]	a = 10.04	14	–	

[a] Ratio of Si/Al of original B = ~1.5
[b] Unit cell contents of original P = $Na_{5.7}(AlO_2)_{5.7}(SiO_2)_{10.3} \cdot 12 H_2O$

of tetragonal P has nearly an identical x-ray powder pattern with that of garronite, although the chemical composition is different. The mineral contains both Ca^{2+} and Na^+ ions.

A comparison of the models of the structures of harmotome, phillipsite, and gismondine shows close similarities. Many interplanar spacings in the x-ray powder patterns should be identical, leading to very similar x-ray powder patterns. Within a given volume, the number of tetrahedra is the same (framework density) resulting in similar densities and void volumes. Variations observed in the various cation-exchanged forms of P due to dehydration, etc. are explained by the flexibility of the framework structure.

The lattice constants of the P zeolites vary with cation content and cation species so that these constants lose significance as a means for differentiating between the various structures (Table 2.9). Some of the reported preparations may be mixtures. The orthorhombic P phase reported may be a mixture of P and a chabazite-like phase (based on the published x-ray powder data and electron micrographs).

Garronite (Table 2.33)

The structure of garronite has not been determined although the x-ray powder pattern is nearly identical to that of the tetragonal form of zeolite P (Table 4.75) and 3.15). It can be assumed that garronite has the same basic structure as zeolite P. Considerable complexity has been observed in the x-ray powder patterns of the zeolites of this

74 STRUCTURE OF ZEOLITES

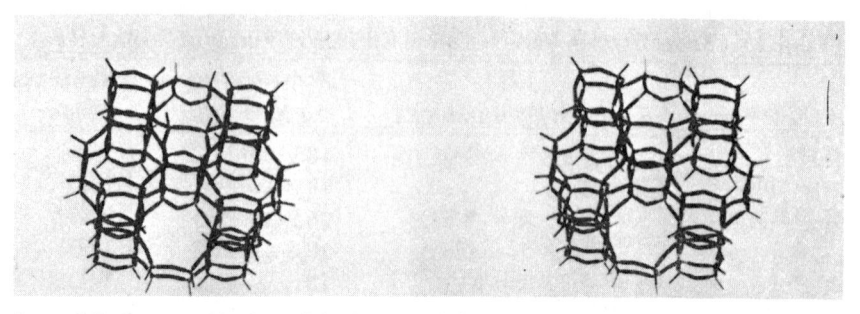

Stereo 2.7 Stereoscopic view of the framework structure once proposed for zeolite P showing the body centered array of D4R units and the main 8-ring channels. Note the close similarity to gismondine, Stereo 2.6.

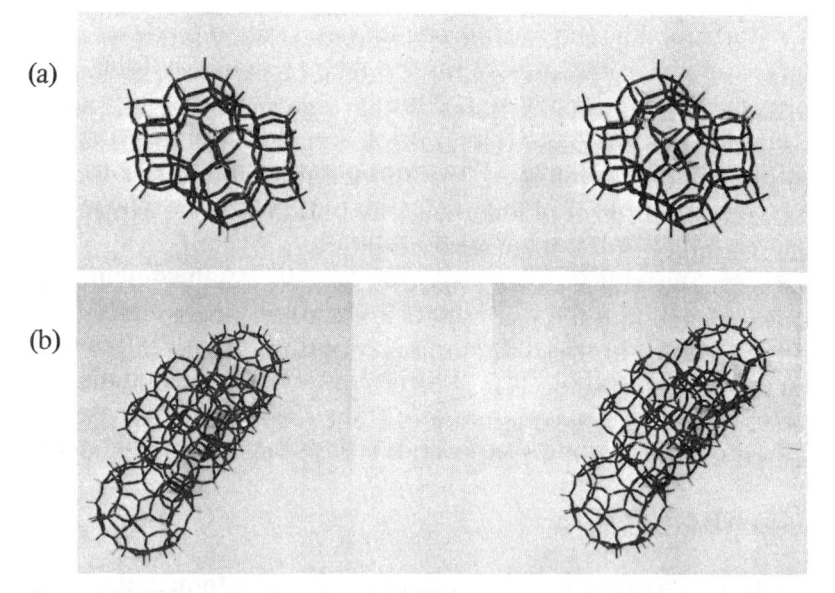

(a)

(b)

Stereo 2.8 Stereoscopic view of the framework structure of paulingite.(a) a portion of the structure of paulingite which has been termed the "c"-cage shown in an oblique view (b) the sequence of the cages α, δ, and γ along the unit cell edge or cubic axis. The α-cages are evident at each end. The sequences of these different cages along the cubic axes are $\alpha, \gamma, \alpha, \gamma$ . . in ZK-5, and $\alpha, \delta, \gamma, \delta, \gamma, \delta, \alpha$ . . in paulingite.

type. This complexity may be due to unrecognized species with other closely related structures.

Paulingite (Table 2.55) Stereo 2.8; 2.9

The framework structure of paulingite is complex and contains 45 crystallographically nonequivalent atoms (81). There are three types of cages present in the framework: the α-cage of the zeolite A and ZK-5 structures; double octagonal rings, or δ-cages, which are present in zeolite ZK-5; and the γ-cage which consists of two planar and four nonplanar 8-rings. The arrangement of the cages as found in the frame framework structure of paulingite is shown in Stereo 2.8 and 2.9. The free aperture of the channels is consistent with the results of adsorption measurements. During dehydration, the crystal structure appears to undergo some degree of distortion as evidenced by the decrepitation of paulingite crystals upon rehydration. The structure of dehydrated paulingite has not been studied.

Laumontite (Table 2.43) Stereo 2.10

Common in occurrence and known since 1801, the structure of laumontite has recently been determined (82, 83). The aluminosilicate framework contains 4-, 6- and 10-rings. The Ca^{2+} ions and H_2O molecules are located in the channels; the H_2O positions appear to be only partially occupied. The dehydration of laumontite occurs in a stepwise fashion and the partially dehydrated form is commonly referred to as leonhardite. The aluminosilicate framework of laumontite is shown in Stereo 2.10. It cannot be definitely associated with any of the known zeolite groups but does bear some topologic resemblance to the framework of analcime because it consists of 4- and 6-rings. The twisted 6-ring can also be considered an SBU.

Yugawaralite (Table 2.65) Stereo 2.11

The mineral yugawaralite does not clearly fit any of the zeolite structure groups; it is best characterized in terms of the 4-rings in the bc plane. We therefore have listed it in group 1. Linkages of 4-rings with those of adjacent layers produce 5-rings (Stereo 2.11)(84 - 87). The structure of the dehydrated form is unknown and adsorption properties are unreported. The synthetic zeolite Sr-Q is apparently structurally related. The two-dimensional channel system lies in the

76 STRUCTURE OF ZEOLITES

(a)

(b)

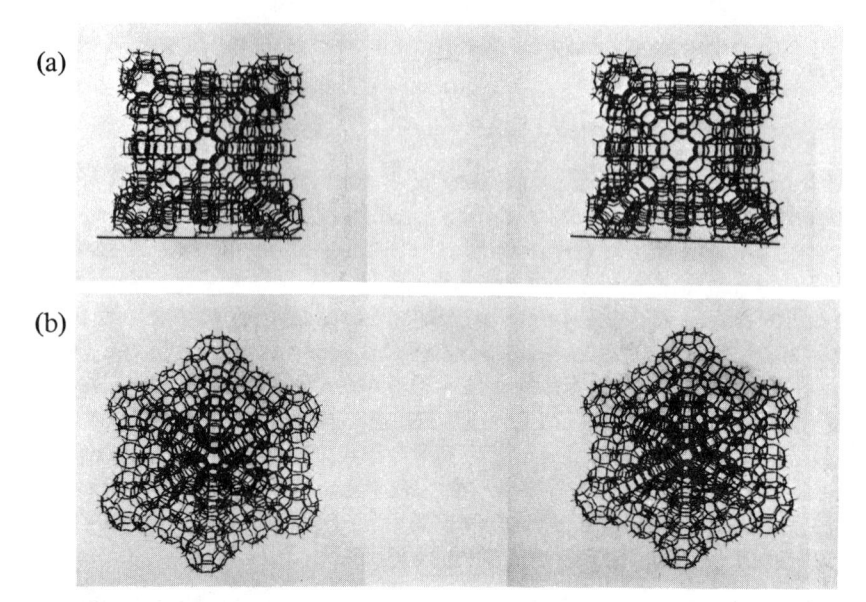

Stereo 2.9 Stereoscopic views of a unit cell of paulingite (a) a view parallel to the cubic axis and (b) a view in the three-fold or [111] direction.

A toroidal channel is produced around each octagonal prism. A continuous three-dimensional channel system in which the α-cages lie at the corners of the cubic unit cell, runs parallel to the cubic axes. A second channel system runs parallel to the first but has the α-cages at the body centers of the unit cell. There is no access from one channel system to the other. The free diameter of 3.8 Å is similar to the channels in zeolites A and ZK-5.

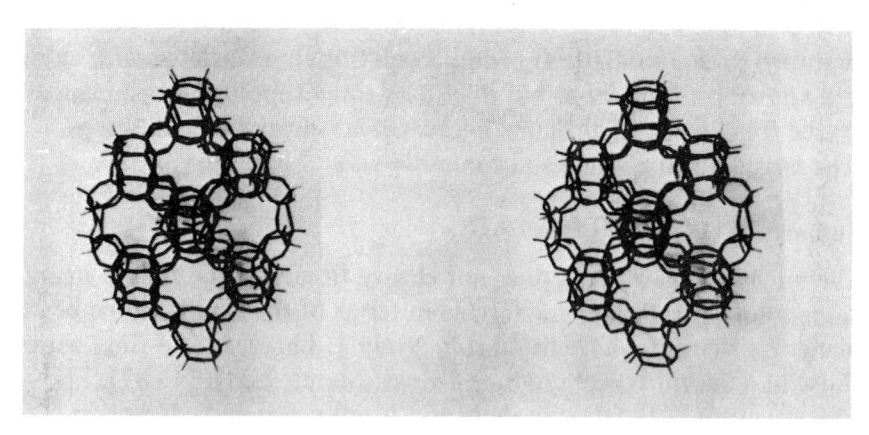

Stereo 2.10 Framework of laumontite viewed parallel to a and showing the main channels found by distorted 10-rings.

ac-plane with 8-rings as apertures of about 3.5 A in free diameter. Calcium ions are located on the wall of the cavity formed at the channel intersections and are coordinated with four framework oxygens and four water molecules.

I. ZEOLITES OF GROUP 2

Group 2 includes erionite, offretite, zeolite T, levynite, omega[a], and HS, a hydrated hydroxysodalite with no interstitial NaOH. The aluminosilicate frameworks exhibit the same SBU—the S6R unit or hexagonal $(Al,Si)_6O_{12}$ rings.

Erionite (Table 2.28) Stereo 2.12

Erionite has a hexagonal structure consisting of parallel arrangements of the $(Al, Si)_6O_{12}$ rings. The erionite framework is shown in Stereo 2.12 and a space-filling model showing the oxygens is in Fig. 2.31.

The erionite framework can also be considered in terms of the cancrinite or ϵ-cages linked by D6R units in the c-direction. In order to diffuse from one cavity to another, molecules must pass through an 8-ring aperture into an adjacent cavity and then through another 8-ring aperture into an adjacent cavity in the original column (Stereo 2.12) (88, 89, 90, 91, 92). Continuous diffusion paths are available for molecules of appropriate size. Erionite readily adsorbs normal hydrocarbons which have a minimum kinetic diameter of 4.3 A.

Typical crystals of erionite are fibrous. Unlike other fibrous zeolites, however, dehydrated erionite has a very stable framework structure. Specimens of erionite exposed to H_2O vapor at 375°C over a long period of time showed essentially no change in adsorption capacity, which confirms a high degree of structural stability.

The chemical composition of erionite is complex and contains appreciable quantities of K^+ and Ca^{2+} as the main cation species. In addition, in sedimentary varieties of erionite (Chap. 3), the potassium ion shows considerable resistance to ion exchange indicating that it is locked within the structure in positions in which it is not free to move during exchange. This position is probably in the ϵ-cage. Sites are available also in the double 6-rings; two are available in the

---

[a]Based on proposed structure for omega in ref. 98. More recent structure of mazzite reported to be isostructural with omega is based on single 4-rings and would be classified in Group 1. See refs. 222 and 223.

78 STRUCTURE OF ZEOLITES

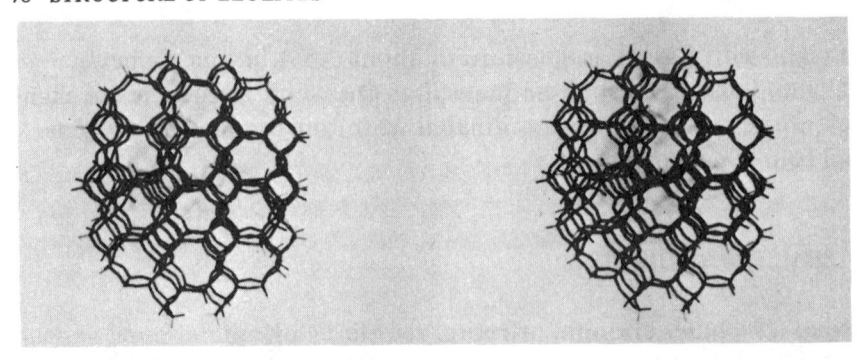

Stereo 2.11 Stereo view of yugawaralite framework in c-direction showing 4-rings and 8-rings. The latter form the c-direction channels, 3.6 x 2.8 A. Channels in the a-direction intersect the c channels. Calcium ions lie in intersections coordinated to 4 oxygen atoms and 4 water molecules.

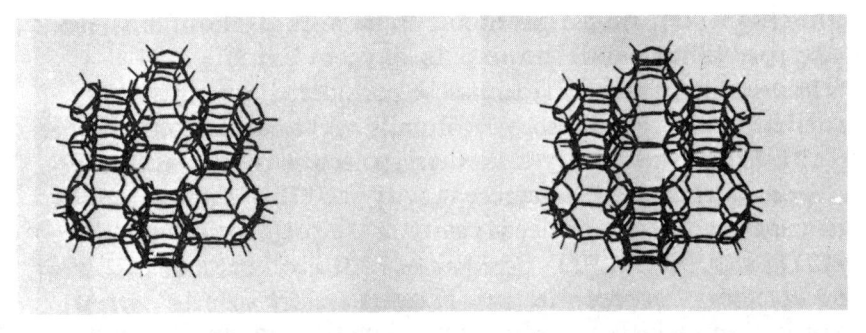

Stereo 2.12 Stereoscopic view of the framework of erionite. The view is parallel to the c-axis and shows the stacking sequence of the 6-rings. The sequence is AAB, AAC. The framework consists of double 6-rings, D6R units, and S6R units which are arranged in parallel planes perpendicular to the hexagonal axis. Because of this stacking sequence the c-dimension is 15 A, i.e., 6 times the 2.5 A spacing of a single 6-ring.

The main cavities have internal dimensions of 15.1 A by 6.3 A. Each of the cavities has a single 6-ring at the top and bottom, which it shares with like cavities above and below. The aperture between these cavities in the c-direction has a free diameter of 2.5 A and is too small to permit diffusion of most molecules. However, six 8-rings form apertures into any single cavity and these have free dimensions of 3.6 by 5.2 A. Three are arranged in the upper half of the cavity and three below. Each window or aperture, formed by these 8-rings is common to two cavities.

The ϵ- or cancrinite-type cages are linked in the c direction by D6R units (hexagonal prisms) in the configuration ϵ-D6R-ϵ-D6R ... These columns are crosslinked by single 6-rings perpendicular to c. (See zeolite L.) In erionite the ϵ-cages are not symmetrically placed across the D6R units.

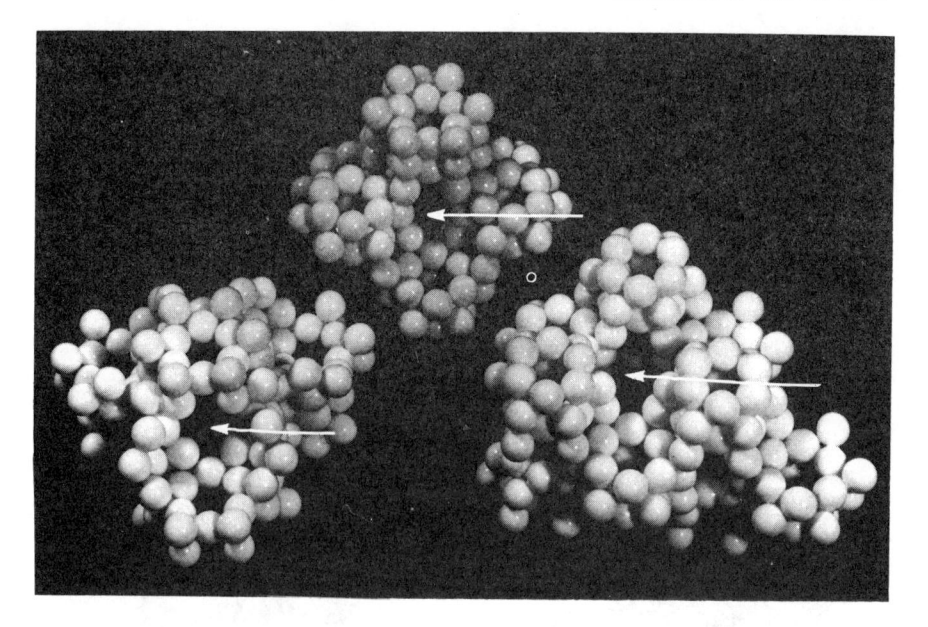

Figure 2.31 Space-filling models of (a) erionite, (b) chabazite, (c) gmelinite. The 8-ring aperture entering the channels is shown by the arrow.

single 6-rings, and six are in the distorted 6-rings of the ϵ-cages. Erionite is one of the more siliceous zeolites with a Si/Al ratio of 3; it is likely that this is related to Si-Al ordering and cation population.

Offretite (Table 2.52) Stereo 2.13

The offretite structure is very closely related to erionite. In fact, at one time, offretite was thought to be identical with erionite. However, it was observed that the c-spacing is half that of erionite and all x-ray reflections with l odd for erionite are absent in the offretite x-ray patterns (90). The framework structure of offretite consists of the AABAAB sequence of 6-rings, compared to AABAAC for erionite. This provides for a 6.4 A c-axis channel which connects with intersecting channels in the a-direction formed by 8-rings into gmelinite-type units (Stereo 2.13). Some of the potassium ions are located within the ϵ-cages. The remaining ions are accounted for by partially occupied sites located on axes parallel to c of the single 6-rings and large channels (91, 93). The relationship of offretite to erionite is similar to that of gmelinite to chabazite.

80 STRUCTURE OF ZEOLITES

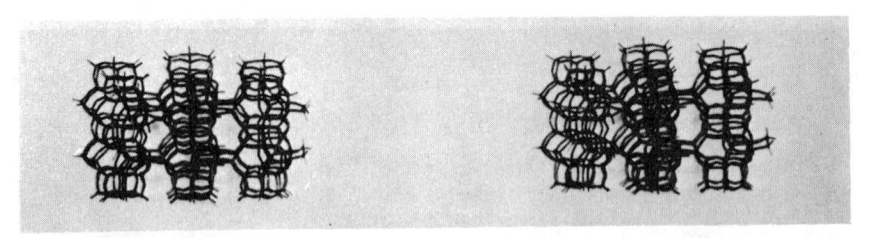

Stereo 2.13 Stereoscopic view of the offretite framework looking perpendicular to c along a. This view shows the stacking of D6R's and €-cages in the c-direction along with the gmelinite, 14-hedron cages. The main c axis channels are not shown in this view but resemble the channels in gmelinite and are 6.4 A in diameter.

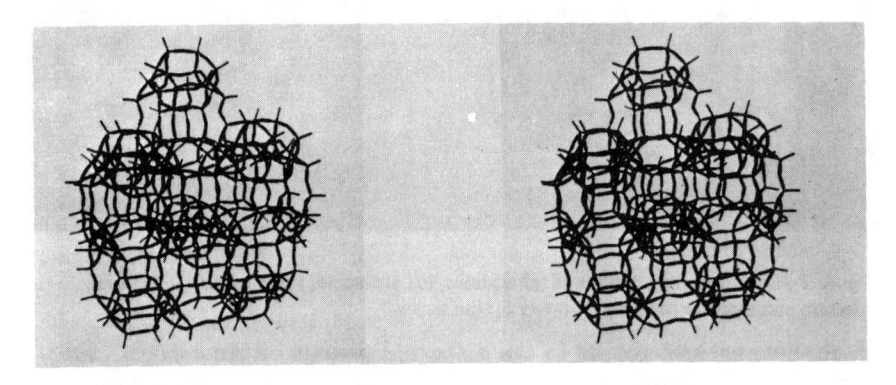

Stereo 2.14 Stereoscopic view of the framework structure of levynite. Nine layers of 6-rings are stacked in the sequence AABCCABBC.

The cavities have a free diameter of about 8 A and each is connected by elliptical 8-rings to three similar cavities. The apertures are not regular. The network of channels lies in the plane perpendicular to the c-axis and spaced at intervals of 7.5 A. There is no channel which connects one system to another above or below it, except through 6-rings which have a smaller free diameter of 2.6 A.

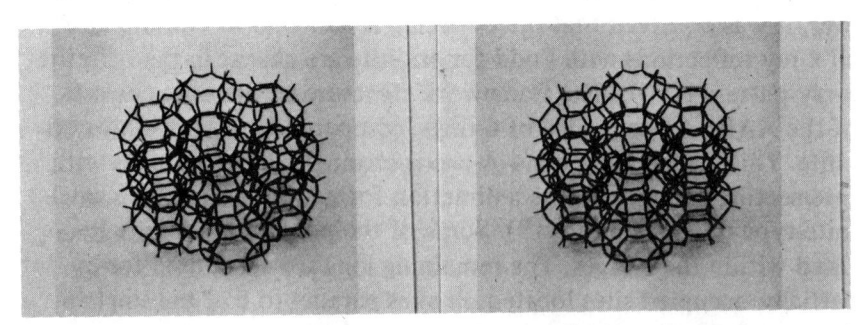

Stereo 2.15 Framework structure proposed for zeolite omega[a]. This is an oblique view showing the array of gmelinite units (14-hedra) in the c-direction surrounding the main central channel.

[a]Proposed structure of ref. 98. See p. 81 and refs. 222 and 223.

Zeolite O or TMA (tetramethylammonium) offretite is a synthetic zeolite with the offretite-type structure (Table 2.51).

Zeolite T (Table 2.59)

Zeolite T is related to both offretite and erionite, but more closely to offretite (89, 94). Current evidence from electron diffraction studies indicates that it is a disordered intergrowth of the two types of structures. X-ray powder patterns of zeolite T show specific differences. The x-ray reflections with l odd of erionite are very weak in zeolite T. Positions for the cations in zeolite T have been proposed (95, 96). These positions are located in the D6R units, the ϵ-cages, and the single 6-rings within the main cavities.

Levynite (Table 2.44) Stereo 2.14

Levynite is the least studied of the five zeolites in the chabazite group. Although an aluminosilicate framework structure has been proposed, the x-ray structural investigation has not been completed (97). The proposed structure is based upon a 9-layer stacking sequence of hexagonal rings. (See Table 2.6.) This results in the framework as shown in Stereo 2.14. Physical adsorption measurements have shown that the largest molecular species adsorbed include O_2 and N_2 with diameters of about 3.6 A. Carbon dioxide is readily adsorbed, but not argon. At present, nothing is known of the positions of cations and water molecules in this zeolite. A synthetic zeolite ZK-20 is reported to have the levynite-type structure (38).

Omega (Table 2.53) Stereo 2.15

A framework structure of the hexagonal, synthetic zeolite omega has been proposed but confirmatory work is necessary (98)[a]. The structure consists of gmelinite-type cages (14-hedra of type II) which are linked in columns parallel to the c-axis, by joining the 6-rings, and laterally by oxygen bridges. This produces small cavities between the cages which may contain Na^+ ions. The main channels run parallel to c and are formed by 12-rings. A second two-dimensional system of channels in the plane perpendicular to c is formed by 14-hedra linked through distorted 8-rings and is not accessible to the main c-channel.

[a]Subsequently Galli, et al (223) reported a different framework structure for the zeolite mineral mazzite and reported it to be isostructural with omega. See Table 2.53, p. 167, for new structural data.

82 STRUCTURE OF ZEOLITES

Access to this two-dimensional system is available only at the surface. These two-dimensional systems do not connect in the c-direction except through small 5-rings. The omega zeolite is synthesized containing TMA cations which may be located within the 14-hedron cages.

Zeolite HS or Zh (Table 2.40) Stereo 2.2

Synthetic zeolite HS or Zh is reported by Zhdanov to be a hydrated sodalite. The framework structure is the same as that of sodalite and consists of the cubic array of β-cages (Fig. 2.16).

Zeolite HS is reported to adsorb 18 wt% water and ethanol in small amounts (99,100). Compositions containing varying amounts of intercalated sodium hydroxide have been prepared.

Losod (Table 2.45)

A new zeolite framework comprised of 6-rings arranged in the ABAC sequence has been synthesized. Because the apertures are 6-rings, it adsorbs only water (101).

Figure 2.32 The cubic unit of 8 tetrahedra as found in the framework structures of the group 3 zeolites. The photo on the left shows a model based on solid tetrahedra and on the right, the space-filling model showing the positions of individual oxygen atoms. This is referred to as the double 4-ring or D4R unit (103).

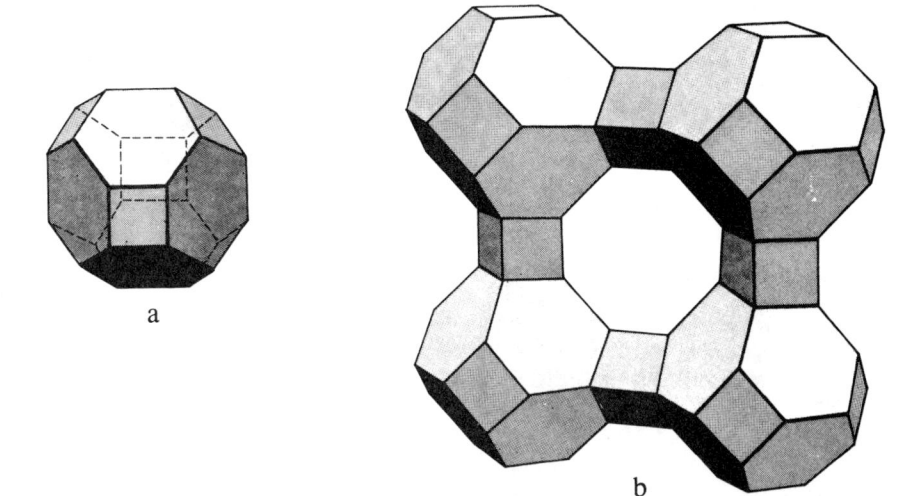

Figure 2.33 The truncated octahedron (a) and (b) the array of truncated octahedra in the framework of zeolite A. The linkage is shown via the double 4-rings (see Fig. 2.16).

J. ZEOLITES OF GROUP 3

The structural unit common to framework structures of the zeolites of group 3 is the double 4-ring, D4R (Fig. 2.32). The framework of the synthetic zeolite A is based upon this unit.

Zeolite A (Table 2.18) Stereo 2.16

Framework

The aluminosilicate framework of zeolite A can be described in terms of two types of polyhedra; one is a simple cubic arrangement of eight tetrahedra, the D4R (Fig. 2.32); the other is the truncated octahedron of 24 tetrahedra or β-cage as previously described for the sodalite-type minerals (Fig. 2.15). The aluminosilicate framework of zeolite A is generated by placing the cubic D4R units ($Al_4 Si_4 O_{16}$) in the centers of the edges of a cube of edge 12.3 A. This arrangement produces truncated octahedral units centered at the corners of the cube (see Fig. 2.33)(102,103). Each corner of the cube is occupied by a truncated octahedron (β-cage) enclosing a cavity with a free diame-

84 STRUCTURE OF ZEOLITES

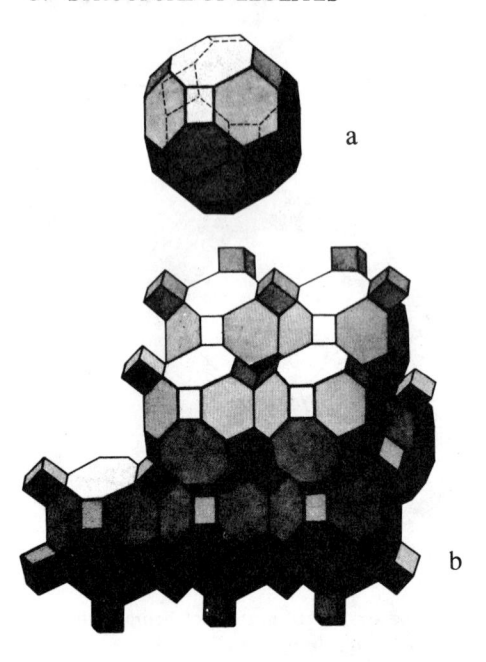

Figure 2.34 (a) The truncated cuboctahedron and (b) the arrangement in the framework of zeolite A (104).

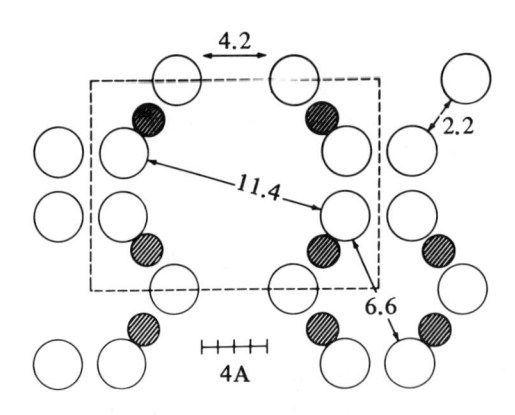

Figure 2.35 A diagram of the (110) section of zeolite A illustrating relative positions and dimensions of the large α-cages, 11.4 A in diameter and the small β-cages, 6.6 A in diameter. The shaded circle represents Na$^+$ ions in site I. There are 2 interconnecting, 3-dimensional channel systems; one consisting of connected α-cages, 11.4 A in diameter, separated by 4.2 A circular apertures; the other system consists of the β-cages, alternating with the α-cages and separated by 2.2 A apertures. The first system runs in the direction of the cubic axis, and the second system in the direction of the 3-fold axis (103).

ter of 6.6 A. The center of the unit cell is a large cavity which has a free diameter of 11.4 A. The tetrahedra centers around this large cavity, referred to as the α-cage, occupy the apices of a truncated cuboctahedron (Table 2.2) (Fig. 2.34) (104). A cross section of the structure in the (110) plane is shown in Fig. 2.35. The skeletal model of the framework appears in Stereo 2.16 and a space-filling model in Fig. 2.36.

ZEOLITES OF GROUP 3   85

(a)

(b)

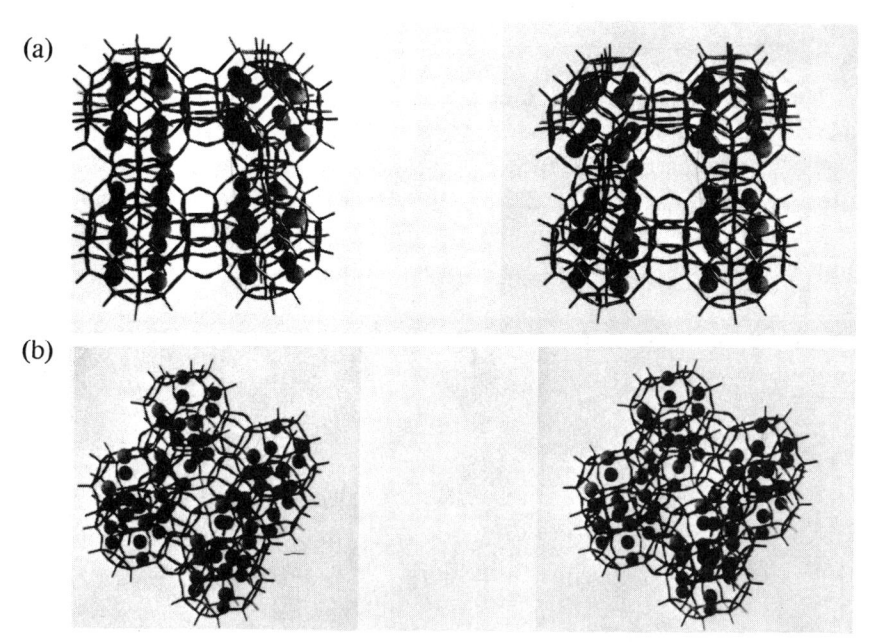

Stereo 2.16 Stereoscopic view of the framework of zeolite A including the positions of typical univalent ions (a) view parallel to the cubic axis showing the large central and regular 8-ring, (b) an oblique view.

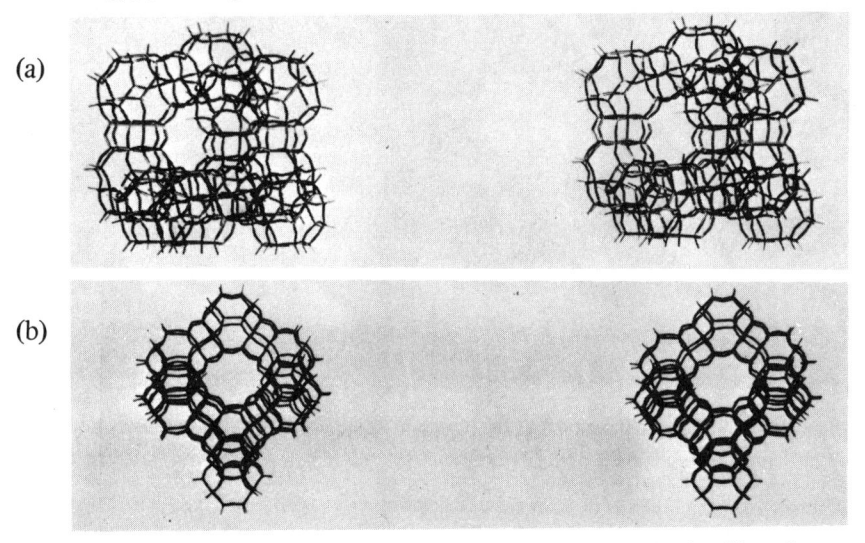

(a)

(b)

Stereo 2.17 Stereoscopic photograph of a model showing the basic aluminosilicate framework of zeolites X, Y, and faujasite. This model illustrates only the spatial arrangement of the tetrahedra and approximate (Si, Al) positions; both views are parallel to [110].

Composition

The unit cell of zeolite A contains 24 tetrahedra, 12 AlO_4 and 12 SiO_4. When fully hydrated, there are 27 water molecules. The electrostatic valence rule, as modified by Loewenstein, requires a rigorous alternation of the AlO_4 and SiO_4 tetrahedra, because the Si/Al ratio is 1:1 (23). In order to achieve this, the lattice constant of the true unit cell of zeolite A must be 24.6 Å and must contain 192 tetra-

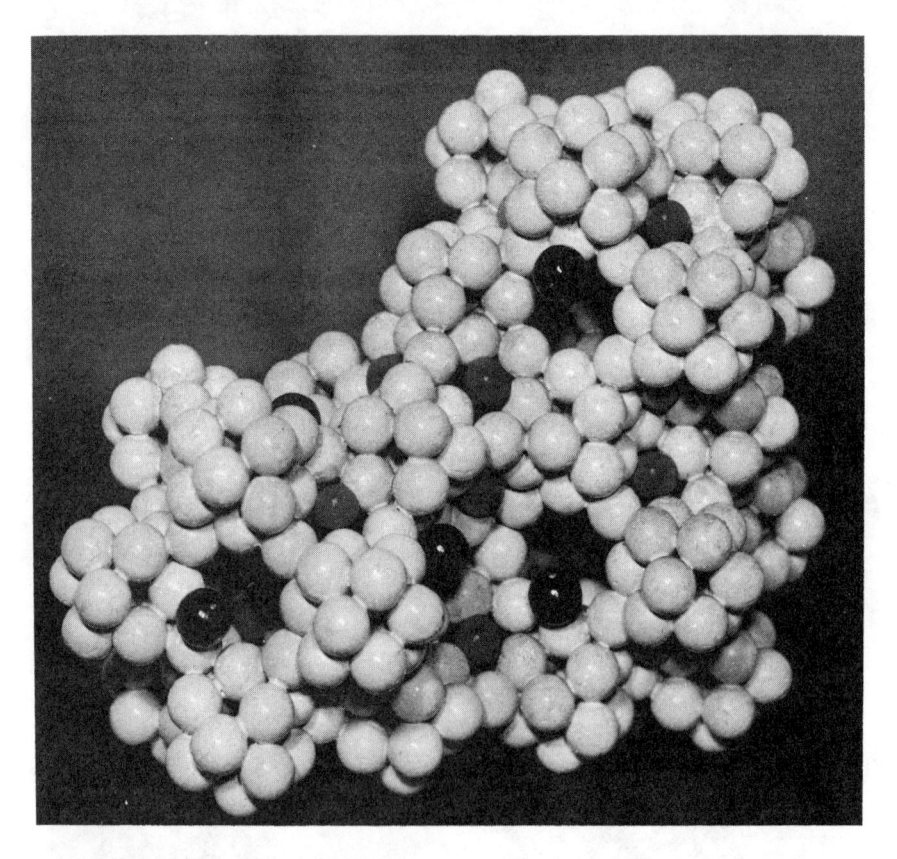

Figure 2.36 A space-filling model of the structure of zeolite A showing the packing of oxygen. Three unit cells are shown and the typical locations of cations in site I and II. The site I cations are indicated by the gray spheres and the site II cations by the black spheres. A ring of oxygens lies in each cubic face. Six apertures open into each α-cage with a free diameter of 4.2 Å. Each α-cage is connected to 8 β-cages by distorted 6-rings with a free diameter of 2.2 Å. The surface of the α-cage is characterized by 8 monovalent ions coordinated in a planar configuration with 6 oxygens of the framework and four ions in a one-sided coordination with at least 4 framework oxygens.

hedra. Ordering of Si/Al has been confirmed by substituting germanium for silicon in a synthetic form of zeolite A (105), and by structural refinement (106). However, it is more convenient to use the pseudo-cell of 12.3 A. In many preparations chemical analyses of zeolite A have indicated a Si/Al ratio of slightly less than 1. This would appear to be a variation from the Loewenstein rule. However, the occlusion of some $NaAlO_2$ within the β-cages, similar to feldspathoid behavior, has been postulated (105). Occlusion of one $NaAlO_2$ would give a SiO_2/Al_2O_3 ratio of 1.85. The usual ratio is 1.92 (102). The starting reactant mixture for zeolite A synthesis may be alumina-rich. (See Chap. 4.) However starting mixtures in which $SiO_2/Al_2O_3 = 2$ yield a stoichiometric zeolite A. There is no crystallographic evidence for $NaAlO_2$ occlusion.

Normally, zeolite A is synthesized in the Na-form. Other cationic forms are readily prepared by ion exchange in aqueous solution. Table 2.10 lists some cation exchanged forms with lattice constants and densities. As expected, density increases with the atomic weight of the cation. Except for lithium ion, cation exchange has little effect on the unit cell constant. When dehydrated, the unit cell constant of NaA decreases by only about 0.02 A, thus confirming the very rigid nature of the aluminosilicate framework.

Hydrated Zeolite A

Of the 12 sodium ions in the hydrated zeolite A, 8 are located near the center of the 6-rings on the 3-fold axis inside the α-cage (107). This

Table 2.10 Cation Forms of Zeolite A (102, 103).

Cation Content/Unit Cell	Hydrated Density (g/cc)	a (A)	No. H_2O's
Na_{12}	1.99	12.32	27
K_{12}	2.08	12.31	24
$Na_{4.2}Li_{7.8}$	1.91	12.04	24
$Na_{8.2}Cs_{3.8}$	2.26	12.30	–
$Na_{2.4}Tl_{9.6}$	3.36	12.33	20
Na_4Mg_4	2.04	12.29	–
Ca_6	2.05	12.26	30
$NaSr_{5.5}$	–	12.32	–
Ag_{12}	2.76	12.38	24

position is referred to as site I. The remaining 4 Na$^+$ ions appear to be located with water molecules in the 8-rings. Water molecules form a pentagonal dodecahedron in the α-cage (106).

Studies of hydrated calcium exchanged zeolite A (Ca/Na =1) containing 28 water molecules indicate that 16 water molecules coordinate with the cations inside the β-cage and α-cage (108,109). Twelve water molecules form a distorted dodecahedral arrangement in the large α-cage; 3 more water molecules may link through the 8-rings between the dodecahedral arrangements of water. Within the small β-cages, the water molecules apparently bond to the framework oxygens. These results are not completely consistent with nmr results which show that water in the larger cavities of zeolite A gives the same pattern as the isolated liquid. For comparison, single crystal studies of the calcium form of chabazite indicate clustering of water molecules around the cation (see Table 2.10).

In zeolites with small cavities, the cations are in contact with few water molecules. As the cavities increase in size, the cations may be surrounded by water molecules, and in zeolites with large cavities, the water may behave essentially as an isolated liquid phase. The pentagonal dodecahedral arrangement of water molecules is not unusual; this configuration is observed in clathrate compounds (hydrates that are formed between water and various guest species such as hydrocarbons, chloroform, and inert gases).

Dehydrated Zeolite A

In dehydrated zeolite A, 8 Na$^+$, Na$_I$, are displaced 0.4 A into the α-cage from the center of the 6-rings (103, 110). Three Na$^+$, Na$_{II}$, are located in the 8-rings displaced about 1.2 A from the center (111). The Na$_{II}$ cations, by partial blocking of the aperture, influence the adsorption of gases and vapors and regulate the pore size. The remaining sodium, Na$_{III}$, has been located opposite the 4-ring. When exchanged by Ca^{2+} so that there are 4 Ca^{2+} and 4 Na$^+$ in each unit cell, the 8 site I positions are occupied and the site II positions are vacant. Consequently the apertures are completely open and capable of admitting molecules with diameters of ~4.3 A. In dehydrated Tl$^+$ exchanged A, the larger Tl$^+$ ions are located off the plane of the 6-rings near site I and extend 1.12 A from the plane of the ring into the α-cage (103). Some are also positioned inside the β-cages. The re-

remaining 4 Tl⁺ ions are near the center of the 8-rings (112). In K⁺ exchanged A, potassium ions in site II restrict the free opening of the 8-rings so that molecules with diameters of >3 A are excluded.

From the unit cell volume and void space, the framework volume and the equivalent volume occupied by each tetrahedron can be calculated. In zeolite A this corresponds to 43 A³ as compared to 38 A³ for α-quartz. Therefore the packing of tetrahedra in the zeolite framework is about as dense as in crystalline quartz. The interstices of the zeolite crystal are characterized as consisting of relatively large cavities surrounded by densely packed oxygens which are contained in a crystalline three-dimensional polyanion. The cations are not well shielded and a molecule within the cavity is acted on by overlapping potential fields from opposite walls.

The volume of water is accounted for by assuming that water molecules occupy both the α- and β-cages, 930 A³ per unit cell. The volume of argon, nitrogen, and oxygen suggests that these species occupy only the large α-cages, 775 A³. For a molecule to enter the β-cages, it must be small enough to pass through the 2.2 A aperture, and also be able to displace the sodium ion which occupies the center of the ring. Through dipole-cation interaction, this occurs with water but does not occur with oxygen, argon, or nitrogen. (See Chap 5.)

Cobalt ions, in dehydrated partially exchanged zeolite A, are located in nearly perfect trigonal coordination with oxygens of the 6-ring. The zeolite studied by single crystal methods contained 4 cobalt ions and 4 sodium ions per unit cell. In the hydrated zeolite, the cobalt ions are located in two positions: one at the center of the β-cage, and three off the 6-rings well into the large α-cage on the 3-fold axes (113).

Since the total void space must remain constant within the rigid zeolite crystal, the volume occupied by various cations will affect the number of water molecules that can subsequently be added. The water content increases with decreasing ionic radius of the cation, from 22 for Tl⁺ (r = 1.49 A) to 27 for Na⁺ (r = 0.98 A). The highest value of 30 is found in CaA zeolite which contains ½ the number of ions. The steric effect imposed on cation exchange is similar to that which occurs during adsorption; that is, the structure exhibits an ion-sieve effect which depends on the size of the exchanging ion. (See Chap. 7.) Many cation forms of the zeolite can be prepared by exchange in aqueous solution. Exchange of lithium and magnesium ions occurs with considerable difficulty. Exchange of sodium ions with barium ions can be achieved but subsequent dehydration destroys the crystal structure, probably because of the size and charge of the barium ion. Similarly, cesium exchanges only partially. Large organic cations such as TMA are completely excluded as would be expected from size considerations (105).

Electrostatic Potential in Zeolite A

Methods for calculating the electrostatic potential in zeolite A have been proposed. The potential at various cation sites is consistent with the structural analysis. Within the cavities, the potential is determined by the total electrical charge of the aluminosilicate framework (114, 115).

Zeolite N-A (Table 2.19)

Zeolite N-A is a silica-rich zeolite with the type A structure but containing TMA ion as the major cation. It is more siliceous than zeolite A or ZK-4 (see the next section); the Si/Al ratio varies from 1.25 to 3.75 (116). As a result cations in site II are probably missing and the dehydrated zeolite exhibits adsorption properties comparable with the 4.2 Å aperture, similar to calcium exchanged A. Since the Si/Al ratio is greater than 1, the average Si, Al–O distance is less than that of zeolite A. In framework structures the average Si, Al–O distance varies from 1.61 Å in those containing no aluminum cations, to 1.75 Å (117); the lattice constant varies with the relative number of AlO_4 and SiO_4 tetrahedra. In zeolite A, the lattice constant is 12.32 Å as compared to 12.12 Å for N-A. The unit cell of N-A contains typically 5 less AlO_4 tetrahedra and 5 fewer cations. As synthesized, N-A contains the large TMA cation which results in a higher Si/Al ratio in the framework due to spatial limitations that necessitate a lower cation population. Vacuum activation accompanied by decomposition of the TMA ion produces a stable decationized form. Ion exchange studies of zeolite N-A have shown that the TMA ions are not replaceable by sodium or calcium. Removal of the TMA ions is accomplished only by thermal decomposition.

Zeolite ZK-4 (Table 2.66)

Zeolite ZK-4 is also a siliceous analog of zeolite A (118). Similar

ZEOLITES OF GROUP 3 91

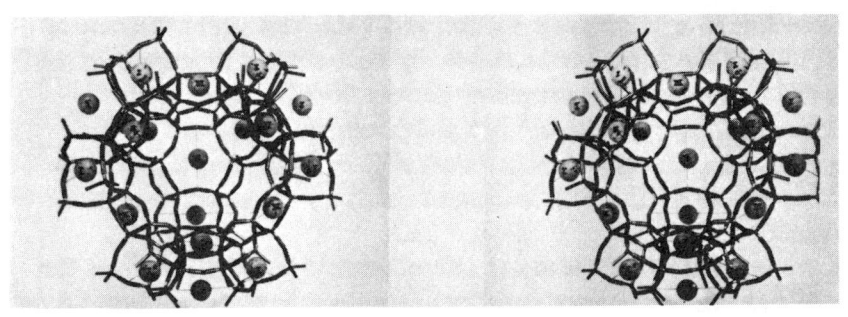

Stereo 2.18 Stereoscopic view of zeolite X showing cation sites surrounding the super cage. The view is perpendicular to [111]. Sites I, II, and III are shown. The nonplanar shape of the 12-ring is evident.

Stereo 2.19 Stereoscopic view of the framework structure of chabazite. This view is approximately paralled to the rhombohedral axis. The horizontal 6-rings are linked through tilted 4-rings to other horizontal hexagonal rings that are displaced both vertically and laterally. The centers of the 6-rings can be seen to lie in about the same pattern as the packing of spheres in simple crystals. The cubic and hexagonal close packing of spheres is generally represented as ABC or AB where A, B, and C represent the three possible horizontal projections. The same terminology can be applied if we designate the center of the hexagonal ring by the alphabetical symbol. When adjacent layers have the same letter designation, i.e., AABBCC the structure contains D6R units. If the adjacent layers have different symbols hexagonal rings are linked by tilted 4-rings. There are many possible structures that can be built hypothetically from paralled 6-rings in this way.

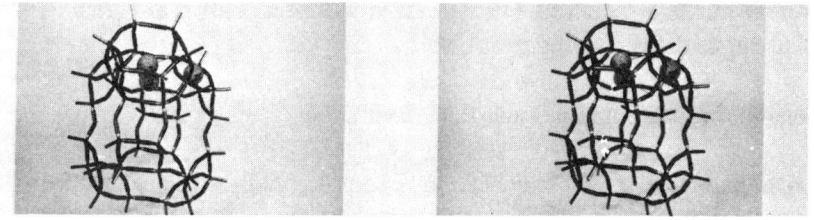

Stereo 2.20 A stereoscopic view of the framework of chabazite illustrating the main adsorption cavity or cage and the approximate location of the three cation sites. Site I in the center of the double 6-ring is at the origin of the unit cell and provides for a near octahedron of oxygen atoms. There is one S_I per unit cell. Site II is located near the center of the 6-ring but displaced away from S_I into the main cavity. There are two site II per unit cell. Site III occurs in 6 pairs of positions, each pair about 1 A apart and near the tilted 4-ring. The locations indicated in the figure is approximate.

to zeolite N-A, it contains sodium and TMA ions. As in the case of N-A, the TMA ions may be thermally decomposed to leave protons in in the framework. (Hydrogen or proton forms of zeolites are discussed in Chap. 6.) The sodium and potassium forms can be prepared by treatment with aqueous NaOH or KOH after thermal decomposition of the TMA ion, presumably by exchange of protonic sites for the sodium or potassium ions.

As in zeolite N-A the lower cation content in ZK-4 increases the effective size of the apertures for adsorption; in the dehydrated form, adsorption is equivalent to that of the Ca-form of zeolite A. The Na-form is able to adsorb n-hexane which is normally excluded by zeolite NaA. If the sodium ion population of the two types of sites in the A framework structure is the same, then ZK-4 contains on the average one monovalent type II ion as compared to four in zeolite A.

Two other synthetic zeolites, ZK-21 and ZK-22, having the A-type structure contain intercalated phosphate in amounts up to one phosphate in each β-cage (40).

K. ZEOLITES OF GROUP 4

The framework structures of the zeolites of group 4 are characterized by the double 6-ring, D6R, as the secondary building unit in their structural frameworks (Fig. 2.18).

These include the minerals chabazite, gmelinite, and R (related to chabazite); S (related to gmelinite); and zeolites X and Y (related to faujasite). (See Table 2.4.) Zeolites L and ZK-5 have no known mineral relatives. Considerable information on the structures of these zeolites has been reported.

Faujasite-Type Structures (Table 2.30, 2.63, 2.64)

Zeolites X and Y and faujasite have topologically similar aluminosilicate framework structures, although they are distinct zeolite species with characteristic differences. The unit cells are cubic with a large cell dimension of nearly 25 A and contain 192 $(Si,Al)O_4$ tetrahedra. The remarkably stable and rigid framework structure contains the largest void space of any known zeolite and amounts to about 50 vol% of the dehydrated crystal.

Chemical Composition

The chemical composition of faujasite is given in Table 2.30. The published data indicate a complex cation distribution including calcium, sodium and variable amounts of potassium and magnesium.

The chemical compositions of zeolites X and Y are related to the synthesis method. (See Chap. 4.) The zeolites are distinguished on the basis of chemical composition, structure, and their related physical and chemical properties (59). Differences are found in the cation composition and distribution, the Si/Al ratio and possible Si-Al ordering in tetrahedral sites.

In zeolites X and Y the relation between the number of tetrahedral Al atoms, N_{Al}, and the Si/Al ratio is:

$$N_{Al} = \frac{192}{(1+R)}$$

where $R = N_{Si}/N_{Al}$.

The aluminum ions (N_{Al}) in the unit cell of zeolite X vary from 96 to about 77. In zeolite Y N_{Al} is about 76 to 48. The value of R varies from 1 to 1.5 for zeolite X and from greater than 1.5 to about 3 for zeolite Y. For the hydrated zeolites as synthesized in the sodium form, a_0 increases from about 24.6 to 25.0 A. (See Fig. 2.37.) The cell constant of faujasite, 24.67 A ($N_{Al} = 59$) does not fit this relation. It is about 0.04 A too small.

A similar plot has been interpreted as indicating Si/Al ordering at particular compositions since discontinuities were indicated at Al contents of 80, 64, and 52 atoms per unit cell (119). This was further supported by similar data obtained on zeolite X with gallium substituted for aluminum (120). The plot of lattice parameter versus Si/Ga ratio shows a break at a Si/Ga ratio of about 2 and another at 1.4. It has been pointed out, however, that the statistical reliability of the data appear better if it is treated in three populations rather than a single population (121). No estimate of the accuracy of the Si/Al ratios in the series of sodium zeolites was given.

Framework Structure (Stereo 2.17)

The aluminosilicate framework consists of a diamond-like array of linked octahedra which are joined tetrahedrally through the 6-rings. The linkage between adjoining truncated octahedra is a double 6-

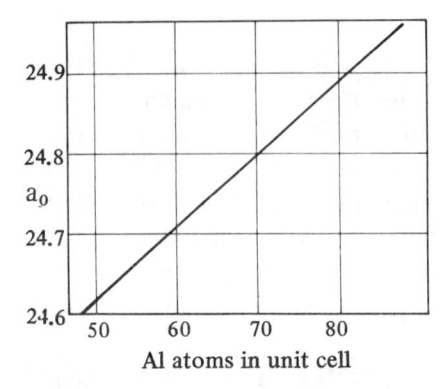

Figure 2.37 Relation between the cell constant, a_0, and aluminum atom density in number/unit cell for hydrated sodium X and Y zeolites. The linear relation is $a_0 = 0.00868\ N_{Al} + 24.191$ which was established by least squares treatment of 37 experimental points. The Si/Al ratio was determined by wet chemical analysis of 37 pure preparations. Analytical error ± 0.5 wt% in SiO_2 and Al_2O_3. Error in a_0 determination estimated to be less than ± 0.005.

$$a_0 = \frac{192b}{1 + (N_{Si}/N_{Al})} + c$$

where b and c are the slope (0.00868) and intercept (24.191), respectively, of the linear relation (59).

ring (D6R) or hexagonal prism, containing 12 (Si,Al)O$_4$ tetrahedra. The framework structure can also be described in terms of the D6R units. The aluminosilicate framework as illustrated by a model using skeletal tetrahedra is shown in Stereo 2.17. Stereo 2.18 shows the supercage and some of the sodium ion locations in zeolite X.

Models illustrate more clearly the nature of the internal surface of the zeolite (Fig. 2.38). The dense packing of oxygens (39 Å^3 per tetrahedron as compared to 38 in α-quartz) has been largely overlooked in the calculation of dispersion interactions between adsorbed molecules and the internal zeolite surface (Chap. 8).

Six cation sites have been defined as shown in Fig. 2.39. Approximate locations of sites I, II, and III are also shown in the model in Fig. 2.38. During initial crystal structure studies on hydrated zeolite X, 48 out of 80 Na$^+$ ions (per unit cell) were located in site I and II positions (107). A later study on a NaX single crystal assigned Na atoms to sites I, I', and II (122). Considerable evidence suggests that the remaining cations and the water molecules behave as a strong electrolyte solution and float freely through the framework. This concept is supported by the infrared spectrum which shows the stretch and deformation frequencies of normal water molecules, the heat of adorption of water which in zeolite X is comparable to the heat of vaporization of water, self-diffusion studies which show that the sodium ions undergo exchange very rapidly, and electrical conductivity studies which have shown that the cations in hydrated zeolite X behave in the same way as they do in salt solutions. The cations

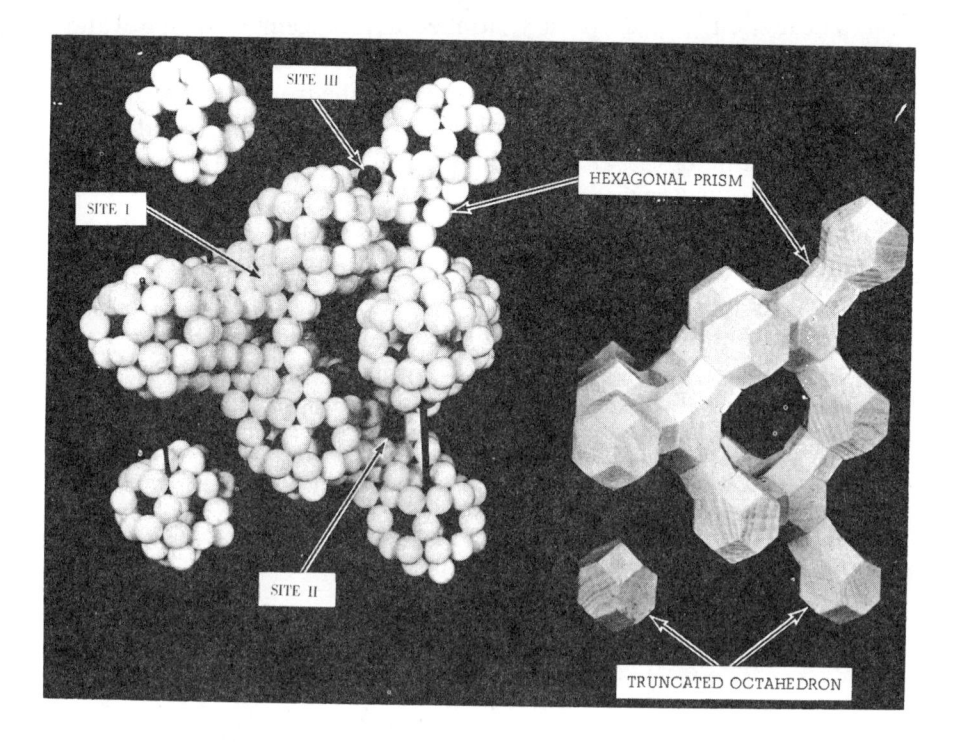

Figure 2.38 The structure of zeolites X, Y, and faujasite as depicted by a model showing the spacial arrangement of truncated octahedral units in a diamond-type array. The space-filling model showing the approximate location of oxygen atoms in the framework is also pictured. The arrangement of truncated octahedra in one unit cell is shown. The large 12-ring is visible as well as the smaller 6-rings which form apertures into the β-cages. The cation positions typical of zeolite X are illustrated as site I within the hexagonal prism unit, site II adjacent to the single 6-rings, and site III within the main cavities (104).

must move upon dehydration from positions where they are coordinated with water positions near framework oxygens. In sodium-enriched faujasite crystals, the site I positions are vacant.

Cation Positions in Hydrated Faujasite-Type Zeolites (Tables 2.11-2.13)

Faujasite, when fully hydrated, contains about 235 water molecules distributed in the large supercages and the small β-cages. Structural investigations on hydrated faujasite of undetermined cationic composition, but assuming a total of 43 sodium and calcium ions per unit cell, have shown that 17 cations are located within the β-

96 STRUCTURE OF ZEOLITES

cages near site I (Fig. 2.38) in a tetrahedral arrangement with 4 water molecules (123). There are four such cation positions within each cage. Each cation is surrounded by a distorted octahedron of three framework oxygens and three water molecules. The remaining 26 cations were not definately located, but may occupy sites next to the single 6-rings inside the β-cages.

The difference in positions of sodium ions should be noted; site I in faujasite is not occupied by cations; in hydrated NaX, site I is occupied by nine Na$^+$ ions.

The location of Ca^{2+} ions in hydrated Ca exchanged faujasite are

Figure 2.39 The cation sites and their designation in zeolites X, Y, and faujasite. Starting at the center of symmetry and proceeding along the 3-fold axis toward the center of the unit cell, site I is the 16-fold site located in the center of the double 6-ring (hexagonal prism). Site I' is on the inside of the β-cage adjacent to the D6R. Site II' is on the inside of the sodalite unit adjacent to the single 6-ring. Site II approaches the single 6-ring outside of the β-cage and lies within the large cavity opposite site II'. Site III refers to positions in the wall of the large cavity, on the 4-fold axis in the large 12-ring aperture. The 4 different types of oxygens O(1), O(2), O(3), O(4) are also indicated in their relative positions.

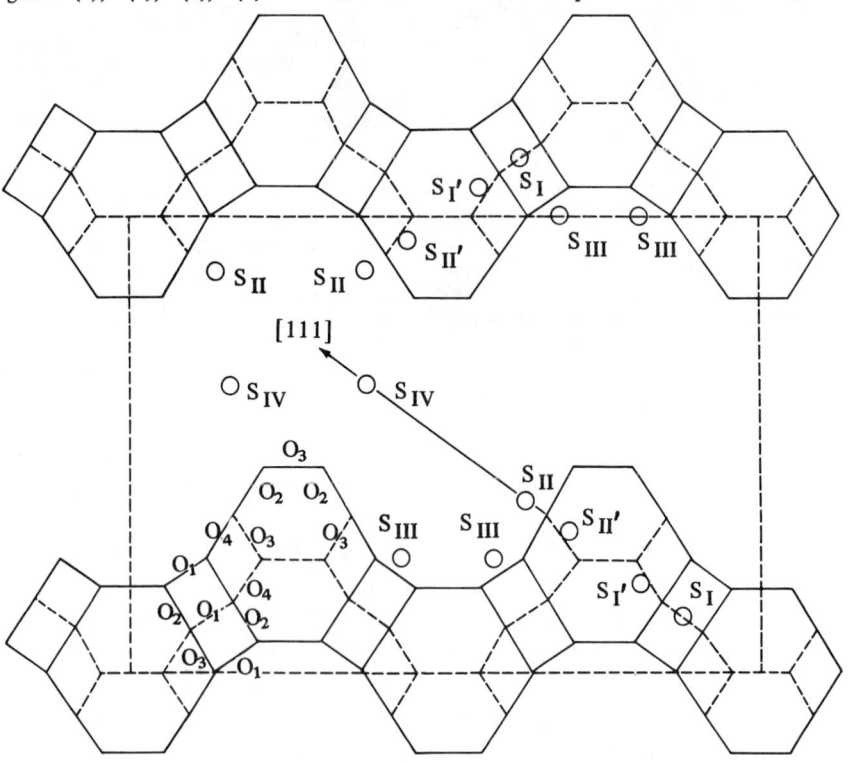

given in Table 2.11. Nearly all of the Ca^{2+} ions are located in the β-cages in site I' and site II'. Some appear to sit in the 12-rings at site IV. Distributions of other cations in the hydrated zeolites are given in Tables 2.11, 2.12, and 2.13.

Cation Positions in Dehydrated Zeolites (Tables 2.11, 2.12, 2.13)

In dehydrated zeolite X, the proposed 48 site III provide for a total of up to 96 monovalent cation positions corresponding to an Si/Al ratio of 1. The occupation of site III by Na^+ ions in dehydrated X was confirmed by the esr spectrum of the Na_6^{5+} complex (124).

The Na^+ ions in dehydrated zeolite Y occupy three sites. On an average, 7.5 Na^+ ions were found in site I, 30 Na^+ ions in site II, and 20 Na^+ ions in site I'. Similar site occupancies in K exchanged and Ag exchanged zeolite Y were found. X-ray diffraction studies on faujasite single crystals which were treated with aqueous Ca^{2+} salt solutions and then dehydrated, have shown that the Ca^{2+} ions occupy site I in preference to site II (125). La^{3+} ions of the dehydrated La exchanged zeolite Y which are in site I at ambient temperature move to site II at 700°C. This effect is reversible (126).

Since only 16 calcium ions are available to occupy the 32 sites II in zeolite Y, half of the sites will have a full charge of +1 centered on the cation while the other half have a net negative charge of -1 distributed on the oxygen ions surrounding the site. Similarly, CaX with a total population of 48 divalent ions would have all sites occupied and no localization of a charge deficiency. Calcium exchanged zeolite Y, Si/Al ratio = 2, has 32 divalent ions per unit cell, presumably 16 at site I and 16 at site II. Since the 32 sites II have a total of 32 units of negative charge localized on these sites, each must have only one unit of charge on the average, only half enough to neutralize the calcium ion. On the other hand, CaX (Si/Al = 1) with 48 calcium ions per unit cell may have all sites occupied with localized charge deficiency.

From Tables 2.11, 2.12, and 2.13 it is apparent that the distribution of cations in the faujasite-type zeolites is dependent upon the presence of adsorbed water or residual OH⁻ groups. For example, the distribution of sodium changes from location in site I' and site II' to site I and site II upon hydration (Table 2.11). With polyvalent ions the situation is more complex and site I is preferred by calcium

Table 2.11 Cation Distribution in Faujasite

Site		Na^{+a}	K^{+g}	Ca^{2+b}	Ba^{2+h}	Ni^{2+c}	Ce^{3+d}	La^{3+e}	Ca^{2+f}, Na^+
16 I	De	10	8.6	14.2	7.3	10.6	3.4 Na^+	11.8	-
	Hy	-	-	-	-	-	-	-	-
32 I'	De	9	12.9	2.6	5.0	3.2	11.5	2.6	-
	Hy	14	-	9.7	-	-	18 Na^+	3.3	17
32 II'	De	-	-	-	-	1.9	16, O_x	1.4, O_x	-
	Hy	15	-	11.5	-	-	32, O_x	28.3, O_x	32, O_x
32 II	De	32	31.7	11.4	11.3	6.4	10.7, Na^+	1.5	-
	Hy	11	-	23.4, O_x	-	-	26, O_x	14.2, O_x	11, O_x
48 III	De	-	-	-	-	-	-	-	-
	Hy	-	-	-	-	-	-	-	-
16 IV	De	-	-	-	-	-	-	-	-
	Hy	-	-	2.2	-	-	6	10.3	-
Reference		129	121	125	121	131	132	133	123
				130	130				

[Diagram showing sites: S_{II}, $S_{II'}$, $S_{I'}$, S_I, S_{III}, S_{IV} along [111] direction]

[a] Data based on Na- and Ca- enriched faujasite from Kaiserstuhl. Dehydration *in vacuo* at 350°C. Si/Al = 2.27. Degree of exchange assumed to be > 80%.

[b] Data on Ca^{2+} form from a completely exchanged crystal; composition: $Ca_{27}[(AlO_2)_{56}(SiO_2)_{133}] \cdot xH_2O$. Dehydration *in vacuo*. Temp. raised to 475°C over 4-days and held at 475°C for 12 hr before cooling to RT.

[c] Ni^{2+} faujasite obtained by treatment with 1.0 M $NiCl_2$ at 90°C. Dehydrated *in vacuo* at 400°C for 7 hr, 1 x 10^{-6} torr. Intensity data collected at RT. Composition: $Ni_2Ca_4[(AlO_2)_{58}(SiO_2)_{134}]$. Two site I' peaks assigned to Ni. Location of site II is further from the 6-ring than usual site I.

[d] Ce^{3+} faujasite uc contents given as: $Ce_{12}Ca_{7.6}Na_{7.8}[(AlO_2)_{59}(SiO_2)_{133}]$ · 270 H_2O. Data on hydrated crystal at RT. Dehydration *in vacuo* at 350°C. Six Ce^{3+} in hydrated crystal (site IV) located at random in supercages. Residual H_2O or OH shown as O_x.

[e] La^{3+} faujasite completely exchanged except for possible hydrolysis. Residual Na, Mg, Ca < 0.1%. Dehydrated *in vacuo* at 475°C for 7 hr. Data collected at RT. 19 La atoms expected but 15.9 found. The La in site I is displaced along the 3-fold axis, half-atoms 0.17 A from site I, Structure (420°C) showed same distribution of La^{3+} 11.7 site I, 2.5 site I', 1.4 site II.

[f] Structure determined on hydrated crystal of mineral; Na-Ca cations assumed to be dominant; 17 located. H_2O molecules shown by O_x.

[g] 33 K^+ ions per unit cell.

[h] 23.1 Ba^{2+} ions per unit cell.

Table 2.12 Cation Distribution in Zeolite X

Site		Na$^+$	Ca^{2+}	K$^+$	Sr^{2+}	La$^{3+}_{25°}$	La$^{3+}_{25°}$	La$^{3+}_{425°}$	La$^{3+}_{735°}$	Ce^{3+}
16 I	De	4	13.3	8	11.2	—	32.1, O$_x$	5.0	5.2	4.2
	Hy	9	—	8.9	2.1	—	—	—	—	11.4, O$_x$
32 I′	De	32	5.0	—	7.0	30	13.8	15.2	14.1	22.8
		8	13							
	Hy	12 H$_2$O	16 H$_2$O	7.2	11.1	12	—	—	—	9
32 II′	De	—	6.0	—	4.2; 5.4, O$_x$	32, O$_x$	24.3, O$_x$	—	—	—
			26, O$_x$							
	Hy	26 H$_2$O	—	—	32, O$_x$	32, O$_x$	—	—	—	32, O$_x$
32 II	De	32	25	—	19.5	—	13.2	4.9	6.3	—
	Hy	24,	27	23.2	15	17	—	—	—	21
		8 H$_2$O	H$_2$O							
48 III	De	4	—	—	—	—	—	—	—	—
16 IV	De	—	—	—	—	—	3.4	—	—	2.6
	Hy	32	—	—	—	4	—	—	—	—
8 U	De	—	—	—	—	—	8, O$_x$	3.3, O$_x$	—	—
Reference		104, 107	134, 135	136	131, 137	132	138	138	138	139
		124, 122								
		130								
Footnote		a	b	g	e	c	d	d	d	f

[diagram of sites: S$_{II}$, S$_{II'}$, S$_I$, S$_{I'}$, S$_{III}$, S$_{IV}$, S$_U$ along [111] direction]

[a] NaX of composition Na$_{80}$[(AlO$_2$)$_{80}$(SiO$_2$)$_{112}$]·xH$_2$O. 32 Na$^+$ ions in hydrated zeolite not located. Assignment in dehydrated zeolite postulated as most probable. Probably space group Fd3 for Na$_2$X single crystal; 47 Na$^+$ not located.
[b] Single crystal, composition Ca$_{40}$[(AlO$_2$)$_{80}$(SiO$_2$)$_{112}$]·235 H$_2$O activated by heating in vacuum at 500°C.
[c] Data for LaX on powder; uc contents: La$_{29}$Na$_{04}$[(AlO$_2$)$_{86.8}$(SiO$_2$)$_{105.1}$]· 270 H$_2$O. Reported to be calcined (specific conditions not published) and "equilibrated to atmospheric conditions." Readsorption of H$_2$O from atmosphere may have occurred.
[d] This data obtained on powder; uc contents: Na$_{46}$La$_{26.4}$Al$_{82}$Si$_{110}$O$_{385}$∼ 260 H$_2$O. Dehydration by heating in dry He at 100 μ.
[e] Data for hydrated SrX obtained on hydrated powder; uc contents: Sr$_{42}$Na[(AlO$_2$)$_{85}$(SiO$_2$)$_{107}$]·2 H$_2$O.
[f] CeX powder, composition Ce$_{26}$Na$_{0.6}$[(AlO$_2$)$_{88}$(SiO$_2$)$_{104}$]·n H$_2$O. Activation by heating in N$_2$ at 540° for 2 hr. 9 deficient cations.
[g] KX powder, composition: K$_{84.5}$Na$_2$[(AlO$_2$)$_{86.5}$(SiO$_2$)$_{105.5}$] activated by heating *in vacuo* at 500°C.

Table 2.13 Cation Distribution in Zeolite Y

Site		Na^+	K^+	Ag^+	Cu^{2+}	La^{3+}_{rt}	Ni^{2+}	$La^{3+}_{725°}$
16 I	De	7.5	12	16	3.4	13.1,O_x	12	5.2
	Hy	—	—	—	—	—	3.5	—
32 I'	De	19.5	14.2	10.7	10.4	16.0	1.1	8.9
	Hy	—	13.6	—	—	—	6	—
32 II'	De	—	—	—	—	29.1,O_x	—	—
	Hy	—	—	—	—	—	1.9	—
32 II	De	30	30	28.3	21 Na	—	21.3 Na	5.5
	Hy	—	17.8	—	—	—	23.6 Na	—
8 U	De	—	—	—	4.6 OH	—	—	—
	Hy	—	—	—	—	—	—	—
Reference		140	136, 140, 141	140	142	126	128	126
Footnote		a	a,c	a	d	b	e	b

[111] ←
S_{II}
S_{III}
$S_{II'}$
U
$S_{I'}$
S_I
$S_{I'}$
U
$S_{II'}$
S_{III}
S_{II}
→
S_{IV}

[a] Na, K, Ag, Y. Composition of NaY powder given as $Na_{57}[(AlO_2)_{57}(SiO_2)_{135}]$. The K^+ and Ag^+ forms prepared by cation exchange; complete exchange assumed. Dehydration at 350°C to 10^{-5} torr. Data are at room temperature.

[b] LaY. Data on powder sample of composition: $Na_{13.4}La_{16.3}(Al_{55}Si_{137}O_{388})$ 270 H_2O, (Na + 3 La) = 62.3 which is 7.3 equivalents in excess of the 55 Al. Presence of some La OH^{2+} is indicated. Dehydration in vacuum at 350°C, data collected at room temperature and at 725°C with sample under dry He. The U site is in the center of the β-cage and is occupied by residual oxygen (H_2O or OH). This is removed at 725°C with movement of La^{3+} in site I' to site I and site II. This effect appears to be reversible, i.e., on cooling to room temperature, La^{3+} in site I and site II returns to site I' and site II'. One La is equivalent to 6.5 Na or O_x. Assignment of Na not made.

[c] KY powder, composition: $K_{47.5}Na_{0.7}[(AlO_2)_{48.2}(SiO_2)_{143.8}]$. References include structure data on intermediate compositions of Si/Al ratio = 2.51 and 1.75 (69.8 and 54.7 K^+ ions per unit cell).

[d] Cu, NaY powder, composition: $Cu_{16}Na_{24}[(AlO_2)_{56}(SiO_2)_{136}]$. Activated in vacuum at 500°C.

[e] NiY powder, composition: $Ni_{14}Na_{23}H_5[(AlO_2)_{56}(SiO_2)_{136}]$. Activation by heating O_2 at 600°C and vacuum for 12 hr.

Table 2.14 Sites Occupied by Mn^{2+} in Various Cation Forms of Zeolite X and Y (144, 145)

Specimen	No. of Exchanged Cations per Unit Cell	Residual Sodium Ions per Unit Cell	Hydrated	200°C	300°C	400°C	600°C
X/Li$^+$ Mn^{2+}	72	12	I, A	+, I	+, I		
Y/Li$^+$ Mn^{2+}	32.4	21.6	A	I	I	I	
X/Na$^+$ Mn^{2+}		84	A	+, II	+, II		
Y/Na$^+$ Mn^{2+}		54 (Si/Al=2.45)	A	i, +	II, +	II	II
X/K$^+$ Mn^{2+}	74	10	A	II, +	II, +		
Y/K$^+$ Mn^{2+}	49.7	4.3	A	i, +	i, +	II, i	
X/Cs$^+$ Mn^{2+}	47	37	A, I	II	II		
Y/Cs$^+$ Mn^{2+}	37.3	16.7	A	I, ii	I, ii	I	
X/Mg^{2+} Mn^{2+}	27	30	I, A	II	II		
Y/Mg^{2+} Mn^{2+}	20.3	13.4	I, A	I, II	II	II	II
X/Ca^{2+} Mn^{2+}	37	10	A	ii	I, II	II, I	
Y/Ca^{2+} Mn^{2+}	20.5	13.0	A	ii	II, ii	II	II, I
X/Zn^{2+} Mn^{2+}	31	22	I	I	I	I	
Y/Zn^{2+} Mn^{2+}	18.0	18.0	A	I	I	I	
X/La^{3+} Mn^{2+}	27	3	A, I	I		I	
Y/La^{3+} Mn^{2+}	13.4	13.8	I	I	I	I	I, II

A, hydrated sites; + other sites. First site listed has higher population.
  i covalent sites

Composition of zeolite X: Na$_{84.4}$[(AlO$_2$)$_{84}$(SiO$_2$)$_{108}$] 250 H$_2$O
  zeolite Y: Na$_{54}$[(AlO$_2$)$_{55.7}$(SiO$_2$)$_{136.3}$] 249 H$_2$O

and nickel ions. Lanthanum ions move from site IV, a location within the large aperture leading to the supercage, to the preferred location in site I. The distribution of lanthanum ions in dehydrated faujasite seems to differ markedly from the cerium ion distribution. In zeolite X Ca^{2+} and Sr^{2+} ions prefer locations at site I, site I', and site II with no population in the higher sites. Site II is the main functional surface cation. The distribution of La^{3+} in dehydrated zeolite X at temperatures ranging from ambient to 735°C is of interest. The La^{3+} ions change with increasing temperature from site II to site I leaving a decreased population of site II. As the residual oxygen is removed with increasing temperature of activation, the La^{3+} ions prefer site I and site I'. In Table 2.13, data for the univalent cations Na^+, K^+, and Ag^+ in zeolite Y are very similar, with preferred population in site II, site I', and site I. A considerable number of surface cations are present in site II. There is no distribution in the higher sites, site III and site IV. In zeolite Y, lanthanum ions redistribute in the dehydrated zeolite as the temperature increases to 725°C. This change is reversible. Lanthanum ions located in site I' redistribute into site I and site II as the residual oxygen, presumably from water or hydroxyl is removed. During the process of ion exchange, the lanthanum ions hydrolyze to form hydroxy cation complexes such as $LaOH^{2+}$. The presence of hydroxyl groups in LaY has been studied extensively by infrared (see Chap. 6). The hydroxyl groups are introduced with the partially hydrolyzed cation during exchange. On dehydration, a hydroxyl ion is left which may link two lanthanum ions within the β-cage located in site I'. With removal of hydroxyl groups, part of the cations shift to a better coordination in site I.
At room temperature there is a marked difference in the distribution of lanthanum ions in faujasite and zeolite Y. In faujasite nearly all of the lanthanum ions occupy site I whereas in zeolite Y the preferred location appears to be site I'. Lanthanum ions in faujasite are displaced from site I along the 3-fold axis about 0.17 A from site I. Apparently 3 oxygens in one ring are closer than the 3 oxygens in the opposite 6-ring. This may be caused by the presence of more aluminum atoms, due to ordering, in one ring than in the other. It is possible to locate 16 lanthanum ions in site I and up to 32 in site II. It is also possible to place up to 22 lanthanum ions, 7 A from each other, in site I, site I', and site II if the total population in site I' is less than

8. Site I is preferable electrostatically, unless extra framework species such as hydroxyl groups are available. The trivalent ions preferentially occupy site I and to some degree site II and site I' when the zeolite is fully dehydrated and dehydroxylated.

The occupancy of cation sites by divalent cations in zeolite X and faujasite depends on the number of residual water molecules (127). Since the polarizing power of the cation determines the amount of water retained, then smaller cations retain more water molecules than larger ions. The approximate rule was proposed that $N_I = 2(16 - N_I)$ for cation occupancies in site I and I'. This depends, however, on the residual water content. The site I' and II' occupancies are related to the number of residual water molecules in each unit cell. In site II', for example, the number of cations is about equal to the water content. At zero water content sites I' and II' should be empty.

Another structural study of nickel exchanged zeolite Y (14 Ni, 10 Ni, and 19 Ni ions per unit cell) showed that during dehydration the nickel ions move into site I as water molecules are removed. The maximum site I occupancy was 12 per unit cell, many less than the full occupancy allows. In comparison, calcium ions prefer sites within the β-cage (128).

Structurally Related Physical Properties

The vapor phase adsorption of a critically sized molecule, triethylamine, $(C_2H_5)_3N$, with an estimated kinetic diameter of 7.8 Å was studied on calcium exchanged zeolites of typical X and Y compositions as shown in Fig. 2.40 (59).

If we consider the Ca-form of zeolite Y with 64 aluminum atoms per unit cell, corresponding to a Si/Al ratio of 2, a Ca^{2+} ion at site I is

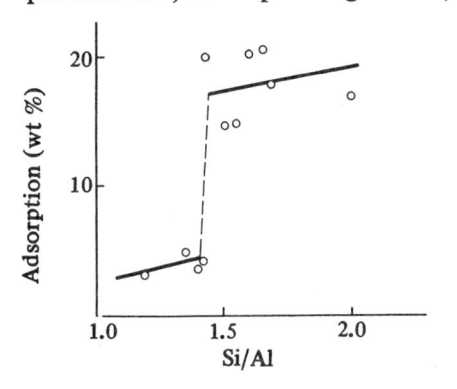

Figure 2.40 Adsorption of triethylamine on calcium X and Y. The adsorption of the amine in wt% of the weight of activated zeolite was measured at 28°C and at a total pressure of 2.2 torr (p/p_0 = 0.03). The dashed line indicates the sharp discontinuity at Si/Al of 1.5. The degree of calcium exchange, that is, the ratio of equivalents of calcium to aluminum, is greater than 0.85 for all of the zeolite samples (59).

surrounded by 12 tetrahedra of the D6R unit. Four of these tetrahedra will on an average contain Al^{3+} ions and carry a negative charge of one. The single Ca^{2+} ion will balance two of the four negative charges leaving two additional charges to be balanced by the cations in site II. All of the tetrahedra are related equivalently to site I and site II. In site II the calcium ion is surrounded by 6 tetrahedra (on an average, two AlO_4 and four SiO_4) having one negative charge.

The $(C_2H_5)_3N$ was rejected by the dehydrated zeolite CaX (Si/Al ratio less than 1.5) and readily adsorbed at room temperature by the CaY (Si/Al ratio greater than 1.5). A proposed explanation is that calcium ions in site II distort the framework of the zeolite slightly if the population of the calcium ions in site II is greater than 75%. When the total divalent cation population is 40 or greater (5 Ca^{2+} ions in each of the eight β-cages, 1 ion in each of the two site I and three in site II) distortion may occur and restrict the main aperture to the point where the critically sized molecule is not readily adsorbed. When the cation population is less than 40 (less than 5 Ca^{2+} ions in each truncated octahedron or β-cage), the effect on the framework is diminished and distortion of the apertures does not result. This is confirmed by studies of the variation in the lattice constant, a_0, with the Si/Al ratio (or Al^{3+} ion population) for a series of dehydrated Ca exchanged zeolites X and Y as shown in Fig. 2.41. A discontinuity in the lattice constant near 77 and 62 Al atoms per unit cell is observed. This may be caused by the framework distortion due to movement of calcium ions with dehydration into more suitable sites such as was observed in the zeolite chabazite (143). The second discontinuity corresponds to a distribution of about four Al and eight Si in the hexagonal prism.

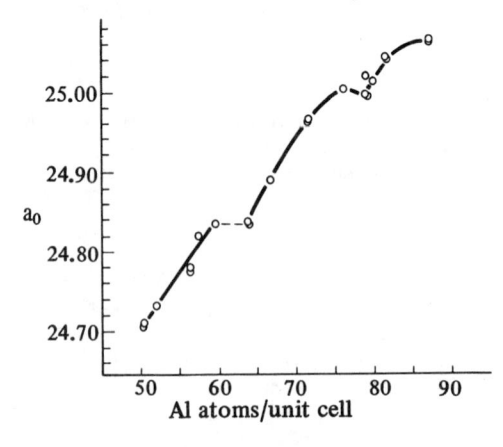

Figure 2.41 The relation between the cell constant, a_0, and the Al/ unit cell of dehydrated calcium-exchanged X and Y zeolites. The degree of calcium exchange is greater than 0.85 for all zeolite samples. Dehydration was carried out by heating in dry air at 400°C for 16 hrs (59).

In another study of the relationships of unit cell constants to the aluminum content of sodium X and Y zeolites, discontinuities in the curve were observed near 77 and 62 aluminum atoms per unit cell. These correspond to those observed in Fig. 2.41. It was suggested that these discontinuities correspond to phase changes which occur along with changes in the Si-Al ordering at these compositions (119). These results were supported by a similar study on a series of gallium substituted zeolites (120, 121). No estimate of the accuracy of the determination of the Si/Al ratio is given. In a comparison of the two sets of data, however, the curves are clearly parallel (121).

The cation distribution in dehydrated zeolites X and Y is consistent with studies of the adsorption of gases and measurements of isosteric heats of adsorption (Chapter 8). Adsorbate interaction energies and heats of adsorption are related to the occupancy of the site I positions by calcium ions in zeolite Y and correlate with the occupancy of the site II positions. The calcium ion in site I is well shielded and does not interact with gas molecules. Occupancy of site II is essential for molecule-zeolite interaction.

Limited information is available on the physical, chemical, and gas adsorption properties of faujasite. This is largely due to the unavailability of faujasite (it is an extremely rare mineral and specimens of any quantity are not available). Adsorption studies for perfluorotriethylamine, $(C_2F_5)_3N$, indicate that faujasite has a smaller aperture size than Na-Y (Chapter 8).

The chemical analyses of faujasite show the presence of four cations, Na^+, Ca^{2+}, Mg^{2+}, and K^+. Experience with ion exchange in other zeolite minerals indicate that Mg^{2+} cations in zeolite minerals are not readily exchangeable. The adsorption capacity of faujasite is generally less than that of zeolite Y. (See Chap. 8.)

Cation Distribution by Electron Spin Resonance

An electron spin resonance (esr) technique, using Mn^{2+} as a probe cation and monitoring its location, has been utilized to determine the distribution of univalent alkali metal cations, divalent alkaline earth cations and Zn^{2+} and La^{3+} in zeolite X and zeolite Y, hydrated and dehydrated at elevated temperatures (144, 145). The zeolite, after exchange to a high degree with the major cation was then treated to introduce about 0.27 atoms of Mn^{2+} per unit cell. The observed esr spectra were related to Mn^{2+} in five environments designated as follows:

A–	Hydrated Mn^{2+}
I–	Mn^{2+} in site I
II–	Mn^{2+} in site II
i–	Mn^{2+} in more covalent centers
ii–	Mn^{2+} in intermediate centers

The results are summarized in Table 2.14. In the hydrated zeolites, the high field strength ions, Li^+, Mg^{2+}, Zn^{2+}, and La^{3+} occupy the larger cavities in the structure while low field strength ions, Na^+ and K^+, are found in site I; that is, Mn^{2+} ions were observed in site I in the Li, Mg, Zn, and La exchanged forms and not in the Na and K exchanged zeolites. In dehydrated Na and K zeolites, Mn^{2+} does not displace the univalent ion from site I and instead occupies site II. The order of preference of site I in divalent ions was determined to be Ca^{2+}, Mg^{2+} > Mn^{2+} > Zn^{2+}. La^{3+} ions in X and Y behaved similarly; Mn^{2+} occupies site I even at 400°C. This is explained by the presence of residual OH. The covalent character of Mn^{2+} as shown by the spectra of type i was explained as due to MnO and $MnOH^+$ groups formed by hydrolysis during ion exchange. These groups may stabilize the location of di- and trivalent ions within the β-cages as indicated by x-ray evidence (124). With decomposition of the groups by severe dehydration conditions, more cations occupy sites I and II.

The esr spectra of a dehydrated copper exchanged zeolite X suggest that Cu^{2+} ions are distributed in two different sites and it is postulated that they occupy sites II' and I'; that is, they are located in the β-cages. Some of these, located in Site II', were observed to interact by complexing with adsorbed NH_3 or pyridine (146). In a separate x-ray diffraction study, pyridine was shown to cause migration of copper ions into the large supercages (142).

Electrostatic Potential and Field

The electrostatic potential and field near site II have been calculated in purely ionic models of calcium X (Si/Al=1) and calcium Y (Si/Al=2) (147, 148). For these calculations it was assumed that the zeolite is fully ionic and covalent bonding was neglected. Every ion was given its full ionic charge. Repulsive interactions were neglected. Calculations of the variation in electrostatic energy for hydrated zeolite

X and Y in the sodium and calcium form showed that based upon the ionic model, the lowest electrostatic energy was obtained in calcium Y by putting all of the cations outside the β-cage. For the sodium exchanged Y three configurations have closely similar energies and it makes little difference in the electrostatic energy if all of the Na$^+$ ions are outside the β-cage or if one or two Na$^+$ in the β-cages moved from site II to site II'. Calculations of the field for X and Y zeolites containing Na$^+$ and Ca^{2+} were made. The field was calculated as a function of the distance from the site II cation along the 3-fold axis. The variation of the magnitude of the field vector proceeding on a straight line from the center of the D6R unit, site I, to one of the 12-rings that outline the large pore of the zeolite is shown in Table 2.15. Starting from an occupied site II the field magnitude is 6.3 V/A at a distance of 2 A from site II (1A from the surface of a Ca$^{2+}_{II}$). For an unoccupied site II it is 0.46 V/A. The field is stronger near an occupied site than it is near an unoccupied site. For a given cation, the field is stronger in zeolite Y than in X; site III fields are higher than site II fields and the Na$^+$ ions are poorly shielded.

No repulsive components are considered in this treatment. A partially covalent character of the crystal would result in smaller fields at both positions than those that were determined for the purely ionic models. Electrostatic fields of this magnitude should cause shifts of the bonding electrons in adsorbed molecules and consequently this is of interest in adsorption and catalysis (147).

Since the oxygen ions in zeolites are in positions of low symmetry, effects due to induced dipoles are important and make an appreciable contribution to the values of the total field at different positions in the zeolites. For zeolite X (Si/Al = 1, Na$^+$ and/or Ca^{2+} ions), the dipole contributions reduce the total calculated fields near the site II ions by as much as 20% at distances 1.5 A from the cation (148).

Chabazite (Table 2.23) Stereo 2.19

A crystal structure of chabazite was first reported in 1933 but was later shown to be incorrect (149, 150). The structure of chabazite has been determined in detail for the dehydrated and hydrated forms and for the zeolite containing adsorbed Cl$_2$ molecules. The aluminosilicate framework of chabazite consists of D6R units arranged in layers in the sequence ABCABC (Stereo 2.19). The D6R units are linked by tilted 4-rings. The resulting framework contains large, ellipsoidal cavities each of which is entered by six apertures that are formed by 8-rings. When dehydrated, the zeolite contains a three-dimensional channel system consisting of the cavities joined through the 8-rings. Detailed studies of a calcium exchanged form of chabazite in the dehydrated and hydrated forms have shown that the positions of the calcium ions depend on the presence or absence of water molecules. A space-filling model of chabazite is shown in Fig. 2.31.

Hydrated Chabazite

In zeolites with large cavities, the cation is surrounded by the water molecules; in those with small cavities, such as analcime, the cation may be in contact with as few as two water molecules. In chabazite the five nearest neighbors of each calcium ion are incompletely coordinated water molecules. The water molecules are not linked directly to the framework, and the calcium ions are also isolated. Although dehydration of this zeolite is continuous and the framework has an inherent rigidity which does not collapse during dehydration, the calcium ions move from the cavities to sites I and II. The free aperture of the 6-ring is 2.6 A and most small ions are capable of exchange. This is accompanied by a slight distortion of the framework. The shape of the aperture in hydrated chabazite is more nearly planar than in the dehydrated chabazite. During the adsorption of water molecules, which coordinate with the calcium ions, the aperture shape changes with movement of the calcium ions into other positions. The rhombohedral angle, α, changes with dehydration from 94° 28' to 92° 01' for the dehydrated form. The location of adsorbed molecules of Cl_2 has been reported.

The study of zeolitic protons by measuring dielectric relaxation has provided interesting information on the nature of the water molecules. The relaxation time, which is about 10^{-2} sec for the movement of calcium ions in the hydrated crystals at ordinary temperatures decreases with dehydration and then increases considerably, indicating that the calcium ions are tightly held and relatively immobile in fully hydrated crystals. Nmr studies of the hydrated zeolites are also interpreted in terms of the contents of the cavity behaving similar to a concentrated electrolyte (151).

Dehydrated Chabazite

The calcium ions occur in three sites, site I, at the center of the hexagonal (di-trigonal) prisms, site II, near the 6-ring displaced away from site I, and site III, near the tilted 4-ring in the cavity (Stereo 2.20). Each unit cell contains one site I, two site II, and 12 site III. The occupancy by Ca^{2+} is 0.6 in site I, 0.35 in each site II, and 1/16 in each site III for a total of 2 Ca^{2+} per unit cell. The site II cation positions are available for all cations; the larger ones would be displaced away from the ring towards the cavity center. Univalent ions such as Cs^+, Tl^+, and Rb^+ with radii of 1.67, 1.47 and 1.47 A, respectively, would probably not exchange for cations in the site I positions. For the case of 4 univalent ions per unit cell, 2 may locate at the 6-rings, while the other 2 will probably lie on the internal surface of the cavity near the 8-rings, similar to the Na^+ ions in zeolite A. It has been shown in adsorption studies on ion-exchanged chabazites that the adsorptive capacity for gases such as oxygen and argon falls sharply when there are more than 3 cations per unit cell. Since 3 cations in each unit cell correspond to occupancy of one-third of the apertures, the total number of cations is sufficient to block one-third of the apertures and reduce the free diffusion of molecules within the zeolite structure. The reduction in adsorption of critically sized molecules occurs at a cation density of 3 to 4 per unit cell.

This blocking occurs only for molecules which do not interact strongly with the cations. Water molecules, which coordinate with the cations and move them into the cavities are not influenced by cations in the apertures. This is similar to the situation in zeolite A. Structural studies on the sodium exchanged form of chabazite are not complete but a reasonable extrapolation would indicate that two of the four sodium ions will occupy the 6-rings and two will occupy positions near or in the 8-rings.

The poor local charge balance in dehydrated chabazite results in an electrically active surface around each of the cavities. This surface is responsible for interesting adsorption properties. If in each 6-ring there are two Al^{3+} and four Si^{4+} ions, the charge distribution is equally balanced about the center of the pseudo-hexagonal prism. However, if the (Si, Al) distribution is unbalanced so that there are, for example, three aluminum ions in one ring and one aluminum ion in the other, then the electric field is unbalanced and the calcium ion would be nearer nearer the ring containing the three aluminum ions (143).

Composition Variations

The ordering of the Si^{4+} and Al^{3+} atoms in the tetrahedral positions will have control over the Si/Al ratio. In chabazite, Si/Al ratios of 1:1, 2:1, and 5:1 are reasonable. This would require 3 Ca^{2+} or 6 Na^+, 2 Ca^{2+} or 4 Na^+, and 1 Ca^{2+} or 2 Na^+, as the balancing cations (152). The requirement of 3, however, is incompatible with the available cation sites (2 per unit cell). The other two combinations are reasonable inasmuch as a sufficient number of cation sites per unit cell are available.

Observed compositions of chabazite indicate an excess of Si over the 2:1 ratio and an excess of cations over the proposed ideal value of 2. Synthetic zeolites K-G, D, and R are structurally related to chabazite but their detailed structures are not known.

Herschelite

Herschelite appears to have a framework isostructural with that of chabazite. It is listed as a separate species based upon the differences in chemical composition and physical properties. In addition to crystal habit, which by itself is not a basis for distinguishing one zeolite species over another, herschelite differs from the typical chabazites in terms of its mean refractive index and its chemical composition (See Chap. 3, Table 3.20). The unit cell of herschelite contains more cations (generally four or more sodium ions per unit cell) than chabazite. The adsorption character of the dehydrated mineral is quite different from that of chabazite. Herschelite does not adsorb molecules such as oxygen at low temperatures. The largest molecule adsorbed by herschelite is ammonia with a kinetic diameter of 2.6 Å. The higher cation population in the herschelite unit cell is also reflected in the lower water content. Presumably each cavity has 4 Na^+ and about 10 H_2O as opposed to 2 Ca^{2+} and 12 H_2O in chabazite.

Gmelinite (Table 2.35) Stereo 2.21

The aluminosilicate framework of gmelinite consists of the parallel stacking of hexagonal rings in the sequence AABBAABB or D6R units in the sequence ABAB. As in chabazite, the framework is produced by joining the D6R units at the edges through tilted 4-rings

(153).The relationship between the AB or hexagonal close packing of D6R units in gmelinite and the ABC or cubic close packing in chabazite is shown in Fig. 2.22. A model of the framework structure is shown in Stereo 2.21 and a space-filling model in Fig. 2.31. It has been found from studies of gas adsorption on dehydrated crystals of gmelinite that the adsorptive properties are equivalent to those of chabazite (154). A free aperture dimension of $\sim$ 4 A is indicated which is due to stacking faults that interrupt the large 7.0 A channels (155). The main path for diffusion occurs in the intersecting channels that run perpendicular to the c-axis. The positions of H_2O molecules and cations in hydrated gmelinite are not entirely known. The unit cell of gmelinite provides two cation sites within the D6R units. Six monovalent cations would have to be located near the 8-rings. Synthetic zeolite S is structurally related to gmelinite but a complete structural analysis has not been made.

Zeolite ZK-5 (Table 2.67) Stereo 2.22

The framework structure of zeolite ZK-5 results by joining truncated cuboctahedra, α-cages, through double 6-rings in a body-centered array as shown in Stereo 2.22. There are two truncated cuboctahedra per unit cell, a total of 30 AlO_4 and 66 SiO_4 tetrahedra (155). The main cavities are the α-cages as found in zeolite A. Due to the same aperture size, ZK-5 compares to zeolite A in adsorption properties. At room temperature 16 of the 30 sodium ions in the hydrated zeolite are located in the α-cages near the center of the 6-rings on the 3-fold axis. Positions of water and the other sodium ions are not known. Upon heating to 150°C, the sodium ions near the 6-rings move to unknown sites.

Table 2.15 Electrostatic Field in Zeolite X and Y (148)

	Field at Indicated Distance from Cation Surface (V/A)			
	site II		site III	
	1 A	2.5 A	1 A	2.5 A
NaX, Si/Al = 1	1.8	.09	3.1	.43
CaX, Si/Al = 1	6.1	1.3	–	–
NaY, Si/Al = 2	2.4	0.43	3.2	.67
CaY, Si/Al = 2	6.3	1.8	–	–

112 STRUCTURE OF ZEOLITES

(a)

(b)

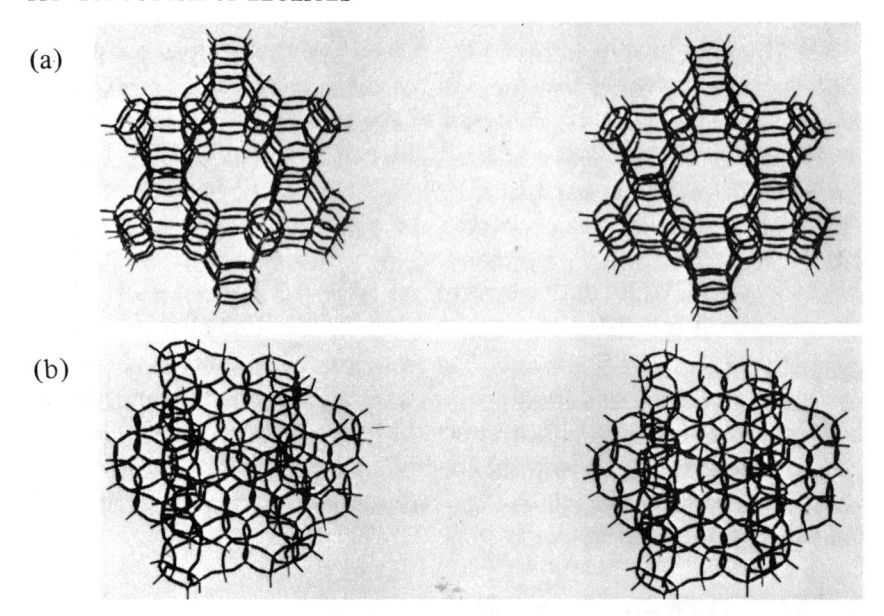

Stereo 2.21 Stereoscopic view of the framework structure of gmelinite. (a) view nearly parallel to the c-axis which shows the large 12-ring and (b) an oblique view which shows the ABAB arrangement of D6R units. Gmelinite has a nonintersecting system of wide channels which run parallel to the c-axis. Since these channels are formed by 12-rings, the free diameter is about 7.0 A. Smaller channels run perpendicular to the main c-axis channels; the apertures are formed by distorted 8-rings, which have free dimensions of about 3.6 x 3.9 A. The dimensions of the cavities formed by the intersection of these two types of channels are about 7.8 A - 6.5 A. Each cavity has three 8-ring apertures of the dimensions given above. Thus a planar network of intersecting channels runs in planes perpendicular to the c-axis and spaced at intervals of about 5 A apart in the c-direction.

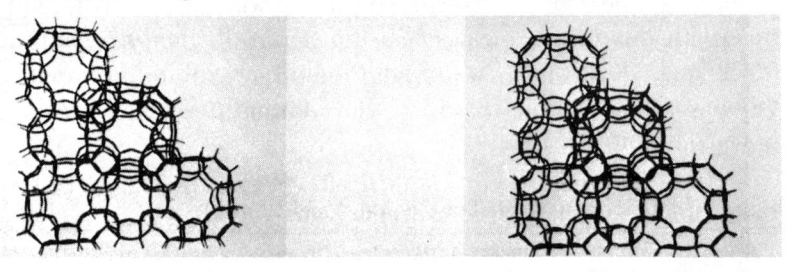

Stereo 2.22 Stereoscopic view of the framework structure of zeolite ZK-5 parallel to the cubic axis. The body centered arrangement of α-cages is shown; the 8-ring apertures form the main channel system.

There are also smaller cavities, called γ-cages, or 18-hedra, which are entered by two, near planar 8-rings and four, nonplanar 8-rings (Fig. 2.18b). The main channels run parallel to the cubic axes and are formed by the planar 8-rings with a free diameter of 3.9 A. There are two sets of three-dimensional channel systems which parallel each other but do not intersect.

The aluminosilicate framework of ZK-5 can alternately be considered as a cubic array of D6R units [$(Al, Si)_{12}O_{24}$], about the centers of the α-cages. There are 8 per unit cell. Synthesis of ZK-5 requires the large divalent organic cation:

$$[H_3C-N\bigcirc N-CH_3]^{2+}$$

Two synthetic aluminosilicates prepared by Barrer in 1948 contained intercalated $BaCl_2$ (Species P) and $BaBr_2$ (Species Q). In this book these are referred to as P-[Cl] and Q-[Br] respectively. These materials have been shown to have the same framework structure as ZK-5. The nonframework atoms and molecules, Ba^{2+}, Br^-, Cl^- and H_2O are located in the 18-hedral γ-cages as well as the large α-cages. The unit cell of P-[Cl] for example, contains about 13 $BaCl_2$ and 35 H_2O molecules (156).

Zeolite L (Table 2.41) Stereo 2.23

The crystal structure of zeolite L is based on the 18 tetrahedra unit (ε-cage) found in the feldspathoid cancrinite and erionite (Fig. 2.42) (61). Zeolite L is classified in group 4 because the main structural unit is the D6R unit; however, additional oxygen bridges are required.

In erionite the ε-cages are separated by double 6-ring units in the c-direction. These units are linked laterally by a single 6-ring connecting the ε-cages. In zeolite L the ε-cages are symmetrically placed across the D6R unit, as shown in Fig. 2.42, Stereo 2.23. Other structures are possible, as yet unknown, based upon different configurations of these units.

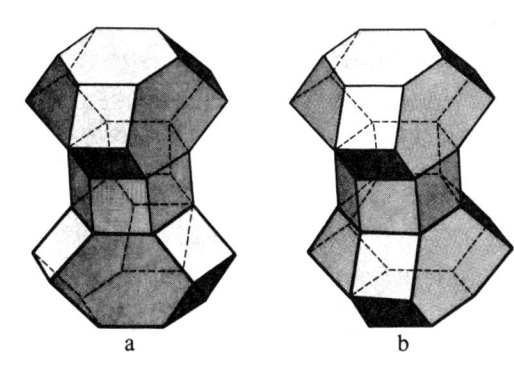

Figure 2.42 (a) The nonsymmetrical configuration of cancrinite units, ε-cages, in erionite, (b) The symmetrical configuration of ε-cages in zeolite L.

114 STRUCTURE OF ZEOLITES

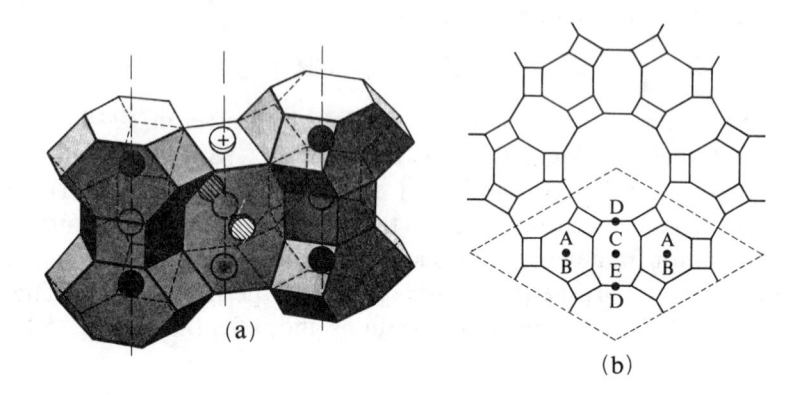

Figure 2.43 (a) Section of the framework of zeolite L showing linkage of ε-D6R-ε units and cation sites as follows: site A ○; B ●; C ⊕; D ⊘; E ○; (b) Projection of framework of zeolite L parallel to c showing the main c-channels 7.1-7.8 A in free diameter (61).

The structure of zeolite L consists of columns of symmetrical ε-D6R-ε units crosslinked to others by single oxygen bridges; planar 12-membered rings produce wide channels parallel to the c-axis (See Fig. 2.43).

Fully hydrated, the structure has four cation positions with occupancies as shown in Table 2.16, Fig. 2.43. The preferred Si/Al ratio in zeolite L is 3.0 which indicates an ordered Si, Al distribution. The calculated value of 9.7 is based on partial occupancy of site A by Na$^+$. If K$^+$ is used, a value of 9.1 is obtained in good agreement with 9 as found by analysis. This structure agrees with physical properties such as adsorption, void volume, density, etc. The cations in site D appear to be the only exchangeable cations at room temperature.

During dehydration, cations in site D probably withdraw from the channel walls to a fifth site, site E, which is located between the A sites. In site D, the cations are coordinated with two H$_2$O molecules in the channels. The cations in site D are readily exchangeable at room temperature. Ordering of Si and Al, indicated by the preferred Si/Al ratio of 3.0, requires lowering the symmetry of a doubled c con-

ZEOLITES OF GROUP 4 115

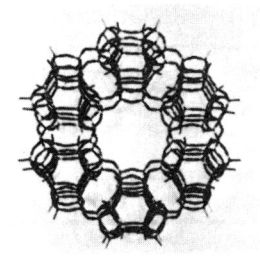

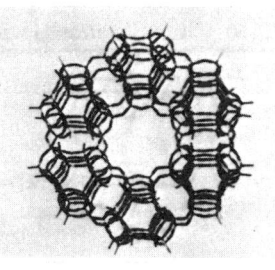

Stereo 2.23 A stereoscopic view of a model of the framework structure of zeolite L. This view shows the main channel, 7.1 Å, parallel to the c-direction formed by 12-rings.

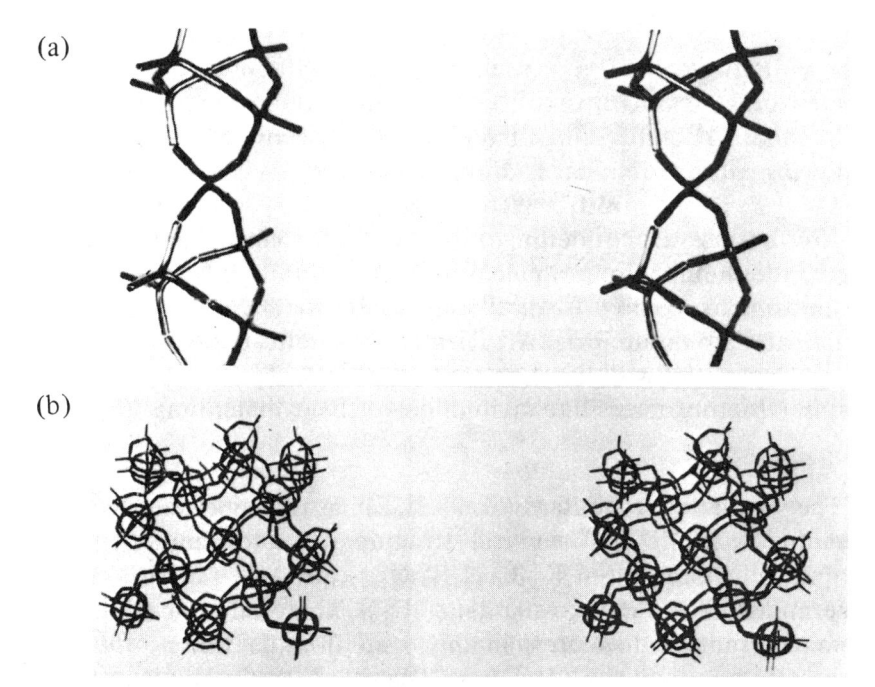

(a)

(b)

Stereo 2.24 (a) Stereoscopic view of a model of the chain of T_5O_{10} units of tetrahedra as present in the zeolites of group 5, (b) stereoscopic view of the framework structure of natrolite. The view is nearly parallel to the c-axis. The tetrahedra chain has a fundamental unit length of 6.6 Å so that one dimension of the unit cell will be 6.6 Å or a multiple.

Table 2.16 Cation Positions in Zeolite L (61)

Site	Location	Number of Sites/uc	Occupancy Observed
A	Center of D6R	2	1.4
B	Center of ε-cage	2	2.0
C	Between adjacent ε-cages	3	2.7
D	Wall of main channel	6	3.6
	Total	13	9.7 [a]

[a] The calculated value of 9.7 is based on partial occupancy of site A by Na^+. If K^+ is used, a value of 9.1 is obtained in good agreement with 9 as found by analysis.

stant. The ratio Si/Al = 3 cannot be obtained in a single ε-unit of 18 tetrahedra.

Zeolite P-L (Table 2.42)

The synthetic zeolite P-L contains PO_4 tetrahedra substituted in the framework. A comparison of the typical unit cell contents given for L in Table 2.41 and P-L in Table 2.42a quite clearly shows that the following substitution occurs during synthesis:

$$AlO_2^- + PO_2^+ \rightarrow 2\ SiO_2$$

This imposes strict ordering of the Al, Si, P atoms since the "avoidance" rule should also be applied to Al-Al and P-P. However, sharing of the same oxygen by Al and P must be allowed because 2/3 of the tetrahedral atoms are P and Al. There are 4 ε-units in the unit cell of P-L and each unit will therefore contain about 6 Si, 9 Al, and 3 P atoms. Ordering necessitates a doubling of the c dimensions (37).

Zeolite Ba-G

The synthetic zeolite Ba-G (Table 2.32) has the same framework structure as zeolite L. The crystal structure of a potassium, barium-G of unit cell composition $K_{2.7}Ba_{7.65}[(AlO_2)_{18}(SiO_2)_{18}] \cdot 23\ H_2O$ was determined from x-ray powder data (157). Three cation positions were determined; one corresponding to site B in the center of the ε-cage is fully occupied by Ba^{2+} ions. The site C has fractional occupancy. Potassium ions are located, not in site A, but in site D on the wall of the main channel. Remaining cations appear to be distributed in the main channel.

L. ZEOLITES OF GROUP 5

The structural characteristics of the group 5 zeolites were established 30 years ago by Pauling (25) and Taylor (62). The detailed structures (in terms of cation and water positions) of most of the members of this group are still not known. The zeolites of this group are called the fibrous zeolites; this morphology is explained by the prominent chains in the framework structure. All of the structures in this group are based upon these crosslinked chains of tetrahedra. An individual chain is comprised of linked units of five tetrahedra as shown in Stereo 2.24a. The SBU is designated by Meier as the 4-1 unit, Fig. 2.18. The five tetrahedra unit is similar to a unit of five tetrahedra in the structure of zunyite. The three possible ways of crosslinking the chains produce the three types of framework structures of the group 5 zeolites. One mode is characteristic of the framework of natrolite, scolecite, and mesolite. A second mode results in the characteristic structure of thomsonite and gonnardite. The third mode of linking is found in edingtonite. Although the chain of tetrahedra is in itself a rigid assembly, the crosslinking between adjacent chains by sharing of oxygen atoms is not rigid and the chains may rotate with dehydration or in order to accommodate different interstitial cations. The three modes of linkage are described in terms of the location of the linking oxygen atom. If all of the linking oxygen atoms lie on reflection planes, then the simplest framework, that of edingtonite, results. If half of the linking oxygen atoms lie on reflection planes and the other half on rotation axes the framework produced is that of thomsonite. If all of the linking oxygen atoms lie on rotation axes then the natrolite framework is produced. The characteristic fibrous habit of the zeolite crystals is therefore parallel to the c-axis. Two types of channels run through the structure which produce small cavities at the intersections. These cavities contain the water molecules and exchangeable cations (Table 2.17). It has always been assumed that the main channels which run perpendicular to the c-axis between adjacent chains have a minimum free diameter of 2.60 A as compared to 2.08 A for the c-parallel channels.

Natrolite, Scolecite and Mesolite (Tables 2.46, 2.50, 2.57) Stereo 2.24b

Although they differ in unit cell composition and symmetry, natro-

Table 2.17 Cation and H_2O Contents in Group 5 Zeolites (158)

	Cations	H_2O	Total
Natrolite	16	16	32
Scolecite	8	24	32
Mesolite	32/3	64/3	32
Gonnardite	12	24	36
Thomsonite	12	24	36
Edingtonite	8	32	40

Based on 40 tetrahedra.

lite, scolecite and mesolite all possess the same type of framework structure, consisting of the linked chains of tetrahedra as shown in Stereo 2.24. Natrolite and scolecite are, respectively, the Na- and Ca- forms of the same type of framework structure. Mesolite is the intermediate member with a Na/Ca ratio of 1. As the cation population in the zeolite varies from 16 Na^+ to 8 Ca^{2+}, the number of H_2O molecules increases accordingly from 16 to 24 (158). The total number of the cations and water molecules in the same unit cell column in each case is 32 (Table 2.17). In these zeolites, the channel dimensions are small and the space occupied by cations is very important. Because of the structural changes associated with dehydration and ion exchange, the zeolites of this group are not important molecular sieve adsorbents.

The Si, Al ordering in the chains of the natrolite structure is indicated in Fig. 2.44. The Na^+ ions and Ca^{2+} ions, along with the H_2O molecules, are arranged in the channel in clusters; that is, the water molecules are clustered around the cations. In the case of natrolite, each Na^+ ion has as nearest neighbors four framework oxygens and two H_2O molecules. In scolecite, the Ca^{2+} ions are in 7-fold coordination with four oxygens of the framework and three H_2O molecules; it is further concluded that in mesolite the cation-water molecule combinations are similar. Dehydration of these zeolites causes a substantial disturbance in the local charge balance, and it is likely that the cations tend to move into different sites for coordination with the framework oxygens. As a result, the lattice dimensions change and the framework collapses if dehydrated at too high a temperature. In all cases, the dehydrated zeolites have shown no tendency to readsorb molecules other than polar molecules such as water and ammonia.

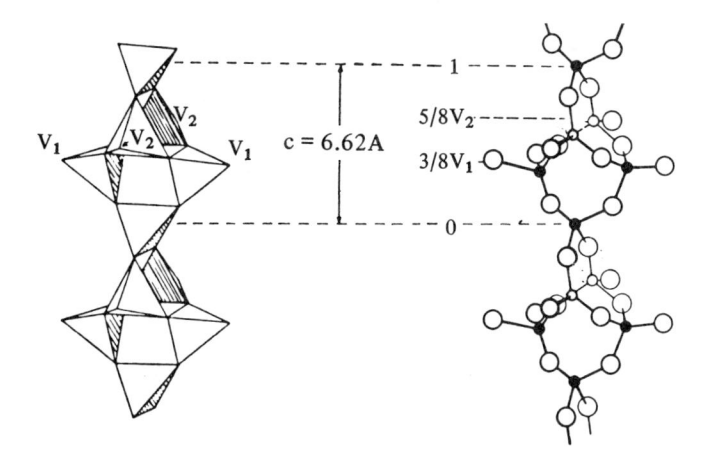

Figure 2.44 Drawing of the tetrahedron chain as found in natrolite showing the ordering of aluminum and silicon (1). Small open circles, Al; filled circles, Si; large open circles, O.

Thomsonite (Table 2.60) Stereo 2.25

The framework structure of thomsonite is based upon the second type of crosslinking of the chains of 4-1 units (Stereo 2.25). Since the Si/Al ratio is 1, ordering of the (Si, Al)-O$_4$ tetrahedra is spread over more than one unit of five tetrahedra; therefore the c-dimension is double the fundamental 6.6 A. Like the other zeolites of group 5, thomsonite occurs as fibrous crystals with water molecules located in the channels in double zigzag chains.

Gonnardite (Table 2.36)

Gonnardite has the same aluminosilicate framework structure as thomsonite but an Si/Al ratio of 1.5. Consequently the ordering of Si, Al atoms in each 4-1 unit is the same as that in natrolite. The difference between gonnardite and thomsonite is due to the ratio Si/Al and a corresponding difference in crystallographic symmetry. Specific locations of the cations in gonnardite are not known.

Edingtonite (Table 2.26) Stereo 2.26

The simplest method of crosslinking the chains of 4-1 units of tetra-

120 STRUCTURE OF ZEOLITES

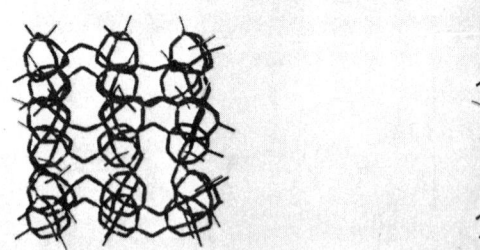

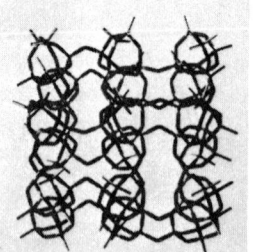

Stereo 2.25 Stereoscopic view of the framework structure of thomsonite nearly parallel to the c-axis. The main channels in thomsonite are perpendicular to the fiber or c-axis. Eight of the 12 cations in the hydrated zeolite are found in coordination with seven oxygens, four of which are framework oxygens and three of which are water molecules. The remaining four cations apparently are in 8-fold coordination. Si, Al ordering due to Si/Al = 1 requires c ≈ 2 x 6.6 Å or two 4 - 1 units.

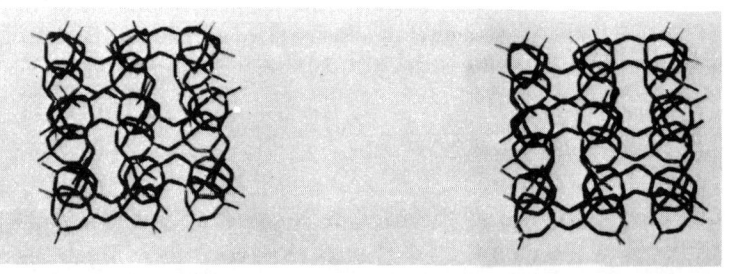

Stereo 2.26 Stereoscopic view of the framework structure of edingtonite parallel to the c-axis.

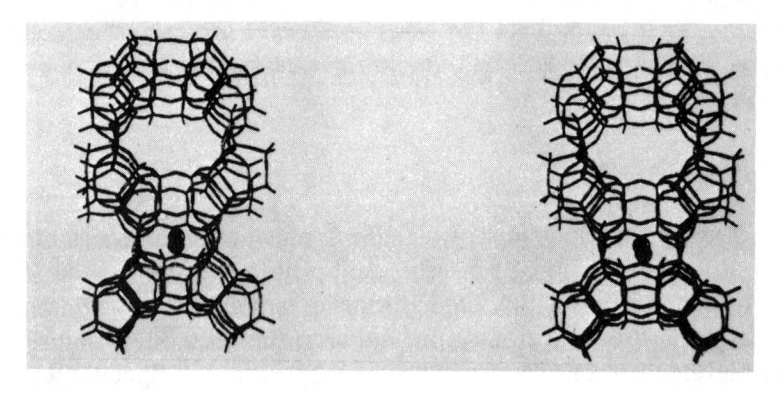

Stereo 2.27 Stereoscopic view of the framework structure of mordenite parallel to the main c-axis showing the large 12-ring and the position of four of the sodium ions which are located as shown in the constrictions with a minimum diameter of 2.8 Å. The remaining four sodium ions per unit cell probably occupy some of the 8- and 12-fold positions at random.

ZEOLITES OF GROUP 5   121

Figure 2.45   A space-filling model of edingtonite shown perpendicular to the c-axis and in the direction of the main channels. Each alternate cavity produced by the intersection of the two types of channels, one parallel to c and the other perpendicular to c is occupied by a barium ion. The barium ion is coordinated with 6 oxygens and 2 water molecules for a total 8-fold coordination.

hedra is found in edingtonite (Stereo 2.26). A space-filling model is shown in Fig. 2.45.

The Si/Al ratio of 1.5 and the c-dimension of 6.5 are consistent with Si-Al ordering of the type found in natrolite; each unit of five tetrahedra has a Si/Al ratio of 3:2 (Fig. 2.44).

The channels in thomsonite and edingtonite are probably more open because the total number of cations and water molecules in each

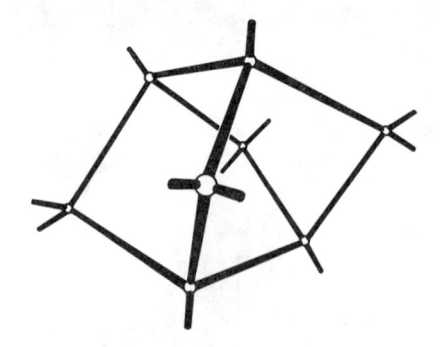

Figure 2.46 Diagram showing the configuration of the T_8O_{16} units of tetrahedra as found in the framework structures of the group 6 zeolites (31).

equivalent unit cell of 40 tetrahedra is larger than that found in the natrolite-type zeolites (Table 2.17).

M. ZEOLITES OF GROUP 6

The common structural element in the framework structures of the zeolites of this group (except bikitaite) is a special configuration of 5-rings (Fig. 2.46). The SBU of six tetrahedra is called the 5-1 unit (Fig. 2.18).

These units form complex chains which are linked to each other in various ways. Six different modes are possible (56). The framework structures of mordenite and dachiardite are based on two arrangements of the chains (Fig. 2.47). The structural element of Fig. 2.47 is also found in epistilbite and ferrierite. The framework of bikitaite is closely related as shown by a projection of the framework structure along the b-axis.

Mordenite (Tables 2.47, 2.48) Stereo 2.27

Mordenite, with a nearly constant Si/Al ratio of 5, is the most siliceous zeolite mineral. This constant ratio indicates an ordered distribution of Si and Al in the framework structure.

The structure consists of chains, such as those shown in Fig. 2.47,

Figure 2.47 The chain of tetrahedra arranged in linked 5-rings as found in mordenite (159).

crosslinked by the sharing of neighboring oxygens (159). Each tetrahedron belongs to one or more 5-rings in the framework. The high degree of thermal stability shown by mordenite is probably due to the large number of 5-rings which are energetically favored in terms of stability. The aluminosilicate framework of mordenite is shown in Stereo 2.27. For the diffusion of small molecules, the dehydrated zeolite has a two-dimensional channel system but for larger molecules the channel system is one dimensional and may be subject to diffusion blocks produced by crystal stacking faults in the c-direction or the presence of amorphous material or cations in the channels. Gases such as nitrogen and oxygen are rapidly adsorbed by dehydrated mordenite whereas hydrocarbons such as CH_4 or ethane are slowly adsorbed. This is inconsistent with the channel dimension of 6.7 Å. Because the main channel system is in one direction the number of diffusion blocks needed to limit the adsorption process is not great (Fig. 2.48). This is similar to the results obtained in adsorption studies on gmelinite where experience has shown that stacking faults apparently block the main channels. Although electron diffraction studies indicate stacking faults in gmelinite, no evidence was found for the presence of stacking faults in mordenite (160).

The positions of 4 of the 8 sodium ions in the hydrated crystal have been determined. Locations of the remaining 4 sodium ions and

Figure 2.48 Packing drawing of mordenite showing oxygen atoms and channels parallel to c and parallel to b. The large continuous channels parallel to c are elliptical with dimensions of 6.7 x 7.0 A. The channels parallel to b consist of pockets, with dimensions of 2.9 x 5.7 A separated by 2.8 A restrictions (159, 161).

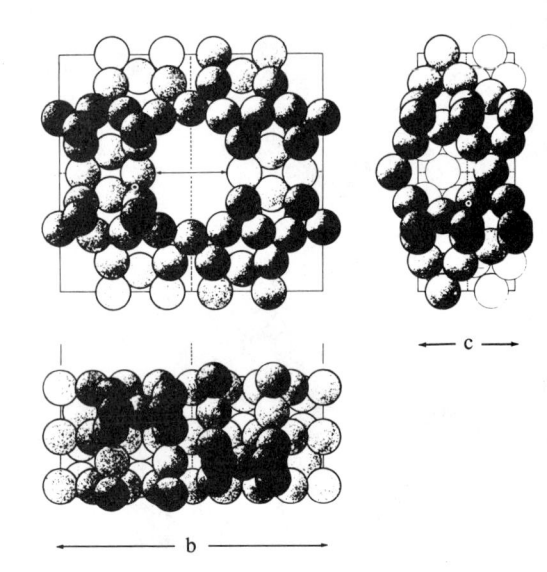

water molecules are not known. Large cations such as cesium cannot occupy the positions determined for sodium (161). It appears doubtful that water molecules are located in fixed positions at room temperature. The fibrous nature and prismatic cleavage of mordenite crystals is explained on the basis of the structure.

A synthetic type of mordenite, known as "large-port" mordenite (Table 2.48) has been prepared which exhibits the adsorption characteristics expected for the free diffusion of molecules in the main 6.7 A diameter channels (Chapter 4) (160). The extraneous matter which is assumed to block the main channels in the mineral is apparently not present in this synthetic form (Table 2.47). Synthetic types of mordenite of the small-port variety can be partially converted to the large-port (Table 2.48) by acid leaching. This treatment presumably removes the extraneous matter and opens up the main channels.

Dachiardite (Table 2.25) Stereo 2.28

The aluminosilicate framework of dachiardite is also based upon a chain structure which parallels the c-axis (162). As shown in Stereo 2.28, the framework involves a different type of linkage of the chains containing 5-rings. The dachiardite structure contains a two-dimensional system of channels. There is no confirmatory data on the mol-

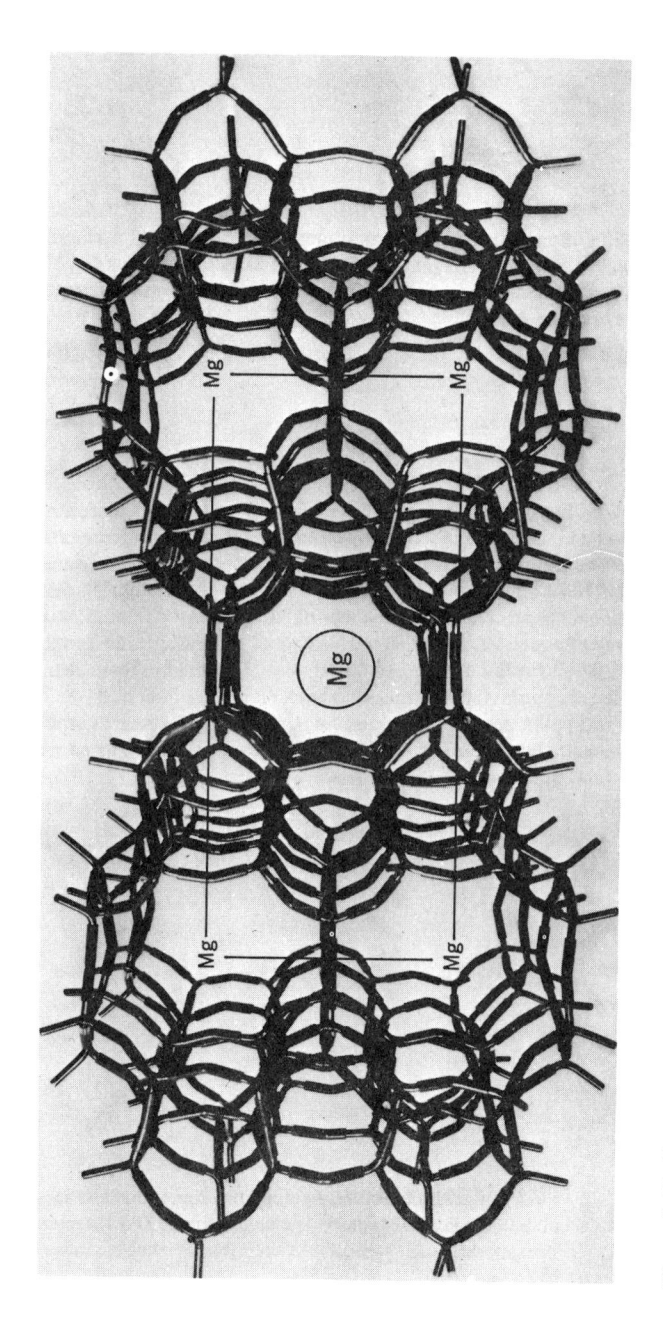

Figure 2.49 Ferrierite parallel to b axis showing location of magnesium ions.

126 STRUCTURE OF ZEOLITES

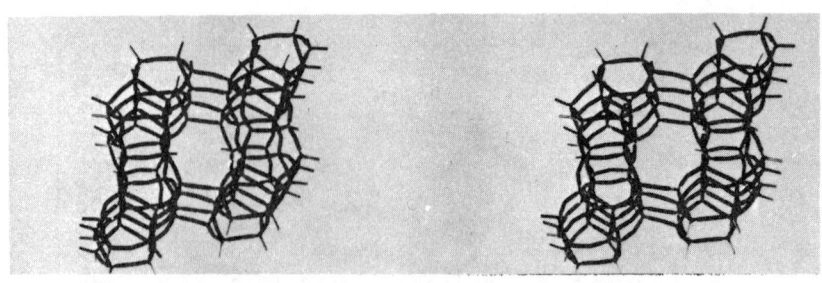

Stereo 2.28 Stereoscopic view of the framework structure of dachiardite shown parallel to the main c-channel. The main channel runs parallel to the b-axis [010] and intersects channels which parallel the c-axis. The minimum free aperture diameter of both of these channels is about 4 A.

Stereo 2.29 Stereoscopic view of the framework structure of ferrierite parallel to the c-axis. The main c-axis channels are shown in cross section. Two intersecting systems of channels run through the ferrierite structure parallel to the b- and c-axes. There is no channel in the a-direction. This is also the case with other zeolites in this group. The c-axis channel has an elliptical cross-section; dimensions = 4.3 x 5.5 A and a cross-sectional area of about 18 A. The second system of channels, parallel to the b-axis, is formed by 8-rings with diameters of 3.4 x 4.8 A. Located in these channels are cavities which are approximately spherical with a diameter of about 7 A. There is only one diffusion path available to a molecule of any moderate size; in order to move from one major channel to another, the diffusing species must move through the smaller 8-ring aperture.

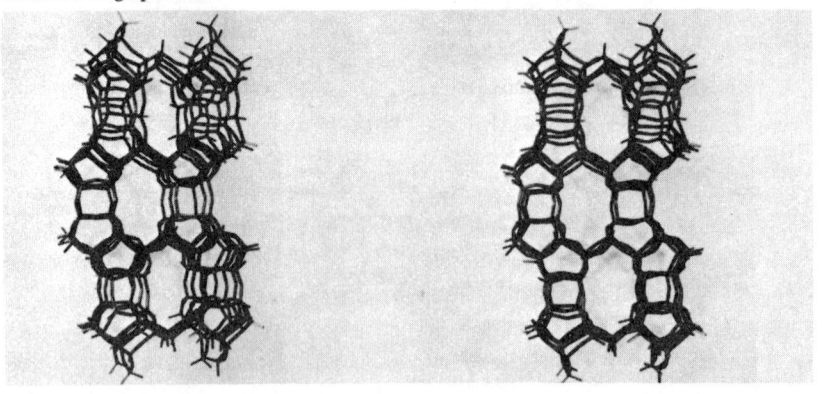

Stereo 2.30 View of epistilbite structure along the a-direction showing the 3.2 x 5.3 A apertures. Channels also intersect these in the c-direction. The c-direction channels are restricted by 8-ring apertures 3.7 x 4.4 A. At the intersections, cavities are present 3.7 x 8.4 A in size.

ecular sieving properties of this zeolite. The positions of the cations and water molecules are not known. The structural information was obtained on crystals of the mineral which contain several different types of cations. In this particular case the cation composition included K^+, Na^+, Ca^{2+}, and Mg^{2+}.

Ferrierite (Table 2.31) Stereo 2.29

The zeolite mineral ferrierite is quite rare and has been found in only one location in British Columbia. A sedimentary occurrence in Nevada has also been reported. A structural analysis was made of the hydrated mineral. The aluminosilicate framework (Stereo 2.29) consists of chains of 5-rings parallel to the orthorhombic c-axis of the crystal and cross-linked by 6-rings, similar to the mordenite and dachiardite structures (163, 164).

In the hydrated zeolite the cavities contain hydrated Mg^{2+} as shown in Fig. 2.49. There are two similar positions in the unit cell. (The Mg ion is surrounded by a regular coordination sphere of 6 H_2O molecules. The remaining univalent Na^+ and K^+ ions have not been located along with the H_2O molecules which are not present in the hydration sphere of the Mg^{2+} ion. The Mg^{2+} ions are tenaciously held within the hydrated zeolite and are not removed by treating the zeolite with hot acid.

The synthetic zeolite type Sr-D is structurally related to ferrierite (85).

Epistilbite (Table 2.27) Stereo 2.30

The aluminosilicate framework of epistilbite consists of another variation of linked chains of 5-rings (165, 166). The framework structure is illustrated in Stereo 2.30. This structure explains the morphology and cleavage of epistilbite and the similarity in the x-ray powder pattern to that of dachiardite. The cations, Ca^{2+} and Na^+, occupy sites near the periphery of the cavity and are coordinated with some of the framework oxygen atoms as well as H_2O molecules in the channels. Preferred occupancy of these sites is due to occupancy of adjacent tetrahedral sites by Al^{3+}.

When subjected to dehydration in vacuum, epistilbite is structurally stable to at least 250°C. It will readsorb water vapor but not other gases or vapors. Although the channel system should permit the diffusion of larger molecules, it is apparent that when dehydrated the cations block the diffusion paths.

Bikitaite (Table 2.21) Stereo 2.31

Although a structural investigation of bikitaite has been reported, no detailed account has been published (167). Bikitaite is a recently discovered and rather unusual zeolite in that the cation composition is essentially all lithium. It is classified in group 6 because a description of the framework structure indicates that it is based upon 5-rings and the 5-1 units (Stereo 2.31). Adsorption measurements have shown that when dehydrated at 230°C, bikitaite exhibits no adsorptive properties. After dehydration at 150°C, zeolitic adsorption of H_2O vapor did not occur. The Li^+ ions in bikitaite do not exchange with NH_4^+.

N. ZEOLITES OF GROUP 7

The zeolites classified in group 7 include the morphologically lamellar zeolites; heulandite, stilbite, and possibly clinoptilolite.

Although the commonly occurring, sedimentary mineral clinoptilolite is generally considered to be an isostructural variant of heulandite, there is not sufficient evidence for this conclusion; accordingly clinoptilolite is listed as a separate species in group 7. Resolution of this question must await a structural investigation of clinoptilolite. The mineral stellerite, considered to be a variant of stilbite, may be a separate species due to compositional and physical property differences.

The common structural feature in the framework structures of group 7 zeolites is the special configuration of tetrahedra shown in Fig. 2.50. Each tetrahedron belongs to one of these elements which contain 4- and 5-rings. These are arranged in sheetlike arrays (Fig. 2.51) which accounts for the cleavage properties of these zeolites. The SBU for group 7 is the 4-4-1 unit proposed by Meier (32), Fig. 2.18.

Heulandite (Table 2.39) Stereo 2.32

The arrangement of the $T_{10}O_{20}$ units in the framework of heulandite

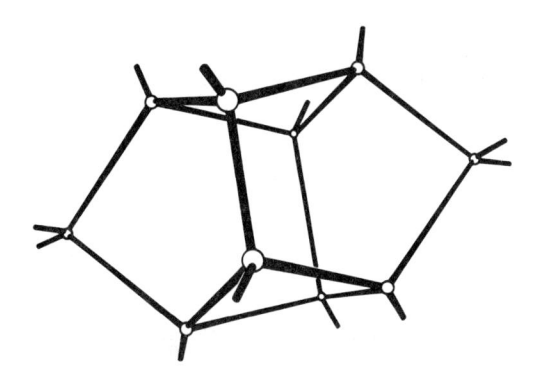

Figure 2.50 Configuration of the $T_{10}O_{20}$ units of tetrahedra in the framework structures of group 7 zeolites (32).

is shown in Fig. 2.51 and Stereo 2.32. The low bond density between the layers is apparent (168). Because of the low bond strength in one direction, heulandite changes structurally during dehydration. If dehydrated at moderate temperatures, below 130°C, heulandite will absorb H_2O and NH_3. After dehydration at higher temperatures no adsorption occurs.

Clinoptilolite (Table 2.24)

In sharp contrast to heulandite, clinoptilolite is very stable towards dehydration and readily readsorbs H_2O and CO_2. Some varieties adsorb O_2 and N_2. The chemical composition differs significantly from that of heulandite in Si/Al ratio and the exchangeable cation. The thermal stability, 700°C in air, is also considerably greater than the stability of heulandite.

X-ray powder data show significant differences. In view of the absence of specific structural data, clinoptilolite is listed as a distinct zeolite. Structurally, it may even resemble mordenite.

Stilbite (Table 2.58) Stereo 2.33

The aluminosilicate framework of stilbite is shown in Stereo 2.33. Structural details of dehydrated stilbite are unknown but it seems evident that, although the channels should be large enough to permit molecular sieve behavior, dehydration (a) distorts the structure or (b) leaves cations blocking the apertures. In hydrated stilbite, the Ca^{2+} ions are located coordinated with H_2O molecules at the cavities formed at channel intersections (169, 170).

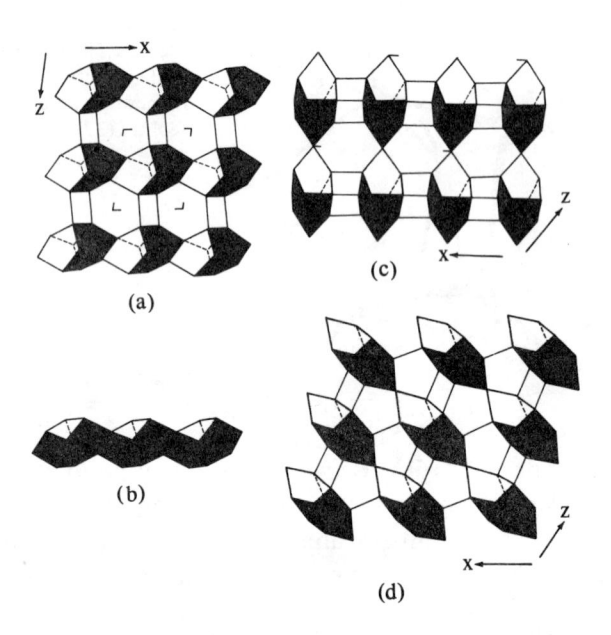

Figure 2.51 Arrangement of the $T_{10}O_{20}$ units of tetrahedra in the framework structures of (a) brewsterite, (b) brewsterite chain of 4-rings, (c) stilbite, and (d) heulandite. The sheets are all parallel to (010) and are connected by relatively few oxygen bridges which result in two-dimensional channel systems. The bond densities across these connecting bridges are very low, 1.7 (Si, Al)-O bonds per 100 A^2 in heulandite and stilbite as compared to 4.2 in analcime. This accounts for the lamellar habit of the crystals. The larger channels, parallel to a, are formed by rings of 10 tetrahedra (32).

Stellerite

Although reported to be a variety of stilbite, stellerite exhibits compositional and property differences. Stellerite is stable to dehydration in vacuum at 250°C, shows no change in the x-ray powder pattern, and adsorbs CO_2 readily. Morphologically, stellerite is found as fibrous aggregates and available analyses indicate essentially no content of sodium.

Brewsterite (Table 2.22) Stereo 2.34

Brewsterite bears no simple structural relationship to the zeolites in group 1 but might be classed with them because it does contain 4-rings as the smallest units in the framework (171). As shown in Stereo 2.34, the framework structure is complex. The structure of hydrated brewsterite has been studied (108). The predominant Sr^{2+} ions occupy po-

ZEOLITES OF GROUP 7   131

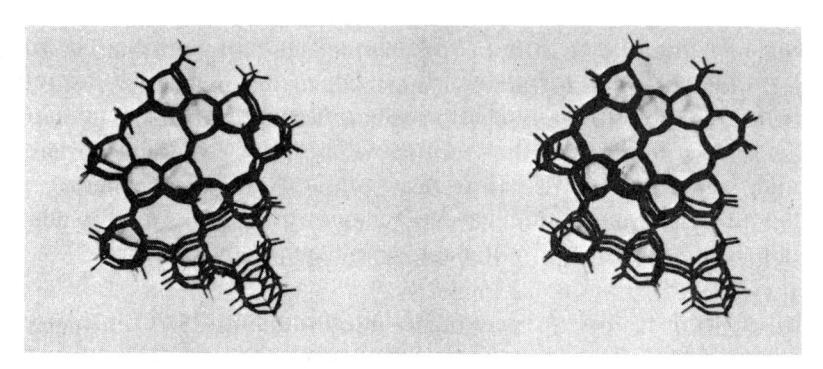

Stereo 2.31   Structure of bikitaite viewed in the b-direction showing one-dimensional channel system 3.2 x 4.9 A in size.

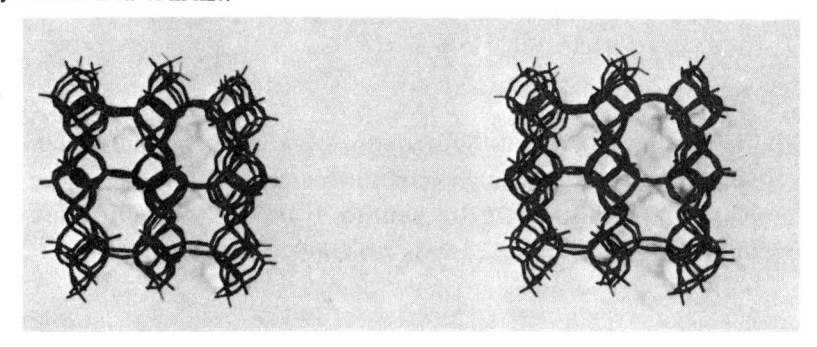

Stereo 2.32   Stereoscopic view of model of the framework structure of heulandite. The channel system is two dimensional; one set parallel to the a-axis formed by 8-rings, 4.0 x 5.5 A in free aperture, and two sets parallel to the c-axis formed by 8- and 10-rings with free apertures of 4.1 x 4.7 and 4.4 x 7.2 A. The cross-section parallel to the ab plane is similar to that of mordenite.

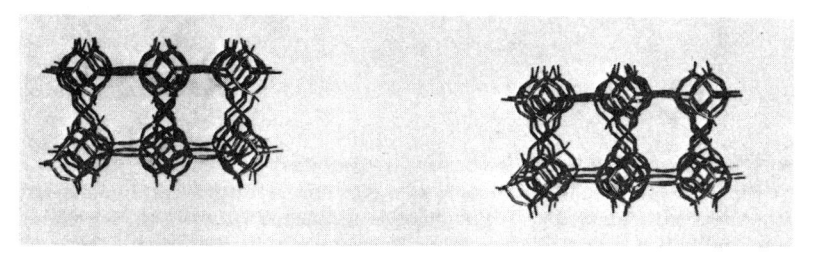

Stereo 2.33   Stereoscopic view of a model of the framework structure of stilbite parallel to the a-axis. The units of 4- and 5-rings are sheetlike and form a two-dimensional channel system in the ac plane with free apertures due to 10-rings, 4.1 x 6.2 A, and 8-rings 2.7 x 5.7 A.

sitions near the intersection of the channels and are coordinated with 5 H$_2$O molecules and 4 framework oxygen atoms in an irregular polyhedron. An incomplete hydration sphere formed about the exchangeable cation is similar to other zeolites where the cavities are not large enough to permit the formation of a complete hydration sphere.

The platy character and cleavage behavior of brewsterite crystals is explained by the weak bonding across the symmetry planes (1.9 Si, Al-O bonds/100 A^2).

Adsorption studies on brewsterite dehydrated at 250°C *in vacuo* have shown that the only molecule readily adsorbed is H$_2$O with a diameter of 2.6 A. The quantity of water adsorbed is not equivalent to the quantity lost upon activation (12% vs. 15% by weight respectively).

O. OTHER ZEOLITES

Synthetic zeolites for which the structure is not known and which are not obviously related to known structures are listed in Table 2.4b. Data related to structures for the zeolites F, H, W, P-W, and N, are summarized in Tables 2.29, 2.37, 2.61, 2.62, and 2.49 respectively.

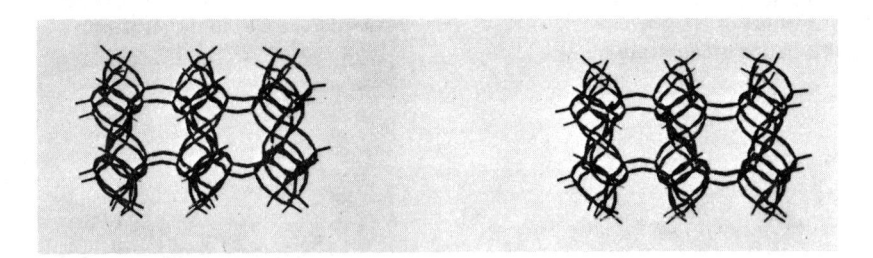

Stereo 2.34 Stereoscopic view of the framework structure of brewsterite illustrating the main c-parallel channel and the low density of oxygen bridges which imparts the laminar or platy character to this zeolite. The aluminosilicate framework contains 4-, 6-, and 8-rings and, in addition, 5-rings which crosslink the networks of the other rings. There are two sets of intersecting channels which run parallel to the a- and c-axes of the crystal. The channels are controlled by 8-rings as apertures; those perpendicular to the a-axis are elliptical with the free aperture size of 2.3 x 5.0 A; the aperture perpendicular to the c-axis is 2.7 x 4.1 A.

Table 2.18 Zeolite: A

Structure Group:	3
Reference:	102, 103, 105–112, 115, 172

Chemical Composition

Typical Oxide Formula:	$Na_2O \cdot Al_2O_3 \cdot 2\ SiO_2 \cdot 4.5\ H_2O$
Typical Unit Cell Contents:	$Na_{12}[(AlO_2)_{12}(SiO_2)_{12}] \cdot 27\ H_2O$, pseudo cell and 8X for true cell
Variations:	Si/Al = ~0.7 to 1.2; occlusion of $NaAlO_2$ in β-cages

Crystallographic Data

Symmetry:	Cubic	Density:	1.99 g/cc
Space Group:	Pm3m	Unit Cell Volume:	1870 A^3
	(Fm3c for true cell)		pseudo cell
Unit Cell Constants:	a = 12.32 A, pseudo cell	X-Ray Powder Data:	Table 4.26
	a = 24.64 A for true cell		

Structural Properties

Framework:	Stereo 2.16	Cubic array of β-cages linked by D4R units
	SBU: D4R	Void volume: 0.47 cc/cc
	Cage type: α, β	Framework density: 1.27 g/cc
	(one each)	
Channel System:	Fig. 2.28a	Three-dimensional, ∥ to [100]; 4.2 A and ∥ to [111]; 2.2 A minimum diameter

Hydrated–
- Free Apertures: 2.2 A into β-cage and 4.2 A into α-cage
- Cation Locations: 8 S_I on 6-rings, 4 cations with H_2O in the 8-rings

Dehydrated–
- Free Apertures: 4.2 A
- Cation Locations: 8 S_I in 6-rings, 3 S_{II} in 8-rings, 1 S_{III} at the 4-ring
- Effect of Dehydration: None on framework, 4 cations move to S_{II}
- Location of H_2O Molecules: Dodecahedral arrangement in α-cage 4 molecules in β-cage.

Largest Molecule Adsorbed: C_2H_4 at RT, O_2 at $-183°C$
Kinetic Diameter, σ, A: 3.9 and 3.6

Table 2.19 Zeolite: N-A

Structure Group:	3
Reference:	116
Chemical Composition	
Typical Oxide Formula:	$(Na, TMA)_2O \cdot Al_2O_3 \cdot 4.8\ SiO_2 \cdot 7\ H_2O$
Typical Unit Cell Contents:	$Na_4TMA_3[(AlO_2)_7(SiO_2)_{17}] \cdot 21\ H_2O$
Variations:	Na/Al up to 0.9; Si/Al = 1.25 - 3.75
Crystallographic Data	
Symmetry:	Cubic Density:
Space Group:	Pm3m Unit Cell Volume: 1780 A^3
Unit Cell Constants:	a = 12.12 A X-Ray Powder Data: Table 4.27
Structural Properties	
Framework: Stereo 2.16	Same as zeolite A
SBU:	D4R Void volume: 0.5 cc/cc
Cage type:	α, β Framework density: 1.3 g/cc
Channel System:	Fig. 2.28a Three-dimensional, ‖ to [100]
Hydrated—	
Free Apertures:	4.2 A into α-cage
Cation Locations:	S_I and S_{II}; TMA ions not exchangeable under usual conditions.
Dehydrated—	
Cation Locations:	Probably in S_I
Effect of Dehydration:	Stable to 700°C in air; calcination removes TMA
Location of H_2O Molecules:	Removal of TMA by decomposition increases void volume by 0.15 cc/g (0.3 to 0.5 cc/cc)
Largest Molecule Adsorbed:	n-paraffin hydrocarbons
Kinetic Diameter, σ, A:	4.3

TMA = tetramethylammonium, $(CH_3)_4N^+$

Table 2.20 Zeolite: Analcime

Structure Group:	1
Reference:	62, 67, 173–176
Chemical Composition	
Typical Oxide Formula:	$Na_2O \cdot Al_2O_3 \cdot 4\, SiO_2 \cdot 2\, H_2O$
Typical Unit Cell Contents:	$Na_{16}[(AlO_2)_{16}(SiO_2)_{32}] \cdot 16\, H_2O$
Variations:	Si/Al = 1.8–2.8; H_2O = 14–18
Crystallographic Data	
Symmetry:	Cubic Density: 2.25 g/cc
Space Group:	Ia3d Unit Cell Volume: 2590 A^3
Unit Cell Constants:	a = 13.72 A X-Ray Powder Data: Table 3.4
Structural Properties	
Framework: Stereo 2.4	S4R of type UDUD connected to form regular S6R which lie ∥ to (111). Also very distorted 8-rings.
SBU:	S4R Void volume: 0.18 cc/cc
Cage type:	None specific Framework density: 1.85 g/cc
Channel System: Fig. 2.26	One-dimensional, ∥ to [111]
Hydrated—	
Free Apertures:	6-ring, 2.6 A
Cation Locations:	16 in 24 sites; in distorted octahedra with 4 cations and 2 H_2O molecules
Dehydrated—	
Free Apertures:	Unknown
Cation Locations:	Unknown
Effect of Dehydration:	Continuous and reversible; stable to 700°C
Location of H_2O Molecules:	On 3-fold axes in the channels
Largest Molecule Adsorbed:	NH_3
Kinetic Diameter, σ, A:	2.6

Table 2.21 Zeolite: Bikitaite

Structure Group:	6
Reference:	32, 167, 177

Chemical Composition

Typical Oxide Formula:	$Li_2O \cdot Al_2O_3 \cdot 4\ SiO_2 \cdot 2\ H_2O$
Typical Unit Cell Contents:	$Li_2[(AlO_2)_2(SiO_2)_4] \cdot 2\ H_2O$
Variations:	Other cations K, Na, Mg, mostly Li

Crystallographic Data

Symmetry:	Monoclinic	Density:	2.29 g/cc
Space Group:	$P2_1$	Unit Cell Volume:	297 A^3
Unit Cell Constants:	a = 8.61 A	X-Ray Powder Data:	Table 3.5
	b = 4.96 A		
	c = 7.61 A		
	$\beta = 114°26'$		

Structural Properties

Framework: Stereo 2.31	Chains of 5-rings linked by AlO_4 tetrahedra
SBU:	Unit of 5-1 Void volume: 0.23 cc/cc
	Framework density: 2.02 g/cc
Channel System:	One-dimensional ‖ to b-axis
Hydrated—	
Free Apertures:	8-ring, 3.2 x 4.9 A
Dehydrated—	
Effect of Dehydration:	Supposed to be stable at 500°C
	Loses 7.5% H_2O at 230°C

TABLES OF ZEOLITE DATA   137

Table 2.22 Zeolite: Brewsterite

Structure Group: 7
Reference: 171
Chemical Composition
Typical Oxide Formula: (Sr,Ba,Ca)O·Al$_2$O$_3$·6 SiO$_2$·5 H$_2$O
Typical Unit Cell Contents: (Sr,Ba,Ca)$_2$ [(AlO$_2$)$_4$(SiO$_2$)$_{12}$]·10 H$_2$O
Variations: Sr > Ba > Ca; Si/Al = 2.57–3.03
Crystallographic Data

Symmetry:	Monoclinic	Density:	2.45 g/cc
Space Group:	P2$_1$/m	Unit Cell Volume:	941 A^3
Unit Cell Constants:	a = 6.77 A	X-Ray Powder Data:	Table 3.6
	b = 17.51 A		
	c = 7.74 A		
	β = 94°18'		

Structural Properties

Framework: Stereo 2.34 Complex, sheet-type structure of 4-4-1 units or chains of parallel 4-rings in [100] direction

SBU: 4-4-1 or S4R   Void volume: 0.26 cc/cc
Framework density: 1.77 g/cc

Channel System: Fig. 2.27 Two-dimensional ∥ to a and c
Hydrated—
 Free Apertures: 8-rings 2.3 × 5.0 A ⊥ to a; 8-rings 2.7 × 4.1 A ⊥ to c
 Cation Locations: In channels coordinated to 5 H$_2$O and 4 framework oxygens

Dehydrated—
 Free Apertures: Not known
Location of H$_2$O Molecules: In channels
Largest Molecule Adsorbed: H$_2$O
Kinetic Diameter, σ, A: 2.6

Table 2.23 Zeolite: Chabazite

Structure Group:	4
Reference:	143, 150, 152, 178

Chemical Composition

Typical Oxide Formula:	$CaO \cdot Al_2O_3 \cdot 4\ SiO_2 \cdot 6.5\ H_2O$
Typical Unit Cell Contents:	$Ca_2[(AlO_2)_4(SiO_2)_8] \cdot 13\ H_2O$
Variations:	Na,K; Si/Al = 1.6–3

Crystallographic Data

Symmetry:	Rhombohedral	Density:	2.05–2.10 g/cc
Space Group:	$R\bar{3}m$	Unit Cell Volume:	822 A^3
Unit Cell Constants:	a = 9.42 A	X-Ray Powder Data:	Table 3.7
	$\beta = 94°28'$		
Hexagonal,	a = 13.78; c = 15.06		

Structural Properties

Framework: Stereo 2.19   Configuration of D6R units in sequence ABCABC linked by tilted 4-rings

SBU:	D6R	Void volume:	0.47 cc/cc
Cage type:	Ellipsoidal, 6.7 x 10 A	Framework density:	1.45 g/cc

Channel System: Fig. 2.29a   Three-dimensional

Hydrated—

Free Apertures:	8-rings, 3.7 x 4.2 A
	6-rings, 2.6 A
Cation Locations:	Coordinated with 4 H_2O in the cavity

Dehydrated—

Free Apertures:	3.1 x 4.4 A
Cation Locations:	0.6 Ca^{2+} in S_I; 0.35 Ca^{2+} in S_{II}; 1/16 Ca^{2+} in S_{III}
Effect of Dehydration:	Some framework distortion
Location of H_2O Molecules:	In cavities, 5 per Ca^{2+} ion
Largest Molecule Adsorbed:	n-paraffin hydrocarbons
Kinetic Diameter, σ, A:	4.3

Table 2.24 Zeolite: Clinoptilolite

Structure Group:	7
Reference:	179–181

Chemical Composition

Typical Oxide Formula:	$(Na_2,K_2)O \cdot Al_2O_3 \cdot 10\ SiO_2 \cdot 8\ H_2O$
Typical Unit Cell Contents:	$Na_6\ [(AlO_2)_6(SiO_2)_{30}] \cdot 24\ H_2O$
Variations:	Ca, K, Mg also present; Na, K $\gg$ Ca
	Si/Al, 4.25 to 5.25

Crystallographic Data

Symmetry:	Monoclinic	Density:	2.16 g/cc
Space Group:	I 2/m	Unit Cell Volume:	2100 A^3
Unit Cell Constants:	a = 7.41 A	X-Ray Powder Data:	Table 3.8
	b = 17.89 A		
	c = 15.85 A		
	$\beta = 91°29'$		

Structural Properties

Framework:	Possibly related to heulandite but not determined
Void volume: 0.34 cc/cc	Framework density: 1.71 g/cc
Dehydrated—	
Effect of Dehydration:	Very stable—in air to 700°C
Largest Molecule Adsorbed:	O_2
Kinetic Diameter, σ, A:	3.5

Table 2.25 Zeolite: Dachiardite

Structure Group:	6
Reference:	162, 182

Chemical Composition

Typical Oxide Formula:	$Na_2O \cdot Al_2O_3 \cdot 7.6\ SiO_2 \cdot 4.8\ H_2O$
Typical Unit Cell Contents:	$Na_5[(AlO_2)_5(SiO_2)_{19}] \cdot 12\ H_2O$
Variations:	Other cations Na, K, Ca, Mg

Crystallographic Data

Symmetry:	Monoclinic	Density:	2.16 g/cc
Space Group:	C2/m	Unit Cell Volume:	1384 A^3
Unit Cell Constants:	a = 18.73 A	X-Ray Powder Data:	Table 3.9
	b = 7.54 A		
	c = 10.30 A		
	$\beta = 107°54'$		

Structural Properties

Framework: Stereo 2.28	Complex chains of 5-rings; crosslinked by 4-rings	
SBU:	Unit of 5-1	Void volume: 0.32 cc/cc
		Framework density: 1.72 g/cc
Channel System: Fig. 2.27	Two-dimensional, ∥ to c-axis and b-axis.	

Hydrated—
- Free Apertures: 8-ring, 3.7 x 6.7 A
  10-ring, 3.6 x 4.8 A

Dehydrated—
- Effect of Dehydration: Probably stable

Table 2.26 Zeolite: Edingtonite

Structure Group:	5
Reference:	183–186
Chemical Composition	
Typical Oxide Formula:	$BaO \cdot Al_2O_3 \cdot 3\ SiO_2 \cdot 4\ H_2O$
Typical Unit Cell Contents:	$Ba_2[(AlO_2)_4(SiO_2)_6] \cdot 8\ H_2O$
Variations:	May have some Ca, Na, K
Crystallographic Data	
Symmetry:	Orthorhombic Density: 2.75 g/cc
Space Group:	P222 Unit Cell Volume: 598 A^3
Unit Cell Constants:	a = 9.54 A X-Ray Powder Data: Table 3.10
	b = 9.65 A
	c = 6.50 A
Structural Properties	
Framework: Stereo 2.26	Crosslinked chains of 4-1 units, Si/Al = 1.5
	Al, Si ordered
SBU:	Unit of 4-1 Void volume: 0.36 cc/cc
	$[Al_2Si_3O_{10}]$ Framework density: 1.68 g/cc
Channel System: Fig. 2.27	Two-dimensional $\perp$ to c
Hydrated–	
Free Apertures:	8-ring, 3.5 x 3.9
Cation Locations:	In alternate cavities, 8-coordinated with 6 oxygens and 2 H_2O
Dehydrated–	
Effect of Dehydration:	Stable at 250°C
Location of H_2O Molecules:	Double chains in each channel
Largest Molecule Adsorbed:	H_2O
Kinetic Diameter, σ, A:	2.6

Table 2.27 Zeolite: Epistilbite

Structure Group:	6
Reference:	165, 166

Chemical Composition

Typical Oxide Formula:	$CaO \cdot Al_2O_3 \cdot 6\ SiO_2 \cdot 5\ 1/3\ H_2O$
Typical Unit Cell Contents:	$Ca_3\,[(AlO_2)_6(SiO_2)_{18}] \cdot 16\ H_2O$
Variations:	Also Na, K; Si/Al 2.58–3.02

Crystallographic Data

Symmetry:	Monoclinic	Density:	2.21 g/cc
Space Group:	C2/m	Unit Cell Volume:	1354 A^3
Unit Cell Constants:	a = 9.08 A	X-Ray Powder Data:	Table 3.11
	b = 17.74 A		
	c = 10.25 A		
	β = 124.54°		

Structural Properties

Framework: Stereo 2.30		Arrangement of chains of 5-rings in a-direction. Tetrahedra are in sheets ∥ to the (010) plane. Partial ordering of Al in joining tetrahedra
SBU:	Unit 5-1	Void volume: 0.25 cc/cc
		Framework density: 1.76 g/cc
Channel System: Fig. 2.27		Two-dimensional, ∥ to a-axis and c-axis
		Ellipsoidal voids at intersections, 3.7 x 8.4 A

Hydrated—

Free Apertures:	10-ring, ⊥ to a, 3.2 x 5.3 A
	8-ring, ⊥ to c, 3.7 x 4.4 A
Cation Locations:	Near periphery of cavity coordinated with framework and water oxygen atoms

Dehydrated—

Free Apertures:	Unknown
Cation Locations:	Unknown
Effect of Dehydration:	Stable to vacuum at 250°C
Location of H_2O Molecules:	Central part of cavities
Largest Molecule Adsorbed:	H_2O
Kinetic Diameter, σ, A:	2.6

Table 2.28 Zeolite: Erionite

Structure Group:	2
Reference:	88, 89, 90, 91, 92

Chemical Composition

Typical Oxide Formula: $(Ca,Mg,Na_2,K_2)O \cdot Al_2O_3 \cdot 6\ SiO_2 \cdot 6\ H_2O$

Typical Unit Cell Contents: $(Ca,Mg, Na_2, K_2)_{4.5} [(AlO_2)_9 (SiO_2)_{27}] \cdot 27\ H_2O$

Variations: Si/Al = 3–3.5. Alkali ions > alkaline earths. Possibly Fe^3 in tetrahedral sites.

Crystallographic Data

Symmetry:	Hexagonal	Density:	2.02 g/cc
Space Group:	$P6_3/mmc$	Unit Cell Volume:	2300 A^3
Unit Cell Constants:	a = 13.26 A	X-Ray Powder Data:	Table 3.12
	c = 15.12 A		

Structural Properties

Framework: Stereo 2.12 Units of S6R arranged ∥ in stacking sequence of AABAAC. Columns of D6R and ϵ-cages in c direction in sequence ϵ-D6R-ϵ linked by 6-rings

SBU:	S6R and D6R	Void volume:	0.35 cc/cc
Cage type:	ϵ, 23-hedron	Framework density:	1.51 g/cc

Channel System: Fig. 2.29c Three-dimensional, ⊥ to c one-dimensional, 2.5 A, ∥ to c

Hydrated—
  Free Apertures: 8-ring, 3.6 x 5.2 A
  Cation Locations: In ϵ-cages
Dehydrated—
  Effect of Dehydration: Stable
  Largest Molecule Adsorbed: n-paraffin hydrocarbons
  Kinetic Diameter, σ, A: 4.3

Table 2.29 Zeolite: F (Linde)

Structure Group:	Unknown
Reference:	41

Chemical Composition

Typical Oxide Formula:	$K_2O \cdot Al_2O_3 \cdot 2\ SiO_2 \cdot 2.9\ H_2O$
Typical Unit Cell Contents:	$K_{11}\,[(AlO_2)_{11}(SiO_2)_{11}] \cdot 16\ H_2O$

Crystallographic Data

Symmetry:	Tetragonal	Density:	2.28 g/cc
Unit Cell Constants:	a = 10.4 A	Unit Cell Volume:	1490 A^3
	c = 13.9 A	X-Ray Powder Data:	Table 4.43

Structural Properties

Void volume:	0.31 cc/cc	Framework density:	1.47 g/cc

Dehydrated—

Effect of Dehydration:	Stable 350°C
Largest Molecule Adsorbed:	SO_2
Kinetic Diameter, σ, A:	3.6

Table 2.30 Zeolite: Faujasite

Structure Group:	4
Reference:	59, 123, 125, 131–133, 187–190

Chemical Composition

Typical Oxide Formula:	$(Na_2,Ca,Mg,K_2)O \cdot Al_2O_3 \cdot 4.5\ SiO_2 \cdot 7\ H_2O$
Typical Unit Cell Contents:	$Na_{12}Ca_{12}Mg_{11}[(AlO_2)_{59}(SiO_2)_{133}] \cdot 235\ H_2O$
Variations:	K observed in variable amounts
	Mg observed in variable amounts

Crystallographic Data

Symmetry:	Cubic	Density:	1.91 g/cc
Space Group:	Fd3m	Unit Cell Volume:	15,014 A^3
Unit Cell Constants:	a = 24.67 A	X-Ray Powder Data:	Table 3.13

Structural Properties

Framework: Stereo 2.17	Truncated octahedra β-cages, linked tetrahedrally through D6R's in arrangement like carbon atoms in diamond. Contains eight cavities ~ 13 A in diameter in each unit cell.
SBU:	D6R, 16/uc Void volume: 0.47 cc/cc
Cage type:	β, 8/uc, 26-hedron (II). Framework density: 1.27g/cc
Channel System: Fig. 2.30	Three-dimensional, ∥ to [110]
Hydrated–	
Free Apertures:	12-ring 7.4 A; 6-ring 2.2 A
Cation Locations:	See Table 2.11
Dehydrated–	
Cation Locations:	See Table 2.11
Effect of Dehydration:	Stable and reversible
Location of H_2O Molecules:	4 in each β-cage
Largest Molecule Adsorbed:	$(C_2F_5)_3N$
Kinetic Diameter, σ, A:	8.0

Table 2.31 Zeolite: Ferrierite

Structure Group:	6
Reference:	163, 164, 181

Chemical Composition

Typical Oxide Formula:	$(Na_2,Mg)O \cdot Al_2O_3 \cdot 11.1\ SiO_2 \cdot 6.5\ H_2O$
Typical Unit Cell Contents:	$Na_{1.5}Mg_2[(AlO_2)_{5.5}(SiO_2)_{30.5}] \cdot 18\ H_2O$
Variations:	Si/Al to 3.8; some Fe^{3+}; also K and Ca; H_2O to 23

Crystallographic Data

Symmetry:	Orthorhombic	Density:	2.13 - 2.14 g/cc
Space Group:	Immm	Unit Cell Volume:	2027 A^3
Unit Cell Constants:	a = 19.16 A	X-Ray Powder Data:	Table 3.14
	b = 14.13 A		
	c = 7.49 A		

Structural Properties

Framework:	Stereo 2.29	Complex chains of 5-rings which ‖ c-axis are cross-linked by 6-rings
	SBU:	Unit of 5-1 Void volume: 0.28 cc/cc
		Framework density: 1.76 g/cc
Channel System: Fig. 2.27		Two-dimensional ‖ to c-axis, and b-axis

Hydrated—

Free Apertures:	10-ring, 4.3 x 5.5 A; 8-ring, 3.4 x 4.8 A
Cation Locations:	Mg^{2+} as $Mg(H_2O)_6^{2+}$ ions in center of cavities

Dehydrated—

Cation Locations:	Unknown
Effect of Dehydration:	Stable
Location of H_2O Molecules:	In hydration sphere of Mg^{2+}
Largest Molecule Adsorbed:	C_2H_4
Kinetic Diameter, σ, A:	3.9

Table 2.32   Zeolite: K, Ba-G

Structure Group:	4
Reference:	157
Chemical Composition	
Typical Oxide Formula:	0.15 K$_2$O, 0.85 BaO·Al$_2$O$_3$·2.14 SiO$_2$·2.6 H$_2$O
Typical Unit Cell Contents:	K$_{2.7}$Ba$_{7.65}$[(AlO$_2$)$_{18}$(SiO$_2$)$_{18}$]·23 H$_2$O
Variations:	Si/Al to ~ 2.7
Crystallographic Data	
Symmetry:	Hexagonal Density: 2.46 g/cc
Unit Cell Constants:	a = 18.70 A Unit Cell Volume: 2271 A^3
	c = 7.50 A X-Ray Powder Data Table 4.49
Structural Properties	
Framework: Stereo 2.23	Same as zeolite L
SBU:	D6R and S4R Void volume: 0.29 cc/cc
Cage type:	ε Framework density: 1.57 g/cc
Channel System: Fig. 2.29b	One-dimensional ∥ to c-axis
Hydrated—	
Free Apertures:	12-rings; 7.1 x 7.8 A
Cation Locations:	In ε-cage, between ε-cages, and on channel
Dehydrated—	
Effect of Dehydration:	Stable to 800°C
Location of H$_2$O Molecules:	In channels
Largest Molecule Adsorbed:	At least neopentane
Kinetic Diameter, σ, A:	6.2

Table 2.33 Zeolite: Garronite

Structure Group:	1
Reference:	78, 191

Chemical Composition

Typical Oxide Formula:	$(Ca,Na_2)O \cdot Al_2O_3 \cdot 3.3\ SiO_2 \cdot 4.7\ H_2O$
Typical Unit Cell Contents:	$NaCa_{2.5}[(AlO_2)_6(SiO_2)_{10}] \cdot 14\ H_2O$
Variations:	Small amounts K and Ba; Si/Al = 1.5 to 1.67

Crystallographic Data

Symmetry:	Tetragonal	Density:	2.13–2.17 g/cc
Unit Cell Constants:	a = 10.0 A	Unit Cell Volume:	990 A^3
	c = 9.9 A	X-Ray Powder Data:	Table 3.15

Structural Properties

Framework: Stereo 2.6   Same as gismondine and P

Structural analysis not made

SBU:	D4R	Void volume:	0.40 cc/cc
		Framework density:	1.62 g/cc

Channel System: Fig. 2.30b   Three-dimensional, ∥ to a and c

Hydrated—
  Free Apertures: 8-ring, 3.5 A
  Cation Locations: Probably in channels

Dehydrated—
  Effect of Dehydration: Probably distorts

Largest Molecule Adsorbed: Probably like zeolite P—not studied

Table 2.34 Zeolite: Gismondine

Structure Group:	1
Reference:	75, 76

Chemical Composition

Typical Oxide Formula:	$CaO \cdot Al_2O_3 \cdot 2\, SiO_2 \cdot 4\, H_2O$
Typical Unit Cell Contents:	$Ca_4\,[(AlO_2)_8(SiO_2)_8]\cdot 16\, H_2O$
Variations:	Si/Al = 1.12–1.49; some K

Crystallographic Data

Symmetry:	Monoclinic	Density:	2.27 g/cc
Space Group:	$P2_1/C$	Unit Cell Volume:	1046 A^3
Unit Cell Constants:	a = 9.84 A	X-Ray Powder Data:	Table 3.16
	b = 10.02 A		
	c = 10.62 A β = 92°25'		

Structural Properties

Framework: Stereo 2.6	4-rings, crosslinked so that UUDD chains ∥ a and b; Si-Al ordered		
SBU:	S4R	Void volume:	0.46 cc/cc
Cage type:	nonspecific	Framework density:	1.52 g/cc
Channel System: Fig. 2.30b	Three-dimensional; ∥ to a, ∥ to b Tube bundle type		

Hydrated–

Free Apertures:	8-rings; 3.1 x 4.4 A ⊥ to a, 2.8 x 4.9 A ⊥ to b
Cation Locations:	At channel intersection coordinated to 4 H_2O and 2 cations.

Dehydrated–

Free Apertures:	Unknown
Cation Locations:	Unknown
Effect of Dehydration:	Stable to 250°C
Location of H_2O Molecules:	Coordinated to Ca^{2+}; H_2O/Ca^{2+} = 4
Largest Molecule Adsorbed:	H_2O
Kinetic Diameter, σ, A:	2.6

Table 2.35 Zeolite: Gmelinite

Structure Group:	4
Other Designation:	Groddeckite
Reference:	57, 97, 153, 188, 192, 193
Chemical Composition	
Oxide Formula:	$Na_2O \cdot Al_2O_3 \cdot 4\, SiO_2 \cdot 6\, H_2O$
Typical Unit Cell Contents:	$Na_8\,[(AlO_2)_8(SiO_2)_{16}] \cdot 24\, H_2O$
Variations:	Na, Ca ≫ K; Si/Al = 1.9–2.46
Crystallogrpahic Data	
Symmetry: Hexagonal	Density: 2.03 g/cc
Space Group: $P6_3/mmc$	Unit Cell Volume: 1647 A^3
Unit Cell Constants: a = 13.75 A	X-Ray Powder Data: Table 3.17
c = 10.05 A	

Structural Properties

Framework:	Stereo 2.21	D6R units in ∥, hexagonal, "close" packed array ABAB... linked by tilted four rings	
	SBU:	D6R	Void volume: 0.44 cc/cc
	Cage type:	14-hedron (II)	Framework density: 1.46 g/cc

Channel System: Fig. 2.29b	Three-dimensional ∥ to c, two ∥ to a; stacking faults apparently block the c channel partially
Hydrated–	
Free Apertures:	12-rings, ⊥ to c-axis, 7.0 A, ⊥ to a, 8-rings, 3.6 x 3.9 A
Cation Locations:	Two/uc in center of D6R
Dehydrated–	
Free Apertures:	Unknown
Cation Locations:	Unknown
Effect of Dehydration:	Stable
Location of H_2O Molecules:	Probably in channels
Largest Molecule Adsorbed:	C_3H_8
Kinetic Diameter, σ, A:	4.3

Table 2.36 Zeolite: Gonnardite

Structure Group:	5
Other Designation:	Ranite
Reference:	183, 194

Chemical Composition

Typical Oxide Formula:	$0.5\ Na_2O, 0.5\ CaO \cdot Al_2O_3 \cdot 3\ SiO_2 \cdot 3.5\ H_2O$
Typical Unit Cell Contents:	$Na_4Ca_2[(AlO_2)_8(SiO_2)_{12}] \cdot 14\ H_2O$
Variations:	Some Mg, K

Crystallographic Data

Symmetry:	Orthorhombic	Density:	2.25 g/cc
Unit Cell Constants:	a = 13.19 A	X-Ray Powder Data:	Table 3.18
	b = 13.32 A		
	c = 6.55 A		

Structural Properties

Framework:	Stereo 2.25	Same as thomsonite but Si, Al distribution is the same as that in natrolite in the 4-1 units	
	SBU:	Unit of 4-1	Void volume: 0.31 cc/cc
		$[Al_2Si_3O_{10}]$	Framework density: 1.74 g/cc
Channel System: Fig. 2.27		Two-dimensional and ⊥ to c	
Hydrated—			
Free Apertures:		Same as thomsonite	
Cation Locations:		Unknown	
Dehydrated—			
Free Apertures:		Unknown	
Location of H_2O Molecules:		Probably with cations in channels	

Table 2.37 Zeolite: H

Structure Group:	Unknown
Other Designation:	K-I
Reference:	43 |их

Chemical Composition

Typical Oxide Formula:	$K_2O \cdot Al_2O_3 \cdot 2\ SiO_2 \cdot 4\ H_2O$
Typical Unit Cell Contents:	$K_{14}[(AlO_2)_{14}(SiO_2)_{14}] \cdot 28\ H_2O$

Crystallographic Data

Symmetry:	Hexagonal	Density:	2.18 g/cc
Unit Cell Constants:	a = 13.47 A	Unit Cell Volume:	2082 A^3
	c = 13.25 A	X-Ray Powder Data:	Table 4.50

Structural Properties

Void volume:	0.39 cc/cc	Framework density:	1.35 g/cc

Dehydrated—
Effect of Dehydration: Unstable at 200°C

TABLES OF ZEOLITE DATA   153

Table 2.38   Zeolite: Harmotome

Structure Group:	1
Reference:	73, 195

Chemical Composition

Typical Oxide Formula:	$BaO \cdot Al_2O_3 \cdot 6\,SiO_2 \cdot 6\,H_2O$
Typical Unit Cell Contents:	$Ba_2\,[(AlO_2)_4\,(SiO_2)_{12}] \cdot 12\,H_2O$
Variations:	$Ba \gg K$ or Na; $Si/Al = 2.3-2.5$

Crystallographic Data

Symmetry:	Monoclinic	Density:	2.35 g/cc
Space Group:	$P2_1$	Unit Cell Volume:	999 A^3
Unit Cell Constants:	a = 9.87 A	X-Ray Powder Data:	Table 3.19
	b = 14.14 A		
	c = 9.72 A		
	$\beta = 124°50'$		

Structural Properties

Framework: Stereo 2.5	Same as phillipsite		
SBU:	S4R	Void volume:	0.31 cc/cc
		Framework density:	1.59 g/cc
Channel System:	Three-dimensional; ‖ to a, ‖ to b, ‖ to c		

Hydrated—

Free Apertures:	8-rings, 4.2 x 4.4, 2.8 x 4.8 A, and 3.3 A
Cation Locations:	Coordinated to 6 O's, 4 H_2O

Dehydrated—

Free Apertures:	Unknown
Effect of Dehydration:	Stable
Location of H_2O Molecules:	8-coordinated to Ba^{2+}
Largest Molecule Adsorbed:	NH_3
Kinetic Diameter, σ, A:	2.6

154 STRUCTURE OF ZEOLITES

Table 2.39   Zeolite: Heulandite

Structure Group:	7
Reference:	168
Chemical Composition	
Typical Oxide Formula:	$CaO \cdot Al_2O_3 \cdot 7\ SiO_2 \cdot 6\ H_2O$
Typical Unit Cell Contents:	$Ca_4[(AlO_2)_8(SiO_2)_{28}] \cdot 24\ H_2O$
Variations:	Also K and Sr; Si/Al 2.47–3.73; water rich variety has 30 H_2O

Crystallographic Data

Symmetry:	Monoclinic	Density:	2.198 g/cc
Space Group:	Cm	Unit Cell Volume:	2103 A^3
Unit Cell Constants:	a = 17.73 A	X-Ray Powder Data:	Table 3.21
	b = 17.82 A		
	c = 7.43 A		
	$\beta = 116°20'$		

Structural Properties

Framework:	Stereo 2.32	Special configuration of tetrahedra in 4- and 5-rings arranged in sheets ‖ to (010)
SBU:	Unit 4-4-1	Void volume: 0.39 cc/cc
		Framework density: 1.69 g/cc
Channel System:	Fig. 2.27	Two-dimensional, consisting of three channels, ‖ to a-axis and c-axis and at 50° to a-axis

Hydrated—
 Free Apertures: 8-ring, 4.0 x 5.5 A ⊥ to a; 10-ring, 4.4 x 7.2 A ⊥ to c; 8-ring, 4.1 x 4.7 A ⊥ to c
 Cation Locations: Three types located in open channels. Ca_1 and Ca_2 in eight-fold coordination with 3 framework oxygens and 5 water molecules near channel walls. Ca_3 is similarly coordinated and located at intersection of 8-ring channels.

Dehydrated—
 Free Apertures: Unknown
 Effect of Dehydration: Structure changes at 215° to Heulandite "B", structure unknown

Location of H_2O Molecules: Coordinated to calcium ions in the channels
Largest Molecule Adsorbed: NH_3 if partially dehydrated at 130°
Kinetic Diameter, σ, A: 2.6

Table 2.40 Zeolite: HS, Hydrated Sodalite

Structure Group:	2
Other Designation:	Sodalite Hydrate, Hydroxy Sodalite, G
Reference:	99. 100, 196

Chemical Composition

Typical Oxide Formula:	$Na_2O \cdot Al_2O_3 \cdot 2\ SiO_2 \cdot 2.5\ H_2O$
Typical Unit Cell Contents:	$Na_6[(AlO_2)_6(SiO_2)_6]{\sim}7.5\ H_2O$
Variations:	May contain varying amounts of NaOH

Crystallographic Data

Symmetry:	Cubic	Density:	2.03 g/cc
Space Group:	$P\bar{4}3n$	Unit Cell Volume:	702 A^3
Unit Cell Constants:	a = 8.86 A	X-Ray Powder Data:	Table 4.28

Structural Properties

Framework:	Stereo 2.22	\multicolumn{2}{l}{Units of S6R ‖ and connected in sequence ABCABC. Close packed truncated octahedral cages or β-cages}
SBU:	S6R	Void volume:	0.35 cc/cc
Cage type:	β	Framework density:	1.72 g/cc
Channel System:	\multicolumn{3}{l}{Three-dimensional, interconnected β-cages}

Hydrated—

Free Apertures:	6-rings, 2.2 A
Cation Locations:	Near the 6-rings but in the cavity of the β-cage

Dehydrated—

Cation Locations:	Probably on the 6-rings
Effect of Dehydration:	None to 800°C
Location of H_2O Molecules:	Probably 4 in each β-cage
Largest Molecule Adsorbed:	H_2O
Kinetic Diameter, σ, A:	2.6

Table 2.41 Zeolite: L

Structure Group:	4
Reference:	59, 61, 157, 197

Chemical Composition

Typical Oxide Formula:	$(K_2, Na_2)O \cdot Al_2O_3 \cdot 6\ SiO_2 \cdot 5\ H_2O$
Typical Unit Cell Contents:	$K_9[(AlO_2)_9(SiO_2)_{27}] \cdot 22\ H_2O$
Variations:	$K \gg Na$; $Si/Al = 2.6-3.5$

Crystallographic Data

Symmetry:	Hexagonal	Density:	2.11 g/cc
Space Group:	P6/mmm	Unit Cell Volume:	2205 A^3
Unit Cell Constants:	a = 18.4 A	X-Ray Powder Data:	Table 4.60
	c = 7.5 A		

Structural Properties

Framework: Stereo 2.23   Cancrinite-type or ϵ-cages linked by D6R's in columns and crosslinked by O bridges to form planar 12-rings

SBU:	D6R and S4R	Void volume:	0.32 cc/cc
Cage type:	ϵ	Framework density:	1.61 g/cc

Channel System: Fig. 2.29b   One-dimensional ∥ to c-axis, no stacking faults

Hydrated—
    Free Apertures: 12-rings, 7.1 A
    Cation Locations: Four types: A in D6R, B in ϵ-cage, C between ϵ-cages, D on channel wall

Dehydrated—
    Free Apertures: Same
    Cation Locations: Unknown, D cations move inward to E-site
    Effect of Dehydration: Stable in air to 850°C
Location of H_2O Molecules: Loosely bound in clusters in channel
Largest Molecule Adsorbed: $(C_4H_9)_3N$ and $(C_4F_9)_3N$ slowly at 50°
Kinetic Diameter, σ, A: 8.1

Table 2.42 Zeolite: P-L

Structure Group:	4		
Reference:	37		
Chemical Composition			
Typical Unit Cell Contents:	$K_{23}[(AlO_2)_{33}(SiO_2)_{26}(PO_2)_{13}] \cdot 42\ H_2O$		
Crystallographic Data			
Symmetry:	Hexagonal	Density:	2.21 g/cc
Space Group:	P6cc or P6/mcc	Unit Cell Volume:	4576 A^3
Unit Cell Constants:	a = 18.75 A	X-Ray Powder Data:	Table 4.62
	c = 15.03 A		
Structural Properties			
Framework: Stereo 2.23	Same as zeolite L with P and Al ordering		
SBU:	S4R and D6R		
Cage type:	ϵ	Framework density:	1.57 g/cc
Channel System: Fig. 2.29b	∥ to c		
Largest Molecule Adsorbed:	Isobutane		
Kinetic Diameter, σ, A:	5		

Table 2.43 Zeolite: Laumontite

Structure Group:	1
Other Designation:	Leonhardite (partially dehydrated)
Reference:	32, 82, 83

Chemical Composition

Typical Oxide Formula:	$CaO \cdot Al_2O_3 \cdot 4\ SiO_2 \cdot 4\ H_2O$
Typical Unit Cell Contents:	$Ca_4\ [(AlO_2)_8(SiO_2)_{16}] \cdot 16\ H_2O$
Variations:	K, Na; Si/Al = 1.75–2.28

Crystallographic Data

Symmetry:	Monoclinic	Density:	2.30 g/cc
Space Group:	Am	Unit Cell Volume:	1390 Å^3
Unit Cell Constants:	a = 14.90 A	X-Ray Powder Data:	Table 3.23
	b = 13.17 A		
	c = 7.50 A		
	β = 111°30′		

Structural Properties

Framework:	Stereo 2.10	4-rings (Si_4O_{12}) in TTTT configuration ∥ to (120) and linked by AlO_4 tetrahedra
	SBU:	S4R Void volume: 0.34 cc/cc
		Framework density: 1.77 g/cc
Channel System:		One-dimensional, ∥ to a; 4.6 x 6.3 A
Hydrated—		
Free Apertures:		10-ring, distorted 4.6 x 6.3 A
Cation Locations:		4 O and 2 H_2O in 6-fold coordination
Dehydrated—		
Free Apertures:		
Cation Locations:		
Effect of Dehydration:		Stepwise dehydration
Largest Molecule Adsorbed:		H_2O, NH_3 if dehydrated at 200°
Kinetic Diameter, σ, A:		2.6

Table 2.44 Zeolite: Levynite

Structure Group:	2
Reference:	97, 192

Chemical Composition

Typical Oxide Formula:	$CaO \cdot Al_2O_3 \cdot 4\ SiO_2 \cdot 6\ H_2O$
Typical Unit Cell Contents:	$Ca_3\ [(AlO_2)_6(SiO_2)_{12}] \cdot 18\ H_2O$
Variations:	$Ca \gg Na, K$; Si/Al − 1.59–1.83

Crystallographic Data

Symmetry:	Rhombohedral	Density:	2.14 g/cc
Space Group:	R3m	Unit Cell Volume:	$1150_{rh}\ A^3$,
Unit Cell Constants:	a = 10.75 A		$3450_{hex}\ A^3$
	$\alpha = 76°25'A$	X-Ray Powder Data: Table 3.24	
Hexagonal	a = 13.32 A		
	c = 22.51 A		

Structural Properties

Framework: Stereo 2.14 Units of S6R arranged ∥ in stacking sequence of AABCCABBC..

SBU:	S6R, D6R Void volume: 0.40 cc/cc
Cage type:	Ellipsoidal, 7–9A Framework density: 1.54 g/cc
Channel System: Fig. 2.29d	Two-dimensional ⊥ to c, 3.2 x 5.1 A and ∥ to c, 2.6 A

Hydrated—

Free Apertures:	8-rings, 3.2–5.1 A
Cation Locations:	Unknown

Dehydrated—

Free Apertures:	By adsorption ~ 3.3
Cation Locations:	Unknown
Effect of Dehydration:	Stable to dehydration and reversible
Location of H_2O Molecules:	Unknown
Largest Molecule Adsorbed:	N_2, O_2
Kinetic Diameter, σ, A:	3.6

Table 2.45 Zeolite: LOSOD

Structure Group:	2		
Reference:	101		

Chemical Composition

Typical Oxide Formula: $Na_2O \cdot Al_2O_3 \cdot 2\ SiO_2 \cdot 3.13\ H_2O$

Typical Unit Cell Contents: $Na_{12}[(AlO_2)_{12}(SiO_2)_{12}] \cdot 19\ H_2O$

Crystallographic Data

Symmetry:	Hexagonal	Density:	2.154 g/cc
Space Group:	$P6_3/mmc$	Unit Cell Volume:	1520.6 A^3
Unit Cell Constants:	a = 12.906 A	X-Ray Powder Data:	Table 4.63
	c = 10.541 A		

Structural Properties

Framework:	ABAC—stacking of 6-rings		
SBU:	S6R	Void volume:	0.33 cc/cc
Cage type:	ϵ, and a 17-hedron	Framework density:	1.58 g/cc

Hydrated—

Free Apertures: 6-rings, ~2.2 A

Largest Molecule Adsorbed: H_2O

Table 2.46 Zeolite: Mesolite

Structure Group:	5
Reference:	158, 198, 201
Chemical Composition	
Typical Oxide Formula:	$(Na_2, CaO) \cdot Al_2O_3 \cdot 3\ SiO_2 \cdot 2.7\ H_2O$
Typical Unit Cell Contents:	$Na_{16}Ca_{16}[(AlO_2)_{48}(SiO_2)_{72}] \cdot 64\ H_2O$
Variations:	$Ca/Na \geqslant 1$; $Si/Al = 1.44-1.58$
Crystallographic Data	
Symmetry:	Orthorhombic Density: 2.26 g/cc
Space Group:	Fdd2 Unit Cell Volume: 6,814 A^3
Unit Cell Constants:	a = 56.7 A X-Ray Powder Data: Table 3.25
	b = 6.54 A
	c = 18.44 A
Structural Properties	
Framework: Stereo 2.24b	Same as natrolite; Al and Si are ordered
SBU:	Unit of 4-1 Void volume: 0.30 cc/cc
	Framework density: 1.75 g/cc
Channel System: Fig. 2.27	Same as natrolite
Hydrated—	
Free Apertures:	8-rings, 2.6 x 3.9 A
Cation Locations:	In channels with H_2O
Dehydrated—	
Effect of Dehydration:	Unstable—new structure at 300°C
Largest Molecule Adsorbed:	H_2O
Kinetic Diameter, σ, A:	2.6

Table 2.47 Zeolite: Mordenite

Structure Group:	6
Other Designation:	Ptilolite, Arduinite, Flokite, Deeckite
Reference:	89, 159, 161
Chemical Composition	
Typical Oxide Formula:	$Na_2O \cdot Al_2O_3 \cdot 10\ SiO_2 \cdot 6\ H_2O$
Typical Unit Cell Contents:	$Na_8\ [(AlO_2)_8(SiO_2)_{40}] \cdot 24\ H_2O$
Variations:	Si/Al = 4.17–5.0; Na, Ca > K
Crystallographic Data	
Symmetry:	Orthorhombic Density: 2.13 g/cc
Space Group:	Cmcm Unit Cell Volume: 2794 $Å^3$
Unit Cell Constants:	a = 18.13 X-Ray Powder Data: Table 3.26
	b = 20.49
	c = 7.52
Structural Properties	
Framework: Stereo 2.27	Complex chains of 5-rings crosslinked by 4-rings. Chains consist of 5-rings of SiO_4 tetrahedra and single AlO_4 tetrahedra
SBU:	Unit 5-1 Void volume: 0.28 cc/cc
	Framework density: 1.70 g/cc
Channel System: Fig. 2.27a	Main system ∥ to c; channels with 2.8 Å restrictions ∥ to b
Hydrated—	
Free Apertures:	12-rings ⊥ to c-axis, 6.7 x 7.0 Å
	8-rings, 2.9 x 5.7 Å, ⊥ to b
Cation Locations:	Four Na^+ in the restrictions with minimum dimension of 2.8 Å in channels ⊥ to b
Dehydrated—	
Free Apertures:	Probably no change
Cation Locations:	Unknown
Effect of Dehydration:	Very stable
Location of H_2O Molecules:	Unknown
Largest Molecule Adsorbed:	C_2H_4
Kinetic Diameter, σ, Å:	3.9

Table 2.48 Zeolite: Mordenite (Large Port)

Structure Group:	6
Other Designation:	Zeolon
Reference:	160
Chemical Composition	
Typical Oxide Formula:	$Na_2O \cdot Al_2O_3 \cdot 9\text{-}10\ SiO_2 \cdot 6\ H_2O$
Typical Unit Cell Contents:	$Na_{8.7}[(AlO_2)_{8.7}(SiO_2)_{39.3}] \cdot 24\ H_2O$
Variations:	Si/Al = 4.5 to 5
Crystallographic Data	
	X-Ray Powder Data: Table 4.68
Structural Properties	
Framework: Stereo 2.27	Same as mordenite; large channels open
SBU:	Unit of 5-1 Void volume: 0.28 cc/cc
	Framework density: 1.7 g/cc
Channel System: Fig. 2.27	Two-dimensional ∥ to b- and c
Hydrated—	
Free Apertures:	12-rings ⊥ to c; 6.7 x 7.0 A
Largest Molecule Adsorbed:	C_6H_6
Kinetic Diameter, σ, A:	6.2

Table 2.49 Zeolite: N

Structure Group:	Unknown	
Reference:	33, 48	
Chemical Composition		
Typical Oxide Formula:	$(Na, TMA)_2O \cdot Al_2O_3 \cdot 2\ SiO_2 \cdot 2.7\ H_2O$	
Variations:	$SiO_2/Al_2O_3 = 2.0 \pm 0.2$	
Crystallographic Data		
Symmetry:	Cubic	Unit Cell Volume: 51557 A^3
	a = 37.2 A	X-Ray Powder Data: Table 4.70
Structural Properties		
Framework:	Unknown	
Largest Molecule Adsorbed:	H_2O	

Table 2.50 Zeolite: Natrolite

Structure Group:	5
Other Designation:	Elagite
Reference:	158, 198, 200, 201
Chemical Composition	
Typical Oxide Formula:	$Na_2O \cdot Al_2O_3 \cdot 3\ SiO_2 \cdot 2\ H_2O$
Typical Unit Cell Contents:	$Na_{16}[(AlO_2)_{16}(SiO_2)_{24}] \cdot 16\ H_2O$
Variations:	Si/Al = 1.44–1.58; H_2O/Na = 1; Ca, K very small

Crystallographic Data

Symmetry:	Orthorhombic	Density:	2.23 g/cc
Space Group:	Fdd2	Unit Cell Volume:	2250 A^3
Unit Cell Constants:	a = 18.30 A	X-Ray Powder Data:	Table 3.27
	b = 18.63 A		
	c = 6.60 A		

Structural Properties

Framework: Stereo 2.24b	Crosslinked chains of 4-1 units. Al and Si atoms are ordered.
SBU:	Unit of 4-1 Void volume: 0.23 cc/cc
	4-1 $[Al_2Si_3O_{10}]$ Framework density: 1.76 g/cc
Channel System: Fig. 2.27	Two-dimensional $\perp$ to c
Hydrated–	
Free Apertures:	8-rings, 2.6 x 3.9 A
Cation Locations:	In channels coordinated to 2 H_2O and 4 framework O atoms
Dehydrated–	
Free Apertures:	Unknown
Cation Locations:	Unknown
Effect of Dehydration:	Framework shrinks due to rotation of chains
Location of H_2O Molecules:	In channels coordinated to oxygen in framework and sodium ions
Largest Molecule Adsorbed:	NH_3
Kinetic Diameter, σ, A:	2.6

Table 2.51 Zeolite: O

Structure Group:	2
Other Designation:	TMA offretite
Reference:	202, 203

Chemical Composition

Typical Oxide Formula:	$(TMA_2, K_2, Na_2)O \cdot Al_2O_3 \cdot 7\ SiO_2 \cdot 3.5\ H_2O$
Typical Unit Cell Contents:	$(TMA, K, Na)_4\ [(AlO_2)_4(SiO_2)_{14}] \cdot 7\ H_2O$
Variations:	SiO_2/Al_2O_3 = 6.8 to 8.5

Crystallographic Data

Symmetry:	Hexagonal	Density:	1.99 (calc) g/cc
Unit Cell Constants:	a = 13.31 A	Unit Cell Volume:	1160 A^3
	c = 7.59 A	X-Ray Powder Data:	Table 4.73

Structural Properties

Framework:	Stereo 2.13	Same as offretite	
	SBU:	D6R, S6R	Void volume: 0.43 cc/cc†
	Cage type:	Same as offretite	Framework density: 1.55 g/cc
Channel System:	Fig. 2.29b	Same as offretite	

Hydrated—
    Cation Locations: TMA located in the 14-hedron (II) cage
Dehydrated—
    Effect of Dehydration: Stable to > 600°C
Largest Molecule Adsorbed: Cyclohexane
Kinetic Diameter, σ, A: 6

†Based on H_2O adsorption on air calcined zeolite

Table 2.52 Zeolite: Offretite

Structure Group:	2
Reference:	90, 91, 93, 192, 204

Chemical Composition

Typical Oxide Formula:	0.28 MgO, 0.41 CaO, 0.26 $K_2O \cdot Al_2O_3 \cdot$ 4.98 $SiO_2 \cdot 5.9$ H_2O
Typical Unit Cell Contents:	$(K_2,Ca,Mg)_{2.5}[(AlO_2)_5(SiO_2)_{13}] \cdot 15$ H_2O
Variations:	Alkaline earths > Alkali metals

Crystallographic Data

Symmetry:	Hexagonal	Density:	2.13 g/cc
Space Group:	P6m2	Unit Cell Volume:	1160 A^3
Unit Cell Constants:	a = 13.291 ± .002	X-Ray Powder Data: Table 3.28	
	c = 7.582 ± .006		

Structural Properties

Framework:	Stereo 2.13	Units of S6R arranged ∥ to c in sequence AABAAB ... Columns of ε-cages and D6R in c-direction	
	SBU:	D6R, S6R	Void volume: 0.40 cc/cc
	Cage Type	ε, and 14-hedron II	Framework density: 1.55 g/cc
Channel System: Fig. 2.29b		∥ to c, 6.4 A; ∥ to a, 3.6 x 5.2 A	

Hydrated—

Free Apertures:	3.6 x 5.2 A; 6.4 A
Cation Locations:	K^+ in ε-cages, and not exchangeable; Ca^{2+} is in D6R; Mg^{2+} is in the 14-hedron cage

Dehydrated—

Free Apertures:	Unknown
Location of H_2O Molecules:	10 are coordinated to cations; 5 mobile in large channels
Largest Molecule Adsorbed:	Cyclohexane
Kinetic Diameter, σ, A:	6

Table 2.53 Zeolite: Omega, Ω[a]

Structure Group:	1
Reference:	98, 205, 223
Chemical Composition	
Typical Oxide Formula:	(Na, TMA)$_2$O · Al$_2$O$_3$ · 7 SiO$_2$ · 5 H$_2$O
Typical Unit Cell Contents:	Na$_{6.8}$TMA$_{1.6}$[(AlO$_2$)$_8$(SiO$_2$)$_{28}$] · 21 H$_2$O
Variations:	Si/Al = 2.5–6.0
	Na/Al = 0.5–1.5 TMA/Al = 0–0.7
Crystallographic Data	
Symmetry:	Hexagonal Density: 2.16 g/cc
Space Group:	P6$_3$/mmc Unit Cell Volume: 2160 Å^3
Unit Cell Constants:	a = 18.15 Å X-Ray Powder Data: Table 4.72
	c = 7.59 Å
Structural Properties	
Framework: Stereo 2.15	14-hedron of gmelinite-type linked by oxygen bridges in columns ∥ to c-axis
SBU:	S4R and Unit 5-1 Void volume: 0.38 cc/cc
Cage type:	14-hedron (II) Framework density: 1.65 g/cc
Channel System:	One-dimensional ∥ to c, 7.5 Å
Hydrated—	
Free Apertures:	12 rings, 7.5 Å
Cation Locations:[b]	TMA in 14-hedron cages; Na$^+$ at 0, 1/2, 1/4 between cages
Dehydrated—	
Effect of Dehydration:	Stable 750°C
Largest Molecule Adsorbed:	(C$_4$F$_9$)$_3$N
Kinetic Diameter, σ, Å:	10

[a]See p. 81. Data from ref. 205 and 223 except where noted.
[b]From proposed structure of ref. 98.

Table 2.54 Zeolite: P

Structure Group:	1
Reference:	77–80, 206
Chemical Composition	
Typical Oxide Formula:	$Na_2O \cdot Al_2O_3 \cdot 2.0\text{-}5.0\ SiO_2 \cdot 5\ H_2O$
Typical Unit Cell Contents:	$Na_6\ [(AlO_2)_6(SiO_2)_{10}] \cdot 15\ H_2O$
Variations:	Si/Al = 1.1 to 2.5
Crystallographic Data	
Symmetry:	Near Cubic
Space Group:	I4/amd
Unit Cell Constants:	a = 10.05 A
Density:	2.01 g/cc
Unit Cell Volume:	1015 A^3
X-Ray Powder Data:	Tables 4.74, 4.75
Structural Properties	
Framework: Stereo 2.6	Same as gismondine
SBU:	S4R
Void volume:	0.41 cc/cc
Framework density:	1.57 g/cc
Channel System: Fig. 2.30b	Three-dimensional, ‖ to a and ‖ to b
Hydrated—	
Free Apertures:	Distorted 8-rings, 3.1 x 4.4 A and 2.8 x 4.9 A
Cation Locations:	Coordinated to H_2O at channel intersections
Dehydrated—	
Free Apertures:	~2.6
Cation Locations:	Unknown—probably blocking channels
Effect of Dehydration:	Framework shrinks to a = 9.6 A, volume decreases 31%
Largest Molecule Adsorbed:	H_2O
Kinetic Diameter, σ, A:	2.6

Table 2.55 Zeolite: **Paulingite**

Structure Group: 1
Reference: 81, 207
Chemical Composition
Typical Unit Cell Contents: $(K_2,Na_2,Ca,Ba)_{76}[(AlO_2)_{152}(SiO_2)_{525}]\cdot 700\ H_2O$
Crystallographic Data
Symmetry:	Cubic	Density:	2.209 g/cc
Space Group:	Im3m	Unit Cell Volume:	43,218 A^3
Unit Cell Constants:	a = 35.09 A	X-Ray Powder Data:	Table 3.29

Structural Properties
Framework: Stereo 2.8, 2.9 Polyhedral units—α, δ, and γ cages arrayed in sequence $\alpha, \delta, \gamma, \delta, \gamma, \delta, \alpha \ldots$ along cubic axis
SBU: S4R   Void volume: 0.49 cc/cc
Cage type: α, γ, and δ Framework density: 1.54 g/cc
Channel System: Fig. 2.28a Three-dimensional ∥ to a
Hydrated—
 Free Apertures: 8-ring, 3.9 A
Dehydrated—
 Cation Locations: U
 Effect of Dehydration: Stable to at least 250°, crystals break up
Location of H_2O Molecules: U
Largest Molecule Adsorbed: Kr
Kinetic, Diameter, σ, A: 3.6

Table 2.56 Zeolite: Phillipsite

Structure Group:	1
Other Designation:	Wellsite
Reference:	74, 208

Chemical Composition

Typical Oxide Formula:	$(Ca,Na_2,K_2)O \cdot Al_2O_3 \cdot 4.4\ SiO_2 \cdot 4\ H_2O$
Typical Unit Cell Contents:	$(Ca,K_2,Na_2)_5[(AlO_2)_{10}(SiO_2)_{22}] \cdot 20\ H_2O$
Variations:	Si/Al = 1.7 to 2.4; Ba found in wellsite; K found in all cases

Crystallographic Data

Symmetry:	Orthorhombic	Density:	2.15 g/cc
Space Group:	B2mb	Unit Cell Volume:	2022 A^3
Unit Cell Constants:	a = 9.96 A	X-Ray Powder Data:	Table 3.30
	b = 14.25 A		
	c = 14.25 A		

Structural Properties

Framework:	Stereo 2.5	4-rings of type UUDD crosslinked so that chains of UUDD units run ∥ to a axis	
SBU:	S4R	Void volume:	0.31 cc/cc
		Framework density:	1.58 g/cc

Channel System:	Three-dimensional; ∥ to a, ∥ to b, ∥ to c
Hydrated—	
Free Apertures:	8-rings, 4.2 x 4.4 A ⊥ to a, 2.8 x 4.8 A ⊥ to b, 3.3 A ⊥ to c
Cation Locations:	In channels, coordinated to H_2O and framework
Dehydrated—	
Free Apertures:	Unknown
Effect of Dehydration:	Structure degrades at 200°C
Location of H_2O Molecules:	Coordinated to cations
Largest Molecule Adsorbed:	H_2O
Kinetic Diameter, σ, A:	2.6

Table 2.57 Zeolite: Scolecite

Structure Group:	5
Reference:	158, 198, 199, 209

Chemical Composition

Oxide Formula:	$CaO \cdot Al_2O_3 \cdot 3\,SiO_2 \cdot 3\,H_2O$
Typical Unit Cell Contents:	$Ca_8[(AlO_2)_{16}(SiO_2)_{24}] \cdot 24\,H_2O$

Crystallographic Data

Symmetry:	Monoclinic		Density: 2.27 g/cc
Space Group:	Cc	Aa	Unit Cell Volume: 2308 A^3
Unit Cell Constants:	a = 18.53 A	9.85	X-Ray Powder Data:
	b = 18.99 A	18.98	Table 3.31
	c = 6.56 A	6.52	
	$\beta = 90°39'$	$110°6'$	

Structural Properties

Framework: Stereo 2.24b   Same as natrolite; Al and Si ordered
        SBU: Unit of 4-1   Void volume: 0.31 cc/cc
                [$Al_2Si_3O_{10}$]   Framework density: 1.75 g/cc

Channel System: Fig. 2.27   Same as natrolite

Hydrated—
    Free Apertures: 8-rings, 2.6 x 3.9 A
    Cation Locations: Coordinated with 3 H_2O molecules and 4 framework O atoms

Dehydrated—
    Effect of Dehydration: Unstable, structure shrinks
    Location of H_2O Molecules: In channels; 8 occupy vacant cation positions; 16 are same as in natrolite

Largest Molecule Adsorbed: H_2O
Kinetic Diameter, σ, A: 2.6

Table 2.58 Zeolite: Stilbite

Structure Group:	7
Other Designation:	Foresite, Epidesmine, Desmine
Reference:	169, 170, 210, 211

Chemical Composition

Typical Oxide Formula:	$(Na_2Ca)O \cdot Al_2O_3 \cdot 5.2\ SiO_2 \cdot 7\ H_2O$
Typical Unit Cell Contents:	$Na_2Ca_4[(AlO_2)_{10}(SiO_2)_{26}] \cdot 28\ H_2O$
Variations:	May contain Na, K; Si/Al 2.47–3.85
	Stellerite: Si/Al = 3.4–3.5

Crystallographic Data

Symmetry:	Monoclinic	Density:	2.16 g/cc
Space Group:	C 2/m	Unit Cell Volume:	2179 A^3
Unit Cell Constants:	a = 13.64 A	X-Ray Powder Data: Table 3.32	
	b = 18.24 A		
	c = 11.27 A		
	$\beta = 128°$		

Structural Properties

Framework:	Stereo 2.33	Special configuration of tetrahedra in 4- and 5-rings arranged in sheets
SBU:	Unit of 4-4-1	Void volume: 0.39 cc/cc
		Framework density: 1.64 g/cc
Channel System: Fig. 2.27	Two-dimensional; ∥ to a-axis, c-axis	

Hydrated—

Free Apertures:	10-ring, irregular, 4.1 x 6.2 A ⊥ to a
	8-ring, 2.7 x 5.7 A ⊥ to c
Cation Locations:	Ca^{2+} at channel intersections coordinated to 8 H_2O molecules, Na^+ ions with 4 H_2O and 2 framework O

Dehydrated—

Free Apertures:	Unknown
Effect of Dehydration:	Structure changes at 120°C
Location of H_2O Molecules:	Coordinated with Ca^{2+} ions at channel intersections
Largest Molecule Adsorbed:	H_2O, NH_3 if dehydrated 200°C
Kinetic Diameter, σ, A:	2.6

Stellerite stable at 250°C; adsorbs CO_2, $\sigma = 3.3$ A

Table 2.59 Zeolite: T

Structure Group:	2
Reference:	89, 91, 94–96
Chemical Composition	
Typical Oxide Formula:	$0.3\ Na_2O \cdot 0.7\ K_2O \cdot Al_2O_3 \cdot 6.9\ SiO_2 \cdot 7.2\ H_2O$
Typical Unit Cell Contents:	$Na_{1.2}K_{2.8}[(AlO_2)_4(SiO_2)_{14}] \cdot 14.4\ H_2O$
Variations:	Si/Al = 3.2–3.7; Na/Al = 0.2–0.4; K/Al = 0.5–0.9

Crystallographic Data

Symmetry:	Hexagonal	Density:	2.12 g/cc
Space Group:	P6m2	Unit Cell Volume:	1145 A^3
Unit Cell Constants:	a = 13.21 A	X-Ray Powder Data:	Table 4.34
	c = 7.58 A		

Structural Properties

Framework:	Stereos 2.12, 2.13	Similar to offretite and erionite; may be disordered intergrowths
SBU:	D6R, S6R	Void volume: 0.40 cc/cc
Cage type:	ϵ, 14-hedron (II)	Framework density: 1.50 g/cc

Channel System:	Three-dimensional
Hydrated—	
Free Apertures:	Probably like erionite
Dehydrated—	
Effect of Dehyration:	Stable
Largest Molecule Adsorbed:	n-paraffin hydrocarbons
Kinetic Diameter, σ, A:	4.3

Table 2.60 Zeolite: Thomsonite

Structure Group:	5
Other Designation:	Faroeite
Reference:	183, 194, 198, 212
Chemical Composition	
Typical Oxide Formula:	$(Na_2, Ca)O \cdot Al_2O_3 \cdot 2\, SiO_2 \cdot 2.4\, H_2O$
Typical Unit Cell Contents:	$Na_4Ca_8[(AlO_2)_{20}(SiO_2)_{20}] \cdot 24\, H_2O$
Variations:	Si/Al = 1.0–1.1; Gonnardite, Si/Al = 1.5
	Faroeite is Na/Ca = 1
Crystallographic Data	
Symmetry:	Orthorhombic	Density:	2.3 g/cc
Space Group:	Pnn2	Unit Cell Volume:	2253 A^3
Unit Cell Constants:	a = 13.07 A	X-Ray Powder Data: Table 3.33
	b = 13.08 A	
	c = 13.18 A	
Structural Properties	
Framework: Stereo 2.25	Crosslinked chains of 4-1 units Si/Al = 1 and Al, Si are ordered
SBU:	Unit of 4-1 Void volume: 0.32 cc/cc
	$[Al_{2.5}Si_{2.5}O_{10}]$ Framework density: 1.76 g/cc
Channel System: Fig. 2.27	Two-dimensional and ⊥ to c
Hydrated—	
Free Apertures:	8-ring, 2.6 x 3.9 A
Cation Locations:	8 coordinated with 4 cations and 3 H_2O's
	Remainder in 8-fold coordination
Location of H_2O Molecules:	Zigzag chains in the channels
Largest Molecule Adsorbed:	NH_3
Kinetic Diameter, σ, A:	2.6

Table 2.61 Zeolite: W

Structure Group:	1
Other Designation:	K-M
Reference:	59, 213
Chemical Composition	
Typical Oxide Formula:	$K_2O \cdot Al_2O_3 \cdot 3.6\ SiO_2 \cdot 5\ H_2O$
Typical Unit Cell Contents:	$K_{42}[(AlO_2)_{42}(SiO_2)_{76}] \cdot 107\ H_2O$
Crystallographic Data	
Symmetry: Cubic	Density: 2.18 g/cc
Unit Cell Constants: a = 20.1 A	Unit Cell Volume: 8072 A^3
	X-Ray Powder Data: Table 4.86
Structural Properties	
Framework:	Phillipsite-related
Void volume: 0.22 cc/cc	Framework density: 1.45 g/cc
Largest Molecule Adsorbed:	SO_2 not O_2
Kinetic Diameter, σ, A:	3.6

Table 2.62 Zeolite: P-W

Structure Group:	1
Reference:	37
Chemical Composition	
Typical Oxide Formula:	$K_2O \cdot 1.8\ Al_2O_3 \cdot 3.1\ SiO_2 \cdot 0.63\ P_2O_5 \cdot 6.9\ H_2O$
Typical Unit Cell Contents:	$K_{16}[(AlO_2)_{29}(SiO_2)_{25}(PO_2)_{10}] \cdot 55\ H_2O$
Variations:	up to 20% P_2O_5
Crystallographic Data	
Symmetry: Tetragonal	Density: 2.22 g/cc
Unit Cell Constants: a = 20.17 A	Unit Cell Volume: 4080 A^3
c = 10.03 A	X-Ray Powder Data: Table 4.87
Structural Properties	
Framework:	Phillipsite-related
Void volume: 0.35 cc/cc	Framework density: 1.57 g/cc
Largest Molecule Adsorbed:	N_2
Kinetic Diameter, σ, A:	3.6

Table 2.63 Zeolite: X

Structure Group:	4
Reference:	59, 78, 104, 107, 124, 138, 214, 215

Chemical Composition

Typical Oxide Formula:	$Na_2O \cdot Al_2O_3 \cdot 2.5\ SiO_2 \cdot 6\ H_2O$
Typical Unit Cell Contents:	$Na_{86}\ [(AlO_2)_{86}(SiO_2)_{106}] \cdot 264\ H_2O$
Variations:	Ga substitution for Al; Si/Al = 1 to 1.5
	Na/Al = 0.7 to 1.1

Crystallographic Data

Symmetry:	Cubic	Density:	1.93 g/cc
Space Group:	Fd3m	Unit Cell Volume:	15,362– 15,670 A^3
Unit Cell Constants:	a = 25.02–24.86 A		
		X-Ray Powder Data:	Table 4.88

Structural Properties

Framework:	Stereo 2.17	Truncated octahedra, β-cages, linked tetrahedrally through D6R's in arrangement like carbon atoms in diamond. Contains eight cavities, ~13 A in diameter in each unit cell
SBU:	D6R,	Void volume: 0.50 cc/cc
Cage type:	β, 26-hedron (II)	Framework density: 1.31 g/cc
Channel System:	Fig. 2.30	Three-dimensional, ∥ to [110]

Hydrated–
Free Apertures:	12-ring, 7.4 A, 6-ring, 2.2 A
Cation Locations:	Table 2.12

Dehydrated–
Free Apertures:	7.4 A
Cation Locations:	Table 2.12
Effect of Dehydration:	Stable and reversible
Location of H_2O Molecules:	See Table 2.12
Largest Molecule Adsorbed:	$(C_4H_9)_3N$
Kinetic Diameter, σ, A:	8.1

Table 2.64 Zeolite: Y

Structure Group:	4
Reference:	59, 124, 128, 140, 216
Chemical Composition	
Typical Oxide Formula:	$Na_2O \cdot Al_2O_3 \cdot 4.8\ SiO_2 \cdot 8.9\ H_2O$
Typical Unit Cell Contents:	$Na_{56}\ [(AlO_2)_{56}(SiO_2)_{136}] \cdot 250\ H_2O$
Variations:	Na/Al 0.7 to 1.1; Si/Al => 1.5 to about 3
Crystallographic Data	
Symmetry:	Cubic Density: 1.92 g/cc
Space Group:	Fd3m Unit Cell Volume: 14,901 to 15,347 A^3
Unit Cell Constants:	a = 24.85–24.61 A
	X-Ray Powder Data: Table 4.90
Structural Properties	
Framework: Stereo 2.17	Truncated octahedra, β-cages, linked tetrahedrally through D6R's in arrangement like carbon atoms in diamond. Contains eight cavities ~ 13 A in diameter in each unit cell.
SBU:	D6R Void volume: 0.48 cc/cc
Cage type:	β, 26-hedron (II) Framework density: 1.25–1.29 g/cc
Channel System: Fig. 2.30	Three-dimensional, ‖ to [110]
Hydrated–	
Free Apertures:	12-ring, 7.4 A; 6-ring, 2.2 A
Cation Locations:	Table 2.13
Dehydrated–	
Free Apertures:	~ 7.4
Cation Locations:	Table 2.13
Effect of Dehydration:	Stable and reversible
Location of H_2O Molecules:	Not specifically located
Largest Molecule Adsorbed:	$(C_4H_9)_3N$
Kinetic Diameter, σ, A:	8.1

Table 2.65 Zeolite: Yugawaralite

Structure Group:	1
Reference:	84, 85, 86, 87

Chemical Composition

Typical Oxide Formula:	$CaO \cdot Al_2O_3 \cdot 6\ SiO_2 \cdot 4\ H_2O$
Typical Unit Cell Contents:	$Ca_2\ [(AlO_2)_4(SiO_2)_{12}] \cdot 8\ H_2O$
Variations:	Some Na

Crystallographic Data

Symmetry:	Monoclinic	Density:	2.20 g/cc
Space Group:	P_c	Unit Cell Volume:	882 A^3
Unit Cell Constants:	a = 6.73 A	X-Ray Powder Data:	Table 3.36
	b = 13.95 A		
	c = 10.03 A		
	$\beta = 111°31'$		

Structural Properties

Framework:	Stereo 2.11	4-rings linked to form 5- and 8-rings
	SBU:	4-ring Void volume: 0.27 cc/cc
		Framework density: 1.81 g/cc
Channel System: Fig. 2.27		Two-dimensional; ‖ a and ‖ c

Hydrated—

Free Apertures:	8-rings, 3.6 x 2.8 A
Cation Locations:	On cavity wall at channel intersections, coordinated to 40 atoms, 4 H_2O molecules

Dehydrated—

Free Apertures:	Unknown
Cation Locations:	Unknown
Effect of Dehydration:	Forms new structure at 400°C
Location of H_2O Molecules:	In channel intersections
Largest Molecule Adsorbed:	Unknown
Kinetic Diameter, σ, A:	Unknown

Table 2.66 Zeolite: ZK-4

Structure Group:	3
Reference:	118, 217–219
Chemical Composition	
Typical Oxide Formula:	$0.85\,Na_2O \cdot 0.15(TMA)_2O \cdot Al_2O_3 \cdot 3.3\,SiO_2 \cdot 6\,H_2O$
Typical Unit Cell Contents:	$Na_8\,TMA\,[(AlO_2)_9(SiO_2)_{15}] \cdot 28\,H_2O$
Variations:	Si/Al = 1.25 - 2.0; Na/Al = 0.7–1.0; TMA/Al = 0.1–0.3

Crystallographic Data

Symmetry:	Cubic	Density:	
Space Group:	Pm3m	Unit Cell Volume:	1798 A^3
Unit Cell Constants:	a = 12.16 A	X-Ray Powder Data; Table 4.94	

Structural Properties

Framework:	Stereo 2.16	Same as zeolite A	
SBU:	D4R	Void volume:	0.47 cc/cc
Cage type:	α, β	Framework density:	1.32 g/cc
Channel System: Fig. 2.28a	Three-dimensional, ∥ to [100]		

Hydrated—
 Free Apertures: 2.2 into β-cage, 4.2 into α-cage
 Cation Locations: 8 S_I in 6-rings
Dehydrated—
 Free Apertures: 4.2
 Cation Locations: 8 S_I, 1 S_{II}
 Effect of Dehydration: Stable; calcination removes TMA and H_2O
Largest Molecule Adsorbed: n-hexane
Kinetic Diameter, σ, A: 4.3

Table 2.67 Zeolite: ZK-5

Structure Group:	4
Reference:	155, 156, 220, 221

Chemical Composition

Oxide Formula:	$(R, Na_2)O \cdot Al_2O_3 \cdot 4.0\text{-}6.0\ SiO_2 \cdot 6\ H_2O$
Typical Unit Cell Contents:	$(R,Na_2)_{15}[(AlO_2)_{30}(SiO_2)_{66}] \cdot 98\ H_2O$
Variations:	Si/Al = 1.75–2.55; Na/Al = 0.3–0.7; R/Al$_2$ = 0.3–0.7

Crystallographic Data

Symmetry:	Cubic	Density:	2.07 g/cc
Space Group:	Im3m	Unit Cell Volume:	6539 A^3
Unit Cell Constants:	a = 18.7 A	X-Ray Powder Data:	Table 4.43

Structural Properties

Framework:	Stereo 2.63	Truncated cuboctahedral units, α-cages, linked through D6R's in body centered array
SBU:	D6R	Void volume: 0.44 cc/cc
Cage type:	α and γ Fig. 2.21	Framework density: 1.46 g/cc
Channel System:	Fig. 2.30c	Two independent, three-dimensional systems ‖ to [100] restricted by 8-rings

Hydrated—
  Free Apertures: 8-rings, 3.9 A
  Cation Locations: 16 in the α-cages near center of 6-rings

Dehydrated—
  Cation Locations: Unknown
  Effect of Dehydration: At 150°, cation positions change; stable 600°C
  Location of H$_2$O Molecules: Unknown
  Largest Molecule Adsorbed: n-paraffin hydrocarbons
  Kinetic Diameter, σ, A: 4.3

R = [1,4-dimethyl-1,4-diazoniabicyclo (2,2,2) octane]$^{2+}$

REFERENCES

1. L. Bragg and G. F. Claringbull, "The Crystalline State," *Crystal Structure of Minerals*, Vol. IV, Cornell Univ. Press, Ithaca, 1965.
2. L. Pauling, The Nature of the Chemical Bond, 3rd Ed., Cornell Univ. Press, Ithaca, 1960.
3. W. M. Meier, *Z. Kristallogr.*, **114**:478(1960).
4. P. B. Jamieson and L. S. Dent Glasser, *Acta Crystallogr.*, **20**:688(1966).
5. R.W.G. Wyckoff, Crystal Structures, Vol. III, Wiley-Interscience, New York, 1960.
6. G. A. Barclay and E. G. Cox, *Z. Kristallogr.*, **113**:23(1960).
7. A. Preisinger, *Naturwissenschaften*, **15**:345(1962).
8. W. B. Kamb, *Acta Crystallogr.*, **11**:437(1958).
9. W. H. Zachariasen, *Z. Kristallogr.*, **74**:139(1930).
10. G. V. Gibbs, D. W. Breck, and E. P. Meagher, *Lithos*, **1**:275(1968).
11. B. E. Warren and W. L. Bragg, *Z. Kristallogr.*, **69**:168(1928).
12. B. E. Warren, *Z. Kristallogr.*, **72**:42(1929).
13. F. Liebau, *Naturwissenschaften*, **24**:481(1962).
14. E. J. W. Whittaker, *Acta Crystallogr.*, **9**:855(1956); **10**:149(1957).
15. G. Brown, The X-Ray Identification and Crystal Structures of Clay Minerals, Mineralogical Society, London, 1961.
16. R. M. Barrer and D. M. McLeod, *Trans. Faraday Soc.*, **51**:1290(1955).
17. R. B. Sosman, *The Phases of Silica*, Rutgers Univ. Press, New Brunswick, N.J., 1965.
18. J. V. Smith and F. Rinaldi, *Mineral. Mag.*, **33**:202(1962).
19. J. V. Smith, *Mineral. Mag.*, **36**:640(1968).
20. F. M. Jaeger, *Trans. Faraday Soc.*, **25**:320(1929).
21. L. Pauling, *Z. Kristallogr.*, **74**:213(1930).
22. V. I. Ponomarev. D. M. Kheiker, and N. V. Belov, *Kristallografia*, **15**:918(1970).
23. W. Loewenstein, *Amer. Mineral.*, **39**: 92 (1954).
24. C. Baerlocher and W. M. Meier, *Helv. Chim. Acta*, **52**: 1853 (1969).
25. L. Pauling, *Proc. Nat. Acad. Sci.*, **16**: 453 (1930).
26. O. Jarchow, *Z. Kristallogr.*, **122**: 407 (1965).
27. R. M. Barrer and E. A. D. White, *J. Chem. Soc.*: 1561 (1952).
28. T. Hahn and M. J. Buerger, *Z. Kristallogr.*, **106**: 308 (1955).
29. J. J. Papikse and T. Zoltai, *Amer. Mineral.*, **50**: 641 (1965).
30. J. V. Smith, *Mineral. Soc. Amer.*, Spec. Pap. No. 1, 1963.
31. K. F. Fischer and W. M. Meier, *Fortschr. Mineral.*, **42**: 50 (1965).
32. W. M. Meier, Molecular Sieves, Society of Chemical Industry, London, 1968, p. 10.
33. D. W. Breck, Molecular Sieve Zeolites, *Advan. Chem. Ser. 101*, American Chemical Society, Washington, D. C., 1971, p.l.
34. R. M. Barrer, *Chem. Ind.*, 1203 (1968).
35. G. H. Kuehl, *Amer. Mineral.*, **54**: 1607 (1969).
36. R. M. Barrer and E. A. D. White, *J. Chem. Soc.*, 1167 (1951).
37. E. M. Flanigen and R. W. Grose, Molecular Sieve Zeolites, *Advan. Chem. Ser. 101*, American Chemical Society, Washington, D.C., 1971, p. 155.
38. G. T. Kerr, *U. S. Pat. 3,459,676* (1969).
39. R. L. Wadlinger, E. J. Rosinski, and C. J. Plank, *U. S. Pat. 3,375,205* (1968).
40. G. H. Kuehl, *Inorg. Chem.*, **10**: 2488 (1971).
41. R. M. Milton, *U. S. Pat. 2,996,358* (1961).
42. R. M. Barrer and J. W. Baynham, *U. S. Pat. 2,972,516* (1961).
43. R. M. Milton, *U. S. Pat. 3,010,789* (1961).

44. D. W. Breck and N. A. Acara, *U. S. Pat. 3,011,869* (1961).
45. D. W. Breck and N. A. Acara, *U. S. Pat. 2,950,952* (1960).
46. D. W. Breck and N. A. Acara, *U. S. Pat. 2,995,423* (1961).
47. D. W. Breck and N. A. Acara, *U. S. Pat. 2,991,151* (1961).
48. N. A. Acara, *U. S. Pat. 3,414,602* (1968).
49. J. Ciric, *U. S. Pat. 3,411,874* (1968).
50. G. T. Kokotailo and J. Ciric, Molecular Sieve Zeolites, *Advan. Chem. Ser. 101*, American Chemical Society, Washington, D.C., 1971, p. 109.
51. *Brit. Pat. 1,117,568* (1968).
52. R. J. Argauer, and G. R. Landolt, *U. S. Pat. 3,702,886* (1972).
53. J. Ciric, *U. S. Pat. 3,692,470* (1972).
54. R. D. Wadlinger, G. T. Kerr, and E. J. Rosinski, *U. S. Pat. 3,308,069* (1967).
55. H. C. Duecker, A. Weiss, and C. R. Guerra, *Ger. Offen. 1,935,861* (1970).
56. I. S. Kerr, *Nature*, **197**: 1194 (1963).
57. G. T. Kokotailo and S. I. Lawton, *Nature*, **203**: 621 (1964).
58. P. B. Moore and J. V. Smith, *Mineral. Mag.*, **33**: 1008 (1964).
59. D. W. Breck and E. M. Flanigen, Molecular Sieves, Society of Chemical Industry, London, 1968, p.47.
60. S. I. Louisnathan and G. V. Gibbs, *Amer. Mineral.*, **57**: 1089 (1972).
61. R. M. Barrer and H. Villiger, *Z. Kristallogr.*, **128**: 352 (1969).
62. W. H. Taylor, *Z. Kristallogr.*, **74**: 1 (1930); **99**: 283 (1938).
63. W. D. Balgord and R. Roy, Molecular Sieve Zeolites, *Advan. Chem. Ser. 101*, American Chemical Society, Washington, D.C. 1971, p. 140
64. R. M. Barrer and J. W. Baynham, *J. Chem. Soc.*, 2882 (1956).
65. R. M. Barrer and D. A. Ibbitson, *Trans. Faraday Soc.*, **40**: 195 (1944).
66. G. Ferraris, D. W. Jones, and J. Yerkess, *Z. Kristallogr.*, **135**: 240 (1972).
67. D. S. Coombs, *Mineral. Mag.*, **30**: 699 (1955).
68. A. Steiner, *Mineral. Mag.*, **30**: 691 (1955).
69. R. M. Barrer and P. J. Denny, *J. Chem. Soc.*, 983 (1961).
70. M. Koizumi and R. Roy, *J. Geol.*, **68**: 41 (1960).
71. D. McConnell, *Amer. Mineral.*, **37**: 609 (1952).
72. D. McConnell, *Mineral. Mag.*, **33**: 799 (1964).
73. R. Sadanaga, F. Marumo and Y. Takeuchi, *Acta Crystallogr.*, **14**: 1153 (1961).
74. M. Steinfink, *Acta Crystallogr.*, **15**: 644 (1962).
75. K. Fischer and H. Kuzel, *Naturwissenschaften*, **45**: 488 (1958).
76. K. Fischer, *Amer. Mineral.*, **48**: 664 (1963).
77. Ch. Baerlocher and W. M. Meier, *Helv. Chim. Acta*, **53**: 1285 (1970).
78. R. M. Barrer, F. W. Bultitude, and I. S. Kerr, *J. Chem. Soc.*, 1521 (1959).
79. R. M. Barrer, J. W. Baynham, F. W. Bultitude and W. M. Meier, *J. Chem. Soc.*, 195 (1959).
80. R. M. Milton, *U. S. Pat. 3,008,803* (1961).
81. E. K. Gordon, S. Samson, and W. B. Kamb, *Science*, **154**: 1004 (1966).
82. H. Bartl and K. F. Fischer, *Neues Jahrb. Mineral Monat.*, 33, (1967).
83. V. Schramm, and K. F. Fischer, Molecular Sieve Zeolites, *Advan. Chem. Series 101*, American Chemical Society Washington, D.C., 1971, p. 259.
84. I. S. Kerr and D. J. Williams, *Z. Kristallogr.*, **125**: 220 (1967).
85. R. M. Barrer and D. J. Marshall, *Amer. Mineral.*, **50**: 484 (1965).
86. I. S. Kerr and D. J. Williams, *Acta Crystallogr. B*, **25**: 1183 (1969).
87. H. W. Lerner and M. Slaughter, *Z. Kristallogr.*, **130**: 88 (1969).
88. L. W. Staples and J. A. Gard, *Mineral. Mag.*, **32**: 261 (1959).
89. J. M. Bennett, Ph.D. Thesis, Univ. of Aberdeen, 1966.

90. J. M. Bennett and J. A. Gard, *Nature*, **214**: 1005 (1967).
91. J. A. Gard, and J. M. Tait, Molecular Sieve Zeolites, *Advan. Chem. Series. 101*, American Chemical Society Washington, D.C., 1971, p. 230.
92. A. Kawahara and H. Curien, *Bull. Soc. Fr. Mineral. Cristallogr.*, **92**: 250 (1969).
93. J. A. Gard and J. M. Tait, *Acta Crystallogr., B*, **28**: 825 (1972).
94. D. W. Breck and N. A. Acara, *U. S. Pat. 2,950,952*, (1960).
95. O. A. Glonti and N. A. Shishakov, *Izv. Akad. Nauk SSR, Ser. Khim.*, **7**: 1275 (1965).
96. M. M. Dubinin, V. Ya. Nikolina, L. I. Piguzova, and T. N. Shishakova, *Izv. Akad. Nauk SSR, Ser. Khim.*, **6**: 1116 (1965).
97. R. M. Barrer and I. S. Kerr, *Trans. Faraday Soc.*, **55**: 1915 (1959).
98. R. M. Barrer and H. Villiger, *J. Chem. Soc.*, **D**: 659 (1969).
99. S. P. Zhdanov and N. N. Buntar, *Dokl. Akad. Nauk SSR*, **147**: 1118 (1962).
100. S. P. Zhdanov. N. N. Buntar, and E. N. Egarova, *Dokl. Adad. Nauk SSR*, **154**: 419 (1964).
101. W. Sieber, Ph.D. Thesis, Eidgenossichem Technischen Hochschule, Zurich, 1972.
102. D. W. Breck, W. G. Eversole, R. M. Milton, T. B. Reed, and T. L. Thomas, *J. Amer. Chem. Soc.*, **78**: 5963 (1956).
103. T. B. Reed and D. W. Breck, *J. Amer. Chem. Soc.*, **78**: 5972 (1956).
104. D. W. Breck, *J. Chem. Educ.*, **41**: 678 (1964).
105. R. M. Barrer and W. M. Meier, *Trans. Faraday Soc.*, **54**: 1074 (1958).
106. V. Gramlich and W. M. Meier, *Z. Kristallogr.*, **133**: 134 (1971).
107. L. Broussard and D. P. Shoemaker, *J. Amer. Chem. Soc.*, **82**: 1041 (1960).
108. K. Seff, Ph.D. Thesis, Massachussetts Institute of Technology, 1964.
109. K. Seff and D. P. Shoemaker, *Acta Crystallogr.*, **22**:162 (1967).
110. J. V. Smith and L. G. Dowell, *Z. Kristallogr.*, **126**: 135 (1968).
111. R. Y. Yanagida, A. A. Amaro, and K. Seff., *J. Phys. Chem.*, **77**: 805 (1973).
112. P. E. Riley, K. Seff, and D. P. Shoemaker, *J. Phys. Chem.*, **76**: 2593 (1972).
113. P. E. Riley and K. Seff, *J. Chem. Soc. Chem. Comm.*, 1287 (1972).
114. V. A. Bakaev and M. M. Dubinin, *Izv. Akad. Nauk SSR, Ser. Khim.*, 2156 (1967).
115. P. A. Howell, *Acta Crystallogr.*, **13**: 737 (1960).
116. R. M. Barrer, P. J. Denny, and E. M. Flanigen, *U. S. Pat. 3,306,922* (1967).
117. J. V. Smith and S. W. Bailey, *Acta Crystallogr.*, **16**: 801 (1963).
118. G. T. Kerr, *Inorg. Chem.*, **5**: 1537 (1966).
119. E. Dempsey, G. H. Kuehl, and D. H. Olson, *J. Phys. Chem.*, **73**: 387 (1969).
120. G. H. Kuhl, *J. Inorg. Nucl. Chem.*, **33**: 3261 (1971).
121. J. V. Smith, Molecular Sieve Zeolites, *Advan. Chem. Ser. 101*, American Chemical Society Washington, D. C., 1971, p. 171.
122. D. H. Olson, *J. Phys. Chem.*, **74**: 2758 (1970).
123. W. Baur, *Amer. Mineral.*, **49**: 697 (1964).
124. J. A. Rabo, C. L. Angell, P. H. Kasai, and V. Schomaker, *Discuss. Faraday Soc.*, **41**: 328 (1966).
125. J. M. Bennett and J. V. Smith, *Mater. Res Bull.*, **3**: 633 (1968); **3**: 933 (1968).
126. J. V. Smith, J. M. Bennett, and E. M. Flanigen, *Nature*, **215**: 241 (1967).
127. E. Dempsey and D. H. Olson, *J. Phys. Chem.*, **74**: 305 (1970).
128. P. Gallezot, Y. Ben Taarit, and B. Imelik, *J. Catal.*, **26**: 481 (1972).
129. R. P. Dodge, *Unpublished Results*.
130. J. Pluth and V. Schomaker, *Private Communication*.
131. D. H. Olson, *J. Phys. Chem.*, **72**: 1400 (1968); *ibid*, 4366.
132. D. H. Olson, G. T. Kokotailo, and J. F. Charnell, *Nature*, **215**: 270 (1967); *J. Colloid Interface Sci.*, **28**: 305 (1968).

133. J. M. Bennett and J. V. Smith, *Mater. Res. Bull.*, **3**: 865 (1968); **4**: 7 (1969); **4**: 343 (1969).
134. J. J. Pluth and J. V. Smith, *Mater. Res. Bull.*, **7**: 1311 (1972).
135. J. J. Pluth and J. V. Smith, *Mater. Res. Bull.*, **8**: 459 (1973).
136. W. J. Mortier and H. J. Bosmans, *J. Phys. Chem.*, **75**: 3327 (1971).
137. D. H. Olson and H. S. Sherry, *J. Phys. Chem.* **72**: 4095 (1968).
138. J. M. Bennett, J. V. Smith and C. L. Angell, *Mater. Res. Bull.*, **4**: 77 (1969).
139. F. D. Hunter and J. Scherzer, *J. Catal.*, **20**: 246 (1971).
140. G. R. Eulenberger, D. P. Shoemaker, and J. G. Keil, *J. Phys. Chem.*, **71**: 1812 (1967).
141. W. J. Mortier, H. J. Bosmans, and J. B. Uytterhoeven, *J. Phys. Chem.*, **76**: 650 (1972): *J. Catalysis*, **26**: 295 (1972).
142. P. Gallezot, Y. Ben Taarit, and B. Imelik, *C. R. Acad. Sci.*, Ser. C, **272**: 261 (1971).
143. J. V. Smith, *Acta Crystallogr.*, **15**: 835 (1962).
144. T. I. Barry and L. A. Lay, *J. Phys. Chem. Solids*, **27**: 1821 (1966).
145. T. I. Barry and L. A. Lay, *J. Phys. Chem. Solids*, **29**: 1395 (1968).
146. I. R. Leith and H. F. Leach, *Proc. Roy. Soc. Lond. A*, **330**: 247 (1972).
147. P. E. Pickert, J. A. Rabo, E. Dempsey, and V. Schomaker, *Proc. Third Int. Congr. Catal.*, 714 (1965).
148. E. Dempsey, Molecular Sieves, Society of Chemical Industry, London, 1968, p. 293.
149. J. Wyart, *Bull. Soc. Fr. Mineral. Cristallogr.*, **56**: 81 (1933).
150. J. V. Smith, F. Rinaldi, and L. S. Dent-Glasser, *Acta Crystallogr.*, **16**: 45 (1963).
151. P. Ducros, *Bull. Soc. Fr. Mineral. Cristallogr.*, **83**: 85 (1960).
152. J. V. Smith, *J. Chem. Soc.*, 3759 (1964).
153. K. Fischer, *Fortschr. Mineral.*, **38**: 201 (1960); *Neues Jahrb. Mineral Monat.*, 1 (1966).
154. R. M. Barrer, *Trans. Faraday Soc.*, **40**: 555 (1944).
155. W. M. Meier and G. T. Kokotailo, *Z. Kristallogr.*, **121**: 211 (1965).
156. R. M. Barrer and D. J. Robinson, *Z. Kristallogr.*, **155**: 374 (1972).
157. Ch. Baerlocher and R. M. Barrer, *Z. Kristallogr.*, **136**: 245 (1972).
158. M. D. Foster, *Geol. Surv. Prof. Paper*, 504-D (1965).
159. W. M. Meier, *Z. Kristallogr.*, **115**: 439 (1961).
160. L. B. Sand, Molecular Sieves, Society of Chemical Industry, London, 1968, p. 71.
161. L. V. C. Rees and A. Rao, *Trans. Faraday Soc.*, **62**: 2103 (1966).
162. G. Gottardi and W. M. Meier, *Z. Kristallogr.*, **119**: 53 (1963).
163. P. A. Vaughan, *Acta Crystallogr.*, **21**: 983 (1966).
164. I. S. Kerr, *Nature*, **210**: 294 (1966).
165. I. S. Kerr, *Nature*, **202**: 589 (1964).
166. A. J. Perotta, *Mineral. Mag.*, **36**: 480 (1967).
167. D. E. Appleman, *Acta Crystallogr.*, **13**: 1002 (1960).
168. A. B. Merkle and M. Slaughter, *Amer. Mineral.*, **52**: 273 (1967); **53**: 1120 (1968).
169. B. Galli, *Acta Crystallogr. B*, **27**: 833 (1971).
170. M. Slaughter, *Amer. Mineral.*, **55**: 387 (1970).
171. A. J. Perotta and J. V. Smith, *Acta Crystallogr.*, **17**: 857 (1964).
172. R. M. Milton, *U. S. Pat.* 2,882,243 (1959).
173. I. R. Beattie, *Acta Crystallogr.*, **7**: 357 (1954).
174. C. R. Knowles, F. F. Rinaldi, and J. V. Smith, *Indian Mineral.*, **6**: 127 (1965).
175. P. Saha, *Amer. Mineral.*, **44**: 300 (1959).
176. P. Saha, *Amer. Mineral.*, **46**: 859 (1961).
177. W. C. Phinney and D. B. Stewart, *U. S. Geol. Surv. Prof. Paper*, **424-D**, **353-7** (1961).
178. J. V. Smith, C. R. Knowles, and F. Rinaldi, *Acta Crystallogr.*, **17**: 374 (1964).

179. F. A. Mumpton, *Amer. Mineral*, **45**: 351 (1960).
180. A. O. Shepard and H. C. Starkey, *U. S. Geol. Surv. Prof. Paper*, **475**-D: 89-92 (1963).
181. W. S. Wise, W. J. Nokleberg, and M. Kokinos, *Amer. Mineral*, **54**: 887 (1969).
182. S. Bonatti and G. Gottardi, *Period. Mineral.* (Rome), **29**: 103 (1960).
183. M. D. Foster, *Geol. Surv. Prof. Paper*, 504-E (1965).
184. M. H. Hey, *Mineral. Mag.*, **23**: 483 (1934).
185. W. H. Taylor and R. Jackson, *Z. Kristallogr.*, **86**: 53 (1933).
186. W. H. Taylor, *Mineral. Mag.*, **24**: 208 (1935).
187. H. Strunz, *Naturwissenschaften*, **42**: 485 (1955).
188. K. Fischer and H. O'Daniel, *Naturwissenschaften*, **43**: 348 (1956).
189. G. Bergerhoff, H. Koyama, and W. Nowacki, *Experentia*, **12**: 418 (1956).
190. G. Bergerhoff, W. H. Baur, and W. Nowacki, *Neues Jahrb. Mineral Monat.*, **9**: 193 (1958).
191. G. P. L. Walker, *Mineral. Mag.*, **33**: 173 (1962).
192. H. Strunz, *Neues Jahrb. Mineral. Monat.*, **11**: 250 (1956).
193. L. S. Dent and J. V. Smith, *Nature*, **181**: 1794 (1958).
194. M. H. Hey and F. A. Bannister, *Mineral. Mag.*, **23**: 51 (1932).
195. Th. G. Sahama and M. Lehtinen, *Mineral. Mag.*, **36**: 444 (1967).
196. T. N. Shishakova and M. M. Dubinin, *Izv. Ada. Nauk, Ser. Khim.*, **7**: 1303 (1965).
197. D. W. Breck and N. A. Acara, *U. S. Pat. 3,216,789*, (1965).
198. W. H. Taylor, C. A. Meek and W. W. Jackson, *Z. Kristallogr.*, **84**: 373 (1933).
199. M. H. Hey, *Mineral. Mag.*, **24**: 227 (1936).
200. W. Meier, *Z. Kristallogr.*, **113**: 430 (1960).
201. M. H. Hey, *Mineral. Mag.*, **23**: 421 (1933).
202. A. Aiello and R. M. Barrer, *J. Chem. Soc.* **A**: 1029 (1970).
203. T. E. Whyte, E. L. Wu, G. T. Kerr, and P. B. Venuto, *J. Catalysis*, **20**: 88 (1971).
204. R. A. Sheppard and A. J. Gude, *Amer. Mineral.*, **54**: 875 (1969).
205. E. M. Flanigen and E. R. Kellberg, *Neth. Pat. 6,710,729* (1968), U. S. Pat. 4,241,036 (1980).
206. Ch. Baerlocher and W. M. Meier, *Z. Kristallogr.*, **135**: 339 (1972).
207. W. B. Kamb and W. C. Oke, *Amer. Mineral.*, **45**: 79 (1960).
208. E. Galli and G. L. Ghittoni, *Amer. Mineral.*, **57**: 1125 (1972).
209. G. W. Smith and R. Walls, *Mineral. Mag.*, **38**: 72 (1971).
210. E. Galli and G. Gottardi, *Mineral. Petrogr. Acta*, **12**: 1-10 (1966).
211. K. Harada and K. Tomita, *Amer. Mineral.*, **52**: 1438 (1967).
212. W. O. Milligan and H. B. Weiser, *J. Phys. Chem.*, **41**: 1029(1937).
213. R. M. Milton, *U. S. Pat. 3,012,853* (1961).
214. R. M. Milton, *U. S. Pat. 2,882,244* (1959).
215. D. W. Breck, W. G. Eversole, and R. M. Milton, *J. Amer. Chem. Soc.*, **78**: 2338 (1956).
216. D. W. Breck, *U. S. Pat. 3,130, 007* (1964).
217. G. T. Kerr, *J. Phys. Chem.*, **66**: 2271 (1962).
218. G. T. Kerr and G. T. Kokotailo, *J. Amer. Chem. Soc.*, **83**: 4675 (1961).
219. G. T. Kerr, *U. S. Pat. 3,314, 752* (1967).
220. G. T. Kerr, *Science*, **140**: 1412 (1963).
221. G. T. Kerr, *Inorg. Chem.*, **5**: 1539 (1966).
222. W. M. Meier and D. H. Olson, Atlas of Zeolite Structure Types, Intl. Zeolite Association, 1978, Distributed by Polycrystal Book Service, P.O. Box 11567, Pittsburgh, PA 15238.
223. E. Galli, E. Passaglia, and D. Pongiluppi, *Contr. Mineral and Petrol.*, **45**: 99 (1974).
224. D. H. Olson, G. T. Kokotailo, S. L. Lawton, and W. M. Meier, *J. Phys. Chem.*, **85**: 2238 (1981).

Chapter Three

MINERAL ZEOLITES

Zeolite minerals have been used in many investigations of zeolite structure and properties (adsorption, ion exchange, etc.). The first experiments involving ion-exchange properties of zeolites and selective adsorption of gases were conducted on zeolite minerals (Chapter 1). Such studies have contributed considerably to the present state of knowledge about zeolites. A discussion of the zeolite minerals, their classification, occurrence, formation, and properties is important to our understanding of the synthesis and properties of zeolites. The zeolite minerals have not as yet found extensive application as adsorbents or catalysts even though some types occur in large quantities; some synthetic analogs, however, are used commercially.

The occurrence of zeolites is somewhat paradoxical. Some minerals which occur extensively are not suitable for most known commercial applications (for example, analcime and phillipsite). Some synthetic zeolites which have extensive commercial use do not have mineral relatives. Other zeolite minerals occur in miniscule quantities; for research purposes, investigators have sometimes been required to work with milligram quantities.

The naming of minerals is thoroughly discussed by Fleischer (1) and McConnell (2). In general, the rules used in Dana's System of Mineralogy have been followed in this book (3). Old names and synonyms have been eliminated. There may be some question concerning a few minerals such as herschelite (as distinguished from chabazite) and stellerite (as distinguished from stilbite). Because of property differences (chemical composition, dehydration behavior, gas adsorption) we have chosen to designate herschelite and stellerite as separate minerals.

Until recently, the nomenclature for zeolite minerals has been confused. Early identifications were based on chemical composition, optical properties, morphology, etc. With the advent of x-ray crystallography, some zeolite names have been discredited. The list of 34 zeolite minerals in Table 3.1 is based on defining properties such as structure and composition, which were discussed in Chapter 2. The minerals are listed in alphabetical order for convenience; also included are the structural group, extent of occurrence, and year of discovery. Tables 3.4 through 3.36 at the end of this chapter provide detailed information on each mineral.

It is not within the scope of this book to discuss in detail the genesis and mineralogy of zeolites. There are extensive publications on this subject (3, 4, 5, 6). Some relevant aspects of the occurrence and formation of zeolite minerals which may be related to zeolite synthesis (Chap. 4) are reviewed.

A. TYPES OF OCCURRENCES

The mineral combinations in which zeolites are found are widespread, and recognition of a special mineral facies, the zeolite facies, has been proposed (7). Until about twenty years ago, zeolite minerals were considered as typically occurring in cavities of basaltic and volcanic rocks. Dating from the discovery of stilbite by Cronstedt in 1756, numerous publications describe the occurrence and mineralogy of zeolite mineral assemblages in basaltic igneous rocks. During the last 50 years, a number of brief descriptions of zeolite minerals found in sediments and as alteration products of volcanic ash and other pyroclastic materials have appeared. Ross (8) speculated in 1928 that analcime in the Wickieup, Arizona lake sediments was formed by the action of saline lake water on volcanic ash. More recently, the use of x-ray diffraction for the examination of very fine-grained submicroscopic particles that occur in sediments has resulted in the identification of several zeolite minerals which formed during diagenesis* of volcanic ash particles in alkaline environments (5). The mineral phillipsite was found in the walls of Roman baths by Daubree at Plombieres, France, in 1879 (3). Drill cores from wells in Wairakei, New Zealand revealed an extensive zone of zeolite formation (9).

In 1891, the famous Challenger expedition found phillipsite in red

---

* A brief glossary has been included in Section F.

Table 3.1 A List of Zeolite Minerals

Name	Structure Group	Year Discovered	Typical Occurrence in Igneous Rocks	Examples of Occurrence in Sedimentary Rocks
Analcime	1	1784	Ireland, New Jersey	Extensive; Wyoming, etc. Deep sea floor
Bikitaite	6	1957	Rhodesia	
Brewsterite	7	1822	Scotland	
Chabazite	4	1772	Nova Scotia, Ireland	Arizona, Nevada, Italy
Clinoptilolite	7	1890	Wyoming	Extensive; Western U.S., Deep sea floor
Dachiardite	6	1905	Elba, Italy	
Edingtonite	5	1825	Scotland	
Epistilbite	6	1823	Iceland	
Erionite	2	1890	Rare, Oregon	Nevada, Oregon, U.S.S.R.
Faujasite	4	1842	Rare, Germany	
Ferrierite	6	1918	Rare, British Columbia, Italy	Utah, Nevada
Garronite	1	1962	Ireland, Iceland	
Gismondine	1	1816	Rare, Italy	
Gmelinite	4	1807	Nova Scotia	
Gonnardite	5	1896	France, Italy	
Harmotome	1	1775	Scotland	
Herschelite	4	1825	Sicily	Arizona
Heulandite	7	1801	Iceland	New Zealand
Kehoeite	1	1893	Rare, South Dakota	
Laumontite	1	1785	Nova Scotia, Faroe Islands	Extensive; New Zealand, U.S.S.R.
Levynite	2	1825	Iceland	
Mesolite	5	1813	Nova Scotia	
Mordenite	6	1864	Nova Scotia	U.S.S.R., Japan, Western U.S.
Natrolite	5	1758	Ireland, New Jersey	
Offretite	2	1890	Rare, France	
Paulingite	1	1960	Rare, Washington	
Phillipsite	1	1824	Ireland, Sicily	Extensive, Western U.S., Africa, Deep sea floor
Scolecite	5	1801	Iceland, Colorado	
Stellerite	7	1909		
Stilbite	7	1756	Iceland, Ireland, Scotland	
Thomsonite	5	1801	Scotland, Colorado	
Viseite	1	1942	Rare, Belgium	
Wairakite	1	1955	New Zealand	
Yugawaralite	1	1952	Japan	

clay on the bottom of the central Pacific Ocean south of the Sandwich Islands (10). Today it is believed that phillipsite is one of the more abundant mineral species on the earth. Over large areas of the Pacific Ocean, sediments have been found that contain over 50% phillipsite (11).

The metastable formation of zeolites in sedimentary environments probably takes place by a natural process which is similar to the process used in laboratory synthesis (see Chap. 4). Studies such as those of Coombs, Ellis, Fyfe, and Taylor (12) and of Hay (5) of zeolite mineral assemblages in New Zealand and other locations, support the concept of a separate zeolite facies. Interpretation (13) has shown that the zeolite facies is not a necessary step between diagenesis and metamorphism. The formation of zeolites, in this case the calcium zeolites laumontite and heulandite, was considered to be dependent on the chemical potential of water relative to CO_2, in addition to temperature and total pressure. The formation of zeolites during early diagenesis is favored by devitrification of volcanic glass, unusually saline water, and a slow sedimentation rate (14).

Although all of the zeolite minerals known have been found in igneous rocks, only a few occur in sedimentary environments. These two major types of occurrence are discussed separately. Table 3.1 list some typical occurrences of zeolite minerals. Detailed summaries are given by Hay (5) and Sheppard (6).

B. IGNEOUS ZEOLITES

The igneous zeolites usually occur as well-developed crystals, sometimes as large as 1 inch in diameter (Fig. 3.1). They are assumed to have crystallized in cavities and along fractures in basalt from the aqueous solutions corresponding to the last stages of magmatic activity. In most cavities, successive growth of several different zeolite minerals is observed and is usually accompanied by other minerals such as calcite, calcium carbonate, quartz, and other hydrous minerals. In Nova Scotia basalt, mordenite is reported to comprise 20% of the rock in some localities (15). However, it is rare that the fraction of zeolite mineral in the parent rock exceeds about 5% by volume. Coombs pointed out that the appearance of zeolites in igneous rocks is difficult to interpret on a physicochemical basis. Although some authors have published the paragenetic sequence, the significance of such

190 MINERAL ZEOLITES

Figure 3.1 Crystals of Stilbite. This specimen from Oregon is 10 cm in length (16).

conclusions is doubtful. The observed sequence of formation of minerals in a cavity may be a function either of the changing composition of the solution from which the mineral is formed, or of other conditions such as the temperature and the pressure (12).

The distribution of zeolite minerals in terms of the relationship between their silica content and the availability of free silica has been used as a basis for correlating occurrence and environment (12). Typical zeolites in veins of quartz-bearing rocks are stilbite, heulandite, laumontite, and analcime. The sequence of silica-rich zeolites, mordenite, heulandite, chabazite and stilbite, is reported to occur with quartz or opal in southern Brazil. The occurrence of mordenite is typical in thermal areas where the silica activity is high. In contrast, the main calcium zeolites of the basalts of northern Ireland, such as chabazite, levynite, thomsonite, natrolite, mesolite, and phillipsite all tend to be low in silica and occur in rocks or environments where the silica activity is low (12). Figure 3.2 illustrates a proposed grouping of zeolites by environment.

An extensive study of a typical zeolite occurrence in the Antrim basalts of Ireland has been reported (17). The Antrim zeolite minerals occur in flat-lying zones that are nearly parallel to the lava strata. The evidence indicates that the formation of zeolites occurred long after the eruption of the individual lava flows in which the zeolites are found and probably after at least 1000 feet of lava had been erupted. Zeolite formation occurred after a critical thickness of lavas had accumulated, permitting the buildup of heat in the lower flows. The formation of zeolites from the lava took place as a result of the reaction of meteoric water heated from below and partly from exothermic hydration reactions. A bore hole in the Antrim basalt to a depth of 2590 feet showed that chabazite is distributed throughout and is probably the most abundant zeolite present. Other minerals are distributed uniformly and include thomsonite. Phillipsite occurs to a depth of 1000 feet and levynite to a depth of 1300 feet. Stilbite was encountered first at 600 and is abundant at below 1000 feet. It is significant that the crystal habit of chabazite was observed to be influenced by the environment under which it crystallized.

The igneous zeolite localities in Nova Scotia are typical. These zeolite occurrences in the basalt along the Bay of Fundy are probably the most important of this type in North America. Along the north shore, the Triassic basalt is exposed for a distance of about 40 miles. Along the southern shore, it is continuously exposed for approximately 125 miles. Zeolites are abundant along these shores. The zeolites fill

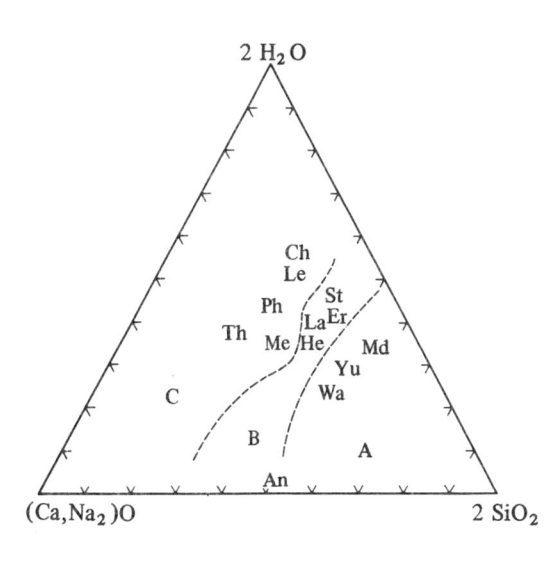

Figure 3.2 Composition in molar proportions of calcium-rich zeolites and of certain other Ca-Al silicates. For the zeolite and the feldspar anorthite, $(Ca,Na_2)O$ is numerically equal to Al_2O_3. A. Field of phases favored by supersaturation in silica. B. Field of phases which can commonly coexist with quartz (erionite coexists with opal). C. Field of phases favored by a silica-deficient environment. From Coombs, et. al. (12).
Abbreviations:
Me = mesolite; Le = levynite
Ch = chabazite; Ph = phillipsite;
La = laumontite; Th = thomsonite; St = stilbite; Md = mordenite;
Wa = wairakite; Yu = yugawaralite; He = heulandite; Er = erionite.

amygdales with their richest concentration not exceeding 20% volume of the mass of the basalt. Mordenite appears to be most common; it is names after the locality of Morden, Nova Scotia. An unusual occurrence of mordenite is in the tubular amygdaloid (18). These are cylindrical columns with diameters that range from a few inches to a foot. In some cases, they are several feet long. They may be filled throughout with zeolite amygdales, or they may have a concentric ring of amygdales surrounding a barren zone. Due to weathering of the basalt, the zeolite minerals are also found as beach pebbles along the shore. Mordenite pebbles up to 3 inches in diameter have been observed.

C. SEDIMENTARY ZEOLITES

Zeolite minerals known to occur in sedimentary rocks are analcime, clinoptilolite, mordenite, phillipsite, erionite, laumontite, chabazite, wairakite and ferrierite (5, 6). A complete summary of zeolites in sedimentary rocks is given in the excellent review by Hay (5). Another recent detailed review by Sheppard deals with zeolites in sedimentary deposits in the United States (6).

Sedimentary deposits of zeolites are subdivided by Hay (5) into two main groups on the basis of depositional environment.

1. Saline, alkaline nonmarine deposits — zeolites are widespread in these deposits and may form relatively pure beds.
2. Fresh-water and marine deposits — the zeolites occur as alteration products of volcanic glass in thick accumulations of tuffaceous sediments.

Some deposits are difficult to classify into these categories. Although the zeolitic reactions in each case have common features, the main difference is due to the high pH, about 9.4, which is common in saline lakes. Selected examples are tabulated in Table 3.2.

Over the last 40 years, a number of occurrences of zeolite minerals, especially analcime and clinoptilolite, have been reported as alteration products of volcanic ash and other pyroclastic materials and as ground mass material in bentonite clay (19). The alteration of a volcanic glass to mordenite was reported (20). The zeolite, comprising about 1% of the parent rock, was identified as a mordenite by x-ray powder data. Zeolites may occur commonly as intermediate products in a breakdown

Table 3.2 Sedimentary Zeolite Deposits (5)

Age	Location	Zeolites	Depositional Environment
1 Some typical zeolites in deposits of saline, alkaline nonmarine environments			
Recent sediments	Lake Natron, Tanzania	Analcime	Sodium carbonate lake of high salinity
Recent-late Pleistocene	Teels Marsh, Nevada	Phillipsite, clinoptilolite, analcime	Sodium carbonate lake
Late Pleistocene	Owens Lake, California	Phillipsite, clinoptilolite, erionite, analcime	Saline, alkaline lake
Late Pleistocene	Olduvai Gorge, Tanzania	Analcime, chabazite, phillipsite	Playa Lake
Middle Pliocene to Middle Pleistocene	North Central Nevada	Erionite, phillipsite	Saline, alkaline lake
Pliocene	Central Nevada	Erionite, clinoptilolite, phillipsite	Saline, alkaline lake
Early and Middle Eocene	Wyoming	Analcime	Saline, sodium carbonate lake
Triassic	New Jersey	Analcime	Saline lake, soda-rich
Early Carboniferous	Tuva, Siberia	Analcime, laumontite	Saline and alkaline lake or lagoon
2 Zeolites deposited in marine and fresh-water environments			
Recent sediments	Gulf of Naples	Analcime	Shallow marine
Recent and Pleistocene	Pacific and Indian Oceans	Phillipsite, harmotome clinoptilolite, natrolite	Deep-sea floor
Recent and late Pleistocene	Atlantic Ocean	Clinoptilolite	Deep-sea floor
Late Miocene and Pliocene	Central Nevada	Clinoptilolite	Lacustrine
Oligocene to early Pliocene	North Central Nevada	Clinoptilolite	Lacustrine
Miocene	Northern Honshu	Clinoptilolite, mordenite	Marine and lacustrine
Oligocene	South Dakota and Wyoming	Clinoptilolite, erionite	Fluvial
Cretaceous	Ural Mountains	Mordenite	Marine
Cretaceous	New Guinea	Laumontite	Marine
Early to late Triassic	New Zealand	Analcime, heulandite, clinoptilolite, laumontite	Marine
Carboniferous	England	Analcime	Land surface
Early Paleozoic or Precambrian	Georgia, USA	Laumontite	Subaerial and subaqueous

of volcanic material. The chemical composition of mordenite, based upon the unit cell, was $(Mg_{0.2} Ca_{0.8} Na_{3.7} K_{1.8})Al_{7.7} Si_{40.3} O_{96} \cdot 22.9 H_2O$. The chemical composition of the original glass is very nearly the same.

Zeolite alteration of Triassic pyroclastic beds in the vicinity of Wairakei in New Zealand has produced zones up to several hundred feet which have been altered almost entirely to heulandite, analcime, wairakite, and laumontite (8). Clinoptilolite occurs in abundance in altered tuff beds of Tertiary age associated with the hectorite deposit at Hector, California (21). The zeolites are formed by the alteration of devitrification products from volcanic ash, usually of Tertiary or early Pleistocene age. Volcanic ash particles carried by the wind probably fell into the many intermountain lakes (which were common in western United States) to form lacustrine tuff deposits. After burial, the ash was probably altered by the action of alkaline lakes and ground waters and transformed into the siliceous zeolites. Commonly the zeolite beds thus formed are mixed with small to major quantities of quartz, feldspar, montmorillonite, and unaltered ash particles. Numerous clay beds of montmorillonite along with fresh ash beds are commonly associated with the zeolitic beds. The zeolite beds are usually light in color and low in density and reflect the volcanic ash genesis of the bed. (See Fig. 3.5.) Analcime, mordenite, and laumontite have been reported in a wide variety of sediments in areas of Russia and Siberia (22). Mordenite and clinoptilolite have been found in Japan in glassy volcanic tuffs (23). Phillipsite, chabazite, and herschelite have been reported in Italy.

The Great Basin area of the western United States has a history of nearly continuous sedimentation. In the areas of Jersey Valley and Pine Valley, Nevada, evidence from the textures of the relics and from other minerals present in these sedimentary beds indicates that the zeolites are alteration products of the vitric ash (24,25). Further evidence of the lacustrine origin is furnished by the presence of fish fossils in bedding planes and ripple marks characteristic of a lake bottom (Fig. 3.4). Figure 3.5 shows an air view of a sedimentary zeolite bed outcrop; Fig. 3.6 is a close-up view showing the horizontal bed exposed, and Fig. 3.7 is a view of a tilted zeolite bed in Nevada.

A possible mechanism has been proposed. Initially, water extracts sodium and potassium from the glass by hydrolysis, making the water alkaline and increasing the dissolution of silica. The rates

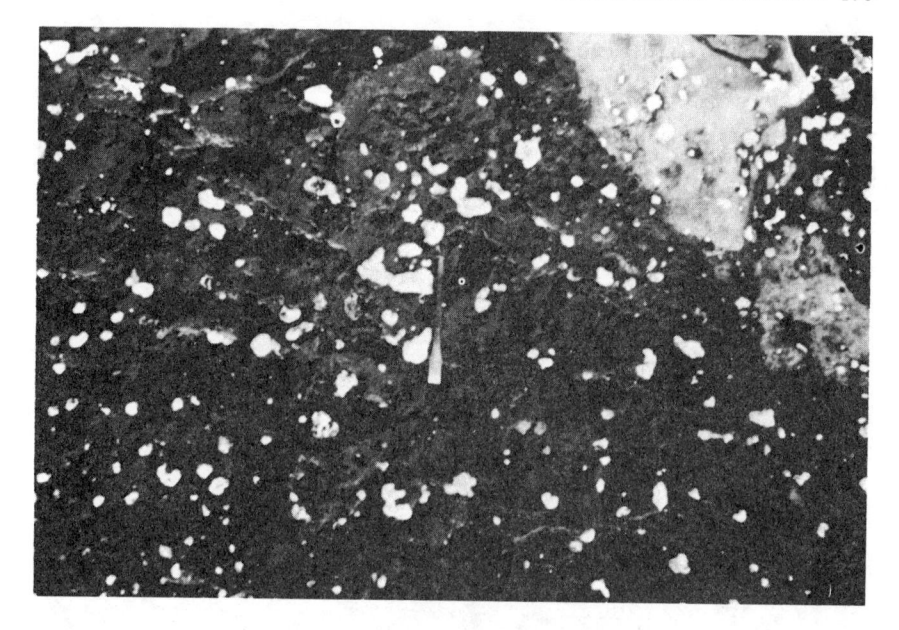

Figure 3.3 Mordenite amygdales in basalt near Canada Creek, Nova Scotia. For scale, note size of rock chisel.

of solution of ordinary commercial glass range from 1 to 100 µg per 100 cm² per day (24). Glass pieces of the size range encountered in most sediments would dissolve entirely in 30-3000 years. Therefore, it is likely that the zeolites are not formed from an internal devitrification process, but from the solution of the material from the glass surface and recrystallization into crystalline zeolites. This is similar to mechanisms that have been suggested for the formation of synthetic zeolites from various amorphous substrates (Chapter 4).

The temperatures of occurrence of some zeolite minerals are shown in Table 3.3. Until recently, the formation of minerals at low temperatures and low pressures was largely neglected in the studies of mineral formation. Mineral formation is probably slowly taking place today under such moderate conditions in the outer portion of the earth's crust.

It is probable that zeolites are being continually formed on the bottom of the oceans. Recent work by the Scripps Institute of Oceanography has indicated that phillipsite may be one of the most abundant mineral species of the earth. It has been found in concentrations

196 MINERAL ZEOLITES

Figure 3.4 This specimen of erionite from Nevada shows the tail of a fish fossil preserved in a bedding plane. The tail is about 14-mm wide (16).

of greater than 50% ranging over huge areas of the floor of the Pacific Ocean (11, 26). Phillipsite appears to have formed by the action of sea water on partially devitrified volcanic glass resulting from submarine volcanic eruptions, occurring in association with nodules of iron and manganese oxides. Rapid chilling of hot lava in the water ap-

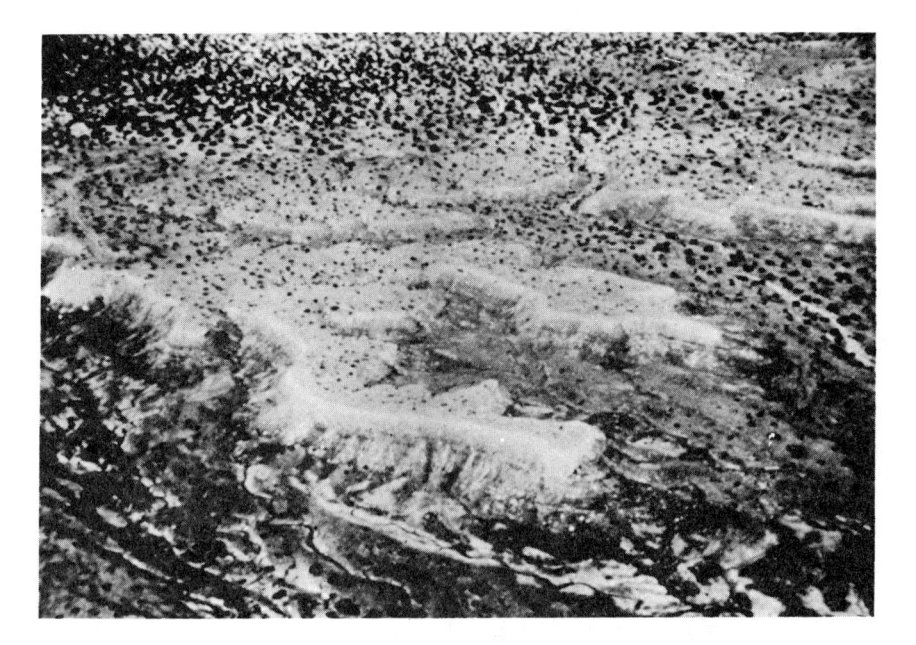

Figure 3.5 Air view of zeolite outcrop in the San Simon Valley, Cochise County, Arizona. The light-colored zeolite bed is evident at the surface where it is exposed by the erosion of an ancient stream. The original lake in which the volcanic ash was altered to zeolite dried up and became eroded by the large streams.

pears to have caused pulverization; the pulverized lava reacted with water to form palagonite. Subsequently, the palagonite gradually altered to phillipsite. Clinoptilolite has also been found on the ocean floor of the equatorial region of the mid-Atlantic ridge (27).

The barium-rich zeolite, harmotome (phillipsite-type structure) is reported to form on the ocean floor by the interaction of sea water and volcanic glass. Subsequent growth of phillipsite engulfs the harmotome crystallite nuclei (28).

The formation of zeolites in active thermal areas has been known for many years. An extensive zone of zeolitization was observed at Wairakei, New Zealand (29). The upper part is largely mordenite; wairakite is prevalent in the lower section, filling veins and cavities as the result of crystallization from solutions. Laumontite was formed in a zone between mordenite and wairakite where temperatures range from 195 to 220°C. As indicated in Table 3.3, clinoptilolite and analcime are observed in drill cores from the Upper Basin, Yellowstone National Park. Wairakite has also been observed in the geysers in California.

198 MINERAL ZEOLITES

Figure 3.6. View of horizontal zeolite bed outcrop in the San Simon Valley, Arizona. The zeolite bed caps the eroded bank at the surface.

Figure 3.7 Erionite Horizon, Jersey Valley Nevada. Note the thick gray ash bed between the two light erionite beds.

Table 3.3 Temperatures of Occurrence of Some Zeolites and Related Minerals (12)

Mineral	Occurrence	Temperature (°C)	Depth (m)
Phillipsite	Deep-sea sediments	0	4000 - 5000
Chabazite, phillipsite, natrolite	Masonry, Roman baths	40 - 70	Surface
Clinoptilolite, analcime	Diagenesis	Low	?
Stilbite	Hunters, Boulder Hot Springs	64, 73	Surface
Clinoptilolite	Yellowstone	125	19 - 26
Analcime	Yellowstone	125 - 155	26 - 60
Mordenite	Wairakei	150 - 230	73 - 300
Heulandite	Wairakei	Within range of mordenite	
Laumontite	Wairakei	195 - 220	150 - 275
Wairakite	Wairakei	200 - 250	180 - 600
Prehnite	Wairakei	200	100
Albite	Wairakei	160 - 240	100 - 600
Adularia	Wairakei	230 - 250	385 - 650
Zeolite and adularia	Steam Boat Springs	170	52

Note: The maximum temperatures recorded at Wairakei boreholes are 250 - 260°C.

The recently discovered mineral, yugawaralite, is found in Japan. Formation by hot spring action on a volcanic tuff was deduced (30). Other associated minerals include laumontite, chabazite and mordenite.

The occurrence of analcime, heulandite, and stilbite in fossils found in Oregon is the result of deposition from hydrothermal solutions which resulted from a basaltic intrusion of the rock strata containing the fossils (31). A volcanic glass crystallized to a mordenite-type zeolite when lowered 1045 feet into a drill hole at Wairakei and exposed for 17 days to the natural hydrothermal solutions at 230°C. (32) The calculated pH of the solution was 5.7, indicating that the zeolite crystallized from acidic solution. This is the only report of this type, however, and it has not been confirmed.

A white-colored vitric tuff (volcanic ash), consisting entirely of glassy fragments, was pulverized to a fine powder and heated with sodium chloride and sodium hydroxide solutions at 100°C (33). A mixture of a sodalite-type phase, and another unidentified material was formed. The published x-ray powder data resembles the synthetic zeolite X and zeolite Y (Chap. 4).

The formation of clinoptilolite in volcanic rock of the John Day formation of Oregon has been interpreted in some detail by Hay (34). From chemical and petrographic studies, Hay has shown that most of the clinoptilolite replaces the glass particles by precipitation within the cavities from which glass had been dissolved. The formation of clinoptilolite from the glass involves principally a gain in water and calcium and a loss of silicon, sodium, potassium, and iron. The water responsible for the formation of the zeolite originated as meteoric water which acquired an increasing pH and increasing concentration of dissolved species while passing downward and laterally through beds of the formation. Most of the zeolitic formations occur between 1250 and 4000 feet deep, and represent temperatures of 27.5°C and 55°C. The hydrostatic pressures at these depths would be 35 to 120 atm. The formation of the siliceous zeolite, clinoptilolite, requires a high Si/Al ratio in the original glass. The coprecipitation of forms of amorphous silica along with the clinoptilolite showed that the zeolite was crystallized from these solutions supersaturated with respect to crystalline forms of silica. The solubility of SiO_2 and Al_2O_3 at room temperature increases above a pH of 9; therefore the solubility of the glass should increase rapidly at higher pH. The water slurry of a finely ground glass was observed to have a pH of 8.5, which indicates hydrolysis of the metallic cations at the surface of the glass particles.

Clinoptilolite probably crystallized near the base of the formation where the pH and the concentration of alkali ions would be the highest. Once the crystallization was initiated, the solution would become increasingly undersaturated with respect to the glass, and additional glass would then dissolve at an accelerated rate. The clinoptilolite would then continue to crystallize until all the glass had dissolved.

D. ORIGIN OF ZEOLITE MINERALS

The mineral makeup of zeolite sedimentary rocks has been correlated by Hay with the chemical composition of the host rock, the water chemistry of the depositional and post-depositional environment, the age, and burial depth (5). The chemical correlations are generally better for zeolites found as the alteration products of volcanic ash in marine and fresh water deposits that have not been deeply buried. High-silica, and alkali-rich zeolites, such as clinoptilolite, predominate in the high-silica rocks; the lower silica-containing zeolites, such as phillipsite and analcime, are characteristic of the more basic or low silica rocks.

The mineral content is related to some extent to the pH, salinity and the dissolved ion composition of the water, which were characteristic of the environment at the time of deposition of the sediments. The pH, for example, seems to be related to the rate of the zeolite reaction and to the types of minerals that formed. In marine environment where the water pH is 7.5 to 8.1, the silicic glass is commonly preserved unaltered for millions of years. In the more basic sodium-carbonate lakes with a pH of 9.1 to 9.9, it appears to have altered within a few tens of thousands of years.

The salinity or salt content of the water also favors the conversion of volcanic glass to form zeolites. The type of metal cation and the relative proportions present also relate to some degree to the type of zeolite formed. (See Chap. 4.) The calcium zeolites do not seem to have formed in deposits in sodium carbonate lakes. On the other hand the ratio of the common alkali ions, sodium and potassium, in a lake water do not seem to correlate with the type of zeolite. An example is the occurrence of phillipsite, primarily a potassium zeolite, in silicic, volcanic tuffs that were deposited in lakes where the sodium to potassium ratio ranged from 6 to 150.

Vertical zoning of zeolite minerals occurs in thick sequences of marine and fresh-water sedimentary rocks. Generally, the zoning is such that the more hydrous and less dense zeolites are found near the surface. With increasing depth of burial the zeolites appear to become replaced by anhydrous framework silicates such as the feldspars. In general, the zoning is from the most hydrous and silica-rich zeolites near the surface to the least hydrous and silica-poor, at increasing depth. In thick deposits of marine and fresh-water volcanic tuffs, the uppermost zone (1000 to 3000 feet) contains fresh glass and lacks any real amount of zeolite minerals. In the uppermost zeolite rocks below the glass zone, the minerals mordenite and clinoptilolite are characteristic. As the depth increases, analcime is found followed by laumontite. The ratio of analcime to clinoptilolite, mordenite, erionite, and phillipsite increases with age in the sedimentary rocks which have been buried to shallow depths (Fig. 3.8). The crystal size of analcime found in deposits in saline alkaline lakes increases with increasing age from

202 MINERAL ZEOLITES

an average of 0.005 mm in recent sediments to 1-2 mm in Eocene rocks. This suggests that the zeolite crystals have continued to grow over a period of a few million years after they were originally formed. In New Zealand the mineral laumontite which is found in the lower half of a 30,000-ft sequence is pseudomorphic after analcime, indicating that the transformation analcime to laumontite occurred. The vertical zoning of the combination of the zeolite and minerals has been interpreted in terms of a pressure-temperature gradient although Hay points out that differences in salinity can produce the same effect. The salinity of the water increases with increasing depth and would favor the same type of vertical zoning as does the variation in pressure and temperatures.

Pressure and temperature

The zeolites are hydrated, low density, aluminosilicates and consequently the environmental pressure and temperature influences the particular mineral formed. The less hydrous and denser zeolites such as laumontite and analcime are more stable at higher temperatures and higher pressures as compared to the more hydrous and less dense types such as chabazite and stilbite. With increasing pressure, as a result of increasing burial depth, the less dense zeolites become unstable toward the more dense types which eventually are unstable toward anhydrous aluminosilicate minerals such as the feldspars. In some examples when burial temperatures are higher than 150°C, the

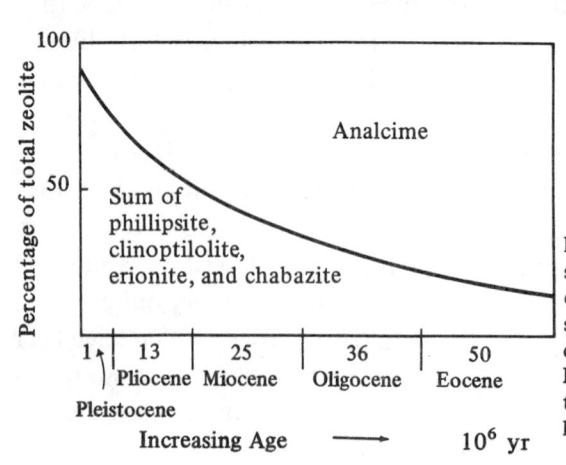

Figure 3.8 Schematic diagram showing the abundance of analcime relative to the sum of phillipsite, clinoptilolite, erionite, and chabazite as a function of age in Eocene to late Pleistocene silicic tuffs deposited in saline, alkaline lakes (5).

zeolites are replaced by feldspars. This is confirmed by experimental work in which zeolites transform to anhydrous aluminosilicates at high temperature and pressure. (See Chap. 6.)

Composition

Since the zeolite minerals appear to have formed by the reaction of a solid phase with water solution (in natural environments) a correlation between the composition of the solid and the solution phases with the zeolite species found would be expected. Such correlations are empirically observed in zeolite synthesis (see Chap. 4). The alkali zeolites such as mordenite or clinoptilolite, for example, might be expected to form from a volcanic glass. The volcanic glass is more reactive in zeolite mineral formation than are crystalline materials at low temperatures because it is relatively more soluble and has a higher free energy. (This is discussed in more detail in Chap. 4 where it is shown that the type of reactant influences the nature of the zeolite formed). The chemical activity of the metal cations, silica, and water, influence the zeolite species which crystallizes from solution; a zeolite requiring a high ratio of base cation will be favored by a high pH. If present, hydrogen ion competes with base cations. The activity ratio will determine whether or not a framework zeolite or a layer silicate such as a clay mineral will form.

Concentration of Silica

The activity of the silica in solution influences the zeolite formed. Thus, mordenite is characteristically formed from solutions with a high silica activity. The presence of hydroxyl ion influences the concentration or silica activity because it independently catalyzes the crystallization of amorphous silica to quartz, consequently diminishing the availability of silica.

The solubility of a silica-rich glass has been recently determined as a function of increasing alkalinity of the solution (35). Although the solubility increases with increasing pH, the ratio Si/Al in the solution decreases. It was proposed that zeolites with a relatively low Si/Al ratio are formed by increasing alkalinity of the solution in contact with the glass during mineralization processes. Synthesis experiments showed that the synthetic type of phillipsite varies in composition with a variation in alkalinity. Increasing the pH from 9 to 11.5 decreased the Si/Al ratio in the zeolite from 3.4 to 2.2. This was in agreement with the ratio of $Si(OH)_4$ concentration to the $Al(OH)_4^-$ concentration in the solution which undergoes a similar variation with pH (35).

It is shown in Chap. 4 that zeolites synthesized in the laboratory are formed metastably. The same possibility exists for zeolites formed in nature. The metastable zeolite materials are most likely to form at low temperatures and pressures on the earth's surface where the kinetic factors are important in determining the particular zeolite species. Since volcanic glass has a higher free energy, the metastable zeolites form from this material at atmospheric pressures. The mechanism for zeolite mineral formation from natural volcanic glass is thought to involve the reaction of glass with the surrounding solution to form a type of gel which subsequently crystallizes to the zeolite. The glass-solution reaction must first involve hydration followed by hydrolysis. The silica concentrations in zeolite alkaline lakes which have a pH of greater than 9 may be as high as 125 ppm. The time required may be a few hundred years versus tens of millions of years depending on the composition and the nature of the chemical environment. Hydrolysis of the glass to form a layer silicate mineral such as montmorillonite may also occur. Formation of the clay produces the alkaline pH and high silica concentration in solution necessary to zeolite formation.

Zeolite minerals may also form by the reaction of other primary silicate minerals. Clay minerals in general do not react to form zeolites in natural environments although they may be converted to zeolites in the laboratory. Halloysite, because of its small particle size and disordered structure, is probably the most reactive of the clay minerals. Analcime may be formed as the result of hydration and exchange of sodium for potassium in leucite.

Formation of zeolite minerals by chemical sedimentation has also been suggested. Analcime in the Lockatong formation in New Jersey appears to have formed as a result of direct precipitation. It was estimated that the silica content and Al_2O_3 content of river waters is such that they are sufficient for the formation of analcime by direct crystallization from solution. There is, however, no evidence of the formation of zeolites by sedimentation on ocean floor bottoms as once was suggested.

E. PHYSICAL PROPERTIES

Pure zeolites are colorless; some mineral specimens may be colored because of the presence of finely divided oxides of iron or similar impurities. Synthetic zeolites (see Chap. 5) may be colored due to the introduction of certain types of transition metal ions such as Co^{2+} by ion exchange. The densities of the zeolites range generally between 2 and 2.3 g/cc. The exchange cation does, of course, change the density which primarily depends on the basic framework structure, that is, the openness and void volume. Zeolites which are rich in barium may have higher densities of 2.5 to 2.8 g/cc. Optically, the refractive indices vary between 1.47 and 1.52 with birefringence of 0 to about 0.015. Optical and physical properties of zeolite minerals are summarized in Tables 3.4 through 3.36. These properties have been selected as representative properties. In some cases, it is difficult to assign a particular x-ray powder diffraction pattern to a certain species because variations do occur in the x-ray pattern as a result of variation in the exchange cation composition. The table for each mineral includes x-ray powder data which have been obtained from selected sources including the Powder Diffraction File. The intensities of the invididual x-ray reflections are those given in the source. In some instances the intensities are given on a relative scale, such as vs, s, ms, m, w, vw and vvw, corresponding to very strong, strong, medium strong, medium, weak, very weak and very very weak.

For the identification of zeolites which occur as fine-grained crystals in sedimentary rocks, x-ray diffraction is the primary tool. Routine diffractometer analysis is generally sufficient for identification of zeolites which occur in rock in concentrations of more than 10%. In order to identify zeolites which are present in less than this amount, some method of concentration is necessary. Although special optical data may be reasonably sufficient to identify a larger zeolite crystal of the igneous zeolites, it generally is a supplement to the modern x-ray method. The refractive index, for example, has been used for distinguishing clinoptilolite from heulandite.

The problem of assigning a specific set of x-ray powder data to a zeolite is exemplified in Fig. 3.9. The x-ray powder patterns are shown for three phillipsite specimens from three different localities. Two of these are in igneous rock and the third is a deep-sea specimen from the Sylvania sea mount. Also included are the x-ray powder pat-

206 MINERAL ZEOLITES

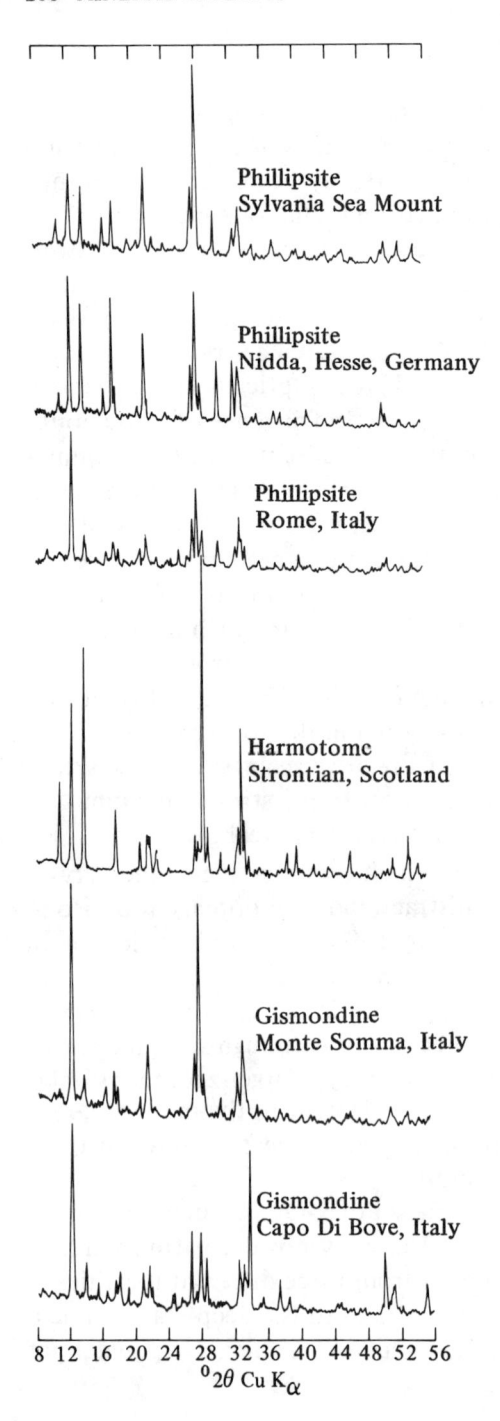

Figure 3.9 X-ray diffraction patterns of phillipsite, harmotome, and gismondine zeolites.

terns for structurally related zeolites, harmotome and gismondine. It appears superficially that harmotome is no less similar to phillipsite than is gismondine, although gismondine has a different framework structure.

Various other methods have been suggested for identifying zeolites. One is based upon contacting the zeolite with its exchangeable cations with a solution of a salt whose cation is excluded by the zeolite due to size. The salt used is tetrabutylammonium chloride. The zeolite is first subjected to exchange with silver using silver nitrate. Then the silver exchanged zeolite is treated with a solution of alkylammonium salt. Because the alkylammonium ion is excluded, the silver ion is removed by precipitation of the insoluble silver chloride (see Chap. 7). In this event, hydrolysis of the zeolite cation occurs and the solution becomes alkaline. The presence of a zeolite is detected by an increase in pH from about 6 to >9. With other likely ion exchanging minerals such as clays, the pH varies between 6.5 and 7.5 since the large organic cation is exchanged without hydrolysis (36).

F. GLOSSARY OF TERMS

A summary of terms that are used in this chapter which may be unfamiliar to the reader is included in this section (37).

Acidic	>66% SiO_2
Amygdale	A gas cavity in volcanic rocks which has become filled with secondary minerals such as zeolites or quartz.
Authigenic	Minerals which are formed *in situ* before burial and consolidation of the sediment to rocks.
Basalt	An extrusive rock, dark-colored and fine-grained, occurring as a lava and composed primarily of calcium feldspar and pyroxene.
Detrital	Refers to minerals occurring in sedimentary rocks which were derived from pre-existing igneous, sedimentary, or metamorphic rocks.
Diagenesis	The chemical and physical changes that sediments undergo during and after their accumulation but before consolidation takes place. These changes may be due to bacterial action, to the solution and redisposition by permeating water, or to chemical replacement.

Epigenitic	Term applied to mineral deposits of later origin than the enclosing rocks or to the formation of secondary minerals by alteration.
Facies	The "aspect" belonging to a geological unit of sedimentation, including mineral composition, type of bedding, fossil content, etc.
Graywacke	A type of sandstone
Lacustrine	Produced by or belonging to lakes
Mafic	"basic" - subsilicic or low in silica, $< 50 - 55\%$
Metasomatism	Replacement process by which one mineral is replaced by another of different chemical composition owing to reactions set up by the introduction of material from different sources.
Mineraloid	A term used to designate materials that are commonly considered to be minerals but which are amorphous, e.g., allophane.
Palagonite	A yellow or orange isotropic mineraloid formed by hydration and other alteration of basaltic glass.
Paragenesis	A general term for the order of formation of associated minerals in time succession, one after another.
Pyroclastic	General term applied to detrital volcanic materials that have been explosively or aerially ejected from a volcanic vent. The term also is used to apply to the rocks made up of these materials.
Shard	A curved spiculelike fragment of volcanic glass.
Syngenetic	A term applied to mineral deposits which are formed contemporaneously with the enclosing rocks as contrasted with *epigenetic* rocks which are of later origin than the enclosing rocks.
Tuff	A rock formed of compacted volanic fragments, generally smaller than 4 mm in diameter.

Table 3.4 Analcime

Structure Group:	1
Typical Unit Cell Contents:	$Na_{16}[(AlO_2)_{16}(SiO_2)_{32}] \cdot 16\,H_2O$
Variations:	$Si/Al = 1.8\text{-}2.8; H_2O/Al_2O_3 = 2\text{-}2.6$
Occurrence:	Igneous and sedimentary rocks; Iceland, Nova Scotia, New Jersey, Wyoming, Arizona, Utah, California, Nevada
System:	Cubic, a = 13.72 A
Habit:	Icositetrahedra, radiating aggregates
Twinning:	[001], [110] lamellar
Cleavage:	[001] poor
Density:	2.24-2.29
Hardness:	5½
Optical Properties:	Isotropic, $n = 1.479\text{-}1.493$; birefringence, δ = slight, < 0.001
Reference:	3-6, 38-40

X-Ray Powder Data (40)

hkl	d(A)	I	hkl	d(A)	I	hkl	d(A)	I
200[a]	6.87	< 10	631	2.022	10	842	1.498	20
211	5.61	80	543	1.940	< 10	761	1.480	20
220	4.86	40	640	1.903	50	664	1.463	10
321	3.67	20	633	1.867	40	754	1.447	10
400	3.43	100	642	1.833	< 10	932,	1.415	40
332	2.925	80	732,	1.743	60	763		
422	2.801	20	651			941,	1.386	<10
431	2.693	50	800	1.716	30	853		
521	2.505	50	741	1.689	40	860	1.372	10
440	2.426	30	820	1.664	10	1011	1.359	40
611,	2.226	40	822,	1.618	20	1031	1.308	10
532			660			871	1.285	20
620	2.168	< 10	831,	1.596	30	1033	1.263	20
541	2.115	< 10	743			963	1.220	30

[a] (200) not in agreement with space group Ia3d

Table 3.5 Bikitaite

Structure Group:	6
Typical Unit Cell Contents:	$Li_2[(AlO_2)_2(SiO_2)_4] \cdot 2 H_2O$
Variations:	Deficient in cations, Li^+ is dominant
Occurrence:	Bikita, Southern Rhodesia
System:	Monoclinic, a = 8.61, b = 4.96, c = 7.61 $\beta = 114°26'$
Habit:	Pseudo-orthorhombic
Cleavage:	[001], [100], good
Density:	2.29
Hardness:	6
Optical Properties:	Biaxial (−), $\alpha = 1.510, \beta = 1.521, \gamma = 1.523$ r < V, $\gamma = 6$, c = 28°; birefringence, 2V = 45°
Reference:	41 - 43

X-Ray Powder Data (41)

hk*l*	d(A)	I	hk*l*	d(A)	I	hk*l*	d(A)	I
100	7.865	80	111	3.284	40	120, 312	2.364	20
001	6.930	50	2̄11	3.215	40			
1̄01	6.732	30	210	3.076	40	021	2.337	10
101	4.374	40	1̄12	3.023	10	1̄21	2.323	10
2̄01	4.265	10	201	2.930	10	310	2.316	10
110	4.195	90	3̄01	2.870	20	1̄13	2.240	10
001	4.022	20	2̄12	2.794	10	301	2.167	10
1̄11	3.991	10	102	2.739	10	2̄21	2.141	10
200	3.926	10	300	2.629	10	220	2.097	10
1̄02	3.806	30	211	2.523	20	013	2.094	10
002	3.462	100	020	2.479	90	1̄22	2.077	20
2̄02	3.371	100	112	2.423	10	022	2.012	10
						103	2.005	10

Table 3.6 Brewsterite

Structure Group:	7
Typical Unit Cell Contents:	$(Sr,Ba,Ca)_2 [(AlO_2)_4 (SiO_2)_{12}] \cdot 10\, H_2O$
Variations:	$Sr > Ba > Ca$; $Si/Al = 2.57 - 3.03$
Occurrence:	Strontian, Argyllshire, Scotland
System:	Monoclinic, a = 6.77 b = 17.51 c = 7.74 $\beta = 94°18'$
Habit:	Platy
Cleavage:	[010]
Density:	2.453
Hardness:	5
Optical Properties:	Biaxial (+), $\alpha = 1.510, \beta = 1.512, \gamma = 1.523$; birefringence, $\delta = 0.013$, $2V\gamma = 47°$
Reference:	3, 4, 44

X-Ray Powder Data (44)

hk*l*	d(A)	I	hk*l*	d(A)	I	hk*l*	d(A)	I
100	6.81	30	0̲60, 1̲51	2.885	90		1.824	10
110	6.15	90					1.771	10
1̄11, 101	4.98	40	1̲60, 240	2.667	30		1.728	10
							1.642	40
1̄21	4.53	100	0̲03, 2̲42	2.549	30		1.595	20
1̄31	3.87	70					1.542	20
140, 131	3.71	10		2.442	30		1.514	10
				2.309	30		1.471	10
050	3.48	10		2.243	30		1.435	20
1̲41, 1̄12	3.35	10		2.191	20		1.386	20
				2.103	30		1.359	20
112	3.21	80		1.989	40		1.324	20
201̄	3.02	10		1.933	20		1.302	20
				1.866	10		1.272	20

Table 3.7 Chabazite

Structure Group:	4
Other Designation:	Acadialite, Haydenite, Phacolite, Glottalite
Typical Unit Cell Contents:	$Ca_2 [(AlO_2)_4(SiO_2)_8] \cdot 13 H_2O$
Variations:	Na and K; Ca > Na, K; Si/Al = 1.6-3; Si/Al = 3.2-3.8 reported in sedimentary mineral
Occurrence:	Ireland, Nova Scotia, Colorado; sedimentary in Italy, Africa, Arizona, California, Nevada
System:	Trigonal, a = 9.42 α = 94°28'
	Hexagonal, a = 13.78, c = 15.06
Habit:	Rhombohedral
Twinning:	[0001] penetration
Cleavage:	[1011]
Density:	2.05 - 2.10
Hardness:	4½
Optical Properties:	Uniaxial (−), ϵ = 1.48-1.50, ω = 1.480-1.50; birefringence, δ = 0.002-0.005
	mean index 1.46-1.47 in sedimentary chabazite
Reference:	3, 4, 6, 38, 45-48

X-Ray Powder Data (46)

hk*l*	d(A)	I	hk*l*	d(A)	I	hk*l*	d(A)	I
101	9.351	50	311	3.235	6	501	2.358	2
110	6.894	10	204	3.190	5	413	2.310	3
102	6.384	5	312	3.033	2	330	2.300	4
201	5.555	9	401	2.925	100	502	2.277	1
003	5.021	30	214	2.890	30	421	2.233	1
202	4.677	6	223	2.842	3	306	2.123	2
211	4.324	76	402	2.776	4	107	2.119	2
113	4.044	1	205	2.690	7	333	2.090	6
300	3.976	2	410	2.605	10	504	2.016	1
212	3.870	28	322	2.574	2	217	1.941	1
104	3.590	23	215	2.507	11	520	1.911	3
220	3.448	13	116	2.361	2	505	1.871	3

Table 3.8 Clinoptilolite

Structure Group:	7
Typical Unit Cell Contents:	$Na_6[(AlO_2)_6(SiO_2)_{30}] \cdot 24\,H_2O$
Variations:	Na, K $\gg$ Ca, Mg; Si/Al = 4.25-5.25; in sedimentary type $Si/(Al + Fe^{3+})$ = 4.1-5.6
Occurrence:	Wyoming; extensive sedimentary occurrences in western U. S.
System:	Monoclinic, a = 7.41 b = 17.89 c = 15.85 $\beta = 91°29'$
Habit:	Tabular, platy
Cleavage:	[010]
Density:	2.16
Optical Properties:	Biaxial (-), α = 1.476, β = 1.479, γ = 1.479
Reference:	6, 38, 49-53

X-Ray Powder Data (52)

hkl	d(A)	I	hkl	d(A)	I	hkl	d(A)	I
020	8.92	100	004	3.964	55	222	3.168	14
002	7.97	3	042	3.897	57	22$\bar{2}$	3.119	15
10$\bar{1}$	6.78	2	14$\bar{1}$	3.74	7	231	3.07	8
031	5.61	2	21$\bar{1}$	3.55	6	044	2.974	80
112	5.15	7	051,	3.48	3	035	2.793	15
130	4.65	14	11$\bar{4}$			12$\bar{5}$	2.793	15
10$\bar{3}$	4.35	2	220	2.419	16	16$\bar{1}$	2.728	33
132	3.964	55	202	3.324	4			

Table 3.9 Dachiardite

Structure Group:	6
Typical Unit Cell Contents:	$K_{0.7}Na_{1.1}Ca_{1.6}Mg_{0.2}[(AlO_2)_5(SiO_2)_{19}] \cdot 12.7\ H_2O$
Occurrence:	Elba, rare
System:	Monoclinic, a = 18.73 b = 7.54 c = 10.30 β = 107°54'
Cleavage:	[100]
Density:	2.165
Hardness:	4½
Optical Properties:	Biaxial (+) α = 1.491, β = 1.496, γ = 1.499; birefringence, $2V_\gamma$ = 65-73
Reference:	38, 54-56

X-Ray Powder Data (56)

hkl	d(A)	I	hkl	d(A)	I	hkl	d(A)	I
010	9.79	vw	121	3.773	w	$\bar{3}31$	3.018	w
200	8.90	m	002	3.750	w	$\bar{3}21$		
$\bar{1}01$	6.91	m	410	3.634	w	$\bar{6}20$	2.964	m
$\bar{1}11$	6.00	mw	012	3.498	w	420	2.862	m
111	5.349	w	202	3.452	s	402		
$\bar{2}20$	4.966	m	$\bar{2}30$	3.396	mw	$\bar{4}22$	2.712	m
020	4.882	m	$\bar{2}12$	3.375	w	222	2.666	m
$\bar{4}10$	4.610	vw	$\bar{5}11$	3.328	mw	412	2.607	vw
400	4.445	vw	501	3.204	s	$\bar{2}40$	2.550	m
$\bar{1}21$	4.228	vw	430,	3.114	vw	$\bar{7}11$	2.517	w
$\bar{4}20$	3.932	m	$\bar{5}21$			103	2.472	w
311	3.848	vw	$\bar{1}31$	3.077	vw	$\bar{7}21$,	2.449	w
220	3.801	m				032		

Table 3.10 Edingtonite

Structure Group:	5
Other Designation:	Glottalite
Typical Unit Cell Contents:	$Ba_2[(AlO_2)_4(SiO_2)_6] \cdot 8 H_2O$
Variations:	Ca, Na, K
Occurrence:	Glasgow, Scotland
System:	Orthorhombic, a = 9.54, b = 9.65, c = 6.50
Habit:	Sphenoids
Cleavage:	[110]
Density:	2.777
Hardness:	4
Optical Properties:	Biaxial (−), α = 1.541, β = 1.553, γ = 1.557; birefringence, δ = 0.015, 2V = 54°
Reference:	3, 4, 57, 58

X-Ray Powder Data (57)

hkl	d(A)	I	hkl	d(A)	I	hkl	d(A)	I
001	6.49	80	311, 131	2.749	100	241, 421	2.035	50
101, 011	5.37	80	230, 320	2.655	40	023, 203	1.974	50
200, 020	4.80	90	212, 122	2.591	90	213, 123	1.933	60
111	4.64	90	231, 321	2.461	50	412, 142	1.889	50B
210, 120	4.29	40	302, 032	2.232	80	431, 341	1.834	70B
201, 021	3.86	20	330	2.260	80		1.795	50
121, 211	3.58	100	141	2.201	60		1.766	50
220	3.39	60	411, 003	2.178	60		1.694	40
002	3.255	50	420, 240	2.143	20		1.678	20
102, 012	3.078	60	331	2.132	70		1.653	60
310, 130	3.010	80	332, 232	2.062	70		1.644	60
112	2.934	70					1.623	40
							1.606	40

216 MINERAL ZEOLITES

Table 3.11 Epistilbite

Structure Group:	6
Typical Unit Cell Contents:	$Ca_3[(AlO_2)_6(SiO_2)_{18}] \cdot 18 H_2O$
Variations:	Si/Al = 2.58-3.02; Ca dominant, also some Na and K
Occurrence:	Berufiord, Iceland, Hawaii
System:	Monoclinic, a = 9.08 b = 17.74 c = 10.25 β = 124.54°
Habit:	Prismatic, sheaflike aggregates
Twinning:	Cruciform interpenetrant
Cleavage:	[010]
Density:	2.21
Hardness:	4-4½
Optical Properties:	Biaxial (−), α = 1.502, β = 1.510, γ = 1.512; birefringence, δ = 0.010-0.014, 2V = 44°
Reference:	3, 4, 38, 57, 59

X-Ray Powder Data (57, 85)

hkl	d(A)	I	hkl	d(A)	I	hkl	d(A)	I
020	8.89	90	220	3.45	100	260	2.322	4
001	8.41	6	2̄03	3.33	14	2̄63	2.214	6
110	6.89	60	113	3.27	25	4̄23	2.187	2
1̄12	4.91	65	150	3.21	90	4̄24	2.060	4
130	4.65	6	2̄42	3.06	8	3̄54	2.004	6
2̄01	4.49	10	060	2.957	12	2̄82	1.964	2
040	4.44	12	1̄52	2.917	60	172	1.939	6
2̄02	4.23	20	240	2.858	25	044	1.907	4
221	4.01	8	061	2.788	6	400	1.871	2
041	3.92	20	3̄32	2.696	12	420	1.838	30
1̄32	3.87	70	023	2.682	16	5̄14	1.778	30
2̄22	3.82	12	2̄04	2.557	12	114	1.775	4
022	3.80	14	310	2.451	4	1̄112	1.538	2
200	3.74	8	171	2.426	16	55̄0	1.377	6

Table 3.12 Erionite

Structure Group:	2
Typical Unit Cell Contents:	$(Ca, Mg, Na_2, K_2)_{4.5}[(AlO_2)_9(SiO_2)_{27}] \cdot 27\ H_2O$
Variations:	$Si/Al = 2.9$-3.7, H_2O to 31. Possibly Fe^{3+} substitution for Al up to 15%
Occurrence:	Oregon, Washington; sedimentary occurrences in Nevada, Oregon, California, USSR
System:	Hexagonal, $a = 13.26$, $c = 15.12$
Habit:	Acicular, fibrous
Density:	2.02-2.08
Optical Properties:	Uniaxial (+), $\epsilon = 1.473$-1.476, $\omega = 1.468$-1.472
Reference:	4, 6, 60-65

X-Ray Powder Data (60)

hkl	d(A)	I	hkl	d(A)	I	hkl	d(A)	I
100	11.41	100	212	3.746	65	401	2.812	52
101	9.07	11	104	3.570	24	304	2.676	15
002	7.51	7	302	3.402	4	402	2.680	12
110	6.61	73	220	3.303	39	205	2.673	8
102	6.28	5	213	3.276	25	410	2.496	20
200	5.72	16	114	3.271	25	224	2.480	17
201	5.34	14	311	3.106	12	330	2.200	11
103	4.595	8	312	2.923	10	107	2.113	6
202	4.551	12	105	2.910	10	422	2.079	5
210	4.322	67	400	2.860	60	512	1.982	4
211	4.156	24	214	2.839	50	430	1.882	6
300	3.813	37				520	1.834	8

Table 3.13 Faujasite

Structure Group:	4
Typical Unit Cell Contents:	$Na_{12}Ca_{12}Mg_{11}[(AlO_2)_{59}(SiO_2)_{133}] \cdot 235\ H_2O$
Variations:	Si/Al = 2.1-2.3, also some K
Occurrence:	Kaiserstuhl, Germany; reported in Oahu, Hawaii
System:	Cubic, a = 24.67
Habit:	Octahedral
Cleavage:	[111]
Density:	1.91
Hardness:	5
Optical Properties:	Isotropic, n = 1.48, 1.471
Reference:	3, 38, 48, 66

X-Ray Powder Data (66)

hk*l*	d(A)	I	hk*l*	d(A)	I	hk*l*	d(A)	I
111	14.418	100	533	3.779	32	751, 555	2.860	22
220	8.784	9	444	3.580	2			
311	7.487	5	711, 551	3.468	5	840	2.767	5
222	7.173	2				911, 753	2.719	3
331	5.695	30	642	3.311	16			
422	5.062	1	731, 553	3.227	5	664	2.641	7
333, 511	4.772	13				931	2.600	4
			733	3.025	6	10,2,2	2.382	5
440	4.387	32	822, 660	2.919	10	666		
620	3.915	4				880	2.189	3

Table 3.14 Ferrierite

Structure Group:	6
Typical Unit Cell Contents:	$Na_{1.5}Mg_2[(AlO_2)_{5.5}(SiO_2)_{30.5}] \cdot 18 H_2O$
Variations:	Na, K in Nevada ferrierite
	Less Mg in Agoura ferrierite
Occurrence:	Kamloops, British Columbia, Vicenza, Italy; Agoura, California; sedimentary in Nevada.
System:	Orthorhombic, a = 19.16 b = 14.13 c = 7.49
Habit:	Blades, needles
Cleavage:	[100]
Density:	2.136-2.21
Hardness:	3 - 3¼
Optical Properties:	Mean index = 1.484 in Na, K ferrierite; biaxial (+), α = 1.478, β = 1.479, γ = 1.482; birefringence, δ = 0.004, $2V\gamma$ = 50°
Reference:	38, 52, 67 - 70

X-Ray Powder Data (67)

hkl	d(A)	I	hkl	d(A)	I	hkl	d(A)	I
110	11.33	20	112, 040	3.54	80	350, 042, 701	2.58	30
200	9.61	100	202	3.49	80		2.49	20
020, 101	7.00	30	501	3.42	20		2.43	20
011	6.61	20	240	3.31	20		2.37	40
310	5.84	50	600	3.20	10		2.32	10
121	4.96	10	141, 312	3.15	30		2.26	10
301, 400	4.80	10	521, 431	3.07	30		2.11	20
130	4.58	10	530	2.97	30		2.04	20
321, 031	3.99	90	620, 132	2.90	20		2.00	30
411	3.88	10	422	2.72	20		1.94	30
330, 510	3.79	20					1.87	30
	3.69	50					1.78	40

Table 3.15 Garronite

Structure Group:	1
Typical Unit Cell Contents:	$NaCa_{2.5}[(AlO_2)_6(SiO_2)_{10}] \cdot 14 H_2O$
Variations:	Low K, trace Ba Si/Al = 1.5-1.67
Occurrence:	Antrim, Ireland; Iceland; USSR
System:	Tetragonal, a = 10.0 c = 9.9
Habit:	Radiating aggregates
Density:	2.13-2.17
Hardness:	4.5
Optical Properties:	Uniaxial, ϵ = 1.502-1.512, ω = 1.500-1.515
Reference:	71-73

X-Ray Powder Data (72)

hkl	d(A)	I	hkl	d(A)	I	hkl	d(A)	I
110	7.15	ms	400	2.54	vw	215	1.805	w
002	4.95	ms	303	2.34	w	440	1.770	mw
211	4.12	s	204	2.22	w	522	1.730	vw
112	4.07	m	323	2.12	vw	433	1.705	mw
311	3.22	m	422	2.05	vw	600	1.665	vw
103	3.14	s	224	2.03	vvw	610	1.645	mw
222	2.88	w	005	1.97	w	116	1.605	mw
213	2.68	vw	314	1.938	w			

Table 3.16 Gismondine

Structure Group:	1
Typical Unit Cell Contents:	$Ca_4[(AlO_2)_8(SiO_2)_8] \cdot 16\,H_2O$
Variations:	Si/Al = 1.12 - 1.49 ; some K
Occurrence:	Italy; Pennsylvania; Hawaii
System:	Monoclinic, a = 10.02 b = 10.62 c = 9.84, $\beta = 92°25'$
Habit:	Tetragonal bipyramids
Cleavage:	None
Density:	2.27
Hardness:	4½ - 5
Optical Properties:	Biaxial (−), β = 1.539; birefringence, 2V = 83°
Reference:	3, 38, 48, 57, 74

X-Ray Powder Data (57, 85)

hk$l$	d(A)	I	hk$l$	d(A)	I	hk$l$	d(A)	I
100	9.99	2	310	3.19	18	042,	2.34	2
110	7.28	16	013	3.13	14	31$\bar{3}$		
11$\bar{1}$	5.93	2	31$\bar{1}$	3.06	2	14$\bar{2}$	2.28	6
111	5.76	4	311	2.99	2	313	2.24	4
020	5.28	2	032	2.87	2	23$\bar{3}$,	2.19	2
200	5.01	4	30$\bar{2}$	2.82	2	12$\bar{4}$		
002	4.91	16	132,	2.74	14	214	2.13	6
021	4.67	4	32$\bar{1}$			15$\bar{1}$	2.04	2
012	4.47	4	21$\bar{3}$	2.70	18	143,	2.01	2
12$\bar{1}$	4.27	35	123	2.66	10	224		
121	4.19	12	312	2.62	2	33$\bar{3}$	1.981	4
211	4.05	4	041,	2.56	2	015,	1.933	2
022	3.61	2	140			34$\bar{2}$		
12$\bar{2}$, 202	3.43	2	23$\bar{2}$, 400	2.51	2	413, 511	1.916	2
130, 300	3.34	100	004	2.46	8	234	1.850	2
			223	2.40	2			

+ means additional indices.

Table 3.17 Gmelinite

Structure Group:	4
Other Designation:	Groddeckite
Typical Unit Cell Contents:	$Na_8 [(AlO_2)_8(SiO_2)_{16}] \cdot 24\ H_2O$
Variations:	Na > Ca, K; Si/Al = 1.92 - 2.46
Occurrence:	Nova Scotia, Ireland, New Jersey
System:	Hexagonal, a = 13.75 c = 10.05
Habit:	Tabular, rhombohedral
Twinning:	Twinning axis, c
Cleavage:	[1010], [0001]
Density:	2.028
Hardness:	4½
Optical Properties:	Uniaxial (−), ϵ = 1.474 - 1.480, ω = 1.476 - 1.494; birefringence, δ = 0.002 - 0.015
Reference:	3, 4, 45, 57, 75

X-Ray Powder Data (57)

hkl	d(A)	I	hkl	d(A)	I	hkl	d(A)	I
100	11.95	70	222, 401	2.849	50		1.631	< 10
101	7.69	50					1,598	< 10
110	6.81	40	213	2.675	50		1.580	< 20
200	5.985	10	402	2.571	10		1.519	10
201	5.067	50		2.292	20		1.465	20
102, 210	4.529	20		2.196	< 10	≃ 1.435		< 10
				2.076	40		1.412	30
211	4.095	80		1.989?	< 10		1.388	20
220	3.440	20		1.941	10		1.37	< 10
221	3.220	50		1.907	10		1.349	< 10
302	3.089	< 10		1.801	40		1.324	40
400, 113	2.959	60		1.722	40		1.296	30
				1.678	20			

Table 3.18 Gonnardite

Structure Group:	5
Other Designation:	Ranite
Typical Unit Cell Contents:	$Na_4Ca_2[(AlO_2)_8(SiO_2)_{12}] \cdot 14\ H_2O$
Variations:	Mg, K; Si/Al = 1.1 - 1.38
Occurrence:	France, Italy, Sicily, Japan
System:	Orthorhombic, a = 13.19 b = 13.32 c = 6.55
Habit:	Spherulites, fibrous
Density:	2.25
Optical Properties:	Biaxial (-), α = 1.497 - 1.506, β = 1.508, γ = 1.499 - 1.508; birefringence, 2V = 50°
Reference:	4, 38, 57, 64, 76 - 78

X-Ray Powder Data (57)

hk$l$	d(A)	I	hk$l$	d(A)	I	hk$l$	d(A)	I
001, 200	6.70	60	311	3.52	5		2.28	20
			410, 321	3.23	50		2.22	40
210, 101	5.93	80					2.16	10B
			112	3.12	40		2.12	10B
111	5.25	20	411, 212	2.92	100		2.07	20
220, 201	4.74	50					1.98	20B
			510, 312	2.61	40		1.895	30
300, 211	4.44	60					1.851	20
			520, 501	2.48	40		1.818	30
310	4.22	30					1.759	30B
320, 301	3.69	10	440	2.36	20		1.697	20
			521	2.33	50		1.647	40

Table 3.19 Harmotome

Structure Group:	1
Typical Unit Cell Contents:	$Ba_2[(AlO_2)_4(SiO_2)_{12}] \cdot 12 H_2O$
Variations:	$Ba \gg K, Na$; $Si/Al = 2.31 - 2.53$
Occurrence:	Strontian, Scotland; New Zealand
System:	Monoclinic, a = 9.87 b = 14.14 c = 8.72 β = 124°50'
Habit:	Prismatic
Twinning:	[001], [021], [110], interpenetrant
Cleavage:	[010]
Density:	2.35 - 2.442
Hardness:	4½
Optical Properties:	Biaxial (+), α = 1.503 - 1.508, β = 1.505 - 1.509, γ = 1.508 - 1.514; birefringence, δ = 0.005 - 0.008, 2V = 80°
Reference:	3, 4, 38, 57, 79 - 81

X-Ray Powder Data (57)

hkl	d(A)	I	hkl	d(A)	I	hkl	d(A)	I
100	8.10	40	3̄01	3.20	10		2.670	70
001	7.16	50	131	3.17	60		2.630	20
011	6.38	100	041				2.527	20
021	5.03	40	3̄12,	3.13	80		2.470	10
1̄02,	4.30	40	3̄11				2.374	20
101			2̄32,	3.08	40		2.315	20
200,	4.08	60	230				2.148	20
2̄02				2.920	20		2.058	20
2̄12,	3.90	30		2.847	20		1.953	10
210				2.751	20		1.713	20
012	3.47	10		2.730	60		1.675	20
140	3.24	60		2.698	60			

Table 3.20 Herschelite

Structure Group:	4
Other Designation:	Seebachite
Typical Unit Cell Contents:	$Na_4[(AlO_2)_4(SiO_2)_8] \cdot 10 - 11\ H_2O$
Variations:	$Na > K, Ca; Si/Al = 3.2$ in sedimentary Herschelite
Occurrence:	Sicily; sedimentary in Arizona
System:	Hexagonal, a = 13.80 c = 15.10
Habit:	Tabular, hexagonal prisms
Twinning:	[0001]
Density:	2.06
Hardness:	4½
Optical Properties:	Mean index = 1.4846; birefringence, δ = 0.002 - 0.004
Reference:	3, 46, 82, 83

X-Ray Powder Data (46)

hk*l*	d(A)	I	hk*l*	d(A)	I	hk*l*	d(A)	I
101	9.361	51	311	3.235	11	116	2.364	2
110	6.894	22	204	3.193	7	413	2.316	7
102	6.379	3	303	3.125	2	330	2.300	5
201	5.555	15	312	3.031	2	315	2.228	1
003	5.032	40	401	2.930	100	422	2.163	2
202	4.679	6	214	2.897	29	306	2.130	5
211	4.322	67	223	2.852	9	511	2.125	4
113	4.109	9	402	2.778	5	333	2.092	15
300	3.976	8	205	2.695	11	512	2.064	2
212	3.877	23	410	2.606	18	504	2.020	1
104	3.600	21	322	2.576	4	217	1.947	2
220	3.448	18	215	2.511	12	520	1.914	4

Table 3.21 Heulandite[a]

Structure Group:	7
Other Designation:	Oryzite
Typical Unit Cell Contents:	$Ca_4[(AlO_2)_8(SiO_2)_{28}] \cdot 24 H_2O$
Variations:	$Ca \gg Na$; $Si/Al = 2.47 - 3.73$
Occurrence:	Berufiord, Iceland; Nova Scotia; sedimentary, New Zealand
System:	Monoclinic, a = 17.73 b = 17.82 c = 7.43 $\beta = 116°20'$
Habit:	Tabular
Cleavage:	[010]
Density:	2.20
Hardness:	3½ - 4
Optical Properties:	Biaxial (+), α = 1.491 - 1.505, β = 1.493 - 1.503, γ = 1.500 - 1.512, 2V = variable, ~ 34°
Reference:	3, 4, 38, 57, 84 - 86, 111

X-Ray Powder Data (86)

hkl	d(A)	I	hkl	d(A)	I	hkl	d(A)	I
020	8.845	80	$\bar{1}52$	2.529	20	$\bar{9}51,$	1.698	20
200	7.796	70	261,	2.430	30	$\bar{1}021$		
001	6.631	60	441,			004	1.662	10
220	5.945	10	$\bar{7}12$			$\bar{5}54,$	1.639	10
$\bar{3}11$	5.277	50	$\bar{2}23$	2.350	10	024		
310	5.096	70	$\bar{6}03$	2.270	10	$\bar{7}13,$	1.608	10
$\bar{1}31$	4.646	60	$\bar{6}23$	2.196	10	$\bar{2}102$		
$\bar{4}01$	4.364	20	730	2.120	20	$\bar{1}43$	1.585	10
421	3.917	100	$\bar{1}72$	2.078	20	$\bar{1}54,$	1.561	10
$\bar{2}41$	3.723	20	$\bar{7}52,$	2.010	20	$\bar{8}81,$		
$\bar{3}21$	3.562	20	$\bar{7}53$			$\bar{8}82$		
$\bar{2}22$	3.420	70	$\bar{8}41,$	1.963	30	$\bar{1}93,$	1.512	10
002	3.320	10	$\bar{5}72$			$\bar{1}131$		
$\bar{4}22$	3.186	50	082	1.850	10	$\bar{5}111,$	1.473	10
510	3.132	40	840	1.814	10	$\bar{1}063$		
350	2.959	90	$\bar{1}02,$	1.770	30	$\bar{7}93,$	1.449	10
530,	2.805	70	243			0121		
$\bar{6}21$			$\bar{2}101,$	1.722	10	$\bar{1}151$	1.431	10
$\bar{5}32$	2.730	40	0101			1060,	1.401	10
042	2.667	10				$\bar{6}103$		
						$\bar{1}105$	1.360	10

[a] Water-rich variety from Nelson Creek, Washington contains 30 H_2O/u.c.

Table 3.22 Kehoeite

Structure Group:	1
Typical Unit Cell Contents:	$Zn_{5.5}Ca_{2.5}[(AlO_2)_{16}(PO_2)_{16}(H_3O_2)_{16}]\cdot 32\ H_2O$
Occurrence:	Rare
System:	Cubic, a = 13.7
Density:	2.34
Optical Properties:	Isotropic, n = 1.52 - 1.54
Reference:	87

X-Ray Powder Data (87)

hkl	d(A)	I	hkl	d(A)	I	hkl	d(A)	I
111	7.63	30	332	2.96	10	543,	1.916	70
211	5.68	30	422	2.816	20	550,		
220	4.85	10	431,	2.708	30	710		
310	4.28	50	510			642	1.816	20
222	3.91	10	532,	2.223	20	651,	1.749	10
321	3.49	30	611			732		
400	3.35	100	541	2.078	10	644,	1.669	10
330, 411	3.13	100	631	1.993	20	820		
						653	1.633	60

Table 3.23 Laumontite

Structure Group:	1
Other Designation:	Leonhardite (partially dehydrated laumontite)
Typical Unit Cell Contents:	$Ca_4[(AlO_2)_8(SiO_2)_{10}] \cdot 16\ H_2O$
Variations:	$Si/Al = 1.75 - 2.28$
Occurrence:	Nova Scotia; sedimentary in New Zealand
System:	Monoclinic, a = 7.55, b = 14.74, c = 13.07, $\beta = 111°9'$
Habit:	Prismatic, fibrous
Twinning:	[100]
Cleavage:	[010], [110]
Density:	2.30
Hardness:	3 - 3½
Optical Properties:	Biaxial (-), $\alpha = 1.502 - 1.514$, $\beta = 1.512 - 1.522$, $\gamma = 1.514 - 1.525$; birefringence, $\delta = 0.01$, $2V_\alpha = 26 - 27°$
Reference:	3, 38, 57, 88, 89

X-Ray Powder Data (57)

hkl	d(A)	I	hkl	d(A)	I	hkl	d(A)	I
110	9.49	100	400	3.411	8		2.575	14
200	6.86	35	131				2.521	4
020	6.54	2	3̄12,	3.367	4		2.463	4
2̄01,	6.19	2	012				2.439	14
011			040	3.272	20		2.361	12
120	5.91	2	3̄31,	3.205	8	600	2.278	2
111	5.052	6	311				2.268	6
220	4.731	20	330	3.152	16		2.217	4
2̄21	4.500	8	420,	3.033	25	060	2.180	6
310	4.314	2	112				2.153	18
130,	4.156	60	240,	2.950	4		2.082	2
201			041				2.060	2
131	3.768	2	511,	2.881	14		2.042	2
4̄01	3.667	14	2̄41				1.991	4
002,	3.510	30		2.798	2		1.955	12
221				2.629	4		1.887	4

Table 3.24 Levynite

Structure Group:	2
Typical Unit Cell Contents:	$Ca_3[(AlO_2)_6(SiO_2)_{12}] \cdot 18\ H_2O$
Variations:	$Ca \gg Na, K; Si/Al = 1.59 - 1.83$
Occurrence:	Antrim, Ireland; Iceland
System:	Hexagonal, a = 13.32 c = 22.51
Habit:	Tabular
Twinning:	Interpenetrant
Cleavage:	[021]
Density:	2.14
Hardness:	4½
Optical Properties:	Uniaxial (-), ϵ = 1.491 - 1.500, ω = 1.496 - 1.505; birefringence, δ = 0.002 - 0.006
Reference:	3, 4, 38, 45, 57, 90

X-Ray Powder Data (57)

hkl	d(A)	I	hkl	d(A)	I	hkl	d(A)	I
101	10.4	35	220	3.35	14	235	2.303	10
012	8.19	65	125	3.17	50	1,0,10	2.256	4
003	7.69	18	312	3.10	20	330	2.234	14
110	6.72	18	401	2.882	10	333, 0,2,10	2.136	18
021	5.64	4	027	2.861	10			
202	5.19	30	042	2.815	80	416	2.113	2
015	4.28	50	306	2.725	6	511	2.072	6
122	4.10	100	217	2.634	40	152,	2.050	4
300	3.87	20	232	2.593	6	244		
006	3.84	6	410	2.534	16	425	1.981	2
205	3.61	6	045	2.453	2	514	1.960	6
214	3.49	16	128, 413	2.406	16	057	1.896	6
303	3.46	6						

Table 3.25 Mesolite

Structure Group:	5
Other Designation:	Metascolecite
Typical Unit Cell Contents:	$Na_{16}Ca_{16}[(AlO_2)_{48}(SiO_2)_{72}] \cdot 64\ H_2O$
Variations:	$Ca/Na \geqslant 1$; $Si/Al = 1.44 - 1.58$
Occurrence:	Nova Scotia, Ireland, Oregon
System:	Orthorhombic, a = 18.43 b = 56.45 c = 6.55
Habit:	Prisms, fibrous aggregates
Twinning:	[100]
Cleavage:	[101], [10$\bar{1}$]
Density:	2.26
Hardness:	5
Optical Properties:	Biaxial (+), $\alpha = 1.504$, $\beta = 1.504 - 1.508$, $\gamma = 1.507$; birefringence, $\delta \leqslant 0.001$, $2V = \sim 80°$
Reference:	3, 4, 57, 76, 91, 92

X-Ray Powder Data (57)

hkl	d(A)	I	hkl	d(A)	I	hkl	d(A)	I
	6.44	40		2.34	10		1.52	10
	5.79	70		2.27	10		1.47	30
	5.46	10		2.19	30		1.43	10
	4.66	30		2.05	10		1.40	10
	4.35	50		1.95	10		1.39	10
	4.16	10		1.86	10		1.35	10
	3.89	10		1.81	30		1.30	10
	3.18	30		1.75	10		1.24	10
	3.08	20		1.72	10		1.21	10
	2.86	100		1.68	10		1.19	10
	2.57	10		1.64	10		1.15	10
	2.47	10		1.59	10		1.15	10
	2.41	10		1.54	10		1.15	10

Table 3.26 Mordenite

Structure Group:	6
Other Designation:	Ptilolite, Arduinite, Ashtonite, Flokite
Typical Unit Cell Contents:	$Na_8[(AlO_2)_8(SiO_2)_{40}] \cdot 24\ H_2O$
Variations:	Na, Ca > K; Si/Al = 4.17 - 5.0; Sr in one variety
Occurrence:	Amygdules in Nova Scotia; sedimentary in USSR, Scotland, California, Nevada
System:	Orthorhombic, a = 18.13 b = 20.49 c = 7.52
Habit:	Laths, fibrous
Cleavage:	[010]
Density:	2.12
Hardness:	3 - 4
Optical Properties:	Biaxial (+) (−), α = 1.472 - 1.483, β = 1.475 - 1.485, γ = 1.477 - 1.487; birefringence, δ = ∼ 0.005, $2V_\alpha$ = 76 - 104°
Reference:	3, 4, 20, 38, 93, 94

X-Ray Powder Data

hkl	d(A)	I	hkl	d(A)	I	hkl	d(A)	I
110	13.52	42	241	3.830	5	171	2.697	33
020	10.19	7	002	3.757	8		2.556	18
200	9.03	100	510	3.563	14		2.519	4
220	6.77	4	151	3.528	3		2.460	4
111	6.57	8	202	3.471	24		2.227	2
130	6.38	37	060	3.410	23		2.130	3
021	6.06	3	350	3.386	44		2.124	2
310	5.78	28	511	3.218	24		2.045	11
330	4.51	42	260	3.198	35		2.041	5
041	4.26	6	261	2.939	3		1.952	5
420	4.133	9	332	2.891	11		1.876	4
150	3.980	100	550	2.710	2		1.808	11

Table 3.27 Natrolite

Structure Group:	5
Other Designation:	Laubanite, Elagite
Typical Unit Cell Contents:	$Na_{16}[(AlO_2)_{16}(SiO_2)_{24}] \cdot 16\ H_2O$
Variations:	Ca, K very small; Si/Al = 1.44 - 1.58
Occurrence:	Nova Scotia, New Jersey, Northern Ireland, New Zealand, etc.
System:	Orthorhombic, a = 18.30, b = 18.63, c = 6.60
Habit:	Needles, prisms, radiating clusters
Twinning:	[110]
Cleavage:	[110], [1$\bar{1}$0]
Density:	2.23 - 2.25
Hardness:	5
Optical Properties:	Biaxial (+), α = 1.473 - 1.483, β = 1.476 - 1.486, γ = 1.485 - 1.496; birefringence, δ = ~ 0.012, 2V = 58 - 64°
Reference:	3, 4, 38, 77, 85, 92, 95, 96

X-Ray Powder Data (77)

hkl	d(A)	I	hkl	d(A)	I	hkl	d(A)	I
220	6.53	s	–	2.67	vvw	153	1.876	mw
111	5.87	vs	242	2.58	m	2,10,0	1.831	vvw
040	4.64	m	–	2.52	vvw	353	1.799	m
311	4.36	vs	171	2.45	m	482	1.754	w
240	4.14	ms	711	2.41	m	4,10,0	1.728	mw
–	3.90	vvw	080	2.33	mw	10,4,0	1.700	w
331	3.64	vvw	062	2.26	w	553	1.676	vw
151	3.17	ms	262	2.19	m	004	1.647	w
202, 022	3.11	m	–	2.12	vvw	733	1.627	mw
			133	2.06	w	10,0,2	1.600	mw
222	2.95	m	191	2.02	vvw	971	1.571	vw
351	2.86	vvs	333	1.96	w			
	2.75	vvw	911	1.93	vvw			

Table 3.28 Offretite

Structure Group:	2
Typical Unit Cell Contents:	$(Mg, Ca, K_2)_{2.5}[(AlO_2)_5(SiO_2)_{13}] \cdot 15\ H_2O$
Variations:	Si/Al = 2.5; Alkaline earths > Alkali metal; no Na
Occurrence:	Montbrison, Loire, France
System:	Hexagonal, a = 13.29 c = 7.58
Density:	2.13
Optical Properties:	Uniaxial (–), ω = 1.489, ϵ = 1.486 birefringence 0.003
Reference:	60, 97, 98

X-Ray Powder Data (60)

hkl	d(A)	I	hkl	d(A)	I	hkl	d(A)	I
100	11.50	100	310	3.190	17	331	2.126	4
110	6.64	20	311	2.942	3	303	2.110	2
200	5.76	35	400	2.880	64	421	2.091	2
201	4.581	4	212	2.858	15	510	2.068	2
210	4.352	59	401	2.693	3	511	1.995	2
300	3.837	43	320	2.642	4	502	1.967	2
211	3.774	11	410	2.510	20	430	1.893	1
102	3.600	3	500	2.300	5	520	1.844	3
301	3.429	2	330	2.214	22	431	1.838	5
220	3.322	22	420	2.177	2			

Table 3.29 Paulingite

Structure Group:	1
Typical Unit Cell Contents:	$K_{68}Na_{13}Ca_{36}Ba_{1.5}[(AlO_2)_{152}(SiO_2)_{520}] \cdot 700\ H_2O$
Occurrence:	Columbia River, Washington
System:	Cubic, a = 35.093
Habit:	Rhombic dodecahedra
Density:	2.21
Hardness:	5
Optical Properties:	Isotropic, $n^{23°} = 1.473 \pm 0.001$
Reference:	62, 85, 99

X-Ray Powder Data (62)

hk*l*	d(A)	I	hk*l*	d(A)	I	hk*l*	d(A)	I
220	12.37	10	642	4.68	30	864, 10,4,0	3.261	90
321	9.45	10	800	4.385	40			
411, 330	8.29	100	820, 644,	4.25	40	954, 873, 11,1,0	3.176	20
420	7.86	10	831, 750,	4.08	40			
422	7.14	20				963, 10,5,1, 11,2,1	3.129	70
510, 431	6.88	100	743 910, 833	3.875	40			
440	6.21	40				970, 11,3,0	3.078	90
600, 442	5.86	50	930, 851,	3.694	20		2.983	80
611, 532	5.70	50	754 844	3.582	80		2.851	20
541	5.42	30	862, 10,2,0	3.440	10		2.789	20
710, 550, 543	4.96	50	952, 265, 10,3,1	3.346	80		2.725	60
							2.615	70
							2.574	10
721, 633, 552	4.78	90					2.520	20
							2.484	10
							2.448	10
							2.046	30

Table 3.30 Phillipsite

Structure Group:	1
Other Designation:	Wellsite
Typical Unit Cell Contents:	$(K,Na)_{10}[(AlO_2)_{10}(SiO_2)_{22}] \cdot 20\,H_2O$
Variations:	$Si/Al = 1.3 - 2.2$; also some Ba, Ca in Wellsite, H_2O to 24; deep-sea phillipsite $Si/(Al + Fe) = 2.4 - 2.8$; sedimentary lacustrine $Si/(Al + Fe) \geq 3.0$
Occurrence:	Ireland, Italy, Sicily; marine sediments in Pacific Ocean; sedimentary in western U.S. (Oregon, Nevada, California)
System:	Orthorhombic, $a = 9.96$, $b = 14.25$, $c = 14.25$
Twinning:	Cruciform penetration
Cleavage:	[010], [100]
Density:	2.15 - 2.19
Hardness:	4 - 4½
Optical Properties:	Biaxial (+), $\alpha = 1.483 - 1.505$, $\beta = 1.484 - 1.511$, $\gamma = 1.486 - 1.514$; birefringence, $\delta = 0.003 - 0.010$, $2V = 60 - 80°$; lacustrine phillipsite, index = 1.44 - 1.48
Reference:	3, 4, 6, 38, 64, 85, 100 - 102

X-Ray Powder Data (64)

hk$l$ [a]	d(A)	I	hk$l$	d(A)	I	hk$l$	d(A)	I
101	8.19	5	024,	3.19	85	343,	2.160	4
020,	7.19	100	042,			305		
002			123			315,	2.136	3
012	6.41	12	311	3.14	34	351		
121	5.37	10	321	2.930	14	262,	2.053	6
022	5.06	25	240,	2.893	6	226		
200	4.98	17	204			344	2.001	3
210	4.69	3	034	2.857	4	064,	1.981	8
103	4.31	10	143,	2.754	21	046,		
131,	4.13	40	105			155,		
113			151,	2.698	34	117,		
202,	4.07	13	115			171		
220			331,	2.667	9	424,	1.964	7
032	3.96	6	313			442		
123	3.70	3	125	2.577	6	361	1.910	3
222	3.54	6	323	2.542	8	264,	1.834	7
014	3.47	6	341	2.389	8	246		
141	3.26	30		2.309	2	452	1.810	3
			062,	2.259	4			
			026					

[a] Based on $a = 9.96$, $b = 14.23$, $c = 14.23$.

Table 3.31 Scolecite

Structure Group:	5
Typical Unit Cell Contents:	$Ca_8[(AlO_2)_{16}(SiO_2)_{24}]\cdot 24\,H_2O$
Variations:	Si/Al constant
Occurrence:	Iceland, Colorado
System:	Monoclinic, a = 9.848, b = 19.978, c = 6.522 $\beta = 110°6'$
Habit:	Prismatic twins, radiating fibers
Twinning:	Common [100]
Cleavage:	[110], [100]
Density:	2.27 - 2.243
Hardness:	5
Optical Properties:	Biaxial (−), α = 1.507 - 1.513, β = 1.516 - 1.520, γ = 1.517 - 1.521; birefringence, δ = −0.007, 2V = 36 - 56°
Reference:	3, 4, 57, 91, 92, 103, 104

X-Ray Powder Data (103)

hkl	d(A)	I	hkl	d(A)	I	hkl	d(A)	I
120	6.590	90	33$\bar{1}$	2.882	100	26$\bar{2}$,		
11$\bar{1}$,	5.848	100	231	2.851	70	360,	2.204	40
011			14$\bar{2}$	2.684	<5	062		
040	4.722	60	260	2.608	<5	42$\bar{2}$	2.191	<5
200	4.608	50	32$\bar{2}$,	2.578	15	222	2.169	5
21$\bar{1}$	4.387	90	042			21$\bar{3}$,	2.143	<5
140	4.208	30	122	2.549	<5	11$\bar{3}$		
220	4.144	20	071	2.474	15	280	2.106	<5
131	3.633	20	41$\bar{1}$	2.440	10	440,	2.076	5
240	3.304	<5	310	2.416	15	37$\bar{1}$		
051	3.221	30	080,	2.366	<5	133	2.039	5
31$\bar{1}$	3.181	30	17$\bar{1}$,			162,	2.030	5
211	3.151	30	400,	2.315	10	013		
20$\bar{2}$,	3.078	20	142			19$\bar{1}$,	1.991	5
122			180,	2.291	10	091		
002	3.071	10	43$\bar{1}$			51$\bar{1}$,	1.954	10
160	2.987	10	16$\bar{2}$	2.267	10	33$\bar{3}$		
22$\bar{2}$	2.929	60	40$\bar{2}$,	2.248	10	411,	1.943	<5
320			420			033		
251	2.900	10						

TABLES OF MINERAL ZEOLITE DATA 237

Table 3.32 Stilbite[a]

Structure Group:	7
Other Designation:	Foresite, Epidesmine, Desmine
Typical Unit Cell Contents:	$Na_2Ca_4[(AlO_2)_{10}(SiO_2)_{26}] \cdot 28\ H_2O$
Variations:	Si/Al = 2.47 - 3.85; also Na, K, sodium-rich variety Na/Ca = 1.3
Occurrence:	Amygdales; Iceland, Scotland, New Jersey, Nova Scotia, Japan
System:	Monoclinic, a = 13.64 b = 18.24 c = 11.27 β = 128°
Habit:	Plates, sheaflike aggregates
Twinning:	Common [001]; interpenetrant
Cleavage:	[010]
Density:	2.16
Hardness:	3½ - 4
Optical Properties:	Biaxial (−), α = 1.484 - 1.500, β = 1.492 - 1.507, γ = 1.494 - 1.513; birefringence, δ = 0.01, 2V = 30 - 49°
Reference:	3, 4, 57, 105, 106

X-Ray Powder Data (57)

hkl	d(A)	I	hkl	d(A)	I	hkl	d(A)	I
001, 020	9.04	100	$\bar{2}$43	2.882	6	380	1.899	6
$\bar{2}$21, 202	5.43	6	241	2.826	4	$\bar{4}$06, 054	1.873	2
$\bar{1}$31	5.30	12	$\bar{3}$14, $\bar{3}$51	2.780	35	$\bar{6}$06	1.826	18
220, $\bar{2}$22	4.65	40	$\bar{4}$04	2.704	6	$\bar{3}$06	1.811	6
$\bar{3}$11, $\bar{3}$12	4.28	14	222, $\bar{4}$41	2.567	12	$\bar{4}$46	1.731	6
041	4.07	95	170	2.516	8	$\bar{5}$75, 204	1.667	4
$\bar{2}$03	3.73	14	071	2.483	6	$\bar{3}$94	1.640	4
$\bar{2}$42	3.49	10	$\bar{2}$63	2.356	10	045	1.626	2
$\bar{1}$13, $\bar{4}$02	3.40	20	360	2.317	4	$\bar{4}$66	1.591	8
$\bar{4}$03	3.19	18	072	2.270	2	$\bar{3}$37	1.555	8
$\bar{1}$52, 330	3.04	70	$\bar{5}$53	2.167	2		1.517	2
			024	2.127	4		1.497	2
			$\bar{1}$73	2.065	8		1.442	4
			$\bar{4}$64	2.029	10		1.412	2

[a] Stellerite is reported to be a Ca-rich stilbite and orthorhombic.

Table 3.33 Thomsonite

Structure Group:	5
Other Designation:	Faroeite, Echellite, Harringtonite, Uigite
Typical Unit Cell Contents:	$Na_4Ca_8[(AlO_2)_{20}(SiO_2)_{20}] \cdot 24\ H_2O$
Variations:	Mg and K; Si/Al = 1.0 - 1.1
Occurrence:	Scotland, Colorado, New Jersey, Faroe Islands, Ireland, Nova Scotia
System:	Orthorhombic, a = 13.07 b = 13.08 c = 13.18
Habit:	Prismatic, radiating aggregates
Twinning:	[110]
Cleavage:	[010], [100]
Density:	2.36
Hardness:	5½
Optical Properties:	Biaxial (+), α = 1.497 - 1.530, β = 1.513 - 1.533, γ = 1.518 - 1.544; birefringence, δ = 0.006 - 0.015, 2V = 42 - 75°
Reference:	3, 4, 57, 76, 77

X-Ray Powder Data (57)

hk*l*	d(A)	I	hk*l*	d(A)	I	hk*l*	d(A)	I
101, 011	9.30	10	223	3.19	45	442, 060	2.18	40
020	6.60	60	204, 024	2.95	70	611, 532	2.12	6
102	5.90	40	142, 412	2.86	100	026, 206	2.09	6
112, 121	5.37	6	332	2.79	10	405	2.06	8
220, 202	4.64	90	242, 422	2.68	80	415, 154	2.03	6
212, 122	4.38	30	314, 134	2.58	25	524	1.956	6
301	4.13	30	423, 243	2.44	18	263	1.876	8
222	3.80	8				064	1.822	10
132, 312	3.51	65	522, 441	2.28	6	640, 460	1.813	20
040, 400	3.27	8	053, 343	2.25	20	721, 712	1.779	6

Table 3.34 Viseite

Structure Group:	1
Typical Unit Cell Contents:	$Na_2Ca_{10}[(AlO_2)_{20}(SiO_2)_6(PO_2)_{10}(H_3O_2)_{12}] \cdot 16\ H_2O$
System:	Cubic, a = 13.65
Density:	2.2
Optical Properties:	Isotropic, $n = 1.53$
Reference:	107

X-Ray Powder Data (107)

hkl	d(A)	I	hkl	d(A)	I	hkl	d(A)	I
211	5.68	40	541	2.11	<10		1.323	10
220	4.98	10	631	2.014	<10		1.282	10
400	3.46	50	640	1.886	30		1.196	20
332	2.92	100	651,	1.740	60		1.155	20
532,	2.20	20	732				1.138	10
611				1.380	10		1.105	10

Table 3.35 Wairakite

Structure Group:	1
Typical Unit Cell Contents:	$Ca_8[(AlO_2)_{16}(SiO_2)_{32}] \cdot 16\ H_2O$
Occurrence:	New Zealand, California, Washington; sedimentary
System:	Monoclinic, pseudocubic, a = 13.69, b = 13.68 c = 13.56, β = 90.5°
Habit:	Microscopic granules
Twinning:	[110]
Density:	2.265
Hardness:	5½ - 6
Optical Properties:	Biaxial (-), α = 1.498, γ = 1.502; birefringence, δ = 0.004, 2Vγ = 70 - 105°
Reference:	9, 40

X-Ray Powder Data (40)

hk$l$ [a]	d(A)	I	hk$l$	d(A)	I	hk$l$	d(A)	I
(200)	6.85	40	440	2.418	30	633,	1.857	30
211	5.57	80	(530),	2.35	<10B	721,		
220	4.84	40	(433)			552		
321	3.64	30	(600),	2.26-2.28	10B	–	1.844	10
400	3.42	60	(422)			642	1.822	<10B
	3.39	100	611,	2.215	40	732,	1.722-1.732	40B
(411),	3.21	<10B	532			651		
(330)			620	2.17	<10	800	1.708	<10
420	3.04-3.06	10B		2.147	10		1.696	<10
322	2.909	50	541	2.115	10	741	1.680	20B
	2.897	30		2.095	<10	820,	1.66	<10B
422	2.783	10	631	1.996	20	(644)		
	2.770	10	543	1.93	<10B	822,	1.612	10B
431	2.680	40	640	1.886-1.895	30B	660		
(510)	2.67	10	–	1.867	10	831	1.595	<10
521	2.50	<10					1.586	20
	2.489	40						

[a] Indices in parentheses are based on pseudocubic unit cell.
B = broad

Table 3.36 Yugawaralite

Structure Group:	1
Typical Unit Cell Contents:	$Ca_4[(AlO_2)_8(SiO_2)_{20}] \cdot 16\ H_2O$
Occurrence:	Yugawara Hot Springs, Japan; Cherra Hot Springs, Alaska
System:	Monoclinic, a = 6.73 b = 13.95 c = 10.03 $\beta = 111°31'$
Habit:	Tabular, lathlike, parallel to [010]
Twinning:	None
Cleavage:	[010]
Density:	2.20
Hardness:	5
Optical Properties:	Biaxial (+), $\alpha = 1.496$, $\beta = 1.497$, $\gamma = 1.504$, 2V = 78°; nearly insoluble in hot HCl, HNO_3
Reference:	30, 85, 108 - 110

X-Ray Powder Data (85)

hkl	d(A)	I	hkl	d(A)	I	hkl	d(A)	I
011	7.79	20	12$\bar{2}$	3.87	5	22$\bar{2}$	2.907	60
020	6.99	60	121	3.78	25	220	2.864	10
100	6.26	25	13$\bar{1}$,	3.75	5	14$\bar{1}$	2.763	30
11$\bar{1}$	5.82	90	130			21$\bar{3}$	2.720	35
021	5.62	5	032	3.30	5	13$\bar{3}$	2.706	5
002	4.68	85	041	3.27	5	051	2.680	25
10$\bar{2}$	4.65	85	131	3.235	55	13$\underline{2}$	2.650	20
01$\bar{2}$	4.45	10	20$\bar{2}$	3.198	10	232	2.638	15
11$\bar{2}$	4.41	30	200	3.135	20	230	2.603	5
111	4.30	65	140	3.056	100	22$\bar{3}$	2.578	15
031	4.18	30	12$\bar{3}$	2.997	10	15$\bar{1}$	2.562	5
022	3.89	15	122	2.937	30	10$\bar{4}$, 221	2.513	15

REFERENCES

1. M. Fleischer, *Amer. Mineral.*, **51**: 1247 (1966).
2. D. McConnell, *Amer. Mineral.*, **43**: 260 (1948).
3. E. S. Dana, System of Mineralogy, 6th Ed., John Wiley & Sons, New York, 1942, p. 570.
4. W. A. Deer, R. A. Howie, and J. Zussman, Rock-Forming Minerals, Vol. 4, Framework Silicates, John Wiley & Sons, New York, 1963, p. 338.
5. R. L. Hay, "Zeolites and Zeolitic Reactions in Sedimentary Rocks," *Spec. Paper No. 85*, Geological Society of America, New York, 1966.
6. R. A. Sheppard, Molecular Sieve Zeolites, *Advan. Chem. Ser.*, *101*, American Chemical Society, Washington, D. C., 1971.
7. W. S. Fyfe, F. J. Turner, and J. Verhoogen, *Geol. Soc. Amer. Mem.*, **73**: 167 (1958).
8. C. S. Ross, *Amer. Mineral.*, **26**: 627 (1931).
9. A. Steiner, *Mineral. Mag.*, **30**: 691 (1955).
10. J. Murray and A. F. Renard, *Deep Sea Deposits; Scientific Results of the Voyage of H. M. S. Challenger, 1873-1876*, Eyre and Spottiswood, London, 1891.
11. E. Bonatti, *Trans. New York Acad. Sci.*, **75**: 938 (1963).
12. D. S. Coombs, A. J. Ellis, W. S. Fyfe, and A. M. Taylor, *Geochim. Cosmochim. Acta*, **17**: 53 (1959).
13. É. Zen, *Amer. J. Sci.*, **259**: 401 (1961).
14. D. S. Coombs, *Australian J. Sci.*, **24**: 203 (1961).
15. G. Klein, M. A. Thesis, Univ. of Kansas, 1957.
16. D. W. Breck, *J. Chem. Educ.*, **41**: 678 (1964).
17. G. P. L. Walker, *Mineral. Mag.*, **32**: 503 (1960).
18. T. L. Walker and A. L. Parsons, *Univ. of Toronto Studies*, Geol. Ser., **14**: 5 (1922).
19. M. N. Bramlette and E. Posnjak, *Amer. Mineral.*, **18**: 167 (1933).
20. P. G. Harris and G. W. Brindley, *Amer. Mineral.*, **39**: 819 (1954).
21. L. L. Ames, Jr., L. B. Sand, and S. S. Goldich, *Econ. Geol.*, **53**: 22 (1958).
22. V. S. Vasil'ev, *Dokl. Akad. Nauk. SSR*, **95**: 149 (1954).
23. T. Sudo, T. Nisiiyama, and K. Chin, *Chishitsugaku Zasshi*, **69**: 1 (1963).
24. K. S. Deffeyes, *J. Sediment. Petrol.*, **29**: 602 (1959).
25. J. Regnier, *Bull. Geol. Soc. Amer.*, **71**: 1189 (1960).
26. E. D. Goldberg and G. Arrhenius, *Geochim. Cosmochim. Acta*, **13**: 153 (1958).
27. J. C. Hathaway and P. L. Sachs, *Amer. Mineral.*, **50**: 852 (1965).
28. G. Arrhenius and E. Bonatti, *Progress in Oceanography*, **3**: 7 (1965).
29. A. Steiner, *Econ. Geol.*, **48**: 1 (1953).
30. K. Sakurai and A. Hayashi, *Sci. Rep. Yokohama, Nat. Univ.*, Sect. II, **1**: 69 (1952).
31. L. W. Staples, *Amer. Mineral.*, **50**: 1796 (1965).
32. A. J. Ellis, *Geochim. Cosmochim. Acta*, **19**: 145 (1960).
33. T. Sudo and M. Matsuoka, *Geochim. Cosmochim. Acta*, **17**: 1 (1959).
34. R. L. Hay, *Publ. Geol. Sci.*, **42**: 199 (1963).
35. R. H. Mariner and R. C. Surdam, *Science*, **170**: 977 (1970).
36. F. Helfferich, *Amer. Mineral.*, **49**: 1752 (1964).
37. J. V. Howell, Chairman, Glossary of Geology and Related Sciences, 2nd Ed. American Geological Institute, Washington, 1960.
38. E. S. Larsen and H. Berman, "Microscopic Determination of the Non-Opaque Minerals," *Geol. Survey Bull.*, 2nd Ed., 848 (1964).
39. P. Saha, *Amer. Mineral.*, **44**: 300 (1959).

40. D. S. Coombs, *Mineral. Mag.*, **30**: 699 (1955).
41. C. S. Hurlburt, Jr., *Amer. Mineral.*, **42**: 792 (1957); **43**: 768 (1958).
42. W. C. Phinney and D. B. Stewart, *U. S. Geol. Survey Prof. Paper*, **424-D**: 353 (1961).
43. P. B. Leavens, C. S. Hurlburt, Jr., and J. A. Nelen, *Amer. Mineral.*, **53**: 1202 (1963).
44. H. Strunz and C. Tennyson, *Neues Jahrb. Mineral. Monat.*, **11** (1956).
45. H. Strunz, *Neues Jahrb. Mineral. Monatsh.*, **11**: 250 (1956).
46. A. J. Gude and R. A. Sheppard, *Amer. Mineral.*, **51**: 909 (1966).
47. E. Passaglia, *Amer. Mineral.*, **55**: 1278 (1970).
48. A. Iijima and K. Harada, *Amer. Mineral.*, **54**: 182 (1969).
49. F. A. Mumpton, *Amer. Mineral.*, **45**: 351 (1960).
50. B. Mason and L. B. Sand, *Amer. Mineral.*, **45**: 341 (1960).
51. W. T. Schaller, *Amer. Mineral.*, **17**: 128 (1932).
52. W. S. Wise, W. S. Nokleberg, and M. Kokinos, *Amer. Mineral.*, **54**: 887 (1969).
53. R. A. Sheppard, *U. S. Geol. Surv. Bull.*, **No. 1332-B** (1971).
54. H. Berman, *Amer. Mineral.*, **10**: 421 (1925).
55. S. Bonatti, *Atti Soc. Toscana Sci. Nat.*, **50**: 14 (1942).
56. E. Galli, *Periodico Mineral.*, **34**: 129 (1965).
57. Powder Diffraction File, Joint Committee on Powder Diffraction Standards, 1972.
58. M. H. Hey, *Mineral. Mag.*, **23**: 483 (1934).
59. D. A. Buckner, Ph.D. Thesis, Univ. of Utah, 1958.
60. R. A. Sheppard and A. J. Gude, III, *Amer. Mineral.*, **54**: 875 (1969).
61. L. W. Staples and J. A. Gard, *Mineral. Mag.*, **32**: 261 (1959).
62. W. B. Kamb and W. C. Oke, *Amer. Mineral.*, **45**: 79 (1960).
63. K. S. Deffeyes, *Amer. Mineral.*, **44**: 501 (1959).
64. K. Harada, S. Iwamoto, and K. Kihara, *Amer. Mineral.*, **52**: 1785 (1967).
65. I. A. Belitskii and G. V. Bukin, *Dokl. Acad. Nauk SSR*, **178**: 169 (1969).
66. D. W. Breck and E. M. Flanigen, Molecular Sieves, Society of Chemical Industry, London, 1968, p. 47.
67. L. W. Staples, *Amer. Mineral.*, **40**: 1095 (1955).
68. R. P. D. Graham, *Trans. Roy. Soc.*, **12**: 185 (1918).
69. L. B. Sand and A. J. Regis, *Abst. Geol. Soc. Amer. Meeting*, 1966; A. J. Regis, *Abst. Geol. Soc. Amer. Meeting*, 1970.
70. A. Alietti, E. Passglia, and G. Scaini, *Amer. Mineral.*, **52**: 1562 (1967).
71. G. P. L. Walker, *Mineral. Mag.*, **33**: 173 (1962).
72. R. M. Barrer, F. W. Bultitude, and I. S. Kerr, *J. Chem. Soc.*, 1521 (1959).
73. G. D. Feoktistov. Z. F. Ushchapovskaya, and T. A. Lakhno, *Dokl. Akad. Nauk SSR*, **188**: 670 (1969).
74. G. P. L. Walker, *Mineral. Mag.*, **33**: 187 (1962).
75. G. P. L. Walker, *Mineral. Mag.*, **32**: 202 (1960).
76. M. H. Hey, *Mineral. Mag.*, **23**: 51 (1932).
77. H. Meixner, M. H. Hey, and A. A. Moss, *Mineral. Mag.*, **31**: 265 (1956).
78. M. D. Foster, *U. S. Geol. Survey Prof. Paper*, **504E** (1965).
79. C. D. Waterson, *Mineral. Mag.*, **30**: 136 (1953).
80. T. G. Sahama and M. Lehtinen, *Mineral. Mag.*, **36**: 444 (1967).
81. P. M. Black, *Mineral. Mag.*, **37**: 453 (1969).
82. B. Mason, *Amer. Mineral.*, **47**: 985 (1962).
83. A. J. Regis and L. B. Sand, Proc. Clay Miner. Conf., Pittsburgh, Pa., 1966; S. W. Bailey, (Ed.), Pergamon, N. Y., 1967, p. 193.
84. W. S. Wise, *Mineral. Mag.*, **64**: 277 (1967).

85. I. Y. Borg and D. K. Smith, "Calculated X-ray Powder Patterns for Silicate Minerals," *Geo. Soc. Amer.*, 122 (1969).
86. A. B. Merkle and M. Slaughter, *Amer. Mineral.*, **53**: 1120 (1968).
87. D. McConnell, *Mineral Mag.*, **33**: 799 (1964).
88. D. S. Coombs, *Amer. Mineral.*, **37**: 812 (1952).
89. M. E. Kaley and R. F. Hanson, *Amer. Mineral.*, **40**: 923 (1955).
90. G. P. L. Walker, *Mineral Mag.*, **29**: 773 (1951).
91. C. J. Peng, *Amer. Mineral.*, **40**: 834 (1955).
92. M. D. Foster, *U. S. Geol. Survey Prof. Paper*, **504D** (1965).
93. B. Stringham, *Amer. Mineral.*, **39**: 819 (1954).
94. A. Reay and D. S. Coombs, *Mineral Mag.*, **38**: 383 (1971).
95. M. H. Hey and F. A. Bannister, *Mineral Mag.*, **23**: 243(1932).
96. A. Pabst, *Amer. Mineral.*, **56**: 560 (1971).
97. M. H. Hey and E. E. Fejer. *Mineral Mag.*, **33**: 66 (1962).
98. J. M. Bennett and J. A. Gard, *Nature*, **214**: 1005 (1967).
99. E. K. Gordon, S. Samson, and B. Kamb, *Science*, **154**: 1004 (1966).
100. R. A. Sheppard, A. J. Gude III, and J. J. Griffin, *Amer. Mineral.*, **55**: 2053 (1970).
101. W. C. Beard, Molecular Sieve Zeolites, *Advan. Chem. Ser. 101*, American Chemical Society, Washington, D.C., 1971, p. 237.
102. E. Galli and A. G. L. Ghittoni, *Amer. Mineral.*, **57**: 1125 (1972).
103. G. W. Smith and R. Walls, *Mineral Mag.*, **38**: 72 (1971).
104. M. H. Hey, *Mineral. Mag.*, **24**: 227 (1936).
105. J. Sekanina and J. Wyart, *Mineral. Cristallogr.*, **59**: 377 (1936).
106. K. Harada and K. Tomita, *Amer. Mineral.*, **52**: 1438 (1967).
107. D. McConnell, *Amer. Mineral.*, **37**: 609 (1952).
108. R. M. Barrer and D. J. Marshall, *Amer. Mineral.*, **50**: 484 (1965).
109. K. Harada and K. I. Sakurai, *Amer. Mineral.*, **54**: 306 (1969).
110. G. D. Eberlein, R. C. Erd, F. Weber, and L. B. Beatty, *Amer. Mineral.*, **56**: 1699 (1971).
111. J. R. Boles, *Amer. Mineral.*, **57**: 1463 (1972).

Chapter Four

THE SYNTHETIC ZEOLITES

Zeolites are hydrated aluminosilicates, formed under hydrothermal conditions. The term hydrothermal is used in a broad sense and includes the crystallization of zeolites from aqueous systems which contain the necessary chemical components. Attempts to synthesize silicates under hydrothermal conditions began with the experiments by Schafhautle, who, in 1845, reported the preparation of quartz by heating a "gel" silica with water in an autoclave. In 1862, St. Claire Deville reported the synthesis of "levynite" by heating aqueous solutions of potassium silicate and sodium aluminate in glass tubes at 170°C. The synthesis of "analcime" was reported in 1882 by de Schulten (1). Additional syntheses of several "zeolites" were reported in the succeeding years; however, few of these reports can be substantiated due to the lack of essential data for identification.

Over a period encompassing many decades, geologists and mineralogists have been extremely interested in reproducing the formation of minerals in the laboratory by methods which were believed to simulate natural processes. In addition, early investigators were interested in the interpretation of the conditions for mineral formation as they were related to geological processes.

A. METHODS OF ZEOLITE SYNTHESIS

Early hydrothermal investigations were confined to temperatures above 200°C and correspondingly elevated pressures. Reaction mixtures were composed of the various components in amounts corresponding to the composition of the desired product. The mixtures were maintained for a period of time at a constant pressure and temperature in the presence of large excesses of water. The water itself was not usually considered a reactant. Although some degree of success may have been achieved in these early experiments, positive identification of the products was not reported. These early experiments were not reproducible by later investigators.

Results of mineral synthesis experiments from multicomponent systems were interpreted in terms of phase equilibrium diagrams (2). This approach has not been meaningful in relating the formation of zeolites to the compositional and environmental factors such as temperature and pressure (3). Most of the synthetic zeolites are produced under nonequilibrium conditions and are considered in a thermodynamic sense as metastable phases. In the four-component system Na_2O-Al_2O_3-SiO_2-H_2O for example, at 1000 atmospheres and excess water, synthetic phases related to albite, analcime, mordenite, hydroxycancrinite, natrolite, nepheline hydrate, hydroxysodalite, and montmorillonite are formed at temperatures ranging from 290°C to 700°C. (2). In the same system, however, from identical or similar chemical compositions, widely different zeolite species may be formed as metastable phases in closed systems at temperatures lower than 200°C. The nature of the starting materials, factors affecting nucleation, and reaction time are important in determining the zeolite species produced from reaction mixtures at low temperatures.

Low Temperature Synthesis

A new approach to zeolite synthesis was initiated by R. M. Milton and associates in the Union Carbide Corporation. This approach was based on starting with very reactive components in closed systems and employing temperatures for crystallization which are more typical of the synthesis of organic compounds than they are of mineral formation (4, 5). Many of the synthetic zeolites are formed at temperatures ranging from about room temperature to the boiling point of water. Extensive empirical data on the relationship between composition, temperature, types of reactants and other factors in the synthesis of zeolite species have been accumulated.

Early attempts to synthesize zeolites were based upon then current ideas concerning their probable mode of formation in basaltic rocks. Recent information on the occurrence and formation of mineral zeolites in sediments at relatively low temperatures and their formation in marine sediments on ocean bottoms is consistent with the relatively mild temperature conditions used in the laboratory synthesis. From an industrial point of view, this is important; the conditions of synthesis of the important zeolites for molecular sieve applications are well suited to manufacture on a large scale.

The synthesis of zeolites by a chemical approach rather than by a phase equilibrium approach is understood in general terms. Conditions for zeolite formation are consistent with various proposals such as those advanced by Fyfe, Goldsmith, and Coombs (6,7,8).

Simplexity Principle of Goldsmith

The crystallization of zeolites is consistent with the simplexity principle proposed by Goldsmith (7) which relates the ease of crystallization to structural "simplexity." Goldsmith defines high simplexity as being "synonomous with disorder, structural simplicity, or high entropy." The disordered form of an aluminosilicate is in a state of higher simplexity and higher entropy than its ordered counterpart. Many substances which can exist in the form of several polymorphs tend to crystallize initially as the high temperature modification in accordance with the rule of Ostwald.

The growth of crystalline aluminosilicates such as zeolites first requires the formation of a nucleus. In a system with high disorder, the principle favors the formation and development of a nucleus with the highest simplexity, which may be the nucleus of a crystal of a metastable phase.

In aluminosilicates, the aluminum ion may exist in either 4-fold or 6-fold coordination as contrasted to silicon, which is always 4-fold coordinated in aluminosilicates. In zeolites, the aluminum is 4-coordinated with oxygen and these structures have a higher ease of crystallization (higher simplexity) than aluminosilicates in which the aluminum ions are located in either an ordered structure or in 6-fold coordination. The nucleus of the disordered structure should be smaller than the corresponding nucleus of the ordered structure. The structurally disordered form is more likely to attain the critical size necessary for growth than is the nucleus for the ordered form. In order to form the complex zeolite structures, crystal nuclei must contain a relatively large number of atoms. In these systems, the nuclei of several metastable phases may develop rather than the nuclei of the ordered, more stable phase (9).

As pointed out in Chap. 2, the smallest structural secondary building units are probably the single 4- or 6-rings. More complex units such as double 4-rings, double 6-rings, or truncated tetrahedra may also be involved in the nucleation (Fig. 2.21). Zeolites contain a large number of atoms in the unit cell. Since the smallest nucleus for growth is smaller than the unit cell, the nucleus of the disordered structure becomes stable for growth at a smaller critical size than that of the ordered form. The formation of smaller nuclei by random fluctuations is favored.

Zeolite Metastability

A mineral zeolite which has existed over long periods of geological time and a synthetic zeolite with a related structure but which was synthesized rapidly in the laboratory may exhibit differences in properties due to the ordering that may occur in the mineral as opposed to the lack of ordering the synthetic structure shows. Various types of analcime have been synthesized which show different types of ordering. Similarly, synthetic types of chabazite show a high degree of disorder as opposed to the mineral. Structural studies of Al-Si ordering in zeolite minerals at present is limited to a few examples (10).

Because of the slow rates of reaction of many of the components, true equilibrium may never be attained. It is probable that many of the synthetic zeolites which do not have mineral relatives are nonequilibrium phases and do not exist when true equilibrium conditions prevail. Many of the known synthetic zeolites are not structurally related to a mineral. This may be due to the metastability of the synthetic zeolite and its ease of conversion to a more stable species. Synthetic relatives of many zeolite minerals and other aluminosilicates are not yet known, although a total of about 100 synthetic zeolites are reported.

In systems where aqueous bases are involved, hydroxyl bonding plays an important part; the hydroxyl can be bonded to the aluminum in preference to the silicon atoms, as is the case with clay minerals where the crystallization also occurs at quite low temperatures.

The crystallization of silicates from reactive gels or glasses can be discussed in terms of the free energy relations (Fig. 4.1). The initial reaction mixture is composed of components which first form phase a, which may convert to b or c in time. In the region where the temperature is such that a or b can both form, the nucleation and growth rates

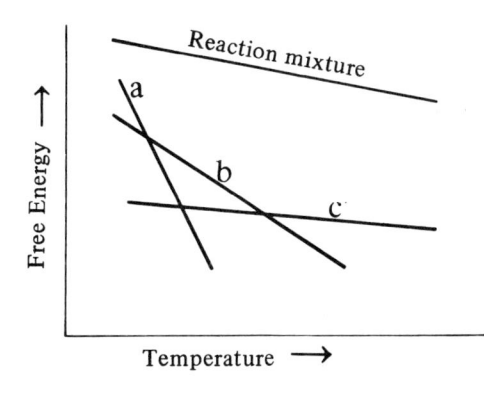

Figure 4.1 Schematic free energy relations between reaction mixtures and various zeolite phases, represented by a, b, c (6).

are important; the metastable phase *a* which preferentially forms and may transform to the more stable phase *b* and in succession to *c* (6).

Reaction Diagrams

The portrayal of equilibria between minerals in equilibrium diagrams is common throughout the earlier work on mineral synthesis and formation of the last few decades. These diagrams aid in interpreting the formation of minerals in nature, but are of less significance under nonequilibrium conditions. The chemist is interested in the formation of synthetic aluminosilicates, such as the zeolites, by reproducible processes and by utilizing the mildest conditions possible. Extensive data in the literature deals with the synthesis of phases and the coexistence of phases; as a result many of the published results are in conflict and not reproducible due to the presence of nonequilibrium conditions.

It is very useful to relate the zeolite compositions with reaction mixture compositions by a diagram. We have used various types of diagrams based on both triangular and rectangular coordinates for this purpose.

Some General Conditions for Synthesis

The conditions generally used in synthesis are: (11, 12)

1. Reactive starting materials such as freshly coprecipitated gels, or amorphous solids.
2. Relatively high pH introduced in the form of an alkali metal hydroxide or other strong base
3. Low temperature hydrothermal conditions with concurrent low autogeneous pressure at saturated water vapor pressure

4. A high degree of supersaturation of the components of the gel leading to the nucleation of a large number of crystals

A gel is defined as a hydrous metal aluminosilicate which is prepared from either aqueous solutions, reactive solids, colloidal sols, or reactive aluminosilicates such as the residue structure of metakaolin (derived from kaolin clay by dehydroxylation) and glasses. The preparation and properties of amorphous silicate gels such as silica gel have been studied for many years. Silica gel has been defined as "a colloidal system of solid character in which the colloidal particles somehow constitute a coherent structure, the latter being interpenetrated by a (usually liquid) system consisting in kinetic units smaller than colloidal particles" (13).

The gels are crystallized in a closed hydrothermal system at temperatures varying generally from room temperature to about 175°C. In some cases, higher temperatures to 300°C are used. The pressure is generally the autogenous pressure approximately equivalent to the saturated vapor pressure (svp) of water at the temperature designated. The time required for crystallization varies from a few hours to several days. When prepared, the aluminosilicate gels differ greatly in appearance — from a stiff translucent nature to opaque gelatinous precipitates to heterogenous mixtures of an amorphous solid dispersed in an aqueous solution. The alkali metals form soluble hydroxides, aluminnates, and silicates. These materials are well suited for the preparation of homogeneous mixtures (11, 14).

The gel preparation and crystallization is represented schematically using the Na_2O-Al_2O_3-SiO_2-H_2O system as an example:

$$NaOH\,(aq) + NaAl(OH)_4\,(aq) + Na_2SiO_3\,(aq)$$
$$\downarrow T_1 \cong 25°C$$

$$[Na_a(AlO_2)_b(SiO_2)_c \cdot NaOH \cdot H_2O]\ \text{gel}$$

$$\downarrow T_2 \cong 25\ \text{to}\ 175°C$$

$$Na_x[(AlO_2)_x(SiO_2)_y] \cdot mH_2O + \text{solution}$$
zeolite crystals

The ease with which the zeolites crystallize is attributed to the high reactivity of the gel, the concentration of the alkali hydroxide, and the high surface activity due to the small particle size of the solid phases concerned. The gel is probably produced by the copolymerization of the individual silicate and aluminate species by a condensation-polymerization mechanism. The gel composition and structure appear to be controlled by the size and structure of the polymerizing species.

B. REVIEW OF EARLY WORK

The hydrothermal alteration and synthesis of silicates as disclosed in 156 papers published from 1845 to 1937 has been reviewed in an extensive paper by Morey and Ingerson (1). Those results encompass the last 100 years and concern the formation of synthetic types of zeolites (Table 4.1). Identification of the various products obtained was based upon chemical analysis or optics. Since in most instances the products are very small in particle size, it has been generally concluded that these results are very much in doubt. For example, Baur reported the synthesis of a potassium "faujasite" which was later shown to be the compound K_2SiF_6 (15). Until identification of polycrystalline materials by means of x-ray powder data became common, positive identification of the products as zeolites was not possible.

Synthesis of Analcime

Synthetic materials with a chemical composition corresponding to that of analcime have been claimed by many investigators (Table 4.1). A summary of these reports has been given by Barrer (16).

In 1911, Baur reported the results of a series of experiments concerned with the hydrothermal synthesis of silicates. Solutions of soluble compounds prepared to produce the necessary ratios of Na_2O, K_2O, Al_2O_3, SiO_2, and CaO were placed in a steel cylinder which was hermetically sealed and heated in a furnace for the desired length of time. The majority of Baur's experiments were conducted at or near 450°C and lasted 12 to 16 hours. The crystalline products obtained were identified by a microscopic examination of their optical characteristics and crystalline structure. From a mixture of SiO_2, $Na_2O \cdot Al_2O_3$, and $Al(OH)_3$, he claimed the preparation of analcime in the form of

Table 4.1 Selected Summary of Unsubstantiated Zeolite Syntheses (1)

Date	Zeolite	Hydrothermal Method	Investigator
1862	Levynite	K silicate + Na aluminate, 170°C	St. Claire Deville
1880	Analcime	Na silicate + Al_2O_3 glass, 180°C	A. de Schulten
1882	Analcime	Na silicate + Na aluminate, 180°C	A. de Schulten
1883	Analcime	SiO_2, NaOH solution, Al_2O_3, 400°C	C. Friedel, E. Sarasin
1885	Analcime	conversion of chabazite, 200°C	J. Lemberg
1887	Analcime	kaolin + Na silicate, 200 - 220°C	J. Lemberg
	Analcime	feldspars + Na_2CO_3, 200°C	
	Natrolite	scolecite + NaCl	
	Chabazite	feldspars + Na_2CO_3, 100°C	
1890	Chabazite	recrystallization, 150 - 170°C	C. Doelter
	Heulandite	anorthite + H_2O + CO_2, 200°C	
	Analcime	Na_2O + Al_2O_3 + SiO_2 + H_2O, 100 - 200°C	
	Scolecite	recrystallization	
1894	Natrolite	anorthite, 174 - 177°C	St. J. Thugutt
1896	Thomsonite	muscovite + NaOH, 200°C	G. Friedel
1906	Analcime	nepheline + Na_2CO_3 + H_2O, 200°C	C. Doelter
	Natrolite	Na_2CO_3 + Al_2O_3 + SiO_2, 90°C	C. Doelter
1911	K Faujasite[a]	K_2O, Al_2O_3, SiO_2, 350°C	E. Baur
1916	Analcime	adularia + $NaAlO_2$, 280°C	E. A. Stephenson
1918	Analcime	Na_2O, Al_2O_3, SiO_2, 300°C	W. J. Müller, J. Konigsberger
1927	Mordenite	feldspars plus carbonates at	R. J. Leonard
	Phillipsite	300 - 500°C in steam at low press.	
1929	Natrolite	paragonite + NaOH, 400°C	E. Gruner
1936	Analcime	Na silicate + Na aluminate, 282°C	F. G. Straub

[a]Shown to be hieratite, K_2SiF_6, (15)

icositetrahedra. From his results, Baur concluded that longer reaction times did not improve yields and results, variation in the temperature over the range 350° to 550°C did not produce a discernible difference, and zeolite synthesis was favored by reaction in an alkaline medium.

The deposition of analcime in steam boilers was reported (17). In 1936, Straub described the preparation of analcime by the hydrothermal treatment at 282°C for 46 hours of a mixture of a sodium silicate solution (70 cc contained 10.8 g of silica) and 8 g of sodium aluminate (93.5% pure) containing 4.6 g of alumina. Identification by x-ray powder pattern was not made (163).

The first synthesis of an analcime-type zeolite substantiated by x-ray diffraction analysis was by Barrer and White (18).

Attempts to Synthesize Chabazite

There are several early, unsubstantiated reports on the synthesis of a chabazite-type zeolite. By heating a potassium silicate and sodium aluminate solution in a sealed glass tube, a product was obtained by Deville in 1862 which he designated as levyne. Doelter claimed the preparation of chabazite in the form of rhombic crystals by the prolonged heating of silicic acid, aluminum hydroxide, calcium hydroxide, and carbonated water at 200°C in a sealed tube. The preparation of sodium chabazite by dissolving silicic acid and aluminum hydroxide in a sodium hydroxide solution was also mentioned. A claim for the preparation of chabazite by digesting a mixture of powdered anorthite, which is a lime-rich feldspar, freshly precipitated silicic acid, and carbonated water for 14 days at 200°C was also reported by J. Lemberg.

Barrer, in a discussion of methods of growing crystalline minerals which emphasized zeolite synthesis, questioned the reliability of the reported chabazite preparations. Attempts to reproduce the results failed; the materials synthesized in each case were definitely not chabazite (19).

Synthesis of Mordenite

In 1927, Leonard reported in his study of the hydrothermal reaction between feldspars and alkali carbonates that mordenite and thomsonite were formed in 7 days at 200°C and at 15 atm pressure (1). He also claimed to have prepared a phillipsite-type zeolite at 350°C from feldspars in the presence of fluorides and superheated water vapor at atmospheric pressure. Although Leonard was one of the first to use x-ray powder data to characterize products, the products were not unambiguously identified.

The first substantiated synthesis of a mordenite-type zeolite was made by Barrer (20). An aqueous sodium aluminate solution which contained sodium carbonate was stirred into an aqueous suspension of silicic acid gel containing some alkali. The mixture, which tended to be a gel, was evaporated to dryness at temperatures below 110°C, thereby producing, when certain proportions were used, a gel of chemical composition similar to that of mordenite. Crystallization was accomplished by heating the gel with water in an autoclave at temperatures up to 300°C. Although an alkaline medium is necessary to crystallize mordenite, in a strongly alkaline solution, the crystals which first precipitated tended to dissolve and convert to other unidentified species when left in contact with the mother liquor at high temperatures. At a pH of between 7 and 8 crystallization was retarded and the crystals which formed were fewer and smaller.

Attempts to Synthesize Faujasite

Baur reported the formation of potassium faujasite and stilbite by hydrothermal synthesis (1). Mixtures of silica and potassium aluminate, or silica, potassium aluminate and aluminum hydroxide gave a product reported to be potassium faujasite. The combination of soda-lime glass and sodium aluminate reportedly gave stilbite. However, Baur states that there is some question as to the certainty of the identification of potassium faujasite. The identification was made on the basis of the low refractive index of the crystal and its octahedral crystalline form. The identification of stilbite could not be made with certainty. Schlaepfer and Niggli later identified the faujasite product as hieratite, K_2SiF_6, which formed from the HF present in the silica gel (15).

Other Zeolite Syntheses

Other zeolites reportedly synthesized include scolecite, thomsonite, phillipsite and various unidentified species (1).

Konigsberger and Müller

Some of the experiments reported by Konigsberger and Müller in 1918 deserve interpretation (21). These investigators conducted a series of hydrothermal experiments at temperatures ranging from 100° to 440°C. As starting materials reactive components such as freshly precipitated aluminum hydroxide, potassium aluminate, and potassium silicate were used. The experiments were conducted in a platinum-iridium lined steel bomb. A summary of the results is given in Table 4.2. The reported synthesis of four zeolite species which they did not identify with known minerals, with the possible exception of analcime, cannot be substantiated due to the lack of x-ray powder data. In some instances, the refractive index, density and chemical composition correspond to a zeolite mineral. The letter designations used by Konigsberger and Müller in these experiments, A, X, Y and Z bear no relationship to the letter designations currently used to identify some well-known synthetic zeolites. The conditions of formation and the properties giv-

Table 4.2 Summary of Some Experiments by Konigsberger and Müller (21)

Exp. No.	Reaction Mixture Composition (moles)					Conditions			Products [b]
	K_2O	Na_2O	Al_2O_3	SiO_2	H_2O	Temp.(°C)	Pressure[a]	Add. Reactants	
80	0.65	0.22	1.0	3.7	136	300/12 hrs	1200	Na_2CO_3	Analcime, Nepheline
1	3.3	–	1.0	19	673	310/3 hrs	1400	CO_2	Zeolite "A", Leucite
52	0.76	0.51	1.0	3.7	194	100/48 hrs	15	$CaCl_2$, CO_2 $CaSO_4$	Zeolite "X", Analcime
53	0.76	0.51	1.0	3.7	194	100/180 hrs	15	same as 52	Zeolite "X"
67	5.7	0.76	1.0	15	265	340/12 hrs	2100	NaCl, CO_2	Zeolite "Y", Nepheline, Pyrophyllite
62	Labradorite feldspar								
	1.1 CaO	0.41	1.0	4.2	excess	400/24 hrs	4000 24% fill	NaCl, K_2CO_3	Zeolite "Z", others

[a] Pressure in psia was estimated from the data given
[b] Product Properties: Analcime, n = 1.465 – 1.478
Zeolite "A," broad needles, parallel extinction, d = 2.3 g/cc does not check with natrolite
Zeolite "X," weakly birefringent, soluble in HCl, n = 1.48 – 1.485, d = 2.17. Chemical analysis gave $(R_2O)_{2.5}(Al_2O_3)_2(SiO_2)_{5.5} \cdot 3\ H_2O$
Zeolite "Y," long needles, soluble in HCl, n = 1.478, d = 2.3 g/cc
Zeolite "Z," biaxial, n = 1.495 – 1.497, d = 2.35 g/cc, thick needles, soluble in HCl, chemical analysis gave composition $3.3\ (CaO + Na_2O + K_2O) \cdot Al_2O_3 \cdot 3.5\ SiO_2 \cdot 3.4\ H_2O$
The experiments 52 and 53 have been repeated. Chloride was added as $CaCl_2$, carbonate as Na_2CO_3 and sulfate as Na_2SO_4. The products were amorphous.

en for zeolite "A" do not correspond to the synthetic zeolite type A discussed in Chap. 3. The zeolites "X" and "Y" do not correspond to the synthetic X and Y zeolites. The conditions of formation and the properties of the product designated by Konigsberger and Müller as "X" are similar to those of the zeolite K-M synthesized by Barrer and Baynham (22).

Synthesis of Chabazite Substitutes

The discovery of a hydrothermal method to convert minerals such as analcime or leucite, or various aluminosilicate gels into zeolites by treatment at high temperature with concentrated solutions of barium chloride, barium bromide or similar salts was reported by Barrer (23, 24). Aluminosilicate gels prepared from sodium silicate and sodium aluminate may be used in this process if the molar ratio of SiO_2/Al_2O_3 is 3 to 5.

> In a typical preparation; the mineral or gel was powdered and mixed well with 0.5 to 1.5 times its volume of solid barium chloride or barium bromide. Water was added in amounts 0.5 - 1.0 of the total solid volume. The mixture was heated in an autoclave at temperatures of 180° to 220°C, preferably for periods of 2 to 6 days. The product was subjected to a series of extractions with water at a temperature between 180 to 270°C. Upon dehydration, a product was obtained that had gas adsorption properties and molecular sieve characteristics similar to those of chabazite.

After a study of the use of various salts as mineralizers, Barrer concluded that their action depended specifically upon the character of both cation and anion. Thus $BaCl_2$, $BaBr_2$, KCl, and KBr were effective; NaCl, $SrCl_2$, $CaCl_2$, KNO_3, KF, and $Ba(NO_3)_2$ were ineffective. Analysis revealed that the "mineralizer" had entered into the composition of the newly prepared products, almost certainly as an interstitial solution. X-ray powder photographs made of the products containing barium chloride or barium bromide, showed some shrinkage of the lattice spacings when the occluded $BaCl_2$ or $BaBr_2$ were removed by extraction. None of the products were like chabazite. (See Section H.)

The preparation of an adsorbent capable of resolving hydrocarbon mixtures was reported by Black (25). The process involves heating a mixture of hydrous aluminum oxide, hydrous silica and an oxide of a group II metal such as CaO with a group II metal salt solution (1 - 10% concentration) at a temperature of 220 - 400°C for 48 to 216 hours. The resulting product, dehydrated in a vacuum, was prepared in the form of a white powder and packed in agglomerates of moderate strength by drying in the form of a moist cake. A crystalline structure was indicated by the x-ray powder pattern.

> In a specific example, calcium hydroxide, alumina gel and silica gel (4:1:4) were mixed in a ball mill. The product was dried at 250°F and formed into granules. Two-hundred ml of this material and two liters of 1% calcium acetate were heated for 84 hr in a bomb at a pressure of 360 lb/sq in. After a distilled water wash and drying at 200°F, the product was dehydrated under vacuum at 500°C.

Variations in experimental conditions modified the results. Temperatures higher than 200°C increased the formation of the product. Salts of group II elements other than the acetates were used successfully. Increasing the concentration of the salt solution from 1 to 5% increased the adsorptive capacity of the product. A beneficial effect was also found for the longer synthesis times used. This material may have been a calcium form of analcime as reported later by Barrer and Denny (26).

Crystalline Permutite of Kurnakov

The preparation of a "crystalline" permutite was described by Kurnakov (27). The material of composition $Na_2O \cdot Al_2O_3 \cdot 2SiO_2 \cdot H_2O$ was formed by mixing dilute solutions of sodium aluminate and sodium silicate at 100°C. The x-ray diffraction pattern indicates crystallinity. In a later paper, Svesknikova and Kurnakov repeated the preparations and tabulated the x-ray powder data. The product was identified as hydroxysodalite (28, 29).

C. FORMATION OF ZEOLITES FROM ALUMINOSILICATE GELS OF ALKALI METALS

Extensive investigations of the formation of zeolites in systems based upon alkali hydroxide-hydrous aluminosilicate gel mixtures have been made (11, 12). Although some of these studies were conducted under conditions which may be near equilibrium, the majority were carried out under nonequilibrium conditions; this approach is more conducive to the formation of the metastable, "stranded" zeolite phases. The following discussion of the synthesis of zeolites from the alkali hydroxide aluminosilicate gels is arranged according to the metal hydroxides used.

Lithium Aluminosilicate Zeolites

One zeolite mineral, bikitaite, is known in which the predominant exchangeable cation is lithium. A synthetic analog of bikitaite has been reported (30). The synthesis of zeolites from compositions in the system Li_2O-Al_2O_3-SiO_2-H_2O was first reported by Barrer and White (18).

> A series of gels Li_2O, Al_2O_3, $nSiO_2$, mH_2O, in which n was varied between 1 and 10, were prepared by mixing as reactants LiOH, freshly prepared hydrous aluminum oxide, $Al_2O_3 \cdot 3H_2O$, and silica gel containing 50 wt% SiO_2 in the desired portions and then evaporating the mixtures to dryness at 120°C. (It was later found that this evaporation diminishes the subsequent reactivity of the gel in terms of its ease of crystallization to zeolite phases.) Portions of the dried gel were mixed with H_2O or LiOH solution in stainless steel autoclaves and heated to temperatures up to 300°C. Crystallization periods ranged from 3 to 4 days for temperatures below 150°C to 36 to 60 hrs at higher temperatures.

In addition to several nonzeolite phases, two new zeolite types were identified by means of their x-ray powder data, optics, and chemical composition. These were designated as zeolite Li-A and zeolite Li-H. Typical synthesis conditions and properties are shown in Table 4.3. They were further characterized as new zeolite species on the basis of their ion exchange and gas adsorption properties. Li-A was formed best at 250°C from a composition of $Li_2O \cdot Al_2O_3 \cdot 2SiO_2$. When dehydrated at temperatures of 220°C, the zeolite was found to adsorb NH_3 at room temperature, but not N_2, thus indicating the presence of small intracrystalline channels. Li-H was formed at 220°C in the presence of an excess of LiOH from a composition of $Li_2O \cdot Al_2O_3 \cdot 8SiO_2$. After dehydration at 320°C, it adsorbed NH_3. The zeolite Li-A has been synthesized at 100°C from freshly prepared LiOH-Al_2O_3-SiO_2 gels (31). (Table 4.3)

Another synthetic lithium zeolite has been prepared by the reaction of LiOH with calcined kaolin (see Section F). The chemical composition is given as $Li_2O \cdot Al_2O_3 \cdot 2SiO_2 \cdot 4H_2O$. Fragmentary characterization

Table 4.3 Synthetic Zeolites Li$_2$O - Al$_2$O$_3$ - SiO$_2$ - H$_2$O System

Zeolite Type	Typical Reactant Comp. (moles/Al$_2$O$_3$) Li$_2$O	SiO$_2$	H$_2$O	Reactants	Conditions Temp. (°C) Time (hr)	Zeolite Comp. (moles/Al$_2$O$_3$) Li$_2$O	SiO$_2$	H$_2$O	Properties	X-Ray Table	Ref.
Li-A	1.0	2.0	xs	silica gel, Al$_2$O$_3$·3 H$_2$O, LiOH	250, 36 - 60 hr	1.0	2.0	4.0	Rods, 30µ long, α = 1.535, γ = 1.525	4.31	18
											30
Li-A	2.0	2.0	100	Same as above	100, 40 hr	1.0	2.0	3.0	—		31
Li-H	>1.0	8.0	xs	silica gel, Al$_2$O$_3$·3 H$_2$O, LiOH	220, 36 - 60 hr, pH = 11[a]	1.0	8.0	5.0	Rods, 30µ, n = 1.480	4.51	18
Li-Zeolite	2.0	8.0	8.5	silica gel, Al$_2$O$_3$·3 H$_2$O, LiOH	254, 625 psi, pH = 7.5	not analyzed, Clinoptilolite-like			—	4.34	33
"Bikitaite"	2.0	4.0	(c)	LiOH, Li$_2$CO$_3$, Li-silicate, Al(OH)$_3$	300 - 350, 2 kilobars	n.a.		—	—		30
ZSM-2	6.0	9.0	(d)	glass (e)	60, 30 days	1.0	3.3 - 4.0	—	—	4.99	34
"Analcime" 1	1	4	34	LiOH, Li$_2$CO$_3$, silica sol, Al(OH)$_3$	152, 67	n.a.			a$_0$ = 13.64-13.69, n = 1.483-1.488, size: 20-80µm	—	9
"Mordenite" 3	3	40	314	LiOH, Li$_2$CO$_3$, silica sol, Al(OH)$_3$	150, 65	contained "phillipsite"			n = 1.470, size: 5 - 25µm	4.67	9
"Phillipsite" 12	12	34.3	270	LiOH, Li$_2$CO$_3$, silica sol, Al(OH)$_3$	150, 370	contained "mordenite"			n = 1.504, size: acicular to 15µm	—	9

[a]Excess LiOH added, pH of final solution; [b]pH of final solution; [c]50 wt% of gel; [d]H$_2$O/glass = 5; [e]glass composition was 6 Li$_2$O·Al$_2$O$_3$·9 SiO$_2$.

and x-ray powder data indicate that the material is a zeolite (32).

A clinoptilolite-like zeolite was synthesized by Ames in the course of an investigation of the crystallization of gels prepared from lithium hydroxide, aluminum hydroxide and silica gel mixtures (33). The crystallization was conducted hydrothermally at temperatures ranging from 250 - 300°C at total pressures of 625 - 1150 psi for 35 days. Since formation of the zeolite phase requires the utilization of lithium introduced as LiOH, the reaction rate could be followed by measuring the pH. The final pH was 7.5 and the resulting zeolite was obtained in a reasonably high purity. Identification by means of x-ray powder data (Table 4.26) indicated a close structural relation to clinoptilolite. Typically clinoptilolite is sodium-rich and contains no lithium; it was rather unexpected that this zeolite would crystallize in the lithium form. Although Li-H reported by Barrer is similar in chemical composition, it does not appear from the x-ray powder data to be related to clinoptilolite.

A lithium zeolite, ZSM-2, is prepared by the reaction of a glass with water at 60°C for about a month. ZSM-2 is reported to be a large-pore zeolite (34).

Synthetic analogs of analcime, phillipsite and mordenite have been prepared in the temperature range 150 to 200°C at autogenous pressure (See Table 4.3). The best results were obtained using colloidal silica sol, aluminum hydroxide, and lithium carbonate as reactants. The analcime-type was prepared pure but the lithium "mordenite" and lithium "phillipsite" did not exceed 75%. The lithium "mordenite" evidenced adsorption characteristic of the small-port variety (9).

Sodium Aluminosilicate Zeolites

Most of the sodium zeolites crystallize from sodium aluminosilicate gels below 150°C. The rate of crystallization and the stability of the sodium zeolite phases are optimum in the vicinity of 100°C. Because of the effect that the various starting reactants and temperatures have upon the zeolite species which crystallize, the sodium zeolites are discussed in terms of the reacting species and synthesis temperatures.

Sodium Zeolites Formed from Aluminosilicate Gels Above 200°C

Based on the phase rule, in the system Na_2O-Al_2O_3-SiO_2-H_2O, two tectosilicates may coexist at equilibrium with the aqueous solution at a given temperature and pressure. In the presence of silica as a separate phase, only one zeolite can theoretically coexist; however, two zeolites have been formed in this system at temperatures above 200°C (8). These are synthetic types of analcime and mordenite. Of these, analcime is the more stable and was one of the first zeolites reported to be synthesized by early investigators. A sodium form of mordenite was first synthesized by Barrer in 1948 (20). Reactant mixtures were prepared by mixing a sodium aluminate solution, silicic acid, and water and drying. This zeolite is designated Na-D.

Barrer and White investigated the hydrothermal crystallization of aluminosilicate gels of composition Na_2O-Al_2O_3-$nSiO_2$-H_2O (35). When n varied from 2 to over 12, quartz and synthetic types of mordenite, Na-D, and analcime, Na-B, were obtained. Identification was made by x-ray powder data. Quartz appeared where n was large, increasing in amount as n increased above 8. The mordenite-type began to appear where n was about 8, diminished as n approached 12, and was produced in highest yields where n had an intermediate value. The analcime-type Na-B, crystallized in high yields for $n = 4$ or 5, but the yield rapidly decreased for $n = 4$ and slowly for $n = 5$. Some Na-B was obtained even in the range where Na-D was the principal species. The results were affected by temperature. Na-B crystallized below 200°C and was still in evidence at 310°; Na-D crystallized best over the narrower temperature range of 265° - 295°C. They did not obtain a natrolite as reported by Thugutt (1). The use of mineralizers such as $BaCl_2$ and $BaBr_2$ shortened the time required for crystal growth. Another decisive factor observed, was the pH of the solution. The Na-D zeolite formed best at a pH of 8 to 10 at temperatures between 265° and 295°C.

The analcime-type was commonly obtained in the form of icositetrahedra; but in the presence of a high concentration of NaF, it was obtained as cubes possessing hemihedral faces (19,20).

Although large crystals of the mineral analcime have been found, synthetic analcime-type zeolite crystals have not exceeded 0.1 mm in diameter. The difficulty in growing large crystals is in sustaining crystal growth after the initial formation of small crystals. When the small synthetic zeolite crystals which initially formed in two days were transferred to a similar mother liquor and heated to the temperature of the first growth for 15 days, there was no observable change in size (19). Figure

Table 4.4 Synthetic Mordenite, Analcime, Natrolite Type Zeolites Na$_2$O - Al$_2$O$_3$ - SiO$_2$ - H$_2$O System

Zeolite Type	Typical Reactant Comp. (moles/Al$_2$O$_3$) Na$_2$O	SiO$_2$	H$_2$O	Reactants	Conditions Temp. (°C) Time	Zeolite Comp. (moles/Al$_2$O$_3$) Na$_2$O	SiO$_2$	H$_2$O	Properties	X-Ray Table	Ref.
Na-D (mordenite)	1	8.2-12.3	xs	NaAlO$_2$, silicic acid	265 - 295, 2 - 3 days, pH = 8 - 10	1	10	6.7[a]	laths, 25µ, n = 1.467	4.37	20, 35, 37
Large Port (mordenite)	6.3	27	61	NaAlO$_2$, diatomite, sodium silicate	100, 158 hr	1	9 - 10	6.7	anhedral, 3µ	4.68	9, 36
Large Port (mordenite)	2.6	15.6	56	NaAlO$_2$, diatomite, sodium silicate	175, 16 days	1	9 - 10	6.7	anhedral, 5µ euhedral	4.68	9, 36
Large Port (mordenite)	1.2	10.7	53	NaAlO$_2$, silicic acid, sodium silicate	260, 24 days	1	9 - 10	6.7	anhedral prismatic, 20µ	—	36
Mordenite	1 + X, X = 20 - 40 mg/g H$_2$O	10	xs	NaAlO$_2$, silica gel, NaOH	150 - 200, 7 days	not analyzed			thin prisms; spherulites to 0.1 mm, n = 1.465 - 1.460	—	39
Mordenite	2	12	xs	NaAlO$_2$, silica sol	300, 1 day, 150, 4-8 days	not analyzed				4.37	37
Na-B (analcime)	>1.0	4.0	xs	Al$_2$O$_3$·3 H$_2$O, NaOH, silicic acid	180	1	4	2	spherulites 20 - 50µ, isotropic, n = 1.486	—	35
Analcime	1.0	2 - 6	xs	Na$_2$O·Al$_2$O$_3$·SiO$_2$ glass	200	1	(2 - 6)	1 - 2[b]	isotropic 0.04 mm	—	40
Analcime	1	2 - 10	xs	silicic acid, NaAlO$_2$	300, 24 hr	not analyzed				—	41

Table 4.4 Synthetic Mordenite, Analcime, Natrolite Type Zeolites Na$_2$O - Al$_2$O$_3$ - SiO$_2$ - H$_2$O System *(continued)*

Zeolite Type	Typical Reactant Comp. (moles/Al$_2$O$_3$) Na$_2$O	SiO$_2$	H$_2$O	Reactants	Conditions Temp (°C) Time	Zeolite Comp. (moles/Al$_2$O$_3$) Na$_2$O	SiO$_2$	H$_2$O	Properties	X-Ray Table	Ref.
Analcime	about 1	2.5 - 13	—	Na$_2$O·Al$_2$O$_3$·SiO$_2$ gels[c] NaOH	250 14 days	1	2.8 - 8.2	—	—	—	42, 43
"Natrolite"	1 - 2	3	xs	seeded gels	120 - 180 19 - 103 days			—	prismatic crystals n = 1.485	—	42

[a]Not analyzed—assumed composition; [b]Not analyzed but optical analysis shows complete conversion; [c]SiO$_2$/Al$_2$O$_3$ varied 2.5 - 13

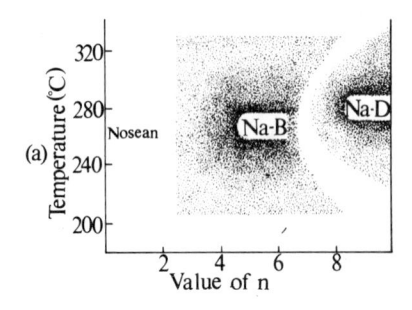

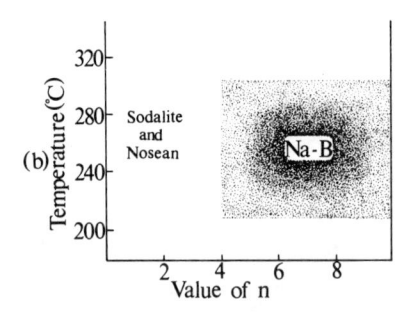

Figure 4.2 Formation of Na-B and Na-D zeolites from gels of composition $Na_2O \cdot Al_2O_3 \cdot n\ SiO_2$ with excess H_2O (35).
(a) No excess NaOH added.
(b) Less than 100% excess NaOH added based on the Na_2O in the initial gel composition.

4.2 relates the reaction mixtures to the zeolite products obtained.

The formation of synthetic types of analcime and mordenite in the Na_2O-Al_2O_3-SiO_2-H_2O system at 250 - 300°C was reported by Ames (33).

Two synthetic types of mordenite have been synthesized by Sand; "large-port" and "small-port" mordenites (36). Studies of the adsorption behavior of mordenite and the first synthetic types indicated that the pore structure was considerably smaller than the structure indicated. These types are referred to as small-port mordenite; they exhibit an adsorption diameter of about 4 A. By varying synthesis conditions a large-port mordenite has been synthesized which possesses the adsorption properties expected for the structure. After activation (dehydration), large–port mordenite adsorbs large molecules such as benzene and cyclohexane which are completely excluded by the small-port variety.

For the synthesis of small-port mordenite the type of starting materials was found to be critical, although the water content could be in large excess. The crystallization temperatures ranged from 275 - 300°C. Small-port mordenite is equivalent to the zeolite Na-D of Barrer.

In the synthesis of large-port mordenite, the water content is critical and excess sodium silicate was used. However, the nature of the starting materials was not as important. Large-port mordenite crystallized as a single phase at 100 - 260°C in 12 to 168 hours depending on the amount of excess sodium silicate (Table 4.4). Large-port mordenite has not been successfully distinguished from small-port mordenite by x-ray diffraction techniques.

The synthesis of both synthetic zeolite forms was accomplished using gels of the composition of mordenite (Fig. 4.3).

Detailed studies of the synthesis of mordenite from sodium aluminosilicate gels and from amorphous sodium aluminosilicates were reported (37) (see Section H). A siliceous mordenite (SiO_2/Al_2O_3 varying from 12.0 to 19.5) has been reported by Whittemore (38). The ad-

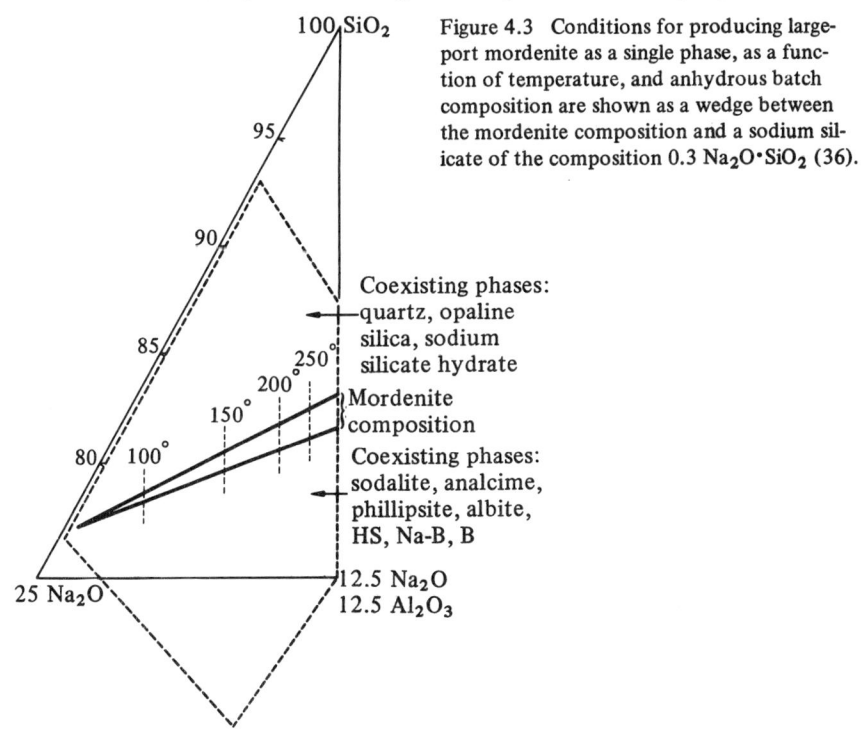

Figure 4.3 Conditions for producing large-port mordenite as a single phase, as a function of temperature, and anhydrous batch composition are shown as a wedge between the mordenite composition and a sodium silicate of the composition 0.3 $Na_2O \cdot SiO_2$ (36).

sorption pore size was not measured.

In a study of the crystallization of sodium aluminosilicate gels, Senderov has reported the formation of analcime, mordenite, and the preparation of a species which resembles natural ferrierite, another member of the mordenite group of zeolites (39). The synthesis of mordenite from an obsidian glass in a natural hydrothermal environment was reported by Ellis. The glass was lowered into a drill hole at Wairakei, New Zealand to a depth of over 1000 feet and left for 17 days. The temperature was 230°C and the pH reported to be 5.7(44).

Synthetic sodium mordenite-type phases have also been reported by Ames and Sand (45), Sand, Roy, and Osborn (2), and Coombs, Ellis, Fyfe and Taylor (8).

The digestion of bauxite in strong, hot NaOH has been reported to form an adherent insulating coating on the autoclave surface. The coating was identified as natrolite (46).

Recent studies of the synthesis of analcime-type phases include those of Saha (40), who prepared synthetic types from glasses ranging in chemical composition from $Na_2O \cdot Al_2O_3 \cdot 2 - 6\ SiO_2$. He also reported the formation of zeolite P and an unknown phase which he referred to as "Y." This "Y" phase appears, from x-ray powder data, to be a mixture of hydroxysodalite and a chabazite-like phase (zeolite R). Guyer, Ineichen and Guyer (41) also prepared synthetic types of analcime at 300°C. The Si/Al ratio was reported to vary from 1 to 5. These compositions are unlikely, and chemical analyses were not given. Although Barrer has reported the preparation of a 5-kilogram lot (19), there is no reported commercial application of synthetic analcime.

Natrolite-type and analcime-type zeolites are formed from amorphous starting materials in the temperature range of 100 to 200°C (42). These zeolites are considered to be thermodynamically stable species under these conditions. According to the simplexity principle, natrolite, due to its ordered structure, should be difficult to nucleate under metastable conditions.

The other metastable zeolites preferentially crystallize from the highly reactive gels. Consequently, seeding was used to nucleate the natrolite-type zeolite from sodium aluminosilicate gels. The analcime phase is a persistent impurity; the data do not suggest that the natrolite-type was crystallized in a pure form (Table 4.4). The composition of the analcime-type varied with the concentration of NaOH in the starting mixture. At high NaOH concentrations, the composition of the resulting zeolite approached the usual stoichiometry of SiO_2/Al_2O_3 equal to 4.

An extensive experimental study (over 900 data points) of the phase equilibria in the system $NaAlSi_3O_8$- $NaAlSiO_4$- H_2O over the pressure-temperature range of 0.5 to 10 kilobars and 159 to 900°C was completed (43). The zeolite phases reported were an analcime and zeolite P. There was no report of the formation of a natrolite-type zeolite. Other hydrated phases formed included the hydrated form of the feldspathoid nepheline. The complete stability field of analcime was mapped out. Below about 200°C, the zeolite P phase appeared to prevail over the entire compositional range, that is, from an Si/Al ratio of 1 to a ratio of 3. In this region, the zeolite was considered to be a metastable phase. The reversibility for the reaction zeolite P = analcime + nepheline hydrate I + water was also established. An invariant pressure-temperature curve for this conversion was demonstrated. Therefore, zeolite P may be an equilibrium phase in this temperature region (200°C).

Synthesis of Sodium Zeolites at Low Temperatures from Sodium Aluminate-Sodium Silicate Gels (Table 4.6)

Typical gels are prepared from aqueous solutions of $NaAlO_2$, $Na_2O \cdot SiO_2$, and $NaOH$. See Table 4.5 for information on the various starting materials used in preparing gels. By this relatively simple method, six synthetic zeolites (A, P, R, S, X, Y in Table 4.6) have been synthesized as pure phases in the $Na_2O-Al_2O_3-SiO_2-H_2O$ system. Zeolite P (Fig. 4.5) refers to the cubic zeolite which is equivalent to the Na-P type described by Barrer (47). Zeolite A is unique (4,48). Zeolite X (5) and zeolite Y are structurally related to faujasite (11); (see Chap. 2); and zeolite S is related structurally to gmelinite (49), and zeolite R to chabazite (50).

From a study of many aluminosilicate gel compositions, relationships between the synthetic zeolite product and the starting reactant mixture composition have been established. These are illustrated by reaction diagrams of the type given in Fig. 4.5 (11,14). The anhydrous compositions of the gels are plotted on a mole percent basis in triangular coordinates. Each diagram is a projection on the $Na_2O-Al_2O_3-SiO_2$ face and cannot show the effect of concentration in detail. Each diagram represents a range of component concentrations. Another approach is

Table 4.5 Some Starting Materials for Zeolite Synthesis

Material	Supplier	Na₂O	K₂O	Al₂O₃	SiO₂	H₂O
Sodium hydroxide (NaOH)	J. T. Baker Co.	1.25				1.25
Potassium hydroxide (KOH)	J. T. Baker Co.		0.89			0.89
Sodium aluminate (NaAlO₂)	Fisher Scientific Co.	0.484		0.431		1.37
	Nalco	0.477		0.438		1.43
	Nalco	0.600		0.443		0.976
Alumina trihydrate (Al₂O₃·3H₂O)	Alcoa			0.64		1.90
Colloidal silica (Ludox)	E. I. duPont de Nemours Co.				0.491	3.92
Sodium silicate (S-35)	Philadelphia Quartz	0.109			0.421	3.77
Potassium silicate (Kasil #6)	Philadelphia Quartz		0.134		0.450	3.36
Silicic acid	J. T. Baker Co.				1.33	1.12

to use rectangular coordinates in which the molar ratios Na_2O/SiO_2, SiO_2/Al_2O_3, H_2O/Na_2O are used as coordinates. Typical diagrams shown in Fig. 4.6 for the Na_2O-Al_2O_3-SiO_2-H_2O system at 25, 100 and 120 - 150°C illustrate the correlation between gel composition and the zeolite produced.

Another method for illustrating diagrammatically the crystallization fields of zeolites uses the Na_2O/Al_2O_3 and SiO_2/Al_2O_3 ratios plotted as the ordinate and abscissa. The plots are shown for each crystallization temperature and for fixed water contents in the mixture (51).

The concentration of the reacting components in the initial mixture, as expressed conversely by the water content in the mixture, is very important in determining the species produced. This parameter was neglected by many early investigators; consequently it is impossible to correlate or duplicate much of the early published data. Other factors include the temperature and the nature of the gel structure as influ-

(a)

(b)

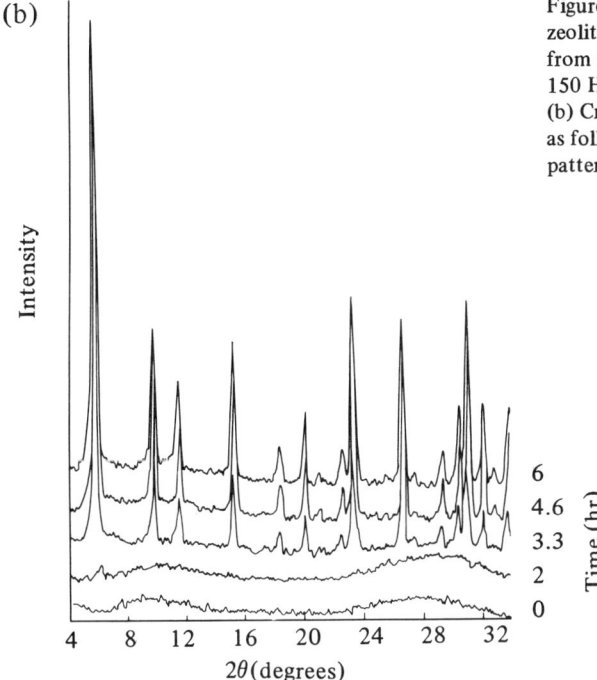

Figure 4.4 (a) The separation of zeolite X during crystallization from a gel (4.2 Na$_2$O · 3 SiO$_2$ · 150 H$_2$O) over a six-hour period. (b) Crystallization of zeolite X as followed by x-ray powder patterns, copper K$_\alpha$ radiation.

enced by the reactants. Zeolite P is readily formed from gels of higher silica content than zeolite A or zeolite X.

Two feldspathoids, basic or hydroxycancrinite and basic or hydroxysodalite crystallize in this system. The sodalite is also referred to a sodalite hydrate and has been designated HS or Zh (47, 35, 56). The ideal unit cell composition for sodalite hydrate is Na$_6$Al$_6$Si$_6$O$_{24}$· 8 H$_2$O.

If NaOH is intercalated during synthesis, the composition varies,

270 THE SYNTHETIC ZEOLITES

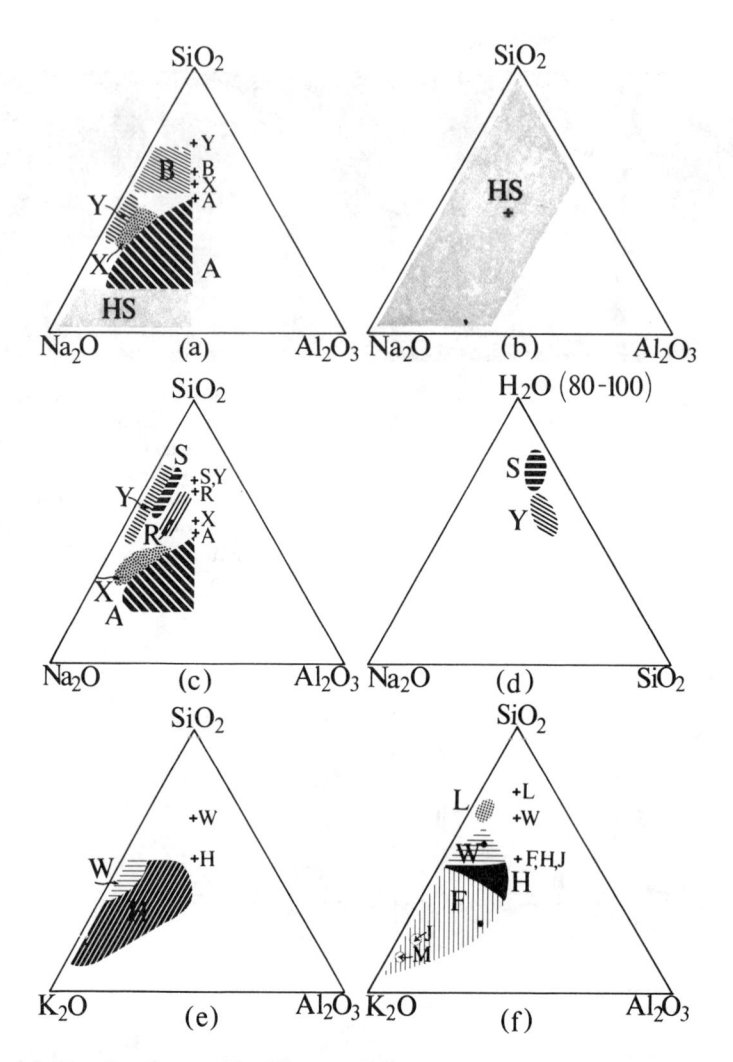

Figure 4.5 Reaction Composition Diagrams (11)
(a) Projection of the Na$_2$O-Al$_2$O$_3$-SiO$_2$-H$_2$O systems at 100°C. H$_2$O content of gels is 90-98 mole %. Areas identified by letters refer to compositions which yield the designated zeolite. The points marked with (+) show typical composition of zeolite phase. Compositions in mole %. Sodium silicate used as a source of SiO$_2$.
(b) Same as (a) with 60-85 mole % H$_2$O in the gel.
(c) Same as (a). Colloidal silica used as source of SiO$_2$.
(d) Effect of water content in gel on synthesis of zeolites Y and S. Colloidal silica employed at 100°C. Al$_2$O$_3$ content is 2-10 mole % of anhydrous gel composition.
(e) Projection of the K$_2$O-Al$_2$O$_3$-SiO$_2$-H$_2$O systems at 100°C. Water content is 95-98 mole %.
(f) Same as (e). Water content is 80-92 mole %.

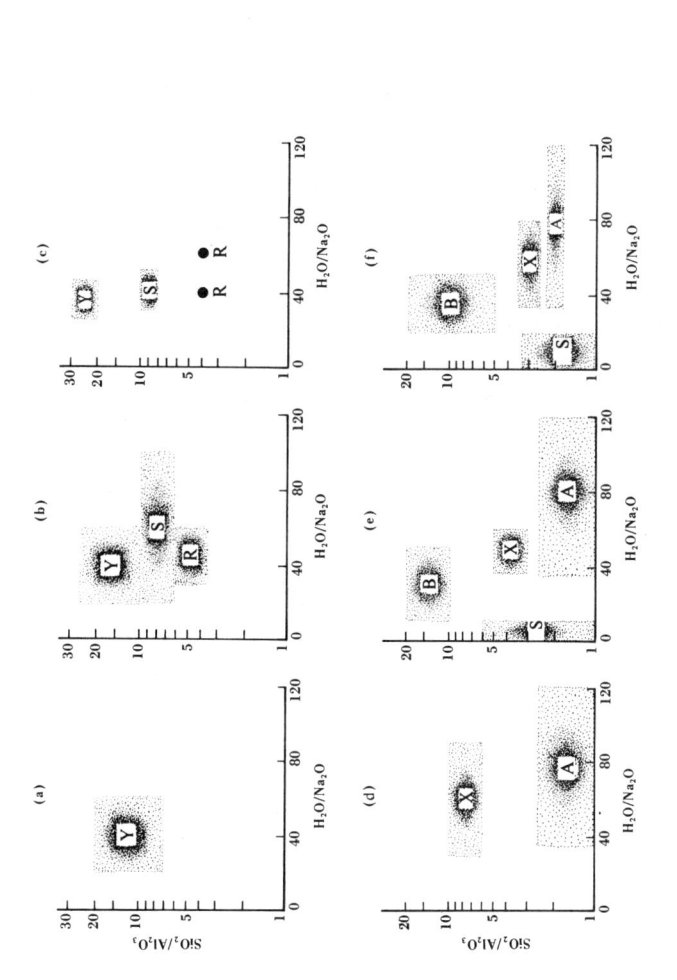

ZEOLITES FROM ALUMINOSILICATE GELS OF ALKALI METALS 271

Figure 4.6 Some typical gel compositions for the synthesis of zeolites in the Na$_2$O·Al$_2$O$_3$·SiO$_2$·H$_2$O system—rectangular coordinates. (a) Colloidal silica, aqueous or solid, is the source of silica. Letters refer to the zeolite formed in the area shown. Temperature = 25°C; Na$_2$O/SiO$_2$ = 0.61 - 0.80. (b) The same as (a) where temperature = 100°C and Na$_2$O/SiO$_2$ = 0.41 - 0.60 for Y; 0.30 - 0.60 for S; and 0.41 - 0.60 for R. (c) The same as (b) where the temperature = 120 - 150°C. (d) Sodium silicate or silicic acid is the source of SiO$_2$. Temperature = 25°C; Na$_2$O/SiO$_2$ = 0.6 - 1.6. (e) The same as (d) where the temperature = 100°C; Na$_2$O/SiO$_2$ = 0.38 - 0.61 for P; 1.2 - 1.5 for X; and 0.8 - 3.0 for A. (f) The same as (d) where the temperature = 120°C; Na$_2$O/SiO$_2$ = 0.6 - 1.3 for P; 1.2 - 1.5 for X and 1 - 2 for A. The feldspathoid-type hydroxysodalite (HS) forms at Na$_2$O/SiO$_2$ = 1 and at low H$_2$O/Na$_2$O ratios.

according to $Na_6Al_6Si_6O_{24} \cdot xNaOH(8-2x) H_2O$ since one NaOH replaces two water molecules. Sodalite hydrate has been observed to adsorb water after dehydration (56, 57).

Basic cancrinite crystallizes at higher temperatures (390°C) with a large excess of NaOH.

The zeolite referred to as LOSOD has been synthesized from a system using mixed bases. However, this zeolite contains only sodium and is a structural type based on the ABAC ... stacking of 6-rings (162). (See Section E.)

From typical reaction compositions as illustrated in Fig. 4.5 the SiO_2/Al_2O_3 molar ratio of the zeolite A is normally two. Silica-rich forms of zeolite A have been prepared by two methods: (1) using gels of a high silica and NaOH content and (2) using an organic cation (e.g., TMA) in place of part of the alkali metal.

From sodium aluminosilicate gels, prepared by the usual combination of a sodium silicate solution and sodium aluminate solution, zeolite A has been prepared which has a silica alumina ratio of 2.5. This would correspond to a unit cell contents of : $Na_{10.7}[(AlO_2)_{10.7}(SiO_2)_{13.3}]$ $\sim 30 H_2O$. Typical starting gel compositions shown in Table 4.6 are high in sodium hydroxide and silica content. If converted to the oxide mole fractions on an anhydrous basis, these reaction mixture-compositions lie in a composition region which normally would be expected to result in the formation of zeolite X or zeolite Y. These compositions are outside of the normal zeolite A composition region. The water content of the gel also is in the range of 91 to 95 mole % (58).

The silica-rich zeolite A appears to crystallize well in a period of 1 hour which is followed by rapid conversion in many instances to hydroxysodalite. This is to be expected from sodium-rich compositions of this type. In these strongly basic gels the zeolite A phase is formed rapidly but has a fleeting existence as it converts to hydroxysodalite very rapidly. The formation of hydroxysodalite was also accelerated when various anions were added to the gel compositions. In the presence of anions, such as chloride or sulfate, no zeolite A phase was formed at all. Uniquely, in the presence of nitrate, the gel crystallized to a cancrinite-type phase.

Another sodium zeolite, Z-21, is formed in this system (59). This zeolite also crystallizes rapidly from sodium aluminosilicate gels prepared from the usual reactants. Apparently, 1 hour is sufficient for the crystallization of this zeolite at 95°C. The reaction mixture compositions are very rich in NaOH, and lie in the Na$_2$O - rich corner of the composition diagram (Fig. 4.5). Water content is 93 mole % of the gel.

Zeolite Z-21 is reported to be cubic with a unit cell dimension of 36.7 A and is supposed to have a pore size of 17 A. However, confirmatory structural and adsorption data are not presented (59).

If after crystallization, zeolite A remains in contact with the mother liquor ($\sim$ 1 N NaOH) recrystallization to zeolite P may occur in 3 days. (See Chap. 6.) With a large excess of NaOH, it converts to hydroxysodalite. Similarly, zeolite X appears to be metastable with respect to zeolite P.

Zeolite A has been crystallized at temperatures ranging from 25 to 150°C with the crystallization time varying from 14 days to 2.5 hours. Zeolite X crystallizes at temperatures ranging from 25 to 120°C with corresponding variations in the time required. The kinetics of zeolite crystallization are discussed in Section I. Zeolite P is stable at higher temperatures and has been formed over a temperature range of 60 to 200°C. In general, at 200°C and svp, the zeolite P, analcime-type and hydroxysodalite or hydroxynosean, are the dominant stable phases.

Senderov (39) has reported the synthesis of a material which resembles ferrierite in x-ray powder data. It was crystallized from Na$_2$O-Al$_2$O$_3$-SiO$_2$-H$_2$O mixtures prepared from silica gel, sodium aluminate and NaOH at 150°C. The reactant composition was Na$_2$O·Al$_2$O$_3$·10SiO$_2$ with a slight excess of Na$_2$O and the total oxides were about 10% of the water content.

Zeolite P appears to be an equilibrium phase. Several structurally related types have been synthesized including a cubic and tetragonal form. The structural relations have been discussed in Chap. 2. An orthorhombic form was also described by Barrer (55).

The synthetic P zeolites (also referred to as B) dominate the synthesis compositon fields in the low temperature range of 100 to 150°C. In the nomenclature adopted here zeolite P refers to all of the synthetic zeolites which have the aluminosilicate framework that is the same as the structure of the cubic zeolite P previously described (gismondine-type). The cubic zeolite P is equivalent to the zeolite B described by Milton (55). For purposes of clarity we adopt the term P_t to refer to the tetragonal variation.

Table 4.6 Synthetic Zeolites Na$_2$O - Al$_2$O$_3$ - SiO$_2$ - H$_2$O System

Zeolite Type	Typical Reactant Comp. (moles/Al$_2$O$_3$) Na$_2$O	SiO$_2$	H$_2$O	Reactants	Typical Conditions Temp. (°C)	Zeolite Comp. (moles/Al$_2$O$_3$) Na$_2$O	SiO$_2$	H$_2$O	Properties	X-Ray Table	Ref.	Other Names, Ref.
A	2	2	35	NaAlO$_2$ sodium silicate NaOH colloidal SiO$_2$	20 - 175	1	2	4.5	cubic, 1 - 2µ, n = 1.46, d = 1.99	4.27	4,48	Q, 47, 53, 54
P$_c$	2	3.8	94	NaAlO$_2$ sodium silicate, NaOH colloidal SiO$_2$	60 - 150	1	2 - 5	5	irregular, n = 1.476, d = 2.01 gismondine-type	4.74	47	53, 55
P$_t$	6	8	a	NaAlO$_2$ sodium silicate NaOH colloidal SiO$_2$	60 - 250	1	3.2 - 5.3	~5	2µ, n = 1.482, d = 2.15 gismondine-type	4.75	47	52
R	3.2	4	260	NaAlO$_2$ sodium silicate, NaOH colloidal SiO$_2$	100	1	3.5	5.7	irregular 0.6 - 7µ d = 1.98 chabazite-type	4.80	50	—
S	2.4	6	80	NaAlO$_2$ sodium silicate NaOH colloidal SiO$_2$	80 - 120	1	4.6 - 5.9	6	spherulites, n = 1.458 gmelinite-type	4.83	49	NaS, 47
X	3.6	3	144	NaAlO$_2$ sodium silicate NaOH colloidal SiO$_2$	20 - 120	1	2.0 - 3.0	6	octahedra, d = 1.94 n = 1.45 - 1.46 faujasite-type	4.88	5, 11	R, 47
Y	8	20	320	NaAlO$_2$ sodium silicate NaOH colloidal SiO$_2$	20 - 175	1	>3.0 - 6.0	9	octahedra, d = 1.92 n = 1.45 faujasite-type	4.90	11, 61	

Table 4.6 Synthetic Zeolites Na$_2$O - Al$_2$O$_3$ - SiO$_2$ - H$_2$O System (continued)

Zeolite Type	Typical Reactant Comp. (moles/Al$_2$O$_3$) Na$_2$O	SiO$_2$	H$_2$O	Reactants	Typical Conditions Temp. (°C)	Zeolite Comp. (moles/Al$_2$O$_3$) Na$_2$O	SiO$_2$	H$_2$O	Properties	X-Ray Table	Ref.	Other Names, Ref.
"Ferrierite"	1	10	b	Silica Gel	150	1	10	?	rectangular, 0.1mm n = 1.469	4.45	39	Z, 8, 84
HS	2.8	3.0	34	Silica Gel	100	1.16	2.1	2.8	—			Hydroxy sodalite, 54
	4	2	xs	Silica Gel	100	2.0	2.1	2.5	spherulitic, n = 1.468 - 1.493	4.53	35, 4	basic sodalite species I
Zh	~7	~2	150	sodium silicate NaAlO$_2$	90	1	2.1	2.7	irregular, n = 1.49		56, 57	G
A (2.5)	7.5-17.8	5-15	165-330	sodium silicate sodium aluminate	92 - 100 1 hr	1	2.5	6	sodalite-type adsorbs n-hexane		58	
Z-21	7	0.3	117	Al$_2$O$_3$ · 3 H$_2$O	100	1	2	~2	cubic, a = 36.7	4.93	59	
	30	2.0	504	NaOH	1 hr							
	43	6.0	720	sodium silicate								
Basic Cancrinite	3	2	xs		390	1	2.4 (0.6 NaOH)		hexagonal up to 500μm n = 1.50		35	

[a]Solids/H$_2$O = 1:6   [b]Some excess NaOH added

It can be concluded from many observations that there is no real relationship between synthesis variables, such as the composition of the gel, and the symmetry of the P phase which results, cubic or tetragonal. This is illustrated by data such as that in Table 4.7. Zeolite P has also been synthesized in the potassium and calcium forms. Ion exchange produces structural variation of the framework as evidenced by the x-ray powder patterns. There are three main structure divisions which depend upon the cation present:

1. A primitive cell in which a is equal to or greater than c.
2. A body-centered tetragonal cell in which a is greater than c.
3. A body-centered tetragonal cell in which c is equal to or greater than a.

The framework structure of zeolite P would permit various types of simple displacive transformation; transformation of the P framework to the harmotome framework cannot occur without a reconstructive transformation. Several types of zeolite P have been identified (60).

When synthetic A is treated for extensive periods of time with NaOH solutions, it converts to the more stable zeolite P(47). In 1 molar NaOH, zeolite A is stable for two days, but in 6 - 10 days, converts to zeolite P.

At temperatures above 100°C, the conversion to zeolite P occurs in

Table 4.7 Composition of Zeolite P Synthesized from Na_2O-Al_2O_3-SiO_2-H_2O Gels

Na_2O	Moles/Al_2O_3 SiO_2	H_2O	Symmetry	Reference
1	2 – 5	6	cubic, tetragonal	55
0.89	3.18	4.6	tetragonal	52
–	3.49	4.2	tetragonal	52
–	3.52	4.7	cubic	52
–	3.68	4.3	tetragonal	47
0.91	3.78	5.1	tetragoral	52
–	3.98	4.7	cubic	47
0.91	4.02	5.6	tetragonal	52
–	4.32	4.7	tetragonal	47
0.94	4.46	5.8	cubic	52
–	5.26	5.7	cubic	47

a shorter period of time. This would tend to confirm the qualitative relationships relative to the stability of the sodium zeolites. In strong NaOH solution, about 10 wt% NaOH, zeolite A converts to a hydroxysodalite phase.

Crystallization of Sodium Zeolites from Gels Based on Colloidal SiO$_2$

When the reaction mixtures are prepared by using a colloidal silica sol or amorphous silica as the silica source, additional zeolites are formed which do not readily crystallize from the homogeneous sodium silicate-aluminosilicate gels. As shown in the reaction diagram of Fig. 4.6a for the crystallization of sodium aluminosilicate mixtures at 100°C, zeolite Y is formed from gel compositions which in the soluble silicate system may result in the formation of zeolite P (11).

When an aqueous colloidal silica sol (Table 4.5) is employed as the major source of silica, the SiO$_2$ content of the starting reaction mixture is higher than that normally employed in the case of the other sod-sodium zeolites. Typically for zeolite Y, the composition ranges within the limits given in Table 4.8, and shown in Figs. 4.5 and 4.6. If the reaction mixture is first digested or aged at ambient or room temperature and then subsequently heated at the higher temperature, usually at 100°C, the Si/Al ratio in the synthetic zeolite Y is higher (Table 4.8b)

It appears that the gel prepared from the colloidal silica sol is heterogeneous on a molecular scale and contains a hydrous aluminosilicate phase together with a solution. After the initial gel formation, the aging step is necessary in order to equilibrate the heterogeneous gel mixture with the solution. Room temperature equilibration or aging reduces the SiO$_2$/Al$_2$O$_3$ ratio in the gel necessary to form zeolite Y. The hydrous gel phase contains relatively large units of silica particles with molecular weights of 100,000 to 500,000 and ranging in size from 10 to 20 millimicrons. The formation of zeolite nuclei must require the solubilization of silicate anions from the colloidal silica particles

Table 4.8 Some Compositions for Synthesis of Zeolite Y (61)

Molar ratio	Non-aged	Aged
Na$_2$O/SiO$_2$	0.4 to 0.6	0.28 – 0.30
SiO$_2$/Al$_2$O$_3$	15 to 25	8 – 10
H$_2$O/Na$_2$O	20 to 50	30 – 50

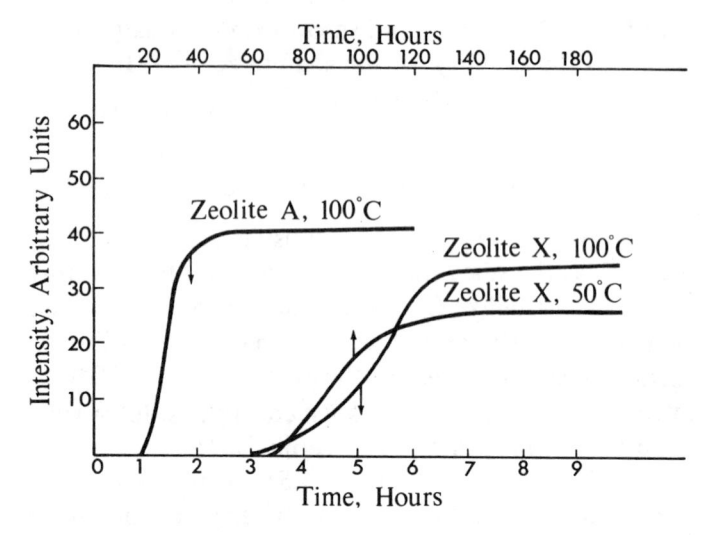

Figure 4.7 Crystallization of zeolites A and X as a function of time. Intensity indicates the degree of crystallization as determined by x-ray powder methods (11).

and their resulting interaction with the aluminate ions present in the solution (see Section I). Another sodium zeolite which is best synthesized from reaction mixtures employing colloidal silica is zeolite R (50).

The activity of SiO_2 in the mixture depends upon the [OH^-]. The higher the [OH^-], the lower the activity; thus synthesis of zeolites with high SiO_2/Al_2O_3 requires lower [OH^-].

The activity of silica during crystallization of aluminosilicates of variable silica content was emphasized by Coombs et al. (8). The use of a more reactive amorphous silica is important in the synthesis of the

Table 4.8b Synthesis of Zeolite Y from Aged Gels (61)

Typical Reactant Composition			Aging Time	Crystallization	Product	
Na_2O/SiO_2	SiO_2/Al_2O_3	H_2O/Na_2O	(RT, hr)	Time at 100°C(hr)	Purity(%)	Si/Al
0.4	20	40	0	72	63	<2.5
0.4	20	40	24	50	92	2.5
0.4	10	40	24	50	92	2.5
0.3	8	40	24	96 – 144	100	2.6
0.3	10	40	24	96 – 144	95	2.8

high silica zeolites (e.g., zeolite Y).

The reaction diagrams exhibit rather definite composition field boundaries with respect to temperature and the alkalinity or concentration of base in the solution. In the Na_2O system at 100°C, the entire field is dominated by the formation of hydroxysodalite (HS) at a water concentration of below 85 mole %. The formation of the zeolites from gels based upon reactive colloidal forms of silica is very dependent upon the base concentration which determines the solubility of the silica in the system.

The effect of temperature on the time required for crystallization of even the most reactive gels is very pronounced. As shown in Figs. 4.7 and 4.20, zeolite X crystallizes in 800 hr at 25°C and ∼ 6 hr at 100°C (11). (See Section I.)

The reaction diagrams given in Fig. 4.6 also show that zeolite P, which forms readily as the equilibrium phase at the elevated temperatures does not crystallize readily at the lower temperatures. At ambient temperature, the composition field is dominated by zeolites Y, A and X. Above 150°C, zeolite P is the dominant phase along with zeolite A and analcime. At high Na_2O concentrations, the synthetic hydroxysodalite is formed.

These results are in accordance with the principle discussed in Section A relative to the free energy-temperature relationship for the metastable phases. At low temperature, the more complex and more easily nucleated zeolite Y or X are formed, whereas the increasing temperature of crystallization favors the formation of the zeolite P followed ultimately by analcime. As shown schematically in Fig. 4.1, zeolite Y corresponds to phase a, zeolite P would correspond to phase b, and analcime to phase c. The free energy curves for b and c would intersect at about 175°C and for a and b at about 100°C. This is consistent with the interesting behavior of the reacting phases during nucleation; for example zeolite Y and zeolite P may form in the same overall composition field, but the type of reactant silica seems to control which species nucleates.

Data on the structures of the synthetic sodium zeolites are summarized in Table 2.4. Data on composition and other properties are listed in Table 4.6. Characteristic x-ray powder data for purposes of identification are given at the end of the chapter.

Sodium-Lithium Aluminosilicate Zeolites

The zeolite ZSM-3 crystallizes in the form of hexagonal plates (62). The crystal structure is based on various stacking sequences of hexagonal layers of linked β-cages. One particular hexagonal form is the hypothetical structure discussed in Chap. 2. The x-ray powder data are indexed in the hexagonal system and the various polytypes appear to have a maximum value of the c-dimension of 129 A. The interlayer distance is 14.3 A.

Crystallization of ZSM-3 (see Table 4.9) is accomplished at relatively low temperatures and utilizes a precursor technique. First, a sodium aluminosilicate solution is prepared and then mixed with additional aqueous sodium silicate and aluminum chloride to form a hydrogel slurry. This slurry is filtered to remove the excess soluble sodium silicate. Lithium is added to the hydrogel as lithium hydroxide solution. Crystallization at 60°C requires 5 days versus 16 hours at 100°C. The structure of the zeolite based on stacking sequences of hexagonal layers of cages accounts for the adsorption of normal hexane. A stacking sequence consisting of nine layers would result in 256 possible structures of this type (63).

Additional studies of this aluminosilicate system at 100°C (64) have resulted in the formation of other types of zeolites or the hydrated feldspathoids which are found in each individual system. In the mixed lithium-sodium aluminosilicate system, nine zeolite species were observed including zeolite A, X, P, sodalite hydrate, cancrinite hydrate, a chabazite-type, the zeolite Li-A and a form of zeolite F. The latter zeolite is normally observed in the potassium aluminosilicate system discussed next. Crystallization sequences were worked out in this study and the products identified by x-ray powder pattern. However, chemical analyses of the zeolites are not reported. For example, although zeolite A is crystallized in this system it was not disclosed whether the zeolite A contains lithium or preferentially incorporates sodium.

A mordenite-type has been synthesized in the sodium-lithium system (9) (Table 4.9). The mordenite-type is formed in many different alkali metal and alkaline earth aluminosilicate systems.

Potassium Aluminosilicate Zeolites

Reaction mixtures (Table 4.10) used in the synthesis of potassium zeo-

Table 4.9 Synthetic Zeolites Li_2O - Na_2O - Al_2O_3 - SiO_2 - H_2O System

Zeolite Type	Typical Reactant Comp. (moles/Al_2O_3) Li_2O Na_2O SiO_2 H_2O	Reactants	Conditions Temp. (°C), Time	Zeolite Comp. (moles/Al_2O_3) Li_2O Na_2O SiO_2 H_2O	Properties	X-Ray Table	Ref.
ZSM-3	6.0 10.6 16.3 448	LiOH sodium aluminate sodium silicate NaOH	60 5 days	0.62 0.43 3.1 —	hexagonal plates 1 mm	4.100	62, 63
"Mordenite"	0.5 0.5 10 50	Al(OH)$_3$ LiOH NaOH silicic acid	200 24 hr	0.35 0.36 10.2 6.6	$n = 1.471$ small port	4.69	9

Table 4.10 Synthetic Zeolites K_2O - Al_2O_3 - SiO_2 - H_2O System

Zeolite Type	Typical Reactant Comp. (moles/Al_2O_3) K_2O	SiO_2	H_2O	Reactants	Typical Conditions Temp. (°C)	Zeolite Comp. (moles/Al_2O_3) K_2O	SiO_2	H_2O	Properties	X-Ray Table	Ref.	Other Names, Ref.
K-E	1	5	ns	$Al_2O_3 \cdot 3\,H_2O$ silica gel colloidal silica KOH, potassium silicate	400 - 450	1	4	1	spherulites, $n = 1.490$ analcime-type	4.41	22	
F	2	1	20	$Al_2O_3 \cdot 3\,H_2O$ silica gel colloidal silica KOH, potassium silicate	100	1	2	2.9	tetragonal, $<1\mu$, $d = 2.28$	4.43	73, 11	
Z	1.7	4	ns	$Al_2O_3 \cdot 3\,H_2O$ silica gel colloidal silica KOH, potassium silicate	120 - 150	1	2	3	rod-like, $n = 1.500$	4.92	22	65,67, K-F
G	2.5	5	ns	$Al_2O_3 \cdot 3\,H_2O$ silica gel colloidal silica KOH, potassium silicate	150	1	3.9	4.6	lenticular aggregates; see Table 4.11 chabazite-type	4.46	22	K-G
H	6	2	150	$Al_2O_3 \cdot 3\,H_2O$ silica gel colloidal silica KOH, potassium silicate	100	1	2.0	4.0	hexagonal plates $2 - 3\mu$, $d = 2.19$ $n = 1.475$	4.50	11, 75	K-I, 66

ns = not specified

Table 4.10 Synthetic Zeolites K$_2$O - Al$_2$O$_3$ - SiO$_2$ - H$_2$O System *(continued)*

Zeolite Type	Typical Reactant Comp. (moles/Al$_2$O$_3$) K$_2$O	SiO$_2$	H$_2$O	Reactants	Typical Conditions Temp. (°C)	Zeolite Comp. (moles/Al$_2$O$_3$) K$_2$O	SiO$_2$	H$_2$O	Properties	X-Ray Table	Ref.	Other Names, Ref.
J	16	4	160	Al$_2$O$_3$·3 H$_2$O silica gel colloidal silica KOH, potassium silicate	100	1	1.9	1.2	rods, 0.2 × 0.1μ	4.56	11, 68	
L	8	20	200	Al$_2$O$_3$·3 H$_2$O silica gel colloidal silica KOH, potassium silicate	100 - 150	1	6	6	d = 2.11 hexagonal	4.60	11, 69	
M	49	10	420	Al$_2$O$_3$·3 H$_2$O silica gel colloidal silica KOH, potassium silicate	100	1	2.1	1.7	sheaflike, 1 - 1.5μ d = 2.34	4.64	11, 70	
Q	16	4	190	Al$_2$O$_3$·3 H$_2$O silica gel colloidal silica KOH, potassium silicate	50	1	2.2	4.3	platelets, d = 2.11	4.77	11, 71	
W	3	5	75	Al$_2$O$_3$·3 H$_2$O silica gel colloidal silica KOH, potassium silicate	100	1	3.6	5.1	sheaflike, 3 - 6μ, d = 2.18 n = 1.490, 1.495	4.86	11, 72	K-M, 22 K-H, 52

lites are typically like those used in the synthesis of the sodium zeolites and are prepared from aqueous potassium silicate solutions, potassium hydroxide and hydrous alumina (Table 4.5). Colloidal silica sols are also employed as a source of SiO_2.

> Generally the alumina is first dissolved in aqueous KOH, and this solution then added to the K_2O-SiO_2 solution, or silica sol, with rapid mixing obtained by using a Waring-type blender. These mixtures were digested at various temperatures, generally in the range of 50 to 150°C.

Ten synthetic zeolites have been prepared in this system in a pure state. The potassium zeolites are identified as types K-E, F, G, H, J, L, M, Q, W, and Z (Fig. 4.5).

The approximate conditions for the synthesis of some of the potassium zeolites above 150°C are given in Fig. 4.8 and Table 4.10.

The synthetic zeolites formed in the K_2O-Al_2O_3-SiO_2 system include two species which crystallize and are stable at temperatures about 200°C. One of these, designated by Barrer as K-E, is somewhat unusual in that it forms at temperatures of 400 to 450°C. It is reported to be a potassium-containing type of analcime (22). Normally, substitution of sodium in analcime by potassium exchange results in the formation of the anhydrous feldspathoid known as leucite, $KAlSi_2O_6$.

The other zeolite species formed above 200°C is designated by Barrer as K-M, here referred to as zeolite W. The optimum reported temperature for the formation of zeolite W is 250°C. However, it crystallizes readily at 100°C, Na_2O/SiO_2 = 0.6 - 0.9 and SiO_2/Al_2O_3 = 4 - 7 (72). Although Barrer reports that zeolite W is very similar to phillipsite, the published x-ray powder data for these zeolites show significant differences. Zeolite W has also been synthesized by Taylor and Roy(52). Most of the synthetic potassium zeolites form at temperatures in the region of 100 to 150°C. Figures 4.5 and 4.8 illustrate the relation between the reaction mixture composition and the synthetic zeolite produced. Two potassium zeolites designated as zeolite F have been reported, one by Milton (73) and one by Barrer (22). Since these are not identical species, as evidenced by the x-ray powder data and chemical composition, the zeolite K-F prepared by Barrer has been redesignated as Z for purposes of clarity (65). The zeolite Z forms best below 150°C from the general composition given in Fig. 4.8 and apparently crystallizes well at 90°C (57).

Zeolite G includes several similar phases of varying composition which are related structurally, as evidenced by their x-ray powder patterns and other data, to chabazite (Table 4.11). The Si/Al ratio in this zeolite varies with the ratio in the starting gel. The species which contain higher amounts of silica are very close to chabazite. Hence, the designation G will include compositions which vary in Si/Al ratio from 1.15 to 2.7. The zeolite G listed in Table 4.10 is a representative composition. X-ray powder data for G are given in Table 4.46.

Zhdanov (14, 56, 57, 74) has reported similar work on the synthesis of potassium zeolites. In addition to zeolite G, W (K-M), Z (K-F), he also reports the formation of a zeolite I. Inspection of the x-ray powder data shows that the zeolite I is identical to zeolite H (75).

It is again obvious that at low temperatures, the relationship between the optimum composition for synthesizing a particular zeolite in a reasonable period of time and the composition of the zeolite which crystallizes is far from stoichiometric. Rather broad areas of reactant composition may lead to the same species (Fig. 4.5e, f). The starting gel

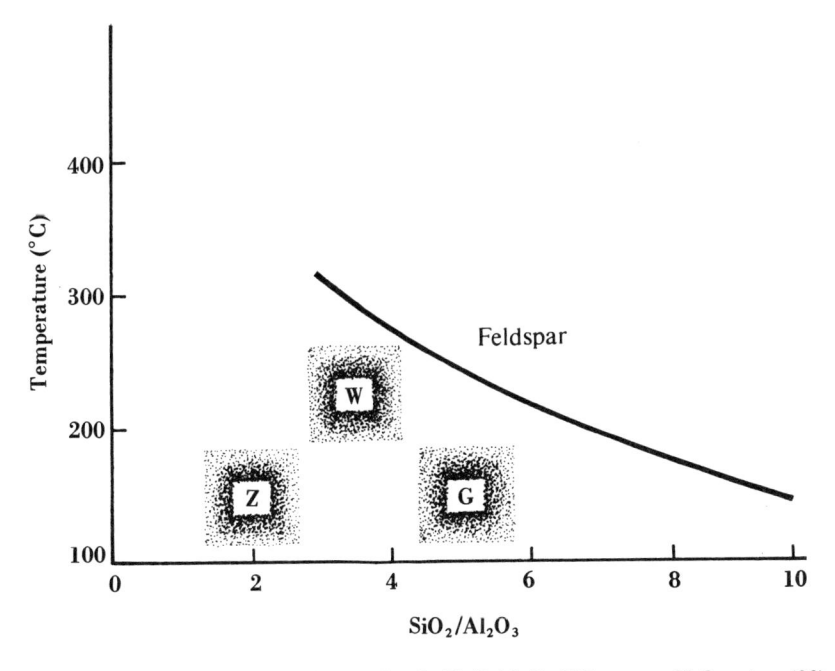

Figure 4.8 Reactant composition diagram for the K_2O-Al_2O_3-SiO_2-excess H_2O system (22). pH > 10.5.

Table 4.11 Synthesis of Zeolite G (22)

Sample No.	n^a	Excess KOH (%)^b	Reactant Compositions (moles/Al$_2$O$_3$) K$_2$O	SiO$_2$	Product Refractive Indices α	γ	Composition (moles/Al$_2$O$_3$) K$_2$O	SiO$_2$	H$_2$O
1	3	150	2.5	3	1.480	1.485	0.92	2.30	3.40
2	4	140	2.4	4	1.490	1.495	1.11	2.56	2.62
3	4	275	3.75	4	1.475	1.480	0.99	2.65	3.72
4	5	310	4.1	5	1.465	1.470	0.95	2.72	3.94
5	5	150	2.5	5	1.460	1.465	1.00	3.91	4.63
6	6	150	2.5	6	1.470	1.475	1.03	4.15	4.38

^{a}Gels of composition K$_2$O·Al$_2$O$_3$·n SiO$_2$ prepared from KOH, aluminum hydroxide, silica gel, or silica sol.
^{b}Added in ratio of 15 ml solution to 0.5 g of gel, K$_2$O·Al$_2$O$_3$·n SiO$_2$.

compositions are controlled to some extent by the possible ranges of composition that can be prepared from the particular starting materials. In the use of aqueous sodium or potassium silicate solutions, the ratio of base to silica in the gel cannot be less than that present in the silicate. To prepare gels or mixtures containing high silica contents, one must use a colloidal silica or silica sol.

Many of the zeolites which form in the sodium system do not crystallize in pure form in the potassium system; this emphasizes the role played by the cation species in zeolite formation. For example, zeolite L (11), is formed instead of zeolite Y when K_2O is substituted for Na_2O in the aluminosilicate reaction mixture. Zeolite Q appears to be an unusually low temperature phase and was only obtained at 50°C. The same gel composition at 100°C produces the zeolite J.

Typical x-ray powder data for the potassium zeolites are given at the end of the chapter.

Sodium-Potassium Aluminosilicate Zeolites

The zeolite species which crystallize from aluminosilicate gels are highly dependent upon the cation species present in the parent gel. The use of two different cations such as Na^+ and K^+ in aluminosilicate gels has a profound effect. Most experimental work has been concerned with the synthesis of zeolites in mixed alkali systems at moderate temperatures, $\sim 100°C$. Some limited work has been reported for temperatures of 150°C and above (47).

In addition to several synthetic zeolites which form in the individual sodium or potassium aluminosilicate systems, three additional synthetic zeolites crystallize from Na, K-aluminosilicate gels. These are the synthetic zeolites designated in Table 4.12 as D, E, T, and ZK-19. Additional zeolites which have been synthesized are A, X, P, F, W, H, L and M. Several nonzeolitic species form in these systems but will not be discussed here.

> In general, the gels were prepared by mixing solutions of the soluble sodium or potassium silicate with the dissolved sodium aluminate, or as in the case of potassium-containing gels, hydrous aluminum oxide. The gels are best prepared by mixing in a blender until homogeneous. Crystallization is generally accomplished by heating at 100°C, although temperatures ranging from as low as 65 to as high as 250°C have been used.

Table 4.12 Synthetic Zeolites Na_2O - K_2O - Al_2O_3 - SiO_2 - H_2O System

Zeolite	Reactant Composition (moles/Al_2O_3) Na_2O	K_2O	SiO_2	H_2O	Reactants	Typical Conditions Temp. (°C)	Zeolite Composition (moles/Al_2O_3) Na_2O	K_2O	SiO_2	H_2O	Properties	X-Ray Table	Ref.	Other Ref.
D	11	2.8	28	565	Sodium silicate, potassium silicate, $NaAlO_2$ $Al_2O_3 \cdot 3 H_2O$ KOH, colloidal SiO_2	100	0.5	0.5	4.8	6.7	Plate-like, 1 - 8μ, d = 2.04 g/cc, chabazite-type	4.35	76	11, 74
E	0.75	0.75	2	30	Sodium silicate, potassium silicate, $NaAlO_2$ $Al_2O_3 \cdot 3 H_2O$ KOH, colloidal SiO_2	100	0.4	0.5	2.0	3.3	Cubic, 2 - 4μ	4.39	77	
T	8	2.8	28	452	Sodium silicate, potassium silicate, $NaAlO_2$ $Al_2O_3 \cdot 3 H_2O$ KOH, colloidal SiO_2	100	0.3	0.7	6.9	7.2	Rod-like, 0.5 x 4 μ, offretite-erionite-type	4.84	78	74
ZK-19	0.3-0.85	0.7-0.15	4-16	—	Sodium metasilicate K_3PO_4, Na_3PO_4		0.1-0.6	0.9-0.4	3-6.25	—	Rods, twinned, phillipsite-type	4.96	79	

Gels which normally produce zeolites A and X in the sodium cation system result in the formation of zeolite F and P, respectively, when potassium is present (Fig. 4.9). The addition of potassium to sodium cation gels which normally would produce zeolite X results ultimately in the formation of zeolite P. The addition of small amounts of potassium produces a mixture of zeolites. The chemical composition of zeolite X, formed from mixed potassium-sodium cation gels, is related to the gel composition in Table 4.13. Similarly the addition of potassium cation to gels which would normally produce zeolite A results in the formation of zeolite F. Zeolite E, which crystallizes only from the sodium, potassium-containing gels, as shown in Figs. 4.9 and 4.10, is formed in a compositional area which would normally be expected to produce zeolite A or zeolite F.

Compositions which produce zeolite Y crystallize to zeolite P or zeolite D, as shown in the reaction diagram of Fig. 4.9. Small amounts of potassium cation in the gel inhibit the formation of zeolite Y, and zeolite P crystallizes instead.

In contrast, zeolite L, which crystallizes in the K_2O system, will incorporate appreciable amounts of sodium. This is illustrated in reaction diagrams Figs. 4.10c and 4.10d. Fig. 4.11a, which is a projection of the anhydrous gel compositions in terms of the Na_2O, K_2O, and SiO_2 components, shows (1) that zeolites Y and L form from the higher silica-containing gels, (2) that zeolite L can form in the presence of a considerable amount of Na_2O in the reaction mixtures, (3) that zeolite P also will form over a wide range of mixed gel compositions and zeolite D and zeolite T form in a limited composition range in this system. Similar results have also been reported by Barrer and coworkers, but they did not determine the effect of the reactant mixture compo-

Table 4.13 Chemical Composition of Zeolite X Crystallized from K-Na Gels
Gel Composition: SiO_2/Al_2O_3 = 3 - 4, H_2O/R_2O = 30 - 45

K_2O/R_2O (gel)	K_2O/R_2O (zeolite)	SiO_2/Al_2O_3 (zeolite)
0.02	0.04	2.65
0.05	0.09	2.58
0.10	0.14	2.44
0.15	0.17	2.39

If K_2O/R_2O is greater than 0.15, zeolite X is mixed with zeolite B. $R_2O = K_2O + Na_2O$.

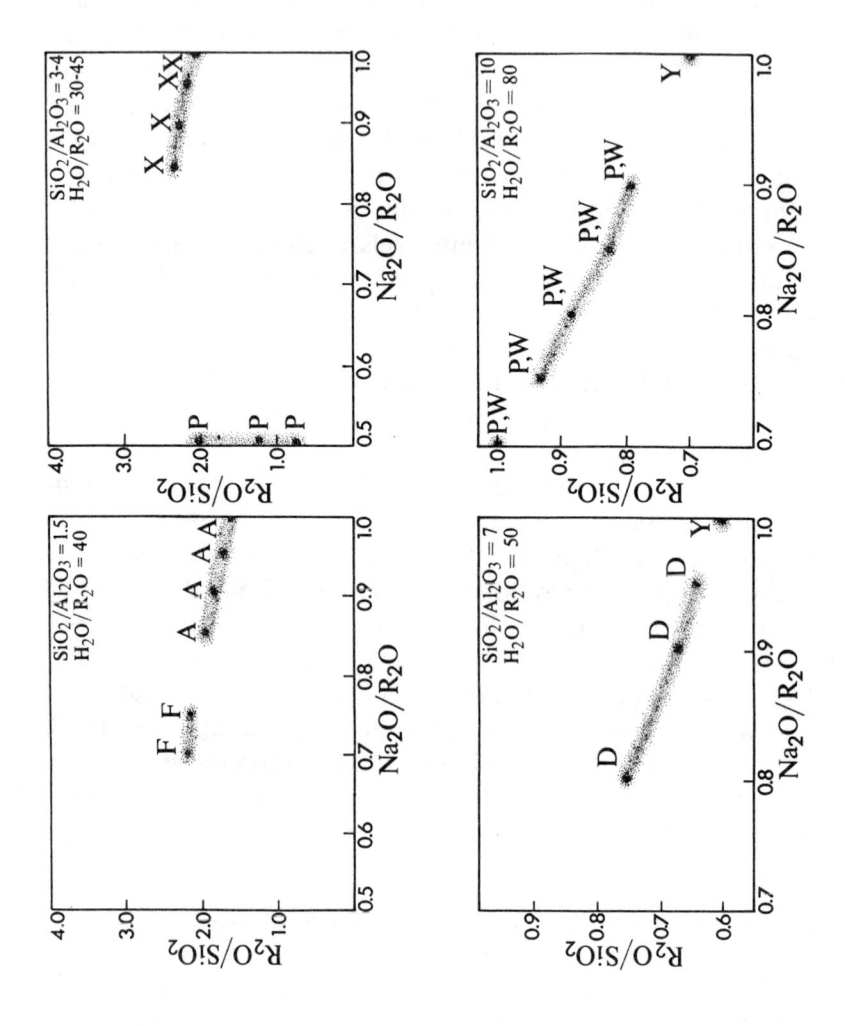

Figure 4.9 Some typical reactant compositions for zeolites crystallized at 100°C from $Na_2O \cdot K_2O \cdot Al_2O_3 \cdot SiO_2 \cdot H_2O$ gels. $R_2O = (K_2O + Na_2O)$

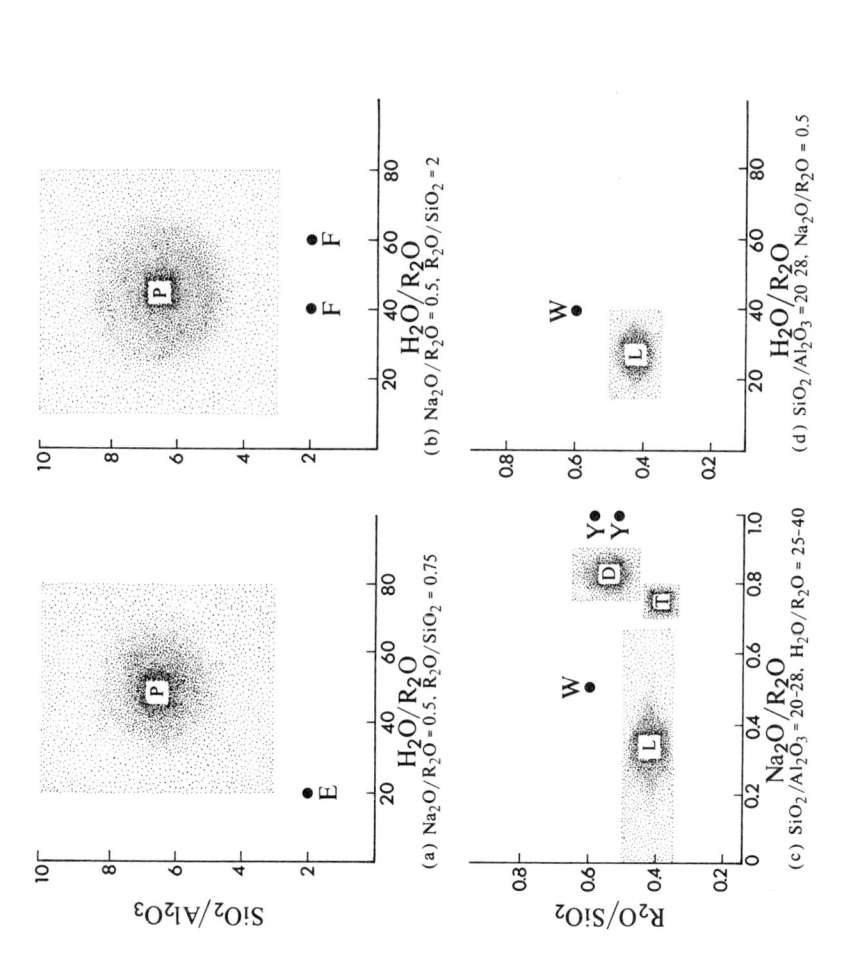

Figure 4.10 Some typical reactant compositions for zeolites crystallized at 100°C from $K_2O \cdot Na_2O \cdot Al_2O_3 \cdot SiO_2 \cdot H_2O$ gels.

292 THE SYNTHETIC ZEOLITES

(a)

(b)

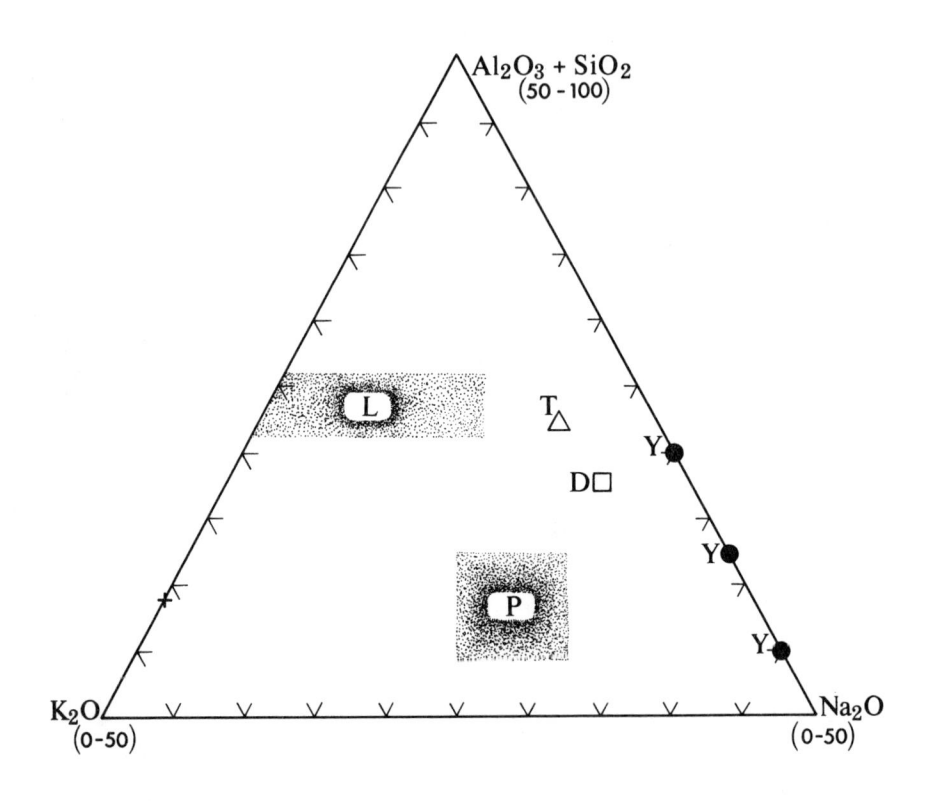

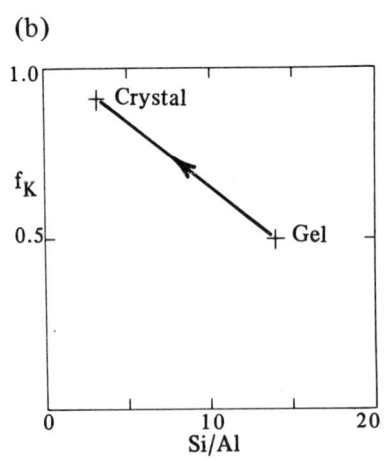

Figure 4.11 (a) Some reactant compositions for zeolites crystallized from Na$_2$O-K$_2$O-Al$_2$O$_3$-SiO$_2$-H$_2$O gels. Compositions are projected on dry basis on Na$_2$O-K$_2$O-(Al$_2$O$_3$ + SiO$_2$) at 100°C. The silica content is 68 - 72 mole % on a dry basis and the water content is 87 - 90 mole % of the total composition. The zeolites formed from the compositions shown are designated by the appropriate letter. (b) As zeolite L crystallizes from a reactant gel of composition f_k = K$_2$O/(K$_2$O + Na$_2$O) = 0.5 and the Si/Al ratio = 14, the zeolite preferentially incorporates K and f_k goes to ∼1 as the Si/Al ratio decreases to 3.

sition upon the composition of the zeolite (47).

Over the temperature range of 60 - 250°C, the predominant type of zeolite found is of the P type. Above 150°C, the analcime type, Na-B, was dominant in the sodium gels. Zeolite G of the chabazite type was found less frequently in mixed gels than from the K_2O gels. Type S (gmelinite) did not form in a mixed base gel.

As shown in Table 4.13, the ratio of K to Na in zeolite X that forms from a mixed gel is approximately the same as that in the starting mixture. Although zeolite L forms from gels containing Na_2O in the ratio $Na_2O/Na_2O + K_2O = 0.5$, the cation content of zeolite L is mostly potassium (Fig. 4.11b). The structure of zeolite L prefers the incorporation of K^+ to Na^+ even though it will tolerate large amounts of the latter cation in the reaction mixtures. Similarly, zeolite T forms from gels rich in Na_2O relative to K_2O, but the composition of the product is much richer in K_2O, again indicating a selectivity during crystallization for potassium over sodium.

Zeolite D appears to incorporate both sodium and potassium ions quite readily. Zeolite D is structurally related to chabazite; synthetic types of chabazite (zeolite R and zeolite G) have been formed in both the Na_2O- and K_2O-Al_2O_3-SiO_2 systems. Therefore it is to be expected that an intermediate member would also form quite readily. Usually, however, the ratio of SiO_2/Al_2O_3 in zeolite D is higher than that of zeolite R.

A zeolite structurally like phillipsite and designated ZK-19 has been synthesized (79). The composition of the zeolite varied over a wide range in terms of silica/alumina ratio depending upon the synthesis conditions. Similarly the proportion of Na_2O and K_2O was varied. The silica/alumina ratio varied with the source of silicate. High silica contents were favored by use of water glass whereas low silica/alumina ratios were favored by the use of sodium metasilicate. In addition, phosphate complexing of the aluminum was employed (see Section G). The x-ray powder pattern for ZK-19 was found to agree very closely with that of phillipsite obtained from a marine environment (Table 4.96).

A summary of the synthesis conditions and properties of zeolites D, E, T and ZK-19 is given in Table 4.12. Characteristic x-ray powder data are in Table 4.35, 4.39, 4.84, and 4.96.

Rb_2O and Cs_2O Systems

Only one cesium aluminosilicate mineral is known, the feldspathoid pollucite, $CsAl_2Si_4O_{12}$, related structurally to analcime. The large cesium atoms occupy spaces normally occupied by H_2O in analcime. No rubidium or cesium zeolite minerals are known. During an extensive study of the hydrothermal crystallization of rubidium and cesium aluminosilicate gels, Barrer and McCallum synthesized an anhydrous analcime-type species as well as a zeolite, Rb-D, which is apparently related to K-F or Z. The x-ray powder pattern is similar to that of the potassium zeolite Z; no Cs zeolite was found (80). (See Section F.)

D. ZEOLITES OF THE ALKALINE EARTHS

The Calcium Aluminosilicate Zeolites

Several calcium zeolites have been synthesized in the $CaO-Al_2O_3-SiO_2-H_2O$ system; all appear to be structurally related to known mineral zeolites. Although hydrous aluminosilicate gels have been employed as reactants, significant crystallization of a zeolite phase does not occur below 225°C. Most of the successful experiments were conducted in the temperature range of about 300 to 400°C at svp. Synthetic zeolites Ca-D, Ca-I, Ca-J, Ca-L and Ca-Q were prepared by Barrer and Denny (26). They employed as reactants two forms of silica; a stable silica sol containing 30 wt% silica, and a finely powdered silica glass of high purity. Calcium aluminosilicate gels do not crystallize readily; long periods of time are required. Further, reactive gels using silica sol were less readily crystallized than mixtures containing the powdered silica glass (Fig. 4.12).

> The alumina was obtained as a freshly prepared, amorphous gel. The calcium introduced as CaO was obtained by the decomposition of pure $CaCO_3$. All of the experiments were conducted in stainless steel autoclaves.

Zeolite Ca-L was referred to as a "calcium harmotome" and was prepared in good yield; this zeolite is a relative of garronite. Zeolite Ca-I is related to thomsonite and was not obtained pure; zeolite Ca-J, related to epistilbite, was obtained in fair yields at 250°C. The synthetic species termed Ca-Q, related to mordenite, was obtained at 390°C. Zeolite Ca-D, related to wairakite, and Ca-E, related to analcime, were very commonly formed in this system. Synthesis conditions are given in Table 4.14. In general, identification was based upon x-ray powder data. No chemical analyses were made. Thermogravimetric characterizations were also employed (See Chap. 6).

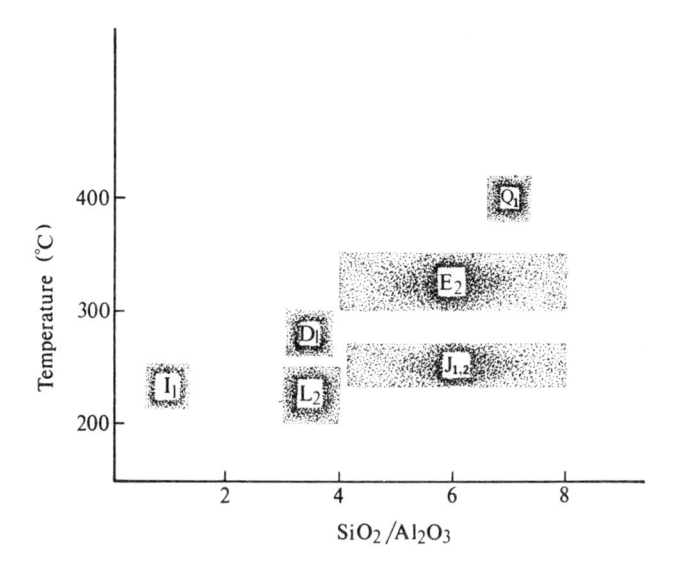

Figure 4.12 Synthesis of zeolites in the CaO-Al$_2$O$_3$-SiO$_2$-H$_2$O system (26). (1) using SiO$_2$ sol, and (2) using powdered SiO$_2$ glass.

Similar experimental studies in this system were made by Koizumi and Roy (81). They reported the preparation of synthetic types of heulandite, wairakite, scolecite, and epistilbite (Ca-J). Identification was based on x-ray powder data. The phase termed CASH-II is a synthetic type of garronite and is probably the Ca-L reported by Barrer and Denny (26).

In an earlier study of this system, Buckner used CaO-Al$_2$O$_3$-SiO$_2$ gels, ranging in compositon from CaO · Al$_2$O$_3$ · 2 SiO$_2$ to CaO · Al$_2$O$_3$ · 10 SiO$_2$ and calcined at 650°C. A synthetic type of epistilbite formed at 250°C while above 300°C the wairakite type (Ca-D) phase resulted. Attempts to synthesize a laumontite-type zeolite failed (82)

Synthetic types of wairakite and mordenite were reported by Ames and Sand (83). The wairakite type was formed from calcined gels of compositions SiO$_2$/Al$_2$O$_3$ ⩾ 5 at 325°C and 1000 atm. The mordenite-

296 THE SYNTHETIC ZEOLITES

Table 4.14 Synthetic Zeolites CaO - Al$_2$O$_3$ - SiO$_2$ - H$_2$O System

Zeolite	Reactant Composition (moles/Al$_2$O$_3$) CaO	SiO$_2$	H$_2$O	Reactants	Typical Conditions Temp. (°C)	Zeolite Composition[d] (moles/Al$_2$O$_3$) CaO	SiO$_2$	H$_2$O	Properties	X-Ray Table	Ref.	Other Ref.
Ca-D	1	4	xs[a]	Al$_2$O$_3$·3 H$_2$O CaO Silica sol	260 24 days pH 5.6	1	4	2	Tetragonal, n = 1.496, wairakite-type	4.36	26	81, 83, 84
Ca-E	1	4	xs[a]	Al$_2$O$_3$·3 H$_2$O CaO Silica glass	350 4 days pH 6.7	1	4	2	n = 1.493, analcime-type	4.40	26	
Ca-I	1	1-2	xs[a]	Al$_2$O$_3$·3 H$_2$O CaO Silica sol	245 34 days	1	2	2	Impure, thomsonite-type	4.54	26	85
Ca-J	1	6-7	xs[a]	Al$_2$O$_3$·3 H$_2$O CaO Silica glass	250 31 days	1	6	5	Lamellar habit, epistilbite-type	4.58	26	81, 82
Ca-L	1	3-4	xs[a]	Al$_2$O$_3$·3 H$_2$O CaO Silica glass	250 31 days	1	3	5	n = 1.510, harmotome-type	4.61	26	CASH II, 81
Ca-Q	1	7	xs[a]	Al$_2$O$_3$·3 H$_2$O CaO Silica sol	390	1	10	6.7	Acicular, mordenite-type	—	26	81, 84
"Clinoptilolite"	1	5-7	xs[e]	Dried gel	340-380 at 15,000 - 24,000 psi					—	84	
"Ferrierite"	1	7	xs[e]	Dried gel	350-370 15,000 - 24,000 psi					—	84	

Table 4.14 Synthetic Zeolites CaO - Al₂O₃ - SiO₂ - H₂O System *(continued)*

Zeolite	Reactant Composition (moles/Al₂O₃) CaO	SiO₂	H₂O	Reactants	Typical Conditions Temp. (°C)	Zeolite Composition (moles/Al₂O₃) CaO	SiO₂	H₂O	Properties	X-Ray Table Ref.	Other Ref.
Heulandite	1	7	xs [b]	Gels dried at 600°C	200 37,000 psi 14 days	1	7	5		4.52	81
Scolecite	1	3	xs [b]	Gels dried at 600°C	230-285 [c] 15,000 psi	1	3	3	$n = 1.513$		81

[a] H₂O not specified in each experiment, but given as 10 cc H₂O to 0.75 g of gel on dry basis. In the case of Ca-D, for example, the mole ratio H₂O/CaO = 300.
[b] H₂O in large excess not specified.
[c] Seed crystals of mineral scolecite used.
[d] No products were analyzed; compositions are assumed.
[e] Syntheses carried out in Morey bombs or "cold-seal" bombs at high pressures (157, 158).

type was produced from a gel of CaO · Al$_2$O$_3$ · 10 SiO$_2$ composition at temperatures above 350°C.

In a rather extensive study of the stability of zeolite phases Coombs et al. (8) also reported the formation of a wairakite-type phase. Most of this work is concerned with the Na$_2$O-CaO-Al$_2$O$_3$-SiO$_2$ system which is discussed later. Because of the high temperatures required and low reactivity of these mixtures the large scale synthesis of calcium aluminosilicate zeolites has not been practiced.

Magnesium in Zeolites

Although magnesium occurs in zeolite minerals (e.g. offretite, faujasite, dachiardite and ferrierite) studies of the hydrothermal treatment of magnesium aluminosilicate glasses or similar oxide mixtures have not resulted in the formation of a magnesium zeolite.

Formation of clay-type phases such as montmorillonite, however, was accelerated by the addition of MgO (86, 87). In experiments simulating the action of sea water on basalt, the presence of magnesium resulted in montmorillonite as the chief product; when no magnesium was present, an analcime-type zeolite formed.

Strontium Aluminosilicate Zeolites

The hydrothermal synthesis of strontium zeolites from hydrous gels of the general composition SrO-Al$_2$O$_3$-nSiO$_2$, where n varied between 1 and 9, was studied by Barrer and Marshall (88). The gels were prepared from recrystallized SrOH·8H$_2$O, freshly prepared Al(OH)$_3$ and a colloidal silica sol. The zeolite products (Table 4.15) all appear to be related structurally to zeolite minerals on the basis of x-ray powder data. These include new synthetic types of ferrierite (Sr-D) and yugawaralite (Sr-Q). The other synthetic species (Sr-G, Sr-F, Sr-I, Sr-M and Sr-R) are related to structures synthesized in other systems. In addition to x-ray powder data the Sr-zeolites were further characterized by differential thermal analysis and thermogravimetric analysis. Some gas adsorption studies were made to confirm the zeolitic character of Sr-F and Sr-D. None of the strontium zeolites were chemically analyzed to confirm their composition. After crystallization the cold solutions were not very basic and had a pH of 7 or a little higher. Figure 4.13 is

Table 4.15 Synthetic Zeolites SrO - Al$_2$O$_3$ - SiO$_2$ - H$_2$O System

Zeolite	Reactant Composition (moles/Al$_2$O$_3$) SrO	SiO$_2$	H$_2$O	Reactants	Typical Conditions Temp. (°C)	Properties[b]	X-Ray Table	Ref.	Other Ref.
Sr-D	1	9	a	Sr(OH)$_2$·8 H$_2$O Al$_2$O$_3$·3 H$_2$O Silica sol	340 10 days	Lathlike crystals, ferrierite-type $\alpha = 1.473, \gamma = 1.488$	4.38	88	89, 84
Sr-G	1	3	a	Sr(OH)$_2$·8 H$_2$O Al$_2$O$_3$·3 H$_2$O Silica sol	150 35 days	Spherulitic-crystals, chabazite-type	4.48	88	
Sr-F	1	4.4	a	Sr(OH)$_2$·8 H$_2$O Al$_2$O$_3$·3 H$_2$O Silica sol	205 6 days	Hexagonal crystals not reproducible, gmelinite-type	4.44	88	
Sr-I	1	3	a	Sr(OH)$_2$·8 H$_2$O Al$_2$O$_3$·3 H$_2$O Silica sol	380 3 days	Impure, not reproducible, spherulitic, analcime-type	4.55	88	84
Sr-M	1	7	a	Sr(OH)$_2$·8 H$_2$O Al$_2$O$_3$·3 H$_2$O Silica sol	300 5 days	Wheat-sheaf habit, mordenite-type	4.66	88	84
Sr-Q	1	8	a	Sr(OH)$_2$·8 H$_2$O Al$_2$O$_3$·3 H$_2$O Silica sol	340 4-7 days	Fibrous crystals, yugawaralite-type	4.79	88	84, 89
Sr-R	1	9	a	Sr(OH)$_2$·8 H$_2$O Al$_2$O$_3$·3 H$_2$O Silica sol	250 23 days	Impure, heulandite-type	4.82	88	
Clinoptilolite	1	10	a	gels and glasses	300-360				84

[a]H$_2$O given as 7 ml/0.5 g gel on dry basis. For Sr-D, this produces a H$_2$O/SrO mole ratio = 485.
[b]No analyses reported.

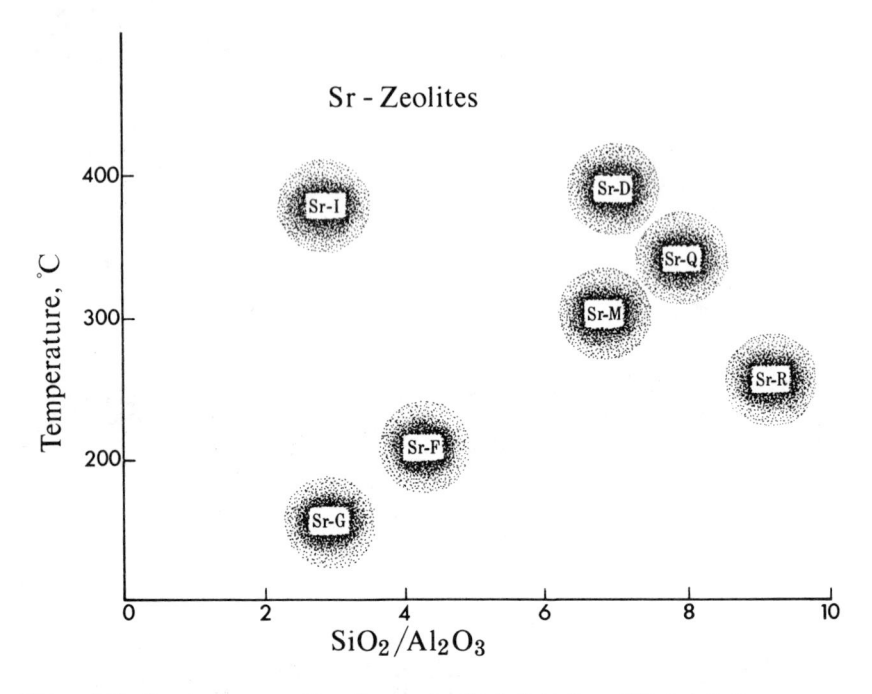

Figure 4.13 Reactant composition diagram for the SrO-Al$_2$O$_3$-n SiO$_2$(xs) H$_2$O system where SrO/Al$_2$O$_3$ = 1 (88).

a reaction diagram relating the silica content of the gel with temperature for the formation of the strontium zeolites.

Synthetic Barium Aluminosilicate Zeolites

Barium-rich zeolite minerals include harmotome, brewsterite and edingtonite. Four synthetic barium zeolites were prepared by Barrer and Marshall (90) from hydrous BaO-Al$_2$O$_3$-nSiO$_2$ gels where n varied between 1 and 9. Temperatures of 110 - 450°C were used for the hydrothermal crystallization. Identification and characterization of the synthetic barium zeolites was based upon differential thermal analysis, thermogravimetric analysis and x-ray powder data. Chemical analyses were not carried out. Only one zeolite could be structurally related to a mineral, the zeolite Ba-M. Ba-G has been related structurally to zeolite L. Data relative to the synthesis conditions are given in Table 4.16 and reaction diagrams are shown in Fig. 4.14.

Synthetic zeolites which contain different alkaline earth cations (Ca^{2+}, Sr^{2+}, or Ba^{2+}) do not, in general, resemble each other. The syn-

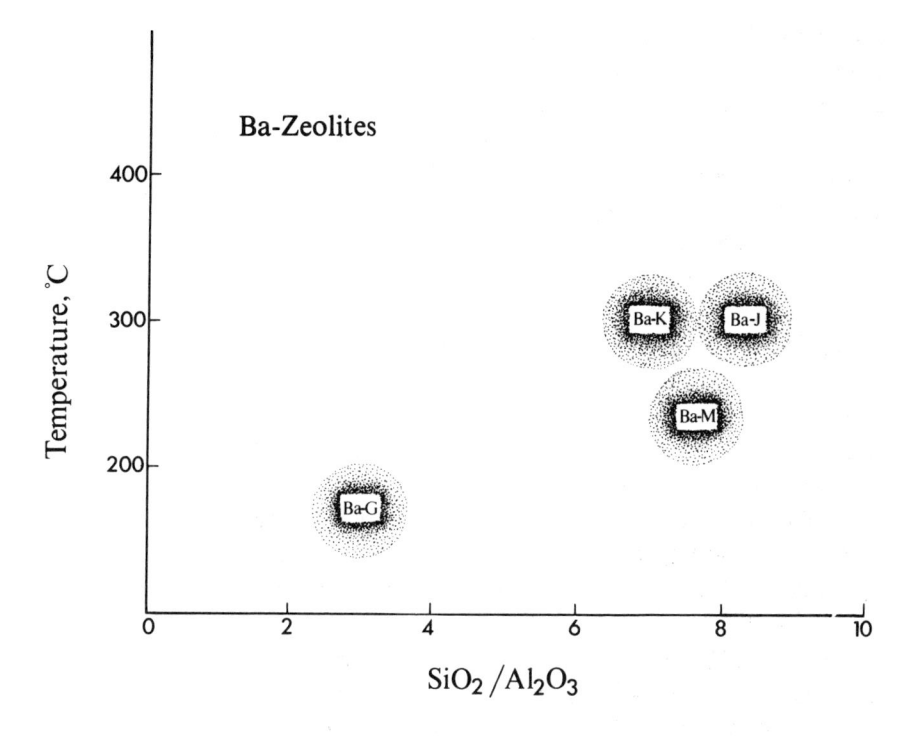

Figure 4.14 Reactant composition diagram for the BaO-Al$_2$O$_3$-n SiO$_2$- excess H$_2$O where BaO/Al$_2$O$_3$ = 1 (90).

thetic harmotome-type, Ba-M is not observed in the calcium or strontium zeolites. Mordenite-types were observed, irreproducibly, in the SrO and CaO systems, but not in the BaO system. It is probable that the reaction conditions of the parent gel are more important in determining the zeolite species than the particular cation species.

Sodium-Calcium Aluminosilicate Zeolites

The hydrothermal crystallization of oxide mixes and glass compositions in the system Na$_2$O-CaO-Al$_2$O$_3$-SiO$_2$-H$_2$O was carried out as part of a study of the stability of zeolite minerals (8). Reaction mixtures were prepared from CaO, NaOH solution, active alumina and analytical reagent grade silicic acid. Reactive gels which would favor the formation of metastable phases were not employed. In most cases, the reaction time was about four weeks. The data for runs at 275°C are shown in the triangular diagram (Fig. 4.15) in terms of the mole % of Na_2O, CaO and ($SiO_2 + Al_2O_3$) on an anhydrous basis. Only the silica-rich portion of the diagram is relevant. Zeolite phases that were synthesized include an analcime-type, a mordenite-type, an epistilbite-type and an unknown "Z-phase." Synthetic types related to thomsonite and heulandite were also reported. The "Z-phase" was obtained only from mixed Na-Ca compositions. X-ray powder data indicate that it resembles ferrierite or zeolite Sr-D reported by Barrer and Marshall (89).

The common and geologically important laumontite-type has not been synthesized. The epistilbite-type was readily formed at the lower temperatures from either the glass compositions (350 - 360°C) or oxide mixes. Heulandite is a common zeolite mineral. A synthetic analog was found as a minor component in these synthesis experiments.

In the presence of quartz, the upper limit for the stability of analcime was established as 280°C at 1000 bars pressure. Mordenite-type appeared at temperatures below 230 - 245°C at 2000 bars pressure. In the $CaO - Al_2O_3 - SiO_2 - H_2O$ system, the synthetic wairakite-type was the most stable zeolite to form, the upper limit varying from 375°C for oxide mixes to 450°C for the glass compositions.

The calcium-rich glass compositions react readily above 300°C. With increasing content of sodium the ease of crystallization diminished. The sodium glass ($Na_2O·Al_2O_3·14.5\ SiO_2$) did not crystallize at 300°C in two weeks. However this result is probably due to the concurrent increase in SiO_2/Al_2O_3 ratio, rather than the replacement of calcium by sodium; i.e., the calcium glass had the composition $CaO-Al_2O_3-6.25\ SiO_2$. This

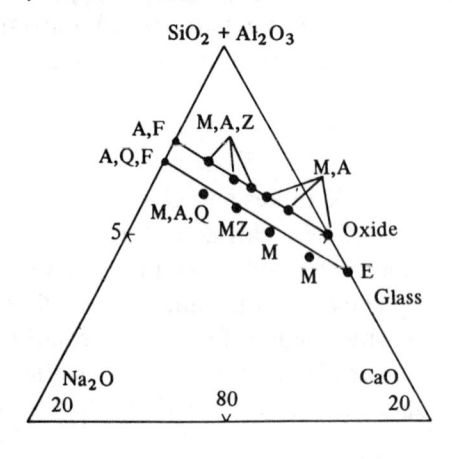

Figure 4.15 Compositions in the $Na_2O-CaO-Al_2O_3-SiO_2-H_2O$ system. Mole % is based on oxide mixes and glasses. The diagram shows Na_2O = 0– 20, CaO = 0 - 20, and ($SiO_2 + Al_2O_3$) = 80 - 100 mole %. $Al_2O_3 = (Na_2O + CaO)$, $t = 275°C$; identified by x-ray powder patterns. Oxide compositions run at 200 atm for 14 days. Glass compositions were run at svp for 28 days. Zeolites and products are indicated according to: Z = "Phase Z" which is probably a ferrierite-type; A = analcime-type; F = feldspar; M = morderite-type; E = epistilbite-type (●).

Table 4.16 Synthetic Zeolites: BaO - Al$_2$O$_3$ - SiO$_2$ - H$_2$O System

Zeolite	Reactant Composition (moles/Al$_2$O$_3$) BaO	SiO$_2$	H$_2$O	Reactants	Typical Conditions Temp. (°C)	Properties	X-Ray Table	Reference
Ba-G	1	3	a	Ba(OH)$_2$·8 H$_2$O Al$_2$O$_3$·3 H$_2$O Silica sol	150-200 3-4 weeks Impure		4.49	90, 91
Ba-J	1	8	a	Ba(OH)$_2$·8 H$_2$O Al$_2$O$_3$·3 H$_2$O Silica sol	300 9-13 days 100% yield	Fibrous	4.57	90
Ba-K	1	8	a	Ba(OH)$_2$·8 H$_2$O Al$_2$O$_3$·3 H$_2$O Silica sol	300 16 days 100% yield	Spherulitic	4.59	90
Ba-M	1	7.5	a	Ba(OH)$_2$·8 H$_2$O Al$_2$O$_3$·3 H$_2$O Silica sol	220-250 3 weeks 100% yield	Cruciform twins, harmotome-like	4.65	90, 91
Clinoptilolite	1	6-8	xs	Calcined kaolin, amorphous silica	250-300			84

[a]Ratio of H$_2$O/gel on dry basis was 7; see Section F on synthesis from metakaolin.
[b]No chemical analyses reported.

is apparent in Fig. 4.15.

E. THE ALKYLAMMONIUM OR NITROGENOUS ZEOLITES

The feldspar mineral, buddingtonite contains 8.3 wt % $(NH_4)_2O$ and synthetic ammonium muscovites have been prepared (92, 93). Ammonium silicates have been proposed by Eugster and Munoz as a primary source of atmospheric nitrogen (94). However, there is no known ammonium or alkylammonium zeolite mineral.

The exchange cations in zeolite minerals are alkali metal or alkaline earth ions. Other cations, including alkylammonium ions may be introduced by ion exchange (Chap. 7).

Although synthesized most readily from alkali hydroxide-aluminosilicate gels, zeolites have been also synthesized from systems containing quaternary ammonium ions (93, 95) in particular, the tetramethylammonium (TMA) cation. The zeolites crystallized from the systems involving alkylammonium ions require two bases. The alkylammonium base is used together with alkali hydroxide in nearly every case. Other types of organic bases have been employed including tetraethylammonium (TEA) and large cations derived from 1,4-diazabicyclo-(2.2.2.) octane (108). Additionally, combinations of two alkali metal bases, that is, sodium hydroxide and potassium hydroxide have been used together with the TMA ion.

The synthesis conditions and important properties of the nitrogenous zeolites are summarized in Table 4.17. Because of the unusual character of these zeolites in terms of their properties and preparation, some additional discussion is warranted.

Initial Investigations (93, 95)

"Nitrogenous"-types of zeolite A (N-A), zeolite X(N-X), zeolite Y(N-Y), zeolite P(N-P) and an ammonium zeolite of the analcime-type were prepared. Identification was made by chemical analysis and x-ray powder data.

The reaction mixtures are prepared from colloidal silica, freshly precipitated aluminum hydroxide and TMA-OH. The $Al(OH)_3$ is dissolved in aqueous TMA hydroxide solution and then added to the colloidal silica. The resulting mixture was heated at 100°C for several days. Typically the TMA counterpart of zeolite A(N-A), was prepared as a pure phase in 8 days from a mixture of compositions: 1.5[TMA_2O]-

Table 4.17 Synthetic Zeolites of Organic Bases

Zeolite	Reactant Composition (moles/Al$_2$O$_3$) $(R_4N)_2O$	Na$_2$O	SiO$_2$	H$_2$O	Reactants	Typical Conditions Temp. (°C)	Zeolite Composition (moles/Al$_2$O$_3$) $(R_4N)_2O$	Na$_2$O	SiO$_2$	H$_2$O	Properties	X-Ray Table	Ref.
N-A	3.5 TMA[a]	—	10	200	Colloidal SiO$_2$ (CH$_3$)$_4$NOH Al(OH)$_3$	100 5 days	0.5 TMA	0.4	4.4	4.9	Cubes, <0.5μ	4.28	93, 95
ZK-4	8.3 TMA[a]	1.46	3.9	322	Silica gel (CH$_3$)$_4$NOH NaAlO$_2$	100 1-3 days	0.15 TMA	0.85	3.2	6	Cubic A-type	4.94	96
N-P	1.5 (CH$_3$)$_2$NH$_2$⁺	—	3.0	200	Colloidal SiO$_2$ (CH$_3$)$_2$NH$_2$OH Al(OH)$_3$	200 6 days	C/N = 1.34		—	—	Euhedral	—	95
N-X	1.5 TMA	—	2.0	190	Colloidal SiO$_2$ (CH$_3$)$_4$NOH Al(OH)$_3$	100 8 days	C/N = 3.44		—	—		4.89	95
N-Y	2.5 TMA	—	3.4	314	Colloidal SiO$_2$ (CH$_3$)$_4$NOH Al(OH)$_3$	100 13 days	0.6 TMA	0.42	3.46	4.5	Octahedrons, 0.5μ	4.91	95
ZK-5	11 DDO[b]	1.5	11	480	Silica gel NaAlO$_2$ C$_8$H$_{18}$N$_2$(OH)$_2$	100 8 days	~0.5 DDO	0.515	4.7	6-10		4.95	98
α	1.52 TMA	1.47	35.3	660	Colloidal SiO$_2$ NaAlO$_2$ TMA-OH	80 6 days	—	0.5	6.08	—	Cubic, A-type adsorbs n-hexane	4.30	97
β	10.5 TEA[c]	0.33	99	1400	Colloidal SiO$_2$ Al and NaAlO$_2$ TEA-OH	100 60 days	—	0.54	75	—	Adsorbs n-hexane	—	99

306   THE SYNTHETIC ZEOLITES

Table 4.17 Synthetic Zeolites of Organic Bases *(continued)*

Zeolite	Reactant Composition (moles/Al$_2$O$_3$) $(R_4N)_2O$	Na$_2$O	SiO$_2$	H$_2$O	Reactants	Typical Conditions Temp. (°C)	Zeolite Composition (moles/Al$_2$O$_3$) $(R_4N)_2O$	Na$_2$O	SiO$_2$	H$_2$O	Properties	X-Ray Table	Ref.
TMAΩ	1.40 TMA	5.60	20	280	Colloidal SiO$_2$ Al(OH)$_3$ NaOH, TMA-OH	100 64 hr	0.36 TMA	0.71	7.3	6.3	Small spherulites, large pore	4.72	107, 160
N	0.4 TMA	1.6	2.0	80	Calcined Kaolin TMA-OH NaOH	Age 24 hr 100 2 hr	0.04 TMA	0.83	2.07	2.86	Small pore adsorbent	4.70	100, 119
ZSM-4	0.35 TMA	6.35	16.4	324	TMA-OH NaOH Na silicate NaAlO$_2$ AlCl$_3$	100 70 hr	0.18 TMA	0.865	6.56		Cubic, a = 22.2 Å, adsorbs cyclohexane	4.101	101, 164
ZSM-5	8.6 TPA^d	10	27.7	453	TPA-OH NaAlO$_2$ SiO$_2$	150 5-8 days		0.89	31.1 (calcined 1000°C)		0.5-2μm, adsorbs n-hexane	4.102	103
ZSM-8	7.14 TEA	1.24	120	1520	TEA-OH NaAlO$_2$ Colloidal SiO$_2$ NaOH	171 6 days	0.57 TMA	0.14	43	—	adsorbs n-hexane	4.103	102
"Gismondine"	Not available				(CH$_3$O)$_4$Si TMA-OH (iso-PRO)$_3$Al	130 8-10 days	TMA	—	6	2	0.4-0.5mm, isometric crystals	—	161
TMA "Offretite"	0.95 TMA	3.6 K$_2$O=6.3	19	300	KOH NaOH TMA-OH Colloidal SiO$_2$ NaAlO$_2$	100 4-8 days	0.28 K$_2$O = 0.63	0.11	7.67 ~8		0.5 × 1.5mm, 4.73 agglomerates of needles		104

Table 4.17 Synthetic Zeolites of Organic Bases *(continued)*

Zeolite	Reactant Composition (moles/Al$_2$O$_3$) (R$_4$N)$_2$O	Na$_2$O	SiO$_2$	H$_2$O	Reactants	Typical Conditions Temp:(°C)	Zeolite Composition (moles/Al$_2$O$_3$) (R$_4$N)$_2$O	Na$_2$O	SiO$_2$	H$_2$O	Properties	X-Ray Table	Ref.
TMA-O	0.9 TMA	K$_2$O=8.1	20	315	KOH TMA-OH Colloidal SiO$_2$	80 7 days	0.33 K$_2$O=0.72	—	8	4.1	Rounded rods offretite-type		107
ZK-20	2.7 9.9 e	3.15 K$_2$O=3.15 1.4	20 10.4	315 —	Al(OH)$_3$ (C$_7$H$_{15}$N$_2$)OH$_2$ Silica gel Sodium aluminate	80-100 5 days	0.47 K$_2$O=0.38 0.1-0.2	0.16 0.8-0.9	7.84 4.5	3.5	Levynite-type	4.97	108
TMA-E	4 TMA	4	15	400	NaOH TMA-OH Al(OH)$_3$ Colloidal SiO$_2$	80 7 days	0.22	0.74	5.78	3.7	Erionite-type	4.42	107
LOSOD	25 BP^f	0.5	2	400		100 20 days		1	2	1.5	Adsorbs H$_2$O	4.63	162
ZSM-10	0.4 DDO	4.5 K$_2$O	15	368		100 10 days	0.3	0.76	7.4		Adsorbs cyclohexane	4.104	109

^{a}TMA = tetramethylammonium
^{b}DDO = [CH$_3$–N◯N–CH$_3$]$^{2+}$
^{c}TEA = tetraethylammonium
^{d}TPA = tetrapropylammonium
^{e}R = CH$_3$–N◯N
^{f}BP = bispyrrolidinium, C$_8$H$_{16}$N$^+$

Al_2O_3 -6 SiO_2 - 200 H_2O. A high purity TMA counterpart of zeolite X (N-X) was prepared in 5 days from a reaction mixture of composition: 4.2[TMA_2O-] Al_2O_3-3 SiO_2 -210 H_2O. The TMA form of zeolite Y (N-Y) was also prepared in 13 days at 100°C from a reactant composition: 2.5[TMA_2O] · Al_2O_3 · 3.4 SiO_2 · 314 H_2O.

Chemical analysis of the N-A and N-Y zeolite products, which were crystallized in glass vessels, showed the presence of significant quantities of sodium ion which was dissolved from the glass container. Even in synthesis experiments conducted in stainless steel vessels, the product compositions contained as much as 5 wt% Na_2O. This is because a small amount of NaOH is used to stabilize the silica sol and is preferentially incorporated into the zeolite. Using ammonium-stabilized silica sol as a reactant in a stainless steel vessel, zeolite N-A could not be crystallized in 11 weeks at 100°C. Traces of sodium help to nucleate crystallization of zeolites N-A and N-Y. The rate of crystallization seems to depend on the amount of sodium present. The unit cell of N-A is significantly smaller than that of NaA. This is due to the difference in cation content and higher SiO_2/Al_2O_3 ratio. Table 4.17 lists pertinent synthesis conditions and chemical compositions determined by analysis of the N-Y, N-X and N-A phases. The Si/Al_2 ratio in zeolite N-A is significantly higher than is found in the sodium counterpart. Variations in the Si/Al_2 ratio in zeolite N-A to 7.0 were found. Analysis of zeolite N-A and N-Y confirmed the C/N ratio of 4 which corresponds to the TMA ion. These zeolites were further characterized by ion-exchange, DTA, TGA, and adsorption measurements.

Zeolite N-A crystallizes as small euhedral cubes and zeolite N-Y as perfect octahedra. The synthetic form of zeolite P, here termed N-P (also referred to as N-L by Barrer and Denny was prepared from gels using dimethylammonium hydroxide at 200°C. With ammonium hydroxide as the base, an ammonium-type analcime, NH_4-B, formed at 450°C, as well as an ammonium type of zeolite P. From TMA-OH as the base, the P-type zeolite was also prepared (93).

Zeolite ZK-4

Using TMA silicate and sodium aluminate as reactants, a similar TMA-containing type of zeolite A has been prepared and named zeolite ZK-4 (96). The reaction mixture was prepared from sodium aluminate and TMA disilicate pentahydrate previously prepared from

TMA-OH and silica gel. The zeolite was crystallized from reaction mixture compositions that fell within the following ranges of composition: $SiO_2/Al_2O_3 = 4\text{-}11$; $Na_2O/Al_2O_3 = 1\text{-}1.5$; $(Na_2O+TMA_2O)/Al_2O_3 = 9\text{-}30$; $H_2O/Al_2O_3 = 100\text{-}350$. Crystallization periods of 24 to 72 hours were usually necessary. Chemical analysis of ZK-4 indicated a high preference for sodium. A typical composition was: $Na_8(TMA)_{12}[(AlO_2)_{9.2}(SiO_2)_{14.8}] \cdot 28H_2O$. The C/N ratio of 3.94 confirmed the presence of the TMA ion in the zeolite. The Si/Al ratio found by chemical analysis is consistent with the smaller unit cell constant, 12.16 A as compared to 12.3 A for zeolite A. Thus ZK-4 is a type of zeolite A with higher Si/Al ratio and sodium as the major cation.

Zeolite alpha (α)

Another zeolite with the type A structure is designated zeolite α (97). The reaction mixture includes TMA-OH and sodium hydroxide as the active bases with compositions which are in the following ranges: $SiO_2/Al_2O_3 = 15\text{-}60$; $Na_2O/(Na_2O+TMA_2O) = 0.1\text{-}0.3$
$H_2O(Na_2O+TMA_2O) = 30\text{-}60$; $(Na_2O+TMA_2O)/SiO_2 = 0.5\text{-}1.0$
Zeolite α is reported to have a pore diameter of 5.5 A in the sodium form and about 5 A in the potassium form. In a typical example, crystallization was carried out at 78°C for 6 days. The chemical composition of the product is reported to be:
$0.2\text{-}0.5\ TMA_2O \cdot 0.5\text{-}0.8\ Na_2O \cdot Al_2O_3 \cdot 4\text{-}7\ SiO_2 \cdot 8H_2O$

Zeolite β

A material, referred to as zeolite β, is crystallized from reaction mixtures containing tetraethylammonium (TEA) hydroxide as the base (99). The composition of zeolite β is given as:
$(TEA,Na)_2O \cdot Al_2O_3 \cdot 5\text{-}100\ SiO_2 \cdot 4H_2O$
The zeolite is characterized by having a very high silica-alumina ratio with values up to 96 being reported. The structure is supposed to be cubic with a unit cell constant of 12.04 A. The x-ray powder data in the Ref. 99 cannot be indexed on a cubic unit cell and characterization is incomplete.

ZK-5

The synthesis of zeolite ZK-5 from sodium aluminosilicate gels was achieved using the dibasic nitrogenous cation (98).

$$CH_3N^+\bigcirc^+NCH_3, \text{ abbreviated as } C_8H_{18}N_2^{2+}$$

This cation was introduced in the form of 1,4-dimethyl-1, 4-diazonia-cyclo (2.2.2) octane silicate, which was first prepared from 1,4- diazabicyclo (2.2.2) octane. The composition of the reaction mixtures for the synthesis of ZK-5 have the following mole ratios:

SiO_2/Al_2O_3:	4 - 11
$Na_2O/Al_2O_3 + C_8H_{18}N_2O/Al_2O_3$:	6 - 19
Na_2O/Al_2O_3:	1.5 - 2.3
H_2O/Al_2O_3:	200 - 700

Alumina was added in the form of sodium aluminate solution, which was then mixed with the quaternary ammonium silicate solution to form a gelatinous amorphous gel. Heating at 100°C for a period of 9 days resulted in the formation of zeolite ZK-5. By analysis the chemical composition of the zeolite is:

$$0.3\text{-}0.7\ Na_2O \cdot 0.3\text{-}0.7\ RO \cdot Al_2O_3 \cdot 4.0\text{-}6.0\ SiO_2 \cdot 6\text{-}10 H_2O$$

where $R = C_{18}H_{18}N_2^{2+}$

The x-ray powder data (Table 4.95) are not like any mineral zeolite. Structural analysis (Chap. 2), confirmed it as a unique zeolite structure.

Omega

The synthetic zeolite designated TMA-Ω (for tetramethylammonium zeolite Ω) is formed at low temperatures (80 - 150°C) in the system (TMA-OH)-NaOH-Al_2O_3-SiO_2-H_2O. The composition of the reaction mixtures in oxide mole ratios should be in the following ranges (160):

$$\frac{Na_2O + (TMA)_2O}{SiO_2} = 0.3 - 0.5 \qquad \frac{SiO_2}{Al_2O_3} = 8 - 20$$

$$\frac{(TMA)_2O}{(TMA)_2O + Na_2O} = 0.2 \qquad \frac{H_2O}{(TMA)_2O + Na_2O} = 15 - 40$$

The most suitable silica sources are the reactive solid silicas as typified by Cab-O-Sil and colloidal silica sol (Ludox). Crystallization requires 2 - 8 days. A typical preparation is given in Table 4.17 and x-ray powder data in Table 4.72. As shown in Chap. 2, zeolite Ω is a thermally stable, large pore zeolite (pore size ∼7.5 A).

Zeolite N (100)

Although zeolite N is prepared from calcined kaolin as the source of silica and alumina, it is included in this section because of the organic base. Zeolite N is cubic with a large unit cell and a lattice constant equal to 37.2 A. This might imply a large channel system. However, the adsorption measurements indicate it is a small pore adsorbent (see Chap. 2).

TMA - Gismondine

A synthetic type of gismondine is crystallized from this system containing no metal cation and only TMA ion (161). The ratio Si/Al in this zeolite is higher than the corresponding mineral, gismondine; 3 as compared to 1 for the latter. This results from incorporation of the large organic cation which, due to spacial limitations, restricts the number of aluminum atoms in tetrahedral positions and, consequently, increases the silica/alumina ratio. This zeolite may be exchanged to the sodium form which is the cubic zeolite P. This confirms the fact that the structure of zeolite P is based upon a gismondine-type aluminosilicate framework.

ZK-20

Zeolite ZK-20 is synthesized utilizing the large organic cation, $C_7H_{15}N_2^+$, mixed with sodium cations. The x-ray powder data for this zeolite corresponds with data for the mineral zeolite levynite indicating that zeolite ZK-20 has the same type of aluminosilicate framework structure (108).

Offretite-type and Erionite-type

The systems involving the mixed bases (TMA, sodium, and potassium hydroxide) produced important zeolite phases including zeolite omega, a zeolite of the offretite-type, zeolite O, and a zeolite of the erionite-type, TMA-E (107). The zeolite O is obtained from starting mixtures containing TMA and potassium and is also obtained from the three base system including sodium. It did not crystallize from gels containing only sodium and TMA. Two typical compositions of offretite-type zeolites obtained from starting mixtures of different compositions are shown in Table 4.17. Ion exchange experiments on zeolite O samples containing different numbers of TMA cations per unit cell showed that one TMA cation in each gmelinite cage of the structure (see Chap. 2) is not exchanged and is effectively trapped. This indicates a templating effect of the TMA ion in zeolite synthesis. The smaller ϵ-type cages must be occupied by the potassium ions as well as with the hexagonal prism, D6R units. The synthesis of offretite-type zeolites from these mixed base systems has been also reported by others (104, 105).

The erionite-like zeolite, TMA-E, was synthesized from gels involving sodium and TMA as the bases (107). The effect of heat treatment on this zeolite is of interest. As the temperature is raised the unit cell lengthens and narrows; at 360°C there is a rapid structural conversion into a sodalite-type structure. This change in structure must involve a change in the stacking sequence of the layers of 6-rings as they are found in erionite (AABAAC) to the simplest sequence, ABC, found in the structure of sodalite (see Chap. 2).

LOSOD

Although this zeolite was synthesized in a mixed base system, sodium and bispyrrolidinium, $C_8H_{16}N^+$, and other organic bases, no organic cation was incorporated. Apparently the organic base serves only as a source of hydroxyl ions (162). The structure was determined and consists of the ABAC . . . stacking of 6-rings (see Chap. 2) (162).

Other Zeolites Containing Organic Cations

Zeolite ZSM-4 has been synthesized in the mixed base system, TMA and sodium (101). It is reported to be cubic, a = 22.2 A. This zeolite adsorbs cyclohexane indicating large pores.

The related zeolites ZSM-5 and ZSM-8 are synthesized in the TPA-sodium, and TEA-sodium mixed base systems, respectively (102, 103). Similar to zeolite β they have unusually high silica/alumina ratios, reported to approach 100, but a distinct x-ray powder pattern. Adsorption of n-hexane and some cyclohexane suggests a pore size near 6 A. No analogy to a previously known structure-type is reported.[a]

Zeolite ZSM-10 is synthesized in the potassium-DDO ($C_8H_{18}N_2^{2+}$) system (109). It adsorbs cyclohexane and the x-ray powder pattern resembles that of zeolite L.

---

[a]Structure of ZSM-5 reported. See Cpt. 2, ref. 224, and Table 4.102, p. 373.

F. THE SYNTHESIS OF ZEOLITES FROM CLAYS

Synthetic zeolites crystallize from reactive amorphous substrates other than aluminosilicate gels. Sudo and Matsuoka (110) have reported the crystallization of hydroxysodalite and a zeolite by treating powdered volcanic glass with NaOH-NaCl solutions. The zeolite, from x-ray powder data, appeared to be type X. It was not prepared pure, however. Ellis was able to convert a volcanic glass into mordenite in a natural hydrothermal system (44). These crystallizations which take place in the laboratory in short periods of time are similar to the formation of zeolites by the diagenesis of sediments of volcanic origin.

The most important process uses minerals of the kaolin group, which may be represented chemically as $Al_2O_3 \cdot 2SiO_2 \cdot 2H_2O$ (see Chap. 2) (111, 112). In order to "activate" the clay for the reaction to occur, it has been found necessary in most instances to convert the kaolin to metakaolin by thermal treatment (calcination) at temperatures of about 600°C. This amorphous material, referred to commonly as metakaolin, is then treated with aqueous, alkali metal hydroxide solutions at a convenient temperature of about 100°C. The zeolite formed depends upon the composition of the reaction mixture. For example, if only sodium hydroxide is added to the metakaolin, then zeolite A is formed according to a reaction scheme represented as follows (113):

a. $2\ Al_2Si_2O_5(OH)_4 \xrightarrow{500 - 600°C} 2\ Al_2Si_2O_7 + 4\ H_2O$

    Kaolin   metakaolin

b. $6\ Al_2Si_2O_7 + 12\ NaOH\ (aq) \xrightarrow{100°C} Na_{12}(AlO_2)_{12}(SiO_2)_{12} \cdot 27\ H_2O + 6\ H_2O$

If raw kaolin, i.e., uncalcined kaolin, is reacted with sodium hydroxide, the product is generally the feldspathoid hydrated or hydroxysodalite. The reaction of clays with sodium hydroxide has been extensively studied.

In 1946, a group at the National Bureau of Standards (114) examined processes for the extraction of alumina from clays and high-silica bauxites. In one step of the process, silica was removed from the sodium aluminate solution by precipitation as a sodalite-type of compound. While most of their products produced in this way corresponded in chemical analysis and x-ray powder pattern to a hydroxysodalite, a few were identified as nepheline-like.

Early work of Nagai

In a series of papers published in the period 1935-1940 Nagai reported the results of the hydrothermal reaction of clay materials with alkali hydroxide bases (115, 116). About 37 different types of clays were heated to aqueous sodium or potassium hydroxide solutions for periods of 1 to 5 hours at pressures ranging from atmospheric to 200 kg/cm^2. The pulverized clay material was mixed with the aqueous base, and then heated in a steam autoclave. The mixture was filtered and the residue and the filtrate were analyzed. Frequently, the solid residue weighed more than the starting clay which indicated that Na$_2$O and H$_2$O were incorporated into the clay. At times, the raw clay materials were calcined at temperatures ranging from 500 to 1260°C before the hydrothermal treatment. Experiments were conducted using NaOH, NH$_4$OH, KOH, and, in some cases, fusion with Na$_2$CO$_3$. On the basis of the chemical analyses and ion exchange studies, Nagai assigned the solid products formulae of the type M$_2$O · Al$_2$O$_3$ mSiO$_2$ · nH$_2$O, where $m = 2 - 4$, $n = 1 - 4.5$, and M is either K or Na. These were referred to variously as "zeolitic silicate, zeolitic compound, products of zeolitic form, nepheline hydrate, alkali aluminosilicate hydrate, alkali kaolin, artificial nepheline hydrate, or artificial felspathic matter of the zeolite group." Since Nagai did not utilize x-ray powder data to identify the reaction products, the formation of crystalline materials was not shown. The use of crystalline clay materials coupled with the chemical composition of the reaction mixture, i.e., the high concentration of NaOH in the solution, would preclude the formation of zeolites. The probable products are the feldspathoids such as hydroxysodalite or basic cancrinite of the type reported by others.

Kaolin Transitions

When heated, kaolin-type clays undergo several transitions in air (117). The first of these [1] takes place at about 550°C, and produces the disordered metakaolin phase by an endothermic dehydroxylation reaction. The metakaolin is then stable to about 925°C where it rearranges to give a defect aluminum-silicon spinel which is also referred to as a gamma-alumina type structure (2). Although considerable controversy

[1] $\quad$ 2 Al$_2$Si$_2$O$_5$(OH)$_4$ $\xrightarrow{550 - 600°C}$ 2 Al$_2$Si$_2$O$_7$ + 4 H$_2$O
$\qquad\qquad$ kaolin $\qquad\qquad\qquad\qquad\quad$ metakaolin

[2] $2\,Al_2Si_2O_7 \xrightarrow{925\text{-}950°C} Si_3Al_4O_{12} + SiO_2$
　　　metakaolin　　　　　　　　spinel

[3] $3\,Si_3Al_4O_{12} \xrightarrow{1050°C} 2\,Si_2Al_6O_{13} + 5\,SiO_2$
　　　spinel　　　　　　　　　mullite　　　cristobalite

has been concerned with the nature of the metakaolin phase, it is now generally concluded that metakaolin is not a simple mixture of amorphous silica and alumina but retains some order which is associated with the hexagonal layers. The so-called γ-Al_2O_3 phase converts to mullite, $3\,Al_2O_3 \cdot 2\,SiO_2$, and/or sillimanite at about 1050°C. Metakaolin is believed to be a defect phase in which the tetrahedral silica layers of the original clay structure are largely retained; adjacent are the AlO_4 tetrahedral units derived from the original octahedral layer. It is known to be more reactive; it is more easily leached by either acids or alkalis.

It is apparent from the stoichiometry that to form zeolite A only the alkali sodium is needed. After forming the initial metakaolin slurry, a low-temperature, aging treatment improves the conversion of the clay to zeolite (111). This is shown in the results given in Table 4.18a for the formation of zeolite A.

In order to form zeolites which have SiO_2/Al_2O_3 ratios greater than 2, additional SiO_2 must be added to the metakaolin. For example, to produce zeolite X, a typical reaction mixture may have a composition of $4\,Na_2O \cdot Al_2O_3 \cdot 4\,SiO_2 \cdot 160\,H_2O$ (112). The additional silica may be added in the form of sodium silicate or other sources such as colloidal silica.

Preformed Shapes

An interesting variation of this method is concerned with the formation of formed shapes or objects by starting with the kaolin in preformed shape. The initial kaolin may be shaped by extruding into pellets or may be shaped into special forms by techniques such as slip casting. Articles of useful shapes such as tubes, cylinders, etc., are formed by precasting the kaolin into an appropriate shape. Subsequent to the forming, the kaolin is dried and calcined at the appropriate temperature, about 700°C, which results in a shaped object consisting of the metakaolin. This shaped object may then be converted *in situ* into zeolites such as zeolite A by aging and digestion in NaOH solution. An-

Table 4.18a Synthetic Zeolites from Metakaolin

Zeolite	Reactant Composition (moles/Al$_2$O$_3$) Na$_2$O	SiO$_2$	H$_2$O	Reactants	Typical Conditions Temp. (°C)	Zeolite Composition (moles/Al$_2$O$_3$) Na$_2$O	SiO$_2$	H$_2$O	Ref.
A	2.4	2	96	Metakaolin, NaOH	80 - 85, aged at RT 16 hr, 8 hr; or RT, 35 days	Same as A			111
P	4	10	120	Metakaolin, NaOH, sodium silicate	100, aged at RT 3 days with stirring, 3 days	Same as P			31
X	4	4	160	Metakaolin, sodium silicate, NaOH	100, aged at RT 3 days, 1 day	1.0	2.5		112
Y	3.5	7	140	Metakaolin, NaOH, sodium silicate, NaCl (NaCl/Al$_2$O$_3$=2)	100, aged at RT 24 hr, 24 hr	1.0	3.5		112
Y	5	10	200	Calcined, acid-leached metakaolin, NaOH	100, aged at RT 24 hr, 3 days	1.0	4.1		118
S	4.4	11	264	Calcined, acid-leached metakaolin, sodium aluminate, SiO$_2$/Al$_2$O$_3$ of leached metakaolin = 143	100, aged at RT 24 hr, 1 - 14 days	0.8	5.1	7.2	31
L	2.2 K$_2$O	2.2	11	176	Calcined, acid-leached metakaolin, NaOH, KOH	100, aged at RT, 24 hr, 4 days	0.9 K$_2$O	5.7	4.7	118

Table 4.18b Synthetic Zeolites from Metakaolin with Some Other Bases

Zeolite [a]	Reactant Composition (moles/Al_2O_3)	Reactants	Conditions Temp (°C)	Remarks	Ref.
K-F	SiO_2 = 2 - 10	Metakaolin Silica gel KOH	80 - 160		120
W(K-M)	SiO_2 = 4 - 10	Metakaolin Silica gel KOH			120
G or K-G [b]	SiO_2 = 2 - 10	Metakaolin Silica gel KOH	80 - 140	SiO_2/Al_2O_3 = 2.1 to 4.5	120
Li-A	SiO_2 = 2 - 10	Metakaolin Silica LiOH	80 - 160		119
Li-H	SiO_2 = 6 - 10	Metakaolin Silica LiOH	120 - 160 7 days		119
Rb-D	SiO_2 = 2	Metakaolin Silica RbOH	80		119
Rb-M [e]	SiO_2 = 4 - 6	Metakaolin Silica RbOH	80 7 days		119
Ba-N [c]	SiO_2 = 2	1.5 M $Ba(OH)_2$ Metakaolin	80 7 days	Table 4.71	91
Ba-T [c]	SiO_2/Al_2O_3 = 2 - 6	2.5 M $Ba(OH)_2$ Metakaolin Silica gel	80	Table 4.85	91
Ba-G, L [d]		Metakaolin Silica gel 1 M $Ba(OH)_2$	100 10 days	Related to Zeolite L	91

[a] Unless otherwise indicated the zeolite type is the same as previously designated.
[b] Zeolite K-G (SiO_2/Al_2O_3 = 4.51) does not adsorb O_2 at 78°K or n-butane at 273°K. The Ca^{2+} and Li^+ exchanged forms do adsorb O_2 and n-butane.
[c] Ba-N and Ba-T are species not previously prepared in gel systems.
[d] The Ba-G, L is reported to be the Ba-zeolite prepared previously from gels (Table 4.16). It appears to be a relative of zeolite L but with larger unit cell dimensions. It was also prepared using mixed bases, $Ba(OH)_2$ and KOH. The Ba-G, L did not adsorb neopentate at 273°K unlike zeolite L due possibly to occluded $Ba(OH)_2$ or silicate.
[e] Rb-M is similar to zeolite W or K-M.

other variation includes forming the shaped object from zeolite powder mixed with raw kaolin. Figure 4.16 illustrates some specific shapes that are essentially 100% zeolite A (112).

Another technique for the synthesis of zeolites with high SiO_2/Al_2O_3 ratios from clays involves increasing the ratio in the starting clay by first leaching alumina from the clay by acid treatment (118). For example, zeolite Y with a $SiO_2/Al_2O_3 = 4.1$ was prepared from acid-leached metakaolin using an overall composition of $4.8\,Na_2O \cdot Al_2O_3 \cdot 9.6\,SiO_2 \cdot 192\,H_2O$. The yield was 95%.

Synthetic zeolite types D, S, T, L, and K-M can be formed using metakaolin with additional silica obtained from soluble sodium silicate solutions. In general, the relationship between the overall composition of the reacting system to the zeolite type formed corresponds with that observed in gel-reactant mixtures.

The conversion of metakaolin to the defect spinel phase (Eq. 2) forms one mole of reactive silica for each two metakaolins. This reactive silica can then be used as the source of additional silica to form the high

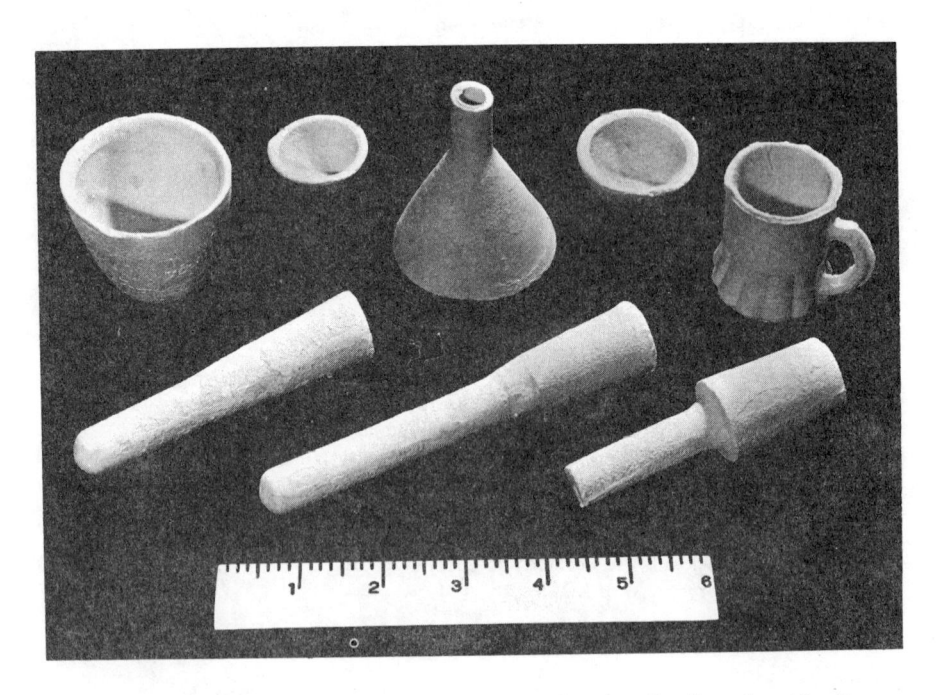

Figure 4.16 Preformed shapes of zeolite A (~90%). Mixtures of kaolin and metakaolin were slip cast prior to crystallization.

silica zeolite such as zeolite Y. This is discussed in more detail in Chap. 9.

Kaolin with Other Bases

The conversion of metakaolin to zeolites by reaction with KOH and Ba(OH)$_2$ at low temperatures has been reported. Zeolites previously synthesized in the hydrous gel systems were crystallized including Z (or K-F), K-G (chabazite-type), W (K-M), and an L-type zeolite. Two additional zeolite phases were prepared: Ba-N and Ba-T. A Ba-GL was found which resembles zeolite L. (See Table 4.18b) (91).

The use of mixed bases (NaOH + LiOH), (NaOH + KOH), (KOH + LiOH), and (NaOH + Me$_4$NOH) was also studied (119). Zeolite phases prepared included additional phases related to Z(K-F), K-G, W(K-M), P(gismondine-type), A, X, S, and N (here referred to as Na, TMA-V).

The conversion of raw kaolinite to zeolites and feldspathoids by reaction with hydroxides of the alkali metals, i.e., LiOH, NaOH, KOH, RbOH, and CsOH has been studied. With LiOH the zeolite Li-A was obtained at temperatures ranging from 130 to 140°C. With KOH the zeolites H, G, and K-F were obtained and with RbOH and CsOH the zeolite products were Rb-D and Cs-D respectively (67).

A series of sodalite hydrates, or basic-sodalites, were also prepared which varied in the amount of water and intercalated sodium hydroxide depending upon the concentration of sodium hydroxide in the initial solution. The solubility of the basic sodalite was studied as a function of sodium hydroxide concentration. In very strong solutions, greater than 30 molar, the Al$_2$O$_3$ and SiO$_2$ remain in solution and no insoluble aluminosilicate was formed. At about 400°C with sodium hydroxide, kaolinite was converted to a basic cancrinite by using about a 300% excess of sodium hydroxide. Similar studies have also resulted in the formation of zeolite P (121).

The morphology of the conversion of kaolin to zeolite A was studied by scanning electron microscopy (SEM). The study involved the conversion of calcined kaolin by a 10 wt% solution of NaOH at 100°C. Periodic samples were photographed by the SEM. The starting material consists of stacks of irregularly-shaped platelets, up to 7 microns across. After 1 hour of crystallization the mixture contains 0.5 to 3.0 micro cubes which, by x-ray powder diffraction analysis, is about 55% zeolite A. After 4 hours the material consists of agglomerated cubes without much change in particle size. (122)

The reaction of kaolin with KOH at 70 to 90°C was studied. The product varied with the concentration of KOH which was varied from 1 N to 9 N. During the first part of the reaction, the SiO_2 and Al_2O_3 content of the solution increased reaching a maximum in 100 to 240 minutes. The time depended upon the alkali hydroxide concentration and temperature. An amorphous aluminosilicate gel then formed. The amorphous aluminosilicate gel then crystallized to a crystalline aluminosilicate with a concurrent decrease in the concentration of SiO_2 and Al_2O_3 in the solution. Various zeolites were obtained including the zeolite H (or I), Z (or K-F), and G (123).

The conversion of different types of kaolin clays to zeolite A depends upon the initial crystallinity of the clay, the concentration of the sodium hydroxide solution, the temperature and time. A well-crystallized variety of kaolin was observed to convert directly to hydroxysodalite whereas a halloysite of low crystallinity formed zeolite A (124). In this study it was observed that 100% zeolite was obtained at 70°C in 16 - 24 hours but if contact with the solution continued, gradual conversion to hydroxysodalite resulted.

The low-crystallinity halloysite may be crystallized to zeolite X if additional silica in the form of aqueous sodium silicate is added. The major impurity in this conversion is zeolite P which results if the SiO_2/Al_2O_3 ratio is low, that is, about 2. From a reaction mixture with SiO_2/Al_2O_3 5, 95% zeolite X was obtained at 90°C in 16 hours. At 100°C, zeolite P formed with zeolite X as a minor component (124).

G. SYNTHETIC ZEOLITES WITH OTHER FRAMEWORK ATOMS

Other atoms which replace aluminum and silicon in oxygen tetrahedra include gallium, Ga^{3+}, phosporus, P^{5+}, and germanium, Ge^{4+}. Iron, Fe^{3+} has also been attributed to tetrahedral sites (151). Phosphorus occurs in tetrahedral coordination with oxygen in the mineral zeolites viseite and kehoeite. Goldsmith showed that the replacement of aluminum by gallium and silicon by germanium occurs in feldspars and in leucite. In addition he reported the synthesis of a thomsonite-type zeolite containing gallium. This was prepared hydrothermally from a gallium-containing glass (125).

Gallium and Germanium

The replacement of aluminum and silicon by gallium and germanium in the A, X and P zeolites was demonstrated by Barrer et. al. (47). Reaction mixtures were prepared from aqueous solutions of sodium gallate, sodium germanate, and aqueous silica sol. The zeolite products were identified by the x-ray powder patterns; types isostructual with zeolites X, A, and P and the mineral thomsonite were found. Zeolite A-type resulted only from sodium aluminogermanate gels. Zeolite X-type formed from sodium gallogermanate and sodium aluminogermanate mixtures. Zeolite P-type was prepared by recrystallization of zeolite A in NaOH. All crystallizations were conducted at 100°C; no chemical analyses were performed on the products.

In similar studies, Selbin and Mason (126) prepared a gallium-analog of zeolite X. The sodium gallosilicate reactant was prepared from tetrachlorogallate solutions which were added to alkaline sodium metasilicate solutions. The composition of the reaction mixture was Na_2O/SiO_2 = 0.5 - 1.0; SiO_2/Ga_2O_3 = 3.0 - 5.0. The crystallization was carried out at 70°C with vigorous stirring for 20 - 22 hours. Chemical analysis of the zeolite product gave the following composition on a dry basis: $Na_2O \cdot Ga_2O_3 \cdot 2.60 SiO_2$. X-ray powder data correspond with data for zeolite X, prepared from typical aluminosilicate gels. Gallium sodalite was prepared but not a gallium zeolite with the type A structure. The gallium analog of zeolite Y has been synthesized (127).

Partial replacement of Al by Ga in chabazite-type and phillipsite-type zeolites was reported by using reactant gels containing equal proportions of Al_2O_3 and Ga_2O_3 (128). The phillipsite-type crystallized from sodium galloaluminosilicate gels. The ratio $Na_2O/(Al_2O_3 + Ga_2O_3)$ in the product zeolite was very low, less than 0.7, which indicated excessive hydrolysis of the sodium. The chabazite-type was prepared from potassium galloaluminosilicate gels. The ratio Al_2O_3/Ga_2O_3 in the product zeolite varied from 5.7 to 23 which shows that the Ga atoms occupied 5 to 15% of the tetrahedral sites. A deficiency in potassium, relative to $Al_2O_3 + Ga_2O_3$, also was observed. Proof of the substitution of Al by Ga (structural, infrared, etc.) was not given.

The distribution of trace elements between clay minerals and zeolites during crystallization by the hydrothermal alteration of synthetic gels and glasses was studied by Hawkins and Roy (87). Trace amounts of boron were incorporated preferentially in the clay (montmorillonite) structure in an undefined state relative to the zeolite (analcime-type) phase. Actual incorporation of boron in a zeolite structure has not been achieved.

Zirconium, Titanium, Chromium

Crystalline zirconosilicate "zeolites" and titanosilicate "zeolites" are recently reported. Although the preparation of the materials is described and some x-ray data given, characterization is insufficient to classify them with certainty as zeolites (129). The chemical composition given for the zircono "zeolite" is: $(D_{2n}O)_x \cdot ZrO_2 \cdot (SiO_2)_y$ where n = valence, D (Me^+, Me^{2+}, NH_4^+, H^+), $x = 1.5 - 4$, $y = 4.5 - 8.5$ and for the titano "zeolite": $(D_{2n}O)_x \cdot TiO_2 \cdot (SiO_2)_y$ where $x = 0.5 - 3$, $y = 1.0 - 3.5$

Zeolites with chromium in the tetrahedral positions have been described (130). The zeolites are claimed to have the zeolite A structure with chromium substituted for silicon or aluminum. A range of materials were prepared containing various amounts of chromium that is described as not being exchangeable and therefore must occupy tetrahedral atom positions. The published x-ray powder data indicate, however, two strong reflections at d = 6.39 and d = 2.94 which are not compatible with the zeolite A structure; these lines cannot be indexed on the basis of the 12.3 A cell. It is likely therefore that the chromium is present as some impurity in insoluble form. The presence of an occluded impurity such as a chromium silicate is confirmed by the nature of the water vapor adsorption isotherm.

Phosphate Zeolites

Proof of the substitution of atoms (other than Al, Si) in tetrahedral sites of a zeolite during synthesis requires detailed and careful characterization both analytically and structurally. In most of the published work it is not possible to conclude from the results presented that substitution by other tetrahedral atoms has been achieved. Because of the presence of phosphorus in tetrahedral PO_4 units in the rare zeolites kehoeite and viseite, attempts to synthesize zeolites containing PO_4 tetrahedra were made. The formation of synthetic analogs of analcime has been reported. However, it was concluded from optical properties such as refractive index that no appreciable substitution of phosphorus for aluminum or silicon occurred (131). Similarly, other attempts to prepare zeolites containing tetrahedral atoms such as phosphorus and boron have been made but the published data do not confirm substitution (132).

Recently, Flanigen and Grose have prepared phosphorus-containing zeolites (also termed phosphate zeolites) and extensively characterized them by chemical analysis, x-ray powder data, electron microprobe analysis and adsorption. The substitution of phosphorus within the framework structure was further confirmed by infrared spectra and crystal structure analysis (133, 134).

Synthetic phosphorus-containing types of analcime (zeolite P-C) phillipsite (zeolite P-W), chabazite (zeolite P-R), zeolite A (P-A), zeolite L (P-L), and zeolite P (P-B), have been prepared. Generally, the synthesis technique involves crystallization from a gel in which phosphorus is first incorporated by a controlled copolymerization and coprecipitation of all of the component oxides in the framework, i.e., aluminate, silicate, and phosphate in the homogeneous gel phase. The crystallization of the zeolite from the gel was carried out at temperatures ranging from 80 to 210°C. Typical synthesis conditions for crystallizing the phosphorus-substituted zeolites are given in Table 4.19. Synthesis of the aluminosilicophosphate zeolite requires a reactive form of phosphorus such as phosphoric acid in the gel structure. Adding a phosphate salt, such as sodium metaphosphate, to the reactant gel does not result in the incorporation of phosphorus in the tetrahedral zeolite framework.

The x-ray powder data for the phosphorus zeolites correspond well with the data for the corresponding aluminosilicate zeolites as exemplified by the data for P-L and P-A (see Tables 4.29, 4.62). The crystals of the phosphate zeolites were generally larger than those produced in the nonphosphorus system and some were nearly 100 microns in size (except in the case of P-A). Consequently, single crystal measurements were possible to establish framework substitution from the changes in unit cell dimensions as well as changes in crystallographic symmetry. The crystals were also used for single crystal electron microprobe analysis. Typical microprobe analyses for four of the phosphorus-containing zeolites are shown in Table 4.20. Comparison with the wet chemical analysis supports framework substitution.

Based upon chemical analyses the unit cell compositions of the phosphate zeolites are given in Table 4.21. Considerable substitution of

324 THE SYNTHETIC ZEOLITES

Table 4.19 Typical Synthesis Conditions for Crystallizing Phosphorus-Substituted Zeolites (133)

Zeolite	Reactant Composition in Moles					Crystallization Temp. (°C)	Crystallization Time (hr)
	Na_2O^a	K_2O^a	Al_2O_3	SiO_2	P_2O_5	H_2O^a	
P-C	0.5	—	1.0	0.6	0.5	55	210	160
P-W	—	0.5	1.0	1.6	0.5	110	150	68
P-G	—	0.5	1.0	1.0	0.5	110	150	116
P-R	1.2	—	1.0	1.8	0.9	110	125	94
P-A	1.8	—	1.0	1.6	1.1	110	125	45
P-L	—	1.0	1.0	1.5	1.0	110	175	166
P-B(P)	0.5	—	1.0	0.6	0.7	110	200	70

[a] Values for Na_2O, K_2O, and H_2O are slightly higher than the values shown and undetermined due to unknown quantities absorbed on the precipitated hydrous aluminophosphate gel.

Table 4.20 Single Crystal Electron Microprobe Analysis of Phosphorus-Substituted Zeolites (133)

Zeolite	Wet Chemical Analysis (Composite Sample) wt% P_2O_5	Electron Probe Analysis (Single Crystal) wt% P_2O_5
P-C	22.0	19.5
P-W	19.8	21.7
P-G	14.1	15.7
P-B	15.6	16.1

Table 4.21 Typical Unit Cell Compositions for Phosphate Zeolites (133) [a]

Phosphorus Zeolite	Unit Cell	Tetrahedra/ Unit Cell	Composition	Charge Deficiency [b]
P-L	Hexagonal, a = 18.75, c = 15.03 [c] d_{meas} = 2.21	72	$K_{22.9}[(AlO_2)_{33.1}(SiO_2)_{26.3}(PO_2)_{2.0}] \cdot 10.7\ H_2O$	−2.6
P-W	Tetragonal, a = 20.17 c = 10.03 [d]	64 [d]	$K_{15.6}[(AlO_2)_{28.8}(SiO_2)_{24.5}(PO_2)_{10.7}] \cdot 55\ H_2O$	+2.5
P-G	Rhombohedral, a = 9.44, α = 94°28'	12	$K_{3.0}[(AlO_2)_{5.6}(SiO_2)_{4.3}(PO_2)_{2.0}] \cdot 10.7\ H_2O$	+0.6
P-R	(chabazite)		$Na_{4.4}[(AlO_2)_{5.5}(SiO_2)_{4.5}(PO_2)_{2.0}] \cdot 10.9\ H_2O$	−0.9
P-C	Cubic, a = 13.73, d_{meas} = 2.30	48	$Na_{15.3}Ca_{0.6}[(AlO_2)_{2.30}(SiO_2)_{18.2}(PO_2)_{6.7}] \cdot 18.6\ H_2O$	−0.2
P-B	Tetragonal, [e] a = 10.1, c = 9.8	16	$Na_{4.7}[(AlO_2)_{7.9}(SiO_2)_{4.3}(PO_2)_{3.8}] \cdot 10.4\ H_2O$	−0.6
P-A	Cubic, a = 12.24 (pseudo cell), d_{meas} = 2.11	24	$Na_{11.5}[(AlO_2)_{11.5}(SiO_2)_{9.7}(PO_2)_{2.8}]OH_{2.8} \cdot 24.8\ H_2O$	0

[a] All calculations were normalized to the appropriate number of tetrahedra/unit cell. Unit cell dimensions are in Å, and densities in g/cc at 25°C.

[b] The charge deficiency values listed represent the additional charge required to balance the unit cell charge. Although the presence of OH^- and $H_3O_2^-$ groups are postulated in some cases in the substitution mechanism, their assignment to balance charge seems arbitrary except in the case of zeolite P-A. Also, the values of charge deficiency in most cases are well within the errors in analysis.

[c] Cell determined from single crystal electron diffraction studies on zeolite P-L.

[d] Unit cell from single crystal x-ray precession studies; tetrahedra/unit cell chosen on the basis of the tetrahedra density in the related harmotome-phillipsite structures. The density was not determined.

[e] Tetragonal cell that of Taylor and Roy (52) for related zeolite P_t.

phosphorus was observed in the case of several of the synthetic zeolites. Optical photomicrographs of several phosphate zeolites are shown in Fig. 4.17.

Observed is an unusual simple cubic habit for the analcime analog and an unusual form of twinning in the zeolites P-W and P-L (see Fig. 5.9). In the synthetic phosphate zeolites the presence of tetrahedral phosphorus was shown by a crystal structure study (134) of the anal-

326 THE SYNTHETIC ZEOLITES

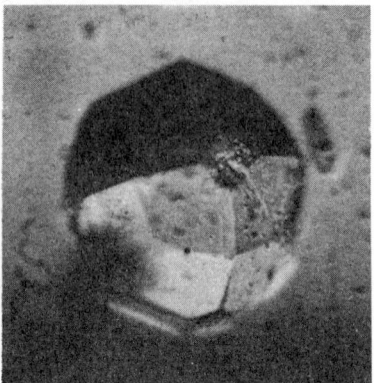

Plate 1

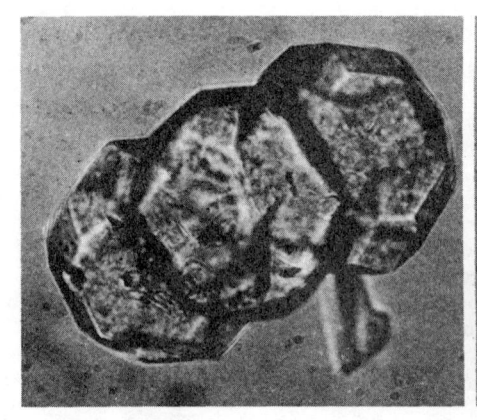

Plate 2

Plate 3

Plate 4

Figure 4.17 Optical photomicrographs of phosphate zeolites.
Plate 1 Zeolite P-C (12.9 wt% P_2O_5) 200X   Plate 3 Zeolite P-B (21.1 wt% P_2O_5) 200X
Plate 2 Zeolite P-C (17.5 wt% P_2O_5) 150X   Plate 4 Zeolite P-W (14.3 wt% P_2O_5) 600X

cime analog, P-C, by the variation in the unit cell dimension with phosphorus content, and by infrared spectra.

In the mid-infrared region, that is, from 200 to 1300 cm^{-1}, the vibrations sensitive to the framework composition were examined. Substitution of aluminum for silicon causes a shift to a *lower* frequency owing to an increase in the Al-O bond distance. Substitution of phosphorus in the framework should show a similar change but, because of a shorter P-O distance, the shift in the main stretch band near 1000 cm^{-1} should move toward a *higher* frequency. As shown in Figs. 4.18 and 4.19, for example, the infrared spectra for zeolites P-A and P-L are compared with the phosphorus-free zeolites. Increased absorption in the higher frequency portion of the asymmetric stretch band is observed due to the presence of PO$_4$ groups in the framework.

Various mechanisms were proposed for phosphorus substitution. These include:

1. $AlO_2^- + PO_2^+ = 2\ SiO_2$
2. $PO_2^+ = (K^+)Na^+ + SiO_2$
3. $PO_2^+ + OH^- = SiO_2$

In most of the phosphate zeolites mechanism 1 seems to occur. There is simultaneous substitution of AlO$_4$ and PO$_4$ for SiO$_4$. This results in

Figure 4.18 Infrared spectra for the sodium form of P-A and zeolite NaA.(133)

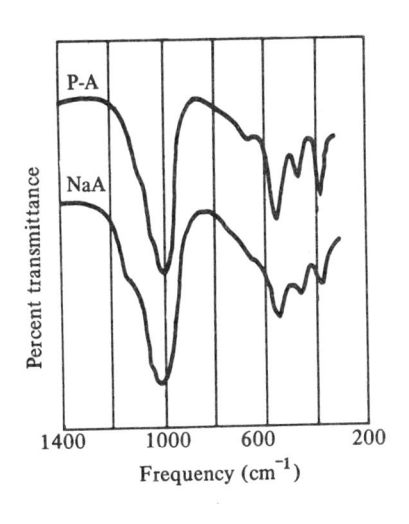

Figure 4.19 Infrared spectra for zeolite P-L and Zeolite KL.(133)

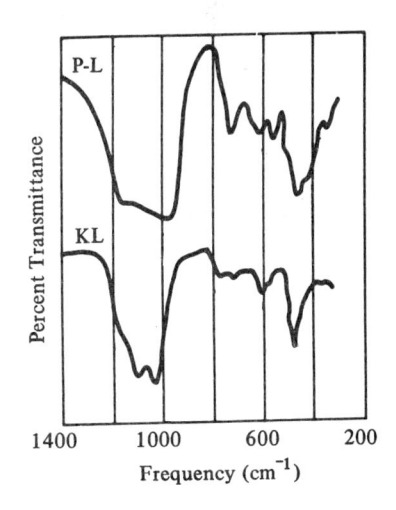

an alumina-rich framework with the corresponding changes in the infrared spectra.

It is possible that more than one mechanism may operate simultaneously. Chemical analysis of the analcime type, P-C, indicates that both mechanism 1 and mechanism 3 occur. If the AlO_2/PO_2 ratio exceeds one, then a counterbalancing anion, OH^- is introduced. Consequently, mechanism 3 accounts for the presence of OH^- groups.

The phosphate zeolites were additionally characterized by adsorption which will be discussed in Chap. 8. The P-A zeolite exhibits adsorption properties similar to the phosphorus-free zeolite A. Substitution of phosphorus in zeolite L reduces the adsorption capacity by about 50% and the adsorption pore size from about 10 to 7 Å. Also, a change in symmetry was observed in the case of zeolite P-L. In zeolite P-L, the unit cell c-dimension is doubled and systematic absences show the presence of two glide planes (133). The phosphate-type P zeolite was not stable thermally. Phosphate has been used in aluminosilicate gels to control aluminum substitution and the Si/Al ratio in the synthesis of zeolite Y (135, 136). Several types of zeolite have been prepared which contain occluded phosphate.

Complexing Agents in Zeolite Crystallization

Certain ions which form stable complexes with aluminum in solution, in particular phosphate, have been used to modify the usual composition of the zeolite as it is influenced by the composition of the initial aluminosilicate gel. A more efficient use of the silica in the crystallization of zeolite Y was reported when phosphate was present in the gel. The phosphate was added in two ways: (1) aluminum phosphate was dissolved in a solution of trisodium phosphate and (2) sodium aluminate was dissolved in disodium hydrogen phosphate (135). Silicate was added as a sodium silicate solution. In addition to zeolite Y, chabazite-type zeolites (also known as ZK-14) were prepared as well as a phillipsite-type (ZK-19). Other complexing agents employed included arsenate and tartrates. The formation of the phosphatoaluminate complex in the gel was represented by the equation:

$$AlPO_4 + M_3PO_4 \rightleftharpoons M_3[Al(PO_4)_2]$$

where $M = NH_4$, Na, K, R_4N.

In the alkaline media present this anion is in equilibrium with the tetra-

hedral aluminate anion:
$$Al(PO_4)_2^{3-} + 4\, OH^- \rightleftharpoons Al(OH)_4^- + 2\, PO_4^{3-}$$

The $Al(OH)_4^-$ ions are primary building units in the formation of the zeolite; the concentration of these ions is lowered as the result of complex formation. The "effective" ratio of silicate to aluminate in the reaction mixture is higher than the overall SiO_2/Al_2O_3 ratio in the gel. As the $Al(OH)_4^-$ is consumed, due to crystallization, it is replaced according to the second equation. The concentration of the $Al(OH)_4^-$ therefore depends on the phosphate and hydroxyl ion concentrations. In order to form high-silica zeolites the concentration of the $Al(OH)_4^-$ should be low. This is brought about by increasing the phosphate concentration and lowering the pH.

The most effective silicate was found to be sodium metasilicate in these experiments since the higher sodium silicate solutions and a silica sol require higher hydroxyl concentrations in order to provide for the depolymerization of the silica.

Gallosilicate zeolites of the Y-type have also been synthesized from phosphate-complexed, aluminate mixtures with the same result; the utilization of the silica is more complete in the presence of phosphate (137).

The effect of the complexing agents is explained on the basis of a polymerization-depolymerization equilibrium. In a study of many complexing agents in the synthesis of zeolite Y, most were found to have some effect but only phosphate and phytate exhibit a buffering action which is also required in order to control the pH and the silica solubility. The various parameters important in zeolite crystallization were discussed. These include (1) the SiO_2/Al_2O_3, (2) pH, (3) type of cation and its ratio, (4) effect of prepolymerization of the silicate, (5) time of crystallization, (6) concentration, and (7) temperature. The SiO_2/Al_2O_3 ratio of the reaction mixture controls the corresponding ratio in the product zeolite. For example, with $SiO_2/Al_2O_3 = 2$, zeolite A can be formed and with a ratio of 4, zeolite X. However, if the mixture is phosphate buffered, then at a ratio of 4 a zeolite of chabazite-type may be formed and, at a ratio of 6, a zeolite similar to S (ZK-15) is formed (136).

The pH influences the polymerization-depolymerization equilibrium of the gel and the nature of the species in solution. It is proposed that the type of precursor species in solution, such as 4-ring and double 6-rings will depend upon the pH; as the pH is lowered the precursor units

Table 4.22 Synthetic Zeolites: Na$_2$O-TMA$_2$O-Al$_2$O$_3$-SiO$_2$-H$_2$O System with Phosphate

Zeolite	Reactant Composition (moles/Al$_2$O$_3$) Na$_2$O	TMA$_2$O	SiO$_2$	P$_2$O$_5$	H$_2$O	Reactants	Conditions Temp. (°C)	Zeolite Composition[b] (moles/Al$_2$O$_3$) Na$_2$O	SiO$_2$	H$_2$O	Properties	X-Ray Table	Ref.
ZK-21	5.72	8.28	4.0	3.99	6.58	Sodium aluminate, sodium metasilicate, TMA-OH, H$_3$PO$_4$	96 5 days	1.01	3.26	a	a = 12.17 Å, adsorbs 14% of n-hexane	4.98	138
ZK-22	6.7	2.7	6.0	4.0	404	Same as ZK-21	90, 74 days 125, 8 days	0.78	5.34	b	a = 12.04 Å adsorbs 14% of n-hexane		138

[a] Contains about 1 phosphate per unit cell in the β-cage. Also contains nitrogenous material which is also intercalated.

[b] Contains about 0.38 phosphorus per unit cell; contains TMA as cation to make up for sodium ion deficiency.

become more complex, that is, change from 4-ring to a cubic unit or a hexagonal prism unit. The type of cation may determine the way in which building units link to form the framework structure (see Section I).

The phosphate complexing technique was used to prepare high-silica forms of zeolite A. These have been designated as ZK-21 and ZK-22 (138). Mixed bases, sodium and tetramethylammonium, TMA, were employed. Both of these zeolites contained intercalated phosphate (up to one phosphate in each β-cage of the structure). The composition of the zeolites, ZK-21 and ZK-22 are given in Table 4.22 which also presents one example of a starting reaction mixture and the conditions of synthesis. Both zeolites have an x-ray diffraction pattern like that of zeolite A but, because of the higher SiO_2/Al_2O_3 ratio, the unit cell constant is smaller. The unit cell constant was found to vary in a linear fashion with the number of aluminum atoms per unit cell. For the zeolite with $SiO_2/Al_2O_3 = 6$, the unit cell dimension was 12.04 A which increased to 12.25 A for a SiO_2/Al_2O_3 ratio of 2.36.

Because of the lower cation density in the higher silica form of the A-type structure, adsorption of a normal hydrocarbon such as n-hexane is expected if the SiO_2/Al_2O_3 ratio is greater than 2.8, which is equivalent to 10 univalent ions per unit cell. The ZK-21 and the ZK-22 were reported to adsorb normal hexane.

The Na^+ ions in ZK-21 are exchangeable by Ca^{2+}. The Na^+ ions in ZK-22 also exchange for Ca^{2+}, but the TMA ions do not exchange. La^{3+} is reported to exchange for sodium in ZK-22 but the examples (see below) show about two La^{3+} ions in each unit cell. In zeolite A, ion exchange with Ce^{3+} does not occur (48). See Chap. 7.

Lanthanum ZK-22

SiO_2/Al_2O_3	Al/unit cell	Equiv. La^{3+}/Al	La^{3+}/unit cell
3.94	8	0.71	1.9
4.06	7.9	0.87	2.3
5.70	6.2	0.47	1.0

H. SYNTHESIS OF SALT-CONTAINING ZEOLITES

Minerals of the feldspathoid group contain intercalated or occluded ions such as Cl^-, CO_3^{2-}, etc. This is typical of the minerals sodalite, cancrinite, and nosean. Synthetic types of the feldspathoids are easily prepared and have been discussed. These include types of sodalite, referred to as hydroxysodalite, and hydroxycancrinite. In addition it has been observed that zeolite A, in the more alumina-rich reaction mixtures, crystallizes with small amounts of occluded aluminate. (See Chap. 6.)

Zeolite-type species were prepared in hydrothermal experiments using salts such as barium chloride and barium bromide. The zeolite-like species synthesized (see Section B) were designated as species, N, O, P, and Q (139). The hydrothermal recrystallization of a synthetic analcime in the presence of an excess of KCl resulted in the species N. The species O was prepared by the use of KBr. All of these phases contain the intercalated salt. The hydrothermal extraction of the occluded salt resulted in the formation of a zeolite-type material which exhibited adsorptive propreties upon activation like those exhibited by the mineral chabazite or, as it is now known, zeolite type A. A reinvestigation of these materials showed that these species (N, O, P, and Q) may be synthesized from hydrogels as well as by the hydrothermal recrystallization of other crystalline aluminosilicates such as a synthetic type of analcime, zeolite X and zeolite Y. The species P and Q referred to in this work are not related to zeolite P and zeolite Q. It was later found that the species P and Q are structurally like zeolite ZK-5 which was previously synthesized using an alkylammonium base (140). The species P and Q were also prepared by using zeolite X as a reactant at 230°C. The x-ray powder data are like data given for ZK-5 but with differences in intensities and other differences in the size of the unit cell constant (See Tables 4.76, 4.78). The unit cell size varies with the nature of the occluded salt. It was also observed that occlusion of a barium salt within the framework of the zeolite increased the thermal stability. X-ray powder patterns for these zeolites showed no change after heating to 1000°C.

The species N and O were formed by the recrystallization of synthetic analcime using KBr and KCl. The species N and O are structurally related to the zeolite K-F or Z, normally synthesized from a gel. Examples of a few of the conversions using salt solutions are given in Table 4.23. These materials therefore lie between a true zeolite and a feldspathoid since they contain both occluded salt and water. Species based on the structure of ZK-5 are more amenable to subsequent extraction of the salt than those based on the tighter structural framework of sodalite.

Table 4.23 Some Syntheses of Salt-Containing Zeolites (140)

Starting Material	Salt	Temp (°C)	Time (d)	Product
Analcime, synthetic	BaBr$_2$	230	2	Q^a-Br
Chabazite, mineral	BaCl$_2$	250	4	P^b-Cl
	BaBr$_2$	250	4	Q -Br
Zeolite Y	BaCl$_2$	250	4	P-Cl
	BaBr$_2$	250	4	Q -Br
SiO$_2$-Al$_2$O$_3$ gel	BaCl$_2$	230	4	P-Cl
with SiO$_2$/Al$_2$O$_3$=4	BaBr$_2$	230	4	Q -Br
SiO$_2$-Al$_2$O$_3$ gel SiO$_2$/Al$_2$O$_3$=2	KCl	100-200	4	K-F (Z)
Zeolite X	KCl or KBr	260	4	K-F (Z)

[a] Species Q has a ZK-5 type structure. Typical composition for preparation from zeolite Y was: BaO·Al$_2$O$_3$·5.1 SiO$_2$·0.74 BaBr$_2$·2 H$_2$O

[b] Species P has a ZK-5 type structure. Typical composition for preparation from zeolite Y was: BaO·Al$_2$O$_3$·4.91 SiO$_2$·0.90 BaCl$_2$·2.5 H$_2$O

I. KINETICS AND MECHANISM OF ZEOLITE CRYSTALLIZATION

Extensive data in the published literature relates the composition of reaction mixtures or gels, the character of the reactant phases, and reaction conditions such as temperature, pressure, and time to the zeolite species or phases which result. Specific information concerned with crystallization mechanism and kinetics is rather limited. Some studies have been made of the rate of crystallization of zeolite A, zeolite X, and mordenite from sodium aluminosilicate gels (Fig. 4.7). These curves are sigmoid in shape and imply that an induction period is necessary during which the initial crystal nuclei grow to a critical size. The effect of temperature on the crystallization of zeolite X is evident since the induction period at 50°C is about 60 hours, as compared to 3 hours at 100°C. After the initial rapid growth, continued crystal growth is difficult to maintain.

The effect of crystallization temperature on the zeolite phase is of interest. The relationship between crystallization time and temperature for the formation of zeolite X from aluminosilicate gels is given in Fig. 4.20. From this type of logarithmic relationship an apparent activation energy can be calculated which is 11 kcal per mole for zeo-

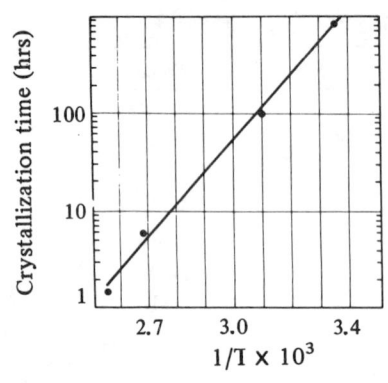

Figure 4.20 Time required for complete crystallization of zeolite X versus the reciprocal of the absolute temperature. Calculated Arrhenius activation energy = 15 kcal/mole of framework units (11).

lite A and 15 kcal per mole for zeolite X. It is difficult to ascribe a physical significance to these values, but it is of interest that Greenburg (141) reports in a study of the rate at which silica is depolymerized in the temperature interval 30 - 60°C that the activation energy for this depolymerization is 21.5 kcal per mole.

The rate of crystallization of zeolite A in a reaction mixture prepared from an amorphous sodium aluminosilicate of composition $NaAlO_2 \cdot 0.82\ SiO_2$ and aqueous NaOH was studied by Kerr (142). At 100°C he found that the rate of formation follows approximately first-order kinetics and is proportional to the quantity of crystalline zeolite present in the system. The formation occurs rapidly after an induction period which was concluded to be due to the formation of nuclei. The induction period was reduced by adding zeolite crystals in the initial mixture. A typical result is shown in Fig. 4.21.

It was concluded that the amorphous solid dissolves rapidly in the alkaline solution to form a soluble active species. The concentration of this species decreases as the amorphous substrate is depleted. The rate of growth increases with concentration of NaOH.

In a further study Ciric (143) used aluminosilicate gels containing excess alumina, i.e., with an SiO_2/Al_2O_3 ratio equal to 1. The effect of changing the concentration of NaOH was interpreted in terms of the transfer of a dissolved species from the gel to the surface of the crystal. It was concluded that the diffusing species contains two negative charges and probably consists of an aluminosilicate dimer or tetramer. The rate of growth of zeolite A from the sodium aluminosilicate gels was found to be represented by:

$$\left(\frac{dW_c}{dt}\right)_T = K_4 (OH)^2$$

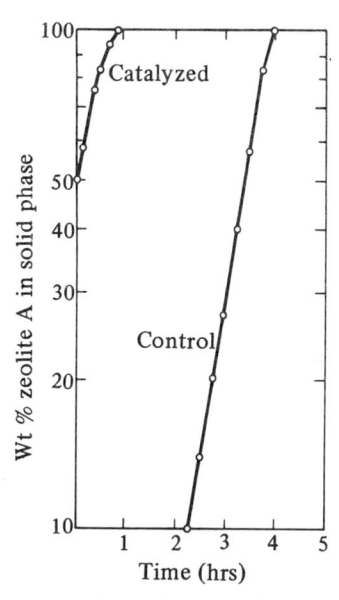

Figure 4.21 An amorphous sodium aluminosilicate of composition $Na_2O \cdot Al_2O_3 \cdot 1.64\ SiO_2$ (dry basis) was first prepared from $NaAlO_2 \cdot NaOH$, and tetraethylorthosilicate. This was used in the initial experiment with 1 M NaOH (35 g of amorphous substrate and 700 ml of 1 M NaOH) and reacted to completion at 100°C in 4 hours. Then, 35 g of the amorphous sodium aluminosilicate was added to the reaction mixture of zeolite A and mother liquor. The time of this addition is zero for the catalyzed curve. Hence, the weight percent zeolite = 50. The straight line portions of the control curve are given by $\ln Z = kt + \ln Z_0$. In this example, $k = 0.7\ hr^{-1}$ (142).

where W_c = degree of conversion. Plots of $\log(dW_c/dt)$ versus $\log(OH^-)$ were linear with a slope that gave the exponent of 2.

Growth curves of the same type as shown in Fig. 4.21 were observed for the crystallization of zeolite X and zeolite P from gels. The same gel composition was used for each zeolite but different mixing times were employed in the gel preparation. The rates of crystallization followed the same equation as for zeolite A. Zeolite P was observed to nucleate more slowly but then grew at a faster rate than zeolite X. The induction period for zeolite P was twice that for zeolite X (144).

Two starting materials were used in studies of the rate of formation of synthetic mordenite: sodium aluminosilicate gels prepared from sodium aluminate and stabilized silica sol, and an amorphous sodium aluminosilicate which was previously prepared from sodium silicate solution and aluminum sulphate (37). The SiO_2/Al_2O_3 ratio varied from 9 to 12. The highest purity synthetic type of mordenite was obtained by using a SiO_2/Al_2O_3 ratio of 12 to 13.

Using the amorphous sodium aluminosilicate, at 300°C the rate of crystallization increased with increasing pH up to 12.8. At a pH of 13.3, the synthetic mordenite apparently converted to an analcime-type as previously observed by Barrer (20). At a pH of 12.6, the induction period varied from one hour at 350°C to 4 weeks at 100°C. Two days were required at 200°C. Crystallization from gels was slower than from

336 THE SYNTHETIC ZEOLITES

the amorphous aluminosilicate.

The chemical composition of the solid and liquid phases of aluminosilicate gels formed from alkaline solutions of sodium and potassium aluminates and silicates were studied by Zhdanov (74). His results, shown in Fig. 4.22, indicate the region of gel formation and the relation between the composition of the liquid phase and the aluminosilicate gel phase. Although the SiO_2/Al_2O_3 ratio in the initial mixture varied widely (36.8 to 0.33) the ratio in the gel phase after washing out the intermicellar liquid varied over a much narrower range of 2.2 to 6.6. Further, it was found that the ratio of Na_2O to Al_2O_3 in the gel phase was about 1. It appears that the Loewenstein rule is operative in the amorphous aluminosilicate gels and that the gel skeleton consists of AlO_4 and SiO_4 tetrahedra with sufficient alkali cations to compensate for the negative charge of the aluminum atoms.

The formation of hydrous ($H_2O/SiO_2 = 7 - 8$) gels of relatively con-

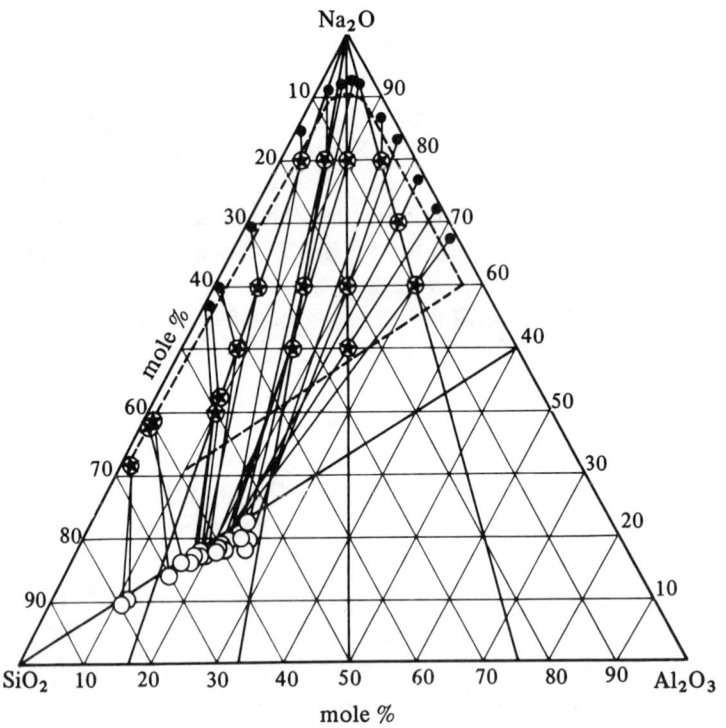

Figure 4.22 Component correlations in initial Na-aluminosilicate gels (starred circles); in gel skeletons washed free of excess alkali (open circles); in gel liquid phases (black circles) in mole percent (74).

Table 4.24 Composition of Amorphous Precipitated Sodium Aluminosilicates (145)

Composition of Initial Solution Moles/Al$_2$O$_3$		Composition of Precipitate Moles/Al$_2$O$_3$	
Na$_2$O	SiO$_2$	Na$_2$O	SiO$_2$
3.3	0.67[b]	1.0	2.5
3.1	0.5[b]	1.0	2.56
3.0	0.5[a]	1.0	2.5
3.2	1.0[a]	1.0	2.5
3.66	1.0[b]	1.1	2.63
3.65	2.0[a]	1.0	2.6
4.0	1.3[b]	1.0	2.63
4.8	2.0[b]	1.0	2.63
6.76	4.0[b]	.95	3.0
6.7	4.0[a]	1.0	3.0
1.6	11.5[b]	1.1	3.16

[a]Prepared from 0.1 M NaH$_3$SiO$_4$ and 0.1 M NaAlO$_2$ solution.
[b]Prepared from 0.1 M Na$_2$H$_2$SiO$_4$ and 0.1 M NaAlO$_2$ solution.

stant composition by precipitation of sodium silicate with sodium aluminate from aqueous solution, was similarly observed by Fahlke, Wieker, and Thilo (145). As shown in Table 4.24, from widely different initial compositions, the precipitated aluminosilicate gel had the composition Na$_2$O·Al$_2$O$_3$·2.5-3.2 SiO$_2$·xH$_2$O. Further, infrared studies of the gel dried to a water content of H$_2$O/SiO$_2$ = 1-1.3 showed a broad 3600 cm^{-1} band indicative of hydroxyl and no indication of free water. A hypothetical structure for this "compound," Na$_4$Al$_4$Si$_6$O$_{18}$(OH)$_4$, was proposed which is based on the cubic unit of eight tetrahedra, D4R (Chap. 2). However, this proposed structure violates the Loewenstein rule by permitting two aluminums to share the same oxygen.

The importance of the composition of the liquid phase in the gels on the formation and growth of zeolite crystals was pointed out (74). For example, it was possible to obtain zeolite A from a washed gel prepared for the crystallization of zeolite X by adding NaOH as a liquid phase, and zeolite X from the washed gel skeleton of a zeolite A composition. In each case, however, the chemical composition corresponds to that required for crystallization of these zeolites.

The crystallization of highly complex zeolite structures at relatively low temperatures from reactive gels requires that the disordered gel phase be converted to a more ordered but metastable zeolite phase. A single mechanism for zeolite crystallization is not known but some qualitative proposals have been made.

These proposals include (1) the formation of a solid amorphous "precursor" which is followed by a solution crystallization mechanism. This has been reviewed by Zhdanov who describes a quasi-equilibrium between the solid and liquid phases in the gel and concludes that nuclei form and grow in the liquid phase (146). The gel material dissolves continuously and the dissolved species from the gel are transported to the nuclei crystals in the liquid phase. This is analogous to the solution mechanism described by Kerr (142). A solution of sodium hydroxide was circulated through an amorphous sodium aluminosilicate (composition $NaAlO_2 \cdot 0.82\ SiO_2$) and then over zeolite A "seed" crystals. At 100°C it was found that all of the aluminosilicate dissolved and converted to zeolite A on the seeds.

It was further shown that zeolite crystallization from solution proceeds through the formation of amorphous lamellae. The solutions used were very dilute (2 to 3 grams of silica and aluminum hydroxide per liter). Subsequently, the lamellae evolve into larger zeolite particles by heterogeneous nucleation (147).

In solution, silica exists as the hydrated monomer, $Si(OH)_4$, or in solutions of high pH, as silicate ions (148). The solubility of silica, particularly amorphous forms, has been extensively studied. Dissolution involves a simultaneous hydration and depolymerization. Dissolution is catalyzed by the presence of a strong base such as sodium hydroxide and when the pH exceeds about 10.5 - 11, silica dissolves extensively. This probably involves depolymerization through hydration to form $Si(OH)_4$, followed by addition of an OH^- ion to form the silicate ion. The dissolution-depolymerization reaction can be represented by

[1] $(SiO_2)_n + 2n\ H_2O = n\ Si(OH)_4$

[2] $Si(OH)_4 + H_2O + OH^- = (H_2O)Si(OH)_5^-$

Assuming that the concentration of the monomer, $Si(OH)_4$, does not change with pH, the increase in soluble silica with increasing pH was explained on the basis of [2] with the equilibrium constant:

$$\frac{(H_2O)Si(OH)_5}{[OH^-][Si(OH)_4^-]} = 1.85 \times 10^4$$

Typical solubility data for amorphous silica in alkaline solution are given in Table 4.25.

When solutions of the aluminate and polysilicate anions are mixed to form the hydrous gel, the aluminate anions and silicate anions undoubtedly undergo a polymerization process. The gel structure thus produced is amorphous and in a state of high simplexity. The composition and structure of this hydrous gel is controlled by the size and structure of the polymerizing species. Since the silicate may vary in chemical composition and molecular weight distribution, different silicate solutions may lead to differences in the gel structure. Therefore gelation controls the nucleation of the zeolite crystallites. The crystal size and morphology of zeolites grown from gels generally appear to support this. The crystals are very small, several microns in size, uniform, and often euhedral. The high degree of supersaturation of the ionic species present in the gel must lead to rapid and heterogeneous nucleation and the formation of a large number of nuclei. Nucleation takes place after an induction period (Fig. 4.7).

The size and charge of the hydrated cation species which serves as a nucleation site for the polyhedral structural unit also influences the nucleation process. The cation is known to have an effect on the silicate and probably on the aluminate species in solution (150). (The more open zeolite structures appear to be crystallized from gels which contain so-

Table 4.25 Solubility of Silica (148)[a] (149)[b]

pH	Solubility (% at 25°C)[a]
1.0	.014
3.0	.015
5.7	.011
7.7	.010
10.26	.049
10.60	.112

$Na_2SO_4(N)$	pH	$H_4SiO_4 (moles/liter \times 10^3)$[b]
0	10.15	4.75
0.005	10.12	4.25
0.01	10.05	4.00
0.03	9.85	3.33
0.06	9.65	2.83
0.08	9.50	2.75

[a]Linde Silica A [b]Mallinckrodt Colloidal Silica, SA = 750 sq in/g

dium as the alkali rather than potassium, the latter leading to less open zeolite structures in general.) During the crystallization of the gel, the aluminate and silicate components must undergo a rearrangement in order to form the crystalline structure. This occurs by depolymerization and solubilization of the gel.

A schematic version of the crystallization of an amorphous aluminosilicate gel to a zeolite is given in Fig. 4.23. The gel structure, represented in two-dimensions, is depolymerized by the hydroxyl ions which produce soluble aluminosilicate species that may regroup to form the nuclei of the ordered zeolite structure. In this version the hydrated cation acts as a template. The representation presented is for the formation of the zeolite X structure based upon the truncated octahedron unit. A similar scheme could be based on the use of other secondary building units such as the double 6-ring.

A second mechanism is based on crystallization occurring in the solid phase by an ordering of the aluminosilicate framework. This is supported by several observations. The crystallization of gel solids in the absence of a liquid phase was observed by Khatami and Flanigen (165). From a typical gel which normally crystallizes to zeolite X, the solids were removed at the end of the induction period, washed free of liquid phase and dried to a free flowing powder of composition

$$1.1Na_2O \cdot Al_2O_3 \cdot 2.7SiO_2 \cdot 4.6H_2O \text{ (20 wt\% water)}$$

After ten days at ambient temperature, the solid contained 2% zeolite X and after 47 days, 20% X. Growing crystals imbedded in gel particles which shrunk and coalesced were observed by Ciric (143).

Phosphorescence and Raman spectroscopy have been used to study the structural chemical aspects of zeolite crystal growth from gels (151). Changes occurred in the solid phase which supported the conclusion that crystal growth takes place in the solid-gel phase by a condensation mechanism.

Changes in the solid phase of the aluminosilicate gels during the crystallization of the zeolite have been studied by means of electron microscopy (11). The morphological changes observed by carbon replica techniques indicate the mode of nucleation and growth of the zeolite crystals. The progressive crystallization of a typical aluminosilicate gel to produce a zeolite A crystal is shown in Fig. 4.24. The initial gel is shown and is followed by the appearance of nuclei and finally the well-developed cubic crystals of the zeolite. Fig. 4.24 illustrates unusual

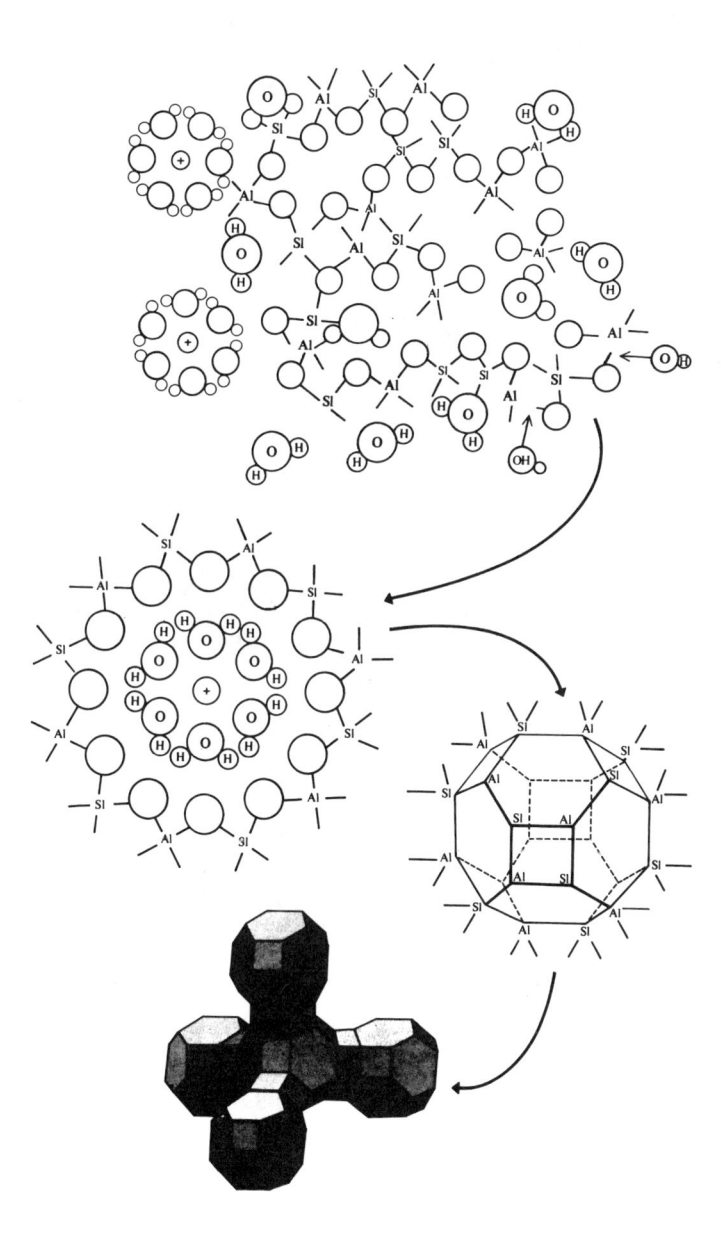

Figure 4.23 Schematic representation of the formation of zeolite crystal nuclei in a hydrous gel. The gel structure, center, is depolymerized by OH⁻ ions. Tetrahedra regroup about hydrated sodium ions to form the basic polyhedral units.

342 THE SYNTHETIC ZEOLITES

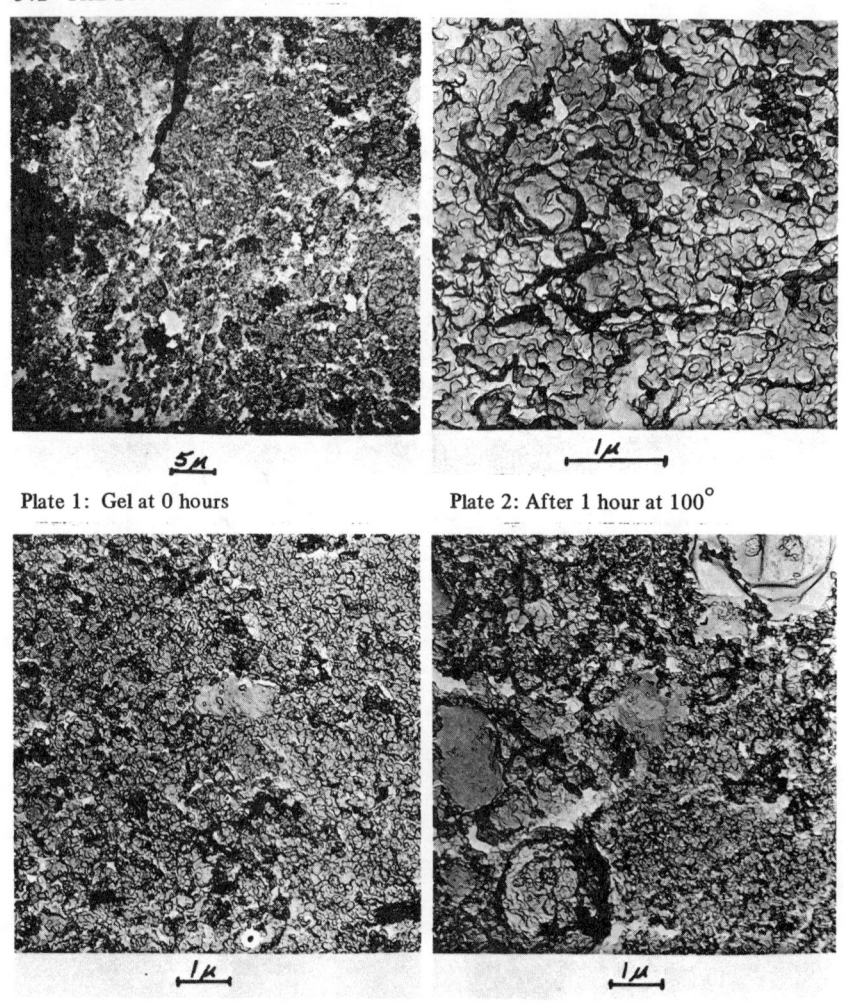

Plate 1: Gel at 0 hours

Plate 2: After 1 hour at 100°

Plate 3: After 1.25 hours at 100°

Plate 4: After 1.5 hours at 100°, some crystals are present.

Figure 4.24 Crystallization of zeolite A from gels as followed by an electron microscope.

CRYSTAL GROWTH 343

Plate 5: After 1.5 hours at 100°, some crystals are present.

Plate 6: After 2 hours, Zeolite A crystals.

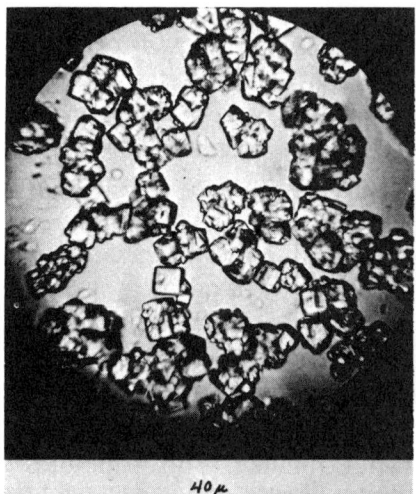

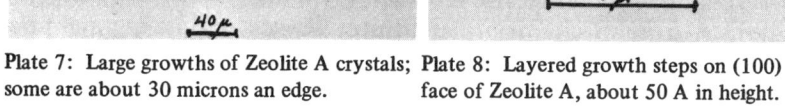

Plate 7: Large growths of Zeolite A crystals; some are about 30 microns an edge.

Plate 8: Layered growth steps on (100) face of Zeolite A, about 50 A in height.

growth of crystals to a reasonably large size. Twinning, which is a common characteristic of zeolite crystals, is evident, as are the layered growth steps on a (100) face of a single zeolite A crystal.

Four subsystems in the crystallization mechanism have been proposed by Tezak (143). These include (1) formation of simple and polynuclear complexes, (2) embryonation as a state of aggregation of comlexes, (3) nucleation as aggregate formation with a crystalline core and formation of micelles and (4) aggregation of primary particles into larger structures through oriented crystalline aggregation.

The formation of zeolite A crystals from a TMA-aluminosilicate sol by the addition of sodium ions (as NaCl) was observed by Acara and Howell (152) (Fig. 4.25). The rate, yield and size of the crystals depends on the amount of NaCl added. The cubic crystallites vary in size from 250 A to 0.3 μm. This is consistent with a mechanism of oriented aggregation.

It has been further suggested that nucleation and transport of species by surface diffusion and crystallization occur at the liquid-solid interface. In concentrated gel systems (typical of most zeolite synthesis) interparticle contact is maximized and processes of aggregation and coalescence may readily occur. The epitaxy observed in the co-crystallization of zeolites L and O (offretite) supports a surface nucleation mechanism (154).

Photographs of large single crystals of zeolite A and X are shown in Figs. 5.3 and 5.5. Sustained hydrothermal growth of single crystals of zeolites such as is possible in the growth of quartz and complex aluminosilicates such as beryl has not been achieved (153).

Intergrowths of zeolite L and O (offretite) resemble hammer-shaped crystals (154). They are due to epitaxial growth of a zeolite L crystal with a flaky habit and zeolite O which is rod-like. Both structures are based on an array of chains of cancrinite, ϵ-cages, and hexagonal prisms (see Chap. 2). The epitaxy is caused by chains of this type passing through the crystal interfaces. These intergrowths are illustrated in Fig. 5.8.

J. CRYSTAL GROWTH

The crystallite size of the zeolite X phase varies with crystallization temperature. At room temperature, the maximum crystallite size obtained is about 0.5 microns, which is sufficient to produce line-broadening of

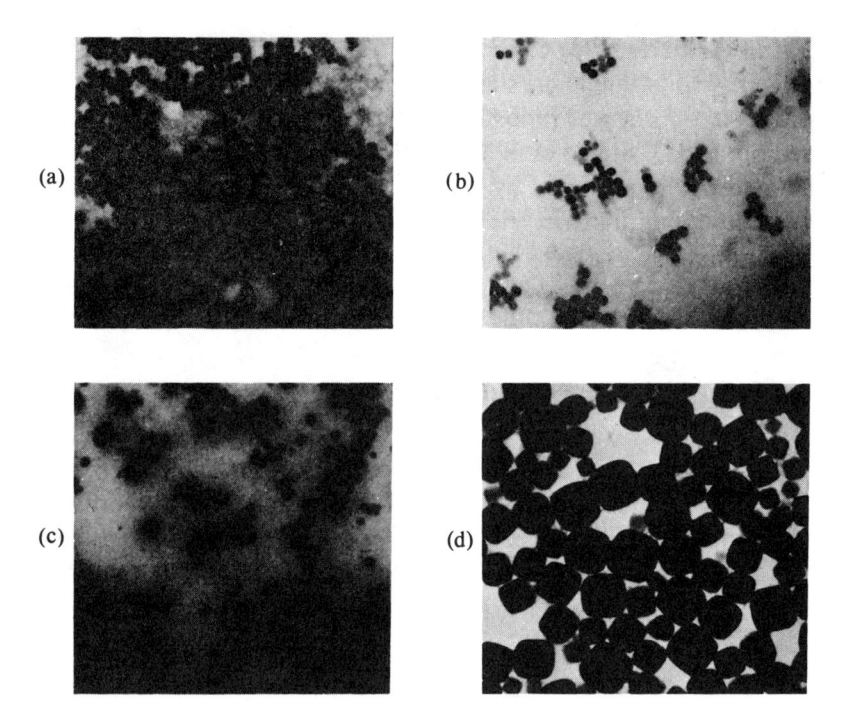

Figure 4.25 Addition of NaCl to a TMA-aluminosilicate sol. Electron photomicrographs of zeolite A crystals recovered after 24 hrs., 100°C. (a) 0.1 NaCl/Al$_2$O$_3$; (b) 0.2 NaCl/Al$_2$O$_3$; (c) 0.5 NaCl/Al$_2$O$_3$; (d) 1.0 NaCl/Al$_2$O$_3$; magnification 20, 240X.

typical x-ray reflections. This is illustrated in Fig. 4.26 which compares typical x-ray powder data of a well-crystallized zeolite X with one grown at room temperature. Spherulitic crystals tend to predominate at the lower temperature. It has been observed that zeolite X crystals, produced at room temperature, do not increase in size when subsequently heated in the same solution at the higher temperature of 100°C.

The growth steps in Fig. 4.24 (Plate 8) indicate that the species in solution contribute to the growth. Aging reaction mixtures prepared from colloidal silica sols adds an equilibration step in order to promote depolymerization of the solid colloidal silica particles. The hydrous gel phase contains large units of silica particles with molecular weights of about 100,000. Formation of zeolite nuclei occurs by the solubilization of silicate anions from the colloidal silica product and interaction with the aluminate anions already present in the solution. Under special conditions, uniform crystallites of zeolites A may be prepared which are about

500 A in diameter. (See Chap. 5, Fig. 5.4.)

The growth of single crystals of zeolite A up to 75 μm in size and zeolite X up to 40 μm was reported by Ciric who used a Carbopol gel as a medium (155). Single crystals of zeolite A and zeolite X 100 - 140 μ in diameter were prepared by Charnell by reacting sodium metasilicate with sodium aluminate in triethanolamine-water solution at 75 to 95°C. The aluminate and silicate solutions were carefully filtered. Crystallization times were 3 - 5 weeks for zeolite X and 2 - 3 weeks for zeolite A (156).

K. SUMMARY LIST OF SYNTHETIC ZEOLITES

A summary list of the synthetic zeolites is given in Table 4.26. This list is presented as an aid in locating information and data concerning any of the synthetic zeolite species. It is believed that this list is as complete as is possible at this time since many of the zeolites are incompletely characterized.

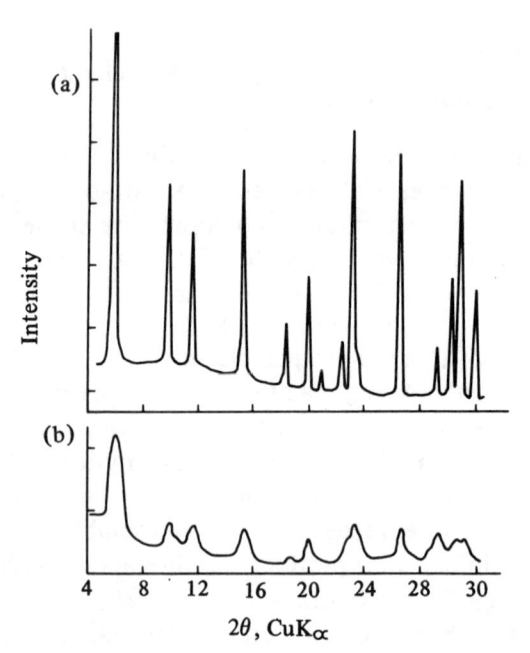

Figure 4.26 Typical x-ray powder patterns for zeolite X. (a) Six hrs, 100°C; (b) 35 days, 25°C.

X-Ray Powder Data for Synthetic Zeolites

The tables are arranged in alphabetical order as in Table 4.26. The data include where possible, the Miller Indices, (hk$l$), the spacing (d), and relative intensity (I). In some cases only visual estimates of intensity are published and these are given as vs(very strong), s (strong), m (medium), w (weak, vw (very weak). Lattice constants are included where available.

The powder data are given for the fully hydrated zeolite unless otherwise indicated. Partial or complete dehydration affects the intensities (48).

Because most zeolites have large unit cells, the x-ray powder patterns contain many reflections at low diffraction angles and consequently overlap of reflections frequently occurs, particularly in mixtures.

348 THE SYNTHETIC ZEOLITES

Table 4.26 Alphabetical List of Synthetic Zeolites

This table is an attempt to sort out the more than 90 species of synthetic zeolites. The zeolites are arranged in alphabetical order either by the letter which indicates the structure type or in some cases by the name of the related mineral where that has been employed in the literature. (See Chap. 1.)

The framework structure-type is given in column 2 if known. The cations shown in column 3 indicate the system from which the zeolite crystallizes. For example, zeolite A is synthesized in the sodium form. (Details are given in the Section C which deals with the $Na_2O \cdot Al_2O_3 \cdot SiO_2 \cdot H_2O$ system.) Zeolite omega, with the two cations—Na^+ and TMA^+—is discussed in Section E. The column labeled "Framework" shows the range in Si/Al ratio—when reported—and other framework atoms such as P, Ge, Ga. Column 5 indicates whether or not the zeolite is structurally stable toward the dehydration necessary to determine adsorption properties. The pore size and pore volumes in columns 6 and 7 are determined from adsorption measurements. For added convenience the table containing x-ray powder data is given in column 8. Other designations for the zeolite are shown in column 9 and pertinent literature references in column 10.

Zeolite	Structure Type	Cations	Framework	Stability	Pore Size (Å)	Pore Vol (cc/g)	X-ray Data Table	Other Des.	Reference
A	A	Na	Si/Al ~ 1, Ge	Stable	4	0.30	4.27		4, 47, 48
N-A	A	Na, TMA	Si/Al = 1.25 - 3	Stable	4.3	0.30	4.28		93, 95
P-A [a]	A	Na	Si-Al-P	Stable	4	0.27	4.29		133
α	A	Na, TMA	Si/Al = 2 - 3.5	Stable	4.3	0.30	4.30		97
Li-A	Unknown	Li	Si/Al = 1	Stable	2.6	0.15	4.31		18, 30, 31
Analcime	Analcime	Li	Si/Al = 1.5 - 3.0	Stable	2.6	0.10	–		9
B	Gismondine	Na	Si/Al = 1 - 2.5	Struct. contr.	–	–	–		55
Na-B	Analcime	Na	Si/Al = 2	Stable	2.6	0.24	4.74	P	35, 40, 41
β	Unknown	Na, TEA	Si/Al = 2.5 - 50	Stable	~6+	0.20	–		99
Bikitaite	Bikitaite	Li	Si-Al	–	–	–	–		30
P-B [a]	Gismondine	Na	Si-Al-P	Unstable	–	–	4.32		133
P-C [a]	Analcime	Na	Si-Al-P	Stable	–	–	4.33		133, 134
Clinoptilolite	Clinoptilolite	Li	Si-Al	–	–	–	4.34		33
		Ca, Sr, Ba	Si-Al	–	–	–	–		84
D	Chabazite	Na, K	Si/Al = 2.3 - 2.5	Stable	4	0.26	4.35		76
Ca-D	Wairakite	Ca	Si/Al = 2	–	–	0.11	4.36		26, 81, 83, 84
NH_4-D	Analcime	NH_4	Si-Al	–	–	–	–		93

SUMMARY LIST OF SYNTHETIC ZEOLITES   349

Table 4.26   Alphabetical List of Synthetic Zeolites *(continued)*

Zeolite	Structure Type	Cations	Framework	Stability	Pore Size (Å)	Pore Vol (cc/g)	X-Ray Data Table	Other Des.	Reference
Na-D	Mordenite	Na	Si/Al = 5	Stable	4	0.13	4.37		20, 35, 37, 39
Rb-D	Z	Rb	Si/Al = 2	—	—	—	—		80
Sr-D	Ferrierite	Sr	Si/Al ~ 3.5	Stable	3.5	0.13	4.38		88, 89, 84
E	Unknown	Na, K	Si/Al = 1 - 2	Unstable	2.8	0.21	4.39		77
Ca-E	Analcime	Ca	Si-Al	—	—	—	4.40		26
K-E	Analcime	K	Si/Al = 2	—	—	—	4.41		22
TMA-E [c]	Erionite	Na, TMA	Si/Al ~ 3	Stable	—	—	4.42		107
F	Unknown	K	Si/Al = 1	Stable	3.6	0.16	4.43		73, 11
K-F	Unknown	K	Si/Al = 1	Stable	2.6	0.13	4.92	Z	22, 65
Sr-F	Gmelinite	Sr	Si-Al	Stable	3.5	0.22	4.44		88
Ferrierite	Ferrierite	Na, Ca	Si-Al	—	—	—	4.45		8, 39, 84
G	Chabazite	K	Si/Al = 1.15 - 2.7, Ga	Stable	—	0.26	4.46		22, 128
P-G [a]	Chabazite	K	Si-Al-P	Stable	—	—	4.47		133
Sr-G	Chabazite	Sr	Si-Al	Stable	—	0.28	4.48		88
Ba-G	L	Ba	Si/Al = 1.5	Stable	3	0.15	4.49		90, 91
Gismondine	Gismondine	TMA	Si/Al = 1.5	—	—	—	—		161
H	Unknown	K	Si/Al = 1	Unstable	2.6	0.22	4.50	I	11, 66, 75
Li-H	Unknown	Li	Si/Al = 4	Stable	2.6	0.16	4.51		18
Heulandite	Heulandite	Ca	—	—	—	—	4.52		81
HS	Sodalite	Na	Si/Al = 1	—	<2.6	—	4.53	G, Zh	35, 56, 47
Ca-I	Thompsonite	Ca	Si-Al	—	—	—	4.54		26, 85
K-I (see H)									
Sr-I	Analcime	Sr	Si-Al	—	—	—	4.55		88, 84
J	Unknown	K	Si/Al = 1 - 1.15	Stable	2.8	0.08	4.56		68, 11
Ba-J	Unknown	Ba	Si/Al = 4	—	—	0.10	4.57		90
Ca-J	Epistilbite	Ca	Si-Al	—	—	0.19	4.58		26, 81, 82

350 THE SYNTHETIC ZEOLITES

Table 4.26 Alphabetical List of Synthetic Zeolites (continued)

Zeolite	Structure Type	Cations	Framework	Stability	Pore Size (Å)	Pore Vol (cc/g)	X-ray Data Table	Other Des.	Reference
Ba-K	Unknown	Ba	Si-Al	~	–	0.08	4.59		90
L	L	K, Na	Si/Al = 2.6 - 3.5	Stable	8	0.20	4.60		69, 11
Ca-L	Harmotome	Ca	Si-Al	–	–	–	4.61		26, 81
P-L [a]	L	K	Si-Al-P	Stable	6 - 7	0.14	4.62		133
Losod	Losod	Na, $C_8H_{16}N^+$	Si-Al	Stable	<2.6	0.18	4.63		162
M	Unknown	K	Si/Al = 1	Stable	2.6	0.10	4.64		70, 11
Ba-M	Harmotome	Ba	Si-Al	–	–	0.13	4.65		90, 91
Sr-M	Mordenite	Sr	Si-Al	–	–	–	4.66		88, 84
Mordenite	Mordenite	Li	Si-Al	Stable	–	–	4.67		9
	Mordenite	Na	Si/Al = 5	Stable	7	0.20	4.68		36
	Large Port	Na, Li	Si/Al = 5	Stable	~4	–	4.69		9
		Ca	–	Stable	–	–	–		83
N	Unknown	Na, TMA	Si/Al = 1	Stable	2.6	0.16	4.70		100, 119
Ba-N	Unknown	Ba	Si/Al = 1	Unstable	–	–	4.71		91
Natrolite	Natrolite	Na	–	–	–	–	–		42
O [c]	Offretite	TMA, K, Na	Si/Al = 3.4 - 4.25	Stable	6	–	–		107
Omega [c]	Omega	Na, TMA	Si/Al = 2.5 - 10	Stable	9	0.20	4.72		107, 160
Offretite [c]	Offretite	TMA, K, Na	Si/Al = 2.5 - 5	Stable	6	0.15	4.73		104
P_c	Gismondine	Na	Si/Al = 1 - 2.5, Ge	Struct. contr.	2.6	0.24	4.74	P-1	47
P_t	Gismondine	–	Si/Al = 1.5 - 2.5	–	2.6	0.22	4.75		52
N-P [c]	Gismondine	Na, TMA	–	–	–	–	–	N-L, N-B	95
P-(Cl) [b]	ZK-5	Ba	Si/Al = 4.4 - 4.9	Stable	–	–	4.76		139, 140
Phillipsite	Phillipsite	Li	–	–	–	–	–		9
Q	Unknown	K	Si/Al = 1 - 1.2	Unstable	2.8	0.23	4.77		71, 11
Q-(Br) [b]	ZK-5		Si/Al = 4.3 - 5.1	Stable	–	–	4.78		139, 140
Ca-Q	Mordenite	Ca	Si/Al = 5	–	–	–	–		26, 81, 84

SUMMARY LIST OF SYNTHETIC ZEOLITES 351

Table 4.26 Alphabetical List of Synthetic Zeolites *(continued)*

Zeolite	Structure Type	Cations	Framework	Stability	Pore Size (Å)	Pore Vol (cc/g)	X-Ray Data Table	Other Des.	Reference
Sr-Q	Yugawaralite	Sr	Si/Al = 3	—	—	0.12	4.79		84, 88, 89
R	Chabazite	Na	Si/Al = 1.7 - 1.8	Stable	3.5	0.26	4.80		50
P-R	Chabazite	Na	Si-Al-P	Stable	3.5	0.19	4.81		133
Sr-R	Heulandite	Sr	Si-Al	—	—	—	4.82		88
S	Gmelinite	Na	Si/Al = 2.3 - 2.95	Stable	3.5	0.23	4.83	E, Na-S	47, 49
T	Offretite, Erionite	Na, K	Si/Al = 3.2 - 3.7	Stable	4	0.23	4.84		78
Ba-T	Unknown	Ba	Si/Al = 1	Unstable	—	0.06	4.85		91
Na-V	Thomsonite	Na	Ga-Si	—	—	—	—		
W	Phillipsite	K	Si/Al = 1.15 - 2.45	Stable	2.8	0.22	4.86	K-M, K-H	11, 72, 22, 52
P-W [a]	W	K	Si-Al-P	Stable	3.8	0.19	4.87		133
X	Faujasite	Na	Si/Al = 1 - 1.5, Ga, Ge	Stable	8	0.36	4.88	R	5, 11
N-X [c]	Faujasite	Na, TMA	Si/Al = 1 - 1.5	Stable	8	0.36	4.89		95
Y	Faujasite	Na	Si/Al = >1.5 - 3	Stable	8	0.34	4.90		11, 61
N-Y [c]	Faujasite	Na, TMA	Si/Al = >1.5 - 3	Stable	8	0.34	4.91		95
Z	Unknown	K	Si/Al = 2	Stable	2.6	0.14	4.92	K-F	22, 65
Z(Cl) [b]	Like Z	K	—	—	—	—	—	N, O	140
Z-21	Unknown	Na	Si/Al = 1.7 - 2.1	—	—	0.14	4.93		59
ZK-4 [c]	A	Na, TMA	Si/Al = 1.25 - 2.0	Stable	4.3	0.25	4.94		96
ZK-5	—	Na, C$_8$H$_{18}$N$_2$$^{2+}$	Si/Al = 2 - 3	Stable	4.3	0.27	4.95		98
ZK-19	Phillipsite	Na, K	Si/Al = 1.5 - 3.1	Stable	2.6	0.14	4.96		79
ZK-20	Levynite	Na, C$_7$H$_{15}$N$_2$$^+$	Si/Al = 2 - 2.5	Stable	3.5		4.97		108
ZK-21	A	Na, TMA	Si/Al = 1 - 2	Stable	4.3	0.25	4.98		138
ZK-22	A	Na, TMA	Si/Al = 2 - 3.5	Stable	4.3	0.25	—		138
ZSM-2	Unknown	Li	Si/Al = 1.65 - 2.0	Stable	6	0.22	4.99		34
ZSM-3	Faujasite	Na, Li	Si/Al = 1.0 - 3.0	Stable	8	0.30	4.100		62, 63

352 THE SYNTHETIC ZEOLITES

Table 4.26 Alphabetical List of Synthetic Zeolites *(continued)*

Zeolite	Structure Type	Cations	Framework	Stability	Pore Size (A)	Pore Vol (cc/g)	X-ray Data Table	Reference
ZSM-4 [c]	Unknown	Na, TMA	Si/Al = 3.0 - 7.5	Stable	6+	0.14	4.101	101, 164
ZSM-5 [d]	Unknown [f]	Na, TPA	Si/Al = 2.5 - 50	Stable	6	0.10	4.102	103
ZSM-8 [e]	Unknown	Na, TEA	Si/Al = 2.5 - 50	Stable	5	0.10	4.103	102
ZSM-10	Unknown	K, $C_8H_{18}N_2^{2+}$	Si/Al = 2.5 - 3.5	Stable	6+	0.14	4.104	109

[a] Phosphate zeolite
[b] Contains a salt, e.g., $BaCl_2$, $BaBr_2$
[c] TMA = tetramethylammonium
[d] TPA = tetrapropylammonium
[e] TEA = tetraethylammonium
[f] Structure reported. See Cpt. 2, ref. 224 and Table 4.102, p. 373.

Table 4.27 Zeolite A (4,48)

hkl	Na$_{12}$A[a] d(A)	I	K$_{12}$A[b] d(A)	I	Ca$_6$A[c] d(A)	I
100	12.29	100	12.31	100	12.24	100
110	8.71	69	8.71	64	8.66	39
111	7.11	35	7.10	30	7.08	32
200	–	–	6.15	4	6.12	12
210	5.51	25	5.50	10	5.48	20
211	5.03	2	5.03	8	5.00	4
220	4.36	6	–	–	–	–
221, 300	4.107	36	4.105	33	4.084	35
310	–	–	3.895	10	3.875	2
311	3.714	53	3.714	62	3.696	34
222	–	–	3.555	5	3.539	4
320	3.417	16	3.414	34	3.398	18
321	3.293	47	3.292	35	3.276	38
400	–	–	3.078	12	–	–
410 322	2.987	55	2.985	80	2.972	32
411 330	2.904	9	2.902	27	2.888	9
420	2.754	12	2.753	65	2.741	7
421	2.688	4	2.687	9	2.676	3
332	2.626	22	2.625	18	2.614	24
422	2.515	5	2.514	28	2.502	7
430, 500	2.464	4	–	–	2.451	7
431, 510	–	–	2.415	4	–	–
511, 333	2.371	3	2.370	9	2.359	3
520, 432	2.289	1	2.287	3	–	–
521	2.249	3	2.248	5	2.238	3
440	2.177	7	2.177	26	2.166	8
441, 522	2.144	10	2.143	12	2.141	8
530, 433	2.113	3	–	–	2.103	5
531	2.083	4	2.081	5	2.074	2
600, 442	2.053	9	2.053	3	2.042	4

[a] Sodium A, Na$_{12}$(A) · 27 H$_2$O where (A) = framework composition of [(AlO$_2$)$_{12}$ (SiO$_2$)$_{12}$] a = 12.32A

[b] Potassium ion exchanged A, K$_{12}$[A] · 24 H$_2$O, a = 12.31A

[c] Calcium ion exchanged A, Ca$_6$[A] · 30 H$_2$O, a = 12.26A

Table 4.28 Zeolite N-A (95)

hkl	d(A)	I
100	12.08	100
110	8.55	60
111	6.99	45
200	6.05	5
210	5.41	8
211	4.94	6
220	4.28	11
221, 300	4.03	59
310	3.83	7
311	3.65	68
320	3.36	24
400	–	–
322, 410	2.93	36
330, 331	2.85	11
	–	–
420	2.710	5
421	2.644	5
332	2.583	15
422	2.474	5
430, 500	2.423	3
431, 510	2.375	0.6
511, 333	2.333	2
520, 432	2.250	0.5
521	2.211	4
440	2.141	6
441, 522	2.111	1.3
530, 433	2.078	2
531	2.049	1
442, 610	2.019	8
611, 532	1.992	0.2
	1.966	1.0
620	–	–
621, 540, 443	1.892	5
541	1.870	5

Table 4.29 P-A (133)

d(A)	I
12.2	100
8.60	89
7.07	57
5.48	35
4.99	5
4.33	16
4.08	57
3.87	8
3.69	81
3.54	5
3.40	32
3.28	68
2.96	92
2.89	22
2.74	16
2.68	11
2.62	49
2.50	11
2.45	8
2.36	5
2.24	3
2.17	14
2.14	8
2.10	5
2.07	5
2.05	14
2.02	2
1.92	12
1.89	7
1.85	5
1.83	5
1.73	18

Table 4.30 Zeolite Alpha (97)

d(A)	I
12.03	68
8.51	69
6.94	70
6.00	8
5.38	26
4.91	5
4.26	38
4.02	86
3.80	9
3.63	100
3.34	63
3.21	64
3.01	2
2.91	87
2.83	25
2.68	10
2.62	10
2.56	35
2.45	8
2.40	2
2.36	4
2.31	7
2.19	5
2.12	11
2.09	3
2.06	4
2.030	3
2.003'	19
1.953	3
1.877	9
1.854	9
1.812	2
1.791	22
1.711	2
1.699	21
1.650	16
1.635	4
1.592	12
1.565	9
1.539	1
1.491	2
1.479	3'
1.470	1
1.463	3
1.453	3

Table 4.31 Li-A (18)

d(Å)	I
6.42	vs
5.21	mw
4.29	vs
4.06	vw
3.27	vw
3.15	vs
3.03	vs
2.490	s
2.392	w
2.326	m
2.243	vw
2.173	mw
2.042	w
1.952	mw
1.868	vw
1.754	m
1.725	mw
1.556	w
1.524	mw
1.474	m
1.445	vw
1.405	mw
1.371	vw
1.348	vw
1.327	vw
1.300	vw
1.270	w
1.244	vw

Table 4.32 Zeolite P-B(P) (133)

d(Å)	I
7.08	58
5.01	58
4.96	45
4.44	7
4.10	57
4.04	17
3.53	5
3.34	6
3.20	100
3.13	57
2.71	37
2.68	38
2.66	22
2.53	8
2.20	5
1.98	6
1.68	7

Table 4.33 P-C (133)

d(Å)	I
5.64	80
4.87	16
3.68	6
3.44	100
3.27	3
2.93	45
2.81	6
2.70	14
2.51	13
2.37	8
1.91	8
1.87	5
1.75	12
1.72	3
1.70	3

Table 4.34 Synthetic Clinoptilolite-type Zeolite (33)

d(Å)	I
8.85	s
7.76	m
3.95	vs
3.41	m
3.12	w
2.96	m
2.80	w
2.73	w
2.56	w
2.43	w

355

356 THE SYNTHETIC ZEOLITES

Table 4.35 Zeolite D (76)

d(Å)	I
9.42	66
6.89	67
5.59	15
5.03	62
4.33	62
3.98	27
3.89	23
3.60	12
3.45	39
3.19	15
2.94	100
2.69	9
2.61	38
2.30	16
2.09	22
1.81	29
1.73	23

Table 4.36 Ca-D[a] (26)

hkl	d(Å)	I
200	6.82	m
211	5.56	vs
220	4.82	m
321	3.63	m
400	3.40	vs
004	3.37	s
332	2.90	s
323	2.89	m
431,	2.66	w
510		
521	2.47	w
440	2.40	w
619,	2.20	vw
532		
640	1.88	vw
633	1.85	vw
—	1.72	vw

[a] $a = 13.62$ Å

Table 4.37 Na-D (37)

d(Å)	I
13.53	40
10.24	—
9.06	50
6.57	55
6.39	—
6.08	—
5.80	15
5.05	—
4.83	—
4.52	25
—	—
4.15	—
4.00	60
3.84	—
3.76	—
—	—
—	—
3.53	—
3.47	100
—	—
3.39	—
3.29	—
3.22	—
3.16	—

Table 4.38 Sr-D[a] (88)

hkl	d(Å)	I
200	9.49	75
020	7.07	20
101	6.96	15
011	6.61	55
310	5.77	15
211	5.43	5
121	4.96	15
400	4.76	15
031	3.99	45
420	3.94	35
411	3.86	25
330	3.78	50
002	3.74	10
510,	3.67	30
231		
112	3.555	10
040	3.536	90
202	3.483	100
501	3.3389	15
240	3.313	20
141	3.142	55
521	3.058	45
530	2.960	25
402	2.938	25
620	2.897	35

[a] Orthorhombic
$a = 19.01$ Å
$b = 14.13$ Å
$c = 7.48$ Å

Table 4.39 Zeolite E (77)

d(Å)	I
9.53	100
7.13	16
5.47	8
4.23	18
3.86	6
3.54	4
3.46	12
3.41	4
3.34	5
3.14	14
3.08	10
3.00	18
2.86	23
2.63	7
2.30	8
2.23	9
1.89	4
1.89	4
1.67	5
1.62	3
1.48	3

Table 4.40 Ca-E[a] (26)

hkl	d(Å)	I
200	6.80	m
211	5.54	vs
220	4.80	m
321	3.63	w
400	3.40	vs
332	2.90	vs
431	2.67	m
521	2.48	m
440	2.40	w
116	2.21	w
640	1.88	w
633	1.85	w
—	1.68	w
—	1.58	w

[a] Tetragonal
a = 13.62 Å
c = 13.56 Å

Table 4.41 Zeolite K-E (22)

hkl	d(Å)	I
211	5.67	vvs
220	4.92	m
400	3.46	vvs
332	2.94	vs
431, 510	2.68	m
521	2.51	ms
440	2.42	m
611, 532	2.22	ms
640	1.90	s
721, 633, 552	1.87	ms
732	1.74	vs
651 932, 763	1.41	s
10·1·1, 772	1.36	s

Table 4.42 (TMA,Na)-E (107)

hkl	d(Å)	I
100	11.58	vw
101	9.21	vs
002	7.64	w
110	6.66	vs
102	6.38	mw
201	5.39	mw
103	4.60	m
211	4.19	ms
300	3.83	m
212	3.78	s
104	3.62	ms
220	3.32	mw
310	3.182	w
311	3.120	vw
222	3.042	vw
312	2.942	mw
214	2.864	m
401	2.825	ms
402	2.687	m
321	2.597	vw
410	2.506	m
323	2.335	vw
404	2.293	vw
413	2.248	vw
330	2.211	mw
324	2.166	vw
306	2.119	m
422	2.086	vw
316	1.988	w
504	1.968	vw

Table 4.43 Zeolite F[a] (73)

hkl	d(Å)	I
002	6.948	100
111	6.506	11
301	3.48	21
302	3.086	56
132	2.964	72
321	2.809	39
403	2.252	8
008	1.738	6
611	1.685	6
620	1.637	5

[a] a = 10.36 Å
c = 13.90 Å

Table 4.44 Sr-F (88)

d(Å)	I
11.91	ms
9.40	w
7.65	vw
6.93	vw
5.95	vvw
5.54	vw
5.14	w
4.99	s
4.53	mw
4.32	m
4.12	s
3.98	m
3.89	vw
3.54	mw
3.46	ms
3.32	w
3.19	w
2.99	s
2.93	ms
2.87	m
2.675	ms
2.608	s
2.302	m
2.091	ms
2.050	vvw
1.947	vvw
1.912	vvw
1.801	m
1.727	s
1.694	w

358 THE SYNTHETIC ZEOLITES

Table 4.45 Synthetic Ferrierite-Type Zeolite (39)

d(Å)	I
13.8	10 a
10.7	10 b
9.57	100
7.12	20
6.70	20
5.68	40
4.54	10
4.00	80
3.76	60
3.66	40
3.57	50
3.49	70
3.40	60
3.33	10
3.22	40
3.15	40
3.06	20
2.91	50
2.69	10
2.57	10

[a] mordenite impurity
[b] broad line

Table 4.46 Zeolite G[a] (22)

K-G (1) d(Å)	I	K-G (4) d(Å)	I	K-G (6) d(Å)	I
9.47	ms	9.45	ms	9.38	ms
				8.02	m
6.90	m	6.94	m	7.00	mw
		5.55	vvw	5.60	mw
5.22	m	5.24	m	5.11	m
		4.78	vvw	4.66	w
4.32	s	4.32	s	4.34	s
3.97	ms	3.95	ms	3.92	ms
3.70	w	3.68	mw	3.61	m
3.46	w	3.45	mw	3.47	mw
3.11	mw	3.18	mw	3.21	mw
2.93	vvs	2.93	vvs	2.93	vvs
2.80	w	2.79	mw		
2.59	s	2.59	ms	2.61	m
2.29	s	2.29	ms	2.31	mw
2.19	ms	2.17	m		
2.09	m	2.10	mw	2.10	mw
1.90	w	1.89	vw	1.86	w
1.84	m	1.84	mw	1.82	m
1.75	vw				
1.71	s	1.71	ms	1.73	m
1.63	m	1.64	m	1.65	mw
1.57	ms	1.57	mw	1.56	mw
		1.52	mw	1.52	mw
1.48	w	1.48	w	1.49	w
1.45	w				

[a] See Table 4.11 for meaning of numbers in ( ).

Table 4.47 P-G (133)

d(Å)	I
9.46	100
6.97	21
5.61	17
5.10	21
4.72	10
4.53	5
4.36	74
4.15	7
4.02	9
3.90	43
3.62	28
3.48	16
3.25	12
3.14	10
2.95	95
2.92	53
2.71	9
2.64	19
2.54	14
2.33	9
2.11	7
1.89	8
1.82	14
1.73	12

X-RAY POWDER DATA FOR SYNTHETIC ZEOLITES 359

Table 4.48 Sr-G (88)

d(A)	I
9.43	m
6.80	m
5.54	m
5.12	m
4.48	vvw
4.37	vw
4.30	ms
3.95	m
3.68	w
3.21	w
3.12	m
2.95	vw
2.91	s
2.82	vvw
2.78	vw
2.74	vvw
2.54	m
2.28	w
2.09	mw
1.81	mw
1.71	mw

Table 4.49 Ba-G[a] (90)

hkl	d(A)	I
100	16.15	s
111	9.35	vvw
102	8.10	vvw
200	7.53	m
210	6.83	mw
103	5.85	vw
220	5.38	w
203	4.85	w
004	4.72	w
204	3.95	vs
322	3.84	w
005	3.75	vw
323	3.49	m
421	3.32	m
403	3.23	m
422	3.19	m
225	3.08	s
206	2.92	s
216	2.874	vw
325	2.817	vw
424	2.753	vw
	2.690	s
	2.635	mw
	2.502	w
	2.447	mw

[a]Tetragonal
a = 15.16 A
c = 18.89 A

Table 4.50 Zeolite H[a] (75, 66)

hkl	d(A)	I
001	13.33	64
100	11.65	100
110	6.86	62
111	6.03	11
201	5.28	38
112	4.75	8
003	4.46	9
210	4.41	8
202	4.31	16
211	4.20	23
300	3.96	48
301	3.72	17
220	3.37	16
310	3.27	18
104	3.17	25
311	3.14	22
222	3.00	31
400	2.92	94
320	2.68	17
402	2.67	25
321	2.63	8
105	2.59	44
410	2.55	15
224	2.34	7
500	2.33	17

[a]Hexagonal
a = 13.47 A
c = 13.25 A

Table 4.51 Li-H (18)

d(A)	I
9.84	m
8.39	mw
8.10	w
6.68	ms
5.83	w
4.88	ms
4.55	vw
4.27	s
3.97	vw
3.78	w
3.58	mw
3.49	w
3.40	mw
3.32	mw
3.08	vw
2.97	w
2.80	w
2.66	w
2.52	m
2.38	vvw
2.02	vw
1.87	w

Table 4.52 Heulandite (81)

d(A)	I
9.03	80
7.97	38
6.84	29
5.25	37
5.13	45
4.65	40
4.37	20
3.97	130
3.91	58
3.72	18
3.56	30
3.46	31
3.42	43
3.40	33
3.32	23
3.17	39
3.12	31
3.07	26
2.97	61
2.80	30
2.73	20
2.53	10
2.43	8

360   THE SYNTHETIC ZEOLITES

Table 4.53 (HS) (35)

hkl	d(Å)	I
110	6.28	80
200	4.44	30
210	3.97	5
211	3.63	100
220	3.13	5
310	2.81	60
222	2.56	80
321	2.37	30
330, 411	2.90	80
420	1.99	5
332	1.888	9
422	1.814	30
501, 431	1.737	40
521	1.623	5
440	1.573	30
530, 433	1.523	30
442, 600	1.480	30
532, 611	1.439	30
620	1.402	5
541	1.371	30

a = 8.86 Å

Table 4.54 Ca-I[a] (26)

hkl	d(Å)	I
022	4.63	m
312	3.50	m
322, 3.19		w
332		
204, 2.94		m
024		
412, 2.85		m
142		
242, 2.67		m
422		
—	2.25	w
—	2.17	w
—	1.76	vw
—	1.62	vw

Orthorhombic

Table 4.55 Sr-I (88)

d(Å)	I
5.60	s
4.87	mw
3.68	vw
3.44	vs
2.93	s
2.806	vw
2.691	m
2.509	m
2.423	m
2.227	m
1.906	ms
1.757	ms
1.719	vw
1.695	vw
1.596	mw
1.415	ms
1.360	ms

Table 4.56 Zeolite J[a] (68)

hkl	d(Å)	I
101	5.862	54
111	5.570	11
200	4.770	32
—	4.725	16
120	4.271	15
112	3.995	51
122	3.231	46
300	3.186	4
103	3.128	93
301	3.033	40
130	3.002	41
113	2.970	36
131	2.890	100
—	2.870	61
—	2.681	14
—	2.662	20
—	2.644	25
—	2.614	11
—	2.580	23
—	2.329	14
—	2.296	8
—	2.188	14
—	2.146	6
—	2.085	4
—	2.043	6

[a] Tetragonal
a = 9.56 Å,
c = 9.92 Å

Table 4.57 Ba-J (90)

d(Å)	I
11.39	s
10.53	vw
10.22	w
7.39	m
7.04	w
6.23	m
6.08	vw
5.84	m
5.73	mw
5.63	vw
4.79	m
4.58	s
4.52	s
4.30	vw
4.19	w
4.08	w
4.02	ms
3.97	m
3.91	ms
3.81	m
3.70	w
3.56	ms
3.51	s
3.45	s
3.28	w
3.20	w
3.16	w
3.12	vs

Table 4.58 Ca-J[a] (26)

hkl	d(A)	I
020	8.89	s
210	6.88	s
	5.54	mw
002	4.89	s
040	4.48	w
022	4.31	mw
240	3.87	s
330		
312	3.44	vs
042	3.33	w
	3.25	w
	3.20	ms
	2.91	mw
	2.85	m
	2.78	w
	2.68	m
	2.50	m
	2.41	w
	2.21	vw
	2.09	vw
	1.86	vw
	1.77	w
	1.77.	w

[a] near-orthorhombic
a = 15.0 Å
b = 17.0 Å
c = 10.25 Å
$\beta \sim 90°$

Table 4.59 Ba-K (90)

d(A)	I
10.59	ms
7.39	mw
6.87	w
6.07	ms
5.58	ms
5.29	w
5.08	w
4.72	m
4.59	ms
4.30	ms
4.21	ms
3.89	w
3.67	vw
3.53	ms
3.49	m
3.45	w
3.34	vw
3.29	mw
3.16	s
3.06	w
3.03	ms
2.94	m
2.88	vw
2.79	mw
2.72	m
2.648	w
2.613	mw
2.553	w
2.493	m

Table 4.60 Zeolite L and Ion-Exchanged Forms[a] (11, 69)

hkl	d(A)	L[a]	BaL	CaL	$Ce^{3+}L$	MgL	NaL
100	15.8	100	100	100	100	100	100
200	7.89	14	38	10	38	12	9
001	7.49	15	62	31	94	24	25
201	5.98	25	56	33	94	29	21
111	5.75	11	31	18	—	16	14
220	4.57	32	69	37	75	33	34
310	4.39	13	38	16	63	12	13
301	4.33	13	38	29	69	22	23
221	3.91	30	56	33	81	39	34
311	3.78	13	13	12	38	14	12
320	3.66	19	50	22	56	20	16
410	3.48	23	62	22	50	24	25
321	3.26	14	25	22	25	20	21
212	3.17	34	100	47	88	51	46
330, 302	3.07	22	50	22	63	29	29
420	3.02	15	38	10	25	12	11
222	2.91	23	62	31	81	29	29
600	2.65	19	44	16	69	22	21
430	2.62	8	31	8	38	14	11

[a] Hexagonal unit cell, a = 18.4 Å, c = 7.5 Å
BaL = Barium exchanged L, degree of exchange = 0.73 equiv. per Al
CaL = Ca^{2+} exchanged L, 0.71 equiv./Al
$Ce^{+3}L$ = Ce^{3+} exchanged L, 0.28 equiv./Al
MgL = Mg^{2+} exchanged L, 0.39 equiv./Al
NaL = Na^{+} exchanged L, 0.41 equiv./Al

Table 4.61 Ca-L[a] (26)

hkl	d(A)	I
110	7.14	vs
002	4.96	ms
211	4.14	s
112	4.07	w
311	3.24	m
103	3.14	vs
213	2.67	ms
400	2.56	vw
303	2.33	vw
323	2.12	vw
422	2.07	vw
005	1.98	vw
314	1.94	vw
215	1.81	vw
440	1.78	mw
522	1.74	vw
433	1.71	w
600	1.67	vw
610	1.65	vw
800	1.26	w

[a] Tetragonal
a = 10.01 Å
c = 9.89 Å

Table 4.62 P-L (133)

d(Å)	I
16.0	100
8.00	4
7.55	8
6.09	18
5.86	4
4.65	29
4.47	9
4.37	6
3.96	28
3.68	9
3.51	19
3.32	15
3.22	31
3.09	24
3.05	5
2.93	28
2.88	4
2.82	4
2.69	24
2.53	9
2.50	4
2.46	4
2.44	5
2.32	3
2.30	3
2.22	9
2.06	3
1.96	2
1.88	6

Table 4.63 Zeolite Losod (162)

hkl	d(Å)	I	hkl	d(Å)	I
100	11.1067	10	231, 401	2.6996	70
101	7.6383	35	321	2.6726	27
110	6.4284	83	004	2.6341	41
200	5.5930	3	321	2.4917	2
201	4.9334	6	402	2.4692	17
102	4.7500	74	114, 410	2.4384	1
210	4.2213	14	204	2.3824	1
112	4.0768	3	313	2.3275	1
211	3.9163	54	214	2.2368	1
202	3.8298	6	403	2.1885	3
300	3.7231	80	304	2.1504	34
301	3.5188	8	502	2.0579	5
103	3.3442	12	404	1.9191	5
212	3.2933	100	215	1.8879	4
220	3.2255	10	512	1.8700	5
310	3.0977	7	324, 430	1.8379	2
203, 311	2.9732	25	423	1.8112	1
400	2.7934	17			

Hexagonal a = 12.906 Å c = 10.541 Å

Table 4.64 Zeolite M[a] (70)

hkl	d(Å)	I
111	7.003	18
200	6.559	6
112	4.513	8
221	4.245	21
301	3.9918	3
003	3.5093	6
203	3.0956	100
213	2.9756	16
421	2.8286	15
500	2.6099	27
530	2.2621	5
314	2.2294	5
611	2.1251	14

[a] Tetragonal a = 13.12 Å c = 10.48 Å

X-RAY POWDER DATA FOR SYNTHETIC ZEOLITES 363

Table 4.65 Ba-M (90)

d(A)	I
8.24	vs
7.56	vw
7.18	s
7.03	w
6.39	vs
6.11	vw
5.02	ms
4.29	ms
4.10	s
4.06	s
4.03	s
3.89	m
3.66	w
3.57	vvw
3.52	w
3.46	mw
3.23	s
3.19	mw
3.16	ms
3.12	vs
3.07	m
2.91	ms
2.838	w
2.745	w
2.724	s
2.674	s
2.668	s
2.624	w
2.557	w

Table 4.66 Sr-M (88)

d(A)	I
13.56	mw
9.10	m
6.63	m
6.43	w
6.07	vw
5.86	m
4.76	mw
4.55	m
4.32	vw
4.18	m
4.00	ms
3.84	vw
3.76	w
3.64	w
3.48	s
3.40	ms
3.23	s
3.04	mw
2.92	s
2.69	mw
2.563	ms
2.523	ms
2.182	w
2.090	w
1.886	w
1.821	w
1.798	vw
1.705	vw
1.638	vw

Table 4.67 Synthetic Li-Mordenite (9)

d(A)	I
13.5	30
10.3	10
9.02	75
6.50	40
6.32	15
5.98	10
5.75	25
5.05	2
4.80	2
4.49	45
3.97	60
3.80	20
3.73	10
3.45	60
3.37	50
3.20	100
3.13	10

Table 4.68 Synthetic Large-Port Na-Mordenite (9, 159)

d(A)	I	hkl	d(A)	I
13.4	40	110	13.59	ms
10.2	10	020	10.25	w
9.02	70	200	9.072	s
6.50	50	111	6.584	s
6.32	30	130	6.371	mw
6.02	10	021	6.056	mw
5.75	20	201	5.782	mw
5.03	2	221	5.039	w
4.84	2	131	4.861	w
4.50	35	330	4.512	ms
4.12	5	041	4.210	vw
3.97	70	420	4.119	vw
3.81	15	150	4.077	vs
3.73	10	241	3.830	m
3.52	10	002	3.746	mw
3.45	100	112	3.640	w
3.37	60	510	3.561	vw
3.28	10	022	3.521	vw
3.21	55	202	3.486	vs
3.13	10	060	3.410	vw
		350	3.382	s
		222	3.272	w
		530	3.150	s

Orthorhombic
a = 18.1 A
b = 20.4 A
c = 7.5 A

364   THE SYNTHETIC ZEOLITES

Table 4.69 Synthetic Na, Li-Mordenite (9)

d(Å)	I
13.5	40
10.2	10
9.02	80
6.51	50
6.32	25
6.01	10
5.75	25
5.03	2
4.82	5
4.50	40
4.11	5
3.96	70
3.80	10
3.72	10
3.52	5
3.44	100
3.37	65
3.28	10
3.20	65
3.13	10

Table 4.70 Zeolite N (100)

hkl	d(Å)	I
111	21.66	41
220	13.17	68
311	11.19	73
400	9.30	50
333	7.14	14
511	—	—
440	6.57	100
531	6.27	36
620	5.86	23
533	5.66	16
553	4.83	27
731	—	—
644	4.52	14
820	—	—
660	4.36	20
822	—	—
842	4.06	89
844	3.78	57
10,0,0; 3.72	80
860	—	—
951	3.57	23
773	—	—
9.5.3	3.48	20
10.4.2	3.38	20
775	3.347	41
11.1.1		
10.6.2	3.139	41

Cubic
a = 37.2 Å

Table 4.71 Zeolite Ba-N (91)

Ba-N d(Å)	I
7.1	m
5.6	ms
5.2	ms
4.5	w
4.35	vw
3.50	m
3.40	s
3.15	mw
2.92	ms
2.84	w
2.79	w
2.69	w
2.59	w
2.57	ms
2.44	m
2.40	w
2.33	w
2.24	m
2.230	w
2.165	w
2.078	mw
2.051	mw
2.020	mw
1.917	w
1.893	m
1.845	mw
1.823	w
1.762	mw

Table 4.72 Zeolite TMA-Ω (160)

hkl	d(Å)	I
100	15.95	20
110	9.09	86
200	7.87	21
101	6.86	27
210	5.94	32
201	5.47	6
300	5.25	8
	5.19	
211	4.695	32
400	3.909	11
002	3.794	58
102	3.708	30
320	3.620	25
112	3.516	53
410	3.456	20
500	3.13	38
302	3.074	21
	3.02	
501	2.911	36
600	2.640	6
103,	2.488	6
601		
332	2.342	17
303,	2.272	6
440		
432	2.139	5

Hexagonal
a = 18.24 Å
c = 7.60 Å

X–RAY POWDER DATA FOR SYNTHETIC ZEOLITES 365

Table 4.73 Zeolite TMA-"Offretite" (104)

d(Å)	I
11.45	100
7.54	16.5
6.63	55.2
6.30	9.9
5.74	15.0
4.57	26.5
4.34	43.3
3.76	89.2
3.59	43.0
3.31	18.6
3.15	17.4
2.93	9.5
2.85	79.7
2.68	19.1
2.51	13.8

Table 4.74 P$_c$ (47)

hkl	d(Å)	I
110	7.10	55
200	5.01	35
211	4.10	55
310	3.16	100
321	2.67	55
400	2.52	5
411, 330	2.36	7
422	2.054	5
510, 431	1.965	10
440	1.771	7
530, 433	1.719	7
600, 422	1.667	7

Table 4.75 P$_t$ (52)

hkl	d(Å)	I
110	7.132	85
101	7.047	83
111	5.776	5
200	5.048	51
002	4.914	26
102	4.420	8
211	4.108	94
112	4.049	22
202	3.527	4
212	3.328	18
310	3.194	100
103	3.117	64
311	3.036	10
113	2.979	5
302	2.776	3
203	2.750	5
321	2.694	46
312	2.679	28
213	2.653	21
400	2.531	6
322	2.435	5
104	2.387	4
420, 402	2.257	2
421	2.206	2
214	2.159	2
422	2.055	1
510	1.982	2
413	1.966	2
105	1.929	1

Table 4.76 Zeolite P-(Cl) (140)

d(Å)	I
13.29	vvs
9.37	mw
4.99	—
4.66	m
4.40	vw
4.17	vs
3.98	w
3.81	s
3.65	m
3.30	vw
3.19	s
3.10	vw
3.02	vvs
2.94	vvw
2.87	vw
2.80	s
2.74	vvw
2.69	m
2.58	vvw
2.53	s
2.49	w
2.45	vvw
2.36	vw
2.32	w

THE SYNTHETIC ZEOLITES

Table 4.77 Zeolite Q[a] (71)

hkl	d(Å)	I
100	13.514	47
003	11.748	100
005	6.9975	3
200	6.7479	7
210	6.0382	11
220	4.7678	4
206	4.4328	8
303	4.2025	17
321	3.7277	17
226	3.6957	5
400	3.3759	16
410	3.2749	21
403	3.2456	8
404	3.1485	26
421	3.0089	39
423	2.9252	27

[a] Tetragonal
a = 13.5 Å
c = 35.2 Å

Table 4.78 Zeolite Q-(Br) (140)

d(Å)	I
13.32	vvs
9.40	m
5.92	vvs
5.41	vvw
5.01	vw
4.41	m
4.19	m
3.82	s
3.67	w
3.31	vw
3.20	vs
3.12	vw
3.03	vvs
2.88	m
2.82	vs
2.70	—
2.70	vw
2.64	s
2.59	vw
2.54	s
2.50	vw
2.45	vw
2.37	vw
2.33	m

Table 4.79 Sr-Q[a] (88)

hkl	d(Å)	I
011	7.80	5
020	6.91	25
200	6.28	15
111	5.85	95
11$\bar{2}$	4.74	80
220	4.65	40
211	4.30	45
031, 4.16	25
13$\bar{1}$		
30$\bar{2}$	3.93	10
131	3.76	30
040	3.48	5
032	3.30	20
041	3.26	70
231	3.231	5
400	3.137	10
330	3.105	35
42$\bar{1}$	3.030	100
222	2.928	60
113, 2.76	20
050		
41$\bar{3}$	2.735	20
232	2.650	25

[a] Monoclinic
a = 13.48 Å
b = 13.86 Å
c = 10.10 Å
$\beta = 111°41'$

Table 4.80 Zeolite R (50)

d(Å)	I
9.51	88
6.97	35
5.75	16
5.61	26
5.10	45
4.75	12
4.37	78
4.13	12
4.02	14
3.92	35
3.80	16
3.63	41
3.48	25
3.34	12
3.21	18
3.13	12
2.95	100
2.89	16
2.80	14
2.71	14
2.66	10
2.62	25
2.53	22
2.39	10
2.14	6
2.10	14
1.93	10
1.89	10
1.82	18

X–RAY POWDER DATA FOR SYNTHETIC ZEOLITES 367

| Table 4.81 P-R (133) || Table 4.82 Sr-R (88) || Table 4.83 Zeolite S (49) || Table 4.84 Zeolite T (78) ||
d(Å)	I	d(Å)	I	d(Å)	I	d(Å)	I
9.46	100	9.04	m	11.88	77	11.45	100
6.94	29	7.99	m	7.73	19	9.18	4
5.59	10	5.28	mw	7.16	100	7.54	13
5.09	32	5.12	ms	5.96	9	6.63	54
4.72	9	4.66	w	5.03	72	6.01	2
—	—	4.37	w	4.50	46	5.74	6
4.35	76	3.98	vs	4.12	79	4.99	2
4.11	5	3.92	w	3.97	20	4.57	8
4.00	8	3.84	vw	3.44	62	4.34	45
3.90	24	3.73	vw	3.305	13	4.16	3
3.63	32	3.58	mw	3.236	23	4.08	2
3.47	16	3.44	w	2.973	80	3.82	16
3.25	10	3.41	w	2.858	47	3.76	56
—	—	3.33	w	2.693	19	3.67	1
2.94	82	3.18	mw	2.603	39	3.59	30
—	—	3.13	mw	2.126	11	3.42	2
2.71	5	3.08	vw	2.089	39	3.31	16
2.62	11	2.97	ms	1.910	12	3.18	12
2.53	8	2.80	m	1.809	40	3.15	18
2.32	5	2.74	m	1.722	32	2.93	11
2.10	8	2.67	vw			2.87	38
1.89	8	2.28	vw			2.85	45
1.82	9	1.96	vw			2.68	11
1.73	6	1.83	vw			2.61	2
						2.51	8
						2.49	13
						2.30	2
						2.21	6
						2.12	5

368 THE SYNTHETIC ZEOLITES

Table 4.85 Zeolite Ba-T (91)

d(Å)	I	d(Å)	I
7.4	w	2.45	m
7.2	w	2.43	ms
5.2	ms	2.38	mw
5.0	w	2.36	mw
4.3	m	2.35	mw
4.24	mw	2.33	m
4.20	mw	2.32	vw
4.16	m	2.31	vw
4.00	vw	2.27	w
3.79	s	2.25	mw
3.70	w	2.24	w
3.68	s	2.22	m
3.52	m	2.18	m
3.37	ms	2.15	w
3.30	mw	2.09	w
3.22	s	2.07	w
3.18	s	2.04	w
3.15	w	2.03	m
3.08	w	2.02	m
3.93	w	2.00	m
2.98	m	1.99	mw
2.96	m	1.97	mw
2.94	vw	1.96	mw
2.92	w	1.95	w
2.88	w	1.92	w
2.84	w	1.89	w
2.64	m	1.83	m
2.63	w	1.78	m

Table 4.86 Zeolite W[a] (72)

hkl	d(Å)	I
200	9.99	20
211	8.17	49
220	7.09	54
321	5.34	28
400	5.01	56
420	4.45	21
332	4.28	35
521	3.64	20
611, 532	3.25	100
620	3.17	75
631	2.95	71
721, 633, 552	2.72	53
642	2.67	12
553	2.60	8
732, 651	2.54	26
644, 820	2.43	9
653	2.40	8
842	2.18	10
10,3,1; 952, 765	1.91	5
953	1.86	7

[a] a = 20.06 Å

Table 4.87 P-W (133)

d(Å)	I
10.2	18
8.3	42
7.2	61
5.40	17
5.07	29
4.51	27
4.31	23
4.11	19
3.68	18
3.25	71
3.19	100
2.96	45
2.80	16
2.75	39
2.69	18
2.57	23
2.46	10
2.20	8
2.08	4
1.79	5
1.78	6
1.73	8

X-RAY POWDER DATA FOR SYNTHETIC ZEOLITES

Table 4.88 Zeolite X[a] (5, 11)

hkl	d(Å)	I
111	14.465	100
220	8.845	18
311	7.538	12
331	5.731	18
333, 511	4.811	5
440	4.419	9
531	4.226	1
620	3.946	4
533	3.808	21
622	3.765	3
444	3.609	1
711, 551	3.500	1
642	3.338	8
731, 553	3.253	1
733	3.051	4
822, 660	2.944	9
751, 555	2.885	19
840	2.794	8
911, 753	2.743	2

[a] Hydrated sodium X, a = 24.93 Å Si/Al = 1.25
[b] Hydrated calcium X, Ca/Al = 0.84 Si/Al = 1.25

Table Ca-X[b] (5, 11)

d(Å)	I
14.371	100
8.792	9
7.506	4
5.709	16
4.793	5
4.405	11
—	—
3.936	2
3.800	20
3.754	2
3.593	2
3.486	2
3.328	12
3.241	3
3.041	4
2.934	8
2.875	6
2.783	7
2.732	4

Table 4.89 N-X (95)

hkl	d(Å)	I
111	14.42	100
220	8.801	25
311	7.512	20
331	5.720	19
333, 511	4.798	13
440	4.407	8
531	4.217	7
620	3.941	11
533	3.800	32
631	3.679	30
642	3.331	31
733	3.045	9
822, 660	2.937	12
751, 555	2.877	31
840	2.784	12
911, 753	2.731	6
664	2.658	14
844	2.543	3
10,0,0; 862	2.492	4
10,2,0; 862	2.442	1

Table 4.90 Zeolite Y[a] (11, 61)

hkl	d(Å)	I
111	14.29	100
220	8.75	9
311	7.46	24
331	5.68	44
333, 511	4.76	23
440	4.38	35
620	3.91	12
533	3.775	47
444	3.573	4
711, 551	3.466	9
642	3.308	37
731, 553	3.222	8
733	3.024	16
822, 660	2.917	21
751, 555	2.858	48
840	2.767	20
911, 753	2.717	7
664	2.638	19
931	2.595	11

[a] Hydrated sodium Y a = 24.73 Å; Si/Al = 2.00

Table 4.91 N-Y (95)

hkl	d(Å)	I
111	14.28	100
220	8.77	18
311	7.48	17
331	5.688	11
333, 511	4.778	10
440	4.385	7
531	4.195	4
600, 442	4.135	1
620	3.924	7
533	3.786	20
631	3.661	9
711, 551	3.474	1
642	3.316	17
733	3.032	6
822, 660	2.923	8
751, 555	2.866	17
840	2.775	6
911, 753	2.721	2
664	2.646	1
844	2.533	2
10,0,0, 860	2.478	1
10,2,0,	2.430	1

370 THE SYNTHETIC ZEOLITES

Table 4.92 Zeolite Z (65)

d(Å)	I
7.45	vvs
4.78	vw
3.98	m
3.47	m
3.29	m
3.09	vs
2.97	s
2.82	vs
2.35	mw
2.20	w
2.11	m
1.85	m
1.74	m
1.68	mw
1.59	m
1.56	m

Table 4.93 Z-21 (59)

d(Å)	I
21.12	50
12.98	75
11.07	70
9.18	35
8.42	10
7.50	3
7.06	10
6.51	100
6.23	50
5.82	20
5.60	10
5.30	5
5.15	15
4.79	15
4.50	10
4.34	20
4.24	3
4.12	15
4.04	70
3.92	10
3.76	45
3.70	75
3.61	5
3.55	10
3.43	10
3.36	15
3.31	1
3.21	15
3.15	2

Table 4.94 Zeolite ZK-4 (96)

hkl	d(Å)	I
100	12.07	100
110	8.57	71
111	7.025	50
210	5.422	23
220	4.275	11
300	4.062	48
311	3.662	59
320	3.390	33
321	3.244	64
410	2.950	60
411	2.862	14
420	2.727	8
421	2.661	4
332	2.593	13
422	2.481	2
430	2.435	1
511	2.341	2
521	2.225	2
440	2.162	2
441	2.120	1
530	2.080	1
531	2.061	1
600	2.033	4
621	1.904	2
541	1.881	1
622	1.885	1
630	1.813	1

Cubic
a = 12.15 Å

Table 4.95 Zeolite ZK-5 (98)

hkl	d(Å)	I	hkl	d(Å)	I
110	13.3	18	651		
200	9.41	100	811,	2.30	2
220	6.62	6	741,		
310	5.93	41	554		
222	5.41	48	822,	2.20	3
321	5.03	2	660		
400	4.69	6	831,	2.17	2
330	4.41	50	750,		
420	4.19	34	743		
332	3.98	22	622	2.14	1
422	3.81	18	910,	2.06	3
510	3.66	6	833		
521	3.41	13	842	2.04	2
530,	3.21	35	921,	2.02	3
433			761,		
611	3.02	28	655		
620	2.94	21	830,	1.97	1/2
541	2.88	2	851,		
622	2.81	26	754		
631	2.75	9	932,	1.93	2
543,	2.64	11	763		
710,			941,	1.89	2
550			853,		
640	2.59	2	770		
721,	2.54	9	10.0.0,	1.87	5
633,			860		
552					
730	2.45	3			
732,	2.37	1			

Cubic
a = 18.7 Å

Table 4.96 ZK-19 (79)

hkl	d(Å)	I
101	8.17	19
002	7.13	88
121	5.37	19
022	5.03 sh	11
200	4.98	34
103	4.29	12
113	4.13 sh	7
220	4.08	26
123	3.68	3
141	3.26 sh	15
301	3.23 sh	13
024, 133	3.18	100
311	3.15 sh	16
321	2.94	30
240	2.89	5
105, 143	2.74	23
224	2.685	31
125	2.56	6
323	2.54	4
044	2.51	6
341	2.39	5
016, 252	2.34	7
244, 422	2.24	5

sh = shoulder

Table 4.97 Zeolite ZK-20(108)

d(Å)	I
14.2	vw
10.4	m
9.5	w
8.2	s
7.7	w
6.7	m
5.2	s
4.3	s
4.1	vs
3.86	m-s
3.62	w
3.48	w
3.34	w
3.18	s
3.10	m
2.87	m
2.81	s-vs
2.64	m
2.59	vw
2.52	w
2.41	m
2.23	w-m
2.18	vw
2.14	m
2.07	vw
2.04	w
1.96	w
1.93	w
1.90	w

Table 4.98 Zeolite ZK-21 (138)

hkl	d(Å)	I	hkl	d(Å)	I
100	12.16	100	511	2.348	6
110	8.65	68	521	2.227	3
111	7.07	63	440	2.154	10
210	5.48	29	441	2.122	4
211	5.02	4	530	2.093	3
220	4.33	15	531	2.061	3
300	4.07	63	600	2.033	9
310	3.86	3	621	1.904	7
311	3.675	83	541	1.881	5
320	3.389	30	622	1.839	2
321	3.264	59	630	1.818	2
410	2.952	77	632	1.740	2
411	2.878	15	710	1.722	14
420	2.732	10	711	1.706	2
421	2.664	5	641	1.674	8
332	2.601	32	721	1.659	2
422	2.491	9	722	1.615	4
430, 500	2.441	4	731	1.586	5
510	2.395	2	650	1.560	5

Table 4.99 Zeolite ZSM-2 (34)

hkl	d(A)	I
002	14.0	83
200	13.8	76
210	12.2	42
310	8.70	47
321	7.34	37
004	7.07	41
400	6.85	8
005	5.63	85
500, 430	5.48	10
510, 501, 431	5.37	16
440, 225	4.86	5
006, 530	4.70	16
600	4.56	7
602	4.34	55
335	4.24	5
425	4.14	35
631	4.04	43
107	3.97	7
227	3.71	11
642	3.66	39
535	3.60	10
650	3.51	4
652	3.40	11
218	3.38	16
733	3.36	8
653 308	3.28	27
715, 555	3.19	8
527	3.15	41
753	3.01	100
920, 760	2.97	16
726	2.94	18
850	2.90	22
930	2.88	3
852	2.84	5
904	2.79	9
816, 746	2.75	10
3,0,10	2.69	23
952	2.61	6

Table 4.100 Zeolite ZSM-3 (63)

d(A)	I	d(A)	I
15.26	62	4.49	5
14.16	124	4.38	51
13.19	26	4.19	14
11.86	4	4.16	6
9.21	2	4.11	13
8.75	44	3.99	1
8.00	3	3.96	2
7.61	9	3.93	7
7.41	32	3.85	14
7.22	7	3.78	19
7.03	15	3.72	23
6.89	2	3.52	1
5.94	4	3.47	10
5.72	37	3.40	10
5.62	25	3.31	42
5.48	6	3.22	16
5.15	4	3.18	17
5.02	7	3.02	60
4.90	3	2.98	11
4.82	2	2.92	31
4.78	2	2.87	10
4.75	18	2.85	8
4.70	1	2.80	9
4.62	3	2.73	11
4.58	1	2.70	11

X–RAY POWDER DATA FOR SYNTHETIC ZEOLITES 373

Table 4.101 Zeolite ZSM-4 (101)

d(Å)	I
16.07	w
9.12	s
7.90	m
6.92	m
5.99	ms
5.37	vs
5.28	w
4.70	ms
4.39	vw
3.93	w
3.79	s
3.71	w
3.62	w
3.52	ms
3.44	w
3.15	m
3.08	w
3.04	w
2.92	m
2.65	w
2.62	w
2.52	w
2.37	w
2.28	w
2.14	vw
2.10	vw
2.08	vw
2.03	vw
1.98	w

Table 4.102 Zeolite ZSM-5 (103)[a]

hkl	d(Å)	I	hkl	d(Å)	I
200	11.36	s	610	3.84	vs
210	10.20	ms	611	3.74	vs
002	9.90	–	540	3.62	s
102	9.14	vw	315	3.50	w
202	7.54	w	630	3.46	w
212	7.17	w	603	3.33	w
311	6.79	vw	632	3.27	vw
302	6.06	w	642	3.07	w
203	5.77	w	513	3.00	m
410	5.63	w			
411	5.42	vw			
420	5.19	vw			
303	5.05	w			
500, 430	4.65	w			
430	4.40	w			
413	4.30	w			
440, 314	4.12	vw			
441	4.04	vw			

Tetragonal a = 23.2Å, c = 19.9Å

[a]Structure reported in Ref. 224 of Cpt. 2: orthorhombic, space group Pnma, a = 20.1, b = 19.9, c = 13.4Å; SBU is complex 5-1 unit, framework density = 17.9T/1000Å³. See also ref. 222 in Cpt. 2.

Table 4.103 ZSM-8 (102)

d(Å)	I
11.1	46
10.0	42
9.7	10
9.0	6
7.42	10
7.06	7
6.69	5
6.35	12
6.04	6
5.97	12
5.69	9
5.56	13
5.36	3
5.12	4
5.01	7
4.60	7
4.45	3
4.35	7
4.25	18
4.07	20
4.00	10
3.85	100
3.82	57
3.75	25
3.71	30
3.64	26
3.59	2

Table 4.104 Zeolite ZSM-10 (109)

d(A)	I
15.85	58
13.92	42
10.22	13
7.87	22
7.55	56
7.04	13
6.29	35
5.96	22
5.46	31
5.25	15
5.06	25
4.50	76
4.41	67
4.32	27
3.87	91
3.64	100
3.54	56
3.47	25
3.42	27
3.32	13
3.22	16
3.16	31
3.10	67
3.04	73
2.89	89
2.73	48
2.69	15
2.57	15

REFERENCES

1. G. W. Morey and E. Ingerson, *Econ. Geol.*, **32**: 607 (1937).
2. L. B. Sand, R. Roy, and E. F. Osborn, *Econ. Geol.*, **52**: 169 (1957).
3. A. J. Regis, L. B. Sand, C. Calman, and M. E. Gilwood, *J. Phys. Chem.*, **64**: 1567 (1960).
4. R. M. Milton, *U. S. Pat. 2,882,243* (1959).
5. R. M. Milton, *U. S. Pat. 2,882,244* (1959).
6. W. F. Fyfe, *J. Geol.*, **68**: 533 (1960).
7. J. R. Goldsmith, *J. Geol.*, **61**: 439 (1953).
8. D. S. Coombs, A. J. Ellis, W. F. Fyfe, and A. M. Taylor, *Geochim. Cosmochim. Acta*, **17**: 53 (1959).
9. M. L. Sand, W. S. Coblenz, and L. B. Sand, Molecular Sieve Zeolites, *Advan. Chem. Ser. 101*, American Chemical Society, Washington, D. C. 1971, p. 127.
10. W. M. Meier, Molecular Sieves, Society of Chemical Industry, London, 1968, p. 100.
11. D. W. Breck and E. M. Flanigen, Molecular Sieves, Society of Chemical Industry, London, 1968, p. 47.

12. R. M. Barrer, Molecular Sieves, Society of Chemical Industry, London, 1968, p. 39.
13. H. G. Bungenberg de Jong, in W. R. Kruyt (Ed.) Colloid Science II, Elsevier, New York, 1949, p. 2.
14. S. P. Zhdanov, Molecular Sieves, Society of Chemical Industry, London, 1968, p. 62.
15. M. Schlaepfer and P. Niggli, *Z. Anorg. Chem.*, **87**: 52 (1914).
16. R. M. Barrer, *Chem. Brit.*, **2**: 9, 380 (1966).
17. S. T. Powell, *Combustion*, **5**: 15 (1933).
18. R. M. Barrer and E. A. D. White, *J. Chem. Soc.*, 1167 (1951).
19. R. M. Barrer, *Discuss. Faraday Soc.*, **5**: 326 (1949).
20. R. M. Barrer, *J. Chem. Soc.*, 2158 (1948).
21. J. Konigsberger and W. J. Muller, *Z. Anorg. Chem.*, **104**: 1 (1918).
22. R. M. Barrer and J. W. Baynham, *J. Chem. Soc.*, 2882 (1956).
23. R. M. Barrer, *Brit. Pat.*, *574,911* (1944).
24. R. M. Barrer, *J. Chem. Soc.*, 127, (1948).
25. J. F. Black, *U. S. Pat.*, *2,442,191* (1948).
26. R. M. Barrer and P. J. Denny, *J. Chem. Soc.*, 983 (1961).
27. N. S. Kurnakov, L. G. Berg, and V. N. Svestnikova, *Izv. Akad. Nauk. SSR, Otd. Mat. Nauk*, 1381 (1937).
28. V. M. Svestnikova and V. G. Kuzneksov, *Izv. Akad. Nauk. SSr, Otd. Khim. Nauk*, 25 (1946).
29. I. E. Sakovov and N. A. Shishakov, *Izv. Akad. SSR, Otd. Khim. Nauk*, 1745(1963).
30. D. J. Drysdale, *Amer. Mineral.*, **56**: 1718 (1971).
31. N. A. Acara, Union Carbide Corporation, *Unpublished Results.*
32. W. L. Haden, Jr. and F. J. Dzierzanowski, *U. S. Pat.*, *3,123,441* (1964).
33. L. L. Ames, Jr., *Amer. Mineral.*, **48**: 1374 (1963).
34. J. Ciric, *U. S. Pat.*, *3,411,874* (1968).
35. R. M. Barrer and E. A. D. White, *J. Chem. Soc.*, 1561 (1952).
36. L. B. Sand, Molecular Sieves, Society of Chemical Industry, London, 1968, p. 71.
37. D. Domine and J. Quobex, Molecular Sieves, Society of Chemical Industry, London, 1968, p. 78.
38. D. J. Whittemore, Jr., *Amer. Mineral.*, **57**: 1146 (1972).
39. E. E. Senderov, *Geokhim.*, **9**: 820 (1963).
40. P. Saha, *Amer. Mineral.*, **46**: 859 (1961), *Amer. Mineral.*, **44**: 300 (1959).
41. A. Guyer, M. Ineichen, and P. Guyer, *Helv. Chim. Acta*, **40**: 1603 (1957).
42. E. E. Senderov and N. I. Khitarov, Molecular Sieve Zeolites, Vol. 101, American Chemical Society, Washington, D. C., 1971, p. 149.
43. K. T. Kim and B. J. Burley, *Can. J. Earth Sci.*, **8**: 311 (1971).
44. A. J. Ellis, *Geochim. Cosmochim. Acta*, **19**: 145 (1960).
45. L. L. Ames, Jr. and J. B. Sand, *Amer. Mineral.*, **43**: 476 (1958).
46. A. Timfoldgyar, *Brit. Pat.*, *939,617* (1963).
47. R. M. Barrer, J. W. Baynham, F. W. Bultitude, and W. M. Meier, *J. Chem. Soc.*, 195 (1959).
48. D.W. Breck, W.G. Eversole, R.M. Milton, T. B. Reed, and T. L. Thomas, *J. Amer. Chem. Soc.*, **78**:5963 (1956).
49. D. W. Breck, *U. S. Pat.*, *3,054,657* (1962).
50. R. M. Milton, *U. S. Pat.*, *3,030,181* (1962), *Brit. Pat.*, *841, 812* (1960).
51. E. E. Senderov and N. I. Khitarov, Zeolites, Their Synthesis and Conditions of Formation in Nature, Nauka, Moscow, 1970.
52. A. M. Taylor and R. Roy, *Amer. Mineral.*, **49**: 656 (1964).
53. A. Pereyron, J. L. Guth and R. Wey, *C. R. Acad. Sci., Ser. D*, **272**: 1331 (1971).

54. I. Y. Borg and D. K. Smith, "Calculated X-Ray Powder Patterns for Silicate Minerals," Geological Society of America Memoir 122, 1969.
55. R. M. Milton, U. S. Pat. 3,008,803 (1961).
56. S. P. Zhdanov and N. N. Buntar, Dokl. Akad. Nauk SSSR, **147**: 1118 (1962).
57. S. P. Zhdanov, N. N. Buntar, and E. N. Egorova, Dokl. Akad. Nauk SSSR, **154**: 419 (1964).
58. J. L. Guth, P. Collins and R. Wey, Bull. Soc. Fr. Mineral. Cristallogr., **93**: 59 (1970).
59. H. C. Duecker, A. Weiss, and C. R. Guerra, U. S. Pat. 3,567,372 (1971).
60. W. C. Beard, "Molecular Sieve Zeolites," Advan. Chem. Ser. 101, American Chemical Society, Washington, D. C., 1971, p. 237
61. D. W. Breck, U. S. Pat., 3,130,007 (1964).
62. J. Ciric, U. S. Pat., 3,415,736 (1968).
63. G. T. Kokatailo and J. Ciric, "Molecular Sieve Zeolites," Advan. Chem. Ser. 101, American Chemical Society, Washington D. C.,1971, p. 109.
64. H. Borer and W. M. Meier, "Molecular Sieve Zeolites," Advan. Chem. Ser. 101, American Chemical Society, Washington D. C. 1971, p. 122.
65. R. M. Barrer and J. W. Baynham, U. S. Pat., 2,972,516 (1961).
66. S. P. Zhdanov and M. E. Ovsepyan, Dokl. Akad. Nauk, SSR, **157**: 913 (1964) .
67. R. M. Barrer, J. F. Cole, and H. Sticher, J. Chem. Soc. (A), 2475 (1968).
68. D. W. Breck and N. A. Acara, U. S. Pat., 3,011,869 (1961).
69. D. W. Breck, U. S. Pat., 3,216,789 (1965).
70. D. W. Breck and N. A. Acara, U. S. Pat., 2,995,423 (1961).
71. D. W. Breck and N. A. Acara, U. S. Pat., 2,991,151 (1961).
72. R. M. Milton, U. S. Pat., 3,012,853 (1961).
73. R. M. Milton, U. S. Pat., 2,996,358 (1961).
74. S. P. Zhdanov, Molecular Sieves, Society of Chemical Industry, London, 1968, p.62.
75. R. M. Milton, U. S. Pat., 3,010,789 (1961).
76. D. W. Breck and N. A. Acara, Brit. Pat., 868,846 (1961).
77. D. W. Breck and N. A. Acara, Brit. Pat., 2,962,355 (1960).
78. D. W. Breck and N. A. Acara, U. S. Pat. 2,950,952 (1960).
79. G. H. Kühl, Amer. Mineral., **54**: 1607 (1969).
80. R. M. Barrer and N. McCallum, J. Chem. Soc., 4029 (1953).
81. M. Koizumi and R. Roy, J. Geol., **68**: 41 (1960).
82. D. A. Buckner, Ph.D. Thesis, Univ. of Utah (1958).
83. L. L. Ames and L. B. Sand, Amer. Mineral., **43**: 476 (1958).
84. D. B. Hawkins, Mat. Res. Bull., **2**: 951 (1967).
85. J. R. Goldsmith and E. G. Ehlers, J. Geol., **60**: 386 (1952).
86. D. Roy and R. Roy, Amer. Mineral., **40**: 147 (1955).
87. D. B. Hawkins and R. Roy, Geochim. Cosmochim. Acta, **27**: 1047 (1963).
88. R. M. Barrer and D. J. Marshall, J. Chem. Soc., **485** (1964).
89. R. M. Barrer and D. J. Marshall, Amer. Mineral., **50**: 484 (1965).
90. R. M. Barrer and D. J. Marshall, J. Chem. Soc., **2296** (1964).
91. R. M. Barrer and D. E. Mainwaring, J. Chem. Soc., Dalton Trans., 1259, (1972).
92. R. C. Erd, D. W. White, J. J. Fabrey, and D. E. See, Amer. Mineral., **49**: 831 (1964).
93. R. M. Barrer and P. J. Denny, J. Chem. Soc., **971**: (1961).
94. H. P. Eugster and J. Mirnoz, Science, **151**: 683 (1966).
95. R. M. Barrer, P. J. Denny, and E. M. Flanigen, U. S. Pat. 3,306,922 (1967).
96. G. T. Kerr, J. Inorg. Chem., **5**: 1537 (1966); U. S. Pat. 3,247, 195 (1966).
97. G. T. Wadlinger, E. J. Rosinski, and C. J. Plank, U. S. Pat. 3,375,205 (1968).
98. G. T. Kerr, Science, **140**: 1412 (1963); J. Inorg. Chem., **5**: 1539 (1966).

99. R.L. Wadlinger, G.T.Kerr, and E.J. Rosinski, *U. S. Pat. 3,308,069* (1967).
100. N. A. Acara, *U. S. Pat. 3,414,602* (1968).
101. Mobil oil Corp., *Brit. Pat. 1,117,568* (1968).
102. Mobil Oil Corp., *Neth. Pat. 7, 014,807* (1971).
103. R. J. Argauer, and G. R. Landolt, *U. S. Pat. 3,702,886* (1972).
104. M. K. Rubin, *Ger. Offen. 1,806,154* (1969).
105. E. E. Jenkins, *U. S. Pat. 3,578,398* (1971).
106. T. E. Whyte, Jr., E. L. Wu, G. T. Kerr and P. B. Venuto, *J. Catal.*, **20**: 88 (1971).
107. R. Aiello and R. M. Barrer, *J. Chem. Soc. A*, 1470 (1970).
108. G. T. Kerr, *U. S. Pat. 3,459,676* (1969).
109. J. Ciric, *U. S. Pat. 3,692,470* (1972).
110. T. Sudo and M. Matsuoka, *Geochim. Cosmochim. Acta*, **17**: 1 (1959).
111. P. A. Howell, *U. S. Pat. 3,114,603* (1963).
112. P. A. Howell and N. A. Acara, *U. S. Pat. 3,119,660* (1964).
113. D. W. Breck, *J. Chem. Educ.*, **41**: 678 (1964).
114. E. P. Flint, W. F. Clarke, E. S. Newman, L. Shartsis, D. L. Bishop, and L. S. Wells, *J. Res. Nat. Bur. Stand.*, **36**: 63 (1946).
115. S. Nagai and T. Suzuki, *J. Soc. Chem. Ind. Japan*, **38**: 371, B (1935); **38**: 732, B (1935); **39**: 7, B (1936); **39**: 45, B (1936); **39**: 96, B (1936); **39**: 252, B (1936); **41**: 105, B (1938); **41**: 167, B (1938); **43**: 155, B (1940); **43**: 362, B (1940).
116. S. Nagai, *Yogyo Kyokai Shi*, **44**: 531 (1936); **45**: 605 (1937); **46**: 77 (1938).
117. G. Brown, X-Ray Identification and Crystal Structures of Clay Minerals, Mineralogical Society, London, 1961, p. 132.
118. P. A. Howell, *U. S. Pat. 3,390,958* (1968).
119. R. M. Barrer, D. E. Mainwaring, *J. Chem. Soc., Dalton Trans*, 2534, (1972).
120. R. M. Barrer and D. E. Mainwaring, *J. Chem. Soc., Dalton Trans.*, 1254, (1972).
121. G. Denk and W. Menzel, *Z. Anorg. Allg. Chem.*, **382**: 209 (1971).
122. J. L. Bass and Hanju Lee, *J. Mater. Sci.*, **6**: 1039 (1971).
123. H. Sticher and R. Bach, *Helv. Chimica Acta*, **52**: 543 (1969).
124. H. Takahashi, Y. Nishimura, *Rep. Inst. Ind. Ser.*, Univ. Tokyo **20**: 85-117 (1970).
125. J. R. Goldsmith, *Mineral. Mag.*, **29**: 952 (1952).
126. J. Selbin and R. B. Mason, *J. Inorg. Nucl. Chem.*, **20**: 222 (1961).
127. R. J. Argauer, *U. S. Pat. 3,431,219* (1969).
128. M. A. Piontkovskaya, G. S. Shameko, and I. E. Niemark, *Izv. Akad. Nauk. SSR, Neorg. Mater.*, **6**: 794 (1970); M. A. Piontkovskaya, G. S. Shameko, and I. E. Niemark, *Izv. Akad. Nauk. SSR, Neorg. Mater.*, **6**: 1151 (1970).
129. D. A. Young, *U. S. Pat. 3,329,480* (1967); *3,329,481* (1967).
130. P. F. Yermolenko and L. V. Pansevich-Kolyada, Second All Union Conference on Zeolites, Leningrad, 1964, p. 171-8 (Published 1965).
131. R. M. Barrer and D. J. Marshall, *J. Chem. Soc.*, **6616** (1965).
132. K. Wacks, *Ger. Pat. 1,204,642* (1965); *Ger. Pat. 1,174,749* (1964).
133. E. M. Flanigen and R. W. Grose, Molecular Sieve Zeolites, Vol 101, American Chemical Society, Washington, D. C., 1971, p. 76.
134. J. D. Birle, C. R. Knowles, and J. V. Smith, (*in press*).
135. G. H. Kuhl, Molecular Sieve Zeolites, Society of Chemical Industry, London, 1967, p. 85.
136. G. H. Kuhl, Molecular Sieve Zeolites, *Advan. Chem. Ser. 101*, American Chemical Society, Washington, D.C., 1971, p.63.
137. G. H. Kuhl, *J. Inorg. Nucl. Chem.*, **33**: 3261 (1971).
138. G. H. Kuhl, *Inorg. Chem.*, **10**: 2488 (1971); G. H. Kuhl, *U. S. Pat. 3,355,246* (1967).

139. R. M. Barrer, L. Hinds, and E. A. White, *J. Chem. Soc.*, 1466 (1953).
140. R. M. Barrer and C. Marcilly, *J. Chem. Soc. A*, 2735 (1970).
141. S. A. Greenberg, *J. Phys. Chem.*, **61**: 960 (1957).
142. G. T. Kerr, *J. Phys. Chem.*, **70**: 1047 (1966).
143. J. Ciric, *J. Colloid Interface Sci.*, **28**: 315 (1968).
144. G. T. Kerr, *J. Phys. Chem.*, **72**: 1385 (1968).
145. B. Fahlke, W. Wieker, and E. Thilo, *Z. Anorg. Allg. Chem.*, **347**: 82 (1966).
146. S. P. Zhdanov, Molecular Sieve Zeolites, *Advan. Chem. Ser.*, *101*, American Chemical Society, Washington, D.C., 1971, p.20.
147. R. Aiello, R. M. Barrer, and I. S. Kerr, Molecular Sieve Zeolites, *Advan. Chem. Ser.*, *101*, American Chemical Society, Washington, D.C., 1971, p. 44.
148. G. G. Alexander, W. M. Heston, Jr., and R. I. Iler, *J. Phys. Chem.*, **58**: 453 (1954).
149. S. A. Greenberg and E. W. Price, *J. Phys. Chem.*, **61**: 1539 (1957).
150. R. K. Iler, The Colloid Chemistry of Silica and Silicates, Cornell Univ. Press, Ithaca, N. Y., 1955, pp. 26-30.
151. B. D. McNicol, G. T. Pott, and K. R. Loos, *J. Phys. Chem.*, **76**: 3388 (1972).
152. N. A. Acara and P. A. Howell, Union Carbide Corporation, *Unpublished Results*.
153. E. M. Flanigen, D. W. Breck, N. R. Mumbach, and A. M. Taylor, *Amer. Mineral.*, **52**: 744 (1967).
154. I. S. Kerr, J. A. Gard, R. M. Barrer, and I. M. Galabova, *Amer. Mineral.*, **55**: 441 (1970).
155. J. Ciric, *Science*, **155**: 689 (1967).
156. J. F. Charnell, *J. Crystal Growth*, **8**: 291 (1971).
157. R. Roy and O. F. Tuttle, Physics and Chemistry of the Earth, Pergamon Press, London, 1956, p. 142.
158. W. Eitel, Silicate Science, Hydrothermal Silicate Systems, Vol. IV, Academic Press, New York, 1966, p. 149.
159. J. M. Bennett, Ph.D. Thesis, Univ. of Aberdeen, 1966.
160. E. M. Flanigen and E. R. Kellberg, *Brit. Pat. 1,178,186* (1970); U. S. Pat. 4,241,036 (1980).
161. Ch. Baerlocher and W. M. Meier, *Helv. Chim. Acta*, **53**: 1285 (1970).
162. W. Sieber, Ph.D. Thesis, Eidgenossichen Technischen Hochschule, Zurich, 1972.
163. F. G. Straub, *Ind. Eng. Chem.*, **28**:113(1936).
164. C. J. Plank, E. J. Rosinski, and M. K. Rubin, *Brit. Pat 1,297,256,* (1972)
165. H. Khatami and E. M. Flanigen, Union Carbide Corporation, *Unpublished Results*.

Chapter Five

PHYSICAL PROPERTIES OF CRYSTALLINE ZEOLITES

A. PHYSICAL PROPERTIES OF ZEOLITE CRYSTALS

Morphology

Zeolite minerals occur in igneous rocks as well-developed, single crystals which may be as large as several millimeters (Fig. 5.1). This type of crystal was studied by early mineralogists. In some instances, the zeolite crystals in igneous rocks may be found as dense polycrystalline aggregates such as the mordenite amygdales in a typical basalt (Chap. 3, Fig. 3.3). These hard aggregates resist weathering and, as the matrix rock disintegrates, the zeolite nodule is left as a pebble.

In recent years zeolite minerals were found to occur in sedimentary deposits. The fine-grained zeolite crystals found in these deposits are not easily identified from their optical properties. Although altered volcanic sediments were known to exist, the presence of zeolite minerals was not discovered until the use of x-ray powder methods for identification.

Synthetic zeolites that are crystallized from the typical aqueous hydrogels or by the conversion of other aluminum silicates such as kaolin are produced as crystalline powders with a typical particle size of a few microns. Under the electron microscope, these crystals may show well-developed faces, or they may appear as irregular, highly twinned aggregates. A stereoscan of the surface of faujasite crystals from Sasbach, Germany is shown in Fig. 5.2. A coating of amorphous material is evident.

Crystals of synthetic zeolite A are shown in Fig. 5.3. The penetration twins are related by rotation of 60°; this is referred to as a fluorspar twin. Figure 5.4 illustrates an unusual morphology of zeolite A; the individual crystallites are cube-shaped, 250 - 500 A in size. The crystallites in this photograph are cubic close-packed. This special preparation of zeolite A is interesting because it physically indicates one aspect of the crystal-

Figure 5.1 A cluster of crystals of stilbite (magnification 1.2×).

lization mechanism (see Chap. 4). This photograph supports one proposed mechanism of zeolite crystal growth that involves direct crystallization from the solid hydrogel phase by the aggregation of the small crystallites which are initially nucleated from the gel.

Spinel-type twins of zeolite X are shown in Fig. 5.5. The twin axis is a 3-fold axis and the twin plane is parallel to an octahedral face. The zeolite X contact twins shown in Fig. 5.5 are related by a 60° rotation about the 3-fold axis. Also present in this field are polycrystalline spheroids of zeolite P.

A closer view of the surface of these spheroids bears marked resemblance to clusters of the zeolite mineral phillipsite as it occurs in igne-

Figure 5.2 Stereoscan of a faujasite crystal (Sasbach). The crystals are coated with an amorphous material (68 ×).

ous rocks. The polycrystalline spheroids of zeolite P comprise a typical morphology for this species; well-defined, single crystals have not been observed (Fig. 5.6). Synthetic zeolite A may recrystallize under certain conditions to the zeolite P (see Chap. 4). An unusual confirmation of this conversion is the spheroid of zeolite P growing from the original cubic crystal of zeolite A in Fig. 6.7.

Electron microscopy is used extensively to study the habit and surface topography of synthetic zeolites. In one study a series of synthetic analcime single crystals were subdivided into polyhedra derived from the icosatetrahedron (24-hedron) and the octahedron. Icosatetrahedra, which have only [211] faces were generally modified by [100] and [110]

382 PHYSICAL PROPERTIES OF CRYSTALLINE ZEOLITES

Figure 5.3 Stereoscan of crystals of zeolite A showing penetration twinning. The spheres are zeolite P (2500X).

truncations to form 30-hedrons and 42-hedrons. Polycrystalline spherulites were also observed. Calcium forms of the synthetic analcime crystals were octahedral (1).

Zeolite intergrowths are common. One unusual type is illustrated in Fig. 5.8 showing the intergrowth of synthetic zeolite L and synthetic zeolite O (2). This intergrowth appears in the shape of hammerhead crystals. The shaft of the hammer is a crystal of zeolite O, while the head is zeolite L. The O-type and T-type (erionite) crystals form as rods, whereas the zeolite L is cone-shaped. Intergrowths of these zeolites have been interpreted on the basis of a common secondary structural unit, the ϵ-cage (see Chap. 2) (3).

Another unusual type of crystal growth is shown in the stereoscans of the synthetic zeolite P-L in Fig. 5.9 (4). This zeolite contains phosphorus in tetrahedral framework positions (Chap. 4). The habit of zeolite P-L is markedly different from the usual zeolite L crystals that are formed from the potassium-sodium-alumina-silica hydrogel system.

PHYSICAL PROPERTIES OF ZEOLITE CRYSTALS 383

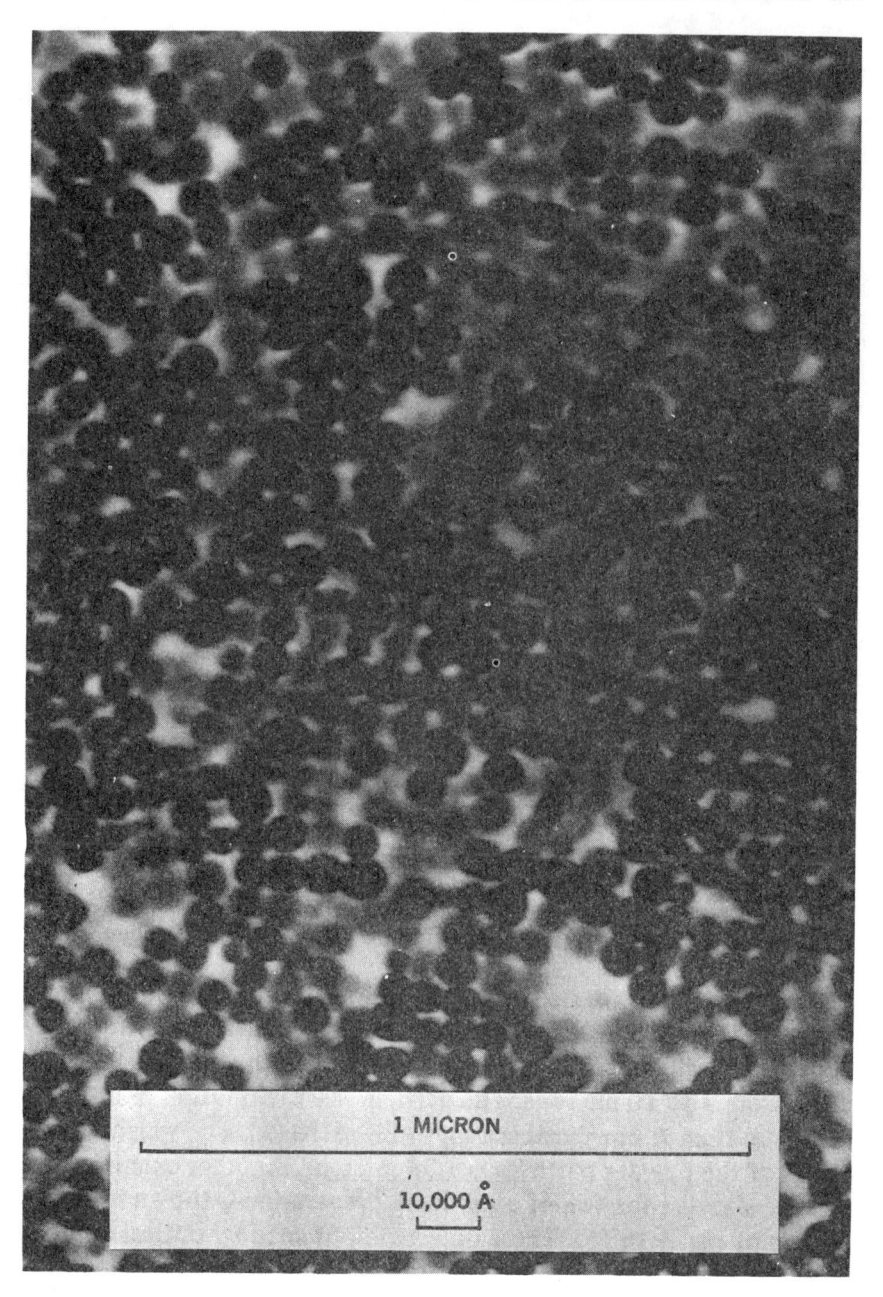

Figure 5.4 Electron micrograph of zeolite A crystallites about 500 A in diameter showing cubic packing (81,000 ×).

384 PHYSICAL PROPERTIES OF CRYSTALLINE ZEOLITES

Figure 5.5 Stereoscan of zeolite X crystal about 50 μm in size showing spinel-type contact twin and spheroids of zeolite P (1380 X).

Particle Size of Synthetic Zeolites

Zeolite crystals are formed under typical synthesis conditions from reactive hydrous gels at low temperatures which contain the various components in high concentrations. Particle sizes of the individual crystals range from 1 to 10 microns. The particle size distribution of a typical zeolite sodium A powder is shown in Fig. 5.10 (5). The weight average radius of the powder particles is 1.39 microns. Another example of the particle size distribution of a zeolite A preparation is shown in the histogram of Fig. 5.11 (6). These data were obtained by optical micrography.

The size distribution of three different batches of zeolite CaA crystals were measured by a photomicrographic method. The mean cube size ranged from 3.05 to 3.91 microns (7, 8).

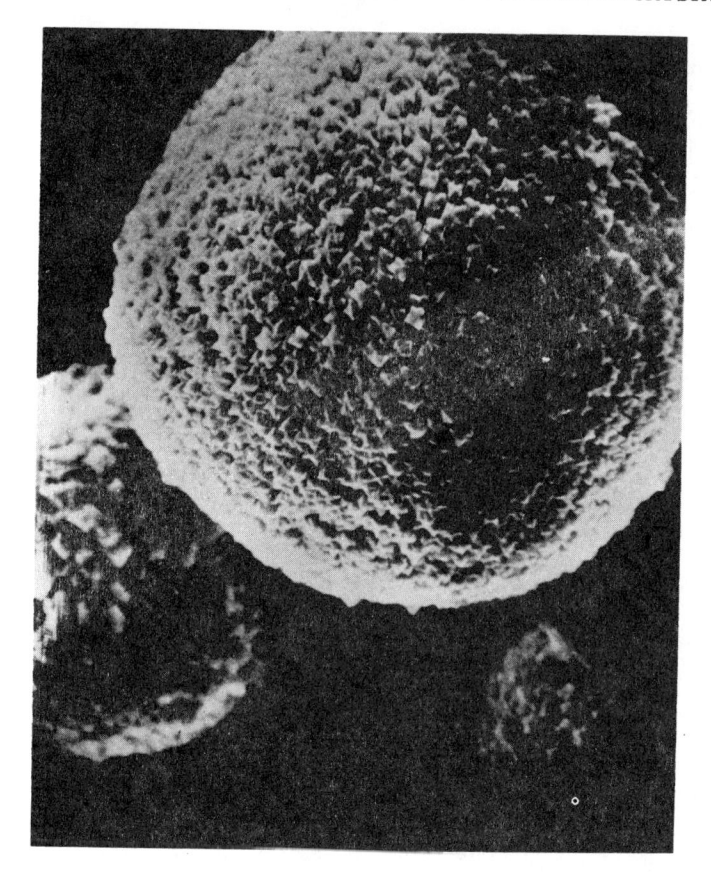

Figure 5.6 Stereoscan of spheroid of zeolite P (2300 X; 1 cm = 4.4 μm).

The very small particle size of most synthetic zeolites is not suitable for use in most applications and, as a result, the crystals must be formed into polycrystalline aggregates in order to be packed into columns and beds for use in adsorption or catalytic processes (see Chap. 9). Although the average particle size of synthetic zeolite crystals is small (1 - 5 microns) the external surface of the crystals is small relative to the internal adsorption void space that becomes available after dehydration of the zeolite. For a spherical particle 1 micron in diameter, the calculated external surface is about 3 square meters per gram of zeolite. Typical zeolites, after dehydration, have equivalent internal surface areas of 800 square meters per gram (see Chap. 8). The largest particle size yet observed for single crystals of synthetic zeolites appears to be about 0.1 mm (100

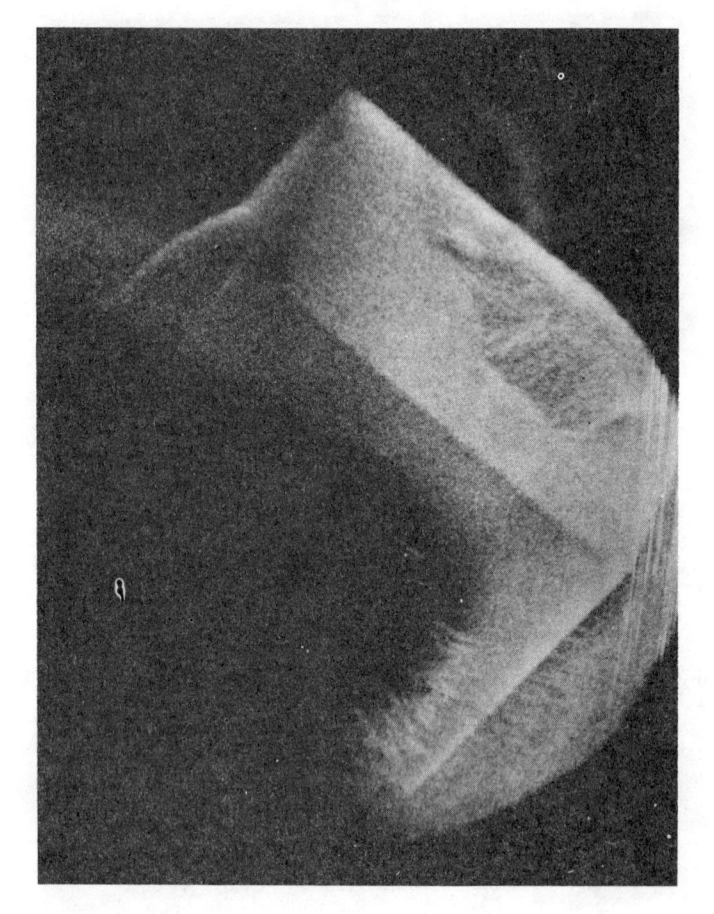

Figure 5.7   Stereoscan of zeolite A crystal partially converted to zeolite P (12,500 X; 1 cm = 0.8 μm).

microns). Earlier preparations of synthetic analcime of this particle size were reported and, more recently, well-developed, single crystals of zeolite X of this size have been prepared. Single crystals of synthetic zeolite A in the range of 25 microns have also been prepared (see Fig. 4.24).

Thermal Expansion

The thermal expansion of fully hydrated zeolite sodium A was measured by the determination of the unit-cell dimension from x-ray patterns over the temperature range of -183°C to 25°C. The thermal expansion coefficient based on the change in unit cell dimension with change in

PHYSICAL PROPERTIES OF ZEOLITE CRYSTALS 387

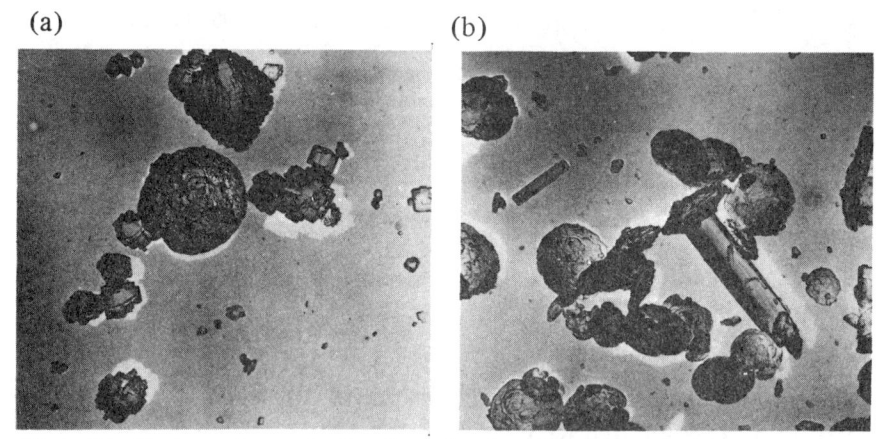

(a)   (b)

Figure 5.8 Electron micrograph of zeolite L showing a "double-cone" habit and a hammer-shaped intergrowth of zeolite L and zeolite T (12,860 X)

(a)   (b)

Figure 5.9 Stereoscan of zeolite P-L showing complex twinning of rod-shaped crystals. Compare this habit with that of zeolite L in Fig. 5.8[(a) 1300X];[(b) 840X].

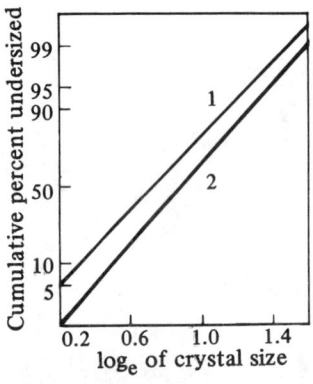

Figure 5.10 Log-normal size distributions for zeolite 4A crystals (5). Determined by optical microscope measurement of 250 particles of each sample. 1—Experimental data. 2—Number distribution.

temperature is about the same as for quartz (9), 6.9×10^{-6}, for hydrated zeolite A, as compared to 5.2×10^{-6} for quartz.

Density

The density of zeolites is low, ranging from about 1.9 to 2.3 g/cc. Cation exchange with heavy ions increases the density; some barium zeolites have densities as high as 2.8. The density depends on the openness of the zeolite structure and the cation. Densities of individual zeolite minerals are given in Tables 3.4 through 3.36. The densities of the synthetic zeolites are given when available in the structural data tables of Chap. 2 and the synthesis data tables in Chap. 4.

Hardness

The hardness of crystals of zeolite minerals ranges from 4 to 5 on the Mohs scale. A hardness of 4 is typical of the mineral fluorite and measures 163 on the Knoop scale. The value of 5 is characteristic of apatite

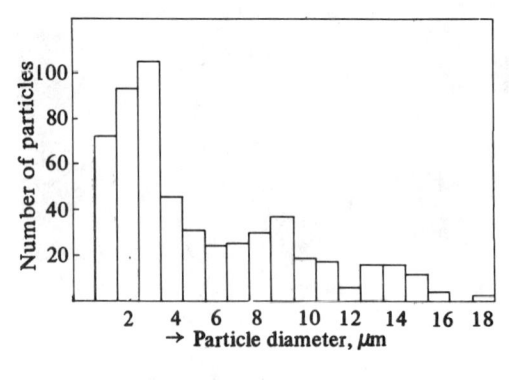

Figure 5.11 Particle size distribution of a zeolite A preparation obtained by measuring and counting 500 particles in optical photomicrographs [After Ciric (6)].

(Knoop 430) and is less than the hardness of feldspar.

B. UNIFORMITY OF COMPOSITION IN SYNTHETIC ZEOLITES

Since most experimental work concerned with the chemical composition, structure, physical, and chemical properties of synthetic zeolites has of necessity been conducted on specimens of the crystalline powders, the question of uniformity of these preparations is important. The chemical analysis of a specimen is the *average* analysis of a relatively large sample containing many small crystals. One gram contains about 1×10^{12} crystals. The composition of the crystallized zeolite generally differs from the composition of the starting mixture. The formation of zeolite crystals from highly concentrated mixtures, supersaturated with respect to the various components, might be expected to produce a range of chemical compositions in a given specimen, particularly if the zeolite itself can exhibit a range in compositions, such as that observed for zeolite Y, G, and others. The chemical composition of a powder specimen might vary from crystal to crystal and possibly within individual crystals. A chemical analysis of the bulk specimen represents an average chemical composition of all of the crystals. A particular physical property, as it is related to composition, could not show any discontinuity since the property would be the average of the bulk specimen.

To determine a particular property of a synthetic zeolite one would preferably work with single crystals which would be expected to have a uniform chemical composition. This has not been possible in most cases because large enough single crystals have not been prepared. In some ion-exchange studies, however, batches of larger crystals have been used (see Chap. 7).

In most instances the properties of zeolite minerals have been determined on a variety of mineral specimens which do not have a well documented chemical composition. Measurements of the properties of sedimentary zeolite minerals are the most in doubt because sedimentary zeolite crystals are very small, and they may contain a relatively large proportion of mineral impurities which are not readily separated. Larger crystals of zeolite minerals of the igneous type may be hand-picked.

Many properties observed in synthetic zeolites such as stability, adsorption phenomena with specific probe molecules, and electrical conductivity, exhibit a discontinuity which is related to specific chemical compositions. (See Fig. 2.58.) This is illustrated schematically in Fig. 5.12. If a powder specimen contains crystals of varying composition, the property measurement will give an average value as illustrated by curve 1. A discontinuity effect, illustrated by curve 2, is indicative of uniform composition. Other data supporting the uniformity of powder specimens are in the form of x-ray powder diffraction patterns. In the synthetic zeolites X and Y, a range of chemical composition in the sample would be evidenced by line broadening in the x-ray powder diffraction pattern; however, in mixtures they are clearly resolved (Fig. 5.13).

Uniformity in the sodium content of zeolite A crystals ranging in size from 30 to 47 microns was demonstrated by measuring the ^{24}Na activity after neutron radiation. The sodium content was shown to be linear with particle volume with a sodium concentration of 0.253 g Na/cc(10).

C. OPTICAL PROPERTIES

Color

Zeolites in a pure state are colorless. Crystals of some mineral specimens are so transparent that it is difficult to see them in a matrix rock. In other instances, small amounts of impurities account for the color of minerals, such as iron in pink chabazite as it is found in Nova Scotia.

If the alkali or alkaline earth cations generally present in synthetic

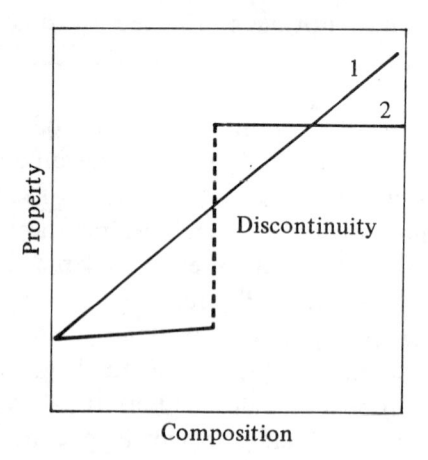

Figure 5.12 1. A composition dependent property does not show a discontinuity when measured on a powder specimen of crystals which encompass a range of chemical compositions. 2. A specimen of uniform composition exhibits the discontinuity in the property-composition relation.

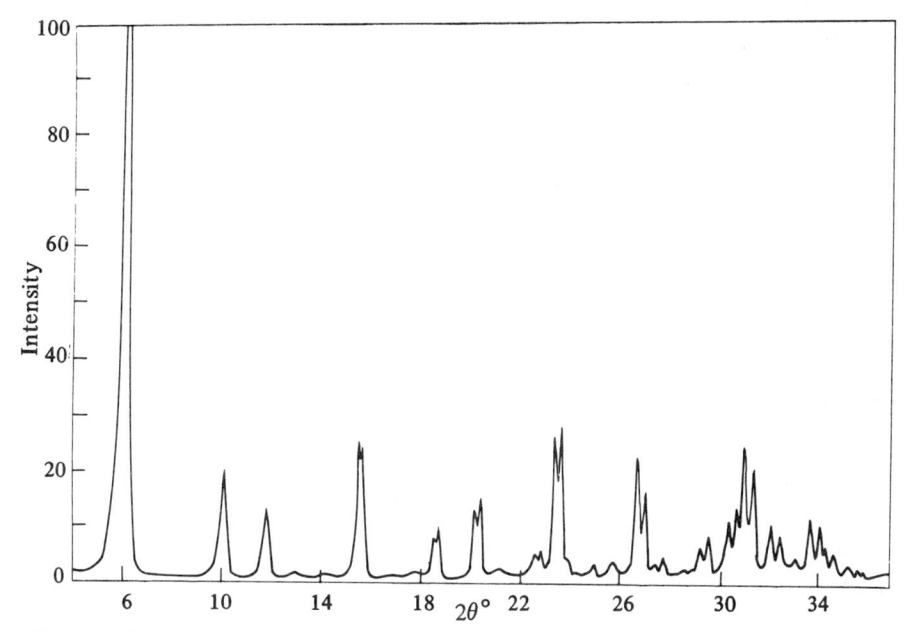

Figure 5.13 An x-ray diffraction pattern of a 1:1 mixture of zeolite X (Si/Al = 1.1) and zeolite Y (Si/Al = 2.4). Peak resolution is clearly evident at $2\theta = 16°$. $\lambda = $ Cu K$_\alpha$

zeolites are exchanged by transition metal ions, the zeolite may have a color (11 - 15). This color will vary with the degree of hydration in many instances as the color of the individual ion varies depending on whether it is hydrated or in an anhydrous state.

When the silver exchanged form of zeolite A, initially white, is dehydrated, the color changes to a deep yellow-red color and ultimately to a bright canary yellow. These color changes occur at partial pressures of water vapor over the zeolite of 3×10^{-2} to 5×10^{-2} torr.

Nickel-exchanged zeolite A changes in color from a lilac to a light-green upon dehydration; the cobalt cation form changes from pink to a blue color. Color change in zeolites has been used as an indicator for the presence of water vapor. Table 5.1 lists some of the color changes associated with the dehydration of nickel-exchanged zeolite A. Reflectance spectra were interpreted on the basis of the nickel ions being located in sites having a trigonal D_{3h} symmetry with the oxygen atoms of the framework structure. Two complexes with the water molecules were identified from the reflectance spectra; one is a pink $Ni^{2+} \cdot H_2O$ and the second is a green $Ni^{2+} \cdot 6H_2O$ complex. The latter complex is observed in

Table 5.1 Color Changes, $Ni^{2+}A$ ($Ni_{1.7}Na_{10.3}$/unit cell) (12)[a]

t(°C)	Moles H_2O/Mole Ni^{2+}	Color
20	15.4	light green
37	2.8	pink
71	1.2	yellowish-pink
149	0.3	yellow
330	0	yellow

[a]Dehydration *in vacuo;* color changes were reversible

the fully hydrated zeolite (11).

Luminescence in Zeolites

Luminescence effects in zeolite minerals were first observed more than fifty years ago. Eight zeolites were found to fluoresce in blue, blue-green, yellow-green, white-yellow, and blue-white when illuminated by 3650 A uv radiation (16). Activator elements such as manganese, lead, silver, and copper can be introduced into the zeolite by cation exchange. Although zeolites exchanged with these activator ions do not exhibit photoluminescence when fully hydrated, dehydration was found to develop cathodoluminescence in the manganese-exchanged forms of chabazite, heulandite, natrolite and stilbite (17). Copper and silver exchanged forms responded to ultraviolet excitation; rehydration reversed this.

Under uv excitation, silver exchanged zeolite X shows six emission bands with three peaks in the blue region (B_1 at 4000, B_2 at 4300, B_3 at 4600 A) and a green 5300 A, yellow 5700 A and red 6700 A peak.

The B_1 green and red bands are due to Ag^+ centers while B_2, B_3 and yellow bands are due to Ag^o centers (18).

D. DIELECTRIC PROPERTIES

The dielectric behavior of zeolite NaA over a range of temperatures and with controlled water contents has been reported. The dependence of the dielectric loss factor ϵ'' on frequency is shown in Fig. 5.14 (19) and dependence on temperature at various water contents is shown in Fig. 5.15 (20). Two types of relaxation processes, denoted as I and II were identified as illustrated in Fig. 5.16. The maximum value of the loss factor ϵ'' increases with increasing water content and is related to relaxation processes of the adsorbed water molecules. The maximum

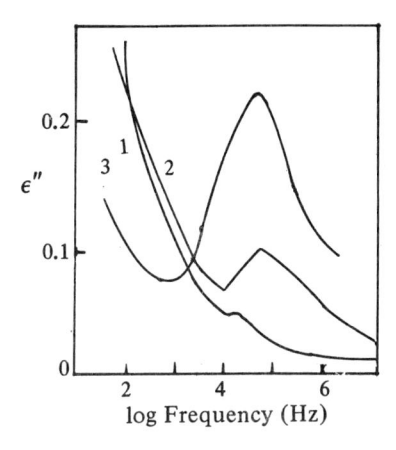

Figure 5.14 Dependence of dielectric loss, ϵ'', of zeolite A on the frequency. The powdered zeolite was contained in a nickel coaxial condenser. Dehydration was in vacuum at 350°C. Curve (1) is for the dehydrated zeolite at 20°C, (2) is the zeolite containing 4.6% H_2O at 22.8°C, and (3) for the zeolite containing 9.3% H_2O at 23.5°C (19).

value of ϵ'' for relaxation process I is greater than the maximum for process II. These two relaxation processes are related to water molecules in two types of adsorption sites. The type I relaxation, characterized by a large value of the activation energy and by being situated in the lower frequency region, is attributed to water molecules which are adsorbed in the large α-cages of zeolite A. The type II relaxation process is assigned to water molecules which are adsorbed in the remainder of the adsorption space.

The dielectric absorption isotherm (21) indicates a break at a water content of 5 - 7 molecules per unit cell which is due to the interaction of water molecules with sodium ions located in site I, the 6-membered oxygen ring (22). The dielectric isotherm shown in Fig. 5.17 for a temperature of 153°K indicates an unusual dependence upon water con-

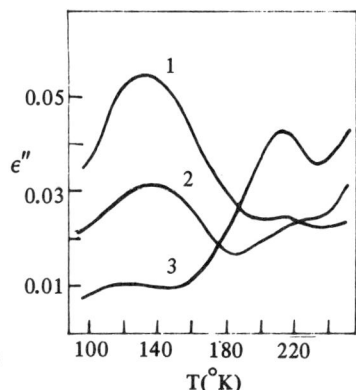

Figure 5.15 Dependence of dielectric loss, ϵ'', of zeolite A on temperature, T°K at 320 kHz. Curve (1) zeolite water content is 1.2% of limiting amount; (2) 4.6%; (3) 9.3 % (20).

Figure 5.16 Dependence of dielectric loss of zeolite A powder on frequency as a function of water content at 20°C. Curve (1) dehydrated zeolite; (2) water content 4.6% of limiting amount adsorbed; (3) 9.3%. The three maxima are evident (20).

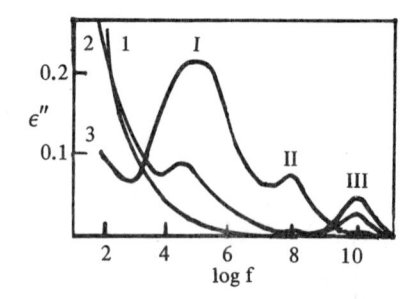

tent. A minimum occurs at a point corresponding to about 12% water. A third relaxation process was postulated for the low temperature region and is associated with the most active adsorption sites (23). The adsorption of water molecules inhibits this third type of relaxation process. The region of temperature independence of the dielectric permeability, ϵ', was observed for fully dehydrated zeolite NaA as shown in Fig. 5.18 (24).

The dielectric properties of zeolite A were studied as a function of time after the adsorption of a small amount of water. At low water loadings, a process of redistribution of the adsorbed water molecules occurs over 3 to 5 days. No change was observed in the behavior of relaxation process III. The change in relaxation processes I and II are attributed to the redistribution of water molecules. After the adsorption of 5% water, changes occurred over a period of as long as 11 days. The redistribution process may be related to the gradual penetration of water molecules into the β-cages which is associated with relaxation process II.

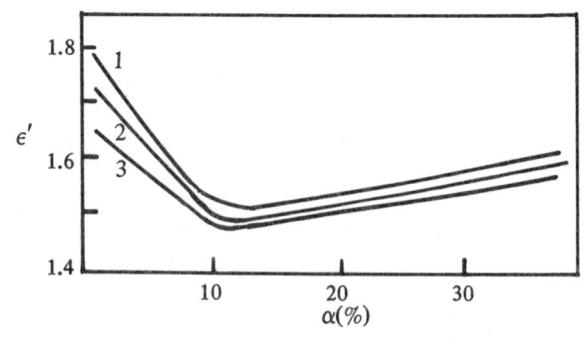

Figure 5.17 Dielectric isotherms for water on zeolite A, 153°K. Curve (1) 3.2 × 10⁵ Hz; (2) 1 × 10⁶ Hz; (3) 3.2 × 10⁶ Hz (23).

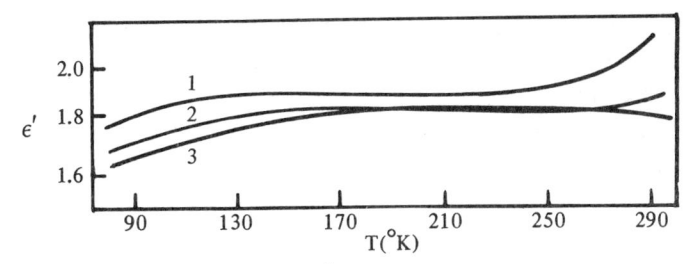

Figure 5.18 Dependence of dielectric permittivity, ϵ', of dehydrated zeolite A on temperature, T°K. Curve (1) 100 Hz; (2) 1000 Hz; (3) 3200 Hz (24).

Additional studies including measurement of the dielectric properties with adsorbed methanol indicate that relaxation process III is caused by the sodium ions in the 6-rings, whereas relaxation process II is associated with the interaction of ions with the water molecules. Dielectric behavior of a magnesium exchanged A showed that the relaxation time of the water molecules in magnesium A is greater than for the sodium A (25).

The dielectric behavior of NaX zeolite revealed two types of relaxation processes and a third type was hypothesized. The dependence of the loss factor on frequency is shown in Fig. 5.19 (26). During hydration, the relaxation maxima are doubled. This may be caused by two overlapping processes, one associated with the sodium ions in the dehydrated crystal and the second with the relaxation of adsorbed water molecules. At low temperatures, another process was observed which was attributed to sodium ions situated in the large apertures.

In contrast, measurements of the dielectric properties of synthetic

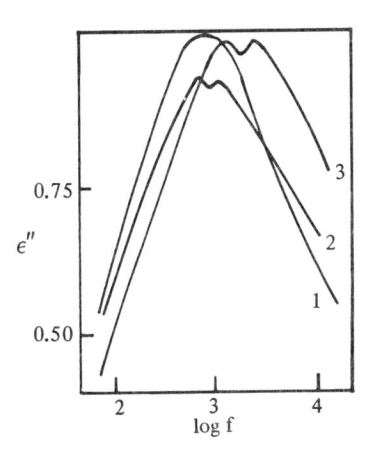

Figure 5.19 Dependence of the loss factor, ϵ'', on frequency, f in Hz, of zeolite X at $-40°C$ curve (1) dehydrated zeolites; (2) zeolite with 5% water; (3) zeolite with 15% water (26).

hydroxysodalite revealed that the relaxation processes characteristic of those found in sodium A are absent. At low temperatures where the relaxation process due to adsorbed water was observed in hydrated zeolite A, the hydroxysodalite showed no appreciable signs of relaxation over the entire range of water contents studied (27).

The apparent permittivity, ϵ', and dielectric loss, ϵ'', of a sodium exchanged chabazite, anhydrous and partially hydrated, were measured over a range of temperatures (28). The permittivity and the loss are high and are frequency-dependent. As in zeolite A, a maximum in the loss was observed as a function of frequency which varies with temperature and the level of hydration. For the loss process, the Arrhenius activation energy was determined to be 7.5 ± 1 kcal per Avogadro number of loss processes. This is comparable to the activation energy for the self-diffusion of sodium and other monovalent ions in fully hydrated chabazite. The loss maximum is not ascribed to cation processes within the structure but may be due to cation jumps on the surface of the chabazite crystals. The interaction of water is more readily explained as being due to the sodium cations. In another study of a hydrated calcium-rich chabazite, two types of loss peaks associated with energy barriers of 16.1 and 8.0 kcal were observed.

The dielectric properties of analcime indicate that the relaxation process in various cation exchanged forms have energy barriers for relaxation in both anhydrous and hydrated forms which are similar in type and in magnitude to the activation energy for conduction and ion self-diffusion (29). A spectrum of relaxation times was observed, the mean varying from 1.8×10^{-5} seconds at 30°C to 2.5×10^{-6} seconds.

Permittivity and dielectric loss factors determined on a calcium ex-

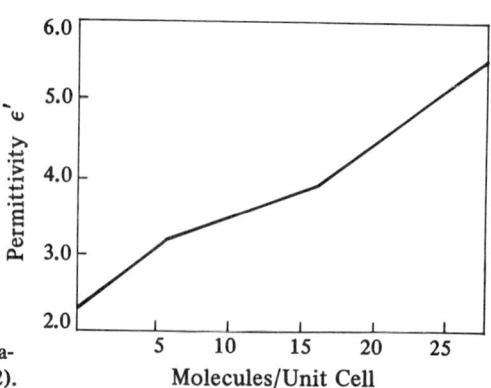

Figure 5.20 Dielectric isotherm for water on calcium exchanged zeolite A (22).

changed zeolite A which varied in water content from anhydrous to full saturation of 30 molecules per unit cell show a well-defined absorption below a frequency of 148 kHz (22). The dielectric isotherm shows a change in slope at 6 H_2O/unit cell and 17 H_2O/unit cell (Fig. 5.20). This corresponds to a discontinuity observed in the increase in electrical conductivity with increasing water content. The value of 6 is equivalent to occupancy of the large 8-rings by one water molecule on each side (i.e., 6/unit cell) which is hydrogen bonded to the ring of oxygen atoms. The second point is due to water in a liquid-like state. A further low-frequency absorption was attributed to adsorbed water.

Values for the conductivity were calculated by the equation:
$$\epsilon'' \times f = 1.8 \times 10^{12} \sigma$$
The conductivity, σ, ranged from $1.3 \times 10^{-13} \, \Omega^{-1} \, cm^{-1}$ for completely dehydrated zeolite, to $68 \times 10^{-12} \, \Omega^{-1} \, cm^{-1}$ for the fully hydrated zeolite. Similarly, calculations of the conductivity from the dielectric loss measured on the sodium zeolite A varied from $1.75 \times 10^{-12} \, \Omega^{-1} \, cm^{-1}$ to $1.19 \times 10^{-10} \, \Omega^{-1} \, cm^{-1}$. These are several orders of magnitude lower than the measured electrical conductivity on zeolite A which varied from 1×10^{-8} for dehydrated zeolite A, to $1 \times 10^{-4} \, \Omega^{-1} \, cm^{-1}$ for the fully hydrated zeolite (see Section E). The reason for this disparity is obscure.

The cation movement, or hopping process, to which the dielectric absorption is ascribed, is interpreted as a jump of a cation in site I (near the 6-rings in the large α-cage) to sites near the 8-ring aperture. A similar behavior was observed with NH_3, SO_2 and CO_2 in the calcium A zeolite (22). At 22°C, the relaxation times for various cation forms of zeolite A, fully hydrated, are given in Table 5.2.

E. ELECTRICAL CONDUCTIVITY

Zeolites contain mobile cations which are located in sites in cavities, on the channel walls, and free within the channels coordinated with water molecules. The electrical conductivity exhibited by zeolites is ionic and

Table 5.2 Relaxation Times for Cation Forms of Zeolite A (22)

Zeolite	Degree of Exchange (%)	H_2O/unit cell	Time (sec)
NaA	—	27	4.9×10^{-7}
K exchanged A	75	27	7.9×10^{-7}
Ag exchanged A	100	24	2.5×10^{-7}
Ca exchanged A	75	30	1.0×10^{-5}

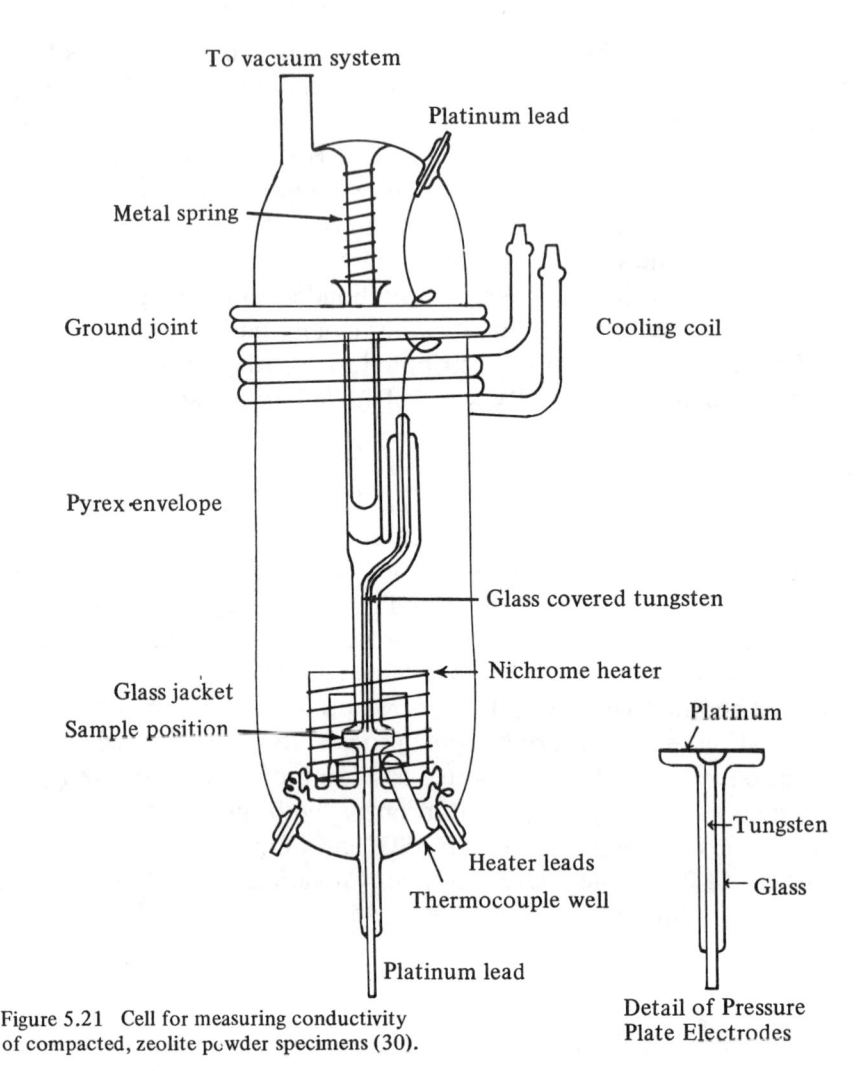

398 PHYSICAL PROPERTIES OF CRYSTALLINE ZEOLITES

Figure 5.21 Cell for measuring conductivity of compacted, zeolite powder specimens (30).

Detail of Pressure Plate Electrodes

arises from the migration of cations through the zeolite framework. The electrical conductivity of various forms of the synthetic zeolites A, X, and Y was measured using polycrystalline compacts contained in a conductivity cell as illustrated in Fig. 5.21 (30). Conductivity data were obtained on different cationic forms of these zeolites as a function of temperature. Typical conductivity data for the zeolites X and Y are given in Fig. 5.22 as Arrhenius plots, log conductivity, σ, versus $1/T$. The activation energy for conduction, ΔH, is shown in Fig. 5.23 as a function

ELECTRICAL CONDUCTIVITY   399

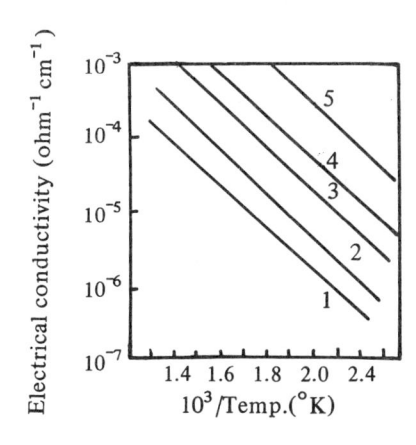

Figure 5.22 Arrhenius-type plots of the electrical conductivity of zeolites X and Y. (30). The SiO_2/Al_2O_3 ratio = (1) 5.20; (2) 4.50; (3) 3.80; (4) 3.05; and (5) 2.40.

of the number of sodium ions per unit cell. The activation energy, ΔH, increases rapidly starting at a cation density of about 78 monovalent ions per unit cell or a Si/Al ratio of 1.5. This indicates two types of cation sites. The ΔH increases as the cation density decreases in zeolite Y, which is characteristic of a more strongly bound cation.

Mass transfer during the conduction process was demonstrated by showing actual cation transfer in silver exchanged zeolite X. The formation of zinc ion at a zinc electrode and silver metal at the opposite gold electrode was observed. The demonstrated mass transfer in the conduction process eliminates cation vacancies as the charge carriers. The measured conductivity is the bulk conductivity because there was no dependence on initial crystallite size.

The conductivity was found to depend strongly on the cation size and the size of the channels within the zeolite structure. The depend-

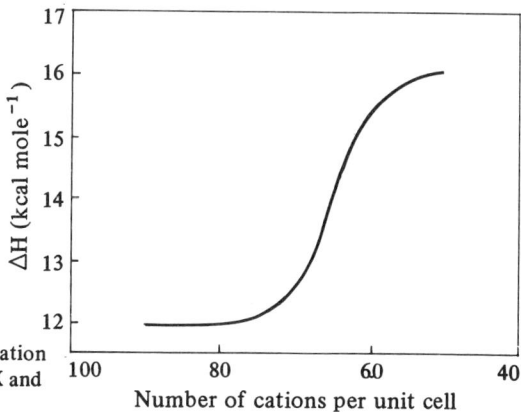

Figure 5.23 Variation of the activation energy with cation density in the X and Y zeolites (30).

Figure 5.25 Activation energies for conduction of the univalent and divalent cation exchanged forms of zeolite Y (30).

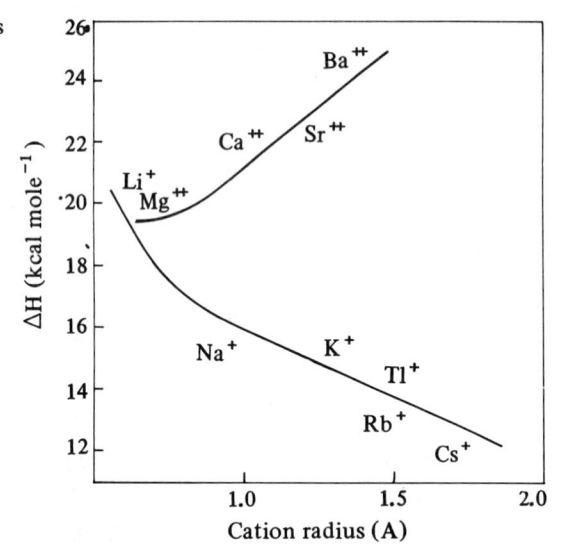

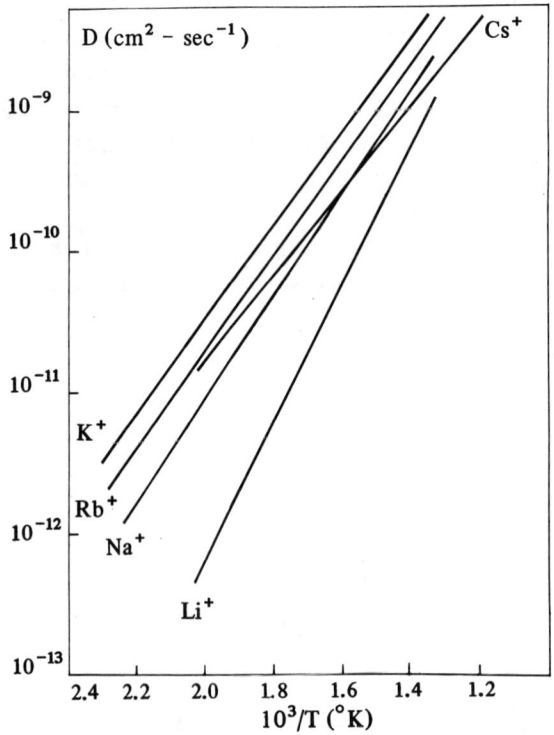

Figure 5.26 Diffusion constant vs. temperature for univalent cation forms of zeolite Y (30).

ence of ΔH on the cation type is shown in Fig. 5.24 for zeolite A and Fig. 5.25 for zeolite Y. The larger cations in zeolite A are sterically restricted; ΔH for cations larger than sodium increases due to the greater energy barrier offered to cation migration by the 8-rings. In contrast in zeolite Y, ΔH decreases with an increase in cation radius of univalent ions and no steric hindrance appears to occur. The activation energy decreases due to a diminishing Coulombic attraction between the cation and the cation site.

Each divalent cation must be associated with two negatively charged sites in the framework structure. The calcium ion should have twice the activation energy for conduction as the sodium ion if only electrostatic interactions are considered and provided that the calcium ion site is completely equivalent to two sodium ion sites. The calcium ion is situated between or near one of the widely separated negatively charged sites in the framework and the bond strength must accordingly be reduced. The sharp increase in the activation energy, ΔH, with increasing size of divalent ions reflects the greater ease of bonding to two separate sites by the larger, more polarizable cations (Fig. 5.25).

Cation self-diffusion coefficients determined from the conductivity data are shown in Fig. 5.26. The high values of D confirm the relatively open zeolite channel structure.

Effect of Water on Electrical Conductivity

The addition of water molecules to the dehydrated zeolite structure

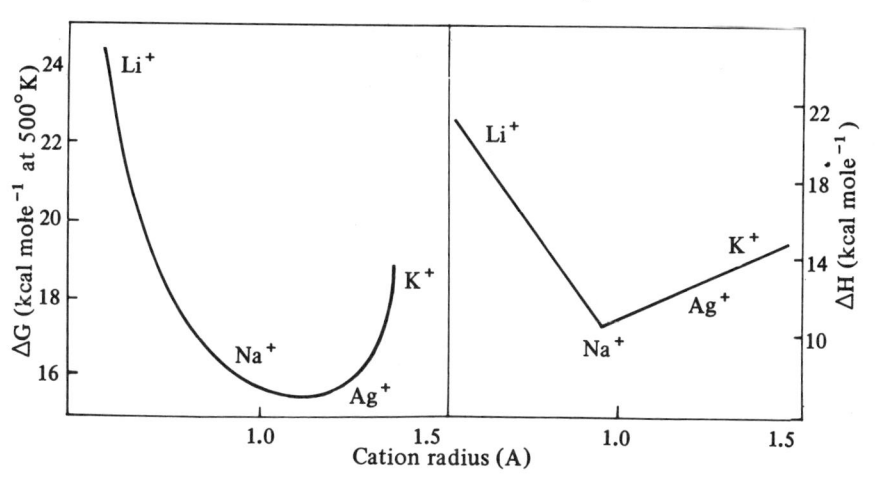

Figure 5.24 Variation in free energy, ΔG, and activation energy, ΔH, for conduction of univalent cation forms of zeolite A (30).

Figure 5.27 Effect of water adsorption on the electrical conductivity of zeolite X at various temperatures (31).

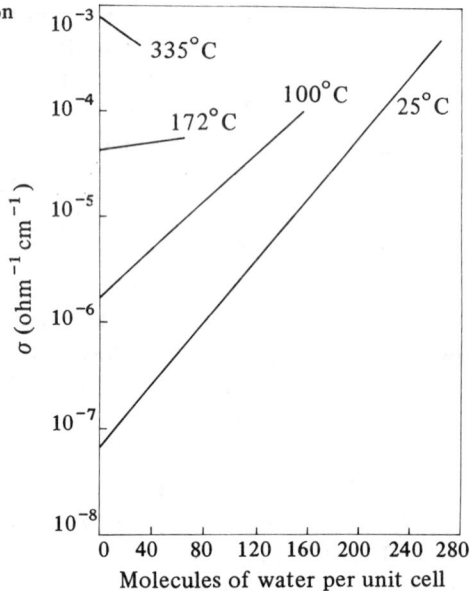

produces a pronounced change in the electrical conductivity as shown in Fig. 5.27 (31). The conductivity of zeolite X increases with increasing water content by a factor of 1×10^4 at 25°C. For zeolite A, (see Fig. 5.28) the conductivity increases nonlinearly. This reflects the preferential hydration of one type of cation in the structure. (Although these measurements were made on compacted powder specimens, the results confirm the uniformity of the structure and composition of the individual zeolite crystallites.)

In zeolite A, conductivity increases with hydration until about five water molecules are present within the cell. This is equivalent to the hydration of the four mobile sodium ions which are located in sites II and III near the oxygen 8-rings that form the apertures into the α-cavities. These sites appear to possess the highest adsorption energy. The sodium water complexes located in these 8-rings effectively block the openings to the α-cages and prevent other moiecules from entering. This has been confirmed by physical adsorption measurements which indicate that traces of water are sufficient to effectively block the adsorption of gases such as oxygen.

After hydration of the four site II cations, water molecules associate

Figure 5.28 The effect of water adsorption on the conductivity of zeolite A (31).

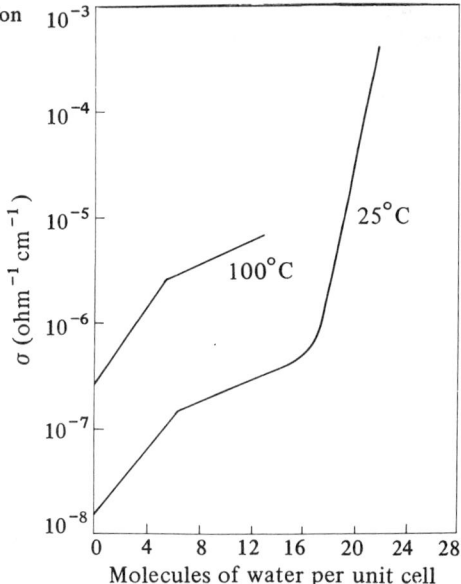

with the site I Na$^+$ ions located in the 6-rings. These provide smaller contributions to the total conductivity. After more than 16 water molecules are present in each unit cell, the water molecules occupy the lowest energy adsorption sites, associating through hydrogen bonding to the anionic zeolite surface. The conductivity then increases as the crystals become saturated, indicating that the channel systems are filled with water molecules. Some of the sodium ions are free to move randomly within the channels. These results are in agreement with the infrared spectra of H$_2$O and D$_2$O and nuclear magnetic resonance experiments (see Sections G and H).

In zeolite X, water molecules diffuse more freely within the structure and can approach both types of cations (31). As shown in Fig.5.29, the self-diffusion constant of Na$^+$ ions in NaX increases regularly with increasing hydration from 1 x 10^{-12} to 1 x 10^{-8} cm^2 sec^{-1}. The activation energy for conduction, ΔH, (Fig. 5.30) shows a regular decrease from 12 kcal/mole to an essentially constant value of 6 kcal/mole when approximately 100 H$_2$O molecules per unit cell (12 H$_2$O's per supercage) are present. This compares with the behavior of salt solutions and indicates that the hydrated zeolite crystal behaves as a solid electrolyte solution. When fully hydrated, zeolite NaX is approximately 18 molal with respect to sodium.

404 PHYSICAL PROPERTIES OF CRYSTALLINE ZEOLITES

Figure 5.29 The effect of water adsorption on the self-diffusion constant of sodium ion, Na$^+$, in zeolite X at 25°C (31).

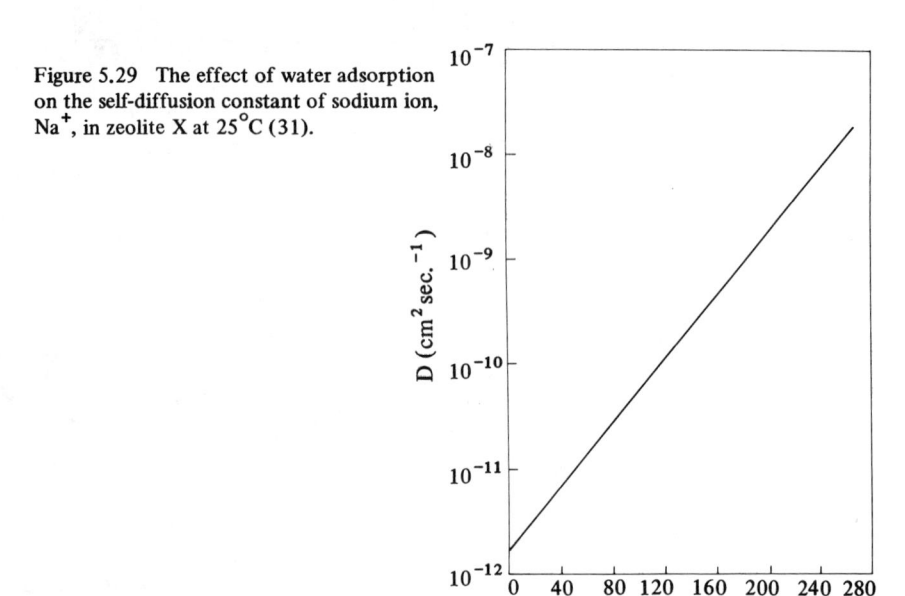

Figure 5.30 The effect of NH$_3$ and water adsorption on the activation energy for conduction, ΔH, in zeolite X (31).

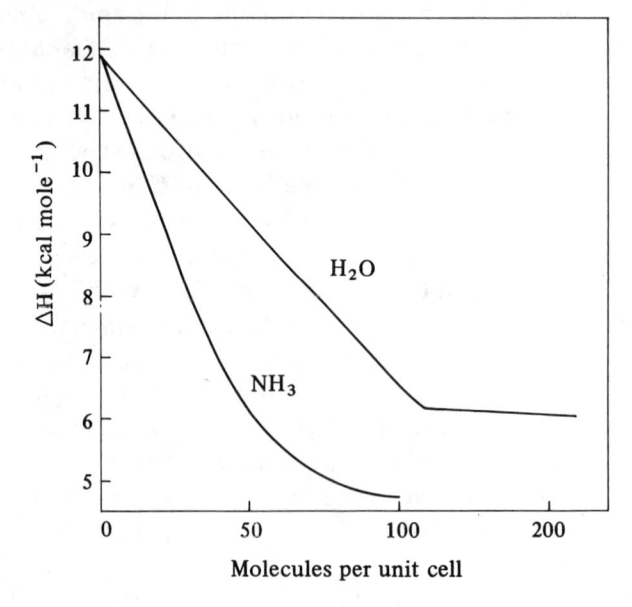

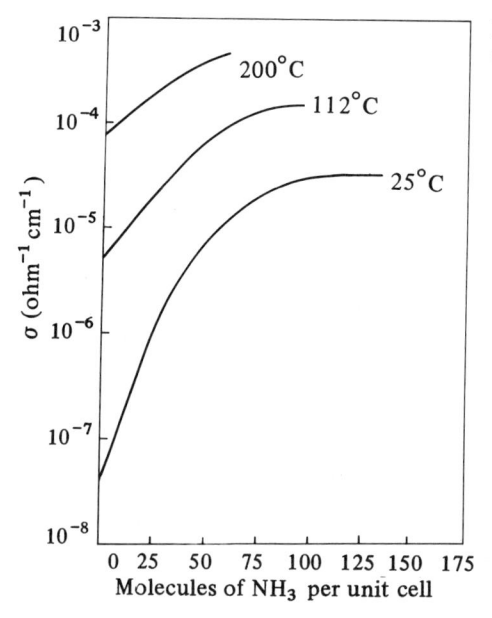

Figure 5.31 The effect of ammonia adsorption on the conductivity of zeolite X (31).

Effect of Ammonia on Electrical Conductivity

Ammonia adsorption in zeolite X shows a different effect upon electrical conductivity (31). Near saturation (Fig. 5.31) the conductivity becomes independent of ammonia content. The ammonia molecules occupy the most energetic adsorption sites associated with cations located in the large cavities. After approximately 45 NH_3 molecules (6 per supercage) are present, the conductivity approaches a constant value. This is the result of two opposing effects; a small increase in the conductivity is counterbalanced by a hindering effect, due to the presence of NH_3 molecules in the migration paths. The steric effect is more pronounced at higher temperatures for water adsorption. The activation energy for conduction for zeolite X with adsorbed ammonia is less than that with water (Fig. 5.30). This suggests that the NH_3 molecules populate the higher energy sites (sites II and III) more selectively than the water molecules; at low NH_3 content, the isosteric heat of adsorption of NH_3 is much higher than for water.

Effect of Nonpolar Molecules on Electrical Conductivity

The change in conductivity with the adsorption of nonpolar molecules in zeolite NaX has also been determined. A higher electrical con-

406 PHYSICAL PROPERTIES OF CRYSTALLINE ZEOLITES

ductivity obtained for the adsorption of nitrogen indicates that the nitrogen molecules are more strongly bound to the structure than oxygen molecules. The adsorption interaction energies of N_2 and O_2 with the zeolite are attributed to cation quadrupole interaction (31).

Mechanisms of Conduction

Two conduction mechanisms were identified from a study of the electrical conductivity and capacitance of a series of dehydrated,[*] univalent (Li^+, Na^+, K^+, Ag^+, Rb^+, Cs^+) cation forms of zeolites X and Y. Decationized forms of Y (prepared from the NH_4^+ exchanged forms) were also measured (32).

One mechanism, called the high temperature (>350°C) mechanism (HT) is ascribed to cation mobility. At 700°C, conductivity is a function of the radius of the exchangeable cations. The specific conductivity, σ, is considerably higher for zeolite X than zeolite Y. In zeolite X, σ decreases with increasing cation radius (Fig. 5.32). Specific conductivity in the decationized samples is a function of residual sodium content (see Fig. 5.33).

Since the high temperature conductivity can be expressed by the Arrhenius equation, an activation energy ΔE for zeolite X was also observed to be lower than ΔE for zeolite Y. For the decationized zeolites,

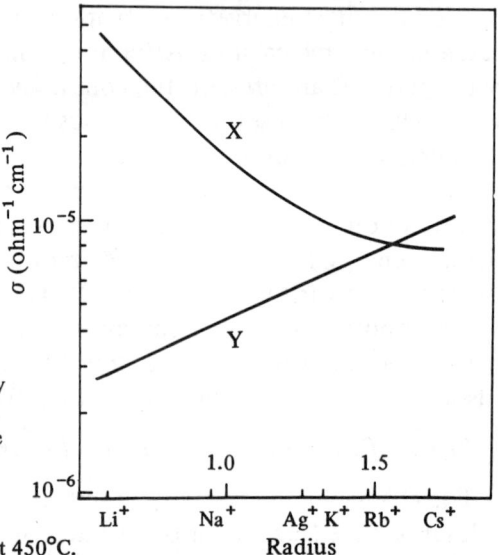

Figure 5.32 The specific conductivity ($ohm^{-1} cm^{-1}$) of zeolite X and Y as a function of the radius of the exchange cation (32). Zeolite X (Si/Al = 1.25); zeolite Y (Si/Al = 2.55).

[*]Ultimate activation in high vacuum at 450°C.

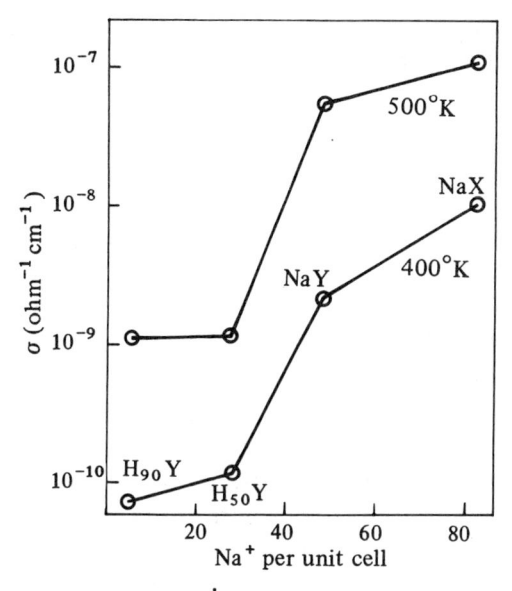

Figure 5.33 The specific conductivity (ohm^{-1}cm^{-1}) of zeolite X and Y as a function of the Na$^+$ content of hydrogen forms (32).

H$_{50}$Y and H$_{90}$Y,[†] ΔE was dependent on residual sodium with values of 25.06 and 20.83 kcal/mole at 19 kHz.

The low temperature mechanism (LT)(<350°C) includes a frequency-dependent process ascribed to motion of cations in the supercages over a range of energy barriers. Assuming that the conduction process involves only cations located in the supercages, ions located on site II would have to migrate over a sequence of three 4-rings to reach another surface site II.

The high temperature (HT) mechanism has as a limiting step the movement of cations from a 6-ring to an interstitial site—the intervening 4-rings. The activation energy ΔE is the energy required to overcome the energy barrier separating site II from site III. All of the X zeolites must contain ions in site III, while the Y zeolites do not.

The difference in the activation energy for conduction, about 4 kcal per mole, has been attributed to the presence of cations in the weakly bound sites or to differences in the electrostatic interactions between cations and the oxygen framework (30). The former interpretation is still preferred—particularly in view of the finding that univalent ions do populate site III.

[†] H$_{50}$Y = 50% decationated Y. Si/Al = 2.5.
H$_{90}$Y = 90% decationated Y.

The use of zeolites as ionic conductors in solid state batteries has been described. High temperature performance at 500°C was reported (33).

Electrical Conductivity of the Interface Between Zeolite Crystals and Salt Solutions

The interfacial conductivity of crystals of the zeolite chabazite in flowing salt solution was studied (34).*

The surface conductivity increases by a factor of 10 in 10^{-3} or 10^{-2} N salt solutions as compared to the value in water. Equilibrium values were obtained in less than 2 minutes showing that the surface conductivity is truly a property of the crystal surface rather than the interior of the crystals. It is suggested that the surface conductivity is due to a skin of "silica-gel" which results when aluminum atoms are removed from tetrahedral positions in the surface of the zeolite as illustrated in Scheme a and b. Apparently this dealumination process occurs readily on the crystal surface. (See also Chap. 6, Section D.)

Surface Reactions on Chabazite

(a)

$$\text{(a)} \quad \begin{array}{c} Na^+ \\ O\ O\ O\ O \\ /\backslash/\backslash/\backslash/\backslash \\ Si\ Al\ Si \\ /\backslash/\backslash/\backslash \end{array} \xrightarrow{H_2O} \begin{array}{c} H\ \ H \\ |\ \ \ | \\ O\ O\ O\ O \\ /\backslash/\backslash/\backslash/\backslash \\ Si\ \ \ \ Si \\ /\backslash\ \ \ /\backslash \end{array} + \begin{array}{c} Al^{3+} \\ Na^+ \\ OH^- \end{array}$$

(b)

$$\text{(b)} \quad \begin{array}{c} H\ H \\ |\ | \\ O\ O\ O\ O \\ /\backslash/\backslash\ \ \backslash/\backslash \\ Si\ \ \ Si \\ /\backslash\ \ /\backslash \end{array} + Al(OH)_4^- Na^+ \longrightarrow \begin{array}{c} H\ \ Na^+\ H \\ \backslash\ \ \ \ / \\ O\ \ \ \ O \\ \backslash\ / \\ Al \\ /\ \backslash \\ O\ O\ \ \ O\ O \\ /\backslash/\ \ \ \backslash/\backslash \\ Si\ \ \ \ Si \\ /\backslash\ \ \ /\backslash \end{array} + 2 H_2O$$

These Si-OH groups produce the surface conductivity by either adsorbing salt from the external solution or by dissociation. The ionic mobilities were assumed to be greater in the leached zeolite crystal surface.

Coherent surface layers of silica gel could be formed by the recon-

---

*The chabazite was ground to 60 - 120 mesh, chemical analysis showed a Si/Al ratio of 2.09. The initial mineral was a calcium-rich chabazite; for some measurements the crystals were converted to the sodium exchanged form. Surface conductivity was measured using a streaming cell containing a plug 1.1 cm in diameter, 3 cm long, weighing about 8 grams.

densation of partially detached chains of polysilicic acid with remaining surface Si-OH groups. The occurrence of this recondensation process was indicated by the observation that the highest conductivity values were obtained from natural crystals which had been ion exchanged with hot sodium chloride solution at pH 6.7. The condensation of the Si-OH groups is favored in this pH range if excess salt is present.

The surface conductivity resulting from the Si-OH groups was drastically reduced when the flowing solution contained aluminate ion. The reaction shown in Scheme b would account for this reduction in surface conductivity by the formation of aluminosilicate links and the elimination of the Si-OH groups. It was first observed in 1936 that the adsorption properties of a zeolite such as chabazite vary with the ion exchange procedures. These variations were in part attributed to the hydrolysis of the external crystal surface to form a gel-like skin which on subsequent dehydration sealed the crystal surface (87).

It was concluded that the surface conductivity of chabazite in sodium chloride solution resulted from the ionization of cations from the fixed surface anions rather than by absorption of salt from the external solution. In solutions of other salts, the salt absorption into the surface gel layer may contribute to the surface conductivity (34).

Variation in the negative charge of chabazite with salt concentration and ion species was attributed to two sources. Ion pairs, M^+X^-, of the salt adsorb onto the hydroxyl groups of the zeolite surface. The surface negative charge is due to the preferential adsorption of the more polarizable anion on the positive end of the hydroxyl dipole. The increase in surface charge with concentration is a result of increased adsorption. Differences in surface charge between salts of the same valence type are due to varying adsorptivities of the anion.

The surface charge may also arise by acidic ionization of the surface Si-OH groups. The charge increases with concentration due to compression of the ionic atmospheres surrounding the surface charges and increasing likelihood of further ionization. For salts with a common anion, the magnitude of the surface charge decreased along the series:

$$Li^+ > Na^+ > Et_4N^+ > K^+ > Mg^{2+} \sim Ba^{2+} > Ca^{2+}$$

In a 10^{-4} M solution of $AlCl_3$, the negative charge of the chabazite crystal surface was reversed. This was attributed to the adsorption of a partially polymerized AlOH species formed by hydrolysis of Al^{3+}.

Figure 5.34 Lowering of the chemical potential, $-\Delta\mu s$, of zeolite X by the adsorption of H$_2$O and NH$_3$. The structural unit is equal to 1/192 part of the unit cell which contains 192 tetrahedra (35). (1) NH$_3$ on NaX; (2) NH$_3$ on NaX; (3) H$_2$O on CaX; (4) H$_2$O on Ca, CsX.

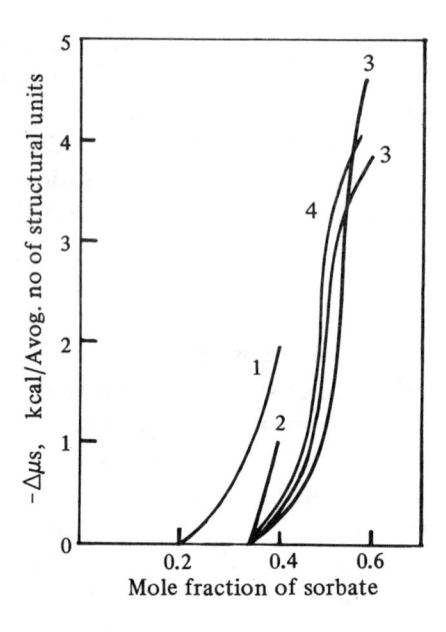

F. THERMOCHEMISTRY

The presence of adsorbed water or ammonia in zeolite NaX has a pronounced stabilization effect because it lowers the chemical potential of the zeolite crystal (35) (Fig. 5.34). The same effect is produced by nonvolatile salts, such as sodium chloride, when occluded in zeolites or in feldspathoids such as sodalite. In hydroxysodalite, the stabilizing effect would be due to the presence of both sodium and hydroxide ions as well as water (36).

The stabilizing effect of water is important in the growth of zeolite crystals. It is known that water or salts, or both, aid in the formation of aluminosilicates. The reported catalytic effects of salts when used as mineralizers may be thermodynamic in nature. Zeolites do not form in the absence of water; that is, it is impossible to synthesize an open zeolite structure equivalent to a dehydrated zeolite under anhydrous conditions. This is explained in part by the stabilization effect of water. Other aspects of mineralizers in the formation of aluminosilicates may be kinetic in nature.

Thermochemical Data

Very little fundamental data and information on the thermochemistry (free energy of formation, heats of formation, etc.) of zeolites are available. Although considerable attention has been given to the thermochemistry of zeolite adsorption and ion exchange, essentially no data exist concerning the basic crystalline zeolites. It should be possible, however, to determine the free energy of formation of zeolites by solubility methods that have been applied to clays (37).

Because of its geological significance, there is some data on the mineral analcime (38). Heat capacity measurements on analcime over the temperature range 51° - 298°K did not show unusual thermal behavior. The entropy of analcime at 298.16°K was determined to be 56.0 cal/deg/mole based on the formula $NaAlSi_2O_6 \cdot H_2O$. For comparison, the entropies of albite, jadeite, and nepheline are 50.2, 31.9, and 29.7 cal/deg/mole. The entropy of analcime is 5.8 to 26.3 e.u. higher which is much larger than the entropy difference for the one mole of zeolitic water (about 9.6 e.u. (39)). This difference is in accord with the more open structure of analcime relative to jadeite. Some thermochemical data for the zeolites analcime, leonhardite, and other aluminosilicates are given in Table 5.3 (40).

Measurements of the thermochemical quantities on several different zeolites should be very informative. An attempt was made by Gruner to arrange the silicates in a reaction energy sequence at low temperatures (41). Silicates which contain hydroxyl or water have low energy indices on this scale, as do those aluminosilicates associated with hydrothermal activity. The energy indices were derived from the electronegativities of the ions concerned.

Zen has calculated the standard free energy of formation and enthalpy of formation of the zeolites laumontite and wairakite from hydrothermal equilibrium data (42)(Table 5.4).

For wairakite the entropy is much greater than the "oxide sum" estimate of 81.2 gibbs/gfw. This is attributed to the mobile state of the water molecules and to less hydrogen bonding between water molecules in the zeolite as compared to liquid water.

A method for computing the bonding energies of silicate minerals was proposed by Keller (43). The energies of formation of various silicates were computed from energies accruing to individual ions and, in

Table 5.3 Thermochemical Data on Certain Aluminosilicates (40)

	Gram Formula Weight	$S°_{298°K}$ (cal/deg/gfw)	Molar Volume (cc)	$\Delta H°_{298°K}$ (kcal/gfw)	$\Delta G°_{298°K}$ (kcal/gfw)	log K_f at 298.15°K
Leonhardite, $Ca_2Al_4Si_8O_{24} \cdot 7 H_2O$	922.867	220.4 ±2.6	404.4 ±2.0	-3397.535 ±2.500	-3146.948 ±2.700	2306.773 ±1.979
Nepheline, $NaAlSiO_4$	142.055	29.72 ±0.30	54.16 ±0.06	-497.029 ±1	-469.664 ±1.010	344.273 ±0.740
Analcime, $NaAlSi_2O_6 \cdot H_2O$	220.155	56.03 ±0.60	97.49 ±0.10	-786.341 ±0.860	-734.262 ±0.880	538.228 ±0.645
Jadeite, $NaAl(SiO_3)_2$	202.140	31.90 ±0.30	60.40 ±0.10	-719.871 ±1	-677.206 ±1.010	496.405 ±0.740
Kaolinite, $Al_2Si_2O_5(OH)_4$	258.161	48.53 ±0.30	99.52 ±0.26	-979.465 ±0.950	-902.868 ±0.960	661.819 ±0.704
Low Albite, $NaAlSi_3O_8$	262.224	50.20 ±0.40	100.07 ±0.13	-937.146 ±0.740	-883.988 ±0.760	647.980 ±0.557

K_f = equilibrium constant of formation.

order to make quantitative comparisons, were normalized to a standard reference cell of 24 oxygen atoms. In this series, only one zeolite mineral, analcime, was included although several hydroxyl-containing minerals were computed. Based on the composition $NaAlSi_2O_6 \cdot H_2O$ the bonding energy of analcime adjusted to 24 oxygen atoms is 32,400 kcal. For nepheline, the energy is 31,860 kcal. Similarly, one can compute for jadeite, which has the same composition as anhydrous analcime, a bonding energy of 33,340 kcal.

G. ZEOLITIC WATER

Nuclear magnetic resonance (nmr) provides information on the state of the water molecules in hydrated zeolites.

The nmr spectrum for chabazite was interpreted as resulting from the interchange of water molecules between sites at a frequency of about 1 x 10^7 times/second (44). The transit time is small compared with the residence time; the probability of occupation of the sites of the same type is, therefore, the same. Nuclear magnetic resonance studies of the protons in hydrated zeolite X and A give a single, sharp line instead of the doublet observed in chabazite.

Water molecules in the larger zeolite cavities exhibit the same pattern as the isolated liquid, thus indicating that molecules in the centers of large cavities do not occupy definite lattice sites. In zeolites with smaller cavities, the molecules of water appear to cluster around the cations. In analcime, the sodium cation is in contact with only two water molecules. In the mineral brewsterite, the Sr^{2+} cation has five associated water molecules, and in chabazite, each calcium ion has five nearest water molecules. Other structures in which cations are linked to water molecules include harmotome, natrolite, and phillipsite. The association of water molecules with cations is consistent with the results of electrical conductivity studies.

During dehydration it appears that the water molecules line the inside of the zeolite structural cages (44). Cation-dipole interactions play an important role in the nature and structure of the zeolitic water.

The proposal has been made that the non-framework water and cations in zeolites behave as a concentrated electrolyte (35). In zeolite X, for example, the intracrystalline phase (86 Na^+ ions and 264 H_2O molecules) corresponds to an 18 molal NaOH solution of 42% NaOH by weight which would have a density of 1.45 g/cc. The nmr measurements show that this water acts as a liquid with a long relaxation time, similar to a viscous liquid. The proton relaxation in hydrated forms of zeolite A containing sodium, calcium, and magnesium cations was determined from the width of the nmr line as a function of the amount of absorbed water (45). This correlation, obtained on several cation-exchanged forms, indicates that the water molecules fill the cavity and are localized near the cations. The widths of the lines for the zeolites A and X at room temperature are very narrow, indicating highly mobile water molecules (46).

At temperatures down to about 200°K, the water molecules retain their mobility; below this temperature the increase in line width as the temperature is lowered suggests that the immobilization of the water molecules occurs gradually and without a sudden phase change. This result agrees with that obtained by Ducros on chabazite at temperatures as low as 150°K (44). At 77°K, the spectra for zeolite A and X are wide and suggest the presence of an icelike structure of water molecules situated in an arrangement resembling the pentagonal dodecahedron.

Proton and ^{23}Na resonance spectra obtained on a dehydrated zeolite X and on the same zeolite containing varying quantities of water showed that the nonlocalized sodium ions form a cation solution in the water with the crystal lattice as the anion (47). No ^{23}Na resonance was observed in zeolite A since there are no double 6-ring units in the structure. Further, the octahedral symmetry of the double 6-ring in zeolite X was removed by the addition of water.

Neutron scattering spectroscopy can be used to measure vibrational and diffusional motion of molecules trapped in solids. From a study of the potassium exchanged zeolite A containing water, heavy water, methyl alcohol, ammonia, and methyl cyanide, it was concluded that the molecule is immobilized in the zeolite (α-cage) for about 10^{-11} sec before jumping to a new orientation in the cage. During this time it may vibrate in resonance with the zeolite lattice (48).

Relaxation times for the protons of water adsorbed to saturation in high-purity zeolite X were measured between 200°K and 500°K. The intracrystalline fluid, i.e., the water within the zeolite cavities containing sodium ions, was shown to be 30 times as viscous as bulk water at room temperature. The density of the zeolite water, however, is the same as liquid water. The time between molecular jumps was employed to measure the fluidity of the intracrystalline water. The jumping rate is 100 times higher for water contained in the pores of charcoal, about 25 A in diameter, than for the zeolitic water (49).

Neutron scattering experiments were conducted on natrolite, chabazite, and analcime (50). Although the H_2O molecules have different environments in each zeolite (in analcime as isolated molecules of two in each cavity, in natrolite isolated individual molecules, and in chabazite groups of "waterlike" molecules), each zeolite showed peaks corresponding to torsional motions of H_2O molecules in the narrow frequency region 480 - 600 cm^{-1}.

In summary, for hydrated zeolites with small and narrow cavities, the positions of the water molecules appear well-defined and coordination occurs with cations in the cavities (for example, in analcime, natrolite, brewsterite, and epistilbite). The cations in the hydrated zeolites

Table 5.4 Thermodynamic Data (42)

Mineral	$S°$ (gibbs/gf)[a]	S_f (gibbs/gf)	$G°_f$ (kcal)	$H°_f$ (kcal)	V/gfw(cal/bar)[b]
Laumontite,	119.3	−439.0	−1598	−1729	4.858
$CaAl_2Si_4O_{12} \cdot 4 H_2O$	± 4.6	± 4.6	± 4	± 5	
Wairakite,	110.4	−335.8	−1477.3	−1577.4	4.558
$CaAl_2Si_4O_{12} \cdot 2 H_2O$	± 1.0	± 1.0	± 1.5	± 1.6	

[a] gibbs/gram formula weight (gf) is the unit of entropy and 1 gibb = 1 calorie/degree.
[b] Volume, V, is in cal/bar where 1 cal/bar = 41.842 cc at 298°K.

seem to be surrounded by as many water molecules as is spatially possible, as long as they do not lie too far away from the aluminosilicate framework and the negative charge distribution.

H. STRUCTURE OF ZEOLITES BY INFRARED SPECTROSCOPY

In recent years a number of aluminosilicate minerals, including tectosilicates and clay minerals have been examined in a phenomenological sense by infrared spectroscopy (ir) (51-53). Some spectra are reported on zeolite minerals. Generally, the reported spectra for the three-dimensional, framework aluminosilicates reach to only 15 microns or about 670 cm^{-1}. Many of the results deal with the behavior of the main (Si, Al-O) band which is found at about 1000 cm^{-1}. An approximate relationship between the frequency of this band and the ratio of Si/Al in the aluminosilicate framework was observed by Milkey (54).

A systematic investigation of the framework structures of many synthetic zeolites has been carried out in the 200-1300 cm^{-1} region (the mid-infrared region) by Flanigen et al. (55). Interpretations of these spectra were based on assignment of the infrared bands to certain structural groups in the various zeolite frameworks. In order to conduct this assignment, it is necessary to know the basic zeolite structure. Hence, this method of investigation is complementary to x-ray structural analysis.

As presented in detail in Chap. 2, the various known zeolite structures can be classified according to common elements of framework topology and the secondary zeolite structure as found in the secondary building units. By an extension of the infrared studies it may be possible to deduce structural information on a new zeolite for which x-ray structural analysis is not complete. The mid-infrared region of the spectrium is useful in this regard since it contains the fundamental vibrations of the framework Al, Si-O$_4$ or (TO$_4$) tetrahedra.

The spectra were determined by the KBr wafer technique (56). Spectral frequencies observed on a series of synthetic zeolites are given in Table 5.5. Each zeolite appears to exhibit a typical ir pattern. The spectra can be grouped into two classes, (1) those due to internal vibrations of the TO$_4$ tetrahedron which is the primary unit of structure and which are not sensitive to other structural variations, and (2) vibrations which may be related to the linkages between tetrahedra. Class 2 vibrations are sensitive to the overall structure and the joining of the individual tetrahedra in secondary structural units, as well as their existence in the larger

416 PHYSICAL PROPERTIES OF CRYSTALLINE ZEOLITES

Table 5.5 Infrared Spectral Data for Synthetic Zeolites (55)

cm^{-1}

Zeolite	$\frac{SiO_2}{Al_2O_3}$	Asymmetric Stretch	Symmetric Stretch	Double Rings	T-O Bend	Pore Opening
A	1.88	1090 1050 995 s vwsh vwsh	660 vw	550 ms	464 m	378 260 ms vwb
Ca^{ex}A	1.9	1130 1055 998 s vwsh vwsh	742 705 665 vwsh vwsh vw	542 ms	460 m	376 m
N-A	3.58	1131 1030 vwsh s	750 675 vwsh vw	572 ms	474 m	385 m
N-A	6.01	1151 1044 vwsh s	750 698 vwsh vw	581 ms	475 m	393 m
X	2.40	1060 971 msh s	746 690 668 m wsh m	560 ms	458 ms	406 365 250 w m vwb
Y	3.42	1135 985 msh s	760 686 m m	564 m	508 460 vwsh ms	372 m
Y	4.87	1130 1005 msh s	784 714 635 m m vw	572 m	500 455 wsh ms	380 260 m vwb
La^{ex}Y	5.0	1135 1006 msh s	790 705 m m	565 m	500 450 wsh ms	382 m
Y	5.63	1130 1017 msh s	789 718 645 m m vw	575 m	504 456 mwsh ms	383 315 m vws
B (P)	2.8	1105 995–1000s mwsh	772 738 670 mwsh mw mw	600 m	435 ms	380 mwsh
Hydroxy-sodalite (HS)	2.0	1096 986 vwsh s	729 701 660 m mw m		461 432 ms ms	282 vwb
Ω	7.7	1130 1024 wsh s	805 722 mw mw	610 mw	451 445 ms m	372 m
ZK-5	6.0	1158 1048 890 wsh s vwb	730 mw	572 m		408 wsh

STRUCTURE OF ZEOLITES BY INFRARED SPECTROSCOPY 417

Table 5.5 Infrared Spectral Data for Synthetic Zeolites (55) (continued)

cm^{-1}

Zeolite	$\frac{SiO_2}{Al_2O_3}$	Asymmetric Stretch	Symmetric Stretch	Double Rings	T-O Bend	Pore Opening
R	3.25	1136 mwsh 1007 s	738 w 678 w	625 m	452 m 426 m	370 vwsh
G	5.44	1138 mwsh 1027 s	720 w 696 w	508 mw 515 m	460 m 408 m	378 vwsh
D	4.62	1184 mwsh 1018 s	755 wsh 711 w	632 m 631 m	459 m 415 m	376 vwsh
S	2.5	1140 wsh 1020 s	770 vwsh 722 mw	513 m 518 m	448 ms 424 m	370 vwsh
T	7.0	1156 1059 1010 wsh s s	771 w 718 mw	623, sh 595 mb 623 575 mb	467 ms 433 ms	366 vwsh
Hydroxy-cancrinite (HC)	2.0	1095 1035 1000 965 mw msh s msh	755 w 680 m	624 mw m w 567 m	498 mw 458 ms 429 ms	410 vwsh 390 mw 353 wb
L	6.0	1160 1080 1015 wsh s s	767 mw 721 mw	606 m 580 wsh	474 ms 435 wsh	375 vwsh
C (Analcime-type)	4.0	1162 1012 952 vwsh s s	740 m 686 wb	642 vwsh 615 m	442 ms 410 msh	
Zeolon	9.95	1216 1180 1046 w vwsh s	795, 715, 772 690 wb wb	vw 621 w 571, 555 w	448 ms	370 vwsh
W	3.6	1128 1006 msh s	786 691 756 mwb mwb	637 mw 590 wb 512 vwsh 483 vwsh	432 ms	375 vwsh

s = strong; ms = medium strong; m = medium; mw = medium weak; w = weak; vw = very weak; sh = shoulder; b = broad.

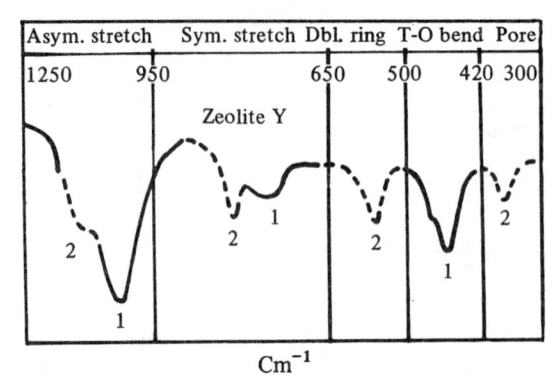

418 PHYSICAL PROPERTIES OF CRYSTALLINE ZEOLITES

Figure 5.35 Infrared assignments illustrated with the spectrum of zeolite Y, Si/Al of 2.5 (55). 1–Internal tetrahedra-structure insensitive; 2–External linkages-structure sensitive.

pore openings.

Individual assignments to specific AlO_4 tetrahedra are not possible. However, the vibrational frequency represents the average composition. A typical spectrum, as illustrated by that of zeolite Y, is given in Fig. 5.35.

The first class of spectra consists of the strongest vibrations found in all zeolites which are assigned to the internal tetrahedron vibrations and are found at 950 - 1250 cm^{-1} and at 420 - 500 cm^{-1}. The assignments are summarized in Table 5.6. The strongest vibration at 950 cm^{-1} is assigned to a T-O stretch. The next strongest band in the 420 - 500 cm^{-1} region is assigned to a T-O bending mode.

Stretching modes involving mainly the tetrahedral atoms are assigned in the region of 650 - 820 cm^{-1} as shown in Table 5.5 and Fig. 5.35. The stretching modes are sensitive to the Si-Al composition of the framework and may shift to a lower frequency with increasing number of tetrahedral aluminum atoms. The bending mode, that is, the 420 - 500 cm^{-1} band is not as sensitive.

The second group of frequencies which are sensitive to the linkages

Table 5.6 Zeolite ir Assignments (in cm^{-1}) (55)

1. Internal tetrahedra	– Asym. stretch	1250 - 950
	– Sym. stretch	720 - 650
	– T-O bend	500 - 420
2. External linkages	– Double ring	650 - 500
	– Pore opening	420 - 300
	– Sym. stretch	750 - 820
	– Asym. stretch	1150 - 1050 sh

between tetrahedra and the topology and mode of arrangement of the secondary units of structure in the zeolite occur in the regions of 500 - 600 and 300 - 420 cm^{-1}. (See Table 5.5 and Fig. 5.35.) A band in the 500 - 650 region is related to the presence of the double rings (D4R and D6R) in the framework structures and is observed in all of the zeolite structures that contain the double 4- and double 6-rings (zeolites X, Y, A, ZK-5, Ω, L, and the chabazite group). The zeolites which do not contain the double rings or larger polyhedral units, such as the β-cage, are typified by zeolite W and Zeolon (mordenite-type). These show only a weak absorption in this region of the spectrum. This was attributed to a stretching mode.

The next main frequency which is assigned to external linkages is in

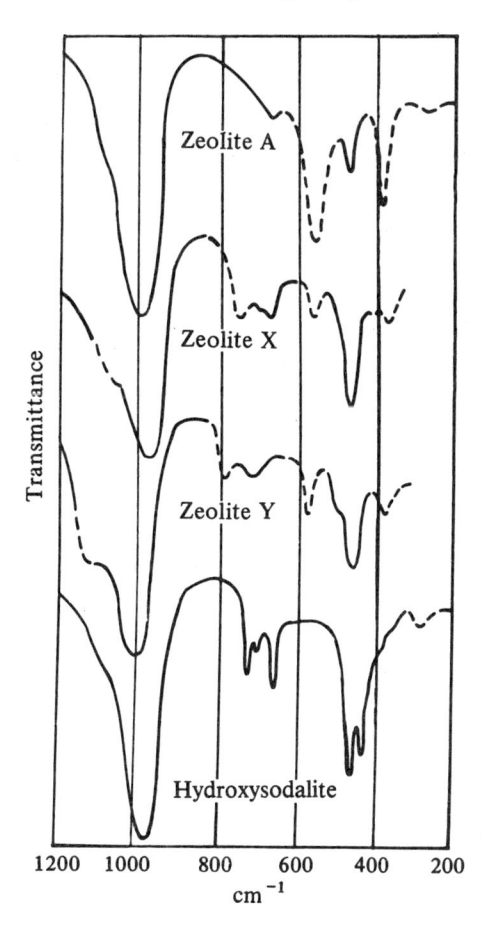

Figure 5.36 Infrared spectra of zeolites A, X, and Y and hydroxysodalite (HS); Si/Al in X is 1 2, and in Y, 2.5 (55).

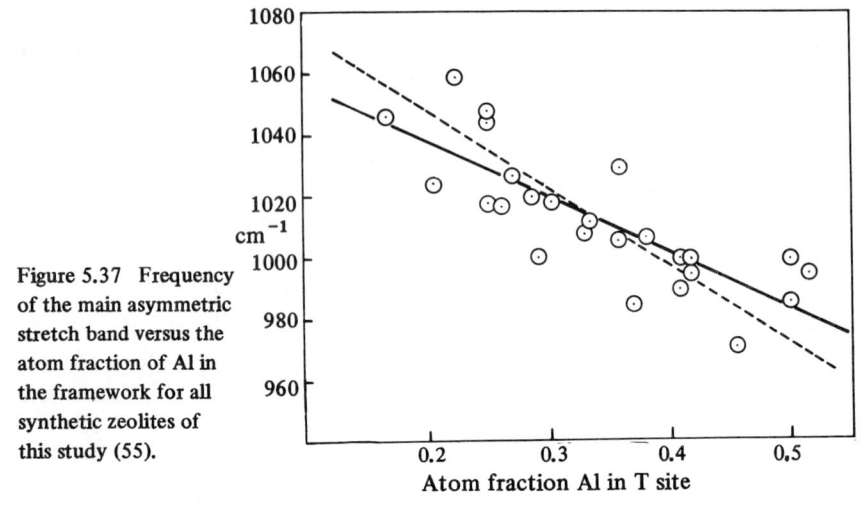

420 PHYSICAL PROPERTIES OF CRYSTALLINE ZEOLITES

Figure 5.37 Frequency of the main asymmetric stretch band versus the atom fraction of Al in the framework for all synthetic zeolites of this study (55).

the 300 - 420 cm^{-1} portion of the spectrum and is related to the pore opening or motion of the tetrahedra rings which form the pore openings in the zeolites. It appears to be more or less observable depending upon the type of zeolite structure. It is a distinct band in the spectra of zeolites A, X, Y and Ω, but is less apparent in the spectra of zeolite P, ZK-5, etc. It seems to be more prominent in cubic structures and decreases in prominence with a decrease in symmetry. This is considered to be a bending mode. As listed in Table 5.5, other bands attributed to the linkages of the primary structural units are found in the 750 - 820 region and the 1050 - 1150 region. The T-O-T angle in zeolite frameworks ranges from 140° to 150°. No correlation was found in the ir with deviations from these values.

Figure 5.38 Frequency versus atom fraction of Al in the framework for zeolites X and Y for several infrared bands (55).

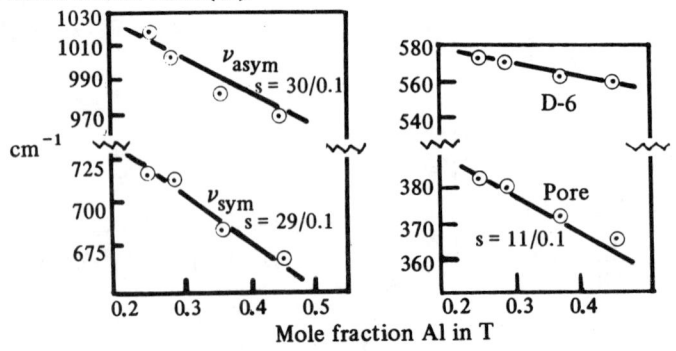

The structures of zeolites A, X, and Y consist of the β-cage (sodalite) type units as a polyhedral unit of secondary structure. One might conclude that there should be a resemblance between the infrared spectra of hydroxysodalite and these zeolites. However, as shown in Fig. 5.36, there are distinct differences in the spectral characteristics of these aluminosilicates. The band attributed to the double 6-ring near 560 cm^{-1} is missing completely in the spectrum of hydroxysodalite. Another example is found in the spectrum of hydroxycancrinite which, like the zeolites T and L, contains the ε-cage as a secondary structural unit.

The main asymmetric stretch of the first type, that is, of the tetrahedra, varies in frequency with the composition of the aluminosilicate framework as expressed by the atom fraction of aluminum tetrahedral atoms. As shown in Fig. 5.37, the frequency varies with the number of Al atoms in the framework structure (55, 57). A similar decrease in frequency with increase of aluminum in the framework tetrahedral sites was observed for other bands, i.e., 970 - 1020 and 670 - 725 cm^{-1} (see Fig. 5.38).

Removal of zeolite water (the water phase itself is not observed in this region of the infrared spectrum) did not change the ir spectra of many of the zeolites that were studied in the synthesized form containing alkali metal ions. Exchange with multivalent cations may change the framework and alter the infrared spectrum. Cation movement and framework distortion is indicated by the ir spectrum of a calcium exchanged zeolite Y (see Fig. 5.39). Dehydration (Chap. 2) causes migra-

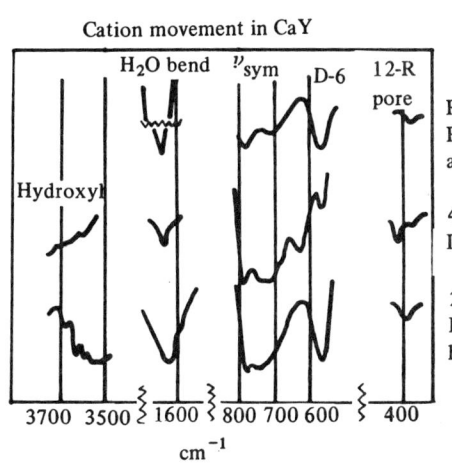

Figure 5.39 Infrared spectra for Ca-exchanged Y zeolite (Si/Al of 2.5) after dehydration, dehydroxylation, and rehydration (55).

tion of the divalent calcium ions from positions within the β-cage to the site I of D6R units. A change in the symmetry of the D6R unit results, as evidenced by a shift in the D6R band (570cm⁻¹) and the band indicative of the pore opening (390 cm⁻¹). These changes are reversible with rehydration. During thermal decomposition of zeolite Y, the structure dependent bands at 1130, 780, 570 and 380 cm⁻¹ decrease with de-

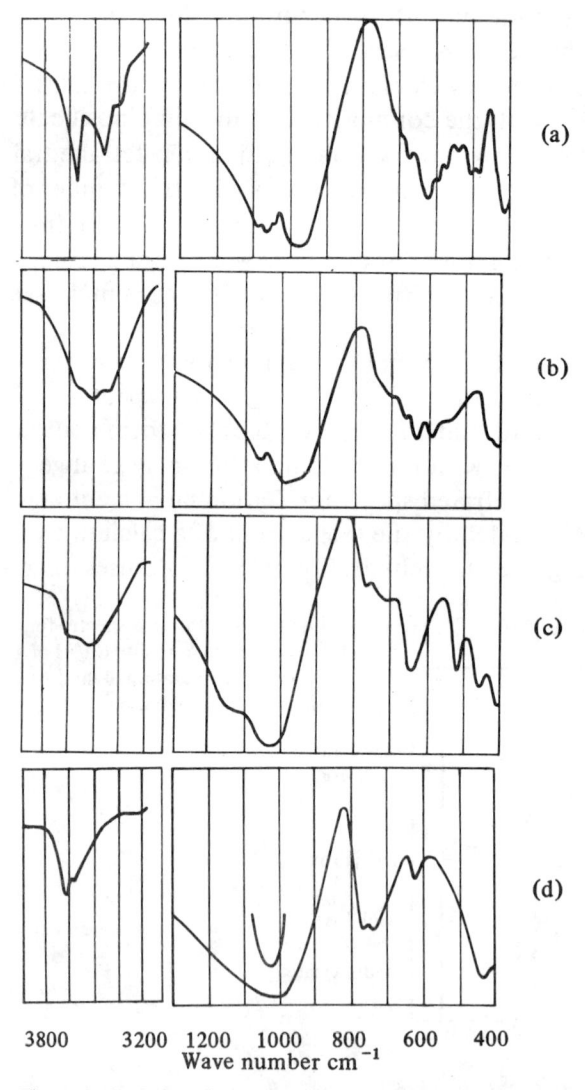

Figure 5.40 Infrared spectra of the samples in the regions of 4000-3000cm⁻¹ and 1300-400 cm⁻¹. a. natrolite; b. thomsonite; c. chabazite; d. analcime; e. stilbite; f. heulandite; g. laumontite; h. mordenite. After Oinuma and Hayashi (58).

creasing x-ray crystallinity. However, the internal T-O vibrations (1000, 710 and 460 cm^{-1}) do not change.

The ir spectra of synthetic zeolites are quite simple and contain several main bands without fine structure. The simplicity seems to increase with increasing openness of the zeolite structure.

The infrared spectra of several zeolite minerals which belong to various structural groups have been reported (58). Some are illustrated in

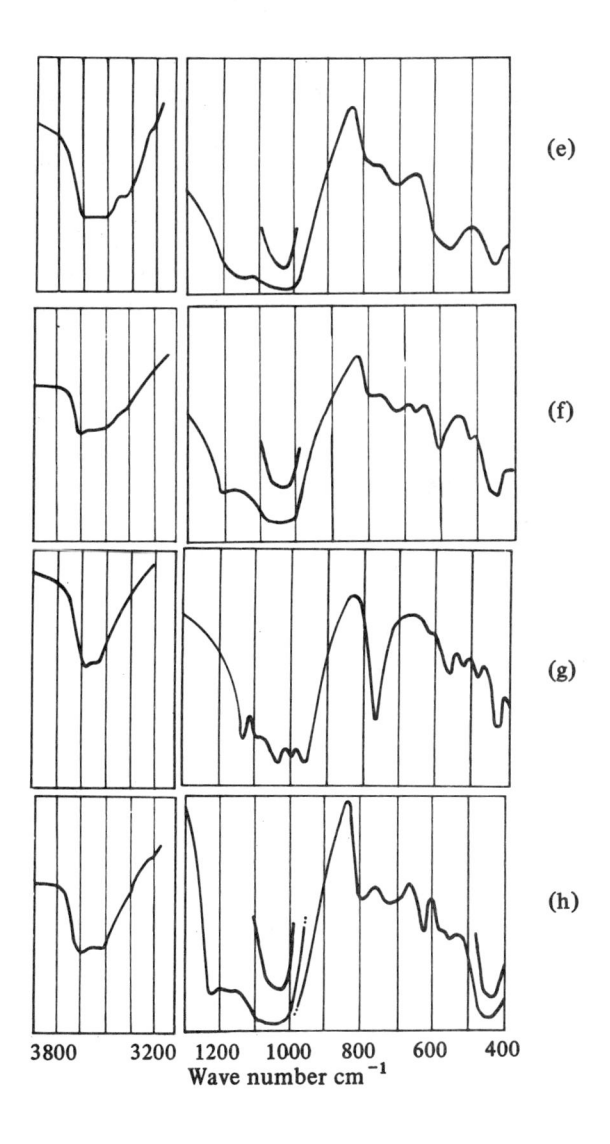

Fig. 5.40. These spectra cover the range in wavelengths of 400 - 4000cm^{-1}. In addition to the bands attributed to water, as discussed previously, other absorption bands are related to the framework structure.

Although they have the same aluminosilicate framework, natrolite and thomsonite exhibit differences in their ir spectra, particularly in the 3200-3500 cm^{-1} region. More fine structure is evident in the 400-700 cm^{-1} region than that found in some other zeolites. This is expected in view of the close water-cation interaction known to exist in the fibrous zeolites.

Similarly, stilbite and heulandite are members of the same structural group, but do not have the same framework structure; however, their infrared spectra do show similarities in the region of 400-800 cm^{-1} and in the main band at 1000 - 1200 cm^{-1}.

The infrared absorption which might be attributed to the pore-opening aspect of the structure was not observed since these characteristic frequencies generally occur below 400 cm^{-1}, beyond the spectral region studied.

Infrared Spectra of Water Adsorbed in Zeolites

The association of the water molecules with the cations and/or framework oxygen ions of a zeolite is dependent upon the openness of the structure. In zeolite NaX, which is representative of the more open structures with larger voids, water molecules are associated with the cation and, to some degree, are hydrogen-bonded to oxygen ions of the framework. This is evidenced by a sharp ir band at 3690 cm^{-1}, the lat-

Figure 5.41 Maximum absorbance as a function of water concentration on NaX: 1, 3695 cm^{-1}; 2, 3400 cm^{-1}; 3, 1650 cm^{-1}. After Bertsch and Habgood (60).

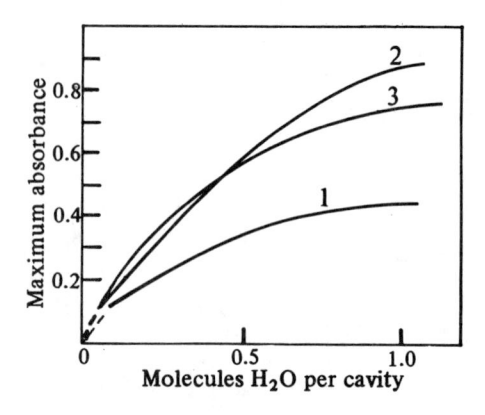

ter being attributed to the hydroxyl of the water molecule bonded to a framework oxygen. In addition, the band at 1645 cm^{-1}, which is characteristic of the bending mode in the water molecule, is present. When the zeolite is fully activated this water bending frequency at 1630 cm^{-1} is absent (59). The marked effect of water on the three ir frequencies at 3690, 3400 and 1645 cm^{-1} is shown in Fig. 5.41 (60). The amount of water in zeolite X when fully hydrated is 30 molecules/large cavity (over 240 molecules/unit cell). The data shown in Fig. 5.41 are for very low coverages, that is, up to about one molecule per cavity or eight molecules in the unit cell. The three typical bands observed are the broad band characteristic of hydrogen-bonded OH at about 3400 cm^{-1}, the sharp band typical of isolated OH at 3700 cm^{-1}, and the usual bending vibration of water at 1645 cm^{-1}.

The isolated OH stretching is attributed to interaction of the water hydroxyl with the cation. The other bands are attributed to the hydrogen bonding of the water molecule to a surface oxygen and to the bending mode of the water (60).

I. PORE VOLUME IN DEHYDRATED ZEOLITES

After dehydration, the zeolite is a crystalline solid permeated by micropores. If major structural distortion occurs on removal of the water molecules, the volume and shape of the micropores may consequently be reduced and/or distorted.

In those zeolites where the framework structure remains essentially unchanged after dehydration, the resulting pore structure and form remains rigid and intact. During the subsequent adsorption or occlusion of various substances, the micropores fill and empty reversibly. Therefore, adsorption is a matter of pore filling and the usual surface area concepts as they are applied to other solid adsorbents are not applicable (61). The pore volume of a dehydrated zeolite and other microporous solids which have type I adsorption isotherms may be calculated by the Gurvitsch rule (62, 63). The adsorption isotherm generally intersects the vertical line at a $p/p_0 \sim 1$ at an angle of nearly 90° (Fig. 5.42). The quantity of material adsorbed, x_s, at this point of saturation is assumed to fill the micropores of the solid as the normal liquid having a density, d_a, of the liquid at that particular temperature. The *total pore volume* in the micropores V_p is given by

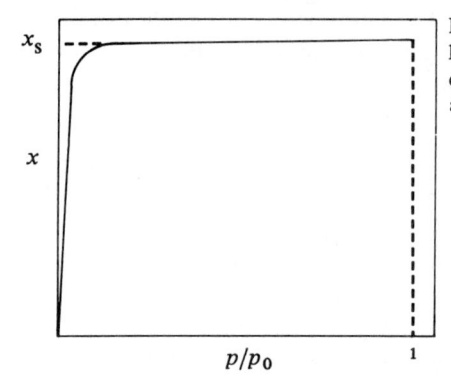

Figure 5.42 The isotherm Type I and relation of the quantity adsorbed at saturation corresponding to complete pore filling, x_s, at $p = p_0$.

$$V_p = \frac{x_s}{d_a} \tag{5.1}$$

where d_a = the density of the liquid adsorbate in g/cc, x_s is in g/g, and V_p in cc/g.

The Gurvitsch rule is obeyed by many different adsorbates on microporous adsorbents such as silica gel and carbon. It is obeyed to a large extent on many of the dehydrated zeolites, but some notable deviations are observed.

The total pore volume of the voids in dehydrated zeolites may be calculated from the amount of adsorbed water on the assumption that the water is present as the normal liquid, with the average density assumed to be that of normal liquid water at the temperature concerned. In zeolites which have more than one type of channel size or void (for example, zeolite A or X) the void spaces available for occupancy by most molecular species are less than the total water volume in the zeolite due to the inaccessibility of the small voids to other molecules. In zeolite A, X and Y, voids of two types are found: (1) the small spherical voids comprised of the β-cage with a diameter of 6.6 A and (2) larger voids, such as the α-cage (truncated cuboctahedron) in zeolite A and the supercages (26-hedron) in zeolite X and Y The diameter of the spherical unit in zeolite A is 11.4 A and in the large cage of the 26-hedron in zeolite X or Y the diameter is about 11.8 A. (See Chap. 2.)

Unlike the amorphous microporous adsorbents, it is possible to calculate the micropore volume of zeolites from the known structure. The measured pore volumes can then be compared with calculated values.

In this section, the pore volumes for zeolites of different structure types as determined from different adsorbates are summarized and compared to calculated pore volumes.

The maximum void volume, V_p, in a zeolite is determined from the water contained in the zeolite at p/p_0 equal to 1 using Eq. (5.1). One can then calculate the void fraction, V_f, by the equation:

$$V_f = x_s \cdot d_c / d_a \qquad (5.2)$$

where V_f = total void fraction in the crystal and
d_c = the density of the dehydrated zeolite crystal.

The total pore volume, V_p', in units of A^3 per unit cell is given by

$$V_p' = V_f \cdot V \qquad (5.3)$$

where V = the unit cell volume in A^3.

The number of adsorbed molecules, N_A, per unit cell can be calculated by

$$N_A = \frac{x_s \cdot M_z}{M_A} \qquad (5.4)$$

where M_z = unit cell formula weight; M_A = molecular weight of the adsorbate molecule.

Zeolite A

Zeolite A contains two types of void spaces: (1) those within the β-cages which are accessible to only a small molecule, such as water, and which involve a volume of 151 A^3; and (2) the large voids in the α-cages of 775 A^3. From the structure one can calculate that the total void volume per unit cell is 926 A^3. The adsorption of gases, such as argon, oxygen, and nitrogen, at a pressure $p = p_0$, is equivalent to the volume of the large α-cage or 755 A^3 per unit cell. Water adsorption in zeolite A is equivalent to a volume of 833 A^3 per unit cell; this rather conclusively indicates that the β-cages must be occupied by water (64) (see Table 5.7).

With the exception of water and nitrogen the observed pore volumes follow the pore-filling theory as far as the large α-cages are concerned. However, the pore volume is greater in the case of water and nitrogen. The amount of adsorbed water and nitrogen cannot be accounted for on the basis that they are normal liquids filling just the large voids. Either

428 PHYSICAL PROPERTIES OF CRYSTALLINE ZEOLITES

these adsorbates occupy the total void volume or the density of the adsorbed phase is considerably greater than the normal liquid density at the temperature concerned. This anomaly on the part of water and nitrogen is observed in several zeolites.

If nitrogen does not occupy the small β-cages, the average density of the adsorbed phase at saturation is 25% greater than the normal liquid density of N_2 at that temperature, $-196°C$. Nitrogen cannot enter the β-cages at low temperatures.

The presence of cations in the zeolite voids may affect the pore volume. In some cases, as typified by zeolite A, a large cation may substantially reduce the pore volume. In Table 5.8, the pore volume of zeolite A, as determined from the water contained in the fully hydrated zeolite, is given for various cation-exchanged forms. Exchange of sodium by thallium, Tl^+, reduces the pore volume by about 250 A^3/unit cell or approximately 30%. From the crystal structure analysis of TlA (65) the cations are located well into the main α-cage on the three-fold axis, and therefore should interfere with the water adsorption.

The exchange of sodium by calcium increases the pore volume since the total cation density (number of cations/unit cell) is reduced by 50%. Exchange by lithium seems to reduce the available pore volume per unit cell. This is in part due to a reduction in the unit cell dimension and consequently, the unit cell volume.

From the Dubinin-Polanyi pore-volume filling theory (See Chap. 8)

Table 5.7 Void Volume in Zeolite A (64, 70)

Zeolite	Adsorbate	$t(°C)$	x_s (g/g)	V_p(cc/g)	V_f(cc/cc)	V'_p(A^3/unit cell)
$Na_{12}A$	H_2O	25	0.289	0.289	0.45	842
	CO_2	-75	0.30	0.252	0.39	725
	O_2	-183	0.242	0.213	0.327	612
Ca_6A	H_2O	25	0.305	0.305	0.48	885
	O_2	-183	0.276	0.242	0.38	700
	Ar	-183	0.358	0.261	0.40	738
	N_2	-196	0.239	0.297	0.465	857
	n-butane	25	0.131	0.226	0.354	655

For $Na_{12}A$ d_c = 1.54 g/cc, V = 1870 A^3, n = 3.58 x 10^{20}/g
For Ca_6A d_c = 1.57 g/cc, V = 1843 A^3, n = 3.64 x 10^{20}/g
n = number of unit cells/g
V_p (calc) = 926 A^3/uc = 0.32 cc/g
V_α = 775 A^3/uc = 0.27 cc/g

Table 5.8 Effect of Cation Exchange on Void Volume in Zeolite A (65)

Unit Cell	Density (g/cc)	$a_o(A)$	$V_p'(A^3/uc)$
$Li_8Na_4(A) \cdot 24 H_2O$	1.91	12.04	735
$Na_{12}(A) \cdot 27 H_2O$	1.99	12.32	833
$Ag_{12}(A) \cdot 24 H_2O$	2.76	12.38	733
$Tl_{9.6}Na_{2.4}(A) \cdot 20 H_2O$	3.36	12.38	584
$Ca_6(A) \cdot 30 H_2O$	2.05	12.26	883

the limiting adsorption volume, W_0, for light hydrocarbons on Ca A is 0.23 cc/g (66). This is in close agreement with the value $V_p = 0.226$ for n-butane in Table 5.7.

Zeolite X

The pore volume of zeolite X, as determined from the adsorption of different types of molecules including water, gases, and hydrocarbons, is shown in Table 5.9. The water pore volume is equivalent to 7908 A^3 per unit cell (2). For most molecules, with the possible exception of water, only the large supercages are occupied. The total pore volume

Table 5.9 Void Volume of Zeolite NaX (Si/Al = 1.25) (2)[a]

Adsorbed Molecule	$t(°C)$	$x_s(g/g)$	$V_p(cc/g)$	$V_f(cc/cc)$	$V_p'(A^3/uc)$	Molecules per Unit cell
H_2O	25	0.355	0.36	0.51	7908	265
CO_2	-78	0.395	0.33	0.48	7360	120
Ar	-183	0.418	0.30	0.44	6757	140
Kr	-183	0.726	0.27	0.38	5950	116
Xe	-78	0.726	0.25	0.36	5622	74
O_2	-183	0.356	0.31	0.45	6923	149
N_2	-196	0.279	0.35	0.50	7680	134
n-Pentane	25	0.184	0.30	0.42	6581	34
Neopentane	25	0.157	0.26	0.38	5860	29
2,2,4-Trimethylpentane	25	0.186	0.27	0.39	6006	22
Benzene	25	0.260	0.30	0.425	6609	45
$(C_4H_9)_3N$	25	0.227	0.29	0.42	6490	

(a) $V = 15531 A^3$ per unit cell; $d_c = 1.43$; d_a from (84); $n = 0.45 \times 10^{20}$ (68)
$V_p(calc) = 7832 A^3/uc = 0.352$ cc/g.
V_p of large voids only = 0.296 cc/g = 6576 A^3/uc

of the large cages was determined from the adsorption of either argon or oxygen and is 6700 Å^3 per unit cell. About 1200 Å^3 (150 Å^3 for each of the eight β-cages) in the unit cell are available only to water. This agrees well with the calculated internal volume of 151 Å^3 for the single β-cage. The volume of each large 26-hedral void in zeolite X has been calculated to be 822 Å^3 (68). Consequently, the total calculated void volume of zeolite X is 7832 Å^3 per unit cell, in good agreement with the observed value of 7908 Å^3 from water adsorption. The total calculated void volume for the large voids is 0.296 cc/g which is consistent with observed pore volumes as determined from all adsorbates, except water and nitrogen.

At a relative pressure (p/p_0) equal to 0.5 the adsorption of paraffin hydrocarbons by zeolite X follows the Gurvitsch rule very well. When expressed in terms of the liquid volume of the adsorbed hydrocarbon using the normal liquid density at the temperature concerned, the void fraction of zeolite X which can be filled is nearly 50% of the total crystal volume (Table 5.10). In each case, with the exception of *iso*-octane, the volume occupied by the adsorbed hydrocarbon is essentially 0.31 cc/g which is equivalent to the large 26-hedral voids only. In the case of *iso*-octane, the adsorption seems anomalous with the volume of the adsorbed hydrocarbon = 0.28 cc/g. This is apparently caused by a packing effect. The more flexible and bulky *iso*-octane molecules do not pack as well within the confinement of the cages of the zeolite as they do in the normal liquid phase. This is counter to the apparent "super-

Table 5.10 Void Volume in Zeolite X (Si/Al = 1.33) for Paraffins and Other Hydrocarbons at p/p_0 = 0.5 (83, 85).

Hydrocarbon	T(°K)	x_s^a(g/g)	V_p (cc/g)	V_f (cc/cc)	Molecules/unit cell
n-Pentane	298	0.193	0.311	0.445	35.8
Isopentane	298	0.190	0.309	0.44	34.8
n-Hexane	298	0.202	0.309	0.44	31.3
n-Heptane	298	0.212	0.311	0.445	28.3
n-Octane	313	0.212	0.308	0.44	24.4
Isooctane	298	0.194	0.282	0.40	22.7
Benzene	303	0.256	0.295	0.42	43.3
Toluene	313	0.256	0.301	0.43	36.8
Cyclopentane	303	0.246	0.334	0.48	45.1
Cyclohexane	313	0.203	0.268	0.38	33.0

[a]Recalculated to a basis of anhydrous zeolite

packing" of nitrogen molecules. The normal paraffin hydrocarbons, however, do not experience this difficulty.

Although the volume of the adsorbed hydrocarbons remains constant, the number of adsorbed molecules per unit cell decreases with increasing complexity of the molecule (see Table 5.10). The void volume occupied by cyclic hydrocarbons in zeolite X is also shown in Table 5.10. The pore volume for cyclopentane is substantially larger than that for cyclohexane. This suggests that adsorbed cyclopentane molecules pack more efficiently than any of the other hydrocarbons used.

Ammonia adsorbed on zeolite X probably occupies the small voids within the β-cages (69). It was observed that ammonia is adsorbed on a hydroxysodalite at 100°C quite rapidly. The structure of hydroxysodalite consists only of β-cages. Adsorption at room temperature was slow.

In a series of publications, Dubinin (61, 70 - 72) has attempted to calculate, from the known crystal structures of the zeolites A and X, the void space and the internal surface area. An attempt was made to relate these to measured values. Based on the structure, the calculated surface areas were 1640 meters^2 per gram for calcium zeolite A and 1400 meters^2 for sodium zeolite X. From the calculated area the adsorption capacity was determined from the ratio to the molecular area of the adsorbate. When compared to the limiting adsorption volumes, V_p, corresponding to complete filling of the voids, it was found that the calculated values exceed the measured values of V_p by factors of 1.3 to 3.3 (61). The adsorption voids as calculated from the structures are, according to Dubinin, 0.278 cc/g for calcium zeolite A and 0.322 cc/g for sodium X. These values are too low compared to measured quantities and do not agree with the calculated values of V_p (see Tables 5.7, 5.9).

Zeolite L

Zeolite L and mordenite have different types of structures: the main adsorption channels run parallel in one direction. In zeolite L, the arrangement of the main adsorption channel parallels the c-direction and the channels are formed by nearly planar 12-rings with a free diameter of 7.4 A (73). These main channels are linked through small apertures so as to form a nonintersecting parallel channel system (see Chap. 2). From the structure it is possible to estimate the void volume for ad-

Table 5.11 Void Volume in Zeolite L[a] (2, 86)

Adsorbate	$t(°C)$	$x_s(g/g)$	$V_p(cc/g)$	$V_f(cc/cc)$	V'_p (A^3/unit cell)
H_2O	25	0.210	0.21	0.36	784
O_2	-183	0.196	0.172	0.29	642
N_2	-196	0.146	0.181	0.31	675
neopentane	25	0.081	0.132	0.22	493
isobutane	25	0.087	0.157	0.27	587

d_c = 1.70 g/cc; V = 2205 A^3/unit cell; n = 2.68 x 10^{20}/g;
V_p (calc) = 614 A^3/uc for large channel = 0.15 cc/g;
V_p (calc) total = 686 A^3/uc = 0.18 cc/g;
[a] NH_4 exchanged L with NH_3 thermally removed; Composition: $K_{24}Al_{90}Si_{27}O_{72}$.

sorption and compare this with the values determined from adsorption measurements. The results are shown in Table 5.11. Water molecules occupy space which is not available to gases such as argon, krypton, or oxygen. However, nitrogen appears to fill voids not available to the other molecules or it must also have a higher average adsorbed density. The additional space available to water must consist of the small voids that are due to the epsilon-cage units that form the channel walls. The calculated free volume of the main channel is 614 A^3/unit cell, which compares well with the pore volume as determined from the adsorption of the permanent gases Ar and O_2, 619 - 642 A^3.

Chabazite

Table 5.12 gives pore volumes as determined from several adsorbates for chabazite. Although calculation of the pore volume from the structure is not easily done, one does observe that the pore volume for N_2 and H_2O is substantially higher than that for the other gases, such as, O_2 and Ar. In this instance, also, it seems that CO_2 fills the voids in chabazite to the same extent as water.

Table 5.12 Void Volume in Chabazite [a] (83)

Adsorbate	$t(°C)$	$x_s(g/g)$	$V_p(cc/g)$	$V_f(cc/cc)$	V'_p(A^3/unit cell)
H_2O	25	0.27	0.270	0.45	370
CO_2	-78	0.298	0.270	0.45	370
O_2	-183	0.248	0.217	0.359	295
Ar	-196	0.30	0.216	0.356	292
N_2	-196	0.206	0.256	0.423	348

[a] Mineral from Nova Scotia; V = 822 A^3/unit cell; d_c = 1.65 g/cc.

Mordenite, Large Port

The total void volume of mordenite has been estimated from the structure (74). Barrer and Peterson observed that *n*-paraffins and *iso*-paraffins were excluded from the main adsorption channels of a sodium mordenite. However, others have shown that the main channels are available to neopentane (Fig. 5.43).

Smaller permanent gas molecules occupy voids in addition to the main channels; these must include the small, niche-type cavities which lie on the side of the main c-axis channels. Replacement of sodium cations by hydrogen opens the channels so that more space is available, to some of the larger hydrocarbon molecules. Consequently, the total pore volume in the mordenite is about 0.21 cc/g with 0.11 cc/g in the main channels only (74).

These results are different from those previously obtained by Dubinin

Figure 5.43 Pore volume of the synthetic forms of mordenite as determined from saturation adsorption values, x_s, (74) in cc/g. Original data are replotted on basis of the molecular kinetic diameter, σ. Solid points are for the sodium mordenite and * points for the hydrogen mordenite. The total void volume is 0.21 cc/g with 0.11 cc/g consisting of the main, c-axis channels. Note that V_p for Ar, O_2, and N_2 *increases* with increasing σ.

The *calculated* volumes of the c-axis channels are given by Dubinin (75) to be 477 A^3 per unit cell or 0.0935 cc/g. The volume contained in the small "niches" is 340 A^3 per unit cell or 0.0666 cc/g. Total V_p (calc) is 0.160 cc/g. (n = 1.96 × 10^{20} unit cells/g) Dubinin found V_p to be 0.16 cc/g for N_2, Ar, and 0.14 cc/g for H_2O. Satterfield and Frabetti give total V_p (calc) = 0.189 and V_p (main channel) = 0.107 cc/g (67).

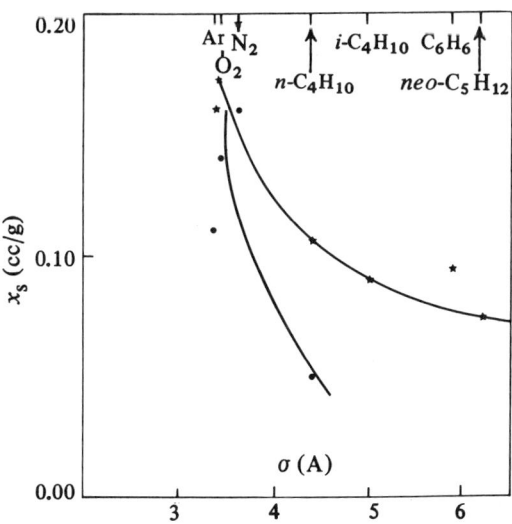

(75), who calculated a total pore volume of 0.16 cc/g and observed the limiting adsorption volumes of 0.16 cc/g for nitrogen, argon, and water.

Pore Volume Correlations

The same type of pore volume calculation can be made for other zeolites, particularly in terms of the void volume as based on the water content of the zeolite when fully hydrated. Values for the void volume in most zeolites expressed as cc/cc or the void fraction, V_f, are given in Table 2.4. In addition, the structure table for each zeolite given in Chap. 2 lists data for the density, unit cell volumes, and framework densities which enable further calculations of this type for other adsorbates.

In all of the examples discussed in this section, as well as in other cases, the behavior of nitrogen at low temperatures appears to be anomalous. Nitrogen seems to fill the voids of zeolites of different structures to the same extent as water. Nitrogen at low temperatures is not adsorbed by zeolite A, and N_2 molecules do not pass the 8-rings although O_2 is rapidly occluded. In comparison, the small 6-ring of an individual sodalite cage has a free diameter of 2.6 A. The diameter of N_2 is 3.64 A as compared to 3.46 A for O_2.

If it is assumed that nitrogen can occupy pore volume accessible to gases such as Ar or O_2, then based on the pore volume of CaA, NaX and chabazite, as examples, the average density of the adsorbed N_2 is about 0.95 g/cc, as compared to 0.80 g/cc for N_2 liquid at the boiling point (76). A density of 0.95 g/cc corresponds to a temperature of 46°K for the liquid. Since at 46°K, liquid N_2 has a vapor pressure of 0.14 mm, this unique behavior corresponds to an increase in density which is achieved by lowering the boiling point 31°K.

The low temperature adsorption of helium, ^{3}He and ^{4}He at 4.2°K, is anomalous in a manner similar to nitrogen. The isotherms of Daunt and Rosen for ^{3}He and ^{4}He yield values of x_s on zeolite NaX of 290 cc NTP/g and 309 cc NTP/g, respectively, after correction for 20% inert binder in the dehydrated zeolite (77). They measured adsorption on a commercial type of zeolite X pellets. Using a liquid density of ^{3}He at 4.2°K of 0.125 g/cc, the values of x_s correspond to 0.415 and 0.44 cc/g for ^{3}He and ^{4}He. It is not probable that helium penetrates into the β-cages of the structure and it must occupy only the supercages. Using the calculated volume of the supercages of 0.30 cc/g, one can

PORE VOLUME IN DEHYDRATED ZEOLITES 435

calculate the mean density of the adsorbed ^{3}He. For ^{3}He the density is 0.173 g/cc and for ^{4}He, 0.184 g/cc. The calculated molar volume of the adsorbed ^{3}He is 23 cc/mole and for ^{4}He, 22 cc/mole. For comparison, the molar volume of He at 4.2°K, liquid, is 32 cc/mole.

Similar results are reported by Monod and Cowen in adsorption and nmr studies of ^{3}He and ^{4}He adsorbed on zeolite X powder at very low temperatures (4.2 to 20.2°K). They conclude that the helium behaves like an inhomogeneous solid. They determine that the molar volume for ^{3}He adsorbed at 1.25°K is 19 cc/mole, corresponding to 30 He atoms in

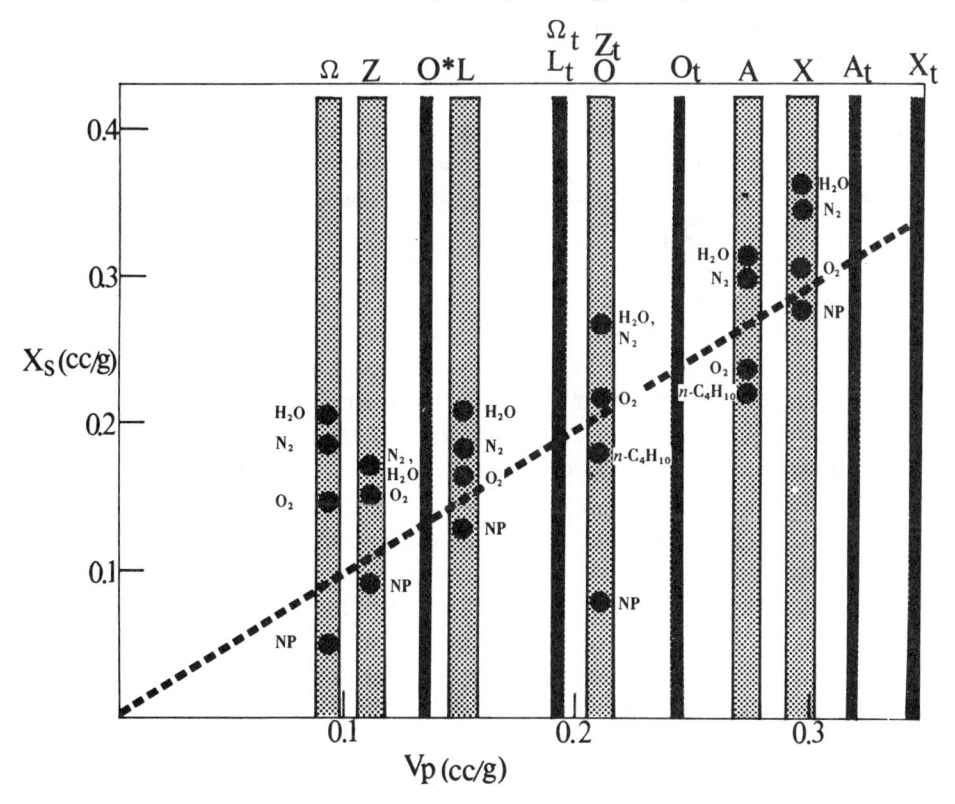

Figure 5.44 Relationship between the measured adsorption volumes, V_p (meas) and calculated void volume V_p of several zeolites. The dashed line corresponds to V_p (meas) = V_p (calc). The symbols represent the zeolites A, X, L, Z (mordenite-Zeolon), omega (Ω), and offretite-type O. Vertical shaded areas containing plotted values of V_p (meas) correspond to calculated values of V_p for the main pore systems. The narrow area, O*, corresponds to the main c-axis void of zeolite O. The value of V_p for $Z_t = V_p$ for zeolite O. Symbols with the subscript t (A_t, X_t, etc.) represent values of V_p for the total void volume shown by narrow shaded areas. The neopentane (NP) volumes lie consistently below the dashed line thus showing a packing effect. In all of these zeolites of varying structure, the H$_2$O and N$_2$ volumes correspond with complete filling of the total voids even though this is not possible in the case of N$_2$ in zeolites A, X, and L.

each supercage (78).

A correlation of measured pore volumes, V_p, versus the calculated values, V_p(calc), is shown in Fig. 5.44 for zeolites A, X, L, Zeolon, omega and zeolite O (offretite).

Dubinin-Polanyi Relation

Adsorption isotherms on microporous adsorbents are described quite satisfactorily by the modified Polanyi relationship of Dubinin (79). This relation has been used to obtain the limiting adsorption volume in zeolites (80 - 82). It is also known as volume filling. (See Chap. 8.) The determination of the limiting adsorption volume by this method for various adsorbates on NaA gave values of 0.264 for H_2O, 0.267 for NH_3, and 0.282 cc/g for SO_2 as compared to a calculated value of 0.290 (81). For zeolite X the calculated value was 0.310 cc/g. These values are low compared to those given previously (0.32 for zeolite A and 0.352 for NaX).

Zeolite Framework Density

The framework density of a zeolite, expressed as the number of structural framework tetrahedra per unit volume of 1000 A^3 has been listed in Table 2.4. The values of the framework density for various zeolites of known structure may be related to the void fraction, V_f. This relationship is shown for various zeolites in Fig. 5.45. Zeolites which contain the larger polyhedral units and double 6-rings (D6R) have void fractions greater than 0.3 and corresponding values of the framework density, d_f, of less than 17 tetrahedra/1000 A^3. In filling space, the packing of the double 6-ring units or double 4-ring units, in conjunction with larger polyhedral units, results in a zeolite framework with a framework density of less than 17. However, within the framework structure, oxygen tetrahedra are densely packed.

The most open known zeolite structures are the zeolites of groups 3 and 4 (Chap. 2). These are based upon the double 6-ring units and truncated octahedral units, or β-cages. Of the zeolites known to date, the largest observed void fraction is 0.5 as found in the zeolites related structurally to faujasite. It has been postulated that zeolites may exist which have a void space fraction approximating 0.6 (73). This was proposed for variations of structures based upon the epsilon

PORE VOLUME IN DEHYDRATED ZEOLITES

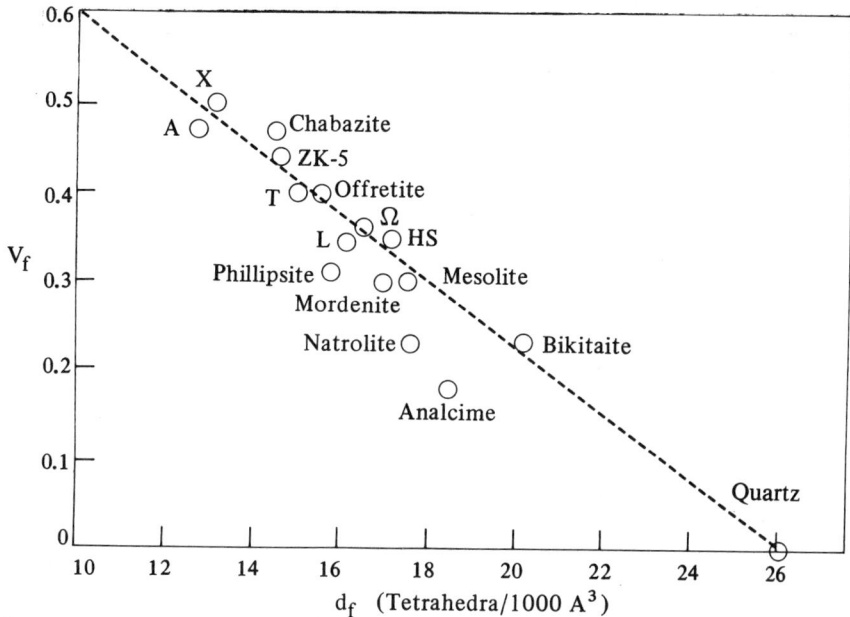

Figure 5.45 Relation between the measured void volume expressed as the void fraction V_f and the framework density, d_f. The dashed line connects the points corresponding to $V_f = 1.0$ at $d_f = 0$ and $V_f = 0$ at $d_f = 26$. The line is therefore expressed by $V_f = 1 - d/26$. The points corresponding to natrolite and analcime deviate because the water molecules in these structures are tightly bound to cations and framework atoms indicating a "feldspathoid" character. The point representing sodalite-hydrate corresponds to no occluded NaOH which is normally present. Typically, synthetic sodalite-hydrate, or basic sodalite, contains occluded NaOH, and the void fraction is accordingly much less, about 0.19 cm^3/cm^3.

cages (Chap. 2) in which the structure may consist of channels formed by 18-rings with apertures of about 15 A in free diameter. These large apertures form the cylindrical channels that parallel the c-axis of these proposed structures. This type of zeolite has a calculated void fraction of approximately 0.6. What are the stability limits for open framework structures of this type? From the known structures, the maximum observed void fraction is about 0.5. Although a value of 0.6 has been postulated, considerations of stability may rule out the likelihood of the formation of such a zeolite. Based on the relation shown in Fig. 5.45, the framework density, d_f, of a zeolite with $V_f = 0.6$ would be 10 tetrahedra/1000 A^3 which corresponds to a density of 1.0 g/cc, or the density of water. Filled with water, the hydrated density of this structure would be 1.6 g/cc.

This simple correlation of V_f with d_f is useful in interpreting unknown structures. If the value of V_f is determined from the measured adsorption of water and d_c, then the framework density can be estimated.

REFERENCES

1. R. M. Barrer and I. S. Kerr, *J. Chem. Soc.*, 434 (1963).
2. D. W. Breck and E. M. Flanigen, Molecular Sieves, Society of Chemical Industry, London, 1968, p. 47
3. I. S. Kerr, J. A. Gard, R. M. Barrer, and I. M. Galabova, *Amer. Mineral.*, **55**: 441 (1970).
4. E. M. Flanigen and R. W. Grose, "Molecular Sieve Zeolites," *Advan. Chem. Ser.*, *101*, American Chemical Society, Washington, D. C., 1971, p. 76.
5. E. F. Kondis and J. S. Dranoff, "Molecular Sieve Zeolites," *Advan. Chem. Ser.*, *102*, American Chemical Society, Washington, D. C., 1971, p. 171.
6. J. Ciric, I. *Colloid Interface Sci.*, **28**: 315 (1968).
7. K. F. Loughlin, R. I. Derrah, and D. M. Ruthven, *Can. J. Chem. Eng.*, **49**: 66 (1971).
8. C. Orr, Jr. and J. M. Dallavalle, Fine Particle Measurement, Macmillan, New York, 1960.
9. R. M. Barrer and W. M. Meier, *Trans. Faraday Soc.*, **54**: 1074 (1958).
10. G. A. Sleater, D. H. Freeman, D. H. Olson, and H. S. Sherry, *Anal. Chem.*, **43**: 1898 (1971).
11. K. Klier and M. Ralek, *J. Phys. Chem. Solids*, **29**: 951 (1968).
12, R. Polak and K. Klier, *J. Phys. Chem. Solids*, **30**: 2231 (1969).
13. G. V. Tsitsishvili, "Zeolites, Their Synthesis, Properties and Applications," Second All-Union Conference on Zeolites, Leningrad, 1964, p. 1-62.
14. M. Ralek, P. Jiru, O. Grubner, and H. Bayer, *Coll. Czech. Chem. Commun.*, **27**: 142 (1962); *ibid.*, **26**: 142 (1961).
15. F. Wolf and H. Fuertig, *Chem. Tech.*, **18**: 339 (1966).
16. E. Engelhard, Dissertation, *Jena* (1912).
17. E. W. Claffy and J. H. Schulman, *Amer. Mineral.*, **36**: 272 (1951).
18. K. Narita, J. Lumin, **4**: 73 (1971).
19. V. M. Federov, B. A. Glazun, I. V. Zhilenkov and M. M. Dubinin, *Izv. Akad. Nauk SSR, Ser. Khim.*, 1930 (1964).
20. B. A. Glazun, V. M. Federov, M. M. Dubinin, and I. V. Zhilenkov, *Izv. Akad. Nauk SSR, Ser. Khim.*, 393 (1966).
21. V. M. Federov, B. A. Glazun, M. M. Dubinin, and I. V. Zhilenkov, *Izv, Akad. Nauk SSR, Ser. Khim*, 1129 (1966).
22. B. Morris, *J. Phys. Chem. Solids*, **30**: 73 (1969); *J. Phys. Chem. Solids*, **30**:89 (1969); *J. Phys. Chem. Solids*, **30**: 103 (1969).
23. P. A. Glazun, V. M. Federov, M. M. Dubinin, and I. V. Zhilenkov, *Izv. Akad. Nauk SSR, Ser. Khim.*, 1297 (1966).
24. B. A. Glazun, V. M. Federov, M. M. Dubinin, and I. V. Zhilenkov, *Izv., Akad., Nauk SSR, Ser., Khim.*, 1301 (1966).
25. B. A. Glazun, M. M. Dubinin, I. V. Zhilenkov, and M. F. Rakityanskaya, *Izv. Akad. Nauk SSR, Ser., Khim.*, 1193 (1967).
26. B. A. Glazun, M. M. Duninin, and I. V. Zhilenkov, *Izv. Akad. Nauk SSR. Ser. Khim.*, 1668 (1966).
27. B. A. Glazun and I. V. Zhilenkov, *Russ. J. Phys. Chem.*, **40**: 734 (1966).
28 R. M. Barrer and P. J. Coen, *Nature*, **199**: 587 (1963).
29. R. M. Barrer and E. A. Saxon Napier, *Trans. Faraday Soc.*, **58**: 156 (1962).

30. D. C. Freeman, Jr. and D. N. Stamires, *J. Chem. Phys.*, **35**: 799 (1961).
31. D. N. Stamires, *J. Chem. Phys.*, **36**: 3174 (1962).
32. R. A. Schoonheydt, Ph.D. Thesis, University of Louvain, March 1970; R. A. Schoonheydt and J. B. Uytterhoeven, "Molecular Sieve Zeolites," *Advan. Chem. Ser.*, *101*, American Chemical Society, Washington, D. C., 1971, p. 456.
33. D. C. Freeman, Jr., *U. S. Pat. 3,106,875* (1965).
34. S. D. James, *J. Phys. Chem.*, **70**: 3447 (1966); A. S. Buchanan and S. D. James, *ibid.*, 34
35. R. M. Barrer and G. C. Bratt, *J. Phys. Chem. Solids*, **12**: 130 (1959); R. M. Barrer, *J. Phys. Chem. Solids*, **16**: 84 (1960).
36. R. M. Barrer and C. Marcilly, *J. Chem. Soc. A*, 2735 (1970).
37. J. A. Kittrick, *Amer. Mineral.*, **51**: 1457 (1966).
38. E. G. King, *J. Amer. Chem. Soc.*, **77**: 2192 (1955).
39. W. S. Fyfe, F. J. Turner and J. Verhoozen, "Metamorphic Reactions and Metamorphic Facies," Geol. Soc. Amer. 73, Geological Society of America, 1958, p. 117.
40. R. A. Robie and D. R. Waldbaum, *U. S. Geol. Soc. Bull.*, 1259 (1968).
41. J. W. Gruner, *Amer. Mineral.*, **35**: 137 (1950).
42. E-an Zen, *Amer. Mineral.*, **57**: 524 (1972).
43. W. D. Keller, *Amer. Mineral.*, **39**: 783 (1954).
44. P. Ducros, *Bull. Soc. Fr. Mineral. Christallogr.*, **83**: 85 (1960).
45. I. V. May
45. I. V. Matyash, M. A. Piontkovaskaya, L. M. Tarasenko, and R. S. Tyutyunnik, *Zh. Strukt. Khim.*, **4**: 106 (1963).
46. S. P. Gabuda and G. M. Mikhailov, *Zh. Strukt, Khim.*, **4**: 446 (1963).
47. A. Knappworst, W. Gunsser and H. Lechert, *Z. Naturforsch*, **21**: 1200 (1966).
48. P. A. Egelstaff, J. S. Downes, and J. W. White, Molecular Sieves, Society of Chemical Industry, London, 1968, p. 306.
49. H. A. Resing and J. K. Thompson, Molecular Sieve Zeolites, *Advan. Chem. Ser.*, **101**, American Chemical Society, Washington D. C., 1971, p. 473.
50. H. Boutin, G. J. Safford and H. R. Danner, *J. Chem. Phys.*, **40**: 2670 (1964).
51. F. Stubican and R. Roy, *Z. Kristallogr.*, **115**: 200 (1961); *J. Amer. Ceram. Soc.*, **44**: 625 (1961); *Amer. Mineral.*, **46**: 32 (1961).
52. P. J. Launer, *Amer. Mineral.*, **37**: 764 (1952).
53. B. D. Saksena, *Trans. Faraday Soc.*, **57**: 242 (1961).
54. R. G. Milkey, *Amer. Mineral.*, **45**: 990 (1960).
55. E. M. Flanigen, H. Khatami, H. A. Szymanski, "Molecular Sieve Zeolites," *Advan. Chem. Ser.*, *101*, American Chemical Society, Washington, D. C., 1971, p. 201.
56. C. N. R. Rao, Chemical Applications of Infrared Spectroscopy, Academic Press, New York, 1963.
57. S. P. Zhdanov, A. V. Kiselev, V. I. Lygin, and T. D. Titova, *Russ. J. Phys. Chem.*, **38**: 1299 (1964).
58. K. Oinuma, and H. Hayashi, *J. Toyo Univ.*, **8**: 1 (1967).
59. C. L. Angell and P. C. Schaffer, *J. Phys. Chem.*, **69**: 3463 (1965).
60. L. Bertsch and H. W. Habgood, *J. Phys. Chem.*, **67**: 1621 (1963).
61. M. M. Dubinin, *Advan. Colloid Interface Sci.*, **2**: 2 (1965).
62. S. Brunauer, The Adsorption of Gases and Vapors, Princeton Univ. Press, Princeton, N. J., 1945, p. 68-82.
63. L. Gurvitsch, *J. Phys. Chem. Soc., Russ.*, **47**: 805 (1915).
64. D. W. Breck, W. G. Eversole, R. M. Milton, T. B. Reed and T. L. Thomas, *J. Amer. Chem. Soc.*, **78**: 5963 (1956).
65. T. B. Reed and D. W. Breck, *J. Amer. Chem. Soc.*, **78**: 5972 (1956).
66. K. F. Loughlin and D. M. Ruthven, *J. Phys. Chem. Solids*, **32**: 2451 (1971).

67. C. N. Satterfield and A. J. Frabetti, Jr., *Amer. Inst. Chem. Eng.*, **13**: 731 (1967).
68. M. M. Dubinin, E. G. Zhukovskaya, and K. O. Murdma, *Izv. Akad. Nauk. SSR, Ser. Khim.*, 760 (1962).
69. R. M. Barrer and R. M.Gibbons, *Trans. Faraday Soc.*, **59**: 2569 (1963).
70. M. M. Dubinin, E. G. Zhukovskaya and K. O. Murdma, *Izv. Akad. Nauk SSR, Ser. Khim*, 220 (1966).
71. M. M. Dubinin, *Izv. Akad. Nauk SSR, Ser. Khim.*, 209 (1964).
72. M. M. Dubinin. S. P. Zhdanov, E. G. Zhukovskaya, K. O. Murdma, E. F. Polstyanov, I. E. Sakavov, and N. A. Shishakov, *Izv. Akad. Nauk SSR, Ser. Khim.*, 1565 (1964).
73. R. M. Barrer and H. Villiger, *Z. Kristallogr.*, **128**: 352 (1969).
74. R. M. Barrer and D. L. Peterson, *Proc. Roy. Soc.*, **280A**: 466 (1964).
75. M. M. Dubinin, E. G. Zhukovskaya, V. M. Lukyanovich, K. O. Murdma, E. F. Polystyanov and E. E. Senderov, *Izv. Akad. Nauk SSR, Ser. Khim.*, 1500, (1965).
76. International Critical Tables, III, pp. 20-21.
77. J. G. Daunt and C. Z. Rosen, *J. Low Temp. Phys.*, **3**: 89 (1970).
78. P. Monod and J. A. Cowen, "Experimental Study of He^3 Adsorbed at Low Temperatures," No. 571, Service de Physique du Solide et de Resonance Magnetique GIF S/Yuette France.
79. M. M. Dubinin, Chemistry and Physics of Carbon
79. M. M. Dubinin, (P. L. Walker, Jr., ed.) Chemistry and Physics of Carbon **2**, Arnold, London, 1966, p. 51; *Chem. Rev.*, **60**: 235; *J. Colloid Interface Sci.*, **21**: 378 (1966).
80. M. M. Dubinin, *J. Colloid Interface Sci.*, **23**: 487 (1967).
81. A. Cointot, J. Cruchaudet, and M. Simonot-Grange, *Bull. Soc. Chim.*, 497 (1970).
82. O. Kadlec and V. Danes, *Coll. Czech. Chem. Commun.*, **32**: 693 (1967).
83 R. M. Barrer and J. W. Sutherland, *Proc. Roy. Soc.*, **237A**: 439 (1956).
84. Timmermans, Physicochemical Constants of Pure Organic Compounds, Elsevier, 1950.
85. R. M. Barrer, F. W. Bultitude, and J. W. Sutherland, *Trans. Faraday Soc.*, **53**: 1111 (1957).
86. R. M. Barrer and J. A. Lee, *Surface Science*, **12**: 341 (1968).
87. E. Rabinovitch and W. C. Wood, *Trans. Faraday, Soc.*, **32**: 947 (1936).

Chapter Six

CHEMICAL PROPERTIES AND REACTIONS OF ZEOLITES

The earliest recognized zeolite transformation was the reversible removal of water when zeolite minerals were subjected to the classical blowpipe analysis. Later, the novel removal of cations from aqueous solutions in contact with zeolites was observed. This observation is the basis for modern ion exchange technology, including important aspects of soil chemistry, plant nutrition, and the whole field of clay mineralogy.

In recent years, new zeolite transformation reactions have been observed and studied by modern techniques. Transformation reactions are the basis for the widespread applications of crystalline zeolites as heterogeneous catalysts, adsorbents, and ion exchangers. (Ion exchange is discussed as a separate subject in Chap. 7.)

Zeolite chemical reactions may be classified in the following categories:

1. Reactions which involve
    a. water as a volatile reacting phase—dehydration and hydrolysis.
    b. volatile phases other than water.
    c. ionic species in solution.
2. Recrystallization reactions after dehydration
3. Formation of structural defects by
    a. decationization and dehydroxylation.
    b. dealumination, that is, the removal of framework aluminum atoms.
    c. hydrothermal stabilization.
    d. metal cation reduction.

These categories of zeolite transformation reactions are discussed in detail in this chapter.

A. DEHYDRATION OF ZEOLITES

Structural Changes

Most zeolites may be dehydrated to some degree without major alteration of their crystal structure; they may subsequently be rehydrated, that is, adsorb water from the vapor or liquid phase. Many zeolites, when *completely* dehydrated, undergo irreversible structural changes and suffer total structural collapse. The early investigation of zeolite minerals was concerned with dehydration behavior and its effect on optical characteristics. Theories concerning the nature of the water in zeolites from a comparison of hydrated zeolites with other crystalline hydrates were postulated.

Early studies of the dehydration of zeolites were based upon observations of the loss in weight with increasing temperature (1). Interpretations were based on a determination of the dehydration isobars and the ability of the dehydrated zeolite to completely rehydrate or adsorb other gases and vapors. Current interpretations of the nature of water in zeolites are based upon x-ray crystal structure analysis, thermal analysis, infrared spectra, nuclear magnetic resonance measurements, and dielectric measurements (see Chap. 5) (2, 3, 4).

In zeolites that undergo dehydration reversibly and continuously, there is no substantial change in the topology of the framework structure. Exchangeable cations that are located in the channels coordinated with water molecules may (in chabazite, faujasite, zeolite X, for example) migrate to different sites located on the channel walls or other positions of coordination.

In zeolites which have several cation sites, such as zeolite X with at least five cation sites, the effect of dehydration or partial dehydration may be pronounced. The water molecules are present in clusters which seem to be joined into a continuous intracrystalline phase. The zeolite is referred to as a *nonstoichiometric hydrate* because the water is present as a guest molecule in the host structure (2). In other zeolites such as analcime, the natrolite-type zeolites, and laminar zeolites such as stilbite, the water molecules are tightly coordinated to the exchangeable cations and the framework oxygen as well as to other water molecules. In some zeolites complete dehydration results in irreversible changes in the framework topology due to weaker linkages in the framework structure in certain directions; exchangeable cations initially located

DEHYDRATION OF ZEOLITES   443

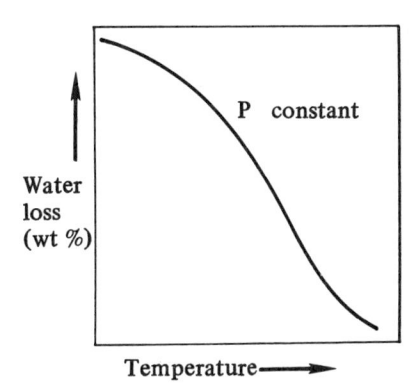

Figure 6.1 Schematic dehydration isobar showing continuous loss of water with increasing temperature. This type of isobar is typical of a stable zeolite structure.

in the channels then may become trapped (see Chap. 2). A dehydration isobar is illustrated schematically in Fig. 6.1.

Generally the thermal analysis curve is similar to Fig. 6.2. The low temperature endotherm represents the loss of water while the higher temperature exotherm represents conversion of the zeolite to another amorphous or crystalline phase.

When the dehydration process of the zeolite is only partially reversible, that is, when the zeolite structure is altered topologically, the dehydration curve may exhibit steps as shown in Fig. 6.3. The differential thermal analysis curves corresponding to these dehydration curves are rather complex (Fig. 6.4) (5, 6, 7).

Early results of dehydration studies on zeolites are inconsistent, largely due to variations in the methods for determining the dehydration curves as well as variations in mineral samples that were being studied. Differential thermal analysis determines the temperatures at which thermal reactions take place within the sample. The degree and nature of these reactions, that is, whether exothermic or endothermic changes are involved, is also shown (8, 9).

Thermogravimetric analysis measures the *loss in weight of the substance* as it is heated to elevated temperatures. The method used in

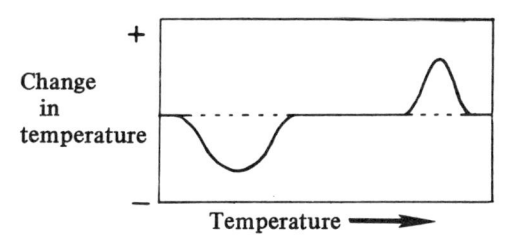

Figure 6.2 Schematic DTA curve for a zeolite showing low temperature endotherm caused by water desorption and high temperature exotherm due to loss of structure.

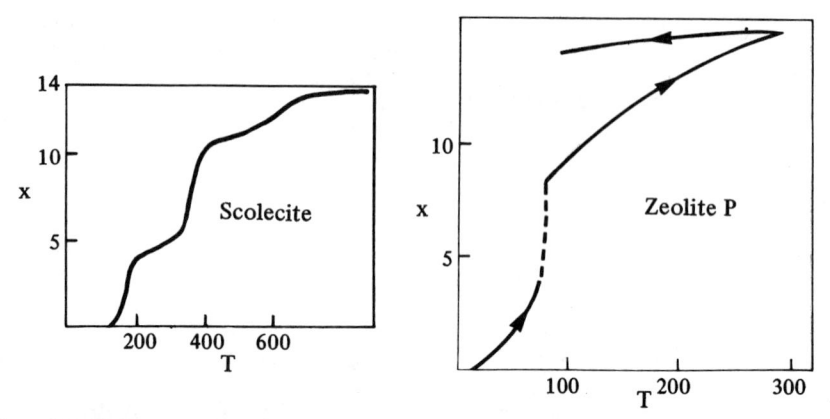

444 CHEMICAL PROPERTIES AND REACTIONS OF ZEOLITES

Figure 6.3 Dehydration curves for scolecite and zeolite P. The loss of water in weight percent is represented by x (5,6,7).

determining the dehydration curve is important; in the isothermal method the specimen may be heated at a given temperature until no further loss in weight occurs and then heated to successively higher temperatures and held until the weight is again constant (5). This process is repeated until no further loss in weight at the desired maximum temperature results. Another method (dynamic) is to heat the specimen continually at a constant rate while recording the loss in weight. In both methods the concentration of water vapor in contact with the specimen is important. Variation in the concentration of water vapor in contact with the sample may greatly affect the result.

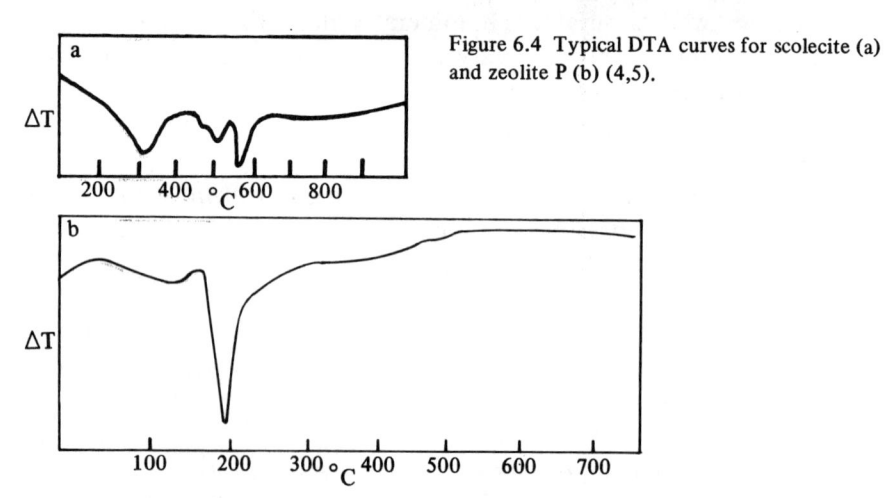

Figure 6.4 Typical DTA curves for scolecite (a) and zeolite P (b) (4,5).

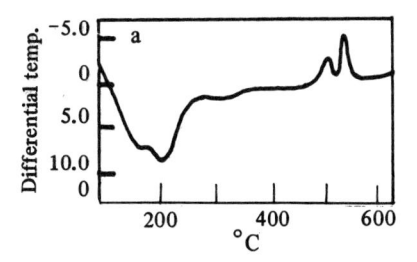

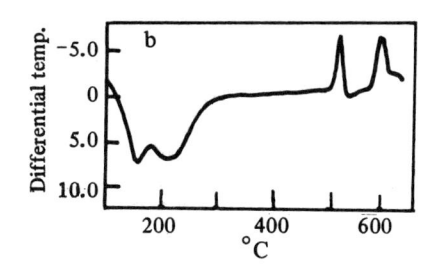

Figure 6.5 DTA curves for (a) zeolite A and (b) calcium exchanged A (4).

Detailed discussions of the theory and techniques of thermal analysis have been published (8, 9). Application of the method to zeolites has been thoroughly described (4). Some of the factors which influence the nature of the differential thermal analysis (DTA) curves include particle size of the specimen, heating rate, and the atmosphere surrounding the specimen. However, these factors cannot be discussed here.

Based on dehydration behavior, zeolites may be classified as (a) those which upon dehydration do not show major structural changes, and

Figure 6.6 DTA curves for a typical zeolite X and zeolite Y. The dehydration endotherm peak for zeolite X is about 40° higher than in zeolite Y.

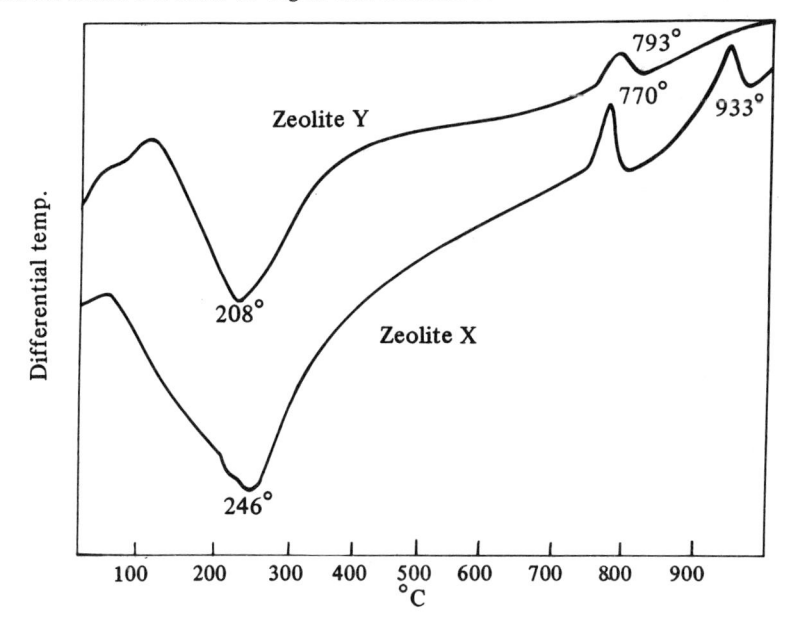

446 CHEMICAL PROPERTIES AND REACTIONS OF ZEOLITES

which exhibit continuous dehydration curves as a function of temperature; and (b) those zeolites which undergo structural change with dehydration and which show steps or other discontinuities in the dehy-

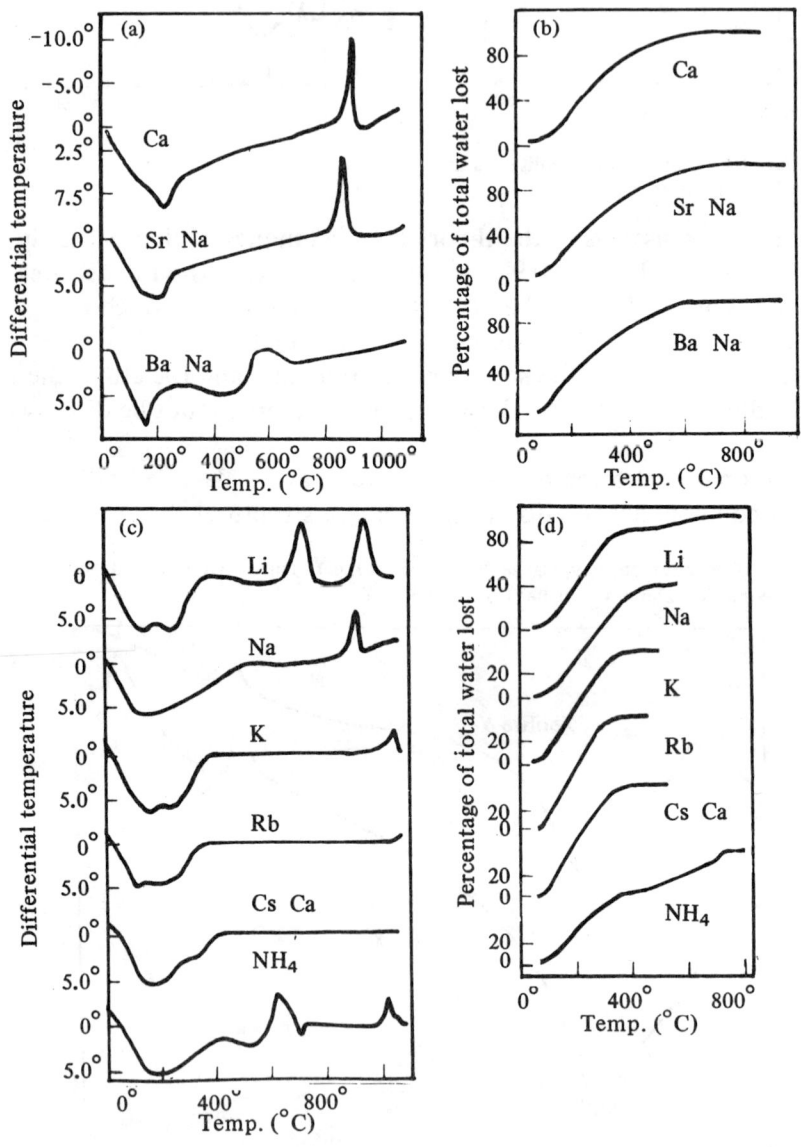

Figure 6.7 (a) DTA and (b) thermogravimetric curves for exchanged chabazites (the cation indicated) (4).

dration curves.

The first type of dehydration and thermal analysis curve is characteristic of zeolite A (Fig. 6.5), zeolite X and Y, (Fig. 6.6), and chabazite (Fig. 6.7), and the second type by the curves for the zeolites natrolite, scolecite, and mesolite (Fig. 6.8). In the latter zeolites, water molecules are arranged in groups within the structure and have different volatilities. The dehydration curves are obviously different from that of chabazite and other rigid, three-dimensional zeolites for which the dehydration curves are continuous and smooth.

Differential thermal analysis curves are useful in characterizing zeolites. This is illustrated by Fig. 6.9 which reproduces the DTA patterns for the structurally similar zeolites heulandite and clinoptilolite (10). Although these two zeolites are sometimes considered to have closely related structures, the DTA patterns indicate that clinoptilolite is much more stable toward dehydration. This is supported by gas adsorption studies. Aside from the broad endotherm which is characteristic of the dehydration of zeolites at the lower temperatures, clinoptilolite exhibits no thermal reactions up to a temperature of about 1000°C and no major structural change upon dehydration. In contrast, heulandite has a sharp endotherm at approximately 300°C. X-ray powder diffraction studies support these conclusions; heulandite does undergo a structural change when heated to about 230°C. The DTA method is a dynamic method; consequently the endotherm which registers on the recorded pattern occurs at a temperature of about 300°C although the beginning of the effect is somewhat below 300°C.

A summary of the dehydration behavior of zeolites as obtained from the literature is given in Table 6.1. In Table 6.1 the dehydration behavior is organized in terms of the structural groups as based upon the structural classification discussed in Chap. 2. Although a detailed treatment of all of the zeolites listed in Table 6.1 will not be undertaken, some discussion of the dehydration behavior of several of these zeolites is necessary to relate to other properties.

Group 1 Zeolites

Analcime is a structurally stable zeolite and may be completely dehydrated. After dehydration, analcime adsorbs small polar molecules. Analcime contains up to 9.1% water and is reportedly unchanged in structure at 700°C (1, 11, 12).

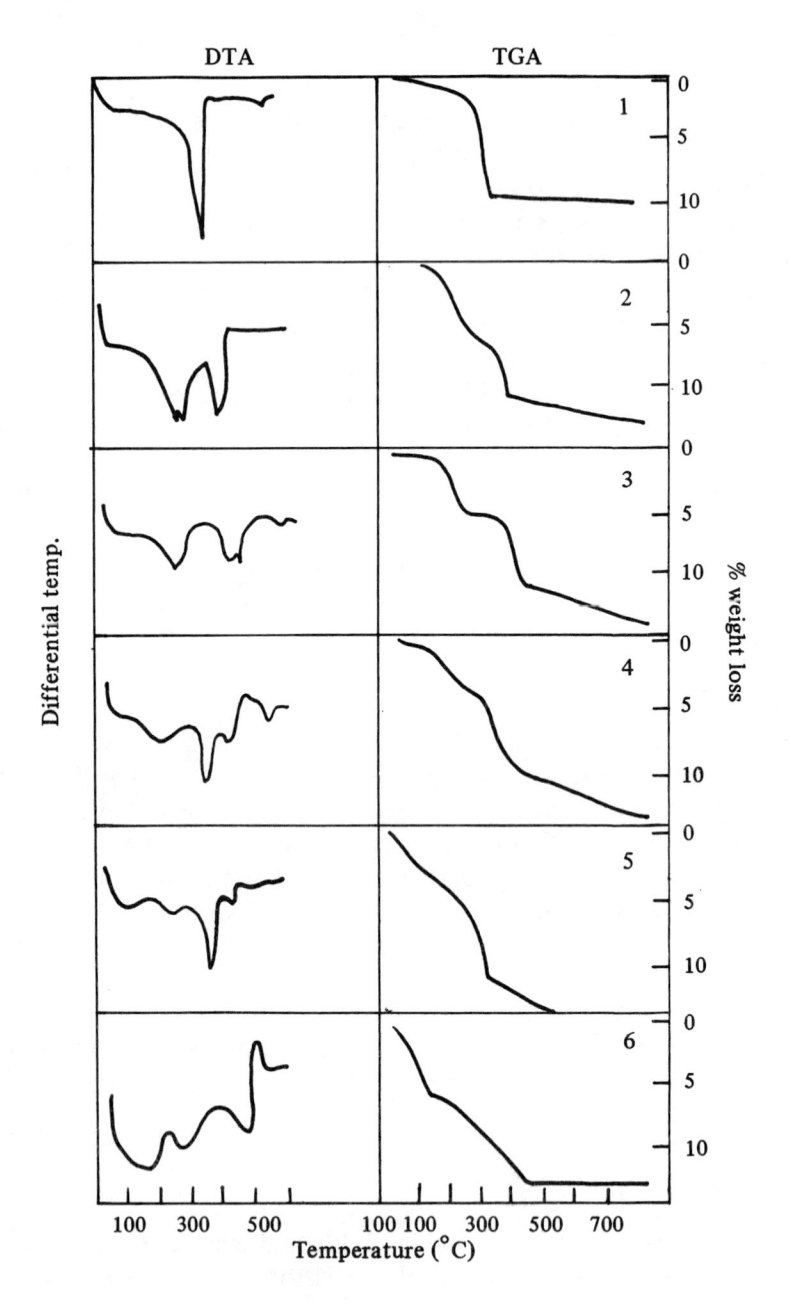

Figure 6.8 DTA and TGA curves for zeolites of the natrolite group. (1) natrolite; (2) mesolite; (3) scolecite; (4) thomsonite; (5) gonnardite; (6) edingtonite (31).

Table 6.1 Summary of Zeolite Dehydration Behavior

Zeolite	TGA [a,b]	DTA [c,d]	Structure	Remarks	Ref.
Group 1					
Analcime	Cont., 8.7% at 400°	End. 200-400°	None up to 700°		1, 11, 12
Harmotome	Step, 170-190°; 15.0% at 300°		New structure at 250° forms celsian 700°		13, 14
Phillipsite	Step, 130°, 18.0% at 300°	End. 100°, 200°, 300°	New structure 160-200°	Rehydrate if not heated > 250°	14, 15
Gismondine	Step, 115-140°; 18.2%	End. 140°, 160°, 190°, 270°	Five metastable phases; feldspar 375°		14, 16
P	Step, 83°; 16.9%	End. 190°	New structure 530°	Structure contracts at 165° by 15.9 vol %	6, 7
Paulingite	— 28%	—	Stable to 250°	Crystals decrepitate	32
Laumontite	Step, 200, 370, 500°; 15%	End. 71, 267, 430, 467°	Structure change at 500°	Rehydrates below 200°	12
Yugawaralite	Step, 200, 400°; 14%	End. 200, 350°; Ex. 540°	New structure 400°	Rehydrates below 400°	39, 40, 41
Group 2					
Erionite	Cont., 14.8%	End. 50-400°; Ex. 920°	Stable 750°	Stable to H_2O at 375°	15, 17
Levynite	Cont., 17.6%			Stable at 340°	42
O (Offretite)	Cont., 9.8%	End. 160, 410, 560°; Ex. 1092°	No change 965°	TMA removed above 350°	18, 19, 20
T	Cont., 9.8%				48, 43
Omega	Cont. to 485°, Step. 485° 22.8%	End. 180, 255° Ex. 410, 510, 675°	Stable to 700°	TMA removed above 400°	18, 44
Sodalite Hydrate	15%	End. 100-300°; Ex. 805°	Stable to 900°	High temp. form reported	26, 45, 46, 47
Group 3					
A	Cont., 22.5%	End. 25-300°; Ex. 860, 910°	Stable to 700°	β-Cristobalite 800°	4, 21, 22
N-A	Cont., 15.0%	End. 100-200°; Ex. 350, 600, 860°	Stable to 700°	TMA removed above 300°	23
ZK-4	25.8%		Stable to 550°		48

450 CHEMICAL PROPERTIES AND REACTIONS OF ZEOLITES

Table 6.1 Summary of Zeolite Dehydration Behavior *(continued)*

Zeolite	TGA [a,b]	DTA [c,d]	Structure	Remarks	Ref.
Group 4					
Faujasite					25
X	Cont., 25%	End. 50-350°; Ex 772, 933°	Stable to 475°	Stability varies with cation	2, 25, 26
Y	Cont., 26.2%	End. 100-400°; Ex. 793°	No change 700°	Stability varies with cation	2, 25, 26
Chabazite	Cont., 26%	End. 25-300°; Ex. 900°	No change 760°	Stability varies with cation	1, 4, 12, 14, 28, 29
Gmelinite	Cont., 23%		No change 700°		14, 28, 29
Zeolite L	Cont., 20%		Structure change at > 300°		25
	Cont., 16.7%		No change 800°		
Group 5					
Natrolite	Step., 240°, 9.7%	End. 350°	New structure 565°; amorphous 785°	Rehydrates to 785°; nepheline at 970-1010°	1, 12, 5, 49, 31
Scolecite	Step., 200, 400°, 14%	End. 225, 410°; Ex. 1000°	Structure decomposes at 490°	No rehydration above 330°	1, 5, 31, 50
Mesolite	Step., 200, 350°. 13%	End. 225, 380°, Ex. 1040°	Structure decomposes at 440-490°	Feldspar at 910°	1, 5, 51
Thomsonite	Step., 150, 300°, 15%	End. 175, 325, 400, 440, 520°	Structure collapse at 520°	No rehydration above 370°	1, 12, 31, 52
Gonnardite	Step., 100, 200, 300°, 14.4%	End. 75; 220, 450°	Structure collapse at 460°	No rehydration above 360°	31, 52
Edingtonite	Step., 100, 250, 400°, 13.1%	End. 160, 270, 450°; Ex. 500°	Recrystallizes to feldspar 500°		31, 53
Group 6					
Mordenite	Cont., 16%	End. 25-300°; Ex.> 1000°	No change 800°		12
Epistilbite	— 15.5%			Stable to at least 250°	32
Bikitaite	Step., 9.8%	End. 190-475°	Goes to β-spodumene at 750°		32, 33
Zeolon-Na	Cont.	End. 25-300°; Ex. > 1000°			

Table 6.1 Summary of Zeolite Dehydration Behavior (continued)

Zeolite	TGA [a,b]	DTA [c,d]	Structure	Remarks	Ref.
Group 7					
Heulandite	Step., 100, 250°, 17%	End. 25-300°; 300°	Heulandite "B" at 250°	Structure collapse at > 360°; cation effects stability	1,11,10, 12, 34, 35, 54
Clinoptilolite	Cont., 14%	End. 125-300°; Ex. > 1000°	No change 750°		10,35,55
Stilbite	Step., 100, 200°, 17%	End. 191, 262°; Ex. 500°	Change at 120°; collapse at 400°	Structure shrinks along b	1, 12, 34,37, 56,57
Stellerite	17%		No change 300°	Rehydrates	32
Brewsterite	12.4%		No change 300°	Rehydrates	32
Group 8					
Li-A	21.2%		Stable 360°		58
F	13.4%		Stable 500°		59
Z	12.5%		Stable > 300°	Forms kaliophilite 600°	60
H	18%		Unstable 100°	Decomposes	61,62
Li-H	13.4%		Stable 350°		58
J	7.7%		Stable 350°	Structure change 430°	63
E	17.4%		Unstable 150°	Some divalent forms stable at 200°	64
M	9.1%		Stable 350°		65
Q	15.5%		Unstable 200°		66
W	18%		Stable 350°		67

[a] Cont. = continuous loss with increasing temperature
[b] Weight percent loss based on original weight
[c] End. = endothermic reaction
[d] Ex. = exothermic reaction

452 CHEMICAL PROPERTIES AND REACTIONS OF ZEOLITES

Figure 6.9 DTA curves for (1) heulandite, Prospect Park, N.J. and (2) clinoptilolite, Hector, California (10).

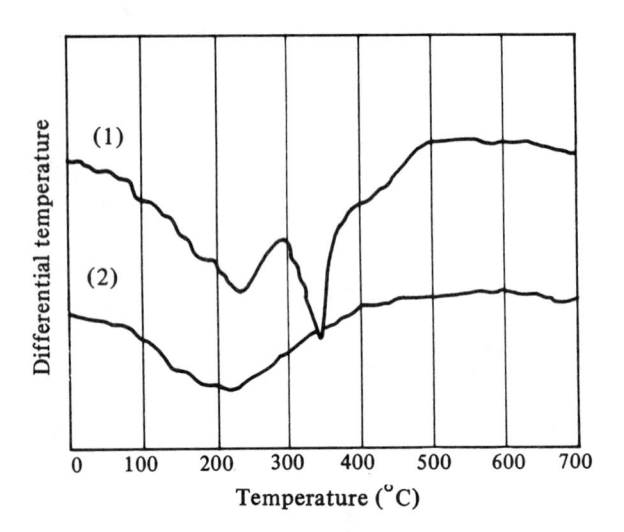

Harmotome exhibits a dehydration curve which has a small step between 170 and 190°C. Parallel high temperature x-ray studies show that the dehydration is accompanied by a slight contraction of the unit cell up to about 130°C. At 150°C a new structural phase is formed which is referred to as meta-harmotome. This meta phase persists to 500°C and converts to a feldspar structure, celsian, at 700°C (13, 14).

Phillipsite shows a similar behavior and forms meta-phillipsite at 200°C. If not heated above 250°C the meta-phillipsite rehydrates reversibly. This structure is unstable above 300°C (14, 15).

Gismondine exhibits two steps in the dehydration curves between 115 and 140°C. Structural changes occur which are similar to those observed for the related zeolite phillipsite (14). When heated in air up to 250°C, gismondine dehydrates reversibly, although the rehydration steps show hysteresis. Five metastable phases were identified. At 330°C, a calcium feldspar forms. About 80% of the total water content can be thermally dehydrated reversibly(16).

Complex structural changes occur during the dehydration of the cubic zeolite P and the tetragonal counterpart (6, 7). The dehydration curve for zeolite P_c exhibits a pronounced discontinuity at 83°C (Fig. 6.4). This discontinuity is due to dehydration of the cubic zeolite (a_o = 10.0 A) resulting in a partially collapsed structure which has a tetragonal unit cell (a = 9.6 A, c = 9.1 A) with a concurrent volume decrease of 16%. A loss of three water molecules per unit cell (out of a total of 15) occurs below the discontinuity and an additional three molecules at the discontinuity. The tetragonal form exhibits a larg step at 80°C, a small step at 112°C, and weak inflections at 60°C and 165°C. A much more complex behavior is indicated.

The reversible transition of the cubic zeolite P to the tetragonal form at 83°C is considered to be an enantiotropic phase transition which results from a displacive rearrangement of the anionic framework. This transition is reversible; the volume change is small and does not produce a loss of water. The two structures of different symmetry have identical chemical compositions.

The potassium exchanged form of tetragonal zeolite P dehydrates continuously to 365°C and returns to its original hydration state at room temperature. X-ray examination shows that a tetragonal structure similar to the room temperature structure exists at temperatures as high as 700°C (6).

Two general types of dehydration reactions are identified in the P zeolites: (1) first-order type which involves the sudden collapse of the zeolite structure in one or several directions when water is released. A displacive movement of the tetrahedral framework results in a reduction in the volume of the unit cell. This type of reaction occurs when the water content is reduced to a fixed number of water molecules per unit cell. However, many of these collapse reactions are reversible. (2) The second-order type of dehydration or collapse reaction involves the continuous loss of water from the zeolite over a short temperature range to form a contracted structure. Some zeolites undergo both types of reaction at the same time.

The zeolite paulingite shows unusual behavior upon dehydration in that the crystals decrepitate after dehydration and rehydration. Although this zeolite may be dehydrated in vacuum at 250°C, upon rehydration the crystals undergo violent decrepitation.

Fully hydrated laumontite, when exposed to the atmosphere, loses approximately 1/8 of its total water to form a phase that has been commonly referred to as leonhardite. Laumontite dehydrates in three steps; between room temperature and 150°C, between 150°C and 300°C, and between 300 and 700°C. The DTA curve correspondingly shows three endothermic peaks. There appears to be little change in crystal structure in the lower two steps, but at 500°C the third step of dehydration produces a marked structural change as observed by x-ray diffraction studies (12).

Group 2 Zeolites

In group 2, erionite undergoes continuous dehydration as exhibited by a broad endotherm in the DTA curve between 50 and 400°C. Up to 14.8 wt% of water is lost continuously up to about 400°C. High-temperature x-ray diffraction studies show that this structure is thermally stable to about 750°C for at least 24 hours. Decomposition occurs at about 950° with the formation of an amorphous phase (15, 17).

Zeolite T (related to the minerals erionite and offretite) is stable toward dehydration and may undergo continuous dehydration and reversible adsorption of water. Its upper limit of thermal stability appears to be about 800°C.

Zeolite O, the synthetic tetramethylammonium (TMA) type of offretite, shows a DTA curve with several transitions. The first endotherm at 160°C corresponds to loss of the adsorbed water and is followed by two endotherms at 410°C and 560°C due to transformations resulting from the decomposition of the organic nitrogen cation. The slight change near 910°C is not due to any significant change in the zeolite structure according to x-ray studies. The high-temperature exotherm at 1092°C is caused by structural recrystallization. The product above this temperature is completely amorphous (18, 19, 20).

Group 3 Zeolites

Zeolite A exhibits a DTA pattern typical of the zeolites with stable three-dimensional structures showing continuous water desorption (Fig. 6.5). At 800°C it recrystallizes to a β-cristobalite-type structure which is probably a "stuffed" derivative of cristobalite (21). Calcium-exchanged zeolite A is very stable and withstands heating to a temperature of over 800°C. The adsorption of H_2O on zeolite A and calcium-exchanged A shows some slight hysteresis which may be related to diffusion of the H_2O molecules into the small β-cages of the zeolite A structure. Hysteresis of this type is pronounced in the adsorption of water by dehydrated sodalite hydrate (HS) (4,22).

The alkylammonium zeolite, N-A, dehydrates with a loss of about 15 wt% water up to a temperature of 300°C. Above this temperature, as evidenced by DTA and thermogravimetric analysis, additional loss in weight up to 28 wt% occurs due to the decomposition of the TMA cation. The mechanism for this decomposition is unknown, but the resulting dehydrated zeolite is stable up to 700°C, as evidenced by adsorption and x-ray diffraction studies. Loss of the alkylammonium cation results in a decationized zeolite (23). See Section D.

Group 4 Zeolites

Because of its scarcity, the mineral faujasite has not been studied. However, outgassing in high vacuum at 475°C produces no gross change in the crystal structure (24).

The differential thermal analysis (DTA) curve of zeolite X indicates a continuous loss of water over a broad range commencing from slightly above room temperature to about 350°C with a maximum at about 250°C (25). Exothermic peaks at 772 and 933°C indicate decomposition followed by recrystallization of the zeolite. After heating for 36 hr at 760°C, the zeolite becomes amorphous in structure and recrystallizes to a carnegeite-like phase at 800°C and to a nepheline-like phase at 1000°C (Fig. 6.6).

A series of cation exchanged forms of zeolite X have been studied by means of thermogravimetry (TGA), DTA, and water adsorption equilibrium measurements. For some cationic forms, a certain amount of hysteresis or irreversible adsorption has been observed, for example in the case of strontium exchanged zeolite X (2,26).

Zeolite Y with a Si/Al ratio of 2.5 has a broad endothermic peak with a maximum at 208°C and an exothermic peak at 790°C. After prolonged heating in air at 760°C no structural change occurs; at 800°C, the structure collapses to an amorphous residue (25, 26) (Fig. 6.6)

The reversible dehydration of chabazite was first observed by Damour in 1840. The dehydration of chabazite was found to be completely reversible up to a temperature of 600°C, but after 800°C it would not adsorb water. The adsorptive properties are not affected by relatively severe treatment such as dehydration at temperatures as high as 500°C. Water in chabazite does not occur as a definite hydrate but is held by adsorptive forces within the channels (1). In adsorption studies on partially dehydrated chabazite, there is no report of the displacement of the remaining water molecules by any adsorbed material except NH_3 (27).

The mineral chabazite and various cation-exchanged forms have been studied extensively in terms of their dehydration behavior and ability to rehydrate. Dehydration is reversible after heating to temperatures of 700°C. Stabilities of ion-exchanged forms of chabazite as studied by thermal analysis and x-ray diffraction have been reported 4, 14, 28). In the mineral with a Si/Al ratio of 2.5, the DTA and thermogravimetric curves (Fig. 6.7) confirm that the nature of the exchanged cation exerts an important influence on the stability and dehydration behavior of the zeolite. As the size of the univalent cation increases, the temperature at which the water is lost (relative ease of water loss) increases. The rubidium and cesium exchanged forms of chabazite are stable to temperatures above 1000°C. This is related to the filling of the void space in the zeolite structure by the cations after the removal of water. Analcime has been observed to behave in a similar fashion. A study of zeolite KG, which has the chabazite-type framework, was conducted in which the Si/Al ratio of the synthetic zeolite varied. The ultimate thermal stability of the structure increased with increasing Si/Al ratio for a given cation-exchanged form (4). The temperature of structural decomposition of the zeolite and recrystallization decreases as the number of tetrahedral aluminum atoms in the framework increases, that is, as the Si/Al ratio decreases. The cation density or the number of cations in the unit cell depends, of course, on the number of Al tetrahedral atoms in the framework and the stability appears to be related to the cation density and the framework charge.

Recrystallization products from the various forms of chabazite vary with the cation and include four types of silica derivatives that are probably the "stuffed" derivatives described by Buerger (165). These include eucryptite, nepheline, kaliophilite, and α-carnegeite. Barrer classified the chabazite zeolites and structural relatives into three types: (1) chabazite, the mineral, (2) synthetic potassium zeolite G with Si/Al ratio greater than 1.5 and (3) the synthetic potassium zeolite G with the Si/Al ratio less than 1.5. These categories are included in the designation KG as described in Chap. 2.

The mineral gmelinite (with a water content of about 21.5 wt%) has been successfully dehydrated by heating it to 350°C *in vacuo* for periods of up to 30 hours. No steps are found in the dehydration curve, however, the sodium and barium exchanged forms show a poorly defined step between 115 and 130°C (7). The structure remains intact up to at least 240°C where 98% of the water is removed. The anhydrous phase is stable to 700°C as evidenced by x-ray studies. If outgassed at 350°C in high vacuum, gmelinite is an excellent adsorbent for gases such as CH_4 (14, 28, 29).

Synthetic zeolite S which is structurally related to gmelinite is similarly stable to dehydration at 350°C *in vacuo* and is capable of adsorbing gases such as O_2 (30).

Zeolite L is structurally stable after dehydration at relatively high temperatures and exhibits no change in the crystal structure as evidenced by x-ray diffraction patterns after prolonged heating in air at 800°C (25).

Group 5 Zeolites

Natrolite, scolecite, and mesolite have the same type of framework topology, but exhibit different behavior upon dehydration (1, 5, 31). Thermal analyses of these zeolites (Fig. 6.8) show that natrolite dehydrates rapidly within a single temperature range, scolecite loses water in two steps, and mesolite in three steps. From this it is concluded and confirmed by x-ray diffraction studies that there is only one type of water molecule in natrolite, while the water in scolecite and mesolite occupies two and three different sites, respectively.

The natrolite structure becomes amorphous at 785°C, whereas scolecite collapses at 560° and mesolite at 490°C (1, 5). The characteristic thermal dehydration behavior of each zeolite is evidenced by the dehydration curves and differential thermal analyses. The arrangement of the cations and water molecules in natrolite has been determined by x-ray structural analysis. Because dehydration produces a change in the local charge balance, the Na^+ ions tend to move towards the framework oxygen atoms which previously were linked by hydrogen bonds to the water molecules. This results in a shrinkage of the zeolite framework structure. All of the group 5 zeolites show contraction of their structure upon dehydration along the a and b axis, and expansion along c. Edingtonite is the only member of the group which does not go to an amorphous phase on heating. It goes directly to another crystalline phase (32).

Group 6 Zeolites

In group 6, mordenite is probably the most stable zeolite. Dehydration is continuous and there is no evidence of any change in the structure at temperatures as high as 800°C (12). The stability is attributed to the prevalence of 5-rings in the mordenite framework structure (see Chap. 2). The relative energies of silicate rings are lowest for 5- and 6-rings.

The low charge density of the mordenite framework (due to the high Si/Al ratio) and the presence of "pockets" in the walls of the channels which provide suitable sites for cations upon dehydration, also contribute to the unusual stability. Although little is known concerning the thermal behavior of other zeolites in this group, it may be assumed that frameworks consisting of linked 5-rings should exhibit high stability. Synthetic types related to mordenite, i.e., Zeolon and zeolite Na-D have equally high stability upon dehydration.

Bikitaite contains 9.9 wt% water (32, 33). The dehydration curve shows that the first loss of water occurs at 160°C, and that final dehydration occurs at 360°C. There are distinct breaks at 180°C and 280°C in the curve. These correspond to ¼ and ¾, respectively, of the water molecules. The stepwise dehydration indicates three energy barriers. The crystal structure determination indicates specific sites for about ½ of the water molecules, but the remainder were not located. The DTA curve indicates a broad endotherm that begins at 190°C and ends at 475°C. Rehydration is reversible but occurs with difficulty.

Group 7

Heulandite and clinoptilolite have been assumed to be structurally related; however, their dehydration and stability behavior are entirely different (1, 10, 34, 35). The DTA curves (Fig. 6.9) are distinctively different. No transitions are observed in clinoptilolite up to about 750°C at which temperature the structure begins to collapse. Heulandite is far less stable and undergoes a structural change at about 230°C. The structure which results is referred to as heulandite "B". Heulandite has a sharp endotherm in the DTA pattern at about 300°C which is indicative of this change. Similarly, the thermogravimetric curve of heulandite shows a stepwise change in the region of 250°C where there is no break in the curve for clinoptilolite. The crystal structure of clinoptilolite is as yet undefined.

From an infrared spectral and x-ray study, water loss from heulandite was shown to be slow between 110°C and 200°C. At 200°C, sudden loss occurs. Lattice contraction was observed up to 180°C and the conversion to heulandite-B appeared sluggish and not complete until about 330°C. The conversion of heulandite to heulandite-B is facilitated by the initial loss of some water (to 5.25%). At 500°C, both heulandite and clinoptilolite retain about 20% of the initial tightly bound water (1 wt% of initial sample weight). For clinoptilolite no change in structure was observed up to 500°C. Infrared spectra of the two zeolites are alike between 2 - 14 μm (35).

The two minerals differ markedly in chemical composition (clinoptilolite is principally a sodium-potassium zeolite with a higher Si/Al ratio than heulandite, principally a calcium zeolite). Replacement of the alkali metal ions in clinoptilolite by exchange with calcium diminishes the thermal stability to some extent, i.e., calcium-exchanged clinoptilolite shows no transition to a B-type phase but loses its structure at 550°C. Exchange of calcium with potassium in heulandite is reported to stabilize the structure of heulandite to higher temperatures. Sodium exchange did not affect the thermal behavior of heulandite. Identical behavior could not be achieved by cation exchange in these two zeolites. The DTA analysis of calcium exchanged clinoptilolite produced a pattern insignificantly different from that of the original alkali metal zeolite and no evidence of transformation to a heulandite-B type phase at low temperatures was indicated (36).

A similar difference in thermal behavior exists between stilbite and stellerite. Dehydration of stilbite to 350°C results in the loss of most of the water with a concurrent shrinkage of the structure in the b direction, i.e., in a direction perpendicular to the layers of 4-4-1 units (see Chap. 2) (34, 37, 38). It was earlier noted that a step occurs in the dehydration isobar for stilbite (1) at about 120°C and an irreversible collapse of structure as evidenced by x-ray diffraction occurs above 200°C with complete breakdown of the structure at about 400°C. After dehydration at 350°C, stilbite may be completely rehydrated. If partially outgassed it readily adsorbs gases such as argon and krypton (34). The DTA curve shows two sharp endothermic peaks. After dehydration in vacuum at 300°C, stilbite readsorbs water but the structure is entirely different from the original.

Group 8 Zeolites

Zeolites of group 8 have unknown structures. Some information on the dehydration behavior of these zeolites is given in Table 6.1.

B. CATION HYDROLYSIS AND STRUCTURAL HYDROXYL GROUPS

The formal description of zeolite crystal structures does not include structural hydroxyl groups as they are found in other aluminosilicates (Chap. 2) such as the clay minerals (e.g., kaolin). The intracrystalline surface of zeolites is not as well defined and orderly as was once believed. Under many conditions, the surface contains defect sites which are important in adsorption and catalysis. The surface may contain hydroxyl groups, replacing the usual metal cations. The presence of these structural hydroxyl groups has been recognized for several years and their nature studied by various means including infrared spectroscopy, thermal analysis, and nmr.

Surface Hydroxyl Groups

Hydroxyl groups are necessary to terminate the faces of a zeolite crystal at positions where bonding would normally occur with adjacent tetrahedral aluminum or silicon ions within the crystal. The number of terminal hydroxyl groups has been estimated to vary as a function of crystal size from 0.15×10^{20} per gram for a 1 micron particle, to 1.4×10^{20} per gram for a 0.1 micron crystal (68, 69). The presence of terminal hydroxyl groups on zeolite crystal surfaces has been confirmed by measurement of the surface electrical conductivity of chabazite crystals (Chap. 5, Section E) and infrared absorption spectra. The hydrogen content of sodium and calcium exchanged forms of zeolite X and Y was measured by deuterium exchange and was found to agree very closely with that calculated from the theoretical number of terminal hydroxyl groups (69).

Structural Hydroxyl Groups

In the hydroxyl-stretching region, three infrared bands attributed to structural OH groups are observed in zeolites of the X and Y type. One, at a frequency of about 3740 cm^{-1} is observed in all cases after dehydration and is attributed to OH groups on the surface, or is possibly related to an amorphous silica impurity. Two others at 3650 cm^{-1} and 3540 cm^{-1} are attributed to protons bonded to framework oxygen atoms, such as O(1) and O(3) (160). They may arise from cation hydrolysis or deammoniation as discussed next. These OH groups are referred to as the high frequency (HF), O(1)H at 3650 cm^{-1}, and the low frequency (LF), O(3)H, at 3540 cm^{-1}. O(1) and O(3) refer to the framework oxygen positions (See Chap. 2).

There is general agreement that, for electrostatic reasons, the protons would preferably bond to an oxygen between two aluminum atoms. Since this situation does not exist, the proton is most likely bonded to a framework oxygen between a silicon and an aluminum, rather than two adjacent silicon atoms (70).

Hydroxyl Groups in Univalent Cation Zeolites

The univalent cation forms of zeolites X and Y do not contain the HF and LF structural hydroxyl groups as do the di- and trivalent ion-exchanged forms. Due to hydrolysis, however, it has been observed that univalent cation forms of some zeolites such as NaX do not contain a total exchange cation equivalency as based on chemical analysis. A deficiency in the metal cation balance has been indicated and confirmed by several investigators. This has been attributed to partial hydrolysis of the cation and replacement by hydronium, H_3O^+ (71, 72). These cation deficient forms may exhibit properties associated with hydroxyl groups. (See Scheme 1.) The pH of most freshly prepared zeolites in water suspension is 9 - 12.

Scheme 1 **Hydrolysis of Univalent Ions.** Limited hydrolysis of cations may result in some cation deficiency and replacement by hydroxyl groups:

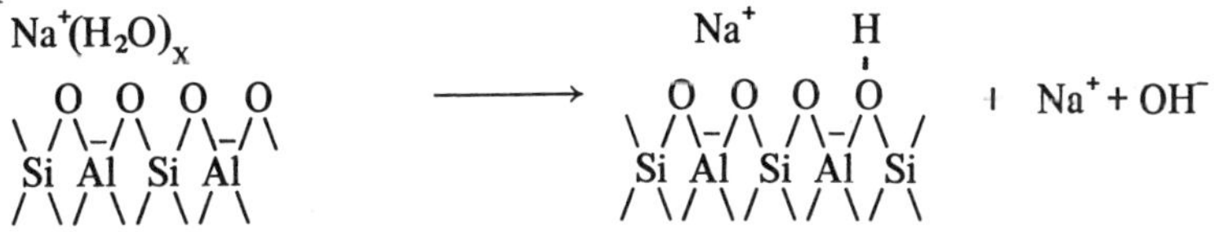

$Na^+(H_2O)_x$

O O O O
\ /\-/\ /\-/\
Si Al Si Al
/ \/ \/ \/ \

⟶

Na⁺   H
      |
O O O O
\ /\-/\ /\-/\ /
Si Al Si Al Si
/ \/ \/ \/ \/ \

+ $Na^+ + OH^-$

Exchangeable cations in some silica-rich zeolites may be replaced by hydrogen or hydronium ions by using a strong acid (see Chap. 7). However, treatment of a zeolite with a strong acid may result in direct attack on the aluminosilicate framework and dealumination. Zeolites with a Si/Al ratio of one are more subject to mild hydrolysis and instability as a result of hydrolysis than the more siliceous zeolites. In general hydrolysis of the exchangeable cation appears to be more likely in the alumina-rich zeolites.

Divalent Cation Hydrolysis

There is general agreement that the formation of hydroxyl groups in the multivalent ion exchanged zeolites is due to hydrolysis of the cation and dissociation of the water molecule by the electrostatic field created by the cation (Scheme 2) (72, 73, 74, 75) (Fig. 6.10). As water is removed, the multivalent cation cannot satisfy the framework change distribution and the cation associated electrostatic field causes dissociation of the coordinated water molecules (76).

Scheme 2 **Divalent Ion Hydrolysis.** Dehydration causes dissociation of H_2O molecules by the electrostatic field of associated cation:

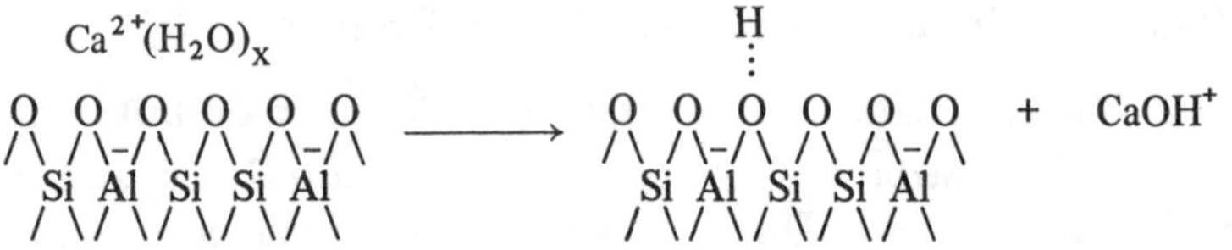

$$Ca^{2+}(H_2O)_x$$

```
                                                         H
                                                         ⋮
  O  O  O  O  O                          O  O  O  O  O            +  CaOH⁺
 /\ /\-/\ /\ /\-/\           ⟶          /\ /\-/\ /\ /\-/\
   Si Al Si Si Al                         Si Al Si Si Al
  /\ /\ /\ /\ /\                         /\ /\ /\ /\ /\
```

The number of hydroxyl groups is related to the polarizing power of the cation (72, 76). Two types of hydroxyl groups in the multivalent cation forms have been found with ir absorption frequencies in the OH stretching region at 3640 and 3530 cm^{-1}. These are attributed to Si-OH groups; another frequency near 3600 cm^{-1} is attributed to the cation-OH interaction. The HF (3640 cm^{-1}) and LF (3540 cm^{-1}) infrared frequencies are also observed in zeolites formed from the deammoniation of an ammonium exchanged form. Because of the importance of the hydroxyl groups in hydrocarbon catalysis and in adsorption, many studies have been carried out to determine the character of the hydroxyl groups and their role in catalytic reactions. Infrared studies have explored the interaction of the hydroxyl groups with adsorbed molecules.

A typical infrared pattern of a divalent cation form of zeolite Y is shown in Fig. 6.11 (77). The band at 3747 cm^{-1} is found characteristically in all of the zeolites studied. With the exception of barium, the hydroxyl groups in divalent cation forms of zeolite X are different from the univalent forms. Barium, however, behaves like a univalent ion and does not produce a strong HF band (76).

Dehydroxylation of Divalent Cation Forms

During dehydration, hydroxyl groups are produced in divalent ion exchanged zeolite Y, as shown in Scheme 2 and Fig. 6.10. At higher temperatures, (above 500°C) dehydroxylation occurs and additional water is formed. The dehydroxylation reaction may involve a framework hydroxyl and a proton of the hydroxycations as shown in Scheme 3 and Fig. 6.10. This results in an oxygen deficient site between a

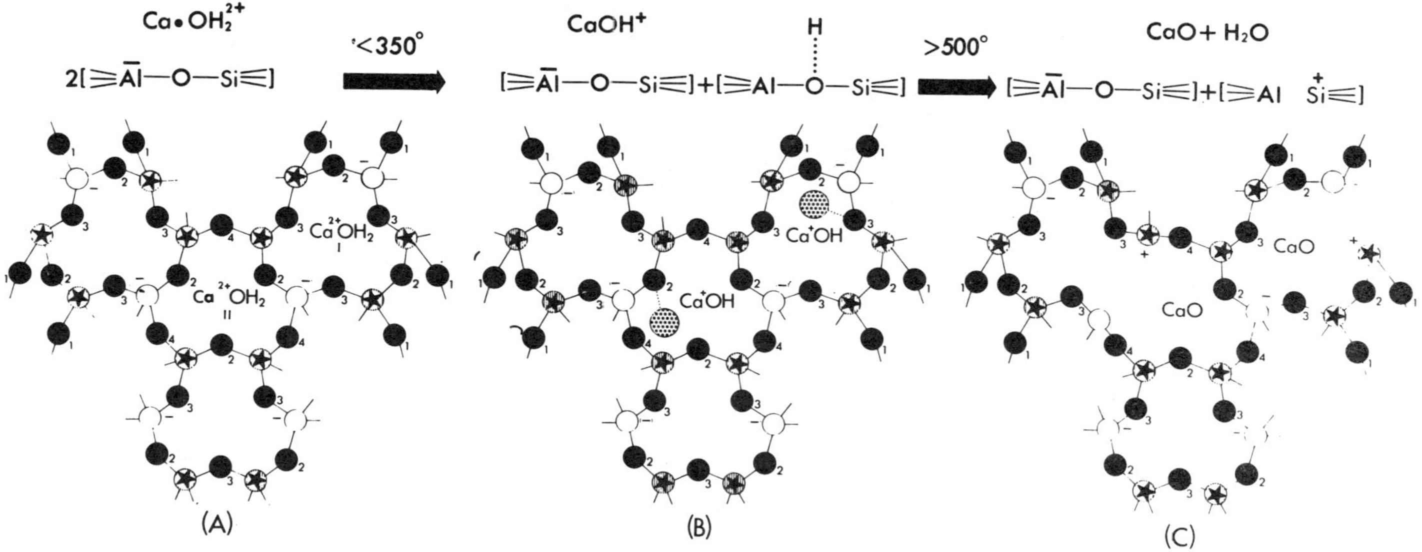

Figure 6.10 Schematic, two-dimensional representations of the formation of structural OH groups in calcium exchanged zeolite Y. A β-cage of the structure is shown folded flat with cation sites shown at site I and site II. Sites I' and II' cannot be differentiated from sites I and II in this drawing. The hydrated calcium ions in A hydrolyse in B to CaOH$^+$ ions with formation of OH groups located at O(2) and O(3). Above 500°C, the dehydroxylation results in formation of CaO interstitially and defect sites, C.

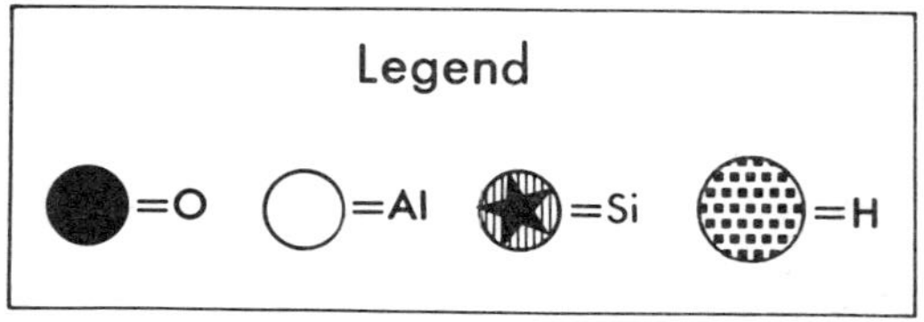

Figure 6.11 Infrared spectra, OH region, (a) calcium exchanged Y and (b) cerium exchanged Y (77). ___Activated at 500°C; ___Activated at 700°C.

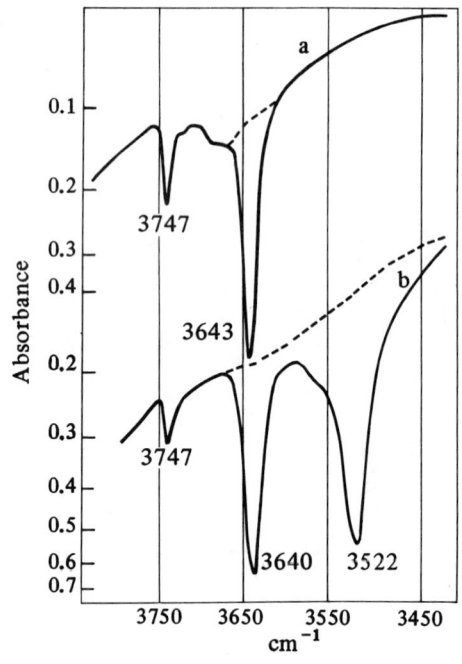

tricoordinate Si and Al (72).

Schemes 3 and 4. Dehydroxylation of the cation-hydroxyl zeolite may occur by (1) removal of a framework oxygen and formation of a defect Lewis acid site (72), scheme 3, or (2) dehydroxylation of the hydroxycation and formation of a bridged cation species, scheme 4 and Fig. 6.12.

Scheme 3:

$$\text{Product of scheme 2} \xrightarrow{-H_2O} \begin{array}{c} \text{CaO} \\ \text{O O O O} + \text{O} \\ /\backslash /\backslash /\backslash /\backslash \quad /\backslash \\ \text{Si Al Si Si Al} \\ /\backslash/\backslash/\backslash/\backslash/\backslash \end{array}$$

Scheme 4 (78,160):

$$\text{Product of scheme 2} \xrightarrow{-H_2O} \begin{array}{c} \quad\quad\quad\quad\quad\quad\quad H \quad\quad\quad Ca^+\text{-O-}Ca^+ \\ \quad\quad\quad\quad\quad\quad\quad\vdots \\ \text{O} \quad \text{O O O O O O O O O} \\ /\backslash \quad /\backslash/\backslash/\backslash/\backslash/\backslash\text{-}/\backslash/\backslash/\backslash\text{-}/\backslash \\ \quad\quad \text{Si Al Si Si Al Si Si Al} \\ \quad\quad /\backslash/\backslash/\backslash/\backslash/\backslash/\backslash/\backslash/\backslash \end{array}$$

The MeOH⁺ cations dehydroxylate at a lower temperature than the Si-OH group as indicated by disappearance of the 3600 cm⁻¹ band assigned to the MeOH⁺ groups (78).

Trivalent Cation Hydrolysis

A similar behavior has been found with trivalent ions (79). Rare earth ions also undergo a hydrolytic reaction and two hydroxyl bands originate in a manner similar to those in the divalent cation form. In a rare earth-exchanged zeolite Y ir absorption bands characteristic of hydroxyl groups are observed at 3640 and about 3520 cm⁻¹(79). The ubiquitous 3740 cm⁻¹ band is also observed (Fig. 6.11) (73,76,79,80,81). The 3640 cm⁻¹ HF band is several times stronger than that observed in the divalent cation form and is attributed to the formation of more associated hydroxyl groups. These hydroxyl groups have been related to the catalytic activity of the polyvalent cation forms of zeolite Y and are similar to the hydroxyl groups observed in the ammoniated zeolites.

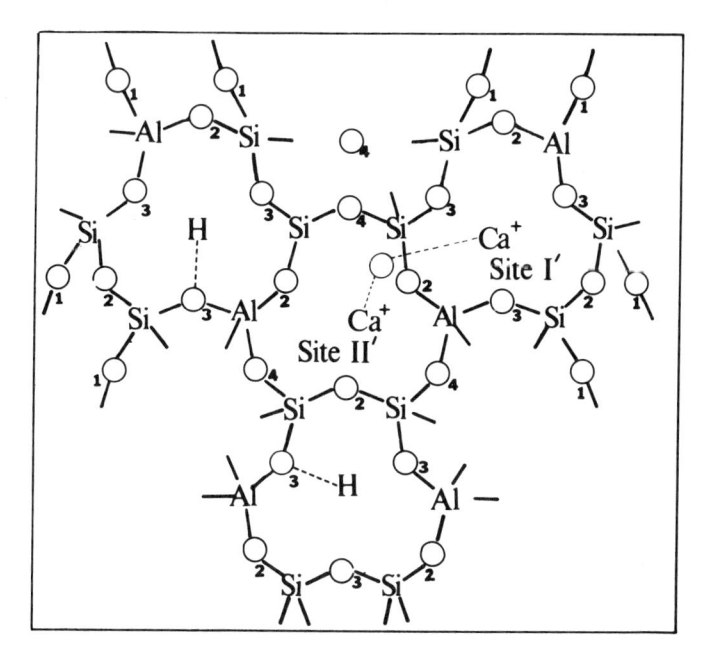

Figure 6.12 Dehydroxylation of the CaOH⁺ zeolite (see Fig. 6.10) forms a bridged cation species and no framework defect by Scheme 4.

466 CHEMICAL PROPERTIES AND REACTIONS OF ZEOLITES

The hydroxyl groups form during dehydration due to a hydrolysis of the rare earth cations (76, 79, 82). The mechanism results in the formation of four hydroxyl groups for every six initial cation exchange sites in the rare earth form of the zeolite (Scheme 5a, 5b, 5c, Fig. 6.13). Six hydroxyl groups are formed by deammoniation of the ammonium exchanged form (See below).

Scheme 5. **Trivalent Ion Hydrolysis.** Dehydration forms a hydroxy cation:

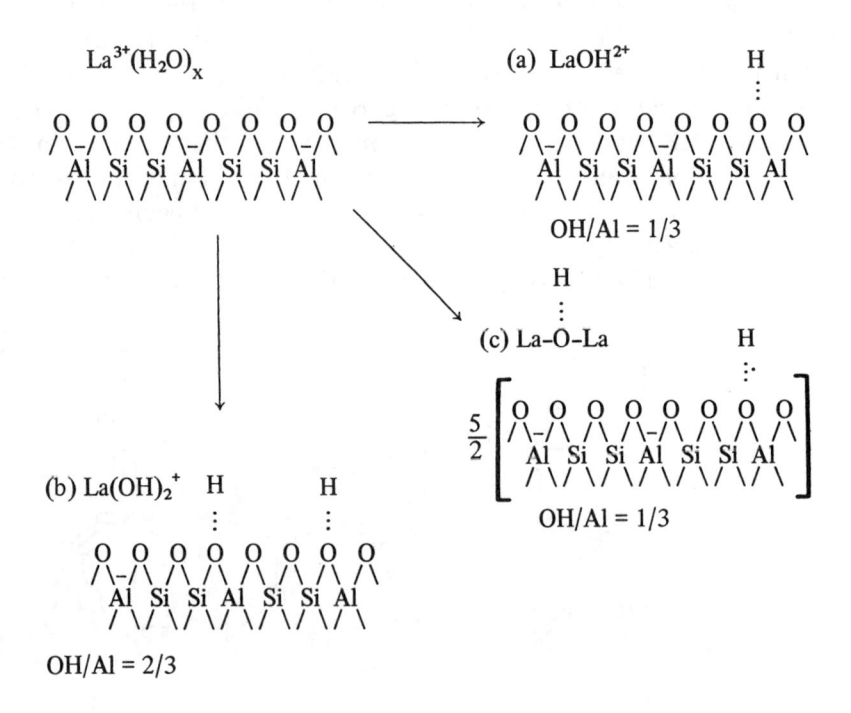

$La^{3+}(H_2O)_x$ → (a) $LaOH^{2+}$

$OH/Al = 1/3$

(c) La-O-La

$\frac{5}{2}$ [...] $OH/Al = 1/3$

(b) $La(OH)_2^+$

$OH/Al = 2/3$

In a recent study, the stoichiometry of rare earth exchanged zeolite Y was obtained by thermal analysis (83). Thermogravimetric data showed that one molecule of hydroxyl water in the zeolite was associated with each rare earth cation. Alternativily the HF band may result from hydrogen ion exchange during ion exchange with a rare earth cation. Hydrogen exchange occurs because of the low pH of the rare earth salt solution. Simultaneous precipitation of a rare earth oxide as an impurity may also occur. This would allow for a cation to aluminum ratio of unity without the actual presence of a cation species. Above an ex-

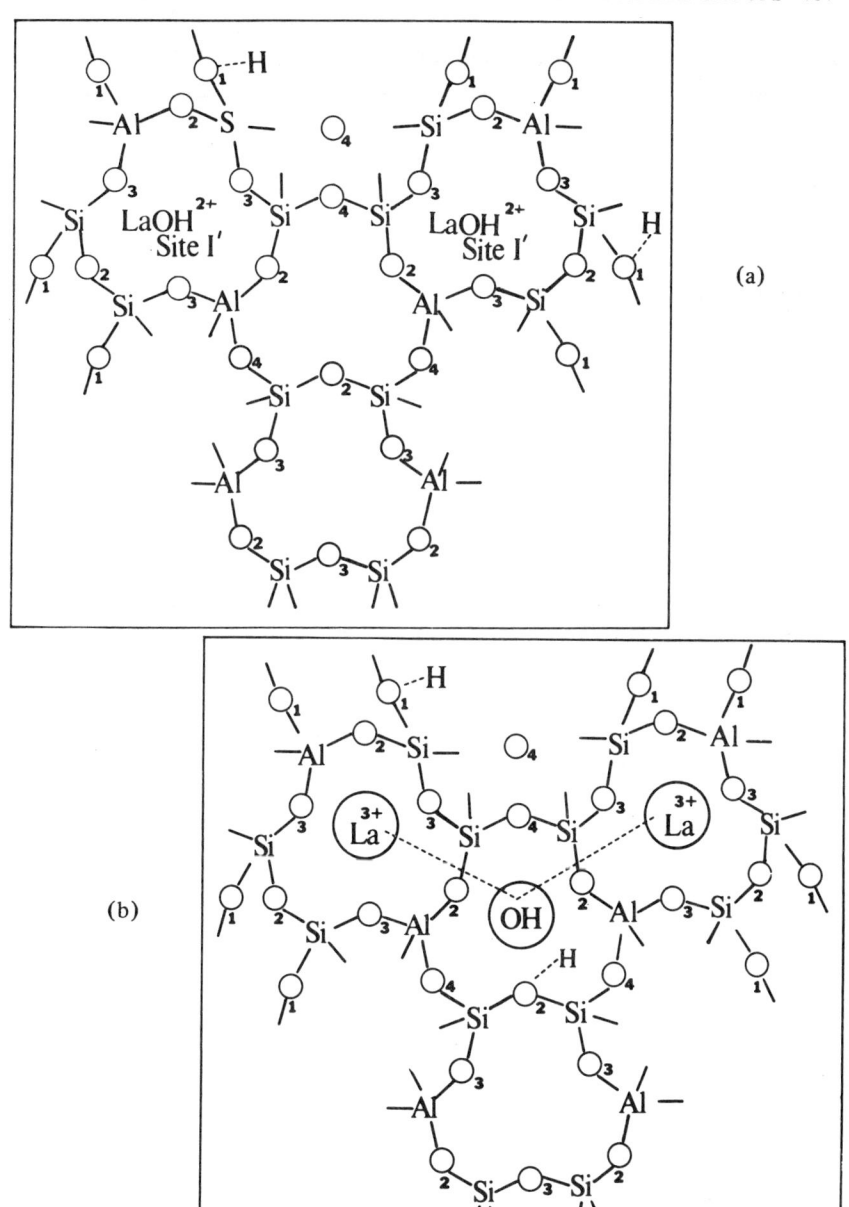

Figure 6.13 (a) Formation of framework hydroxyl groups in lanthanum exchanged zeolite Y by cation hydrolysis during dehydration by the mechanism of Scheme 5A. The ratio of OH/Al = 1/3. Structure indicates lanthanum atoms at site I' with two in each β-cage; (b) Formation of a hydroxyl bridged complex by the mechanism in Scheme 5C.

change level of 75% of the original sodium ions, the more selective rare earth cation may displace the remaining sodium and hydrogen ions. At this point, the cation deficiency is at a minimum and the HF ir band disappears.

Interpretations of the ir spectra, x-ray structural data, and thermal analytical data have been made and several mechanisms for the formation of hydroxyl groups during dehydration of the trivalent form have been postulated.

Dehydroxylation of Trivalent Cation Forms

All of the hydroxyl bands in rare earth-exchanged zeolite Y are removed by calcination at 680°C (Section D). During the high temperature dehydroxylation process, one water molecule is evolved up to 700°C for each initial rare earth cation. Proposed dehydroxylation mechanisms (Scheme 6a and 6b) require a water to rare earth cation ratio of 0.5 or 2 (79, 82). As shown in Scheme 6a it is not necessary to produce a structure with vacant framework oxygen sites. Although structural studies of decationized zeolites have shown the absence of some framework oxygens, x-ray structural analysis of high-temperature calcined lanthanum Y (725°C) showed no framework oxygen deficiency.

Scheme 6 Dehydroxylation of a trivalent zeolite Y may take place in one of two ways:

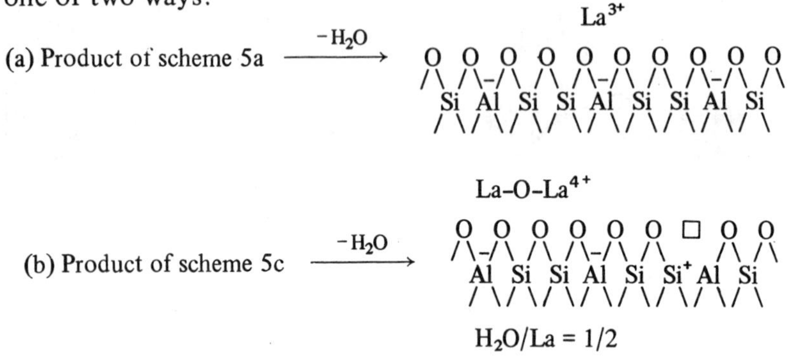

(a) Product of scheme 5a $\xrightarrow{-H_2O}$

La^{3+} over framework: O O O O O O O O O / Si Al Si Si Al Si Si Al Si

(b) Product of scheme 5c $\xrightarrow{-H_2O}$

$La-O-La^{4+}$ over framework: O O O O O O □ O O / Al Si Si Al Si Si$^+$ Al Si

$H_2O/La = 1/2$

Dehydroxylation of 5b is the same as 5a except that the ratio OH/La = 2. Dehydroxylation of 5c results in a bridged cation, formation of an oxygen vacancy, □ , and a trigonal aluminum Lewis acid site.

Structural Studies

Low temperature dehydration (below 400°C), results in partial hydrolysis of the hydrated trivalent lanthanum ion to form the divalent hydroxy complex (See Fig. 6.13). Structural studies indicate that these ions are located in site I' to the extent of 16 per unit cell or 2 in each β-(sodalite) cage. Simultaneously, the proton attaches to a framework oxygen, here pictured as the O(1) oxygen, to form a hydroxyl group. Infrared spectra in the hydroxyl region show the presence of the HF band at 3640 cm^{-1} which corresponds to the hydroxyl group. The 3524 cm^{-1} band and the ratio of the 3524 to the 3640 cm^{-1} band increases in intensity with increasing lanthanum exchange.

After low temperature dehydration at 350°C, in addition to the 16 lanthanum ions located in the site I positions, 12 sodium ions and some 29 water molecules are found in site I' and site II'. Consequently, the OH attached to the lanthanum in LaOH^{2+} located in site I', can occupy the site II' position within the β-cage (84).

As dehydroxylation proceeds with increasing temperature (up to 700°C), loss of water results in the formation of the rare earth cation, La^{3+}, in sites I, II, and I'. (Fig. 6.14b) (84). As originally depicted dehydroxylation should remove a proton plus an oxygen from the framework (Scheme 6b and Fig. 6.14a). Alternatively, dehydroxylation may remove hydroxyl from the rare earth cation and a proton from the framework oxygen thus restoring the ideal zeolite structure. This accounts for the loss of one water molecule per rare earth cation (Scheme 6(a) and Fig. 6.14b).

One argument against this cation distribution is concerned with the population of sites within the β-cages by highly charged cations (79). However, only 5.2 La^{3+} cations are found in site I and 5.5 in site II with 8.9 in site I', or one ion within each β-cage. It has been observed that in Ce^{3+}X (calcined at a high temperature) 23 Ce^{3+} ions are found in site I', which is equivalent to about three within each β-cage (85). An oxygen, or hydroxyl, bridging three Ce^{3+} ions was postulated. Consequently, in a typical rare earth-exchanged Y, 16 univalent ions may occupy site I and site I' without a bridging anion. Upon cooling to room temperature partial redistribution may occur with most of the La^{3+} ions occupying the original site I'.

This proposal, which does not result in loss of framework oxygen, logically accounts for increased thermal stability of the rare earth forms of the zeolite. The presence of oxygen-deficient Lewis sites has been reported in dehydroxylated rare earth zeolite Y (82).

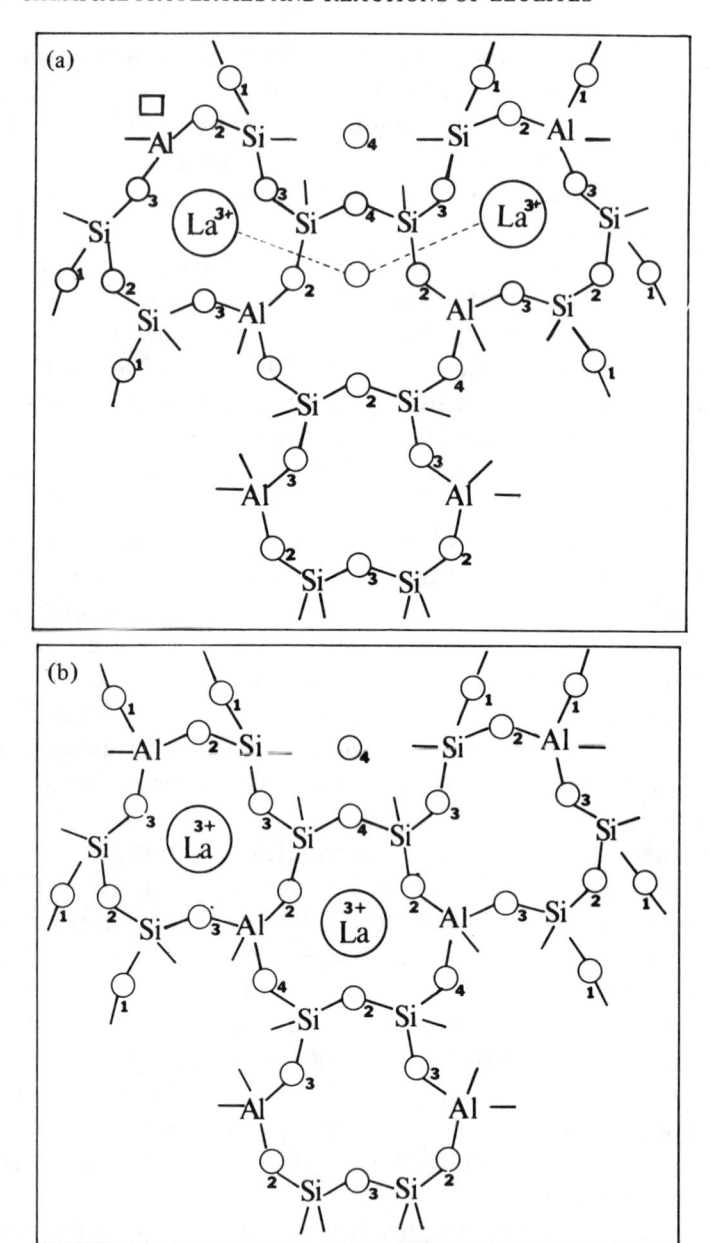

470 CHEMICAL PROPERTIES AND REACTIONS OF ZEOLITES

Figure 6.14 (a) The dehydroxylation of the structures of Fig. 6.13(b) forms an oxygen bridged lanthanum species in the β-cage and an oxygen deficiency in the framework structure. Ratio of dehydroxylation H_2O to La = 1/2; (b) Dehydroxylation of the structure of Fig. 6.13(a) by the mechanism of Scheme 6A forms lanthanum ions in sites I and II with one lanthanum ion in each β-cage and no framework oxygen deficiency. Ratio of hydroxyl/La = 1.

Interaction of Bases with Hydroxyl – Dehydroxylated Zeolites

The interaction of pyridine has been used to measure the concentration of hydroxyl groups (Bronsted sites) and trigonal aluminum (Lewis sites). The formation of a pyridinium ion is evidenced by an infrared absorption band at 1545 cm^{-1}. If pyridine is coordinated to a trigonal aluminum (Lewis) site or to cations, then the band is observed in the 1440 - 1450 cm^{-1} region (72). A band at 1545 cm^{-1} indicates the hydroxyl (Bronsted sites).

Calcination of divalent cation forms of zeolite Y above 650°C produces the Lewis-type absorption band of pyridine at 1451 cm^{-1}. After calcination of a rare earth exchanged zeolite Y at 680°C, a pyridine absorption band was observed and attributed to the presence of oxygen deficient Lewis sites (81); it can also be caused by a cation (e.g., rare earth), pyridine interaction. Below 480°C, pyridine produced a Bronsted-type 1545 cm^{-1} band, as well as a 1451 cm^{-1} band due to coordinated pyridine (82).

In a recent study of rare earth-exchanged forms of zeolite Y by means of ir spectroscopy, a shift in the frequency of the LF 3540 cm^{-1} band was observed as the result of pyridine adsorption (86). This occurred for a series of 11 different rare earth forms of zeolite Y and amounted to a frequency shift of 20 to 30 cm^{-1} to higher values. This shift is opposite to that expected if hydrogen bonding occurs. This effect was attributed to the formation of a pyridinium ion by interaction with accessible hydroxyl groups (the protonated framework oxygens reponsible for the HF 3640 cm^{-1} ir band) and by an inductive effect which strengthens the bond of the inaccessible LF hydroxyl group.

Other Spectral Studies

The diffuse reflectance spectra have been recorded of a rare earth cation exchanged zeolite X and zeolite Y (87). The rare earth forms (chemical analyses were not reported) were prepared by exchange with neodymium, Nd^{3+} and europium, Eu^{3+}. The exchanged zeolites were 1.) heated in air at temperatures of 250°C and 594°C, then in one atmosphere of steam of 675°C for zeolite X and 750°C for zeolites Y; or 2.) vacuum activated at 10^{-6} torr for 4 hours at 700°C (zeolite X) and 800°C (zeolite Y). Reflectance spectra were typical of rare earth cations with an oxygen coordination number of 7 to 9. The activation treatments caused a migration of rare earth cations from the large supercages to sites within the beta cages of the zeolite framework. The degree of cation movement increases with the severity of the activation treatment. These results do not agree with the four coordinate structure previously proposed for a singly bridged lanthanum-oxygen-lanthanum cation species [La-OH-La]$^{5+}$ (Scheme 5c). Instead, a complex involving four bridging hydroxy groups was proposed:

$$\begin{array}{c} O \\ O-La \\ O \end{array} \begin{array}{c} OH \\ OH \\ OH \\ OH \end{array} \begin{array}{c} O \\ La-O \\ O \end{array}$$

During dehydroxylation this complex should evolve water (four H$_2$O/ 2 La^{3+}). If the loss of hydroxyl occurs only from the bridging hydroxyls then the ratio should be one H$_2$O/La^{3+} which would leave a double oxygen bridge:

$$=La\begin{array}{c}O\\O\end{array}La=$$

The 1452 cm^{-1} band, which is attributed to interaction of pyridine with a Lewis site (oxygen deficient aluminum ion) was observed to be the same for the entire series of various rare earth exchanged zeolite Y; it did not vary with type of rare earth cation as did some of the other ir bands. The LF 3540 cm^{-1} band increased in frequency with increasing ionic radius of the rare earth cation in going from yttrium, Y^{3+} to ytterbium, Yb^{3+} (86).

The environment of Fe^{2+} ions in zeolite Y was interpreted from Mossbauer spectroscopy. Changes in the spectra were observed with increasing degree of dehydration in vacuum. Two different environments for the Fe^{2+} cations were established (164). One site I is in the center of the hexagonal prism units; other cations occupy sites which lie off the plane of the single six rings in site II on the trigonal axis. and in tetrahedral coordination with three of the ring-oxygen atoms and one remaining hydroxyl group.

$$\begin{array}{c} H \\ | \\ O \\ | \\ Fe^{2+} \\ O\diagup | \diagdown O \\ Al \\ | \\ O \end{array}$$

Dehydration forms Fe-O-Fe linkages within the β-cage. This bridged group can be formed by dehydroxylation which removes water from two adjacent Fe(OH)$^{2+}$ groups or dehydroxylation of Fe(OH)$^{2+}$ in conjunction with a framework oxygen followed by migration of the Fe^{2+} ion from site I into the β cage in site I' or migration of two Fe^{2+} from site I to site I' with formation of an oxygen bridge from residual water molecules. The adsorption of different gases agrees with these conclusions. Adsorption of oxygen or nitrogen produces no change in the spectrum which indicates no interaction with Fe^{2+}. Ethanol and pyridine removed the spectrum characteristic of Fe^{2+} ions in site II but no interaction with Fe^{2+} ions in site I was observed.

Other Observations

The main evidence for trigonal aluminum (or Lewis sites) is based on the effect of pyridine adsorption on the infrared spectra. This effect may equally be explained by coordination of the pyridine with the polyvalent cation.

Observations by infrared spectroscopy and thermogravimetry of a Ce^{3+} exchanged zeolite Y have shown that the ratio H$_2$O/RE^{3+} = 1 (83, 88). The hydroxyl groups are removed from the Ce^{3+} zeolite by heating in a vacuum at 700°C. The crystal structure of the zeolite is not affected by this dehydroxylation as would be expected if framework oxygens were involved. The OH bands at 3640 and 3524 cm^{-1} which are removed by this treatment may be regenerated by adsorbing small amounts of water at room temperature followed by heating to about 200°C (79).

The addition of water to dehydrated divalent zeolite Y (Ca^{2+} and Mg^{2+}) increases the hydroxyl group concentration (89). Calcination of Ca^{2+}Y at 400°C and treatment with water produced an increase in the acidity as well as a new ir band at 3585 cm^{-1} which was assigned to a CaOH$^+$ group. This indicates that treatment with water increases the formation of framework hydroxyls.

A chemical approach for the determination of acidic OH groups in the rare earth exchanged zeolite Y was based on reaction with lithium aluminum hydride, LiAlH$_4$, and Karl Fischer titration (90). It was concluded that after heating to 300°C the zeolite contains only hydroxyl groups in the supercages, with one OH for each rare earth ion. The method used did not measure OH groups within the β-cages.

Hydroxyl Groups in Zeolites by Deammoniation of NH₄ Zeolites

The usual method for forming structural hydroxyl groups in zeolites is by the thermal decomposition, or deammoniation, of the ammonium ion-exchanged form of the zeolite. By controlling the degree of ammonium exchange the concentration of structural hydroxyl groups in the "hydrogen" zeolite can be controlled. Thermal decomposition of the ammonium exchanged zeolite evolves ammonia and leaves a hydrogen on the framwork as an OH group (Scheme 7 and Fig. 6.15). It has been suggested that protons are initially formed but rearrange to form hydroxyl groups.

The zeolite can contain a maximum of one ammonium ion for each aluminum atom. The electrostatic charge remains neutral and the proton of the hydroxyl group acts as the charge balancing species. Considerable attention has been given to the study of these hydroxyl forms of zeolites and, in particular, those zeolites of interest in catalysis such as the zeolites X, Y, mordenite, and T.

In the earliest work, the product of decomposition of ammonium exchanged zeolites was referred to as a "decationized" zeolite (91). After dehydration and removal of NH_4^+ as NH_3, further thermal treatment results in the loss of hydrogens and framework oxygens as water. This process is now generally called dehydroxylation. The resulting product contains no cations except the original unexchanged sodium ions. However, this original concept has recently been modified (see Section D). It now appears that thermal treatment of the ammonium zeolite at 500°C may leave an unspecified mixture of the hydrogen and dehydroxylated forms.

The thermal decomposition of ammonium exchanged zeolites, such as zeolite Y, may be represented ideally by Scheme 7.

Scheme 7

(a) NH_4^+ ... NH_4^+ on Si–Al–Si–Si–Al–Si framework $\xrightarrow{-NH_3}$ H ... H on Si–Al–Si–Si–Al–Si framework

(b) H ... H on Si–Al–Si–Si–Al–Si framework $\xrightarrow{-H_2O}$ Si–Al ⁺Si–Si–Al–Si framework

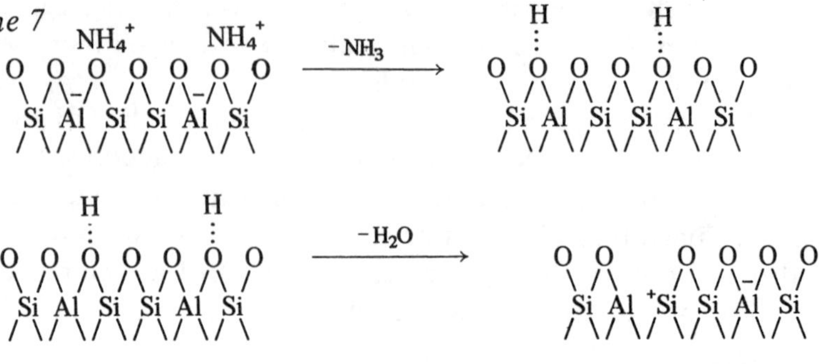

Decationization

The decationization process as originally formulated is illustrated in detail in Fig. 6.15.

Heating of an ammonium exchanged zeolite results in loss of NH_3 at about 400°C. The hydrogen atoms or protons are situated on O(2), near site II, and on O(3) near site I', or possibly site I. Other protons may be located at site II. This picture is not in agreement with a reported crystal structure analysis which locates hydrogen atoms on the bridging oxygen atoms O(1) and O(3) (92).

Further thermal treatment of the zeolite at 500 - 600°C results in the loss of one H_2O molecule from two nearby OH groups or one H_2O for each two tetrahedral aluminum atoms. This causes the loss of oxygens near site I [O(3)], thus producing a lattice vacancy. It is further assumed that the adjacent Al and Si atoms will tend to assume a planar-type sp^2 configuration with the remaining 3 oxygen atoms. After partial dehydroxylation, remaining OH groups on the O(2) near site II may be more strongly acidic due to an inductive interaction from the nearby defect site. Other schemes based on other types of ordering are certainly possible and may involve the O(1) atoms.

The O(4) atoms are least likely to be involved in this scheme although the structural analysis of a decationized form of zeolite Y indicates loss of oxygen from the O(3) and O(4) positions.

Location of Hydroxyl Groups

Hydroxyl groups which are produced in zeolite Y by deammoniation give rise to two ir absorption bands, one at 3640 cm^{-1} referred to as the high-frequency (HF) band, and a second low-frequency (LF) band at 3540 cm^{-1}, (Fig. 6.16). These have been attributed to the location of hydrogen atoms on two different oxygen atoms in the framework. One type, the HF band, is located within the large supercages and is accessible to adsorbed molecules; the second type, represented by the LF band, is less accessible and probably hidden within the β-cages of the structure. One x-ray crystal structure determination has indicated indirectly that the hydroxyl groups involve the O(1) and O(3) framework oxygens (92). Another structural analysis of deammoniated zeolite Y assigns the HF band to an O(2)H group (93).

The locations of hydroxyl groups in a hydrogen faujasite (a calcined, ammonium exchanged mineral faujasite) have been reported (92). The

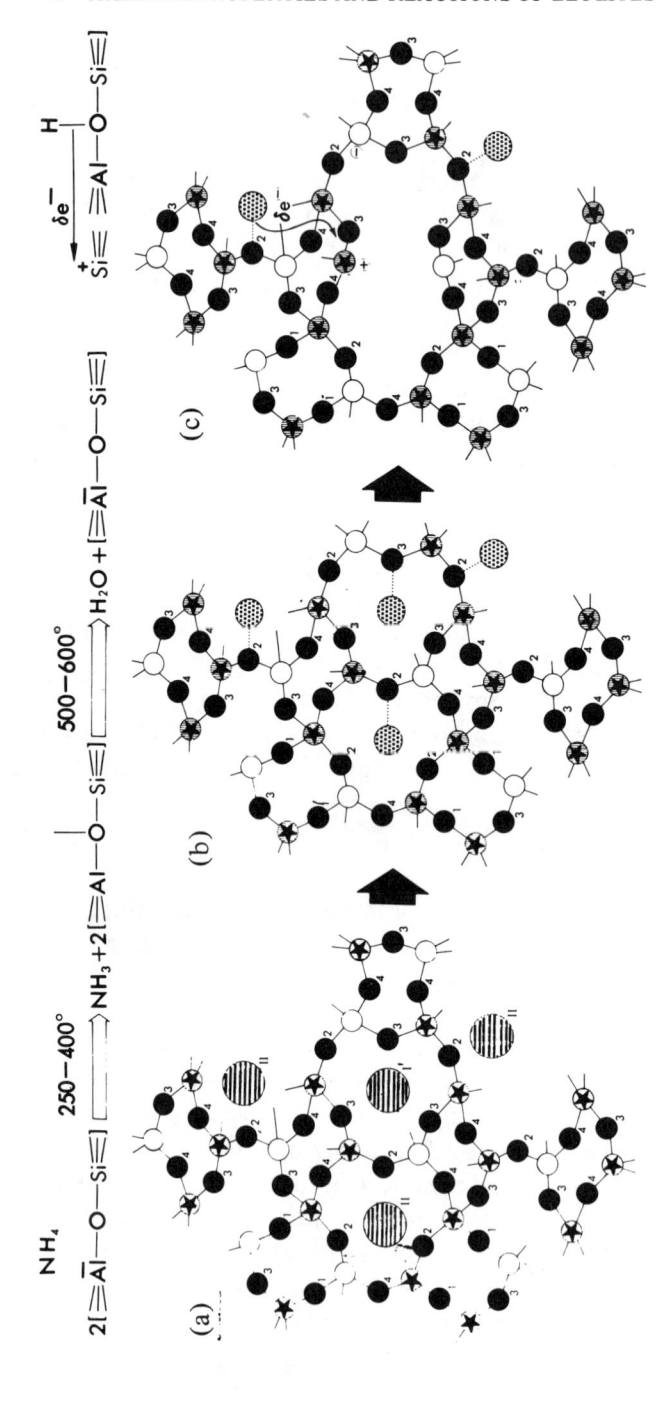

476   CHEMICAL PROPERTIES AND REACTIONS OF ZEOLITES

$$NH_4 \atop 2[\equiv\overline{Al}-O-Si\equiv] \xrightarrow[250-400°]{} NH_3 + 2[\equiv\overline{Al}-O-Si\equiv] \xrightarrow[500-600°]{} H_2O + [\equiv\overline{Al}-O-Si\equiv]$$

$$\overset{\delta e^-}{\underset{Si\equiv}{\longleftarrow}} \overset{H}{\underset{\equiv Al-O-Si\equiv}{|}}$$

Figure 6.15 A schematic representation of the regular decationization process in zeolite Y. (a) View of the β-cage of zeolite Y perpendicular to (100). The cage is unfolded showing one 6-ring of the D6R unit at right and a single 6-ring adjacent to the supercage at left. The Si/Al ratio of the zeolite is assumed to be 2 with 64 tetrahedral Al atoms per unit cell or 8 Al atoms per β-cage. Ammonium ions are assumed to occupy site I' and site II with essentially full occupancy of both sites. Deammoniation of the ammonium exchanged Y at about 400°C leaves hydrogen atoms on O(2), near site II and on O(3), near site I', or possibly site I. Other protons at site II are shown; (c) Dehydroxylation at about 500° or 600°C with the loss of one water from two nearby OH groups removes the O(2) near site II producing a lattice vacancy; the adjacent aluminum and silicon atoms assume a planar configuration with the 3 remaining oxygen atoms. Remaining OH groups on O(2) are acidic due to inductive interaction by the nearby defect site. Other schemes may involve the O(1) atoms.

● = O   ○ = Al   ✳ = Si   ◉ = H   ▦ = NH$_4^+$

positions of the hydrogens were interpreted from the four independent Si(Al)-O distances (Table 6.2). It was concluded that the hydrogen atoms are bonded to two of the four types of oxygen atoms, the O(1) and the O(3). One hydrogen is located within the hexagonal prism in site I and the second hydrogen is on the O(1) bridging oxygen and is accessible to the large supercages. This model agrees with proposed Si, Al ordering schemes (see Chap. 2). Other crystallographic studies based on x-ray powder data obtained on a hydrogen exchanged sodium zeolite Y do not agree with these results.

The crystal structure of zeolite Y at three levels of sodium replacement by hydrogen (from NH_4^+ exchange and decomposition at 400°C *in vacuo*) were determined using x-ray powder data (93). The levels of ammonium exchange were 52%, 75%, and 96%. Residual sodium ions were located at sites I, I', and II. Interpretation of the positions of the hydrogen atoms was based on the tetrahedral atom-oxygen distances. The distances T-O(2) and T-O(3) were found to be larger as shown in Table 6.2 which compares the results on a deammoniated faujasite with a deammoniated Y.

Data on zeolite Y disagree with earlier conclusions obtained on faujasite which were based upon the lengthening of the tetrahedral atom-oxygen distance and located the hydrogen atoms on the O(1) and the O(3). The HF infrared band at 3640 cm^{-1} is assigned to the O(2)H in this work rather than the O(1)H. At present, the specific locations of the hydrogen atoms in a hydrogen zeolite such as zeolite Y appear to be in doubt.

The ir spectra in the hydroxyl stretching region have been resolved into six components. High resolution spectra were obtained on NH_4^+X and NH_4^+Y. In addition to the O(1) and O(3) as preferential

Table 6.2 Summary of T-O Distances in H Zeolites

Distance (Å)	G + I (93)[a]	O + D (92)[b]
T-O (1)	1.622	1.653
T-O (2)	1.645	1.634
T-O (3)	1.643	1.663
T-O (4)	1.618	1.623

[a]Based on $Na_{2.2}H_{53.8}$[Y, Si/Al = 2.42].
[b]From single crystal analysis of a deammoniated, ammonium exchanged faujasite crystal.

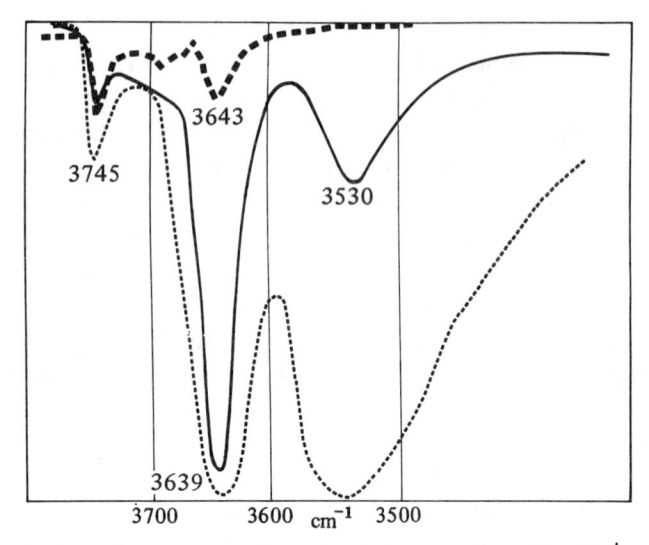

478 CHEMICAL PROPERTIES AND REACTIONS OF ZEOLITES

Figure 6.16 Infrared spectra of hydrogen Y zeolite (71). ----- 7% initial NH_4^+ exchange; ———— 33% initial NH_4^+ exchange; .... 86% initial NH_4^+ exchange.

sites, all of the cage oxygens O(2), O(3), and O(4) were considered to be possible locations (95).

Broadline proton magnetic resonance spectra of the hydrogen form of zeolite Y were interpreted as based on a model consisting of an isolated Al-H pair with an internuclear distance of 2.38 A (94). These results indicate that the hydrogen atoms are located on the O(1) and O(3) atoms with an O-H distance of 1.00 to 1.03 A. The configuration is as follows:

$$\text{Al} \overset{116°}{\diagdown} \text{O} \diagup \text{Si}, \quad \text{with H on O}$$

where Al-O = 1.72 A and Al-H = 2.38 A

The acidity of the OH group is due to the decreased p-character of the O atoms.

The Stoichiometry of Deammoniation-Dehydroxylation

The decomposition of ammonium exchanged zeolite Y and the stoichiometry of the decomposition reaction have been extensively

studied (96 - 98). It is worthwhile to review some of these results. In one study ammonium exchanged zeolite Y had the following chemical composition (97): $Na_5(NH_4)_{46}[(AlO_2)_{51}(SiO_2)_{141}]$.
After deammoniation and dehydroxylation at 500°C the loss in weight corresponded to 22 water molecules for each unit cell (one water molecule for each two tetrahedral Al atoms). This is the reaction stoichiometry that would correspond to the simple picture of the decationization process as shown in Fig. 6.15. After dehydroxylation, it was found that the hydrogen form of the zeolite could not be reconstituted by reaction with water, which was counter to the experience of others (96). After the deammoniation reaction to form the hydrogen zeolite, ammonia was readsorbed (in the ratio of one ammonia molecule for each hydrogen atom) to reconstitute the initial ammonium form. Further, six molecules of ammonia were strongly adsorbed and assigned to specific sites.

From pulsed nmr studies, it was found that from 19.5 to 24 hydrogen atoms are present in each unit cell after dehydroxylation at 500°C. This cannot be explained on the basis of the classical model because the residual proton content is too high. It was further concluded that there are two types of ammonium ions present in the initial zeolite. However, the stoichiometry agreed with the classical picture of decationization.

In another study, an ammonium zeolite of similar composition was used ($Na_2(NH_4)_{53}[(AlO_2)_{55}(SiO_2)_{137}]$). The hydroxyl water that was evolved was found to be 19 molecules per unit cell corresponding to an observed weight loss of 3.0 wt% (96). The calculated loss, based upon the zeolite composition should have been 3.7%. There are consistent reports of the presence of a 500°C exotherm in the DTA pattern. Various interpretations of this exotherm have been made including the rearrangement of a proton with a framework oxygen to form a hydroxyl group (see Section D).

The ion exchange capacity of the zeolite after deammoniation was determined as a function of the dehydroxylation temperature (Fig. 6.17). With increasing dehydroxylation temperature the NH_4^+ exchange capacity drops and levels off at about 50% of the original ion exchange capacity. The curve shows a change in slope which is indicative of a compound phenomenon. This change in slope occurs in the temperature range at which the major dehydroxylation reaction occurs, 500 - 600°C, as evidenced by independent ir studies (96).

Figure 6.17 Re-ion exchange of ammonium exchanged Y (95% exchanged) by aqueous NH₄Cl as a function of the temperature of heating in air. The ion exchange capacity should diminish to zero as dehydroxylation is completed at 600°C. Ion exchange is due to formation of a cationic aluminum species (96).

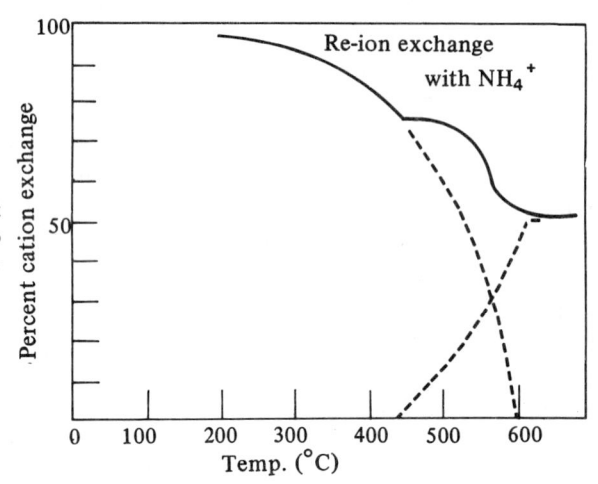

The typical model of decationization as shown in Fig. 6.15 does not account for this type of ion exchange reaction. With complete dehydroxylation there should be no residual ion exchange capacity because the structure is electrostatically neutral; there is no exchangeable cation species present. The potential proton content in a hydrogen zeolite is extremely high. If one assumes that all of the hydrogen ions present in the hydroxyl groups can be exchanged, the H⁺ concentration would correspond to 7.2 molar within each supercage. (The pH of 1 gram of this zeolite in 100 cc of a salt solution should be about 1.6.)

This ion exchange relation indicates that two cocurrent phenomena occur. Above 400°C, a cation species is formed which accounts for the residual ion exchange capacity. This curve is the sum of two individual curves. The formation of another species accounts for the exchange capacity with increasing temperature. Above 400°C, a cationic species of aluminum is formed by removal of aluminum from tetrahedral positions and reaction with the acidic OH groups. (See Section D.) The number of aluminum cations correspond to 50% of the original aluminum atoms present in the zeolite. The formation of hydroxyaluminum cations is discussed in Section D.

The early investigations were interpreted on the basis of the simple mechanisms shown in Fig. 6.15. Characteristic ir spectra for a typical hydroxyl of hydrogen zeolite Y are shown in Fig. 6.16. The intensity of the ir bands of the two types of hydroxyl groups vary with the temperature of treatment after thermal decomposition of the ammonium zeolite. Increasing temperature causes the intensity to increase and remain constant up to 500°C, while above 500°C the intensities rapidly decrease for the HF and LF bands (Fig. 6.18). These data have been confirmed by several investigators. The initial stages of dehydroxylation overlap the final stage of deammoniation at the lower temperatures; there appears to be no clear point where a completely deammoniated form is present with no dehydroxylation.

The DTA curves for ammonium exchanged zeolites X and Y are shown in Fig. 6.19 (96). The first endotherm indicates the loss of zeolitic water by desorption which is complete at about 250°C. At a higher temperature, indicated by a second endotherm, the deammoniation process begins. These results have been confirmed by chemical analysis of the effluent gases. Approximately 90% of the ammonium cations are decomposed below a temperature of 400°C. The exothermic peak that shows up at 500°C in Fig. 6.19 is proportional to the degree of ammonium exchange and, consequently, must be associated with the decomposition of the ammonium cation. Although this peak has been associated with the formation of the hydroxylated species, indicating the rearrangement of the classical Bronsted-type proton to the hydroxyl group (96), it may be caused by reaction of an acidic hydrogen with a cationic aluminum (See Section D).

The small endotherm at 680°C corresponds to the dehydroxylation step as confirmed by thermogravimetric analyses. Accompanying this dehydroxylation step there is a significant contraction in the size of the unit cell.

Formation of a cationic aluminum species has been related to a process known as stabilization. A "super stable" form of zeolite Y, derived by the thermal decomposition of the ammonium exchanged forms in

Figure 6.18 Intensity of the structural hydroxyl group infrared bands as a function of temperature for a 97% ammonium exchanged Y [Si/Al = 2.5]. ● represents the 3540 cm^{-1} band and ★ represents the 3640 cm^{-1} band (96).

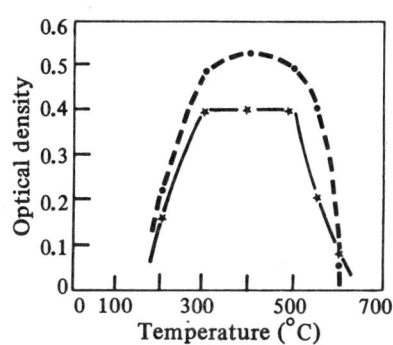

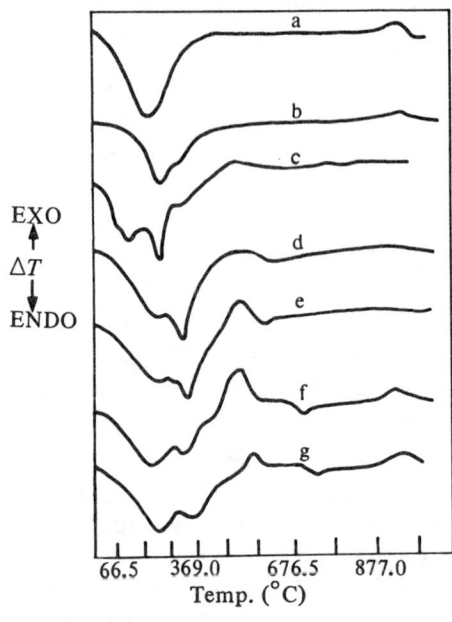

Figure 6.19 Differential thermal analysis of ammonium exchanged zeolites in a helium atmosphere and a sodium Y for comparison (96).
(a) NaY(SiO$_2$/Al$_2$O$_3$ = 5.0)
(b) acid treated NaY(SiO$_2$/Al$_2$O$_3$ = 5.0) (35% Na removed))
(c) NH$_4$X(SiO$_2$/Al$_2$O$_3$ = 2.5)
(d) NH$_4$Y(SiO$_2$/Al$_2$O$_3$ = 3.9)
(e) NH$_4$Y(SiO$_2$/Al$_2$O$_3$ = 4.2)
(f) NH$_4$Y(SiO$_2$/Al$_2$O$_3$ = 5.0)
(g) NH$_4$Y(SiO$_2$/Al$_2$O$_3$ = 5.6)

water vapor exhibits enhanced thermal stability. However, it now appears that the usual process of dehydroxylation as discussed in this section is accompanied by loss of aluminum from the framework and the creation of framework defects which may lead to instability. (See Section D.)

Hydroxyl Groups From Alkylammonium Zeolites

The thermal decomposition of ammonium zeolites to form the protonic or hydrogen form is, in principle, relatively simple. More complex cations, such as (CH$_3$)$_4$N$^+$, in zeolites have been used as proton precursors. The thermal decomposition of alkylammonium exchanged zeolite Y occurs by several pathways. The generation of OH groups as evidenced by ir studies was confirmed by NH$_3$ adsorption since interaction with NH$_3$ produces NH$_4^+$ ions. Interactions of NH$_3$ with the dehydroxylated sites indicated Lewis-type sites. The decomposition products are complex (99).

The decomposition of tetramethylammonium in a synthetic offretite-type zeolite yields three OH bands which interact with NH$_3$. Complete dehydroxylation occurs at 600°C. After dehydroxylation, exposure to NH$_3$ produces ir bands due to NH$_3$ coordinated with Lewis acid sites. This NH$_3$ interaction is weaker than in the Y zeolite (100).

Summary Comments on Hydroxyls in Zeolites

In summary, the detailed study and observation of hydroxyl groups in zeolites has been confined to primarily the synthetic zeolites X and Y. Other synthetic zeolites which have received some treatment include zeolite L and synthetic types related to mordenite. Study of hydroxyl groups has included both the deammoniated and polyvalent cation forms.

Six types of hydroxyl groups have been identified in zeolites. The major interpretation of these hydroxyl groups has been based upon ir absorption spectra and the interaction of the hydroxyl groups in the various zeolites with other types of adsorbed molecules. The latter measurement provides information concerning the location of the hydroxyl group, its functionality and, indirectly, information concerning the location of the cations themselves. Some confirmatory studies have been made by x-ray structure analysis coupled with thermal analysis.

A summary of the information available in the literature on hydroxyl groups in various zeolites is given in Table 6.3. The hydroxyl group indicated by band A observed in nearly every case at 3745 cm^{-1} is associated with occluded impurities within the zeolite structure and does not apparently have any structural function. The bands C and D are the most important of the hydroxyl groups which we have previously referred to as the high frequency, HF, or low frequency, LF, bands, respectively. These are associated directly with framework hydroxyl groups. Other important bands include band E which is associated with trivalent cations in the 3530 cm^{-1} region and is attributed to hydroxyl groups on the cations which result from cation hydrolysis.

C. TRANSFORMATION REACTIONS

Hydrothermal Transformation of Zeolites

The stability of zeolite minerals and synthetic zeolites and their structural changes when exposed to water vapor at elevated temperatures and pressures (i.e., hydrothermal stability) is important in understanding the mechanism and conditions of recrystallization. The transformation of zeolites, or their recrystallization, at elevated temperatures in the presence of water vapor is to be distinguished from the recrystallization reactions or transformations of zeolites to other zeolites or nonzeolite species in solution, such as solutions of strong bases. Because of

484   CHEMICAL PROPERTIES AND REACTIONS OF ZEOLITES

Table 6.3   IR Observations of OH Groups in Zeolites

Zeolite	Cation Form	Activation Treatment (°C)	A	B	C	D	E	F	Other	Ref.
X	NH$_4$	vac, 520°			3660	3570			NH$_3$ adsorption reforms NH$_4^+$	69
Y	NH$_4$				3660	3570				
X, Si/Al = 1.25	NH$_4$, 40%				3650	3556			NH$_3$ removes C and D and reforms NH$_4^+$	104
	NH$_4$, 75%				3650	3556				
Y, Si/Al = 2.4	NH$_4$, 50%				3650	3556				
	NH$_4$, 90%				3650	3556				
X	Ca	vac, variable temps.	3740	3695	3650	3550				78
X	Mg		3740		3650	3550				
X	Ba				3650	3550				
Y	Ca		3740	3690	3650	3550		3605		
Y	Ba, Mg				3650	3550				
Y, Si/Al = 2.45	NH$_4$	vac, 500°	3742		3643	3540			Pyridine reduces A and C	105, 106
		750°	no OH groups							
Y, Si/Al = 2.45	Me$^+$	vac, 500°	Broad band removed by 350°						Pyridine adsorption to characterize B and L sites	72
	Mg	vac, 500°	3739	3690	3645	3530				
	Ca	vac, 500°	3739	3688	3642	3520		3582		
	Sr	vac, 500°	3739	3691	3645	3560		3480		
	Ba	vac, 500°	Broad band removed by 400°							
Y, Si/Al = 2.45	Mg, NH$_4$	vac, 350°	3742		3643	3540			Band C constant to 575°C; pyridine adsorption to characterize B and L sites	107
Y, Si/Al = 2.45	RE	vac, 340°	3740		3640		3522		Pyridine and piperidine interact with A and C	76, 82
Y, Si/Al = 2.45	Na, H variable	vac, 480°	3742		3640	3540			Pyridine to characterize B and L sites	108
			Vary with Na content							

TRANSFORMATION REACTIONS 485

Table 6.3 IR Observations of OH Groups in Zeolites *(continued)*

Zeolite	Cation Form	Activation Treatment	A	B	C	D	E	F	Other	Ref.
Y, Si/Al = 2.4	NH$_4$	540°, steam	3705		3677	3567			Effect of adsorbed gases on B and L sites	91
Y, Si/Al = 2.4	NH$_4$	vac, 500°	3744		3670			3615	Effect of water vapor	109
Y, Si/Al = 2.4	NH$_4$		3748							71
	Na		3745	3688	3636	3544				
	Mg		3746		3652					
	Ca		3748		3643	3540				
	Mn		3746	3682	3645	3545				
	Ni		3748		3644	3544				
	Co		3744	3673	3643	3540				
	Zn		3745		3646	3542				
	Ag				3642	3550				
					3634					
X, Si/Al = 1.25	Li	vac, 600°	3740		3660					101
	Na		3750	3695	3655					
	K		3750	3715	3650					
	Ag		3750	3685	3630					
	Ca		3750	3695	3590					
	Sr		3750	3700	3660			3605		
	Ba		3750	3695	3620					
	Cd		3750	3690				3600		
Y, Si/Al = 2.4	Ba	vac, 250°, ½hr			3645	3550				73
	Mg	vac, 250°	3745	3685	3642	3550				
	Zn	vac, 250°	3745	3675	3645	3550				
	Ce	vac, 250°	3745		3640	3555				
	Ba	vac, 460°, 14 hrs		none						

486 CHEMICAL PROPERTIES AND REACTIONS OF ZEOLITES

Table 6.3 IR Observations of OH Groups in Zeolites (continued)

Zeolite	Cation Form	Activation Treatment	A	B	C	D	E	F	Other	Ref.
Y, Si/Al = 2.4	Mg	vac, 460°	3750	3690	3655	none				73
	Zn	vac, 460°	3750	3680	none	none				
	Ce	vac, 460°	3750		3650		3530		Pyridine removes band C at 70°	
X, Si/Al = 1.25	NH$_4$	vac, 270°, 2 hr	3750		3660	3570			Pyridine reduces band C	102
	Ca	vac, 200°	3750	3700	3650	3570			Pyridine reduces band C	
	Zn	vac, 200°	3750		3650	3570			Pyridine reduces band C	
	La	vac, 300°	3750		3650	3570			Pyridine reduces band C	
Y, Si/Al = 2.5	NH$_4$	vac, 300°			3650	3550	3530		Pyridine removes C Piperidine removes C and D	103
Y, Si/Al = 2.5	Na									
	Ca	vac, 500°	3747		3643				Band C and D removed above 650°C	80
	Ce	vac, 500°	3747		3640		3522			

Table 6.4 Hydrothermal Conversion of Zeolite Minerals at 1000 bars, 10 days (110)

Starting Zeolite	Degree of Exchange[a]	Temp.(°C)	Phase Formed [b]
Chabazite[a]	0.73 CaO	230	wairakite
Li[ex]	0.94	310	poorly crystalline chabazite
Na[ex]	0.98	210	analcime
K[ex]	0.92	300	high sanidine[c]
Ca[ex]	0.90	230	phillipsite
Ba[ex]	0.68	250	unknown
Harmotome	0.77 BaO	330	celsian[d]
Li[ex]	0.86	350	wairakite and montmorillonite
Na[ex]	0.77	250	analcime
K[ex]	0.79	300	high sanidine[c]
Ca[ex]	0.12	320	celsian[d]
Ba[ex]	0.86	320	celsian[d]
Gmelinite	0.99 Na$_2$O	200	analcime
Li[ex]	0.96	200	phillipsite
Na[ex]	0.99	200	analcime
K[ex]	0.93	300	high sanidine[c]
Ca[ex]	0.93	220	phillipsite
Ba[ex]	0.90	250	harmotome
Gismondine	0.80 CaO	235	natrolite
Li[ex]	0.88	200	LASH-1[e]
Na[ex]	0.91	200	analcime
K[ex]	0.53	300	feldspar
Ca[ex]	nd	235	natrolite
Ba[ex]	0.48	220	harmotome
Phillipsite	0.71 CaO	270	wairakite
Li[ex]	0.78	250	montmorillonite
Na[ex]	0.92	230	analcime
K[ex]	0.65	300	high sanidine[c]
Ca[ex]	0.90	260	wairakite
Ba[ex]	0.25	250	harmotome
Stilbite		300	wairakite
Clinoptilolite		360	mordenite

[a] Degree of exchange expressed as $\frac{\text{Sum of exchange cation}}{\text{total exchangeable ions}}$

[b] These were identified by x-ray powder patterns and indicate the structure-type formed

[c] High sanidine is a potassium feldspar, KAlSi$_3$O$_8$

[d] Celsian is a barium feldspar, BaAl$_2$Si$_2$O$_8$

[e] LASH-1 = Zeolite Li-A, Li$_2$O·Al$_2$O$_3$·2SiO$_2$·4H$_2$O

thermodynamic metastability in the hydrothermal magma, a zeolite may transform to another crystalline species over a period of time. Information on stability is important when considering commercial applications of zeolites as molecular sieves in adsorption and catalytic processes where water vapor at elevated temperatures may be encountered.

One experimental method which has been used to study zeolite behavior at elevated temperatures and with an excess of water vapor is based on heating specimens of the zeolite in sealed, gold envelopes in a high pressure vessel at a constant pressure of water at several thousand atmospheres (110). Table 6.4 presents the results of some typical experiments utilizing zeolite minerals in different cation-enriched forms. Different cation forms exhibit different relative stabilities as well as structural conversion to other zeolite and nonzeolite-type products. For example, Ca^{ex} chabazite converts to a phillipsite structure and Na^{ex} chabazite to an analcime.

Transformations of this type are generally dependent on time and, consequently, should be studied as a function of time at constant temperature and pressure. The temperature-time stability limits of several synthetic zeolites and various of their ion-exchanged forms under hydrothermal conditions are summarized in Fig. 6.20 (111). (These diagrams are reaction diagrams and should not be interpreted as equilibrium phase diagrams since the boundary has been approached from only one direction.) Many zeolites do not convert to products of the same chemical composition as the starting material. Since the experiments are conducted with an excess of water present, preferential dissolution of the starting zeolite may occur into the water in the vessel surrounding the sample. For example, synthetic zeolite A converts to an analcime-type zeolite which has less Al_2O_3 and Na_2O, thus indicating preferential solution of these components. The data in the figure is for a hydrostatic pressure of 20,000 psi (Fig. 6.20). Zeolite A is stable for one day at 220°C, but converts after longer periods of time to the zeolite P which finally converts to an analcime-type phase. In contrast, the calcium exchanged form of zeolite A transforms to a scolecite-type phase and then to a feldspar of the anorthite-type.

The zeolite X converts to an analcime-type structure at 240°C, whereas the calcium exchanged form is stable at 300°C and at 280°C for 14 days. The magnesium exchanged X is less stable for it converts to a layer-

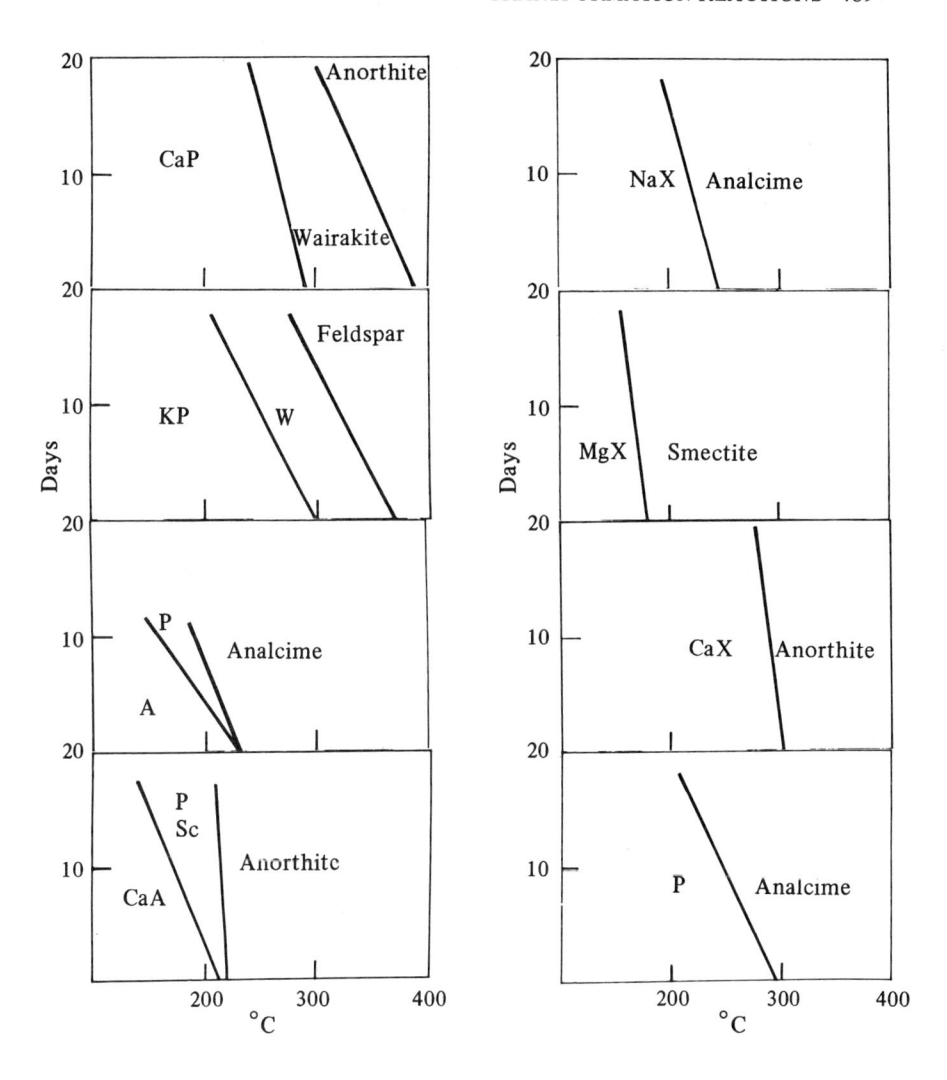

Figure 6.20 Time-temperature diagrams showing hydrothermal transformation of some zeolites at 20,000 psi pressure of H_2O (111). Compositions of starting materieals are as follows (in moles of oxides):

(a) Zeolite X, NaX	$Na_2O \cdot Al_2O_3 \cdot 2.4SiO_2 \cdot 4.3H_2O$
(b) Calcium X	$CaO_{1.0} \cdot Al_2O_3 \cdot 2.4SiO_2 \cdot nH_2O$
(c) Magnesium X	$MgO_{0.81} \cdot Na_2O_{0.19} \cdot Al_2O_3 \cdot 2.4SiO_2 \cdot nH_2O$
(d) Zeolite P	$Na_2O \cdot Al_2O_3 \cdot 5SiO_2 \cdot 4.3H_2O$
(e) Calcium P	$CaO \cdot Al_2O_3 \cdot 4.9SiO_2 \cdot 5.5H_2O$
(f) Zeolite A, NaA	$Na_2O \cdot Al_2O_3 \cdot 1.9SiO_2 \cdot 3.7H_2O$
(g) Potassium P	$K_2O_{0.88}Na_2O_{0.12} \cdot Al_2O_3 \cdot 4.8SiO_2 \cdot 2.8H_2O$
(h) Calcium A	$CaO_{0.96}Na_2O_{0.04} \cdot Al_2O_3 \cdot 1.9SiO_2 \cdot 5H_2O$

type, smectite-type, structure at above 180°C. The zeolite P converts to the analcime-type and the calcium exchanged form of zeolite P to wairakite. The potassium exchanged form of zeolite P converts at 290°C to the zeolite phase W (see Chap. 4). Magnesium exchanged P converts to a smectite-type material below 200°C.

Effect of Water Vapor at Elevated Temperatures on Zeolites

Pore Closure in Zeolite A

Inefficient or slow removal of water vapor in contact with zeolite crystals at elevated temperatures may produce unusual effects. In the presence of water vapor, zeolite NaA may lose adsorption capacity for certain gases, such as oxygen, but without any appreciable effect on the adsorption of a polar molecule such as water.

Coincident with the decrease in gas adsorption, crystallographic changes occur which are not indicative of structural degradation and which are reversible (Table 6.5). The effect of water vapor is evidenced by changes in adsorption of oxygen, adsorption of water, and by changes in the x-ray powder pattern (the intensities of certain low angle x-ray reflections change). These changes are most evident in zeolite A in the (100) and (110) reflections and are illustrated by the intensity ratio. Also shown is the effect of compaction of the zeolite; that is, results obtained on loose zeolite powder are compared with a zeolite pellet formed by cold pressing. Except for a variation in certain intensities, the overall crystallinity of the zeolite is not affected.

This "pore closure" effect is also evidenced by an increase in the pH of a water slurry of about 0.5 pH units, indicating the release of free base from the zeolite. Other variations include the appearance of the (200) reflection in the electron diffraction pattern of the steam-treated zeolite and a decrease in refractive index.

Table 6.5 gives results obtained with zeolite A after partial exchange by calcium. It appears that partial exchange by calcium diminishes the steaming effect; the pore closure does not occur to the same extent. That the chemical composition of zeolite A is probably not related is evidenced by the data in Table 6.6. The nonstoichiometric A which appears to contain occluded sodium and aluminate ions produces about the same results when subjected to high water vapor concentrations at 450°C as does a zeolite with the correct stoichiometric proportions and no occluded aluminate species. In this example, the water vapor atmos-

Table 6.5 Effect of Water Vapor on Zeolite A (112)[a]

Zeolite	Adsorption of O_2[b]	pH[e]	I(110)/I(100)
1 NaA powder	19.6	11.1	1.8
2 NaA pellet[c]	3.5	11.5	1.38
3 NaA powder, untreated	25	10.5	0.7
4 CaA[d]	19.7	11.1	0.7
5 No. 2 after H_2O[f] treatment	24.4		0.7
6 No. 4 after H_2O[f] treatment	25.3		0.7

[a] Heated in steam at 450°C for 7 hours
[b] −183°C, 700 torr
[c] Pressed cold at 10,000 psi
[d] Level of exchange = Ca_z of 0.22
[e] pH measured in slurry containing 3 g/50 cc H_2O
[f] Suspended in H_2O at 25°C

phere over the zeolite powder resulted from water desorption in a constricted glass tube (12 g of zeolite in a 25 x 200 mm tube).

The effect of steam on the adsorption characteristics of zeolite A (pore closure) can in some cases be reversed by suspending the zeolite in water and reactivating or dehydrating the zeolite in vacuum as shown in Table 6.5.

In summary, it appears that exposing zeolite NaA to steam at high temperatures causes cation migration as evidenced by changes in the intensities of certain x-ray reflections. Cation hydrolysis of some of the Na^+ ions (probably those located in the α cage) results in the formation of hydroxyl ion when in a water slurry. Partial replacement of sodium by calcium reduces the hydrolysis and inhibits the cation migration (112, 113).

Prolonged heating of zeolite A at 350°C in dry nitrogen for 10 hours results in an intensity change in the (110) and (100) reflections, but to a lesser extent.

Table 6.6 Effect of Steam on Zeolite A Powder of Varying Composition (112)[a]

Composition, moles anhydrous basis	I(110)/I(100) Before	After	pH[b] Before	After
$Na_2O \cdot Al_2O_3 \cdot 2\ SiO_2$	0.73	1.22	10.6	10.9
$0.96\ Na_2O \cdot Al_2O_3 \cdot 1.84\ SiO_2$	0.68	1.22	11.0	11.5

[a] Heated in restricted tube at 450°C for 3 hr
[b] 3 g/50 cc in H_2O

Stability of Zeolite A to Water Vapor

The interaction of water vapor with zeolite A and its effect on the properties was measured, considering the following parameters: (1) the exchange cation in the zeolite and the degree of ion exchange; (2) the temperature of water vapor treatment; (3) partial pressure of water vapor over the zeolite; and (4) duration of the water vapor treatment (114). The adsorption of water vapor by the zeolite before and after the treatment with water vapor was measured.

The structure change of NaA was found to be highly dependent upon the water vapor partial pressure. The zeolite is sensitive to small amounts of water vapor at a temperature of 600°C. The exchange cation has a pronounced effect. The lithium exchanged form behaves similarly to the sodium cation form, whereas with cesium, which could only be exchanged to a level of 30%, an enhancement of stability was observed. Similarly, potassium exchange increased the hydrothermal stability after a level of 40% exchange was obtained. Exchange with the divalent cations, calcium, magnesium, and strontium, improved the hydrothermal stability, with magnesium having the most effect. It was observed that the stability decreased with increasing ionic radius of the divalent cations. With the univalent ions the reverse occurred, that is, the larger ion, cesium, produced the greater improvement in stability. These results appear to be supported by the x-ray powder patterns of the treated exchanged forms. The concept that the degradation of the structure occurs from the surface inward was supported by electron microscope photographs. Crystal degradation appears to occur by what is described as delamination, that is, layers of the outside of the crystal seem to detach. Beyond a certain level of exchange the cation does not improve stability and very little improvement occurs at higher cation exchanges.

Stability of Zeolites X and Y to Water Vapor

When exposed to steam at 350°C, sodium X zeolite loses structure and adsorption character. Results in Table 6.7 show that partial hydrogen exchange, obtained by treatment with a mineral acid to lower the pH of the zeolite in water to about 6, improves the stability. Loss in adsorptive capacity (oxygen at low temperature) is minimal and the crystal structure, as followed by the (111) and (533) x-ray powder reflections, is retained (Table 6.7).

Table 6.7 Steam Stability of Zeolite X[a] (112)

Sample Zeolite X		X-Ray Intensity [b]		Adsorption [c]
		(111)	(533)	(wt %)
NaX		17	40	5.3
Na, HX[d]		76	35	30.4
NaX untreated		150	30	35.6
		X-Ray	Adsorption	(wt %)
			H_2O [f]	Kr [g]
NaX pellets [e]	before	–	26.5	60.7
	after [h]	50% loss	14.3	24.2

[a] Powder sample 350°C, 8 hours in 100% steam
[b] Arbitrary units
[c] O_2 at −183°C, 700 torr
[d] Titrated in water slurry to pH 5.9 with HCl (0.5N)
[e] Pellets 4 - 8 mesh hot pressed at 200°C, 25,000 psi
[f] 25°C, 4.6 torr
[g] −183°C, 18 torr
[h] 350 - 375°C steam, 14 hr

Further, the stability of zeolite X can be greatly improved by cation exchange of sodium by Ce^{3+}. Exchange with cations such as K^+ and Ca^{2+} has no significant effect on the stability to water vapor. (Table 6.8).

Many zeolites undergo structural degradation when subjected to water vapor at elevated temperatures. This is of considerable practical importance since water vapor at elevated temperatures may contact zeolites being used in adsorptive processes or catalytic processes. Retention of structure under these hostile conditions is desired. It has been observed that zeolites with a high Si/Al ratio in their framework are more stable. The tetrahedral aluminum ions seem to be subject to hydrolytic attack (see Section D).

The synthetic zeolites X and Y, which have the same framework topology, differ in steam stability. As shown in Fig. 6.21, zeolite Y, with Si/Al > 1.5, retains its structure and crystallinity when subjected to water vapor at 410°C (25).

Zeolites such as mordenite and zeolite L resist hydrolytic degradation.

Transformations of Dehydrated Zeolites

When heated to elevated temperatures in air after dehydration, the crystal structure of zeolites ultimately breaks down, resulting in the form-

Table 6.8 Steam Stability of Zeolite X[a] (112)

Cation Form	%Exchange	Structure[b]	Adsorption[c]
K$^+$	77	−60%	−89%
Na$^+$	100	−80%	−84%
Ca^{2+}	84	−60%	−71%
Ce^{3+}	77	no change	−21%

[a] Loose powder (300°C, 8 hr in 100% steam)
[b] Determined from loss in intensity of selected x-ray powder reflections.
[c] As determined from argon adsorption at −183°C and 700 torr

ation of an amorphous solid and recrystallization to some nonzeolitic species. Examples of this type of recrystallization or transformation include the following:

Zeolite A $\xrightarrow{800°C}$ β-Cristobalite [Buerger structure (165)]
Zeolite X $\xrightarrow{1000°C}$ Carnegeite
Mg^{ex}X $\xrightarrow{1500°C}$ Glass $\xrightarrow{1000°C}$ Cordierite
Zeolite Y $\xrightarrow{1000°C}$ Glass

Recrystallization of NaA starts in the crystal exterior and progresses toward the center. Adsorption measurements were conducted on the thermally treated sample; after grinding the water adsorptive capacity of the treated sample was restored. At 790°C, a large proportion of the specimen was found to be low carnegeite (115).

Information on the thermal stability data for the synthetic zeolites A, X, and Y is shown in Table 6.9 and Fig. 6.22 (116).

The crystallinity was determined by averaging the intensities of selected x-ray powder reflections for each zeolite. Each point on the curve was obtained on a new specimen so that no accumulative effect is observed. All of the zeolites studied seem to exhibit the same type of a decomposition curve as evidenced by complete loss of crystalline structure over a narrow temperature interval.

The weight loss curve exhibits a substantial change in weight over the temperature interval in which the crystal structure is destroyed. For univalent ions a slight loss in addition to the loss on dehydration is shown. This is followed by additional loss in weight as the structure is decomposed. For polyvalent ions, such as calcium or lanthanum the weight loss curve has additional steps. These steps may be due to loss of water from hydroxylated cations in the structure (see Section A).

Table 6.9 The Thermal Stability of Zeolites in Air (116) [a]

Zeolite	Si/Al_2	A_z [b]	t_{init} [c]	$t_{0.5}$ [d]
NaA	2	1	660	755
$Li^{ex}A$	2	0.89	660	675
$K^{ex}A$	2	1	800	825
$NH_4^{ex}A$	2	0.5	120	140
$Mg^{ex}A$	2	0.92	620	755
$Ca^{ex}A$	2	0.96	540	830
$Ba^{ex}A$	2	1.0	90	100
$La^{ex}A$	2	0.11	740	800
NaX	2.5	1	660	770
$Li^{ex}X$	2.5	0.89	680	690
	2.5	0.9	700	710
$K^{ex}X$	2.5	1.0	600	720
$NH_4^{ex}X$		0.7	80	100
$Mg^{ex}X$		0.81	700	755
$Ca^{ex}X$		0.93	710	880
$Ba^{ex}X$		1.0	500	780
$La^{ex}X$		0.9	500	820
NaY	3.4	1	710	795
NaY	4.8	1	700	760
NaY	5.0	1	700	780
$NH_4^{ex}Y$	3.4	0.83	400	500
$NH_4^{ex}Y$	4.7	0.91	500	680
$K^{ex}Y$	3.4	1.0	780	810
$Li^{ex}Y$	3.5	0.75	705	730
$Li^{ex}Y$	4.7	0.66	700	720
$K^{ex}Y$	4.7	1.0	780	835
$Ca^{ex}Y$	3.5	0.91	780	890
$Ca^{ex}Y$	4.7	0.85	800	885
$Ba^{ex}Y$	3.5	0.97	660	750
$Ba^{ex}Y$	4.7	0.83	700	800
$La^{ex}Y$	3.5	0.91	800	900
$La^{ex}Y$	4.8	0.80	840	870
$Mg^{ex}Y$	3.5	0.80	780	820
$Mg^{ex}Y$	4.7	0.78	820	880
KL	3.0	1	845	890

[a] Heated for 16 hr in air after a 2-hr dehydration (in air) at 350°C
[b] Equivalent fraction of exchange cation in the zeolite
[c] Temperature at which structural degradation is first observed from the x-ray powder pattern, in °C
[d] Temperature at which the structure is 50% decomposed, in °C

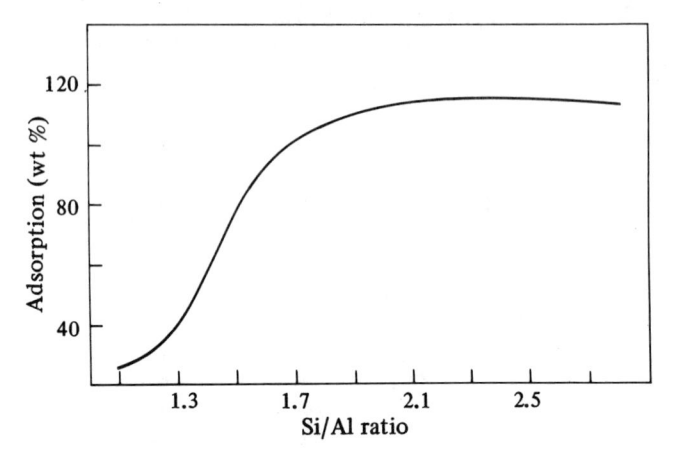

Figure 6.21 Effect of Si-Al composition on hydrolytic stability of zeolites X and Y (25). Zeolites subjected to water vapor at 410° and atmospheric pressure for three hours. Retention of zeolite crystallinity determined from percent retention of oxygen adsorption capacity at 100 torr and −183°C.

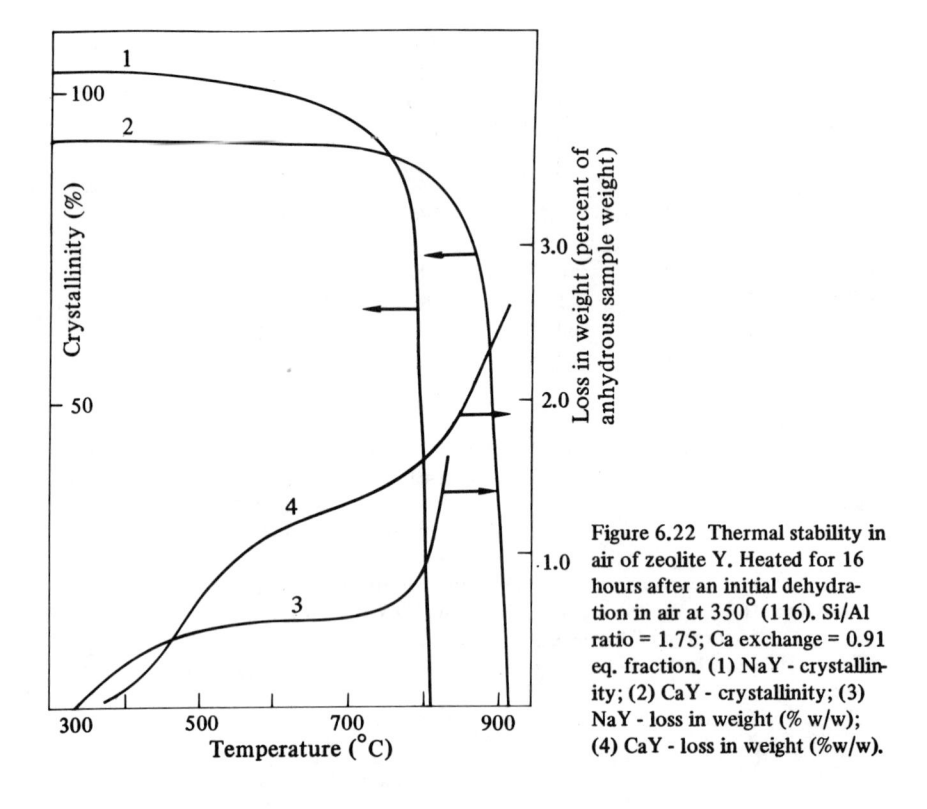

Figure 6.22 Thermal stability in air of zeolite Y. Heated for 16 hours after an initial dehydration in air at 350° (116). Si/Al ratio = 1.75; Ca exchange = 0.91 eq. fraction. (1) NaY - crystallinity; (2) CaY - crystallinity; (3) NaY - loss in weight (% w/w); (4) CaY - loss in weight (%w/w).

Table 6.10 Some Products of the Transformation of Zeolite Y at High Pressure (117)

Temp (°C)	Pressure (kilobars)[a]	Product[b]
300	15	P
400	10	analcime
400	25	jadeite
500	5	analcime
500	20	jadeite
600	5	analcime
600	25	jadeite
700	15	albite

[a]Experiments performed in a uniaxial "simple squeezer" of the Griggs-Kennedy type (118).
[b]Product as identified by x-ray powder pattern.

Transformation of Zeolites at High Pressures

At very high pressures and temperatures, zeolites transform to denser aluminosilicates. This is illustrated by the well-known recrystallization of analcime to jadeite, a pyroxene. Jadeite has a chain structure (see Chap. 2) with a density of 3.3 to 3.4 g/cc. The transformation of synthetic zeolite Y to jadeite has been studied with the results summarized in Table 6.10 (117). Depending upon the conditions, synthetic zeolite Y may transform to the zeolite P, an analcime-type structure, jadeite, or a feldspar of the albite-type. These results are illustrated schematically as follows:

Scheme 8

$Na_{64}[(AlO_2)_{64}(SiO_2)_{128}] \cdot 260\ H_2O$
Zeolite Y, $d_f = 1.25$

$\xrightarrow[15\ \text{kilobars}]{300°C}$ (a) $12[Na_{5.33}(AlO_2)_{5.33}(SiO_2)_{10.66} \cdot 15\ H_2O] + 80\ H_2O$
Zeolite P, $d_f = 1.57$

$\xrightarrow[10\ \text{kilobars}]{400°C}$ (b) $4[Na_{16}(AlO_2)_{16}(SiO_2)_{32}] \cdot 16\ H_2O] + 196\ H_2O$
Analcime, $d_f = 1.85$

$\xrightarrow[20\ \text{kilobars}]{500°C}$ (c) $16[Na_4Al_4(Si_2O_6)_4] + 260\ H_2O$
Jadeite

$\xrightarrow[15\ \text{kilobars}]{700°C}$ (d) $8[Na_4Al_4(Si_3O_8)_4] + 4[Na_8Al_8Si_8O_{32}] + 260\ H_2O$
Albite   Nepheline

At the lower temperature of 300°C, zeolite Y transforms to zeolite P. This is in accordance with the framework densities, (d_f), of the zeolites concerned (1.25 g/cc for zeolite Y and 1.57 g/cc for zeolite P). At higher pressures the zeolite Y converts to an analcime-type with a framework density of 1.85. At still higher pressures, zeolite Y converts to jadeite. The data thus indicate that the zeolite does not convert directly to jadeite but must first pass through either the zeolite P or analcime-type phases. The transformation of analcime to jadeite has been extensively studied and detailed thermodynamic data on this reaction are available (118).

The Transformation of Dehydrated Zeolites by Volatile Phases Other Than Water
Ammonium Chloride

One of the first methods for preparation of a "hydrogen" zeolite was based upon treatment with ammonium chloride vapor at about 300°C (119, 120). When the ammonium exchanged zeolite was subsequently heated in vacuum and treated with oxygen at 350°C, oxidation of the ammonium cation occurred in accordance with Scheme 9.

Scheme 9

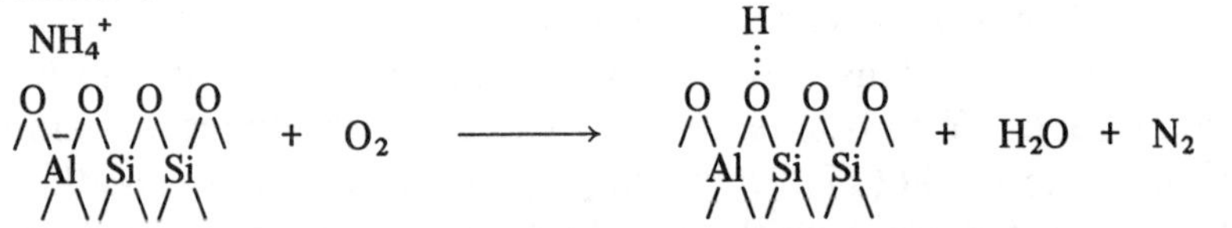

$$NH_4^+ \quad\quad\quad\quad\quad\quad\quad\quad H$$

$$\begin{array}{c} O\ O\ O\ O \\ /\backslash\text{-}/\backslash/\backslash/\backslash \\ Al\ Si\ Si \\ /\ \backslash/\ \backslash/\ \backslash \end{array} + O_2 \longrightarrow \begin{array}{c} O\ O\ O\ O \\ /\backslash/\backslash/\backslash/\backslash \\ Al\ Si\ Si \\ /\ \backslash/\ \backslash/\ \backslash \end{array} + H_2O + N_2$$

This reaction was carried out with chabazite and mordenite. More recently (120), the adsorption of ammonium chloride vapor by several synthetic zeolites has been studied. Since ammonium chloride vapor dissociates into ammonia and hydrogen chloride, the adsorption of ammonia and hydrogen chloride by various zeolites was first examined.

Gaseous hydrogen chloride reacted with various zeolites in the cation form. In some instances, a certain degree of cation exchange occurred with the hydrogen of the hydrogen chloride. In others, where the initial silica to alumina (SiO_2/Al_2O_3) ratio was less than about five, the framework structure of the zeolite was affected. This was attributed to direct attack by the hydrogen chloride on Si-O-Al bonds. The adsorption isotherms for HCl consequently were not totally reversible.

The relative stability of zeolites upon exposure to HCl at different temperatures is given in Table 6.11. The most stable zeolites appear to be sodium Zeolon (mordenite), zeolite Y, and zeolite L. The adsorption of ammonia is reversible except when adsorbed on the hydrogen form of zeolite Y or hydrogen Zeolon. Some degree of chemisorption occurs

Table 6.11 Stability of Zeolites to Anhydrous HCl at 228 Torr (120)

Zeolite	Ratio Si/Al	Temperature (°C) 200°	610°
Na Mordenite	4.75	stable	stable
Na Clinoptilolite	4.65	stable	stable
Zeolite L	3.0	stable	stable
Na Chabazite	2.5	not stable	—
Na Y	2.4	stable	stable
Na X	1.33	not stable	—
Na A	1.0	not stable	—

Stable = no change in x-ray powder diffraction lines.

due to formation of the ammonium cationic form.

When ammonium chloride vapor was adsorbed on hydrogen zeolite Y or hydrogen mordenite, the ammonium chloride dissociation equilibrium was perturbed due to chemisorption of NH_3 by the hydrogen form of the zeolite.

The ammonium forms of zeolite Y and mordenite dissociate by a first order process with a rate constant of about 2×10^{-6} sec^{-1}. The adsorption of ammonium chloride on zeolite L or zeolite NaY follows the isotherm contour of the Brunauer type IV (see Chap. 8).

In zeolite L, ammonium chloride adsorption is reproducible and reversible. Isotherms were measured at temperatures of 245 to 315°C. Similar isotherms were measured on zeolite Y.

The vaporized ammonium chloride is considered as an equimolar mixture of NH_3 and HCl. The solid-vapor equilibrium is subject to perturbation by adsorption on the zeolite. Selective adsorption for one component, such as ammonia, on the hydrogen forms of the zeolites occurred initially, but as the isotherms approached saturation, the balance of each component approached the equimolar ratio in NH_4Cl. This was thought to be due to formation of intracrystalline NH_4Cl.

The decomposition of ammonium exchanged zeolites to form hydrogen and hydroxyl zeolites is discussed in further detail in Sections A and D.

Reaction with NH_3

The hydroxyl groups of a hydrogen zeolite Y, HY, are replaced by H_2N- when the zeolite is exposed to ammonia at 500°C and 1 atm pressure. The ammonia is taken up in a quantity equivalent to conversion of 50% of the hydrogen atoms of the parent zeolite to ammonium ions (121). This material retained the crystal structure of zeolite Y and the ammonia content was observed to be equivalent to the original water content of the hydrogen zeolite. The ammonia is removed by purging the zeolite with an inert gas at temperatures above 600°C. Infrared spectra of the original hydrogen zeolite showed both the HF band (3623 cm^{-1}) and LF band (3533 cm^{-1}) which are the OH stretching frequencies. Upon adsorption of ammonia, bands in the N–H stretching region appeared. The mechanism proposed for this reaction is shown in Scheme 10:

Scheme 10

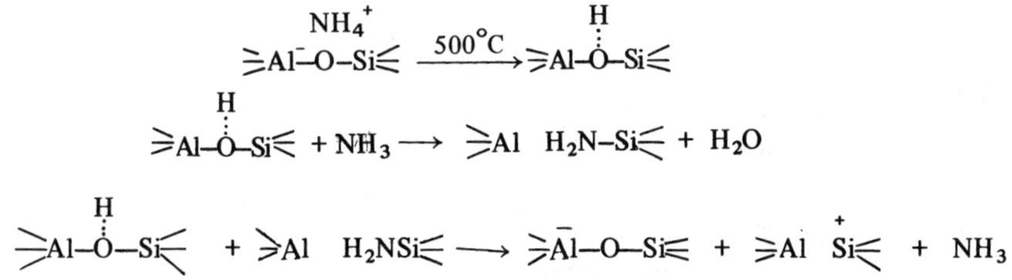

$$\underset{\equiv Al-O-Si\equiv}{NH_4^+} \xrightarrow{500°C} \underset{\equiv Al-O-Si\equiv}{H}$$

$$\underset{\equiv Al-O-Si\equiv}{H} + NH_3 \longrightarrow \equiv Al\ H_2N-Si\equiv + H_2O$$

$$\underset{\equiv Al-O-Si\equiv}{H} + \equiv Al\ H_2NSi\equiv \longrightarrow \equiv \bar{Al}-O-Si\equiv + \equiv Al\ \overset{+}{Si}\equiv + NH_3$$

The amido zeolite reacted with the hydrogen zeolite to form one NH$_3$ molecule for each tetrahedral aluminum in the zeolite structure. One-half of this ammonia was released over the temperature range of 200 - 400°C which is the same as that released from an ammonium exchanged zeolite Y. The formation of the amido zeolite occurs in two steps. Strongly acidic (referred to as type I) sites react with NH$_3$ to form ammonium ions. The second reaction to form the NH$_2$ group then occurs (Scheme 11).

Scheme 11

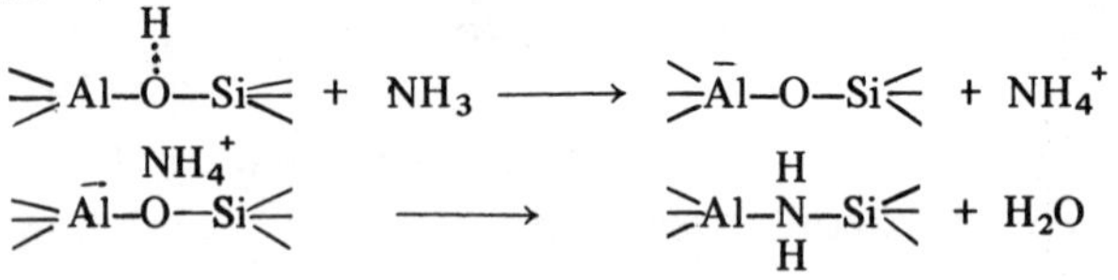

$$\underset{\equiv Al-O-Si\equiv}{H} + NH_3 \longrightarrow \equiv \bar{Al}-O-Si\equiv + NH_4^+$$

$$\underset{\equiv \bar{Al}-O-Si\equiv}{NH_4^+} \longrightarrow \underset{\equiv Al-N-Si\equiv}{\overset{H}{\underset{H}{|}}} + H_2O$$

This suggests that in thermal dehydroxylation (see Section A) the O(1)H hydroxyl groups form protons and the O(3)H groups result in hydroxyl ions. The OH groups in a deammoniated zeolite appear to be removed more readily by the use of NH$_3$ at 500°C than by thermal dehydroxylation at higher temperatures. This reaction may be an alternate route to form a more stable zeolite (see Section D).

Volatile Inorganic Compounds

A dispersed metal phase is produced within zeolite crystal structures by decomposition of adsorbed volatile inorganic compounds. A typical example of this reaction is the decomposition of adsorbed nickel tetracarbonyl, $Ni(CO)_4$. The thermal decomposition of $Ni(CO)_4$ produces a metal dispersion in the zeolite channels although there is some migration to the exterior surfaces; nickel in the form of larger crystallites deposits on the surface of zeolite crystals. The formation of the dispersed metal phase within the zeolite is represented by the following equations (122, 123)

$$NaX + Ni(CO)_4 \longrightarrow NaX \cdot Ni(CO)_4 \qquad (6.1)$$
$$NaX \cdot Ni(CO)_4 \xrightarrow{\Delta} NaX \cdot Ni + 4 CO \qquad (6.2)$$

At room temperature dehydrated zeolite NaX adsorbs 0.30 cc/g of nickel carbonyl. This is equivalent to 32 molecules of $Ni(CO)_4$ per unit cell or 4 molecules in each supercage. By repeated adsorption and decomposition of the carbonyl, the zeolite channels can be nearly filled with nickel.

There are other volatile inorganic compounds which behave in a similar manner such as iron pentacarbonyl, $Fe(CO)_5$, and certain π complexes such as dicyclopentadienyl nickel, $\pi(C_5H_5)_2 Ni$.

Metal phases dispersed in the zeolite in this manner may undergo other reactions. For example, iron when introduced into the zeolite as the carbonyl can be converted by oxidation to small, intracrystalline clusters of alpha iron oxide, Fe_2O_3 (124). Electron photomicrographs showed no extra crystal material. It is reported that this results in iron oxide clusters of controlled particle size within the zeolite. At high temperatures, 900°C, the zeolite decomposes and crystalline αFe_2O_3 is formed. The resulting iron oxide dispersion displays superparamagnetism.

Nonmetallic elements such as sulfur and tellurium may be deposited within the zeolite cavities (NaX) by vapor phase treatment at elevated temperatures. A tellurium-loaded NaX was prepared by ball milling a dry Te-NaX mixture. Structural studies showed that after heating to 538°C in He and H_2 at 482°C five Te atoms per unit cell occupy sites in the supercages coordinated to Na^+ ions in site II and in the β-

cages coordinated to Na⁺ ions in site I'. Additional coordination to a Na⁺ ion in site III was indicated (125). The zeolite containing Te is a dehydrocyclization catalyst.

Some reducible metal cations when exchanged into a zeolite may be reduced upon further contact with a metal vapor such as mercury. In the examples illustrated in Eqs 6.3 and 6.4, the mercury exchanged form of zeolite X, Hg^{2+}, or the silver-exchanged form were contacted with mercury vapor at a temperature of over 200°C. In the first case (Eq. 6.3) disproportionation occurs and the resulting zeolite contains the Hg_2^{2+} ion. In the second case (Eq. 6.4), a dispersed phase of silver remains with Hg^{2+} in the exchange sites (126).

$$Hg^{ex}X + Hg \xrightarrow[300 \text{ torr}]{236°C} Hg_2^{2+}X \qquad (6.3)$$

$$Ag^{ex}X + Hg \xrightarrow[100 \text{ torr}]{213°C} Hg^{2+}X \cdot Ag \qquad (6.4)$$

It has been reported that nitrosyl chloride, NOCl, may react with the silver exchanged form of the zeolite analcime to convert it to a nitrosyl zeolite; silver chloride is formed simultaneously (127).

$$Ag^{ex} \text{ analcime} + NOCl \longrightarrow NO^+ \text{ analcime} + AgCl \qquad (6.5)$$

Reactions may occur between adsorbates within the zeolite. Silicon tetrafluoride, SiF_4, is irreversibly adsorbed to some extent on both zeolite NaA and NaX at 0°C. By maintaining the zeolite at 200°C in the presence of SiF_4, reproducible adsorption isotherms were obtained. It was concluded that reaction occurred between residual hydroxyl groups in the zeolite and SiF_4 since this resulted in some reduction in the usual nitrogen adsorption capacity. Attempts were made to react adsorbed SiF_4 and amines within the zeolite structure by first adsorbing ammonia or methylamine and then treating the zeolite with SiF_4 at -78°C. The ratio of SiF_4 to NH_3 in the NaA zeolite was 1:1. The normal gas phase adduct is $SiF_4 \cdot 2NH_3$. Apparently the 1:1 ratio could not be exceeded within the zeolite. A similar result was obtained with methyl amine. In the case of zeolite X complex the ratio was greater than 1. At higher temperatures the ratio was about 4 in zeolite A and 7 or higher in zeolite X (128).

D. REACTIONS IN SOLUTION

Reaction with Strong Acids

It was early recognized that zeolite minerals are decomposed by acids, many of them with the subsequent formation of gels (129). A detailed classification of silicate minerals that is related to the internal structure was proposed (130). Silicates which are decomposed by treatment with strong acids may be classified into one of two groups: (1) those that separate insoluble silica without the formation of a gel and (2) those that gelatinize upon the acid treatment. The rule generally followed is that those zeolites that have a framework structure silicon/ aluminum ratio of 1.5 or less gelatinize. Zeolites with a silicon/aluminum ratio of greater than 1.5 generally decompose and form a precipitate of hydrous silica. Zeolite A, for example, is readily decomposed by HCl and precipitates a clear gel. Similar results are obtained with zeolite X. On the other hand, zeolite Y with a silicon/aluminum ratio of greater than 1.5, upon treatment with HCl decomposes and precipitates silica. Some zeolites, such as mordenite, resist acid treatment, although the cations may be removed by strong acids to produce a hydrogen form. Hydrogen exchange is further discussed in Chap. 7.

Mordenite, and the synthetic variety, Zeolon, and the mineral erionite, are stable in aqueous acid and may be converted to the hydrogen form through ion exchange of the metal ions by proton or hydronium, H_3O^+. Under hydrothermal conditions these acid-treated or hydrogen zeolites recrystallize to kaolinite (131). Because the zeolites are aluminum deficient relative to kaolin, additional aluminum is added in the form of a salt, such as aluminum chloride, $AlCl_3$. The hydrogen form of the mineral crionite was found to transform into a kaolin-type structure at temperatures as low as 175°C. In the absence of additional aluminum chloride, erionite at 230°C for 10 days in 0.1 N HCl showed little transformation. Erionite pretreated with aluminum chloride transformed in 10 days at 175°C. It was further observed that the synthetic mordenite-type Zeolon did not transform to the same extent as a synthetic, small port-mordenite. Mechanistically it is proposed that the coordination change of aluminum from 4 to 6, due to the presence of hydronium, is the initiating step. This would produce Si^{IV}-O-Al^{VI} units which have a strong polymerization tendency to form the sheet-type structure characteristic of kaolin. The first step of the transformation is the hydration of the aluminum tetrahedra with a change in coordination to octahedral. In the synthesis of layer-type minerals, the formation of these short chains is the structure-determining step rather than the "grafting" of aluminum octahedra onto an already preformed silica sheet.

Table 6.12 Siloxane Derivatives of Zeolites (132)

Zeolite	\multicolumn{5}{c}{Percent of Si Recovered As:}
	SiO_4	Si_2O_7	Si_3O_{10}	$(SiO_3)_4$	Total
Sodalite	76.0	8.7	2.5		87.2
Natrolite	10.0	13.1	67.5		90.6
Laumontite				80.9	80.9

The disintegration of zeolite structures by strong acid is related to the number of aluminum atoms in tetrahedral positions in the framework since they appear to be the site of acid attack, probably by hydrolysis.

The nature of the silicate anion species in sodium silicate solutions and in residues obtained from the acid disintegration of crystalline silicates has been determined by a chemical method (132). The crystalline silicate is disintegrated using a strong acid followed by end-blocking of the silicate species with trimethylsilyl derivatives. The silicate is decomposed in acid solution which contains hexamethyldisiloxane and the siloxane derivative is subsequently identified by chromatography. For example, it was shown that (see Table 6.12) natrolite contains groups of three SiO_4 tetrahedra and laumontite contains 4-rings. These results agree with the known crystal structures of these zeolites.

Direct trimethylsilylation of various crystalline silicates, including laumontite, has been utilized to study their structures. From the zeolite laumontite over 96% of the silica converted to the derivative $[(CH_3)_3Si]_8Si_4O_{12}$; this is consistent with the presence of 4-rings (133).

Reaction with Strong Bases

The stabilities of zeolite minerals in various environments and the structural transformation of the mineral and synthetic zeolites are important to an understanding of the mechanism and the conditions for crystallization either in nature or in the laboratory. As discussed in Chap. 4, many of the synthetic zeolites, although stable under the particular conditions of their synthesis, may undergo conversion to different species with time. Thus zeolite A when exposed to dilute sodium hydroxide solutions for prolonged periods of time converts to the zeolite P phase. In more concentrated solutions, it converts further to the hydroxysodalite hydrate, HS.

Scheme 12

$$5\ Na_{12}[(AlO_2)_{12}(SiO_2)_{12}]\cdot 27\ H_2O$$
Zeolite A

$\downarrow$ NaOH(aq), 100°C

$$6\ Na_6[(AlO_2)_6(SiO_2)_{10}]\cdot 14\ H_2O\ +\ 24\ NaAlO_2$$
Zeolite P

$\downarrow$

$$Na_6[(AlO_2)_6(SiO_2)_6\cdot x\,NaOH]\cdot 4.5\ H_2O$$
Sodalite hydrate $0 < x < 1$

Apparently, resolution of the zeolite A phase is followed by recrystallization of the P phase from solution. Electron micrograph investigations indicate that the P phase in many instances crystallizes on the surface of the A crystals still present in the system. The process is not truly in equilibrium since the P zeolite will not reconvert to zeolite A (see Fig. 5.7 of Chap. 5) and will further transform to a hydrated hydroxysodalite.

Chelating Agents

Aluminum may be removed from the tetrahedral positions in the framework structure of clinoptilolite and mordenite by treatment with strong acids such as HCl; other zeolites are completely destroyed. The removal of aluminum from the tetrahedral sites of zeolite Y by treatment of a zeolite with a chelating agent has also been reported (134). The chelating agent was ethylenediamine tetraacetic acid, abbreviated as H_4EDTA. The mechanism for this reaction is proposed to be as follows:

Scheme 13

(a) Partial conversion to hydrogen zeolite.

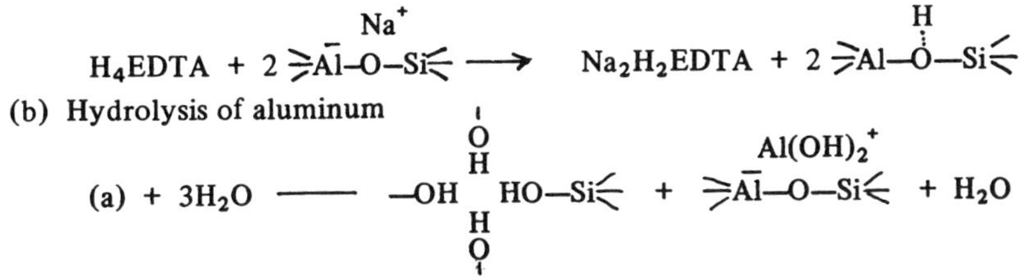

$$H_4EDTA\ +\ 2\ {\geq}Al\text{–}O\text{–}Si{\leq} \xrightarrow{Na^+}\ Na_2H_2EDTA\ +\ 2\ {\geq}Al\overset{H}{\text{–}}O\text{–}Si{\leq}$$

(b) Hydrolysis of aluminum

$$(a)\ +\ 3H_2O\ \longrightarrow\ \text{–OH} \cdots \text{HO–Si}{\leq}\ +\ {\geq}\bar{Al}\text{–}O\text{–}Si{\leq}\ +\ H_2O\ \ (+\ Al(OH)_2^+)$$

(c) Exchange of aluminum cation by sodium ion.

(b) $+ Na_2H_2EDTA \longrightarrow \geq\overline{Al}\text{—}O\text{—}Si\leq \overset{Na^+}{} + NaAlEDTA \cdot H_2O + H_2O$

(d) Overall stoichiometry.

$$xH_4EDTA + Na(AlO_2)(SiO_2)_y$$
$$\downarrow$$
$$xNaAlEDTA \cdot H_2O + (NaAlO_2)_{1-x}(SiO)_y + xH_2O$$

where $x \leq 1$ and $y \geq 2.5$

The complete removal of aluminum from the framework by this method destroys the structure completely. The optimum level of removal for producing a thermally stable product is in the range of 25 to 50%. Starting with zeolite Y, Si/Al = 2.63, removal of 50% of the aluminum atoms raises the Si/Al ratio to about 10.6. This corresponds to the loss of about 27 aluminum ions per unit cell (or an average of two aluminum ions in each double 6-ring). The four hydroxyls that are formed within the tetrahedron are probably attached to a silicon atom and may undergo condensation with the release of water to form new Si-O-Si bonds. This would leave about two aluminum atoms in each double 6-ring as well as two defect sites. An alternative explanation is needed.

Removal of framework aluminum by extraction with acetylacetone has been accomplished. Zeolite X and Y are first exchanged to the ammonium form using an ammonium salt such as NH_4Cl. The NH_4 zeolite is then dehydrated and deammoniated by heating in air at about 550°C for up to three hours. This material was treated with acetylacetone at reflux temperature. In one example the SiO_2/Al_2O_3 ratio of a zeolite Y was increased from 4.94 (55 Al/unit cell) to 6.94 (43 Al/unit cell) starting with a 40% NH_4 exchanged zeolite (135).

Scheme 14

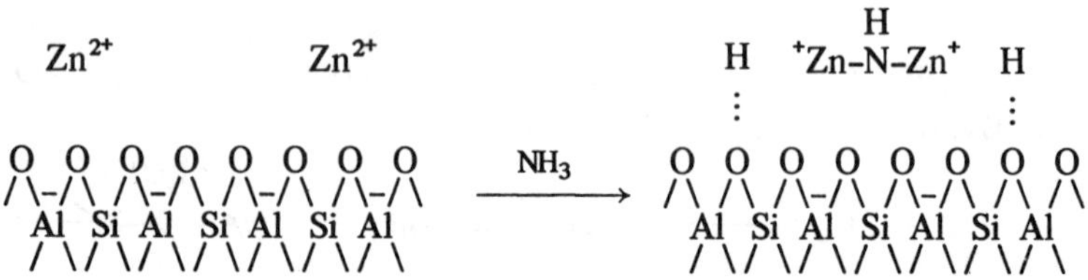

```
                                                              H
    Zn²⁺           Zn²⁺                      H   ⁺Zn-N-Zn⁺   H
                                                 ⋮           ⋮
    O O O O O O O O       NH₃        O O O O O O O O
    /\-/\ /\-/\ /\-/\ /\-/\   ———→   /\ /\ /\-/\ /\-/\ /\ /\
    Al Si Al Si Al Si Al              Al Si Al Si Al Si Al
    /\/ \/ \/ \/ \/ \/ \             /\/ \/ \/ \/ \/ \/ \
```

Zeolite A was exchanged with Zn^{2+} to a level of about 80%, then treated with NH_3 in aqueous suspension and dried at 250°F to 300°F. It is reported that imido bridges form between the Zn^{2+} cations with co-current protonation of a cation site (136) Scheme 14.

This form of zeolite A shows the superstructure lines characteristic of the true ordered cell, a = 24.68 A (see Chap. 2). Thermally it is s stable to 370°C in inert atmospheres but at 480°C the —N— bridges break and the superstructure lines disappear from the pattern. However, if NH_3 is readmitted below 370°C, the superstructure is reformed (136).

E. DEFECT STRUCTURES - STABILIZATION - SUPERSTABLE ZEOLITES

During the first investigations, the dehydration and thermal decomposition of ammonium-exchanged zeolite Y was referred to as "decationization" (91). The decomposition product consists of a crystalline zeolite from which the cations were removed by extensive NH_4^+ exchange followed by dehydration and removal of NH_4^+ ions as NH_3. The crystal structure remained unchanged at temperatures as high as 600°C, with some structural changes occurring at 700°C (91). Thus the decationization process consists of (1) deammoniation — to refer to the removal of ammonium ions as ammonia — and (2) dehydroxylation, which refers to the elimination of hydroxyl groups as water. These two processes may overlap somewhat.

A decationized zeolite Y has been prepared which exhibits improved thermal and hydrothermal stability (137). This form has been referred to as "ultrastable" zeolite Y. In order to avoid confusion this and similar materials are more broadly referred to as superstable zeolites. The crystal structure of this ultrastable Y is retained after heating to temperatures in excess of 1000°C.

The procedure for preparing this ultrastable form of zeolite Y was described in four steps:

1. Sodium zeolite Y (Si/Al = 2.45 to 3.0) was first ion exchanged with ammonium sulfate at 100°C to a level of about 80% (equivalents NH_4^+/total cation equivalents).
2. The ammonium exchanged zeolite, containing residual sodium, was heated at 540°C for three hours; this results in a material that corresponds to the regular decationized zeolite.

3. Additional ammonium exchange was achieved by treatment of the calcined zeolite of step 2 at 100°C in ammonium salt solution resulting in a reduction of the residual Na_2O content to 0.1% by weight. It is proposed that the calcination treatment after the first ion exchange step causes the release of sodium ions from positions within the structure (site I) to sites which are more accessible to the incoming ammonium cations. Thermal activation of the cations is the basis for increasing the facility of this ammonium exchange.
4. The final step is to heat the zeolite after the second exchange to 815°C for 3 hours.

The intermediate phase after step 3 is reported to be metastable. The final product is thermally stable to 1000°C as determined by x-ray powder diffraction studies and adsorption measurements. The usual decomposition temperature is about 800°C for sodium zeolite Y. If the final secondary treatment is performed at 425°C, instead of 815°C, the resulting product has the same chemical composition as the stabilized or ultrastable form, but is not stable.

The removal of the ammonium ion by thermal decomposition and dehydroxylation, followed by conversion to the stabilized form, results in a decrease in the unit cell constant. This decrease is about 1.25%. For example, a unit cell dimension of 24.35 A is observed for the ultrastable form as compared to a typical value of 24.7 A for an ammonium exchanged zeolite Y.

Various modifications of this stabilizing treatment have been disclosed. In one modification the zeolite is first exchanged with ammonium until the residual alkali metal is less than 1 wt %. Stabilization of the resulting zeolite is then achieved by heating at a temperature of 700°C to 1040°C. This eliminates the double exchange treatment and the improved stability seems to be related to the residual level of alkali metal (138).

In one study (139), the thermal stability of zeolite Y was reported to be a function of the degree of ammonium exchange. Based on x-ray powder diffraction analyses, the zeolite crystal structure was retained until the breakdown temperature; beyond this temperature the zeolite crystallized to a mullite-type phase. The mechanism proposed for this decomposition was based upon the interaction of the alkali-metal ion, sodium, with the Al-O bonds. This interaction was said to weaken the bonds and result in aluminum which is coordinated in a trigonal arrangement with the oxygen ions and thus forms some unstable Al-O-Al linkages. The formation of the mullite phase is indicated by the usual exotherm peaks in the DTA patterns. It was further reported that samples which were prepared with equal low residual sodium contents were as stable as those prepared by the ultrastabilization procedure described previously.

The atmosphere surrounding the ammonium exchanged zeolite during the thermal treatment is very important. The hydrogen form of zeolite Y dehydroxylates rapidly when heated to 650°C in an inert gas (140). If water is removed by heating in vacuum the resulting zeolite has relatively poor thermal stability. If the zeolite is calcined in the presence of water vapor, the thermal stability is considerably increased; the crystal structure is retained at 1000°C. Further, it was reported that 25% of the aluminum atoms in tetrahedral positions were converted to a cationic form which could be removed from the structure by ion exchange with sodium (0.1 N NaOH solution).

Scheme 15 Hydrolysis of Framework Aluminum (140).

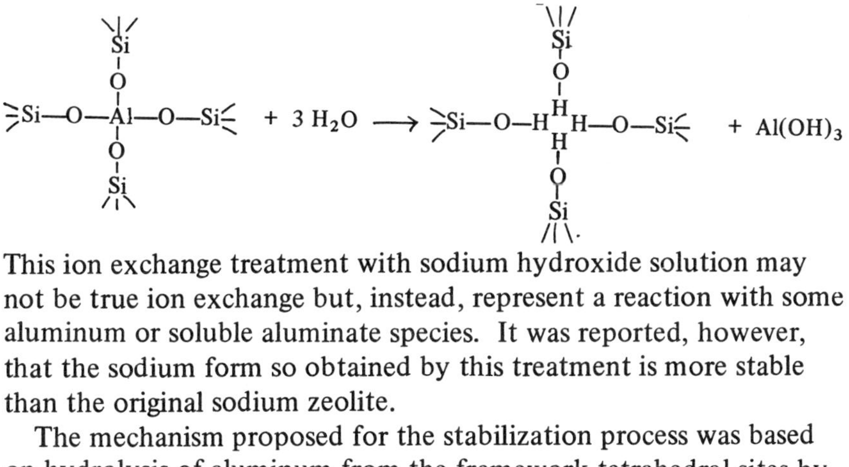

$$\equiv Si-O-\underset{\underset{\underset{Si}{|}}{\overset{|}{O}}}{\overset{\overset{Si}{|}}{\underset{|}{Al}}}-O-Si\equiv \;+\; 3\,H_2O \longrightarrow \;\equiv Si-O-H\overset{H}{\underset{H}{}}H-O-Si\equiv \;+\; Al(OH)_3$$

(with additional Si—O—H...O—Si groups above and below)

This ion exchange treatment with sodium hydroxide solution may not be true ion exchange but, instead, represent a reaction with some aluminum or soluble aluminate species. It was reported, however, that the sodium form so obtained by this treatment is more stable than the original sodium zeolite.

The mechanism proposed for the stabilization process was based on hydrolysis of aluminum from the framework tetrahedral sites by water at elevated temperatures. This is followed by neutralization of the transient aluminum hydroxide phase by the remaining zeolite acidic hydroxyl groups (Scheme 15). The hydroxyl groups produced within the tetrahedra, like those proposed to be formed when a zeolite is treated with a strong acid, are removed during the final calcination procedure and it was speculated that new Si-O-Si linkages are formed. It was further concluded that the loss of aluminum from the tetrahedral positions results in the reduction of the unit cell dimension.

In another variation of this procedure, an ammonium exchanged zeolite Y was calcined at 500°C in a container of varying bed geometry. When the zeolite was calcined in a "shallow" bed, in order to permit the rapid removal of the gaseous decomposition products, ammonia and water, the product lost crystallinity when heated to 850°C and exhibited hydrothermal instability. The zeolite was calcined in a "deep" bed in order to retard the loss of the gaseous products. This material retained crystallinity to a temperature of 1050°C. Furthermore, the deep-bed calcined zeolite exhibited better hydrothermal stability. When treated with sodium hydroxide solution, aluminum in a cationic form was removed (98).

These materials were studied by utilizing ir spectra (141). The hydroxyl region was observed for a "deep-bed" calcined ammonium Y zeolite. In addition to the usual HF and LF bands at 3650 cm^{-1} and 3550 cm^{-1}, intense bands were observed at 3700 and 3600 cm^{-1}. Chemically, it was found that upon treatment of the deep-bed calcined ammonium Y with dilute NaOH aluminum was released corresponding to 31% of the original aluminum. This represents about 17 tetrahedral aluminum atoms per unit cell. The unit cell constant, as determined by x-ray diffraction of the deep-bed calcined Y, also decreased. In these experiments a considerable amount of the original sodium was still present (up to 31% of the original sodium in the zeolite) in the deep-bed calcined sample. This indicates that the residual sodium may not be a major factor in the stabilization process.

From these results it appears that calcination in the presence of water vapor or ammonia, or both, is important to the stabilization process, although the stability may be influenced by some residual sodium content. Partial dehydroxylation in the presence of some cationic hydroxyaluminum species on ion exchange sites appears to be essential. Although the stabilized zeolite has a structure which is deficient in aluminum, vacant tetrahedral sites were not proposed as important to the stabilization mechanism.

Crystal structure analyses, based on x-ray powder data, have been reported on various products obtained from ammonium exchanged zeolite Y by different types of heat treatment (142). These four materials were: (1) a regularly "decationized" zeolite Y, (2) an ammonium-aluminum zeolite Y, (3) a typical stabilized zeolite as prepared by the double-exchange method described previously, and (4) a re-exchanged and heat-treated form of the material of structure 3 (Table 6.13). The results of this study, as shown by the population of framework and nonframework positions, are given in Table 6.14. The structure 1, referred to as regularly decationized zeolite Y, is missing 21 framework O(1) atoms and 15 of the tetrahedral aluminum atoms are converted into cationic aluminum ions located in site I. This occurs without additional water vapor treatment. Structure 2, the ammonium re-exchanged form, shows a restoration of the framework sites but with an excess of cationic aluminum species. Structure 3, which is the stabilized zeolite, has 15 framework aluminum atoms missing along with 25 O(3) and 13 O(4) atoms from the framework. Structure 4, exhibits a mean Si-O bond length of 1.610 A, which indicates that very little framework aluminum is present. The unit cell parameters are summarized in Table 6.13. In structure 4, the occupancy of all of the tetrahedral sites together with the Si-O distance sug-

Table 6.13. Composition and Parameters of Treated Zeolite Y (142)

Structure	Composition	a (A)	Mean Si,Al-O (A)
1	$Na_9[(AlO_2)_{53}(SiO_2)_{139}]$	24.56	1.69
2	$Na_{0.5}[(AlO_2)_{53}(SiO_2)_{139}] \cdot nH_2O$	24.59	1.65
3	$Na_{0.5}[(AlO_2)_{53}(SiO_2)_{139}]$	24.31	1.62
4	High Silica Y	24.24	1.61

Table 6.14 Summary of Occupancy of Framework and Nonframework Position[a] (142)

Position	Number	Structure 1	Structure 2	Structure 3	Structure 4
Si, Al	192	176.6	192.0	176.6	192
O (1)	96	74.9	96	96	96
O (2)	96	96	96	96	96
O (3)	96	96	96	71.0	83.5
O (4)	96	96	96	82.6	96
S I'	32	14.7 Al^{3+}	13.4 Al^{3+}	4.9 AlO(OH)	2.4 AlO(OH)
S II'	32	32 OH^-	15.0 OH^-		
S IV'	16	5.1 Na^+	5.9 NH_4^+		

[a] See Chap. 2 for definiton of cation or nonframework sites

gests that the sites vacated by the aluminum are subsequently reoccupied by silicon. It was observed that the stabilization reactions are performed under hydrothermal conditions which are sufficient to transport silica. However, chemical analyses of the superstable zeolite and the high silica zeolite of structure 4 showed that they have essentially the same SiO_2 and Al_2O_3 content. It was concluded that the thermal treatments which cause removal of aluminum from the framework also result in the formation of an insoluble polymeric aluminum-oxygen compound which cannot be removed by the ammonium sulfate treatment used in the preparation of structure 4.

The products of variously pretreated specimens of NH_4Y have been studied by ir spectroscopy (95, 109). The treatment of NH_4Y with water vapor at four levels (*in vacuo*, moist air, self-steaming and 100% steam) at 540°C and 760°C resulted in a range of products of different stabilities (109). The zeolite Y (Si/Al = 2.1) with the Na_2O level reduced to below 0.2 wt% by the usual process of NH_4^+ exchange, calcination at 540°C and additional NH_4^+ exchange was converted to the stabilized form by either self-steaming (deep bed, static atmosphere) or 100% steam treatment at 820°C. The SiO_2/Al_2O_3 ratio increased and the unit cell constant decreased to 24.24 A. The increased SiO_2/Al_2O_3 ratio was attributed to a recrystallization process which involves the migration of SiO_2 into the tetrahedral vacancies created by dealumination. The silica possibly originates from regions of structurally degraded zeolite.

Infrared spectra of the zeolites obtained after treatments (at the same water vapor levels) at 540°C showed that the 3650 cm^{-1} hydroxyl band increased in intensity under the self-steaming treatments; with 100% steam this band (HF) disappears but new bands at 3600 and 3700 cm^{-1} appear. Heat treatment of this product reduced the intensity of the 3600 cm^{-1} band and this band was attributed to a hydroxyl group close to an aluminum resulting from hydrolytic dealumination.

Similar pretreatments of a partially ammonium exchanged Y were followed by ir spectra of the hydroxyl region (95). The spectra were resolved into several individual bands. Zeolite Y of unit cell composition $Na_{17}(NH_4)_{38}[(AlO_2)_{55}(SiO_2)_{137}]$, was calcined under deep-bed (self-steaming) conditions at 560°C and treated with NH_4Cl solution. This NH_4Cl treatment did not result in aluminum removal. However, treatment of the self-steamed sample with 0.1 N NaOH removed 31%

of the aluminum in confirmation of the earlier results (140). The main difference in the hydroxyl spectra was a decrease in the HF, 3650 cm^{-1}, and increase in the 3600 cm^{-1} band. This 3600 cm^{-1} band was not attributed to an AlOH species since treatment with NaOH did not reduce the intensity. On the other hand it was assigned to an aluminum vacancy in the framework. It was further concluded that the aluminum ions in site I' and site II' are present as a polynuclear cation of the type $Al^{2+}-O-Al^{2+}$. The presence of these cations in the β-cages enhances the thermal stability. Bands at 3690 - 3670 and 3600 cm^{-1} were also observed as the result of hydrolysis of aluminum in LaY and CaY zeolite. These were also assigned to nonacidic hydroxyls in the structure associated with lattice vacancies.

The enhanced stability at high temperatures and under hydrothermal conditions of the stabilized zeolites is of considerable significance since these materials may be used as a base for preparing zeolite catalysts. Considerable information has appeared in the patent literature which is concerned with stabilized forms of zeolites and their methods of preparation.

This increase in thermal and hydrothermal stability of zeolite Y is a complex structural-chemical phenomenon. The product of various treatments probably consists of more than one phase. However, the increase in stability seems to be related to the removal of aluminum from the framework positions by hydrolysis of the ammonium exchanged form. A decrease in the unit cell is consistent with the removal of aluminum atoms from tetrahedral sites and replacement by silicon atoms. The presence of hydroxyl groups in vacant tetrahedra would not result in a reduction in the unit cell dimension.

The silicon which is transported at elevated temperatures in the water vapor atmosphere is probably from a zeolite decomposition product. Reducing the sodium to a low level may not be as important.

It may be concluded that the stabilization process, as it is applied to the ammonium form of the zeolite, is a process of controlled thermal decomposition and dehydroxylation. It seems to be related to the formation of a hydroxy, polyvalent cation in the cation sites. This is similar to the enhancement in the stability of zeolite X by the introduction of polyvalent rare earth ions. It is proposed that dehydration and decomposition of ammonium exchanged zeolites causes some of the framework aluminum to be rejected into sites within the β-cage.

During this removal, as a result of the dehydroxylation process, framework defects may be present. This is borne out by the crystal structure studies. Subsequent heat treatment in a hydrothermal atmosphere may cause silica (from a small portion of degraded zeolite) to take positions in the vacant tetrahedral sites and reconstruct the framework.

The chemistry of the stabilization process in ammonium exchanged zeolites is certainly concerned with the chemistry of NH_3, NH_4^+, aluminum cations (Al^{3+}, $AlOH^{2+}$, etc.), anions and hydrous aluminum oxide. Relevant equilibria concerning the various species that may be involved in the stabilization process are summarized in the accompanying Table 6.15. These data are based upon the known aqueous solution chemistry of aluminum (143). In the zeolite environment, they are not totally applicable.

The relation between pH and the aluminum species in aqueous solution is shown in Fig. 6.23. Over a wide range of pH the stable species is amorphous aluminum hydroxide. When treated with a concentrated ammonium salt solution (pH~4) an aluminum cation species or $Al(OH)_3$ will most likely form in the zeolite. Similarly, the acidic environment within a hydrogen zeolite Y will form cationic aluminum species from the aluminum hydroxide.

In the hydrogen zeolite obtained by the dehydration and decomposition of the hydrated ammonium exchanged zeolite, a cationic $AlOH^{2+}$ is favored. In the ammonium exchanged form of a decationized zeolite and in the presence of water vapor it is likely that $Al(OH)_3$ is favored. It is highly unlikely that anionic aluminate will be present in the hydrogen zeolite.

Table 6.15 Some Equilibria of Aluminum Species in Aqueous Solution (143)

		log K
1	$Al(OH)_3 + 3H^+$ aq = Al^{3+}aq + 3 H_2O amorphous	9.71
2	$Al OH^{2+}$aq + H^+aq = Al^{3+}aq + H_2O	4.9
3	$Al(OH)_3 + 2 H^+$ = $Al OH^{2+}$ aq + $2 H_2O$ amorphous	4.99
4	$Al(OH)_3$ = H^+aq + AlO_2^-aq + H_2O amorphous	−10.66
5	$Al(OH)_3 + H^+$aq = $Al(OH)_2^+$ aq + H_2O amorphous	0.65
6	2 $Al OH^{2+}$ = Al^{3+} + $Al(OH)_2^+$ aq	0.5

Figure 6.23 Concentration of aluminum as Al^{3+} or $AlOH^{2+}$ in aqueous solution in equilibrium with amorphous $Al(OH)_3$ as a function of pH.

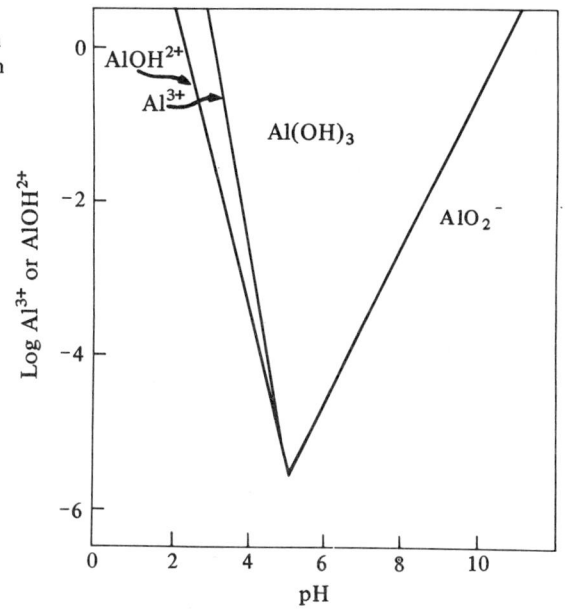

Proposed Mechanism and Chemistry of Destabilization and Stabilization Processes in Zeolites

The decationization and stabilization processes have many chemical aspects in common. The following section summarizes two proposed reaction mechanisms which explain the observed differences. These proposed mechanisms are probably over simplified but are an effort to rationalize the stability differences between superstable and normal decationized zeolite Y.

Any explanation of the decationization and stabilization phenomena observed in several zeolites must make sense chemically, structurally and in terms of the overall reaction stoichiometry. Most published studies have been concerned with reactions involving ammonium exchanged zeolites, in particular ammonium Y. Two general types of phenomena may occur. In the first type, deammoniation and dehydroxylation are conducted *in vacuo* at low pressures or in a purge, either of which removes the volatilized products, ammonia and water. Normally these reactions must overlap to a considerable extent. In the second type, decomposition products are retained; water vapor and ammonia are present in the vapor phase. This is characteristic of the deep-bed or self-steaming calcination reaction described above. Water vapor may also be added in a steaming treatment.

The first type results in destabilization, that is, the formation of a metastable, decationized zeolite containing structural defects. The second type results in a stabilized form or, a structure of increased thermodynamic stability, and with a minimum of structural defects such as lattice oxygen vacancies. The driving force appears to involve the reaction of the cationic, hydroxy-aluminum groups and mobile acidic hydrogens on the zeolite framework oxygen atoms as discussed previously.

In the following discussion reaction schemes are used to illustrate changes in the zeolite stoichiometry based on the zeolite unit cell contents. Although this may appear to be unwieldy and complex, it does enable one to interpret the stoichiometry in terms of the zeolite structure consisting of the tetrahedral atoms, framework oxygen atoms, cation types and positions, etc.

As an example, a typical zeolite Y is exchanged to a level of about 85% by NH_4^+ and consequently will have the following unit cell contents: $Na_8(NH_4)_{48}[Al_{56}Si_{136}O_{384}] \cdot 234H_2O$.

As dehydration of this zeolite proceeds in a vacuum or a purge, a composition is reached where the H_2O/NH_4^+ ratio is about unity. It is proposed that tetrahedral aluminum atoms are removed from the framework by reaction with adjacent acidic protons to form a hydroxy-aluminum cation and a hydrogen on an oxygen atom as a hydroxyl group (Eq. 6.6). The change in unit cell stoichiometry is shown in Scheme 16; reaction 6.7 is exothermic and may account for the exotherm which is observed at 500°C in the DTA of ammonium Y (Fig. 6.19).

Initially the number of aluminum atoms removed from tetrahedral sites is governed by the availability of sites within the sodalite or β-cages, two in each. The approximately 16 aluminum atoms, as the 16 cationic species, account for one-third of the total number of hydrogen atoms available on the framework. As the temperature increases, dehydroxylation occurs leaving a framework with 16 tetrahedral vacancies as well as 32 oxygen vacancies. The hydroxyl water that is removed may vary from 16 to 32 molecules for each 56 aluminum atoms or a ratio of about 1 to 3. This has been observed. The ion exchange capacity, as determined by treatment with an ammonium salt solution, would be in the ratio of 32/56 or a little over 50% of

Scheme 16 Decationization in Inert Purge or Vacuum

(a) $Na_8[NH_4]_{48}[Al_{56}Si_{136}O_{384}] \cdot 234\ H_2O$

$$\downarrow \text{Dehydration, deammoniation} \sim 350°C$$

(b) $Na_8H_{48}[AlOH^{2+}]_{16}[Al_{40}Si_{136}\square_{16}O_{352}] + 48\ NH_3 + 16\ H_2O$

$$\downarrow \sim 500°C$$

(c) $Na_8[AlOH^{2+}]_{16}[Al_{40}Si_{136}\square_{16}O_{352}] + 16\ H_2O$

MECHANISM (a) to (b):

$$\geq Al-O + NH_4^+ \cdot H_2O \longrightarrow \geq Al-OH + H_2O + NH_3 \qquad (6.6a)$$

$$\geq Al-OH + H_2O \longrightarrow Al(OH)_2^+ + HO-\square \qquad (6.6b)$$

$$Al(OH)_2^+ + HO-\square \longrightarrow AlOH^{2+} + \square + H_2O \qquad (6.7)$$

the initial, based upon the initial aluminum content. It is proposed that this product, as described by Scheme 16, represents a typical decationized zeolite. (See Section A.)

In the stabilization reaction additional water vapor is present during the thermal treatment. The mechanism is illustrated by Scheme 17. A tetrahedral aluminum in the presence of excess water is removed by total hydrolysis to form an aluminum hydroxide and a tetrahedral vacancy occupied by four hydrogen atoms as hydroxyl groups. The initial dealumination reaction involves 16 aluminums. After, or simultaneous with the dealumination reaction, the aluminum hydroxide phase reacts further by Eq. 6.9 with the acidic hydrogen to form hydroxyaluminum cations. The overall stoichiometry of this reaction is also illustrated. One noticeable difference is that this does not require the formation of an oxygen deficient framework.

The next step in the stabilization process is the replacement of the H_4 groups in the tetrahedral positions by silicon introduced as $Si(OH)_4$ (Eq. 6.10). This is the species that is responsible for the transport of silica in the water vapor at high temperatures and may originate from two sources. First it may be occluded amorphous silica in the initial preparation. Second, it originates from a small amount of decomposed zeolite. Simple calcuation shows that only 6 volume percent of the zeolite is needed to provide the required silica for this additional substitution. This is equal to a layer of 1×10^{-2} micron thick on the surface of a typical one micron zeolite crystal.

As the substitution reaction occurs at increasing temperature in the presence of water vapor, the tetrahedral vacancies are eliminated and the framework tetrahedral sites are fully occupied, and the Si/Al ratio is increased. The cationic aluminum species may undergo additional dehydroxylation to form two aluminum cations linked by an oxygen bridge (95). Until silicon substitution occurs in the tetrahedral sites, no real shrinkage in the unit cell constant would be observed. When silicon substitution does occur, the unit cell should have a value of about 24.5 A with a Si/Al ratio of 3.8.

The mechanism of framework dealumination has been studied by esr. Fe^{3+} ions, present as a low level impurity occupy tetrahedral sites in the structure of zeolite Y and mordenite. This was confirmed by phosphorescence spectroscopy. Thermal deammoniation and dehydroxylation of $NH_4^{ex}Y$ above 400°C produces a ferromagnetic resonance; no similar effect was observed in the Na^+ form. It was concluded that iron is removed from the framework by hydrolysis of the tetrahedral Fe to $Fe(OH)_3$ as shown for Al in Scheme 17 (144).

Scheme 17 Decationization of Ammonium Exchanged Y in Excess Water Vapor

(a) $Na_8(NH_4)_{48}[Y] + H_2O$

$\downarrow 500°C$

(b) $Na_8H_{32}[Al(OH)_3]_{16}[Al_{40}Si_{136}(H_4)_{16}O_{384}] + 48 NH_3$

MECHANISM:

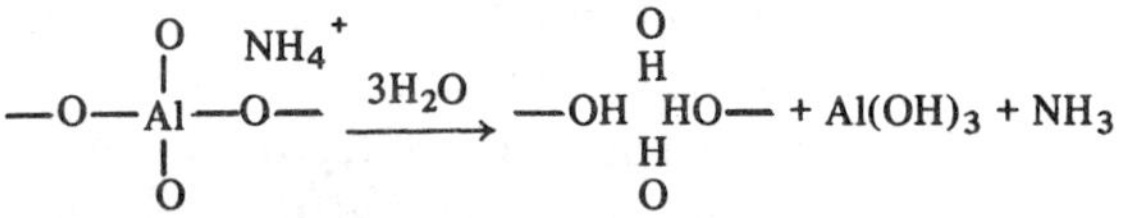

$$-O-\underset{\underset{O}{|}}{\overset{\overset{O}{|}}{\underset{}{Al}}}-O- \;\;\overset{NH_4^+}{} \;\; \xrightarrow{3H_2O} \;\; -OH \;\; \underset{\underset{O}{H}}{\overset{\overset{H}{O}}{HO-}} + Al(OH)_3 + NH_3 \quad (6.8)$$

$$Al(OH)_3 + 2H^+ \longrightarrow AlOH_2^+ + 2 H_2O \quad (6.9)$$

At higher temperatures,

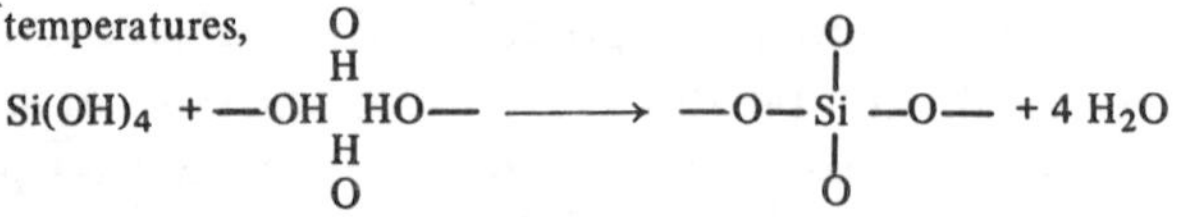

$$Si(OH)_4 + \underset{\underset{O}{H}}{\overset{\overset{O}{H}}{-OH \; HO-}} \longrightarrow -O-\underset{\underset{O}{|}}{\overset{\overset{O}{|}}{Si}}-O- + 4 H_2O \quad (6.10)$$

The final material, stabilized zeolite, is $Na_8[Al^{2+}OH]_{16}[Al_{40}Si_{152}O_{384}]$. The $Al^{2+}OH$ cations may dehydroxylate and form $[Al-O-Al]^{4+}$ cations.

Cationic Exchange by Aluminum and Stabilization

Cation exchange by Al^{3+} and hydronium, for sodium, in zeolite Y has been reported. Zeolites with varying Al cation and proton contents were prepared; exchangeable Al was measured by treatment with NaOH solution. Removal of Al from tetrahedral positions was not indicated and it was concluded that the presence of cationic aluminum "is not the critical factor" in stabilizing the zeolite, but rather the replacement of tetrahedral aluminum atoms is important. The initial ion exchange was carried out with $Al(NO_3)_3$ solution at 70°C and at room temperature (145).

Oxygen Mobility in Zeolite Y

The mobility of the oxygen ions in the framework structure of zeolite Y is not measurable below a temperature of 315°C. Above 315°C, the oxygen mobility, as determined from ^{18}O exchange between O_2 and the zeolite, is enhanced by the presence of Cu^{2+} cations. The exchange activation energy decreases from 45 to 23 kcal/mole when the Cu^{2+} exchange level is 42% (146).

Defect Structures - Removal of Cations by Reduction to the Metal and Formation of Intracrystalline Metal Dispersion.

Charge compensating cations in a zeolite may be removed by (1) replacing with a decomposable cation such as ammonium, as discussed above, or (2) by reducing the cation to the zero valent state with a chemical reductant which leaves the metal dispersed through the zeolite structure as metal atoms or as small agglomerates of metal atoms in the idealized case. Charge compensation in this event must occur by the simultaneous formation of the hydroxyl group on a framework oxygen if the reductant is hydrogen.

An example of this is:

$$Ni^{ex}Y + H_2 \xrightarrow{300-350°C} HY \cdot Ni^° \qquad (6.11)$$

In this example, sodium ions in zeolite Y are first exchanged with divalent nickel to the nickel exchanged form. After dehydration the nickel-zeolite is treated with hydrogen at 350°C which reduces nickel to the metal and forms a hydrogen zeolite (147). The dispersed metal phase may not be very stable in the zeolite structure and may diffuse to the zeolite crystal surface and agglomerate into small crystallites (148, 149). The appearance of the metal crystallites can be observed by x-ray powder methods.

Several combinations of zeolites containing dispersed metal phases have been prepared by this method. Reducible metals, such as lead, copper, and silver are readily dispersed in this manner (150). Other types of chemical reductants include carbon monoxide. In one instance, a copper X was first prepared by ion exchange of NaX with a Cu^{2+} salt. Reduction was accomplished by heating the zeolite in carbon monoxide at 350°C. The color of the zeolite changed with reduction from a light blue to a light purple. The HF and LF hydroxyl groups have been observed in the ir spectra of a hydrogen reduced $Cu^{2+}Y$ (151).

Reduction of the metal ion in the zeolite when suspended in aqueous solution is also possible. Hydrazine hydrochloride has been used as a reductant in a basic solution. During the reduction, nitrogen is evolved and the zeolite color changes to colorless from blue, indicating the formation of Cu^+ ions. The color then changed to a deep red brown.

The question of maintaining charge balance during reduction by carbon monoxide is interesting. For reduction to occur, the carbon monoxide must be oxidized to CO_2. Under completely anhydrous conditions this would have to be accomplished by the removal of an oxygen from the zeolite framework according to the following reaction:

$$CO + Cu^{2+} \text{ Zeolite} \longrightarrow Cu^+ \text{ zeolite} + CO_2 \qquad (6.12)$$

It is also possible that some residual water may produce hydrogen which acts as the reducing agent.

Scheme 18

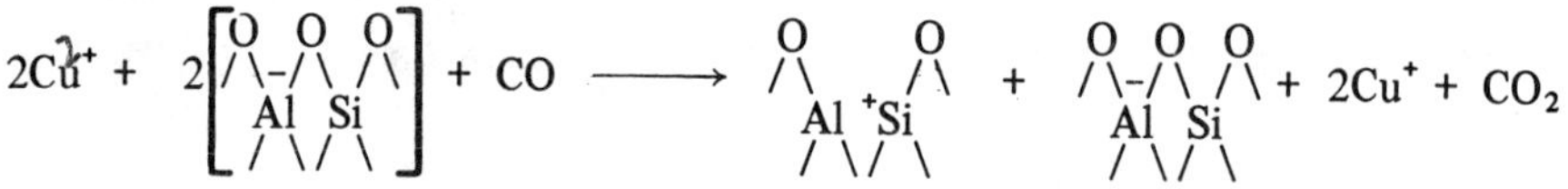

$$2Cu^{2+} + 2\begin{bmatrix} O & O & O \\ \backslash\text{-}/\backslash/\backslash \\ Al & Si \\ /\backslash/\backslash \end{bmatrix} + CO \longrightarrow \begin{array}{c} O O \\ /\backslash/\backslash \\ Al ^+Si \\ /\backslash/\backslash \end{array} + \begin{array}{c} O O O \\ /\backslash\text{-}/\backslash/\backslash \\ Al Si \\ /\backslash/\backslash \end{array} + 2Cu^+ + CO_2$$

This reaction has been proposed for CO reduction but appears unlikely since electrical neutrality is not maintained (151).

More exotic types of metal-zeolite compositions are possible. The platinum group metals may be introduced by cation exchange using an ammine complex followed by reduction with hydrogen (152). The following reaction is typical:

$$Pt(NH_3)_4^{2+} + NaX \longrightarrow Pt(NH_3)_4^{2+}X + Na^+ \qquad (6.13)$$
$$Pt(NH_3)_4^{2+}X + H_2 \longrightarrow Pt°·HX + NH_3 \qquad (6.14)$$

The dispersion of Pt in zeolite is also accomplished by thermal decomposition at 500°C of the complex, $Pt(NH_3)_4^{2+}$, to give Pt^{2+} cations. It is concluded that the Pt^{2+} cation occupies surface sites (site II) in the supercage of zeolite X because it is too large (d = 2.3 Å) to enter the structure to reach site I. Reduction at 300°C of the Pt^{2+} zeolite with H_2 follows Scheme 19 (153).

Scheme 19

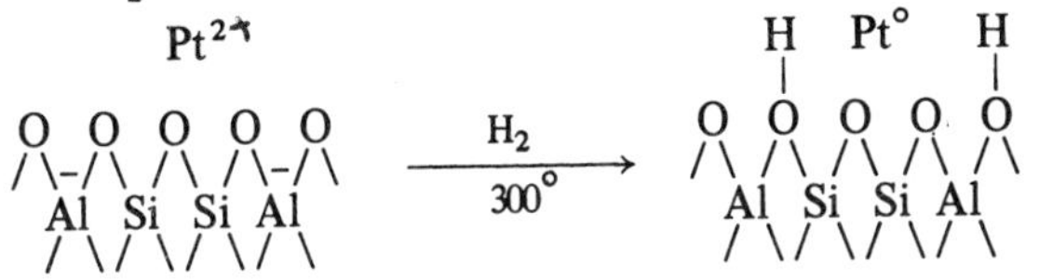

$$Pt^{2+} \begin{array}{c} O\ O\ O\ O\ O \\ /\backslash\text{-}/\backslash\ /\backslash\ /\backslash\text{-}/\backslash \\ Al\ \ Si\ \ Si\ \ Al \\ /\backslash/\backslash/\backslash/\backslash \end{array} \xrightarrow{H_2 \atop 300°} \begin{array}{c} H\ \ Pt°\ \ H \\ |\ \ \ \ \ \ \ \ | \\ O\ O\ O\ O\ O \\ /\backslash\ /\backslash\ /\backslash\ /\backslash\ /\backslash \\ Al\ \ Si\ \ Si\ \ Al \\ /\backslash/\backslash/\backslash/\backslash \end{array}$$

The total hydrogen consumption will then be 4H/Pt; 2 for reduction and 2 for subsequent chemisorption. After degassing at 500°C, the H_2 chemisorption was found to give an H_2/Pt ratio of 2 corresponding to formation of PtH_2

Nickel, copper, and silver exchanged into the synthetic zeolites types T and Y were reduced by H_2 in the temperature range of 280 - 400°C (148). In these cases, the reduced metal is nonvolatile and hydrogen in the form of protons is incorporated into the framework structure. For reduction of a metal ion by hydrogen one can write the following two generalized equations:

$$Me^{z+}Z_z^- + z/2\,H_2g \rightleftharpoons z[zH^+Z^-, Me°] \text{ and } [ZH^+Z^-,Me°] \rightleftharpoons [ZH^+z^-] + [Me] (6.15)$$

Equation 6.16 describes the formation of the hydrogen zeolite containing reduced metal in the single, solid zeolite phase and the formation of an external metal phase as the result of diffusion of the metal atoms to the external surface followed by crystallization.

Reversible uptake of hydrogen on the nickel zeolites was generally observed in the temperature range of 280°C to 400°C. For a highly exchanged NiT (0.79 Ni_z), the reversible hydrogen uptake followed the general relation

$$S_{H_2} = K(P_{H_2})^{1/3} \qquad (6.16)$$

In this relation S is the moles of hydrogen taken up per gram of zeolite. The change in enthalpy, ΔH, was −22 kcal/mole H_2. In the case of nickel T(Ni_z = 0.38) equilibrium was not attained. On a nickel Y, (Ni_z = 0.61), the hydrogen uptake at 350°C was like that of the nickel

T but slower. Diffusion of nickel and formation of nickel crystallites was observed after a long time. With copper exchanged Y (Cu_z = 0.57) the uptake was stoichiometric and followed the following equation:

$$2Cu^{2+} + H_2 \longrightarrow 2Cu^+ + 2H^+ \qquad (6.17)$$

When contacted with O_2 at 360°C, water was released according to the equation:

$$2Cu^+ + 2H^+ + 1/2 O_2 \longrightarrow 2Cu^{2+} + H_2O \qquad (6.18)$$

The silver exchanged Y followed a pattern like copper exchanged Y except that full reduction to elemental silver occurred.

The reducible metal ions, Zn^{2+}, Cd^{2+}, and Hg^{2+} can be vaporized from the zeolites containing them if heated in hydrogen at elevated temperatures (154). The ease of removal increases with increasing vapor pressure of the metal. A mercury exchanged X, $Hg^{2+}X$, lost metal at 200°C, a $Cd^{2+}X$ lost cadmium at 450°C, and $Zn^{2+}X$ lost the metal phase at 600°C. X-ray powder diffraction patterns confirmed the formation of an extra-crystalline metal phase during the reduction of the Ni^{2+} in a nickel exchanged zeolite X.

In a study of the magnetic properties of NiA, NiX and NiY it was proposed that Ni^{2+} ions in dehydrated NiX are tetrahedrally coordinated and are less prone to reduction by H_2 at $\sim 700°K$ than Ni^{2+} ions in octahedral coordination (tetrahedral in the β cages, octahedral in the site I) (155).

The reduced Ni^+ and Na_4^{3+} ions were formed in zeolite Y by means of reduction of the Ni^{2+} or Na^+ by alkali metal vapor (156). Detection of this species was by esr. Reduction of the Na^+ ions in NaY and NaX formed the complexes Na_4^{3+}, which is tetrahedral, and Na_6^{5+} which is octahedral, respectively.

$$Ni^{2+}Z + Na^0 \longrightarrow Na^+Z + Ni^+ \qquad (6.19)$$

Reduction of Ni^{2+} to Ni^+ in zeolite Y occurs with NO. The NO^+ ion is produced (157).

$$Ni^{2+}Y + NO \longrightarrow Ni^+ + NO^+ \qquad (6.20)$$

The formation of a Cu-Ni alloy in zeolite Y by reduction of the cations in H_2 at 550°C was shown by a decrease in the ferromagnetism of Ni (158).

The platinum in a Ca^{ex}Y zeolite (0.5 wt% Pt) was studied by x-ray absorption edge spectroscopy, combined with line broadening and H$_2$ adsorption. Sixty percent of Pt° was concluded to consist of 10 A crystallites, small enough to fit inside the Y zeolite supercages. Forty percent, by x-ray line broadening, consisted of 60 A crystals. The ratio of adsorbed hydrogen to platinum, H/Pt, after exposure to hydrogen at 1 atm for one hour and 100°C or 300°C was found to be 0.5 and was interpreted as hydrogen bonded to Pt since it affected the x-ray absorption edge (159). This is not consistent with other measurements of H$_2$ adsorption from which it was concluded that all the Pt was atomically dispersed (153).

F. RADIATION EFFECTS ON ZEOLITES

The extent of radiation damage produced by neutron irradiation of zeolite X was determined from adsorption measurements. Simultaneous x-ray diffraction patterns were also utilized. At a dose of 6.2 x 10^{17} neutrons/cm^2 damage was first detected; complete collapse of the crystal structure occurred when the dose was increased to 7 x 10^{19} neutrons/cm^2. Changes in the nature of the adsorption sites due to irradiation were followed by krypton adsorption measurements. The x-ray measurements showed that radiation produced a contraction of the crystal structure with displacement of some atoms from the regular lattice positions. This was confirmed by measurements of the density which increased with the level of the radiation dose from 2.0 g/cc for the initial zeolite to 2.4 g/cc for a material which had become completely amorphous (160).

Zeolite X exchanged with ^{141}Pr was subjected to neutron irradiation and the capture product, ^{142}Pr, was preferentially eluted from the zeolite. The eluate contained about 50% of the product and less than 1% of the initial ^{141}Pr. The object of this experiment was to use the Szilard-Chalmers reaction to prepare ^{247}Cm and other isotopes which have considerable value, but exist in very low concentrations in the normal products from nuclear reactors. The ^{141}Pr exchanged zeolite X was subjected to a thermal neutron flux of 4 - 5 X 10^{23} neutrons/cm^2 for time periods varying from a few seconds to 10 minutes. Subsequent elution of the irradiated zeolite with a 10 - 11 M lithium chloride solution removed the neutron capture product. Similar results were obtained with neodymium, erbium, and thulium. Generally 40-60% of the neutron capture product was eluted along with 1% of the original target material. The increase in ease of removal of the capture product was interpreted as caused by the movement of the lanthanide ion from an unavailable site within the β-cage of the structure to a more accessible site. This movement was produced by the recoil energy arising in the course of neutron capture. The conventional irradiation of ^{246}Cm yields about 2% ^{247}Cm at equilibrium. Enrichments obtained by using the zeolite can produce a product containing about 40% of the desirable curium isotope (161).

Superoxide ions, O_2^-, attached to the sodium cations of zeolite Y are formed when the zeolite is irradiated by gamma rays from a ^{60}cobalt source or x-rays in the presence of oxygen. The formation of these centers was followed by the use of esr spectra. When the zeolite was irradiated in vacuum a center was formed by gamma irradiation which was due to an electron trapped within a large zeolite cavity but shared among 4 Na$^+$ cations at the site II positions. This center is responsible for the pink color which is observed when zeolite Y is irradiated *in vacuo* (162, 163), and is the same Na_4^{3+} species described previously, and formed by reduction of NaY zeolite with alkali metal vapor (156).

REFERENCES

1. W. O. Milligan and H. B. Weiser, *J. Phys. Chem.*, **41**: 1029 (1937).
2. R. M. Barrer and G. C. Bratt, *J. Phys. Chem. Solids*, **12**: 130 (1959).
3. J. V. Smith, *J. Chem. Soc.*, 3759 (1964).
4. R. M. Barrer and D. A. Langley, *J. Chem. Soc.*, 3804 (1958); *J. Chem. Soc.*, 3811, 3817 (1958).
5. C. J. Peng, *Amer. Mineral.*, **40**: 834 (1955).
6. A. M. Taylor and R. Roy, *J. Chem. Soc.*, 4028 (1965).
7. R. M. Barrer, R. W. Bultitude, and I. S. Kerr, *J. Chem. Soc.*, 1521 (1959).
8. P. F. Kerr, J. L. Kulp, and P. K. Hamilton, Rep. No. 3, API Project 49, Columbia Univ., New York, 1949.
9. R. E. Grimm, Clay Mineralogy, 2nd Ed., Chap. 9, McGraw-Hill, New York, 1970'
10. F. A. Mumpton, *Amer. Mineral.*, **45**:351 (1960).
11. R. M. Barrer, *Proc. Roy. Soc. A 167,* 393 (1938).
12. M. Koizumi and R. Kiriyama, Scientific Reports, (2), Osaka University, 67 (1953); *Mineral J.*, **1**: 36 (1953).
13. R. M. Barrer, *Annu. Rep. Prog. Chem.*, **41**: 31 (1944).
14. M. Hoss and R. Roy, *Beitr. Mineralogie U. Petrog.*, **7**: 389 (1960).
15. D. L. Peterson, F. Helfferich, and G. C. Blytas, *J. Phys. Chem. Solids*, **26**: 835 (1965).
16. L. P. van Reeuwijk, *Amer. Mineral.*, **56**: 1655 (1971).
17. L. W. Staples and J. A. Gard, *Mineral. Mag.*, **32**: 261 (1959).

18. R. Aiello and R. M. Barrer, *J. Chem. Soc. A*, 1470 (1970).
19. T. E. Whyte, Jr., E. L. Wu, G. T. Kerr and P. B. Venuto, *J. Catalysis*, **20**: 88 (1971).
20. R. Aiello, R. M. Barrer, J. A. Davies, and I. S. Kerr, *Trans. Farad. Soc.*, **66**: 1610 (1970).
21. D. W. Breck, W. G. Eversole, R. M. Milton, T. B. Reed, and T. L. Thomas, *J. Amer. Chem. Soc.*, **78**: 5693 (1956).
22. A. S. Berger and L. K. Yakovlev, *Zh. Prikl. Khim.*, **38**: 1240 (1965).
23. R. M. Barrer, P. J. Denny, and E. M. Flanigen, *U. S. Pat. 3,306,922* (1967).
24. J. M. Bennett and J. V. Smith, *Mat. Res. Bull.*, **3**: 633 (1968).
25. D. W. Breck and E. M. Flanigen, Molecular Sieves, Society of Chemical Industry, London, 1968, p. 47.
26. R. M. Barrer and A. F. Denny, *J. Chem. Soc.*, 4684 (1964).
27. A. B. Lamb, *U. S. Pat. 1,813,174* (1931).
28. R. M. Barrer and B. E. F. Fender, *J. Phys. Chem. Solids*, **21**: 1 (1961).
29. R. M. Barrer, *Trans. Faraday Soc.*, **40**: 555 (1944).
30. D. W. Breck, *U. S. Pat. 3,054,657* (1962).
31. L. P. van Reeuwijk, *Amer. Mineral.*, **57**: 499 (1972).
32. D. W. Breck, *Unpublished Results*.
33. W. C. Phinney and D. B. Stewart, *U. S. Geol. Survey Prof. Paper*, **424-D**: 353 (1960).
34. R. M. Barrer and D. E. W. Vaughan, *Surface Sci.*, **14**: 77 (1969).
35. I. A. Breger, J. C. Chandler, and P. Zubovic, *Amer. Mineral.*, **55**: 825 (1970).
36. A. O. Shepard and H. C. Starkey, *U. S. Geol. Survey Prof. Paper*, **475-D**, 89 (1963).
37. M. H. Simonot-Grange, A. Cointot, J. Cruchaudet, *C. R. Acad. Sci.*, Paris, Ser. C., **267**: 1300 (1958).
38. E. N. Korytkova and A. D. Fedoseev, *Izv. Akad. Nauk, SSR, Ser. Khim.*, **11**: 1925 (1968).
39. K. Sakurai and A. Hayashi, Ser. Rep., Yokohama National University, Sec. II, No. I, 69 (152); *Amer. Mineral.*, **38**: 426 (1953).
40. H. W. Leimer and M. Slaughter, *Z. Kristallogr.*, **130**: 88 (1969).
41. G. D. Eberlein, R. C. Erd, F. Weber, and L. B. Beatty, *Amer. Mineral.*, **56**: 1699 (1971).
42. R. M. Barrer, *Trans. Faraday Soc.*, **45**: 358 (1949).
43. D. W. Breck and N. A. Acara, *U. S. Pat. 2,950,952* (1960).
44. E. Flanigen, *Neth. Pat. 6,710,729* (1968).
45. S. P. Zhdanov, N. N. Buntar, and E. N. Egorova, *Dokl. Akad. Nauk, SSR*, **154**: 419 (1964).
46. T. N. Shishakova and M. M. Dubinin, *Izv. Akad. Nauk, Ser. Khim.*, **11**: 2020 (1968).
47. L. P. Ni, O. B. Khalyapina, and L. G. Romanov, *Russ. J. Inorg. Chem*, **11**: 976 (1966).
48. G. T. Kerr, *Inorg. Chem.*, **5**: 1537 (1966).
49. M. H. Hey, *Mineral. Mag.*, **23**: 243 (1932).
50. M. H. Hey, *Mineral. Mag.*, **23**: 421 (1933); **24**: 227 (1936).
51. M. H. Hey, *Mineral. Mag.*, **23**: 421 (1933).
52. M. H. Hey, *Mineral. Mag.*, **23**: 51 (1932); A. E. Mourant, *Mineral. Mag.*, **23**: 371 (141) (1971).
53. M. H. Hey, *Mineral. Mag.*, **23**: 483 (1932).
54. M. H. Simonot-Grange and A. Cointot, *C. R. Acad. Sci.*, Paris, Ser. C. **264**: 1471 (1967).
55. R. M. Barrer and M. B. Makki, *Can. J. Chem.*, **42**: 1481 (1964).
56. H. B. Weiser, W. O. Milligan, and W. C. Ekholm, *J. Amer. Chem. Soc.*, **85**: 1261 (1936).
57. M. H. Simonot-Grange, A. Cointot, and A. Thrierr-Sorel, *Bull. Soc. Chim. Fr.*, **12**: 4286 (1970).
58. R. M. Barrer and E. A. D. White, *J. Chem. Soc.*, 1167 (1951).
59. R. M. Milton, *U. S. Pat. 2,996,358* (1961).
60. R. M. Barrer, and J. W. Baynham, *J. Chem. Soc.*, 2882 (1956).

61. R. M. Milton, U. S. Pat. *3,010,789* (1961).
62. R. M. Barrer, J. F. Cole, and H. Sticher, *J. Chem. Soc. A*, 475 (1968).
63. D. W. Breck and N. A. Acara, U. S. Pat. *3,011,869* (1961).
64. D. W. Breck and N. A. Acara, U. S. Pat. *2,962, 355* (1960).
65. D. W. Breck and N. A. Acara, U. S. Pat. *2,995, 423* (1961).
66. D. W. Breck and N. A. Acara, U. S. Pat. *2,911,151* (1961).
67. R. M. Milton, U. S. Pat. *3,021,253* (1961).
68. G. T. Kerr, E. Dempsey, and R. J. Mikovsky, *J. Phys. Chem.*, **69**: 4050 (1965).
69. J. Uytterhoeven, L. G. Christner, and W. K. Hall, *J. Phys. Chem.*, **69**: 2117 (1965).
70. J. V. Smith and J. M. Bennett, *Nature*, **219**: 1040 (1968).
71. C. L. Angell and P. C. Schaffer, *J. Phys. Chem.*, **69**: 3463 (1965).
72. J. W. Ward, *J. Catal.*, **10**: 34 (1968).
73. L. G. Christner, B. V. Liengme, W. K. Hall, *Trans. Faraday Soc.*, **64**: 1679 (1968).
74. P. E. Eberly, Jr., *J. Phys. Chem.*, **72**: 1042 (1968).
75. A. E. Hirschler, *J. Catal.*, **2**: 428 (1963).
76. J. W. Ward, *J. Phys. Chem.*, **72**: 4211 (1968).
77. J. A. Rabo, C. L. Angell, P. H. Kasai and V. Schomaker, *Chem. Eng. Progr., Symp. Ser.*, **63**: 31 (1967).
78. J. B. Uytterhoeven, R. Schoonheydt, B. V. Liengme and W. K. Hall, *J. Catal.*, **13**: 425 (1969).
79. J. A. Rabo, C. L. Angell and V. Schomaker, Actes Congr. Intern. Catalysis, 4th, Moscow, 1968.
80. J. A. Rabo, C. L. Angell, P. H. Kasai and V. Schomaker, *Discuss. Faraday Soc.*, **41**: 328 (1966).
81. P. B. Venuto and P. S. Landis, *Advan. Catalysis*, **18**: 259 (1968).
82. J. W. Ward, *J. Catal.*, **13**: 321 (1969).
83. A. P. Bolton, *J. Catal.*, **22**: 9 (1971).
84. J. V. Smith, J. M. Bennett and E. M. Flanigen, *Nature*, **215**: 241 (1967).
85. F. D. Hunter and J. Scherzer, *J. Catal.*, **20**: 246 (1971).
86. P. E. Eberly, Jr., and C. N. Kimberlin, Molecular Sieve Zeolites, *Advan. Chem. Ser.*, *102*, American Chemical Society, Washington, D.C., 1971, p. 374
87. J. R. Feins and P. A. Mullen, "Preprints," Division of Petroleum Chemistry, American Chemical Society Marketing, Houston, Texas, 1970, p. 89.
88. F. W. Kirsch, J. L. Lauer, and J. D. Potts, "Preprints," Division of Petroleum Chemistry, American Chemical Society Meeting, Washington D. C., 1971, p. 374
89. J. W. Ward, *J. Catal.*, **11**: 238 (1968).
90. L. Moscou, "Molecular Sieve Zeolites," *Advan. Chem. Ser.*, *102*, American Chemical Society, Washington, D. C., 1971, p. 337.
91. J. A. Rabo, P. E. Pickert, D. N. Stamires, and J. E. Boyle, Actes Congr. Intern. Catalyse, 2e, Paris, **2**: 2055 (1961).
92. D. H. Olson and E. Dempsey, *J. Catal.*, **13**: 221 (1969).
93. P. Gallezot and B. Imelik, *C. R. Acad. Sci. Ser. B*, **271**: 912 (1970), *J. Chem. Phys. Physicochem. Biol.*, **68**: 816 (1971).
94. R. L. Stevenson, *J. Catal.*, **21**: 113 (1971).
95. P. A. Jacobs and J. B. Uytterhoeven, *J. Chem. Soc. Faraday Trans.*, **69**: 359, 373 (1973).
96. A. P. Bolton and M. A. Lanewala, *J. Catal.*, **18**: 154 (1970).
97. J. Cattanach, E. L. Wu, and P. B. Venuto, *J. Catal.*, **11**: 342 (1968).
98. G. T. Kerr, *J. Catal.*, **15**: 200 (1969).
99. E. L. Wu, G. H. Kuhl, T. E. Whyte, Jr., and P. B. Venuto, "Molecular Sieve Zeolites," *Advan. Chem. Ser.*, *101*, American Chemical Society, Washington, D. C., 1971. p. 490.

REFERENCES 527

101. J. L. Carter, P. J. Lucchesi, and D. J. C. Yates, *J. Phys. Chem.*, **68**: 1385 (1964).
102. H. Hattori and T. Shiba, *J. Catal.*, **12**: 111 (1968).
103. T. R. Hughes and H. M. White, *J. Phys. Chem.*, **71**: 2192 (1967).
104. J. B. Uytterhoeven, P. Jacobs, K. Makay, and R. Schoonheydt, *J. Phys. Chem.*, **72**: 1768 (1968).
105. J. W. Ward, *J. Catal.*, **9**: 225 (1967).
106. J. W. Ward, *J. Catal.*, **9**: 396 (1967).
107. J. W. Ward, *J. Catal.*, **11**: 251 (1968).
108. J. W. Ward and R. C. Hansford, *J. Catal.*, **13**: 364 (1969).
109. J. Scherzer and J. L. Bass, *J. Catal.*, **28**: 101 (1973).
110. H. Hoss and R. Roy, *Beitr., Mineral. U. Petrogr.*, **7**: 389 (1969).
111. F. A. Mumpton, Union Carbide Corporation, *Unpublished Results*.
112. D. W. Breck, Union Carbide Corporation, *Unpublished Results*.
113. C. R. Allenbach and F. M. O'Connor, *U. S. Pat. 3,506,593* (1970).
114. F. Wolf, H. Fuertig, G. Nemitz, *Chem. Tech.*, Leipzig, **19**: 83 (1967).
115. J. L. Thomas, M. Mange, and C. Eyraud, "Molecular Sieve Zeolites," *Advan. Chem. Ser. 101*, American Chemical Society, Washington, D. C., 1971, p. 443.
116. N. R. Mumbach, Union Carbide Corporation, *Unpublished Results*.
117. D. W. Breck and N. R. Mumbach, Union Carbide Corporation. *Unpublished Results*.
118. D. T. Griggs and G. C. Kennedy, *Am. J. Sci.*, **254** 722 (1956).
119. R. M. Barrer, *J. Chem. Soc.*, 2342 (1950), *Nature*, **164**: 112 (1949).
120. R. M. Barrer and A. G. Kanellopoulos, *J. Chem. Soc. (A)*, 765, 775 (1970).
121. G. T. Kerr and G. F. Shipman, *J. Phys, Chem.*, **72**: 3071 (1968).
122. D. W. Breck, *J. Chem. Educ.*, **41**: 678 (1964).
123. C. R. Castor and R. M. Milton, *U. S. Pat., 3,013,987* (1961).
124. D. W. Collins and L. N. Mulay, *Inst. Elec. Electron. Eng. Trans. Mag.*, **8**: 470 (1968).
125. R. J. Mikovsky, A. J. Silvestri, E. Dempsey, and D. H. Olson, *J. Catal.*, **22**: 371 (1971).
126. R. M. Barrer and J. L. Whiteman, *J. Chem. Soc. A*, 19 (1967).
127. I. R. Beattie, *J. Chem. Soc.*, 367 (1957).
128. A. Walton, *Chem. Ind.*, (London), 338 (1970).
129. E. S. Dana, System of Mineralogy, 6th Ed., John Wiley and Sons, New York, N.Y., (1942).
130. K. J. Murata, *Amer. Mineral.*, **28**: 545 (1943).
131. C. R. DeKimpe and J. J. Fripiat, *Amer. Mineral.*, **53**: 216 (1968).
132. C. W. Lentz, *Inorg. Chem.*, **3**: 574 (1964).
133. J. Gotz and C. R. Masson, *J. Chem. Soc., (A)*, 686 (1971).
134. G. T. Kerr, *J. Phys. Chem.*, **72**: 2594 (1968).
135. P. E. Pickert, *U. S. Pat., 3,640,681* (1972).
136. H. B. Jonassen, G. P. Hamner, J. A. Rigney, R. B. Mason, and S. M. Laurent, *U. S. Pat., 3,493,518* (1970).
137. C. V. McDaniel and P. K. Maher, Molecular Sieves, Society of Chemical Industry, London, 1968, p. 186
138. C. V. McDaniel and P. K. Maher, *U. S. Pat., 3,449,070* (1969).
139. W. J. Ambs and W. H. Flank, *J. Catal.*, **14**: 118 (1969).
140. G. T. Kerr, *J. Phys. Chem.*, **71**: 4155 (1967).
141. P. Jacobs and J. B. Uytterhoeven, *J. Catal.*, **22**: 193 (1971).
142. P. K. Maher, F. D. Hunter, and J. Scherzer, "Molecular Sieve Zeolites," *Advan. Chem. Ser.*, *101*, American Chemical Society, Washington, D. C., 1971, p. 266.
143. G. A. Parks, *Amer. Mineral.*, **57**: 1163 (1972).
144. B. D. McNicol and G. T. Pott, *Chem. Comm.*, 438 (1970); *J. Catal.*, **25**: 223 (1972).

145. K. M. Wang and J. H. Lunsford, *J. Catal.*, **24**: 262 (1972).
146. G. V. Antoshin, Kh. M. Minachev, E. N. Sevastjanov, D. A. Kondratjev, and Chan Zui Newy, "Molecular Sieve Zeolites," *Advan. Chem. Ser.*, *101*, American Chemical Society, Washington, D. C., 1971, p. 514.
147. D. W. Breck, C. R. Castor, and R. M. Milton, *U. S. Pat.*, *3,013,990* (1961).
148. L. Riekert, *Ber. Bunsengisell, Phys. Chem.*, **73**: 331 (1969).
149. D. C. J. Yates, *J. Phys. Chem.*, **69**: 1676 (1965).
150. D. W. Breck and R. M. Milton, *U. S. Pat.*, *3,013,982* (1961).
151. C. M. Naccache and Y. Ben Taarit, *J. Catal.*, **22**: 171 (1971).
152. D. W. Breck and S. W. Bukata, *U. S. Pat.*, *3,200,082* (1965).
153. J. A. Rabo, V. Schomaker, and P. E. Pickert, *Proc. Intern. Confr. Catalysis*, 3rd, Amsterdam, **2**: 1264 (1964).
154. D. C. J. Yates, *J. Phys. Chem.*, **69**: 1676 (1965).
155. T. A. Egerton and J. C. Vickerman, *J. Chem. Soc. Faraday Trans.*, **69**: 39 (1973).
156. J. A. Rabo, C. L. Angell, P. H. Kasai, and V. Schomaker, *Discuss. Faraday Soc.*, **41**: 328 (1966).
157. P. H. Kasai and R. J. Bishop, Jr., *J. Amer. Chem. Soc.*, **94**: 5560 (1972).
158. W. G. Reman, A. M. Ali, and G. C. A. Schuit, *J. Catal.*, **20**: 374 (1971).
159. P. H. Lewis, *J. Catal.*, **11**: 162 (1968).
160. L. V. C. Rees and C. J. Williams, *Trans. Faraday Soc.*, **61**: 1481 (1965).
161. D. O. Campbell, *Inorg. Nucl. Chem. Lett.*, **6**: 103 (1970), *Proc. 8th Rare Earth Conf.*, Reno, Nevada, 1970, p. 448.
162. P. H. Kasai, *J. Chem. Phys.*, **43**: 3322 (1965).
163. D. N. Stamires and J. Turkevich, *J. Amer. Chem. Soc.*, **86**: 757 (1964).
164. W. N. Delgass, R. L. Garten, and M. Boudart, *J. Phys. Chem.* **73**: 2970 (1969).
165. M. J. Buerger, *Amer. Mineral.*, **39**: 600 (1954).

Chapter Seven

ION EXCHANGE REACTIONS IN ZEOLITES

The cation exchange property of zeolite minerals was first observed 100 years ago. The ease of cation exchange in zeolites and other minerals led to an early interest in ion exchange materials for use as water softening agents. Synthetic, noncrystalline aluminosilicate materials were primarily used; in more recent years, organic ion exchange resins are used. Crystalline zeolites have not been used commercially as water softeners.

The ion exchange behavior of various inorganic exchangers and other types of crystalline silicates such as clay minerals and feldspathoids has been extensively reviewed (1). Because of their three-dimensional framework structure, most zeolites and feldspathoids do not undergo any appreciable dimensional change with ion exchange; clay minerals, because of their two-dimensional structure, may undergo swelling or shrinking with cation exchange. One application for commonly occurring zeolite minerals (such as clinoptilolite) is in the selective removal of radioactive ions from radioactive waste materials (2).

The cation exchange behavior of zeolites depends upon (1) the nature of the cation species, the cation size, both anhydrous and hydrated, and cation charge; (2) the temperature; (3) the concentration of the cation species in solution; (4) the anion species associated with the cation in solution; (5) the solvent (most exchange has been carried out in aqueous solutions, although some work has been done in organic solvents); and (6) the structural characteristics of the particular zeolite. Cation selectivities in zeolites *do not* follow the typical rules that are evidenced by other inorganic and organic exchangers. Zeolite structures have unique features that lead to unusual types of cation selectivity and sieving. The recent structural analyses of zeolites form a basis for interpreting the variable cation exchange behavior of zeolites.

Cation exchange in zeolites is accompanied by dramatic alteration of stability, adsorption behavior and selectivity, catalytic activity and other important physical properties. Since many of these properties depend upon controlled cation exchange with particular cation species, detailed information on the cation exchange equilibria is important. Extensive studies of the ion exchange processes in some of the more important mineral and synthetic zeolites have been conducted.

A. ION EXCHANGE THEORY

The treatment of ion exchange reactions in zeolites by different investigators seems to be somewhat confused because a uniform system of nomenclature has not generally been employed (3). The ion exchange process may be represented by the following equation:

$$Z_A B^{z_B^+}_{(z)} + Z_B A^{z_A^+}_{(s)} \rightleftharpoons Z_A B^{z_B^+}_{(s)} + Z_B A^{z_A^+}_{(z)} \qquad (7.1)$$

where Z_A, Z_B are the charges of the exchange cations A and B and the subscripts z and s refer to the zeolite and solution, respectively (4).

The equivalent fractions of the exchanging cation in the solution and zeolite are defined by:

$$A_s \equiv \frac{Z_A m_{AY_{Z_A}}}{Z_A m_{AY_{Z_A}} + Z_B m_{BY_{Z_B}}} \equiv \frac{Z_A m_S^A}{Z_A m_S^A + Z_B m_S^B} \qquad (7.2)$$

$$A_z \equiv \frac{\text{no. equivalents of exchanging cation A}}{\text{total equivalents of cations in the zeolite}}$$

where m_S^A and m_S^B are the molalities of the ions A and B, respectively, in the equilibrium solution, also $(A_z + B_z) = 1$ and $(A_s + B_s) = 1$. The ion exchange isotherm is a plot of A_z as a function of A_s at a given total concentration in the equilibrium solution and at constant temperature. The preference of the zeolite for one of two ions is expressed by the separation factor, α_B^A, defined by:

$$\alpha_B^A \equiv \frac{A_z B_s}{B_z A_s} \qquad (7.3)$$

If ion A is preferred, α_B^A is greater than unity (3). The separation factor depends on the total concentration of the solution, the temperature, and A_S. It is not affected by choice of concentration units. The relation between α_B^A and the exchange isotherm is illustrated in Fig. 7.1.

If $\alpha_B^A = 1$, the exchange is ideal and obeys the law of mass action. Normally, however, the isotherm deviates from the diagonal line, represented as $\alpha_B^A = 1$, and thus shows a selectivity for one of the two ions. In most noncrystalline exchangers, the isotherms terminate at the lower left and upper right corners of the diagrams. In zeolites, however, there are many exceptions due to exclusion of the entering ions or trapped cations in the zeolite structure. This is referred to as the ion-sieve effect, that is, the entering ions cannot reach all of the sites occupied by the ions initially in the zeolite. The isotherm then terminates at a point where the degree of exchange, x, is less than 1.

The exchange isotherms for exchange of cations in zeolites may be classified into five types (Fig. 7.2). In curve a, the zeolite exhibits a preference for the entering ion $A(\alpha_B^A > 1)$ and the isotherm lies above the diagonal. In curve c, $\alpha_B^A < 1$, and the isotherm lies below the diagonal. In many cases, the selectivity varies with the degree of exchange and a sigmoidal isotherm results as illustrated by b. In curve d, complete exchange is not attained by the entering ion, $x_{max} < 1$, due to an ion-sieve effect. The curve e represents an unusual case where exchange results in two zeolite phases and produces a hysteresis loop.

Figure 7.1 Derivation of the separation factor α_B^A for the exchange reaction from the isotherm. As illustrated here, α_B^A is given by the ratio of area I/area II (3):

$$\alpha_B^A = \frac{A_z B_s}{B_z A_s} = \frac{I}{II}$$

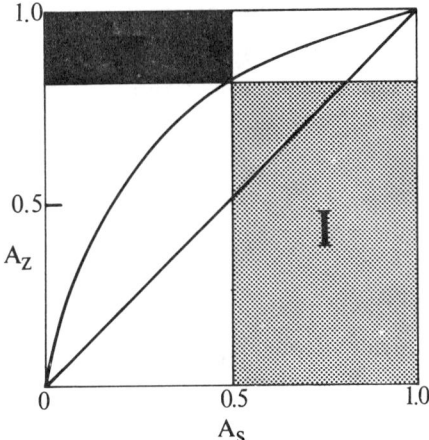

532 ION EXCHANGE REACTIONS IN ZEOLITES

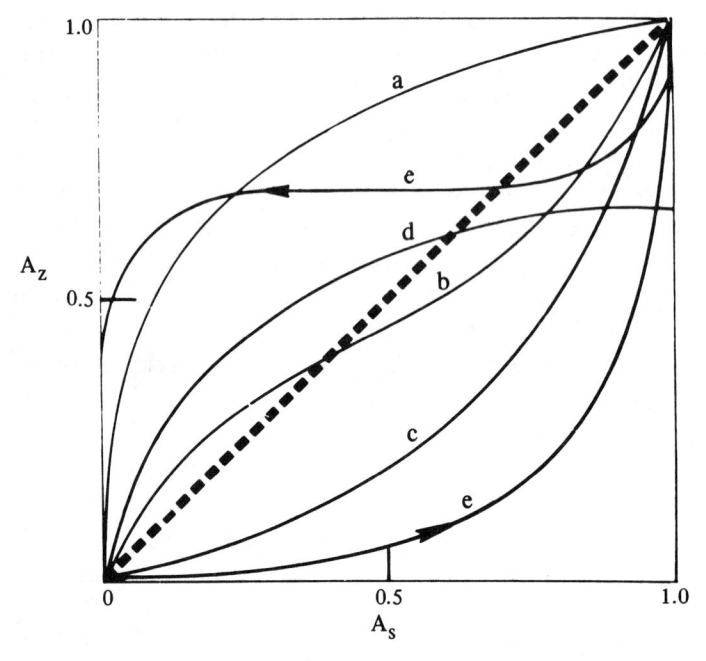

Figure 7.2 Types of ion exchange isotherms for the reaction $A_s + B_z \rightleftharpoons A_z + B_s$. Five types of isotherms are illustrated: (a) selectivity for the entering cation over the entire range of zeolite composition; (b) the entering cation shows a selectivity reversal with increasing equivalent fraction in the zeolite; (c) selectivity for the leaving cation over the entire range of zeolite compositions; (d) exchange does not go to completion although the entering cation is initially preferred. The degree of exchange, $x_{max} < 1$ where x is the ratio of equivalents of entering cation to the gram equiv. of Al in the zeolite; (e) hysteresis effects may result from formation of two zeolite phases.

Families of isotherms can be used to develop an empirical selectivity series. Since the selectivity generally changes with the degree of exchange, such a series is at best qualitative.

The rational selectivity coefficient, K_B^A, includes the charge of the ions, z_A and z_B

$$K_B^A \equiv \frac{A_Z^{z_B} B_S^{z_A}}{B_Z^{z_A} A_S^{z_B}} \qquad (7.4)$$

If the ions are of equal valence ($z_A = z_B$), then $K_B^A = (\alpha_B^A)^{z_A}$. For univalent ions, $K_B^A = \alpha_B^A$.

If $z_A \neq z_B$, then

$$[\alpha_B^A]^{z_A} = K_B^A \left(\frac{A_z}{A_s}\right)^{z_A - z_B} \tag{7.5}$$

The corrected selectivity coefficient, K'^A_B, includes a correction for the activity coefficients of the ions in the equilibrium solution:

$$K'^A_B \equiv \frac{A_z^{z_B} \cdot B_s^{z_A}}{B_z^{z_A} \cdot A_s^{z_B}} \cdot \frac{\gamma_B^{z_A}}{\gamma_A^{z_B}} \tag{7.6}$$

where γ_B, γ_A are mean ionic activity coefficients of the ions in solution. In terms of the dissolved electrolytes:

$$K'^A_B \equiv K_B^A \frac{\left[\gamma_{\pm BY_{z_B}}^{(z_B + 1)}\right]^{z_A}}{\left[\gamma_{\pm AY_{z_A}}^{(z_A + 1)}\right]^{z_B}} \tag{7.7}$$

where $\gamma_{\pm AY_{z_A}}$ and $\gamma_{\pm BY_{z_B}}$ are mean molal activity coefficients of the electrolytes in solution.

Equilibrium Constant

The thermodynamic equilibrium constant K_a is defined by

$$K_a \equiv K'^A_B \frac{f_A^{z_B}(z)}{f_B^{z_A}(z)} \tag{7.8}$$

where $f_{A(z)}$ and $f_{B(z)}$ are the activity coefficients of A and B in the zeolite.

The thermodynamic equilibrium constant is evaluated by using the relation (5,6):

$$\ln K_a = (z_B - z_A) + \int_0^1 \ln K'^A_B \, dA_z \tag{7.9}$$

assuming that changes in the activity of water in the zeolite and solution phases can be neglected. In order to obtain K_a, log K'^A_B is plotted against A_z which is referred to as the Kielland plot (7). The second term is then evaluated graphically from the area under these curves. The first term, $z_B - z_A$, is the difference between the charges of the two ions.

Activity coefficients of the ions A^{z_A+} and B^{z_B+} in the zeolite are given by:

$$\ln f_{A(z)}^{z_B} = -(z_A - z_B) B_z - \int_{K^0}^{K} B_z \cdot d\ln K_B^{'A} \qquad (7.10)$$

$$\ln f_{B(z)}^{z_A} = -(z_A - z_B) A_z + \int_{K^1}^{K} A_z \cdot d\ln K_B^{'A}$$

where $K^0 = K_B^{'A}$ when $B_z = 0$ and $K^1 = K_B^{'A}$ when $B_z = 1$. For uni-univalent exchange ($z_A = z_B = 1$), the log $K_B^{'A}$ versus A_z plots are generally linear and

$$\log K_B^{'A} = 2 C A_z + \log K_B^A \qquad (7.11)$$

The slope 2 C is called the Kielland coefficient. The plots of log $K_B^{'A}$ versus A_z may show three types of behavior which are illustrated in Fig. 7.3 (8,9). If d log $K_B^{'A}$ /d A_z is 0, then the curve 1 is typical and $K_a = K_B^{'A}$ for $f_{A(z)}/f_{B(z)} = 1$. If the relationship is linear as shown by curve 2, d log $K_B^{'A}$/d A_z is a constant, and K_a is given by

$$\log K_a = \log K_B^{'A} + C(1 - 2 A_z) \qquad (7.12)$$

In this case,

$$\log f_{A(z)} = C B_z^2$$
$$\log f_{B(z)} = C A_z^2 \qquad (7.13)$$

and $K_a = K_B^{'A}$ at $A_z = 0.5$.

When d log $K_B^{'A}$/d A_z is not constant, the curve may have many forms such as those illustrated by curve 3. They may show maxima, minima, or

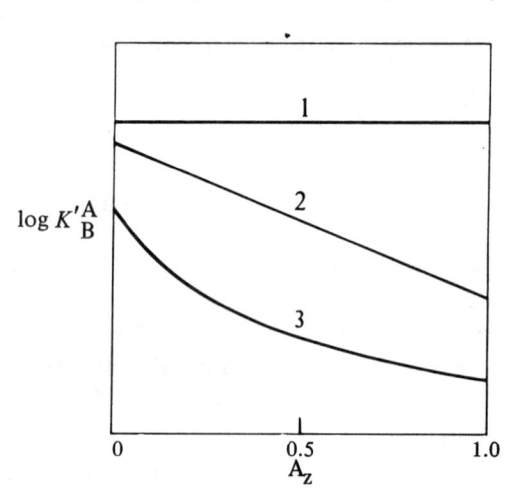

Figure 7.3 Typical curves that result when log $K_B^{'A}$ is plotted versus the entering cation in the zeolite, A_z(1, 8).

inflection points. Kielland plots of this type have been related to the presence of exchange sites of different types (4).

Uni-Univalent Ion Exchange

In uni-univalent exchanges, the isotherm plots of A_z versus A_s are almost independent of ionic strength since

$$K'^A_B = \frac{A_z B_s}{B_z A_s} \cdot \frac{\gamma^2_{\pm BY}}{\gamma^2_{\pm AY}} = \text{constant at constant } A_z \qquad (7.14)$$

The ratio of activities ≈ 1 in dilute solution, and

$$K'^A_B = \frac{A_z B_s}{B_z A_s} = \alpha^A_B \qquad (7.15)$$

In uni-divalent exchanges

$$K'^A_B = \frac{A_z B_s^2}{B_z^2 A_s} \cdot \frac{\gamma^4_{\pm BY}}{\gamma^3_{\pm AY_2}} \qquad (7.16)$$

and

$$K_a = K'^A_B \frac{f_{A(z)}}{f^2_{B(z)}} \qquad (7.17)$$

The ratio $\dfrac{B_s^2 \, \gamma^4_{\pm BY}}{A_s \, \gamma^3_{\pm AY_2}}$ is constant since K'^A_B has a unique value for each A_z(5). If the ratio of activity coefficients in a solution approaches unity, m^2_{BY}/m_{AY_2} is constant and B_s^2/A_s is constant. The exchange isotherms, however, are related to B_s/A_s and depend on the total amount of cation in the solution. In uni-divalent exchanges, the isotherm will vary with the total amount of cation in solution (5).

Most investigators have normalized the ion exchange isotherms in cases where complete exchange is not obtained ($x_{max} < 1$, isotherm type d) for several reasons, such as the inaccessibility of some cations or an ion-sieve effect with large cations. This normalization procedure involves setting the maximum limit of exchange, x_{max}, equal to unity and then multiplying each point of the exchange isotherm by a normalization factor, f_N, such that $f_N = 1/x_{max}$. The usual procedure is then used to obtain the equilibrium constant and the standard free energy of exchange. This procedure has been questioned in a recent publication that shows that these free energy values are controversial (10). Another isotherm normalization procedure was proposed which adopts standard states that allow a comparison between specimens which are exchanged to different extents. The earlier normalization procedure is based on the assumption that the residual ions remaining in the zeolites that are not exchanged do not influence the exchange equilibria. Since cations, as in zeolite X or Y for example, are distributed over different sites (Chap. 2) and the relative distribution varies from one type of ion to another, then the distribution of the ions over different sites is important for minimizing the total free energy of the system. Consequently, the residual ions in the structure which do not exchange may influence the exchange equilibria and should be considered.

Normalization of the selectivity coefficient introduces a correction term into the relation for obtaining the ion activity coefficient in the zeolite (10).

Integration of the log K'^A_B function with respect to dA_Z may be made only after the isotherm is normalized. Some results using this corrected approach are compared to those utilizing the earlier approach in Table 7.1. A sizeable difference exists. Further, the enthalpy values ΔH^0 agree with a sequence of endothermicity of the experimental isotherms.

B. ION EXCHANGE EQUILIBRIA IN AQUEOUS SOLUTION

Early published literature describes the cation exchange behavior of zeolites in a qualitative manner (11). Ion sieve effects were observed in several instances. The mineral analcime was observed to exchange with rubidium (radius = 1.48 A), but excluded cesium (a radius of 1.65 A). Several zeolite minerals have been studied in detail including analcime, chabazite, clinoptilolite, erionite, phillipsite, and stilbite. Synthetic zeolites A, X, and Y as well as the zeolites T, P, and KF have been investigated. Variations in structure, cation sites, cation population and distribution lead to considerable variation in the ion exchange behavior.

Exchange Capacity

The ultimate base exchange capacity of a zeolite depends on the chemical composition; a higher exchange capacity is observed with zeolites of low SiO_2/Al_2O_3 ratio (Table 7.2). In aqueous solution, the relevant capacity is that of the hydrated zeolite. In many cases the measured exchange capacities deviate from these values due to impurities (as in mineral zeolites) or variation in chemical composition. The specific

Table 7.1 Thermodynamic Data for $C_2H_5NH_3^+$ and $C_3H_7NH_3^+$ Exchange in Zeolite Y (10)

Ion	Temp(°K)	$\Delta G°$, (J mole^{-1}) a	$\Delta G°$, (J mole^{-1}) b	$\Delta H°$, (J mole^{-1}) a	$\Delta H°$, (J mole^{-1}) b
$C_2H_5NH_3^+$	277	142	1480	9614	1480
	328	-924	1626	565	-5137
$C_3H_7NH_3^+$	277	706	1643	17849	1643
	328	-1108	1643	3340	-5488

[a] From normalized equation
[b] From the adapted equation

exchange capacities vary with the exchange cation.

The exchange capacity has been measured as part of a characterization procedure in determining ion exchange equilibria. Space does not permit a complete detailed review of all the equilibria data. Tables 7.3 through 7.14 summarize available equilibria data. In these tables K'^A_B (Eq. 7.6) is abbreviated to K'. Selected examples from this tabulation are used for this discussion.

Ion Exchange of Zeolite A

Zeolite A displays a *double ion-sieve action*. Only small cations can penetrate the single 6-rings into the β-cages. Large organic cations (for example, tetramethylammonium) cannot penetrate the 8-rings into the α-cages. In the typical zeolite A, 12 univalent ions are locat-

Table 7.2 Exchange Capacity of Various Zeolites

Zeolite	Si/Al	milliequiv/g Anhydrous Powder	milliequiv/g Anhydrous Pellets with about 20% binder[a]	milliequiv/g Hydrated Powder	milliequiv/g Hydrated Pellets with about 20% binder[a]
Chabazite	2	5		3.9	
Mordenite	5	2.6		2.3	
Erionite	3	3.8		3.1	
Clinoptilolite	4.5	2.6		2.2	
Zeolite A	1	7.0	5.6	5.5	4.4
Zeolite X	1.25	6.4	5.1	4.7	3.8
Zeolite Y	2.0	5.0	4.0	3.7	3.0
Zeolite T	3.5	3.4		2.8	

[a] See Chap. 9

ed in each psuedo-cubic unit cell of 24(Al, Si)O$_4$ tetrahedra (see Chap. 2). In some preparations zeolite A has been reported to occlude up to one sodium ion per β-cage (total of 13 sodium ions per unit cell) along with an additional anion which is probably AlO$_2^-$. Therefore, determination of the true exchange capacity $(12 + x)$ sodium ions per unit cell, where $0 < x < 1$, depends upon the cation (9). Since silver (r = 1.26 A) exchanges with all of the Na$^+$ ions, including those trapped in the β-cages, the total exchange capacity can be determined experimentally. The Tl$^+$ ion (r = 1.40 A) cannot penetrate the single 6-ring into the β-cage. Therefore, Tl$^+$ ion exchanges with a maximum of 12 Na$^+$ ions per unit cell. Excess Na$^+$ ions in the β-cages are generally less than one per cage (9, 12, 13).

Ion exchange equilibria in zeolite A have involved mostly univalent and divalent counter ions. Attempts to exchange with trivalent cerium ions were unsuccessful (14). The equilibria are based upon the 12 cations in the large α-cages per unit cell (Table 7.3). The extra amount of sodium is not considered. Ion exchange isotherms and selectivity coefficients are shown in Fig. 7.4.

Sodium-lithium exchange involves two sets of exchange sites. Up to an exchange level of Li$_z$ = 0.35, the process is independent of temperature. Above Li$_z$ = 0.35, the process is temperature dependent (15). The isotherm exhibits an inflection at Li$_z$ = 0.35, not shown in Fig. 7.4.

The order of decreasing selectivity for univalent ions in zeolite A is (14):

$$Ag > Tl > Na > K > NH_4 > Rb > Li > Cs$$

For divalent ions, the order of decreasing selectivity is:

$$Zn > Sr > Ba > Ca > Co > Ni > Cd > Hg > Mg$$

Barium ion exchange in zeolite A is possible without disruption of the structure if the zeolite is not dehydrated (14, 16).

Selectivity plots for calcium exchange in zeolite A were found by Barrer et al. to exhibit maxima (12). This was not confirmed by Sherry and Walton who attributed their different results to their attainment of true equilibrium (13). In the exchange of strontium ions, the attainment of equilibrium was slow, but no maximum in the selectivity plot was observed if equilibrium was reached.

The calcium ion exchange for sodium shows a nonuniform temperature dependence. It was concluded that calcium ion pairs interact

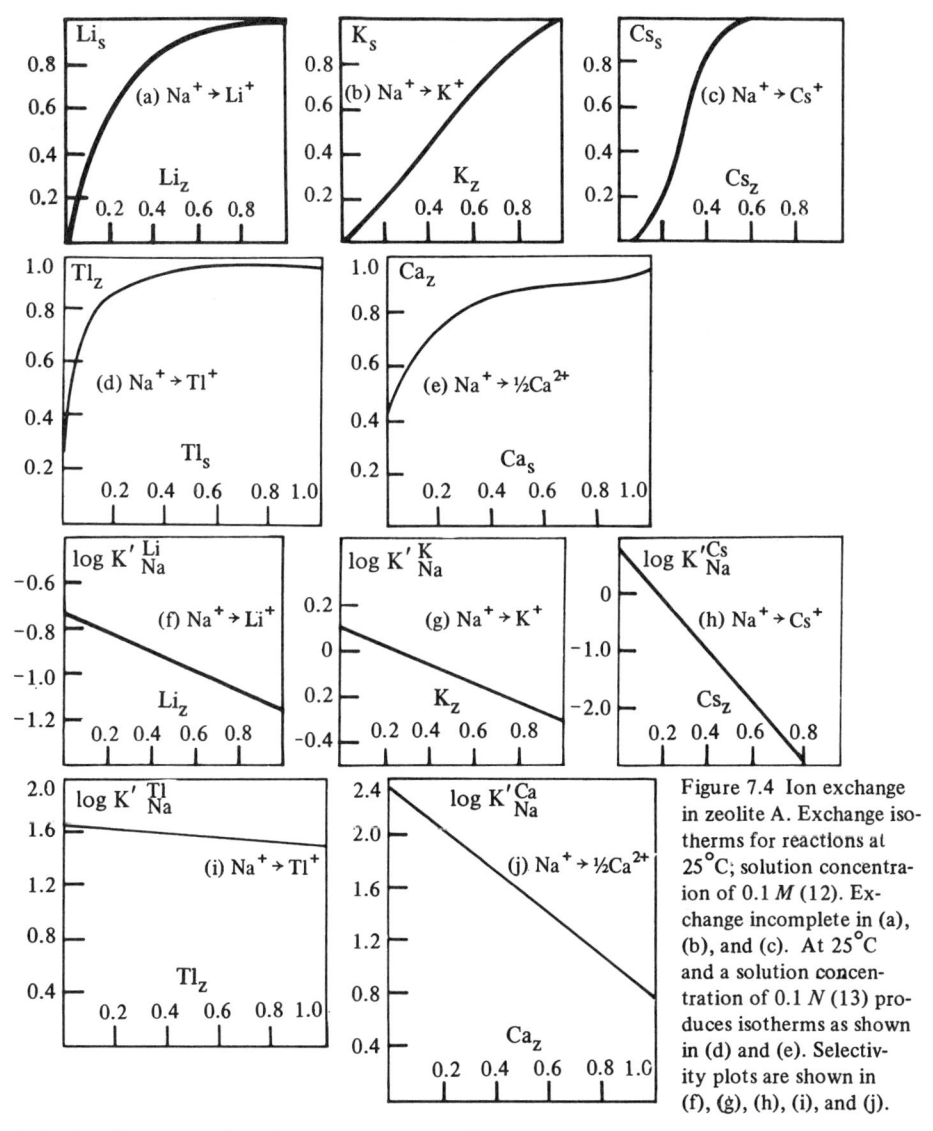

Figure 7.4 Ion exchange in zeolite A. Exchange isotherms for reactions at 25°C; solution concentration of 0.1 M (12). Exchange incomplete in (a), (b), and (c). At 25°C and a solution concentration of 0.1 N (13) produces isotherms as shown in (d) and (e). Selectivity plots are shown in (f), (g), (h), (i), and (j).

with the framework anion sites in the 6-rings. Ion pairing by strontium ions is less whereas the barium ions show no evidence of pair formation; the electrostatic interaction between barium ions and the framework structure is therefore weak. Consequently, the fully hydrated barium ions are in the large α-cage of the structure.

Entropy change for the barium ion exchange is consistent with the conclusion that a net transfer of water occurs from the zeolite to the solution since the barium exchanged zeolite contains fewer water molecules in each α-cage than the original sodium form.

The cobalt exchange isotherms terminated at a value of A_z equal to 0.89 and the nickel isotherms were extrapolated to a value of 0.8 although this extrapolation was not supported by experimental data. The nickel exchanged zeolite A was observed to be unstable and lost its crystal structure at a temperature above 70°C. The other zeolites are thermally stable at 750°C. Relevant data are summarized in Table 7.3. The thermal instability of the nickel exchanged zeolite A was interpreted in terms of a ligand-field effect. After dehydration octahedral coordination for the cation is not favored. The cobalt ion is more stable relative to nickel in a tetrahedral configuration (17).

Ion exchange isotherms for various cation pairs are shown in Fig. 7.4 along with plots of the corrected selectivity coefficients (Eq. 7.10). Thermodynamic quantities corresponding to these data are listed in Table 7.3.

Complex Ion Effects: Zeolite A

Limited data are available on the effect of the anion and of the presence of complexing agents on the exchange equilibria of zeolite A (18). In one study involving calcium, cadmium, and zinc ions from solutions of electrolytes, a method was developed for determining the ion exchange equilibrium constant of the complex ion and the number of ligands in the complexes formed. The selectivity coefficient decreased with the increasing electrolyte concentration. The electrolyte reacted with the exchanging metal ion, such as cadmium, to form an anion complex which greatly reduced the selectivity. The electrolyte cation may also take part in the exchange. The formation of an insoluble salt of a cation, such as silver, has been utilized for preparing hydrogen forms of zeolites.

Ion Exchange of Zeolites X and Y

Zeolites X and Y are ideal materials for studying cation exchange phenomena in zeolites. Exchange with many different cation species including complex cations and large organic cations can take place; even salt imbibement may occur. Differences in structures (cation positions, cation distribution, framework charge, Si/Al ordering) are

Table 7.3 Cation Exchange in Zeolite A

Exchange Reaction	Conc.	t(°C)	Isotherm type[a]	x_{max}[b]	$Log_{10}K'$[d] A_Z=0.2	$Log_{10}K'$[d] A_Z=0.5	$Log_{10}K'$ vs A_Z type	$\Delta G°_{298}$ (cal/g equiv)	$\Delta H°$	$\Delta S°$ (e.u/g equiv)	Ref.	Figure	Other Ref.
Na$^+$ → Li$^+$	0.1 m	25	C	1.0	−0.8	−0.95	linear	1300	2260	3.2	12	7.4a	14,15,20-22
Na$^+$ → K$^+$	0.1 m	25	B	1.0	+0.1	−0.1	linear	+140	−2390	−8.4	12	7.4b	14,20-23
Na$^+$ → NH$_4^+$	0.2 N	25	C	1.0							14		20-22
Na$^+$ → Cs$^+$	0.1 m	25	D	0.45	0	−1.6[c]	linear	+1980[c]	−3800[c]	−19.3	12	7.4c	8,14,23
Na$^+$ → Rb$^+$	0.1 m	25	D	0.68	−0.1	−0.5[c]	linear	+680	−2550[c]	−10.8	12		14,21,22
Na$^+$ → Ag$^+$	0.1 N	25	A	1.0	+3.2	+2.9	linear	−3930	−2780	3.9	13		9.14
Na$^+$ → Tl$^+$	0.1 N	25	A	1.0	+1.6	+1.6	linear	−2320			13	7.4d	9,20
Na$^+$ → ½Ca^{2+}	0.1 N	25	A	1.0	+2.1	+1.6	linear	−733	+2700	11.5	13	7.4e	9,12,14,20
Na$^+$ → ½Sr^{2+}	0.1 N	25	A	1.0	+2.7	+2.6	linear	−1010	+500	5.1	13		20,24
Na$^+$ → ½Ba^{2+}	0.1 N	25	A	1.0	+2.6	+2.1	nonlinear	−1168	0	3.9	13		13,16,20
Na$^+$ → CH$_3$N$^+$H$_3$	0.5 N	100	D	0.43			linear				9		
Na$^+$ → C$_2$H$_5$N$^+$H$_3$	0.5 N	100	D	0.29							9		
Na$^+$ → n-C$_3$H$_7$N$^+$H$_3$	0.5 N	100	D	0.12							9		
Na$^+$ → ½Cd^{2+}	0.1 N	25	A	1.0				−1090	6230	14.3	17		
Na$^+$ → ½Zn^{2+}	0.1 N	25	A	1.0				−871	3680	15.6	17		
Na$^+$ → ½Ni^{2+}	0.1 N	25	A	0.8[e]				70			17		
Na$^+$ → ½Co^{2+}	0.1 N	25	A	0.89				−249	6230	21.7	17		

[a]See Fig. 7.2.
[b]Maximum degree of exchange achieved under the conditions cited,
$$x_{max} = \frac{\text{equiv. of entering cation}}{\text{gram atoms Al in zeolite}}$$
[c]Based on normalized isotherms so that $x/x_{max} = 1$, reduced exchange capacity.
[d] Estimated values from published plots in the References.
[e] Unstable above 70°C; actual degree of exchange, $N_{i_Z} = 0.6$.

evidenced in the differences observed in the cation exchange behavior of these zeolites. The mineral faujasite has not been studied because of the lack of an adequate quantity.

Typically, zeolite X contains about 86 cations per unit cell which may occupy up to five different sites in the hydrated zeolite. Consequently, the selectivity varies with the degree of cation exchange (19). Ion exchange equilibria are summarized in Table 7.4 and Figures 7.5 - 7.9. Below a level of 40% exchange, the selectivity series for univalent ions was observed, in terms of decreasing selectivity, to be:

$$Ag \gg Tl > Cs > Rb > K > Na > Li$$

This series corresponds to occupancy of the most accessible cation sites (site III or IV) within the supercages. At 50% exchange, which includes site II in the 6-rings adjacent to the supercages, the selectivity series was found to be:

$$Ag \gg Tl > Na > K > Rb > Cs > Li$$

Cations in site I in the hexagonal prisms are the most difficult to exchange. Rubidium or cesium ions are excluded due to their size; they are too large under normal conditions to pass through the 2.4 A diameter opening.

The maximum degree of rubidium or cesium ion exchange (x_{max} in Table 7.4) has been reported to be 0.65 - 0.8 (6, 19, 25). The isotherm for the exchange of cesium ion at room temperature terminates at $x = 0.82$ (see Fig. 7.5b) which corresponds to 16 of the 85 sodium ions in the unit cell that are not exchanged. The highly polarizable silver and thallium cations exhibit high selectivity over the entire isotherm.

The barium ion (radius 1.43 A) did not exchange completely at room temperature and thus did not occupy site I. Because of this exclusion phenomena, the free energy of exchange, $\Delta G°$, was calculated on the basis that 16 of the sodium ions are not replaced (6). Complete exchange was obtained with calcium and strontium but exchange with barium was limited to $x = 0.8$ at 25°C (6,30). For divalent ions, the enthalpy of exchange is either a small negative quantity or a small positive quantity. The selectivity shown for the calcium and strontium ions is primarily due to the positive entropy. The dependence of the selectivity coefficients on the zeolite composition indicates that some calcium and barium ions are located in the

ION EXCHANGE EQUILIBRIA 543

Table 7.4 Cation Exchange in Zeolite X

Exchange Reaction		Conc.	$t(°C)$	Isotherm type[b]	x_{max}[a]	$Log_{10}K'$ $A_z = 0.2$	[d] $A_z = 0.5$	$Log_{10}K'$ vs A_z type	$\Delta G°_{298}$ (cal/g equiv)	$\Delta H°$	Ref.	Figure	Other Ref.
Na$^+$ → Li$^+$	(f)	0.1 m	25	C	1.0	−0.7	−1.0	linear	1350	1790	6		25
	(g)	0.1 N	25	C	1.0	−0.75	−0.8	nonlinear	1600		19		
Na$^+$ → K$^+$	(f)	0.1 m	25	B	1.0	+0.5	+0.1	linear	−190	−1250	6		25
	(g)	0.1 N	25	B	1.0	+0.2	−0.05	linear	140		19	7.5a	
Na$^+$ → Rb$^+$	(f)	0.1 m	25	D	0.65	+1.5[c]	+1.0	linear	−1300	−1500	6		25
	(g)	0.1 N	25	D	0.8[h]	+0.6[c]	−0.1	linear	140		19		
Na$^+$ → Cs$^+$	(f)	0.1 m	25	D	0.65	+1.5[c]	+0.5	linear	−780	−1660	6	7.5b	24, 25
	(g)	0.1 N	25	D	0.8	+0.6[c]	+0.2	nonlinear	87		19		
Na$^+$ → Ag$^+$	(g)	0.1 N	25	A	1.0	+1.9	+1.9	nonlinear	−2520		19		25
Na$^+$ → Tl$^+$	(g)	0.1 N	25	A	1.0	+1.9	+1.6	nonlinear	−1840		19	7.5c	25
Na$^+$ → NH$_4^+$	(g)	0.05 N	20	D	0.62						25		26
Na$^+$ → ½Ca^{2+}	(f)	0.1 m	25	A	1.0	+0.8	+1.2	nonlinear	−160		6		
	(g)	0.1 N	25	A	1.0	+1.6	+1.1	nonlinear	−320	1200	30	7.7a	24,27
Na$^+$ → ½Sr^{2+}	(f)	0.1 m	25	A	1.0	+0.09	+2.1	nonlinear	−710		6		24
	(g)	0.1 N	25	special[e]	1.0	+2	+1.5	nonlinear	−740	530	30	7.8a	
Na$^+$ → ½Ba^{2+}	(f)	0.1 m	25	D	0.74	+1.5[c]	+2.0	nonlinear	−1130		6		28
	(g)	0.1 N	25	D	0.8	+3.0[c]	+2.7	nonlinear	−1310	−430	30	7.7c	
Na$^+$ → 1/3Ce^{3+}	(f)	0.5 N	25	D	0.8	+2.3[c]	+1.8	nonlinear	−2300		27		24
Na$^+$ → 1/3La^{3+}	(f)	0.3 N	25	D	0.85						29	7.9	
Na$^+$ → 1/3Y^{3+}	(f)	0.5 N	25	D	0.7				−1100		27		29

[a] Maximum degree of exchange achieved under the conditions cited.

$$x_{max} = \frac{\text{equiv. exchanged cation in zeolite}}{\text{gram atoms Al in zeolite}}$$

[b] See Fig. 7.2.
[c] Based on reduced exchange capacities so that $x/x_{max} = 1$...
[d] Estimated from published plots in the Ref.
[e] Mutual insolubility of end members from $x = 0.71$ to $x = 0.87$.
[f] Zeolite X containing 87 Al/unit cell, Si/Al = 1.21.
[g] Zeolite X containing 85 Al/unit cell, Si/Al = 1.26.
[h] Based on extrapolation of isotherm. Published data indicate x_{max} is actually lower.

544 ION EXCHANGE REACTIONS IN ZEOLITES

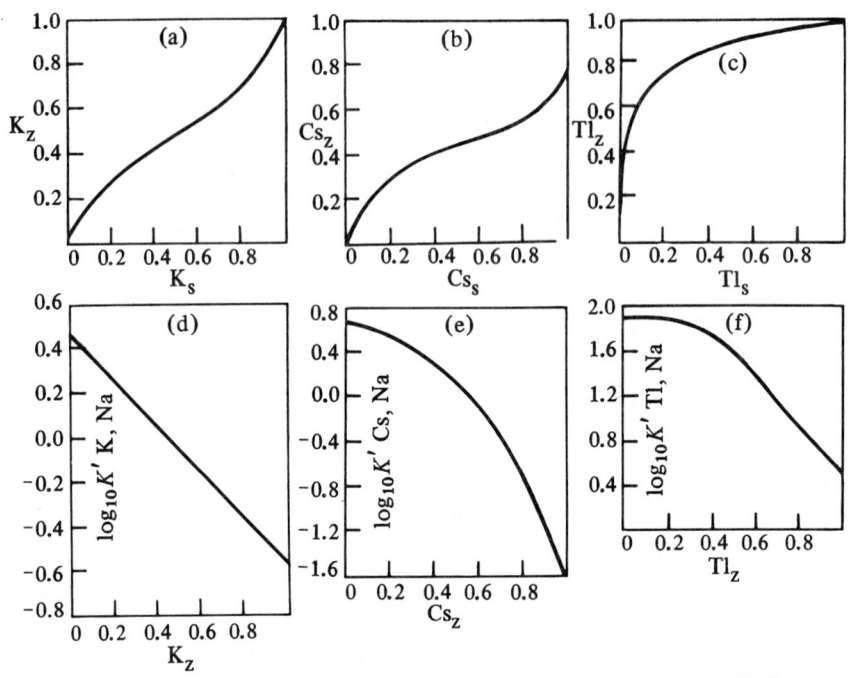

Figure 7.5 Ion exchange isotherms for univalent-univalent ion exchange in zeolite X (Si/Al) = 1.26 at 25°C and 0.1 N total normality (19). (a) Na$^+$ → K$^+$; (b) Na$^+$ → Cs$^+$; (c) Na$^+$ → Tl$^+$.

Plots of corrected selectivity coefficients, $\log_{10} K'$ as a function of the equivalent fraction, A$_z$, in the zeolite for univalent-univalent exchange in zeolite X at 25°C and 0.1 N total normality. (d) Na$^+$ → K$^+$; (e) Na$^+$ → Cs$^+$; (f) Na$^+$ → Tl$^+$.

supercages of zeolite X. (See Chapter 2.) In contrast, relatively few cations are located in the supercages of zeolite Y.

The exchange isotherm for strontium ions in zeolite X exhibits an unusual shape (Fig. 7.8). This is due to the formation of two phases. The region of mutual solid phase insolubility ranges from Sr$_z$ = 0.71 to 0.87 (30).

The formation of two solid phases has been observed in ion exchange reactions in analcime with sodium and potassium ions (31); in lithium exchange in harmotome and phillipsite (32); and in ion exchange in the synthetic zeolite P (B) (33). This behavior has not been observed in zeolite A; the complete solubility of the end members is reported to be due to the high cation mobilities. The self-diffusion coefficient for sodium ion is about 1 x 10^{-8} cm^2/sec. In analcime, the self-diffusion coefficient is about 1 x 10^{-13}. The formation of two solid zeolite

ION EXCHANGE EQUILIBRIA 545

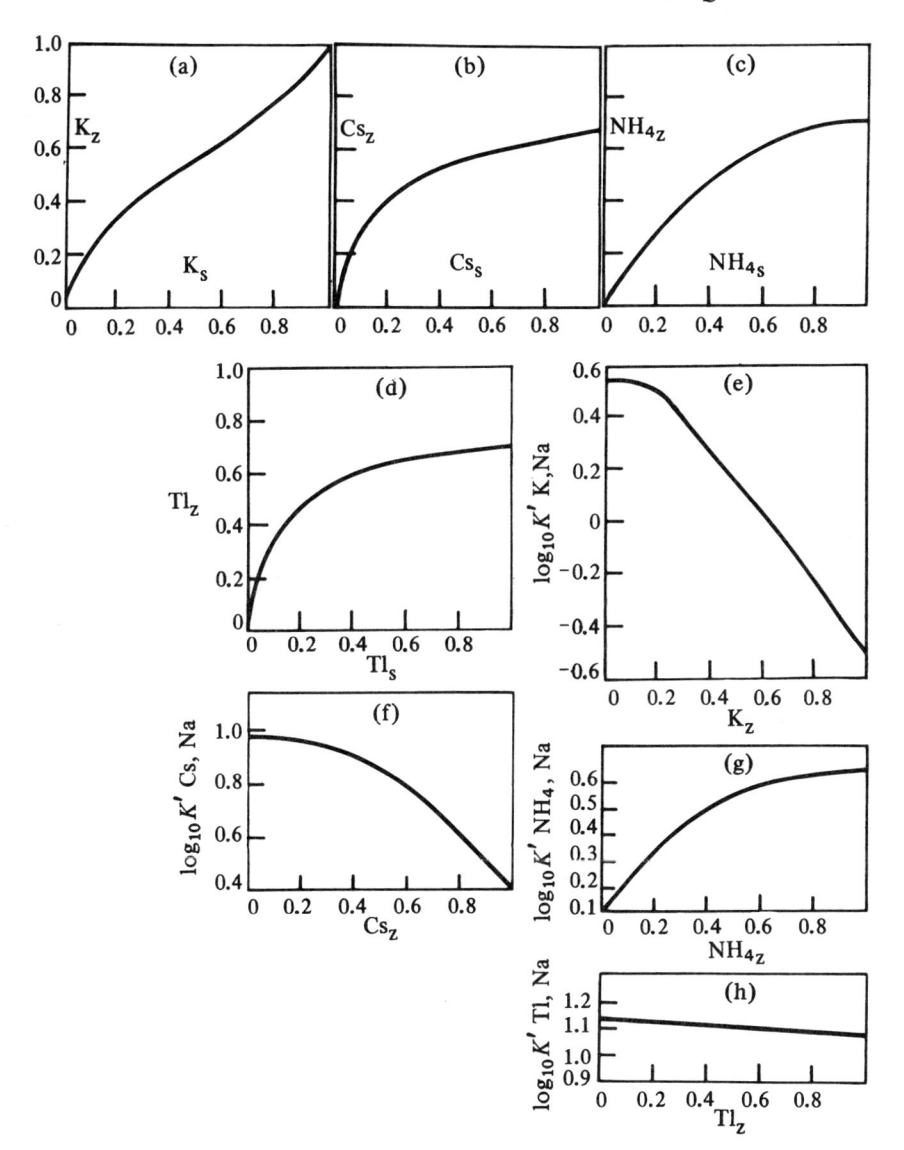

Figure 7.6 Ion exchange isotherms for univalent-univalent ion exchange in zeolite Y (Si/Al = 2.83) at 25°C and 0.1 N total normality (19). (a) Na$^+$ → K$^+$; (b) Na$^+$ → Cs$^+$; (c) Na$^+$ → NH$_4^+$; (d) Na$^+$ → Tl$^+$.

Plots of corrected selectivity coefficients, $\log_{10} K'$ as a function of the equivalent fraction, A$_z$, in the zeolite for univalent-univalent exchange in zeolite Y at 25°C and 0.1 N total normality. (e) Na$^+$ → K$^+$; (f) Na$^+$ → Cs$^+$; (g) Na$^+$ → NH$_4^+$; (h) Na$^+$ → Tl$^+$. In (f), (g), and (h), K' is normalized to $x/x_{max} = 1$.

546 ION EXCHANGE REACTIONS IN ZEOLITES

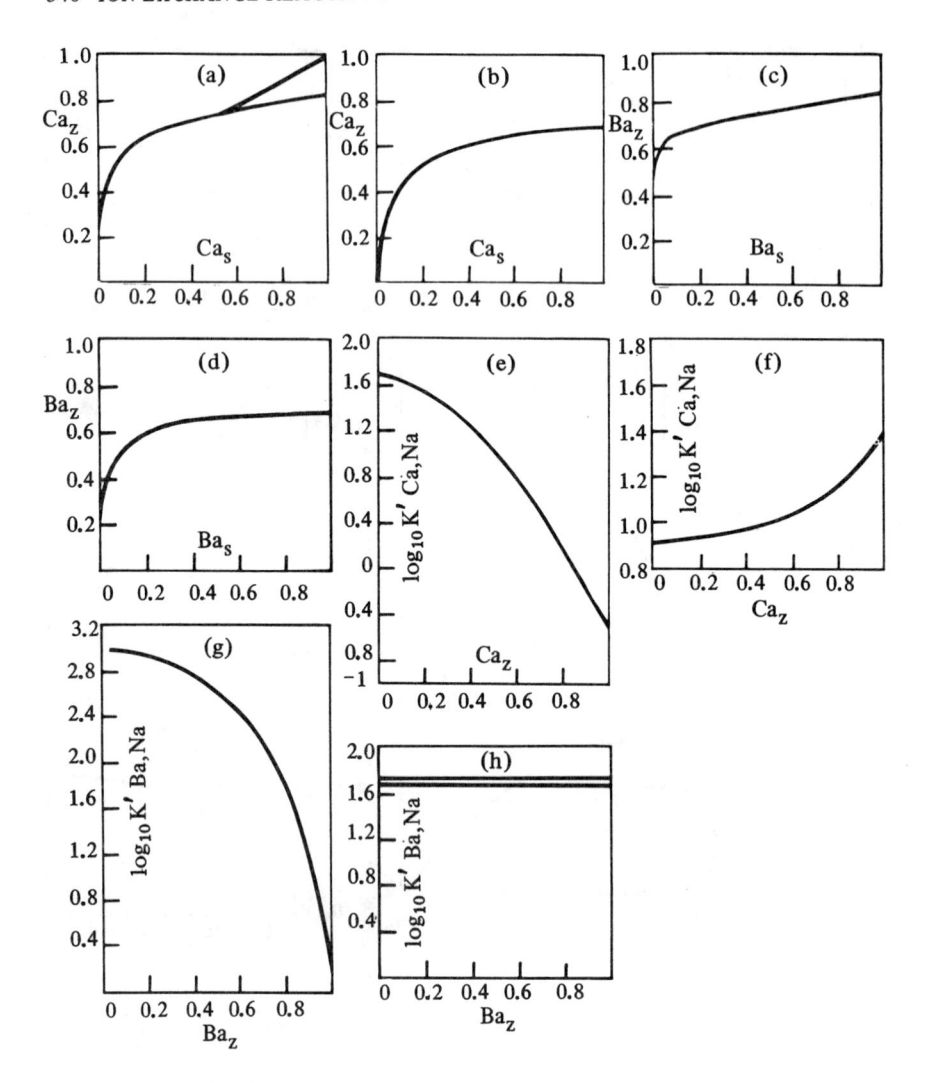

Figure 7.7 Ion exchange isotherms for uni-divalent ion exchange in zeolite X (Si/Al = 1.26) and zeolite Y (Si/Al = 2.76) at 25°C and 0.1 N total normality (30). (a) Na$^+$ → ½Ca^{2+} in zeolite X; (b) Na$^+$ → ½Ca^{2+} in zeolite Y; (c) Na$^+$ → ½Ba^{2+} in zeolite X; (d) Na$^+$ → ½Ba^{2+} in zeolite Y.

Plots of corrected selectivity coefficients, $\log_{10} K'$ as a function of the equivalent fraction in the zeolite, A_z, for univalent-divalent ion exchange in zeolite X (Si/Al = 1.26) and zeolite Y (Si/Al = 2.76) at 25°C and 0.1 N total normality. A_z is normalized to $x/x_{max} = 1$. (e) Na$^+$ → ½Ca^{2+} in zeolite X; (f) Na$^+$ → ½Ca^{2+} in zeolite Y; (g) Na$^+$ → ½Ba^{2+} in zeolite X; (h) Na$^+$ → ½Ba^{2+} in zeolite Y.

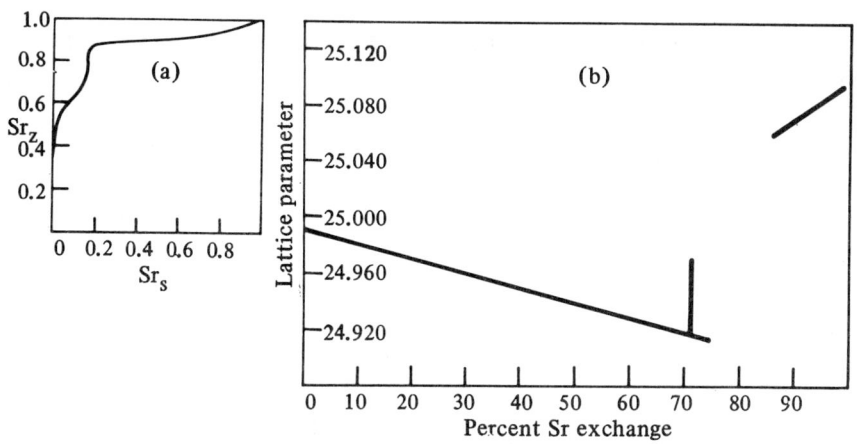

Figure 7.8 (a) ion exchange isotherm for the Na$^+$ → ½Sr^{2+} exchange in zeolite X at 25°C and 0.1 N total normality (30); (b) lattice parameter, a_0 versus degree of Sr^{2+} exchange in zeolite X in percent (34).

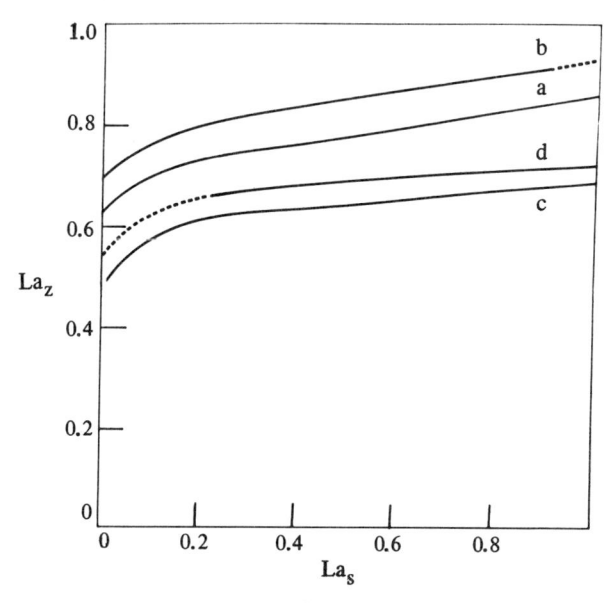

Figure 7.9 Exchange isotherms for La^{3+} exchange in zeolite X and zeolite Y at 25°C, 82°C and 0.3 N total normality (29). (a) Na → 1/3 La^{3+} X, 25°; (b) Na → 1/3 La^{3+} X, 82°.

The X chemical composition is Na$_{85}$[(AlO$_2$)$_{85}$(SiO$_2$)$_{107}$] dry basis; (c) Na → 1/3 La^{3+}, 25°C; (d) Na → 1/3 La^{3+} Y, 82°C.

The Y chemical composition is Na$_{51}$[(AlO$_2$)$_{51}$(SiO$_2$)$_{141}$] dry basis.

phases in the exchange of strontium ions for sodium ions in zeolite X is therefore very unusual.

The nucleation mechanism for the expanded phase involves cation relocation. From x-ray studies the cation positions in the strontium form of zeolite X have been determined, and the unit cell constant, measured as a function of strontium exchange (Fig. 7.8) (34). It was concluded that in the range up to 71% exchange, about half of the strontium ions move into site I' (in the β-cage). The cation distributions are given in Table 7.5. For greater than 71% exchange more sodium ions in the large supercages are replaced by strontium ions. At about 71% exchange, site I loses sodium ions and strontium ions occupy sites I' and II. The movement of sodium ions out of site I produces a new phase which has an expanded unit cell constant (Fig. 7.8). The lattice contracts slightly up to 71% exchange and during the occupancy of site I' by strontium ions.

Optical studies of larger single crystals confirmed the formation of a strontium-rich phase. The crystals cracked and decrepitated as strontium exchange occurred, indicating the formation of an expanded zeolite phase.

At 25°C, the 16 site I cations in zeolite X are apparently not replaced by lanthanum ions (29). At a higher temperature, 82°C, partial replacement takes place over a long period of time. Exchange isotherms of trivalent rare earth ions such as lanthanum in zeolite X at different temperatures are shown in Fig. 7.9. The rate-controlling step in this reaction appears to be removal of water molecules from the rare earth ions in the supercages. Similar results were obtained for exchange with yttrium ions. The exclusion of lanthanum ions from

Table 7.5 Cation Distribution in the NaX-SrX System (34)

Site	$Na_{80}X \cdot n H_2O$ [a]	$Na_{24}Sr_{30.5}X \cdot n H_2O$ [b]	$Sr_{42.0} NaX \cdot n H_2O$ [c]
I	16 Na$^+$	12 Na$^+$	2.1 Sr^{2+}
I'		7.3 Sr^{2+}	11.1 Sr^{2+}
II	32 Na$^+$	11.5 Sr^{2+}	15.0 Sr^{2+}
II'		26 H$_2$O	32 H$_2$O

[a] Cation compositions given for unit cell, based on unit cell of $Na_{80}[(AlO_2)_{80}(SiO_2)_{112}] \cdot n\ H_2O$.
[b] Equivalent of 35 univalent cations not accounted for and presumed to be located with water molecules in the large cages, includes about 19 Sr^{2+} ions.
[c] Equivalent of 29 univalent cations not accounted for and presumed to be located in the large cages, includes about 14 Sr^{2+} ions.

the small cages is due to the large dehydration energy. Apparently the exchange isotherm for the higher temperature of 82°C is not reversible. The exchange of large organic cations (alkylammonium) in zeolite X diminishes with increasing size or with substitution of alkyl groups in the ammonium ion (section F). There are approximately the same number of water molecules per unit cell in zeolite Y as in zeolite X. This provides for 15 water molecules per lanthanum ion in zeolite Y and 9 water molecules per lanthanum ion in zeolite X. Consequently, the dehydration of the lanthanum ions which is necessary for diffusion into β-cages requires more energy.

Cation exchange between mixtures of a zeolite and a fine-grained clay mineral has been studied in one case (35). The rate of exchange between potassium in bentonite and calcium in zeolite X is very rapid; reaction is complete in 40 minutes. The equivalent ratio of calcium to potassium in the zeolite was greater than in the bentonite clay. This was attributed to a higher affinity shown by bentonite for potassium.

Ion Exchange of Zeolite Y

The ion exchange capacity of a typical zeolite Y is lower than that of zeolite X due to the lower framework charge. Ion exchange indicates a different occupancy of cation sites in the hydrated form (5, 19). As shown in Table 7.6, many exchange reactions do not go to completion at normal temperatures. The type D isotherm is characteristic of most exchanges. The isotherms for exchange with the univalent cesium, ammonium, and thallium, all terminate at a cation fraction, A_Z, of about 0.7. About 16 univalent ions in site I are not exchanged (Fig. 7.6) indicating that these cations are too large to penetrate the small β-cages at room temperature. The level of exchange in this instance corresponds to 35 of the 51 univalent ions originally present in the unit cell. The difference, 16 univalent ions, corresponds to the 16 site I ions located within the double 6-rings. Therefore, the selectivity series changes with the degree of exchange. Up to 68% of exchange, the order of decreasing selectivity is

$$Tl > Ag > Cs > Rb > NH_4 > K > Na > Li$$

Lithium, sodium, silver, and potassium ions occupy sites in the β-cages. They exhibit the expected selectivity for a surface that binds the majority of cations tightly. Thallium cations replace all of the cations in zeolite X, but they do not exchange for the 16 site I cations in zeolite

Table 7.6 Cation Exchange in Zeolite Y

Exchange Reaction		Conc.	$t(°C)$	Isotherm type [g]	x_{max} [f]	$Log_{10}K'$ $A_z = 0.2$	$Log_{10}K'$ [i] $A_z = 0.5$	$Log_{10}K'$ vs A_z type	$\Delta G°_{298}$	$\Delta H°$ (cal/g equiv)	Ref.	Fig.
Na$^+$ → Li$^+$	c	n.s.[d]	25	C	1.0	−0.9	−1.2	nonlinear	2860	1350	5, 25	
	a	0.1 N	25	C	1.0	−0.9	−1.4	nonlinear	2700		19	
Na$^+$ → K$^+$	c	n.s.[d]	25	B	1.0	+0.6	+0.3	linear	−370	−1440	5, 25	
	a	0.1 N	25	B	1.0	+0.5	+0.15	nonlinear	−190		19	7.6a
Na$^+$ → Rb$^+$	c	n.s.[d]	25	D	0.7	+0.9 [h]	+0.8	linear	−1060	−1220	5, 25	
	a	0.1 N	25	D	0.68	+0.8 [h]	+0.7	nonlinear	−1300		19	
Na$^+$ → Cs$^+$	c	n.s.[d]	25	D	0.7	+1.5 [h]	+1.0	linear	−1420	−1610	5, 25	
	a	0.1 N	25	D	0.68	+1.0 [h]	+0.9	nonlinear	−1460		19	7.6b
Na$^+$ → NH$_4^+$	a	0.1 N	25	D	0.68	+0.3 [h]	+0.55	nonlinear	−660		19, 25	7.6c
Na$^+$ → Ag$^+$	g	n.s.[d]	25	A	1.0	+0.9	+0.8	linear	−1081	−500	5, 25	
	g	0.1 N	25	A	1.0	+0.8	+0.7	linear	1100			
Na$^+$ → Tl$^+$	c	n.s.[d]	25	D	0.7	+1.1 [h]	+1.0	linear	−1490		5, 25	
	a	0.1 N	25	D	0.68	+1.1 [h]	+1.1	linear	−1630		19	7.6d
Na$^+$ → ½Ca^{2+}	c	n.s.[e]	25	D	0.7	0 [h]	+0.4	nonlinear	170	1090	5	
	a	0.1 N	25	D	0.68	+0.9 [h]	+1.0	nonlinear	−420	649	30	
Na$^+$ → ½Sr^{2+}	c	n.s.[e]	25	D	0.7	+0.3 [h]	+0.7	nonlinear	130	1030	5	7.7b
	b	0.1 N	25	D	0.68	+1.2 [h]	+1.2	linear	−510	1540	30	
Na$^+$ → ½Ba^{2+}	c	n.s.[e]	25	D	0.7	+0.5 [h]	+1.4	nonlinear	−390	−30	5	
	b	0.1 N	25	D	0.68	+1.7 [h]	+1.4	linear	−870	−350	30	7.7d
Na$^+$ → 1/3La^{3+}	b	0.3 N	25	D	0.69						29	7.9

[a] Zeolite Y containing 50 Al atoms/unit cell.
[b] Zeolite Y containing 51 Al atoms/unit cell, Si/Al = 2.77.
[c] Zeolite Y containing 57 Al atoms/unit cell, Si/Al = 2.38.
[d] Concentration not given.
[e] Initial molality varied in exchanging ion salt solution.
[f] $x_{max} = \dfrac{\text{equivalents exchanged}}{\text{gram atoms Al in the zeolite}}$
[g] See Fig. 7.2
[h] K' and thermodynamic data based on reduced exchange capacities, $x/x_{max} = 1$.
[i] Estimated from the plots published in the Ref.

Y. The reason for this is not clear.

Although the results of various studies of ion exchange reactions in zeolite Y are reasonably consistent for univalent exchange, wide differences are reported in the case of divalent exchange (Table 7.6). The major differences appear at low degrees of exchange and may be related to nonequilibrium conditions. None of the divalent or trivalent ions studied in zeolite Y attains complete exchange under normal temperature conditions; and the maximum degree of exchange appears to be $A_z = 0.68$. This indicates nonoccupancy of site I (Fig. 7.7). In the cases where incomplete exchange occurs, the thermodynamic quantities and selectivity coefficient plots were based upon reduced exchange capacities, i.e, $x/x_{max} = 1$. For initial exchange in the zeolite the molality of the exchanging salt solution in the equilibrium solution is very low (10^{-5} to 10^{-6}). Any loss of ions from the zeolite to the solution can significantly modify the selectivity coefficient. Consequently the plots which are used for determining the thermodynamic quantities are affected. With the trivalent lanthanum ion, complete exchange was not accomplished at room temperature and was very slow at high temperatures. In 1 M LaCl$_3$ solution, 13 days were required to prepare a completely exchanged form of LaX at 100°C, and 47 days to achieve 92% exchange in zeolite Y. The difference may be due to the water stripping effect (29) (Fig. 7.9).

Some disagreement exists in the maximum level of ammonium ion exchange in zeolite X (Table 7.4). The isotherm by Sherry (29) shows that the exchange Na$^+\rightarrow$ NH$_4^+$ is complete ($x_{max}= 1$) while that of Theng et al. (25) indicates that $x_{max}= 0.63$. The isotherms for alkylammonium exchange in NaX terminate at values of x_{max} which are lower in zeolite X than zeolite Y. As the cation volume increases, the extent of exchange decreases. The volume required by the supercage cations (total available 0.30 cc/g or 6700 Å^3/uc), determines the maximum degree of exchange; the experimentally determined values of x_{max} (Table 7.7) are generally lower than the calculated values.

The degree of exchange by alkylammonium ions is also limited by the number of sodium ions present in the large cages (sites II, III, and IV). Zeolite X has a total of 85 sodium ions of which 69 should be available in the large cages. Correspondingly, a typical zeolite Y has about 39 ions available. The larger entering alkylammonium ions cause a transfer of water molecules from the hydrated zeolite to the solution

Table 7.7 Exchange of Alkylammonium Ions for Sodium in Zeolite X and Y. Concentration of the Alkylammonium Chloride generally 0.05 N at about 20°C (25)

Exchanging Ion	Cation Volume (A^3)	Zeolite X [a] x_{max}	$\Delta G°$[c] (cal/g equiv)	Zeolite Y [b] x_{max}	$\Delta G°$[c] (cal/g equiv)
NH_4^+	14	0.63 [d]	−670	0.70	−620
Methyl	49	0.52 [d]	−130	0.68	−450
Ethyl	94	0.45 [d]	+550	0.66	−130
Propyl	115	0.35	+590 [e]	0.61	−100 [f]
iso-Propyl		0.33	+920	0.51	+38
Butyl	131	0.24	+650	0.53	−35
$(Methyl)_2$	96	0.30	+700	0.55	+150
$(Ethyl)_2$	134	0.12	+760	0.40	+980
$(Methyl)_3$	129	0.19	+710	0.44	+470
$(Ethyl)_3$	168	0.02		0.26	+1300

[a] Anhydrous unit cell composition $Na_{85}(AlO_2)_{85}(SiO_2)_{107}$, c.e.c. = 6.25 milliequiv/g.
[b] Anhydrous unit cell composition $Na_{55}(AlO_2)_{55}(SiO_2)_{137}$, c.e.c. = 4.34 milliequiv/g.
[c] $\Delta G°$ values all based on reduced exchange capacities so that $x/x_{max} = 1$.
[d] Barrer, et al. (73) report x_{max} = 0.92 for NH_4^+, 0.58 for methyl, 0.37 for $(methyl)_2$, 0.23 for $(methyl)_4$, 0.50 for ethyl.
[e] Revised normalization method gave $\Delta G°$ = +544 cal/mole at 25°C.
[f] Revised normalization method gave $\Delta G°$ = +303 cal/mole at 25°C.

resulting in a "water-transfer" effect in the reaction entropy and an endothermic contribution to the heat of reaction (36). Theoretically, more than 16 sodium cations might occupy sites within the β-cages. Hydrated zeolite NaX contains 9 sodium ions in site I and only 8 sodium ions in site I' (Chap. 2).

A further study was made of the exchange of ammonium and alkylammonium ions on four zeolite X and Y samples (37). Thermochemical values for the propylammonium exchange were derived using the revised procedure. The maximum limit of exchange for the alkylammonium ions was equivalent to 32 ions per unit cell and independent of the nature of the ion. With zeolite Y, interaction with the lattice was concluded to be the most important factor in exchange; in zeolite X, with higher aluminum content in the framework, the exchange is determined by the change in hydration state of the large cavity. Ammonium exchange is exothermic, consequently, over normal temperature ranges, the extent of ammonium exchange should be a maximum at room temperature.

Comparison of Zeolite X and Zeolite Y

There are substantial differences in the exchange of sodium ions in zeolites X and Y by other cations (5, 6, 19, 25, 29, 30, 38). Equilibria (e.g., level of exchange attained), selectivities, and thermodynamic aspects are quite different (Tables 7.4 and 7.6). In some instances, the results from different investigators of the same exchange reaction (e.g., cesium, sodium), in the same zeolite, do not agree (6, 19, 38).

Steric factors based on cation size relative to zeolite aperture diameters may explain the incomplete exchange observed with rubidium (d = 2.98 Å) and cesium (d = 3.38 Å) in zeolites X and Y. These cations cannot enter the hexagonal prisms to site I. This would limit exchange to 80% for zeolite X and 70% for zeolite Y. For rubidium ions and cesium ions, the observed limit in zeolite X is about 65%. This has been attributed to "crowding" by the cesium and rubidium ions exchanged into site III or IV within the supercages. This forces about 14 sodium cations from site II into site I' in the β-cages. In zeolite Y, with low cation population in site III, this crowding does not occur. The thallium ion replaces all of the sodium ions in zeolite X, but only 69% in zeolite Y. Although large, the thallium ion is polarizable and may pass through the 6-rings in NaX. The greater framework charge in zeolite X is sufficient to deform the cation and permit full occupancy of site I.

Electrostatic considerations have been employed to account for the slow but complete exchange reactions of calcium, strontium, and barium ions in zeolite X as compared to the faster but limited exchange observed in zeolite Y.

Using the relation

$$B(Y) + A(X) \rightleftharpoons A(Y) + B(X) \tag{7.18a}$$

to represent the equilibrium for solid phase exchange, where B(Y) represents one equivalent of zeolite Y in the B^+ cation form, A(X) one equivalent of zeolite X in the A^+ form and similar designations for the other phases, a mass action quotient, K' has been represented by the relation (38)

$$K' = \frac{A_y B_x}{B_y A_x} \tag{7.18b}$$

where A_y, B_y, A_x, and B_x are the equivalent fractions of the ions A^+, B^+ in zeolites Y and X.

554 ION EXCHANGE REACTIONS IN ZEOLITES

For the Na⁺ → Li⁺ exchange, the mean value of K' was determined to be 0.61 and for the Na⁺ → K⁺ exchange, 1.35. Thus for the three ions Li⁺, Na⁺, and K⁺, zeolite X, with higher framework charge, exchanges with the smaller ion selectively.

The standard differential heats of exchange, $\partial(\Delta H_x^0)/\partial x$, as a function of the degree of exchange, x, also show differences. The two uni-univalent exchange reactions, (Na⁺ → K⁺ and Na⁺ → Li⁺) in zeolite X show a change in the curve at a composition of x = 0.6. This indicates there are two kinds of sites. In zeolite Y there is no change. For uni-divalent exchanges the differential heats differ considerably (Fig. 7.10). The differential heat of exchange for Na⁺ → ½Ca²⁺ in zeolite Y has a maximum at x = 0.40, which corresponds to 20 sodium ions per unit cell. For the Na⁺ → ½Sr²⁺ exchange, a minimum occurs at this point. Almost half of the 36 exchangeable sodium ions in zeolite Y behave differently from the other sodium ions.

Although the zeolites X and Y have the same framework topology, differences in cation density and distribution cause sizeable differences in the exchange selectivities and thermochemical quantities.

Ion Exchange of Chabazite

In studying the ion exchange behavior of chabazite, various methods for obtaining cationic forms have been employed (11, 39-42). A summary of equilibrium data for uni-univalent and uni-divalent exchange reactions is presented in Table 7.8 (42). Isotherms are illustrated in Fig. 7.11. Due to a sieve effect, attempts to exchange lanthanum ions, tetramethylammonium, and tetraethylammonium ions in chabazite,

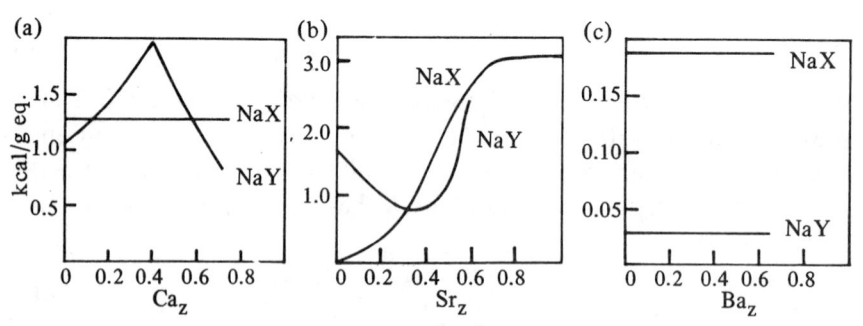

Figure 7.10 Differential heats of exchange, $\partial \Delta H_x^\circ/\partial x$, in kcal/g equiv. as functions of the degree of exchange. (a) Na⁺ → ½Ca²⁺; (b) Na⁺ → ½Sr²⁺; (c) Na⁺ → ½Ba²⁺ for zeolites X and Y as shown. The degree of exchange is based on x_{max} = 1 (38).

Table 7.8 Cation Exchange in Chabazite

Exchange Reaction	Conc.	$t(°C)$	Isotherm type [e]	x_{max} [f]	$Log_{10}K'$ $A_z = 0.2$	$A_z = 0.5$	$Log_{10}K'^g$	$Log_{10}K'$ vs A_z type	$\Delta G°$ (cal/g equiv)	$\Delta H°$	Ref.	Other Ref.
$Na^+ \to Li^+$	a	c	25	B	1	−1.25	−1.25	−1.25	linear	+1720	+3050	42	43
$Na^+ \to K^+$	a	c	25	A	1	+1.3	+1.2	+1.2	linear	−1600	−2610	42	40, 43
$Na^+ \to Rb^+$	a	c	25	A	1	+1.4	+1.0	+1.0	linear	−1340	−2240	42	43
$Na^+ \to Cs^+$	a	c	25	D	0.84	+2.0	+1.9	+1.9	linear	−2500 [h]	−2890	42	40, 43
$Na^+ \to NH_4^+$	a	c	25	A	1	+1.0	+0.75	+0.75	linear	−990	−1230	42	
$Na^+ \to Tl^+$	a	c	25	A	1	+1.9	+2.1	+2.1	nonlinear w. max	−2620	−1900	42	41
$Na^+ \to Ag^+$	a	c	25	A	1	+1.25	+1.1	+1.1	linear	−1440	−440	42	41
$2Na^+ \to Ca^{2+}$	a	d	50	A	1	+0.25	+0.8	+0.8	nonlinear	+90		42	41, 43
$2Na^+ \to Sr^{2+}$	a	d	50	A	1	+0.1	+1.0	+1.0	nonlinear w. max	+40		42	43
$2Na^+ \to Ba^{2+}$	a	d	50	A	1	+0.3	+0.6	+0.6	nonlinear w. max	0	+1760	42	41, 43
$2Na^+ \to Pb^{2+}$	a	d	50	A	1	0	+1.0	+1.0	nonlinear w. max	−170	+1030	42	
$K^+ \to Cs^+$	b	1.0 N	23	D	0.82							40	
$\tfrac{1}{2}Sr^{2+} \to Cs^+$	b	1.0 N	23	A	1.0							40	
$K^+ \to Na^+$	b	1.0 N	23	B	1.0	−1.0	−1.1	−1.1	linear			40	

[a] Chabazite mineral analysis, in moles of oxides 0.15 Na₂O · 0.83 CaO · Al₂O₃ · 4.98 SiO₂ · 6.78 H₂O with c.e.c. of 3.49 meq/g. Sodium exchanged forms were prepared and analyzed to give Na₂O/Al₂O₃ = 0.94 to 1 with slight increase in SiO₂/Al₂O₃ attributed to some surface hydrolysis (42).

[b] Chabazite mineral of composition Na₀.₀₉Ca₁.₇₂Al₃.₄₇Si₉.₅₁O₂₄ in the anhydrous unit cell contents, c.e.c. = 2.59 meq/g.

[c] Concentration not given.
[d] Concentration varied.
[e] See Fig. 7.2.
[f] $x_{max} = \dfrac{\text{equivalents exchanged}}{\text{gram atoms of Al in the zeolite}}$
[g] Estimated from plots in (42).
[h] Based on reduced exchange capacity, $x/x_{max} = 1$.

556 ION EXCHANGE REACTIONS IN ZEOLITES

were unsuccessful. The unit cell constants of chabazite are not greatly altered by exchange (41).

Of the alkali metal ions, only cesium does not replace all of the sodium ions in sodium chabazite; the extent of exchange is 84%. Other ions (Table 7.8) replace all of the sodium ions. The positions of univalent ions in the structure of hydrated chabazite are not known, but they are postulated to occupy sites in the 8-rings, and in the single 6-rings or hexagonal prisms (Chap. 2). The larger rubidium, thallium, and cesium ions would not be expected to replace any sodium ions within the hexagonal prisms. Instead, a mechanism involving migration of sodium ions into the larger cavities during exchange is likely (42). The cations present in these larger cavities result in less space available for water molecules with increasing cation size. A smooth decrease in the number of water molecules per unit cell with increasing cation size was observed (42) (Fig. 7.12).

All of the plots of log K'^A_{Na} versus A_z for the uni-divalent exchanges exhibit maxima, (Fig. 7.11). These have been related to the presence of different types of exchange sites (4). A reversal in selectivity resulted when the degree of exchange reached A_z = 0.8 - 0.9. The selectivity sequence observed is.

$Tl^+ > K^+ > Ag^+ > Rb^+ > NH_4^+ > Pb^{2+} > Na^+ = Ba^{2+} > Sr^{2+} > Ca^{2+} > Li^+$

Exchange reactions with divalent ions are slower than those for univalent ions; the more strongly hydrated divalent ions pass through the 8-rings with more difficulty. For example, a calcium ion must lose many of the strongly coordinated water molecules in order to enter the structure.

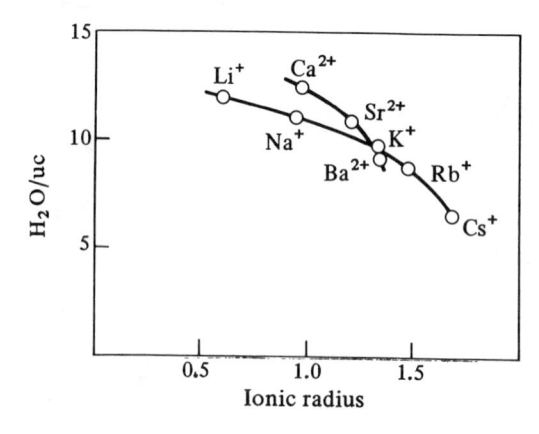

Figure 7.12 Variation of the number of water molecules/rhombohedral unit cell with ionic radius for univalent and divalent cation exchanged forms of chabazite(42).

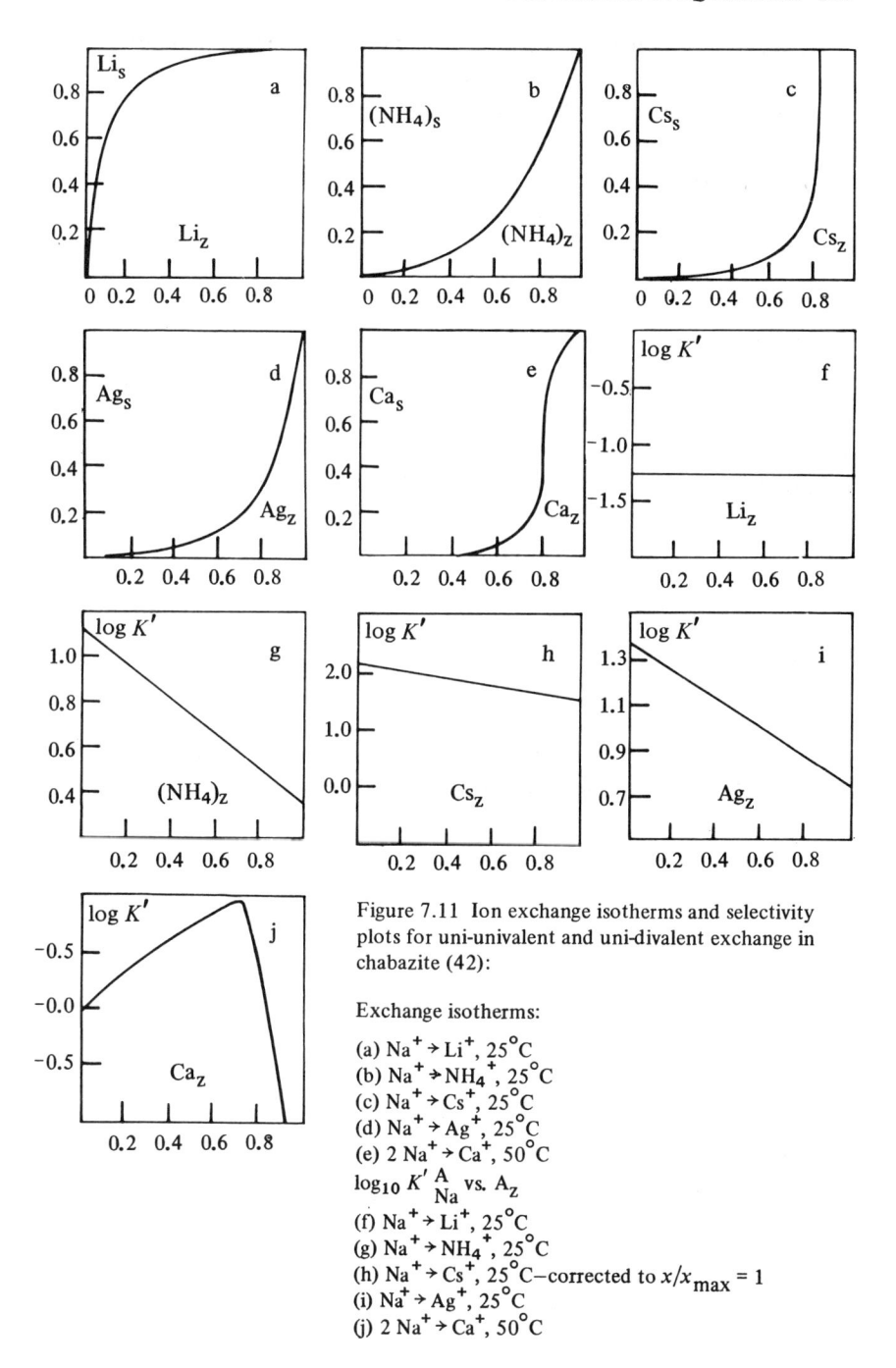

Figure 7.11 Ion exchange isotherms and selectivity plots for uni-univalent and uni-divalent exchange in chabazite (42):

Exchange isotherms:

(a) $Na^+ \rightarrow Li^+$, 25°C
(b) $Na^+ \rightarrow NH_4^+$, 25°C
(c) $Na^+ \rightarrow Cs^+$, 25°C
(d) $Na^+ \rightarrow Ag^+$, 25°C
(e) $2 Na^+ \rightarrow Ca^+$, 50°C

$\log_{10} K' {}^A_{Na}$ vs. A_z

(f) $Na^+ \rightarrow Li^+$, 25°C
(g) $Na^+ \rightarrow NH_4^+$, 25°C
(h) $Na^+ \rightarrow Cs^+$, 25°C—corrected to $x/x_{max} = 1$
(i) $Na^+ \rightarrow Ag^+$, 25°C
(j) $2 Na^+ \rightarrow Ca^+$, 50°C

For cesium exchange, the cesium ions may occupy the sites in the 8-rings (3/unit cell) leaving 1 sodium ion per unit cell within the main ellipsoidal cavity (40). The sample studied contained 3.5 potassium ions per unit cell and occupancy of three sites by three cesium ions corresponds to 85% exchange. Observed results are in agreement.

Standard heats of partial exchange and differential heats have been reported (42). The latter indicate that two thirds of the sodium ions occupy one type of cation site, probably in the intersecting channels.

Ion Exchange of Clinoptilolite

Clinoptilolite displays an ion sieve effect for large organic cations (36). (See Table 7.9.) In a series of substituted alkylammonium ions, the degree of exchange, x_{max}, decreases with increasing size of the cation. Tetramethylammonium and tertiary butylammonium ions were completely excluded. Exchange isotherms are shown in Fig. 7.13a and the selectivity coefficient function in Fig. 7.13b. The number of water molecules displaced by the larger organic cations varies linearly with (1) the degree of exchange and (2) the volume of the organic ion. The ion sieve effect was interpreted on the assumption that the aluminosilicate framework structure of clinoptilolite is isostructural with that of heulandite. (This may not be correct; see Chap. 2). Clinoptilolite is quite selective for ammonium ions as compared to other zeolites. Good agreement with the "triangle rule" was observed; i.e., in terms of K, the rule states:

$$K_B^A \times K_C^B = K_C^A \text{ for ions A, B, and C} \quad (7.19)$$

Ion Exchange of Zeolite T

Univalent and alkaline earth ion exchange equilibria were measured on synthetic zeolite T at 25°C (47). In these exchange studies it was found that 25% of the potassium ions could not be exchanged out of the original zeolite at temperatures below 300°C in dilute solution. The zeolite was found to prefer potassium to the divalent barium and calcium ions. The order of decreasing selectivity for univalent ions was as follows:

$Cs^+ > Rb^+ > Ag^+ > K^+ = NH_4^+ > Na^+ > Li^+$

The equilibria data summarized in Table 7.10 include estimates of selectivity coefficients determined from $\log K'{}_B^A$ vs A_z plots. The exchange behavior of this zeolite was interpreted based on the structure

Table 7.9 Cation Exchange in Clinoptilolite

Exchange Reaction	Conc.	$t(°C)$	Isotherm type [a]	x_{max} [b]	$Log_{10}K'$ [c] $A_z = 0.2$	$Log_{10}K'$ [c] $A_z = 0.5$	$Log_{10}K'$ vs A_z type	$\Delta G°_{298}$ (cal/g equiv)	$\Delta H°$	Ref.	Footnote
$K^+ \to Na^+$	1.0 N	25	C	0.4				+1500		44	d
$K^+ \to Cs^+$	1.0 N	25						−800		44	d
$Na^+ \to Cs^+$	1.0 N	25						−2300	−4000	44	d
$Na^+ \to Cs^+$	0.02 N	30		1.0				−2420	−2700	45, 46	e
$Na^+ \to NH_4^+$	0.02 N	30		1.0				−1290	−880	45	e
$Na^+ \to NH_4^+$	0.018 N	60	A	1.0	+0.9	0.9	linear	−1370		36	f
$½Ca^{2+} \to Na^+$	1.0 N	25	B	1.0				−100		8	d
$Na^+ \to ½Sr^{2+}$	1.0 N	25	D	0.7				−100		8	d
$Ca^{2+} \to Sr^{2+}$	1.0 N	25	B	1.0				−100		8	d
$Na^+ \to EtNH_3^+$	0.018 N	60	A	1.0	+1.3	+1.2	linear	−1900		36	f
$Na^+ \to MeNH_3^+$	0.018	60	A	1.0	+1.1	+1.0	linear	−1400		36	f
$Na^+ \to Me_2NH_2^+$	0.018	60	A	1.0	+1.3	+1.0	linear	−1470		36	f
$Na^+ \to nPrNH_3^+$	0.018	60	B	1.0	+0.9	+0.3	linear	−330		36	f
$Na^+ \to Me_3NH_3^+$	0.018	60	D	0.5	0					36	f
$Na^+ \to nBuNH_3^+$	0.018	60	D	0.3	−0.9					36	f
$Na^+ \to isoPrNH_3^+$	0.018	60	D	0.4	−0.9					36	f

[a] See Fig. 7.2.
[b] $x_{max} = \dfrac{\text{equiv. exchanged}}{\text{g atoms of Al in the zeolite}}$
[c] Estimated from plots in the references.
[d] Clinoptilolite from Hector, California, c.e.c. of 1.7 meq/g.
[e] Original clinoptilolite (locality not given) 10% impurities with c.e.c. of 2.04 meq/g; for the $Na^+ \to Cs^+$ reaction, $T\Delta S° = -280$ and for the $Na^+ \to NH_4^+$ reaction, $T\Delta S° = +410$.
[f] Clinoptilolite from Hector, California of exchange capacity 1.83 meq/g and having the composition after sodium exchanging, in moles,
$K_2O_{0.034}Na_2O_{0.784}CaO_{0.106} \cdot Al_2O_3 \cdot SiO_{2_{9.23}} \cdot H_2O_{6.62}$

560   ION EXCHANGE REACTIONS IN ZEOLITES

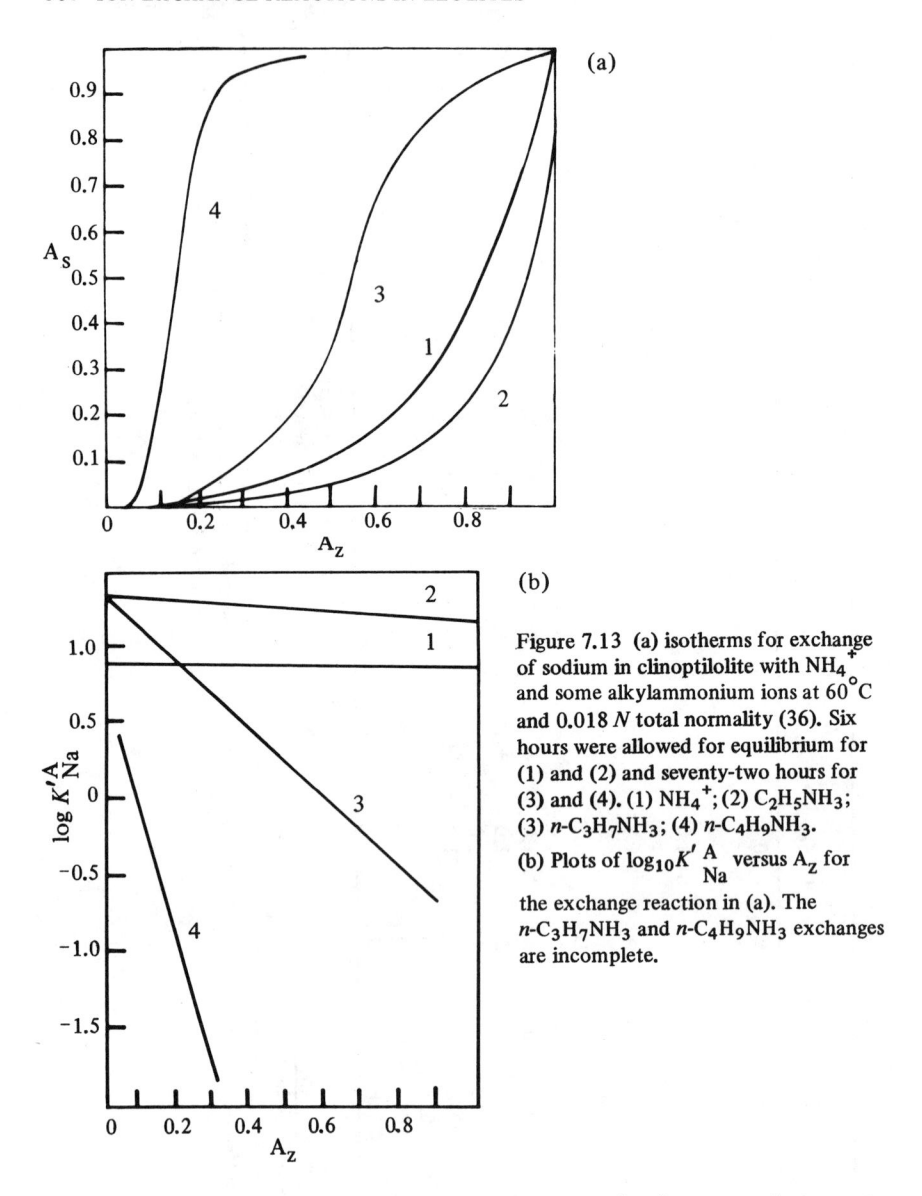

Figure 7.13 (a) isotherms for exchange of sodium in clinoptilolite with NH_4^+ and some alkylammonium ions at 60°C and 0.018 N total normality (36). Six hours were allowed for equilibrium for (1) and (2) and seventy-two hours for (3) and (4). (1) NH_4^+; (2) $C_2H_5NH_3$; (3) n-$C_3H_7NH_3$; (4) n-$C_4H_9NH_3$. (b) Plots of $\log_{10} K'^A_{Na}$ versus A_z for the exchange reaction in (a). The n-$C_3H_7NH_3$ and n-$C_4H_9NH_3$ exchanges are incomplete.

of the related zeolite offretite. Assuming that the framework structure of zeolite T is topologically related that of the zeolite offretite, there would be 18 tetrahedra per unit cell. Each unit cell, consequently, would contain 4 AlO_4 tetrahedra.

Zeolite T shows a specificity for ammonium which is about the

Table 7.10 Cation Exchange in Zeolite T (25°C, 0.1 N) [a] (47)

Exchange Reaction	Isotherm Type	x_{max} [b]	$Log_{10}K'$ $A_z = 0.2$	$Log_{10}K'$ $A_z = 0.5$	$Log_{10}K'$ vs A_z type	ΔG (298°) (Kcal/g equiv)
$Na^+ \rightarrow Li^+$	B	1.0	1.2	1.1	nonlinear	1.41
$Na^+ \rightarrow K^+$	A	1.0	1.0	1.0	linear	-1.30
$Na^+ \rightarrow Rb^+$	A	1.0	1.4	1.3	nonlinear	-1.77
$K^+ \rightarrow Cs^+$	A	1.0	1.0	0.9	nonlinear	-1.30
$K^+ \rightarrow NH_4^+$	C	1.0	-0.1	-0.1	nonlinear	0.23
$Na^+ \rightarrow \frac{1}{2}Ca^{2+}$	A	1.0	0.5	0.2	nonlinear	0.21
$K^+ \rightarrow \frac{1}{2}Ba^{2+}$	C	1.0	0	-0.5	linear	0.04
$Na^+ \rightarrow Ag^+$	A	1.0	1.3	1.2	linear	-1.65

[a] Initial composition of Zeolite T is: $0.713 K_2O, 0.335 Na_2O \cdot Al_2O_3 \cdot 6.76 SiO_2$ expressed as oxides and $K_4(AlO_2)_4(SiO_2)_{14}$ for the offretite-type unit cell.
[b] In this work, 0.25 equivalent fraction of K^+ ions in the zeolite could not be exchanged. Consequently, all data were normalized to the reduced exchange capacity.

same as for potassium ions. The difference in the selectivity shown for various alkali metal ions is due to the small water content and lack of appreciable ion hydration within the zeolite crystals. The zeolite shows a strong preference for potassium over the alkaline earth cations. Utilizing an extension of the Eisenman treatment for predicting ion exchange selectivity, it was shown that univalent cations may be preferred to divalent cations when the Si/Al atom ratio of the zeolite increases; this is observed in the case of zeolite T (50).

Ion Exchange of Phillipsite

The isotherms for the sodium-lithium, sodium-potassium, sodium-rubidium, sodium-cesium, sodium-calcium and sodium-strontium were reversible (Table 7.11) (32). However, barium exchange is irreversible and there is evidence of fixation of barium in the phillipsite structure. In terms of selectivity in the 0.1 N solution there is little preference for calcium or strontium over sodium as compared to the very strong selectivity shown for barium.

Ion Exchange of Zeolite P

The ion exchange equilibria for various univalent and divalent ions exchanging with sodium in the synthetic cubic zeolite P are listed in Table 7.12 (33). Two patterns of exchange behavior were observed as shown by the effect of the exchanged ion on the symmetry of the zeolite crystal. In one pattern a limited range of two forms of zeolite

Table 7.11 Cation Exchange in Phillipsite (25°C, 0.1 N Solution)

Exchange Reaction	Isotherm Type	x_{max}	$Log_{10}K'$ $A_z = 0.2$	$A_z = 0.5$	K_a	$\Delta G°$ (cal/g equiv)	Ref.
$Na^+ \to Li^+$	C	0.5	nonlinear		0.006	3002	32
$Na^+ \to K^+$	A	1.0	1.53	1.11	12.9	−1500	32, 44
$Na^+ \to Rb^+$	A	0.8	1.62	1.02	10.5	−1403	32
$Na^+ \to Cs^+$	A	1.0	nonlinear		24.2	−1900	32, 44
$Na^+ \to \frac{1}{2}Ca^{2+}$	C	1.0	−0.68	−1.02	0.035	+989	32
$Na^+ \to \frac{1}{2}Sr^{2+}$	C	1.0	−0.59	−0.75	0.066	+803	8, 32

Sedimentary Phillipsite, Pine Valley, Nevada, $(Na_{1.42}K_{0.38}Ca_{0.03})O \cdot Al_2O_3 \cdot 5.6\, SiO_2 \cdot 5.7\, H_2O$.

compositions may coexist, whereas in the second pattern the cubic form changes to a tetragonal type at a particular zeolite composition. For example, the exchange with potassium altered the cubic structure to a tetragonal one with an associated volume contraction of 60 Å³ in the unit cell. This is caused by a loss of water during the exchange of sodium for potassium of 12 to 8 water molecules per unit cell. This zeolite is very selective for the heavy alkali metal ions, rubidium and cesium. It also shows considerable selectivity for sodium over lithium. With the divalent cations, strontium and barium, the selectivity was even greater; zeolite P shows such selectivity for barium that up to an exchange level of $Ba_z = 0.5$, barium was removed almost quantitatively from the exchange solution.

Ion Exchange of Zeolite Z (K-F)

Unusual exchange equilibria in the synthetic zeolite Z (K-F) were observed for univalent alkali metal and divalent alkaline earth ion pairs at 25°C(48). Complete reversibility was observed only in the sodium-cesium and potassium-lithium exchange. In some cases the

Table 7.12 Cation Exchange in Zeolite P (25°C, 0.1 N Solution) (33)

Exchange Reaction	Isotherm Type	x_{max}	$Log_{10}K'$ $A_z = 0.2$	$A_z = 0.5$	$Log_{10}K'$ vs A_z	K_a
$Na^+ \to Li^+$	C	0.5	−1.36	−2.0	nonlinear	0.009
$Na^+ \to K^+$	B	1.0	0.2	0.56	linear	3.67
$Na^+ \to Rb^+$	A	0.9	0.73	1.23	nonlinear	3.64
$Na^+ \to Cs^+$	A	1.0	1.13	1.93	nonlinear	3.46
$Na^+ \to \frac{1}{2}Sr^{2+}$	A	1.0	0.94	1.00	linear	3.75
$Na^+ \to \frac{1}{2}Ba^{2+}$	A	1.0	−	−	linear	−

exchange isotherm exhibits hysteresis but is reversible. This is true in the case of sodium-potassium, sodium-lithium, etc. (See Table 7.13.) The univalent exchange reactions were interpreted in terms of two types of exchange site, one type of which may be associated with smaller cavities within the structure. Since the crystal structure of zeolite Z is unknown, further interpretation of the exchange phenomena cannot be made. Like zeolite P, zeolite Z shows a very high selectivity for barium and will remove this ion from solution quantitatively. It is likewise very selective for strontium and calcium. The reversible isotherm for the sodium-cesium exchange is shown in Fig. 7.14a and the irreversible isotherm for the sodium–calcium exchange in Fig. 7.14c.

Ion Exchange Equilibria of Other Zeolite Minerals

Other zeolite minerals for which limited ion exchange equilibria have been determined include erionite (8, 44), stilbite (39), and mordenite (49). The equilibria data are summarized in Table 7.14. Stilbite has a high selectivity for cesium and potassium. The isotherm for $Sr^{2+} \rightarrow Cs^+$ exchange is sigmoidal and the selectivity reverses with increasing degree of exchange by cesium. Increasing the temperature from 23°C to 85°C produced a large drop in selectivity for cesium over sodium. The potas-

Table 7.13 Cation Exchange in Zeolite Z (K-F)[a] (0.1 N Solution) (48)

Exchange Reaction	Isotherm type	x_{max}	$Log_{10}K'$[b] $A_z = 0.2$	$A_z = 0.5$	K_a	Remarks
$Na^+ \rightarrow K^+$	E	1.0	0.9	0.5	4.57	Isotherm shows hysteresis region
$Na^+ \rightarrow Cs^+$	B	1.0	1.5	1.0	3.42	Two types of exchange sites
$Na^+ \rightarrow Li^+$	E	1.0	−0.2	−0.2	1.01	Isotherm has hysteresis region
$Na^+ \rightarrow ½Ca^{2+}$	E	1.0	–	–	–	Hysteresis irreversible
$Na^+ \rightarrow ½Sr^{2+}$	E	1.0	–	–	–	Hysteresis irreversible
$Na^+ \rightarrow ½Ba^{2+}$	E	1.0	–	–	–	Hysteresis irreversible
$Li^+ \rightarrow K^+$	B	1.0	0.75	−0.8	0.51	
$K^+ \rightarrow Cs^+$	E	1.0	–	–	–	
$K^+ \rightarrow ½Ba^{2+}$	E	0.25	–	–	–	Irreversible

[a] Initial zeolite composition $K_2O \cdot Al_2O_3 \cdot 2.3\ SiO_2 \cdot 2.5\ H_2O$
[b] Interpolated values from (48).

Table 7.14 Cation Exchange in Erionite [a]

Exchange Reaction	Conc. (N)	$t(°C)$	Isotherm type	x_{max} [c]	$Log_{10}K'$ $A_z = 0.2$	$Log_{10}K'$ vs A_z $A_z = 0.5$	$\Delta G°$ type	$\Delta H°$ (cal/g equiv)	Ref.
$Na^+ \to Cs^+$	1.0	25	C	[e]			−2100	−2800	44
$K^+ \to Na^+$	1.0	25	C				+1400		44
$Na^+ \to ½Ca^{2+}$	1.0	25	B	1.0			0		8
$Na^+ \to ½Sr^{2+}$	1.0	25	C	1.0			+100		8
$Ca^{2+} \to ½Sr^{2+}$	1.0	25	B	1.0			+200		8

Cation Exchange in Stilbite [b]

$Na^+ \to Cs^+$	1.0	23	A	1.0	+1.52	+0.87	linear	−1156	−4920	39
$K^+ \to Na^+$	1.0	23	C	1.0	−0.65	−0.87	linear	+1204		39
$K^+ \to Cs^+$	1.0	23	B	1.0	+0.22	+0.04	linear	−15		39
$½Sr^{2+} \to Cs^+$	1.0	23	B	1.0				−735		39

[a] Erionite mineral from Pine Valley, Nevada about 90% impurity with a c.e.c. of 2.2 meq/g.
[b] Stilbite mineral from Nova Scotia with chemical composition of $(Na_2O)_{0.18}(CaO)_{0.82} \cdot Al_2O_3 \cdot (SiO_2)_{6.2}$ in moles, anhydrous basis, measured c.e.c. of 1.79 meq/g (for Na^+).
[c] $x_{max} = \dfrac{\text{equiv. exchanged}}{\text{g atoms of Al in the zeolite}}$
[d] Estimated from plots published in Ref. 39.
[e] In view of published data on zeolite T, it is unlikely that $x_{max} = 1$, i.e., all of the K^+ is not exchangeable. Indeed, data in Fig. 6 of Ref. 44 indicate that the isotherm terminates at $x_{max} = 0.6$ to 0.7.

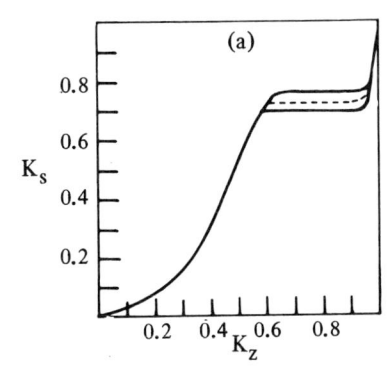

Figure 7.14 Exchange isotherms in zeolite Z(K-F) at 25°C in 0.1 N halide solution (48). (a) exchange of Na$^+$ → K$^+$ shows hysteresis loop; (b) exchange Na$^+$ → Cs$^+$ shows reversibility; (c) exchange Na$^+$ → ½Ca^{2+} shows hysteresis and high selectivity for Ca^{2+}. Ba^{2+} exchange is very similar.

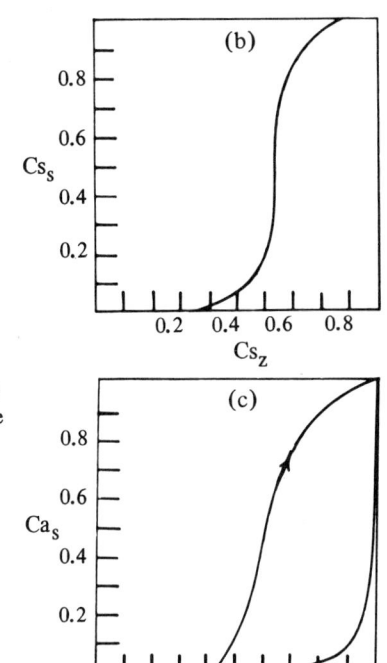

sium in erionite in fact does not exchange fully with sodium as indicated in Table 7.14.

Selectivity coefficients (the separation factor, α) were determined for exchange of sodium ions in a synthetic mordenite by uni- and divalent ions in 0.1 N aqueous solutions at 20°C (49). The synthetic mordenite had the composition 0.86 Na$_2$O · Al$_2$O$_3$ · 10.1 SiO$_2$, on an anhydrous basis with an exchange capacity of 1.9 meq/g.

The isotherms for Na$^+$ → K$^+$, Na$^+$ → Cs$^+$, Na$^+$ → ½ Ba^{2+} terminate at x_{max} = 1. However, the divalent ions, Mg^{2+}, Sr^{2+}, and Ca^{2+} do not exchange completely and the isotherms terminate at x_{max} = 0.5, 0.8, and 0.6 respectively. For divalent ions, the order of decreasing relative selectivity was found to be Ba^{2+} > Sr^{2+} > Ca^{2+} > Mg^{2+}. The separation factor versus the degree of exchange shows a minimum at A_z = 0.9 for the ions K$^+$, Cs$^+$, Ag$^+$, and Ba^{2+}.

C. THERMODYNAMICS OF ION EXCHANGE PROCESSES

For the ion exchange reaction as expressed by Eq. 7.1, the free energy of exchange, ΔG^0 is given by:

$$\Delta G^0 = \frac{-RT}{Z_A Z_B} \ln K_a \qquad (7.20)$$

and the standard entropy, ΔS^0 by:

$$\Delta S^0 = \frac{\Delta H^0 - \Delta G^0}{T} \qquad (7.21)$$

The standard enthalpy is obtained from the variation of K_a with temperature:

$$\frac{d \ln K_a}{dT} = \frac{\Delta H^0}{RT^2} \qquad (7.22)$$

Enthalpy changes in zeolite ion exchange reactions are generally small.

The thermodynamic data shown in Tables 7.3 to 7.13 should be used with caution. In many cases complete exchange is not obtained due to cation sieve effects. Further, in some cases equilibrium may not have been reached where very dilute solutions were employed.

The ion exchange model originally proposed by Eiseman (50) has been extended to account for the variation in ion specificity exhibited by zeolites (26, 47). Interaction of the ion with the zeolite and solution phases is considered. For the uni-univalent ion exchange reaction

$$A_s^+ + B_z^+ \rightleftharpoons A_z^+ + B_s^+ \qquad (7.23)$$

where s and z represent the solution and zeolite phases. The free energy of the reaction is considered to consist of two terms:

$$\Delta G^0 = (\Delta G_z^A - \Delta G_z^B) - (\Delta G_s^A - \Delta G_s^B) \qquad (7.24)$$

The first term in this expression represents the difference between the free energy of A^+ and B^+ in the zeolite while the second term represents the free energy difference of hydration of the ions A^+ and B^+ in solution. The first term is more important if the force fields in the zeolite are very strong (zeolites with a high framework charge and correspondingly low Si/Al ratio) and small ions are preferred. If the fields are weak (as in zeolites with a high Si/Al ratio) the second term is more important and large, weakly hydrated cations are preferred.

The selectivity series

$$Na > K > Rb > Cs > Li$$

is exhibited by zeolites A (Si/Al = 1) and X (Si/Al = 1.2). Zeolite Y

(Si/Al = 2.8) displays a weak field selectivity pattern of :
$$Cs > Rb > K > Na > Li$$
The variation in selectivity coefficient for the sodium-calcium exchange reaction is shown in Fig. 7.15 as it is related to the Si/Al ratio which is used as a measure of field strength. The Si/Al ratio for the synthetic zeolites A, X, Y, and T are plotted.

For uni-divalent ion exchange, the theory predicts that only one selectivity series should exist:
$$Ba > Sr > Ca > Mg$$
When, however, the field strength becomes very weak as is the case in the silica-rich zeolite T (Si/Al = 3.4), sodium ion may be preferred over calcium. In fact, zeolite T does prefer certain univalent ions over divalent ions (Table 7.10).

The model proposed by Sherry (47) considers the entropy change that occurs in the zeolite phase and the corresponding difference in entropy of hydration of the ions in the equilibrium solution. For the uni-univalent exchanges of lithium and cesium for sodium in zeolite A, the difference between the total entropy of the exchange reaction and the hydration entropy of the ions in solution is large. It is concluded that a net transfer of water molecules occurs from the solution to the zeolite when lithium is the exchanging cation and that there is a net loss of water from the zeolite to the solution with cesium as the exchanging ion. Transfer of water to the zeolite decreases the entropy because water molecules adsorbed in the zeolite have fewer degrees of freedom.

The water content of the zeolite decreases with increasing atomic number of the univalent alkali metal ion present. As each univalent

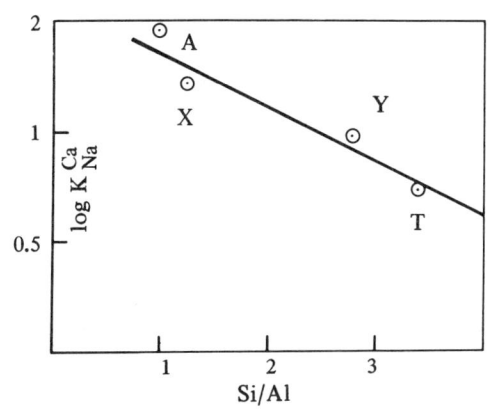

Figure 7.15 The selectivity coefficient, K_{Na}^{Ca}, for Ca-Na exchange in zeolites as a function of Si/Al ratio. The values of K_{Na}^{Ca} were chosen for a loading of $Ca_z = 0.25$ for comparison since calcium population of different cation sites occurs at higher degrees of exchange (26, 47).

ion replaces another univalent ion above it in the series, a decrease occurs in water content in the zeolite phase and consequently an increase in entropy takes place. For uni-divalent ion exchange in zeolite A, entropy changes for the ion exchange reaction indicate extensive ion-pair formation between calcium ions and the negative charge of the aluminosilicate framework as compared to essentially no ion-pair formation in the case of barium. In the latter case, the barium ions appear to be fully hydrated and free within the zeolitic water in large cages such as the α-cages in zeolite A.

The heats of ion exchange between gaseous cations and a second ion in anhydrous zeolites have been calculated from the measured heats of ion exchange of hydrated zeolites in aqueous solutions, heats of immersion of dehydrated zeolites, and the heats of hydration of the gaseous ions. This has been done for the zeolites X, Y, A, and chabazite (51). The numerical values of ΔH for the heat of exchange of gaseous ions with anhydrous zeolites are about ten times greater than the heat of exchange between the ion in aqueous solution and the hydrated zeolite. The magnitudes of ΔH increase in the sequence, chabazite, zeolite A, zeolite X, zeolite Y. The large numerical magnitude of ΔH indicates the high solvation energy of the exchanged cation by the anhydrous zeolite framework. Because the framework oxygen atoms are more rigid in the lattice than are the oxygen atoms in the water, the ion solvation energy is less than the solvation energy in water. The heats of solvation of the gaseous ions by the hydrated zeolite were estimated by using an appropriate heat cycle (Table 7.15).

The individual heats of solvation for a given cation in the hydrated zeolite lie between the values for solvation by the anhydrous zeolite

Table 7.15 Some Solvation Heats of Gaseous Ions by Water and Zeolite Frameworks (51)

Cation	Water	Chabazite Hyd.	Chabazite Anhyd.	Zeolite X Hyd.	Zeolite X Anhyd.	Zeolite Y Hyd.	Zeolite Y Anhyd.	Zeolite A Hyd.	Zeolite A Anhyd.
Na^+	−105.9	−94.6	−68.4	−98.4	−89.5	−99.0	−93.8	−95.0	−94.2
K^+	−85.7	−76.6	−55.4	−79.6	−72.4	−80.1	−75.9	−76.9	−76.2
Rb^+	−79.8	−71.3	−51.5	−74.1	−67.4	−74.6	−70.7	−71.6	
Cs^+	−71.9	−64.2	−46.4	−66.6	−60.8	−67.2	−63.7	−64.5	−64.0

ΔH (kcal/g ion)

and by water. Chabazite exhibits the smallest heat of solvation as compared to the other zeolites. However, the increase in the solvation heat from the anhydrous zeolite to the hydrated zeolite is greatest for chabazite. This is interpreted to support the view that the cations in this zeolite are generally more exposed than in other zeolites. This factor contributes to strong affinity in adsorption and greater electrical fields in the vicinity of the cations. Because the binding energy of the cations is less than in the other zeolites, a larger proportion of the cations are not as well solvated by oxygen anions in the framework.

D. HYDROGEN EXCHANGE IN ZEOLITES

Many of the crystalline zeolites decompose when treated with strong acids (Chap. 6). However "hydrogen" forms have been prepared by acid treatment of the silica-rich zeolites — mordenite and clinoptilolite (52, 53). Synthetic zeolites with a Si/Al ratio $\sim$ 1 in the framework have a high cation density. Cations, such as sodium, that are located within the large cavities exhibit a limited amount of hydrolysis when the zeolite is in water suspension. For example, zeolite NaA when slurried in distilled water produces a pH of 10 - 10.5. This is probably due to hydrolysis of the sodium ions and replacement by hydrated hydrogen ions (14). Direct treatment of silica-rich zeolites with strong acids results in the progressive replacement of the cation by hydronium ion. This is illustrated schematically for the zeolite mordenite by the reaction:

$$Na_8[(AlO_2)_8(SiO_2)_{40}] \cdot 24\ H_2O$$
$$\downarrow n\ H_3O^+ \qquad (7.25)$$
$$Na_{8-n}(H_3O^+)_n[(AlO_2)_8(SiO_2)_{40}] \cdot m\ H_2O + n\ Na^+ + (24-m)\ H_2O$$

Replacement of all of the original cations by hydronium ions has been carried out in mordenite (54) and clinoptilolite (53, 55). Further treatment with strong acid removes the framework aluminum ions, which are probably replaced by groups of 4 hydroxyls (53).

$$\text{Si-O-Al(O-Si)(OSi)-O-Si} + HCl \longrightarrow \text{Si-O(H···O-H)(H-O···H)-O-Si} + Al^{3+}(aq) + 3Cl^- \qquad (7.26)$$

Treatment of clinoptilolite with 2 N HCl yields a completely "dealuminated" form with retained crystallinity (55). The composition of the dealuminated form is reported as:

$$Si_{4.65}O_{7.3}(OH)_{4.0} \cdot nH_2O$$

The existence of hydronium hydrates in minerals such as the $H_9O_4^+$ and $H_{15}O_7^+$ groupings has been proposed (56). They probably exist in the layer structures of smectites (montmorillonite). The existence of tetrahedral H_3O_4 groups in the structure of the zeolite viseite is probable (Chap. 2) (57). Formation of the hydroxyl group in place of the exchange cation in the zeolite structure framework will necessarily result in the disruption of an Si, Al-O bond. This is discussed in Chap. 6.

Hydrogen forms can be prepared in other ways. The simplest method involves exchange of the cation by ammonium from an ammonium salt-water solution. As shown previously, the ammonium ion undergoes exchange readily. Subsequent thermal treatment of the ammonium-exchanged zeolite results in the liberation of NH_3 and the formation of hydroxyl groups on the framework structure. In alumina-rich zeolites, such as zeolite A, partial replacement of the sodium ions is possible. In the silica-rich zeolites, such as zeolite Y, complete exchange of the sodium ions by ammonium followed by thermal decomposition leaves a crystalline zeolite free of cations. In the case of zeolites A and X, attempts to achieve complete hydrogen exchange by this method result in structural disintegration.

Another technique is based upon an ion exclusion effect (41). When a silver exchanged zeolite, in this example chabazite, is heated with an aqueous $(CH_3)_4NBr$ solution at 95°C, AgBr is precipitated. The silver ions are removed and replaced by hydrogen ions. When zeolite A is completely exchanged with silver and treated with tetramethylammonium chloride, silver ions are similarly removed with the formation of silver chloride.

$$Ag_{12}A + n(CH_3)_4NCl + nH_2O \rightarrow Ag_{12-n}H_nA + n(CH_3)_4N^+ + nOH^- + nAgCl \quad (7.27)$$

where $A = (AlO_2)_{12}(SiO_2)_{12} \cdot nH_2O$

Since the $(CH_3)_4N^+$ ion is too large to enter the zeolite channels, the net effect is a proton replacement for silver. A single treatment of silver exchanged zeolite can remove 63% of the silver ions. Similarly, a single treatment of silver exchanged zeolite A with ammonium hydroxide can remove 44% of the silver ions. The loss of silver ion is due to formation of the silver ammonium complex (see reaction scheme below). Hydrogen exchange is thus achieved in a basic solution:

$$Ag_{12}A + 2nNH_3 + H_2O \rightarrow Ag_{12-n}H_nA + nAg(NH_3)^+_2 + nOH^- \quad (7.28)$$

The ion exchange selectivity for silver is affected by complex formation with ammonia as shown in Fig. 7.16. The selectivity of the exchange $Na^+ \rightarrow Ag^+$ in zeolite X is reversed with formation of the NH_3 complex, $Ag(NH_3)_2^+$.

Partial hydrogen (proton) exchange in NaA and CaA by treatment with p-toluenesulfonic acid (p-TsOH) in dioxane was reported (58). Hydrogen exchange occurred by the addition of water or methanol. In the presence of methanol, up to 75% hydrogen exchange in NaA was achieved without significant structure loss. In CaA only residual sodium was exchanged. The stability of the hydrogen zeolites to water vapor adsorption was poor. The adsorption of cyclohexane by a 50% exchanged hydrogen NaA was reported and attributed to an increase in the zeolite pore size. It is more likely, however, that this adsorption is due to the presence of an amorphous phase resulting from degradation of the zeolite structure.

The treatment of hydrogen and ammonium forms of zeolites and the properties of these materials is discussed in Chap. 6.

E. ION EXCHANGE KINETICS AND ION DIFFUSION

In the crystalline zeolites, ion exchange is controlled by diffusion of

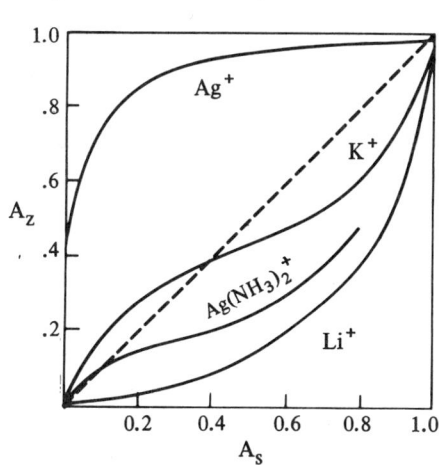

Figure 7.16 Exchange isotherms in zeolite X at 25°C and 0.2 N for $Na^+ \rightarrow Ag^+$ and $Na^+ \rightarrow Ag(NH_3)_2^+$ exchange. Also shown are the $Na^+ \rightarrow K^+$ and $Na^+ \rightarrow Li^+$ isotherms (59).

the ion within the crystal structure. It has been shown that for spherical particles, the extent of exchange follows in the initial steps the relationship (31):

$$\frac{Q_t}{Q_\infty} = \frac{Q_0}{Q_0 - Q_\infty} \cdot \frac{2S}{V}\sqrt{\frac{D^i t}{\pi}} = \frac{6}{r}\sqrt{\frac{D^i t}{\pi}} \qquad (7.29)$$

where Q_t and Q_∞ are the amounts of exchange at time t and equilibrium (∞) respectively; Q_0 = concentration of tracer ion initially present, and S, V, and r represent the surface area, volume, and radius of the exchanger particles. The biggest practical problem in determining the diffusion coefficient is the measurement of the surface area. One method involving gas adsorption and another involving a projection method have been used to determine S (60). A better method would be one which is independent of the surface area. The effective or apparent diffusion coefficient, D^i, is related to interdiffusion of the two exchanging cations, or interchange of cation A and B with the same valence, $D_A^i = D_B^i = D^i$, and is independent of the composition. The apparent diffusion coefficient varies with temperature according to the Arrhenius equation. For zeolites, D^i varies from 1×10^{-8} to 1×10^{-13} cm^2 sec^{-1}. Detailed studies of ion exchange kinetics and diffusion have been made on chabazite (60, 61) and mordenite (62, 63) using radiochemical methods to follow the ion exchange process. In chabazite, the narrowest apertures, about 3.9 A in diameter (Chap. 2), are large enough to permit diffusion of most ions without squeezing. It is probable that replacement of some of the ions found within the hexagonal prism would involve cation diffusion through the 6-rings. The self-diffusion coefficients of certain ions in chabazite follow the Arrhenius equation (see Table 7.16). The energy barriers associated with the self-diffusion process vary from about 7 kcal for the univalent ions to about double that value for calcium and strontium ions. However, the energy barrier was 8.7 kcal for barium ion, the largest of the divalent ions.

The Arrhenius equation is used to determine the activation energy, E:

$$D^i = D_0 \exp(-E/RT) \qquad (7.30)$$

The value of D^i is obtained from a plot of Q_t/Q_∞ versus $\sqrt{t}$ using the slope of the initial portion of the curve. The rates of ion exchange for the ion pairs Ca^{2+}-Sr^{2+}, Ca^{2+}-Ba^{2+}, Sr^{2+}-Ba^{2+}, and Na$^+$-Ba^{2+} in chabazite may be determined radiochemically (Fig. 7.17)(60). Pronounced

ION EXCHANGE KINETICS 573

Table 7.16 Self-Diffusion in Chabazite [a] (T = 298.3°K) (60)

Cation	Energy of Activation (kcal/mole)	Cation Diameter (A)	D^i [b] ($cm^2 sec^{-1}$)	S (Krypton) (cm^2/g)	D_0 ($cm^2 sec^{-1} \times 10^7$)
Na$^+$	6.5	1.90	6.1×10^{-12}	750	4.0
K$^+$	7.0	2.66	6.9×10^{-12}	1120	9.4
Rb$^+$	6.7	2.96	19.5×10^{-12}	503	17.0
Cs$^+$	7.5	3.38	4.9×10^{-13}	1170	1.7
Ca^{2+}	13.8	1.98	4.1×10^{-16}	8100	55.0
Sr^{2+}	14.6	2.26	1.3×10^{-16}	7690	101.0
Ba^{2+}	8.8	2.70	1.3×10^{-13}	817	4.0

[a] Chabazite (Nova Scotia) ground to 18 x 32 mesh except for preparations of pure Ca and Sr forms which were less than 200 mesh. Analysis showed SiO_2/Al_2O_3 varied 4.7 to 6.7.
[b] From $D^i = D_0 \exp(-E/RT)$.

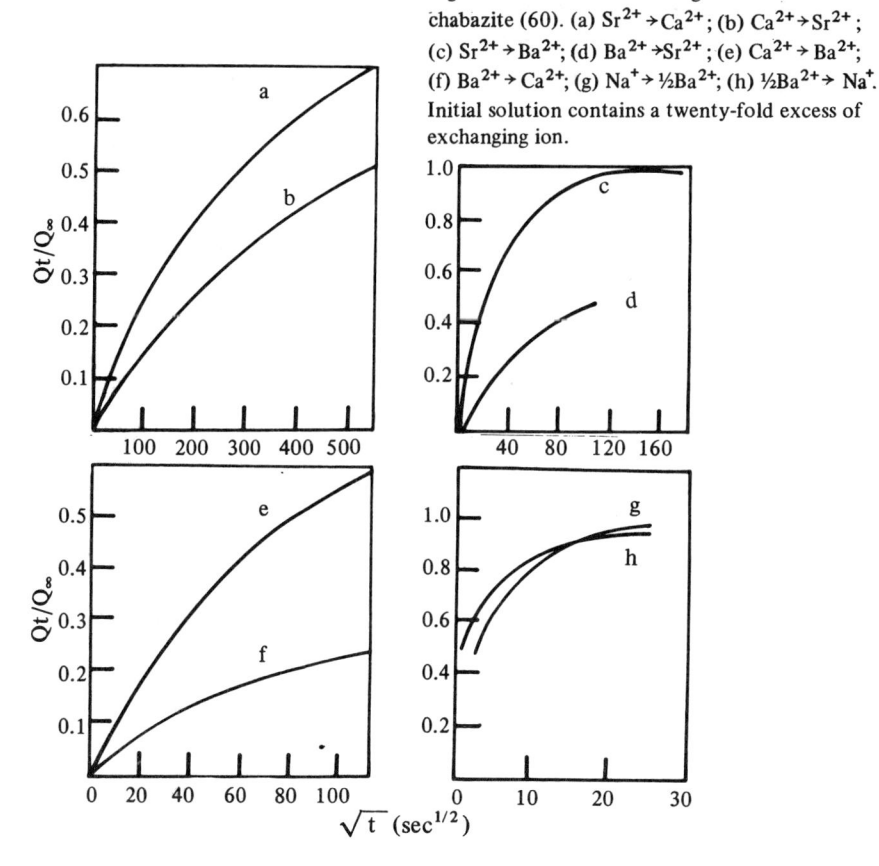

Figure 7.17 Rates of exchange of divalent ions in chabazite (60). (a) $Sr^{2+} \rightarrow Ca^{2+}$; (b) $Ca^{2+} \rightarrow Sr^{2+}$; (c) $Sr^{2+} \rightarrow Ba^{2+}$; (d) $Ba^{2+} \rightarrow Sr^{2+}$; (e) $Ca^{2+} \rightarrow Ba^{2+}$; (f) $Ba^{2+} \rightarrow Ca^{2+}$; (g) $Na^+ \rightarrow \frac{1}{2}Ba^{2+}$; (h) $\frac{1}{2}Ba^{2+} \rightarrow Na^+$. Initial solution contains a twenty-fold excess of exchanging ion.

574 ION EXCHANGE REACTIONS IN ZEOLITES

Table 7.17 Rates of Ion Exchange in Mordenite (60°C)(62)

System	$D\,(cm^2\,sec^{-1})$
$Na^+ \rightarrow K^+$	1.3×10^{-12}
$K^+ \rightarrow Na^+$	1.8×10^{-13}
$Na^+ \rightarrow Rb^+$	7.5×10^{-12}
$Rb^+ \rightarrow Na^+$	8.1×10^{-13}
$Na^+ \rightarrow \tfrac{1}{2}Ca^{2+}$	2.3×10^{-12}
$Cs^+ \rightarrow Na^+$	1.0×10^{-13}

deviation from linearity is evident. Rates of exchange between radioactive and isotopic cations obey the $\sqrt{t}$ law. The rapid Na^+-Ba^{2+} exchange (Fig. 7.17) is of interest. Note the larger size of barium ions relative to strontium and calcium ions. The self-diffusion behavior of barium ions in terms of E and D^i (Table 7.16) is compared to that of strontium and calcium ions. Barium self-diffusion shows a lower activation energy, E, and higher diffusion coefficient, D^i, than strontium and calcium ions.

In chabazite the energy barrier for ion migration is governed by a variation in Coulombic energy while the ions are moving through the 8-ring aperture. This energy barrier should be about twice as large for divalent as for univalent ions. However, the barium ion behaves as if it were univalent.

The rate of exchange of the ions Na^+-K^+, Na^+-Rb^+, and Na^+-Cs^+ were measured on Nova Scotia mordenite (62), using a radiochemical technique (Table 7.17). The self-diffusion of sodium and potassium ions decreased as the ions were gradually introduced. With potassium, the initial decrease was slow, but a rapid decrease occurred as the cation composition approached 50% sodium exchange. With sodium, the decrease in self-diffusion was rapid initially, then changed little over the range of 55-90% sodium exchange.

The specific locations of potassium ions in a potassium exchanged mordenite have not been determined. It is not known, for example, whether they can occupy sites in the small 2.8 A windows. When sodium ions are introduced into the potassium exchanged mordenite, the sodium ions should replace the potassium ions in the large 5. 8 A channels. When the sodium exceeds 50%, sodium ions can enter the side channel. The self-diffusion of potassium would therefore decrease slowly at first, and then more rapidly as the large channels became vacant of potassium ions. The self-diffusion coefficients for univalent cations are greater in

Table 7.18 Self-Diffusion in Mordenite [a](63)

Cation	Cation Diameter (A)	S^b (cm^2/g)	D^i $(cm^2 sec^{-1})$	$t(°C)$	Activation Energy (Kcal/mole)
Na$^+$	1.90	2520	2.5 x 10^{-13}	24.0	8.74
K$^+$	2.66	6370	1.22 x 10^{-13}	20.0	7.14
Rb$^+$	2.96	4980	1.31 x 10^{-13}	22.0	20.26
Cs$^+$	3.38	4980	0.05 x 10^{-13}	28.75	4.31
Ca^{2+}	1.98	9090	1.13 x 10^{-17}	27.0	10.45
Sr^{2+}	2.26	10,600	0.15 x 10^{-15}	20.5	15.97
Ba^{2+}	2.70	10,100	0.45 x 10^{-15}	18.5	10.14

[a] Nova Scotia, SiO_2/Al_2O_3 = 10.5.
[b] by BET method

chabazite than in mordenite. At ambient temperature, the self-diffusion coefficient varies from about 2.5 to 0.05 x 10^{-13} For the univalent ions in mordenite (Table 7.18) (63). For the divalent ions, Ca^{2+}, Sr^{2+}, and Ba^{2+} complete exchange is not possible. A peculiar behavior was evidenced by the self-diffusion coefficients and the energy of activation. In mordenite, cesium diffuses 50 times more slowly than does sodium.

Diffusion of a large cesium cation cannot occur between the adjacent main channels in mordenite. Restriction in the number of diffusion paths decreases the cesium diffusion coefficient in mordenite to a value lower than that found in chabazite. The smaller 2.8 A apertures in mordenite, however, control the diffusion of sodium. The large value of E for rubidium in mordenite (20 kcal/mole) indicates that rubidium must diffuse through the small 2.8 A windows. The lower value of E for cesium indicates that the cesium ions do not diffuse through the smaller apertures, but instead only in the larger channels. It was concluded that four cesium ions per unit cell occupy the sites along the 3.87 A channel. In the case of calcium and barium, only 73% of the sodium ions could be exchanged: the unit cell therefore contains three divalent ions and two of the original sodium ions. Occupancy of the 2.8 A windows by the sodium ions prevents diffusion of the divalent ions. Behavior of the divalent ions is then similar to the behavior of cesium. The activation energy E for the divalent ions should be about double that of cesium.

Information on ion migration in anhydrous zeolites A, X, and Y has been obtained by electrical conductivity measurements. (See Chap. 5.)

Table 7.19 Self-Diffusion Parameters in Synthetic Zeolites (65)

Zeolite [a]	Ion	D_0 (m²/sec)	E (kcal/mole)	ΔS (cal/mole/degree)
A	Ca^{2+}	5.10×10^{-6}	16.1	−4.0
A	Sr^{2+}	1.70×10^{-5}	19.6	+2.7
A	Ba^{2+}	1.48×10^{-3}	21.6	+11.6
X	Ca^{2+}	$\sim 1 \times 10^7$	19.9	+20.0
X	Sr^{2+}	$\sim 1 \times 10^7$	20.0	+20.0
X	Ba^{2+}	$\sim 1 \times 10^7$	20.1	+20.0
ZK-4	Ca^{2+}	3.6×10^{-4}	20.1	+8.6
ZK-4	Sr^{2+}	7.35×10^{-3}	22.6	+14.8
ZK-4	Ba^{2+}	7.60×10^{-7}	15.8	−3.4

[a] Particle radius of zeolite A = 2.64 μm
Particle radius of zeolite X = 2.04 μm
Particle radius of zeolite ZK-4 = 1.10 μm

This includes the effect of water and other adsorbed species upon the cation mobility. The kinetics of self-diffusion of divalent ions into hydrated zeolites A, ZK-4, and X have also been studied by radiochemical methods. Activation energies were evaluated, as well as entropy and free energy data (64, 65). The results, summarized in Table 7.19, indicate that the activation energy, E, does not vary with the particular cation type. This is not in agreement with electrical conductivity results. However, since the conductivity measurements were made on *anhydrous* zeolites, the comparison is incorrectly drawn. The effect of adsorbed water on ion migration is pronounced. In hydrated zeolite X, the migrating species is a cation-water complex which results in large positive values of ΔS. In zeolite A, the cation water complex is not as large and there may be Coulombic interaction with the framework oxygens in the apertures.

Using a radiochemical method, self-diffusion of the divalent cations calcium, strontium, and barium were measured in the synthetic zeolite ZK-4. As expected, movement of the cations takes place through the 8-membered oxygen rings. Two mechanisms are possible. One is characteristic of calcium and strontium diffusion in ZK-4 and chabazite. The cation moves from a completely hydrated state through the restriction of the 8-ring losing part of its water of hydration, returning to the hydrated state on the other side. The second mechanism found in zeolite A does not involve movement between positions of complete hydration as evidenced by the anomalous behavior of barium. The activation energy for diffusion of barium, E_a, is less than that of calcium or strontium despite the larger size of the barium ion. In a zeolite with a larger pore size such as zeolite X, the rate-controlling step is the diffusion of the cation-water complex ion and there is no ion size effect noted for the divalent ions. In zeolites with smaller channels, as in zeolite A, the activation energy varies with cation size. Activation energies and diffusion constants for zeolite ZK-4 are listed in Table 7.19.

A model for particle-controlled isotopic exchange of bound and randomly distributed cations in hydrated zeolites has been described (66, 67). The model combines Fick's diffusion equation and first-order kinetics. Kinetics of isotope exchange of sodium in NaX and NaA were determined (68). In order to follow these very fast exchange reactions, large crystals were used. Crystals of NaX with two sizes (95 ± 10 μm in mean length and 77 ± 10μm) and NaA in one size, (53 ± 5μm) were used. Use of the larger crystals increases reaction half-times by up to 1×10^4 times.

Zeolite NaX with both bound and mobile sodium ions in the supercages exhibited a single diffusional process. In both NaX and NaA exchange takes place in two steps, a fast initial step followed by an overlapping slow step. It was concluded that diffusion occurs by the sodium ions in NaX migrating from type II sites into the large cages followed by diffusion to the surface through the large channels. Diffusion through the system of interconnected β-cages does not occur. In NaX the slow step was independent of crystal size and the fast, diffusional step was slower in the larger crystals. Thus, isotope exchange in NaX involves sodium ions located in the supercages that diffuse rapidly through the main channel, and sodium ions in the β-cage that exchange, in a slow step, with mobile ions in the supercages.

$$\text{Bound ions in } \beta\text{-cages} \underset{k_3}{\overset{k_4}{\rightleftarrows}} \text{Ions in supercages} \rightarrow \text{Diffusion to surface}$$

where k_3 = rate constant for migration into the β-cages and k_4 = rate constant for migration into the supercages.

In zeolite A, the model involves only the diffusivity of the mobile cations since the exchange between bound and mobile cations is too fast to be observable.

$$\text{Bound ions in } \alpha\text{-cages} \underset{k_2}{\overset{k_1}{\rightleftarrows}} \text{Mobile ions} \rightarrow \text{Diffusion to surface}$$

and k_1 = rate constant for desorption of sodium from cation site
and k_2 = rate constant for adsorption of sodium on cation site.

In synthetic zeolites X and Y the replacement of the sodium ions by divalent calcium, strontium, and barium can be completely achieved only at elevated temperatures. After essentially complete exchange by the divalent ion, subsequent re-exchange of the divalent cation is difficult. The ease of exchange by divalent cations is governed by the size of the hydrated cation and the size of the bare ion. Exchange appears to be at an optimum in the case of strontium ions, and consequently the sodium in zeolite X is more readily replaced by strontium. In zeolite Y, however, due to the reduced charge on the zeolite aluminosilicate framework the interaction between the cations and the water molecules in the large supercages results in incomplete replacement. This is similar to the case in zeolite A where the structure or semi-structure of water in the large cavities influences the ease with which a divalent ion removes water and enters the smaller sodalite or β-cages (69).

In another study of the diffusion of strontium and barium in synthetic zeolite X over the entire exchange range, it was shown that the diffusion process is not uniform. About 80% of the exchangeable barium ions and only 65% of the exchangeable strontium ions diffuse with a high diffusion coefficient; the remainder diffuse with a low coefficient which differs by 4 to 5 orders of magnitude at room temperature. The heat of activation for this slow process is about twice as great as for the fast exchange process. The behavior is interpreted in terms of two types of cation locations. It is stated that the cations are sited in two types of channels that are separated from each other (70).

Self-diffusion in Nonaqueous Solvents

No significant exchange of the sodium in zeolite A by divalent alkaline earth cations occurs in a nonaqueous solvent (methanol or ethanol). Similarly, self-exchange of the divalent ions was immeasurably slow (Sr^{2+} and Ca^{2+}) (71).

In contrast, the use of a nonaqueous solvent had little effect on the self-diffusion of sodium ions in zeolite A. In zeolites X and Y, the self-diffusion rates of alkaline earth cations were greatly reduced in nonaqueous solvents and the activation energy for diffusion increased with increasing size and decreasing dielectric constant of the solvent molecules.

The self-diffusion of calcium, strontium, and barium ions in zeolite

X and Y was determined in nonaqueous solutions of methyl and ethyl alcohol. The energy of activation, which reflects the ease with which a cation can migrate through the zeolite structure, varied with the solvent. In the series water-methanol-ethanol the value of E_a increased. The diffusing species, therefore, is an ion-solvent combination. The increase in the activation energy is caused by an increase in the size of the solvated ion and its interaction with the zeolite framework (72).

F. CATION SIEVE EFFECTS IN ZEOLITES

The sieving and partial sieving effects of zeolites toward various cations have been attributed to one or more of three possible mechanisms; (1) the cation may be too large to enter smaller channels and cavities within the zeolite structure, or in some zeolites the exchangeable cation is locked in during synthesis, e.g., potassium ions in zeolite L, and cannot be replaced, (2) the distribution of charge on the zeolite structure may be unfavorable for the cation, (3) the size of the hydrated cation in aqueous solution, or solvation in nonaqueous solution, may influence and retard exchange of the cation since an exchange of solvent molecules must occur for the cation to diffuse through apertures which are too small to accommodate the solvated cation (43). If the cavities in the zeolite are too small to contain sufficient entering cations because of their size, ion exchange is restricted.

An ion-sieve effect was observed in the exchange of sodium for cesium (radius = 1.65 A) in analcime while extensive exchange was obtained with rubidium (r = 1.48 A). Similarly, analcime exhibits negligible sodium exchange for divalent ions (Mg^{2+}, Ca^{2+}, Ba^{2+}) (11). Chabazite exhibits complete exclusion for the ions La^{3+}, $(CH_3)_4N^+$, and $(C_2H_5)_4N^+$, due to the large size of these cations (41). Incomplete exchange with cesium (x_{max} = 0.85) has been attributed to a mutual repulsion of the large cesium ions within zeolite cavities (43). Ion sieve effects of the second type are observed in the exchange of zeolite A with trivalent ions such as lanthanum and cerium (14).

In zeolite X exchange with alkylammonium cations is restricted due to the ion size. Since the internal volume in the zeolite is constant, the extent of exchange by a large cation which is capable of entering the channels must be governed by the size or volume of the cation (see Table 7.7). As the size of the alkylammonium ion increases, the degree of exchange decreases (25, 36, 73). Similar results were observed in zeolite A with monosubstituted alkylammonium ions, RNH_3^+, where R is methyl, ethyl, n-propyl, and n-butyl. The degree of exchange decreased with increasing size of the alkyl group from about 0.4 to less than 0.1 (20).

The cation sieve properties of chabazite, mordenite, erionite, and clinoptilolite have been investigated using a flow method (74). Solutions flowed through a column of the zeolite and then were analyzed radiochemically. The effect of other cations which are present in the same solution on the cesium selectivity is illustrated in Fig. 7.18. The typical replacement series of hydrated cations was observed. Of several zeolites tested, clinoptilolite has been found to have a high selectivity for cesium and is of considerable interest in removing cesium from radioactive waste materials (75). Zeolite A with a high strontium selectivity (K_a = 83) is of considerable interest in the removal of radioactive strontium isotopes at low chemical concentration (75)

Subtle differences in the behavior of hydrated transition metal ions can greatly affect exchange selectivity. In Fig. 7.19, the separation by zeolite A of a solution of cobalt and nickel is illustrated. On elution, a cobalt-rich effluent results. The relative stabilities of the aquo complexes of these ions are believed to cause the sharp difference in exchange behavior.

A hypothesis has been advanced that a zeolite should show enhanced ion exchange selectivity for a given cation if the cation is present in synthesis. This proposal was tested for strontium and calcium. A synthetic strontium mordenite and a calcium mordenite were prepared. The strontium-calcium separation factor, α_{Ca}^{Sr}, varied from 3.5 to 8.4 for the strontium zeolite and from 0.5 to about 1.5 for the calcium zeolite. synthetic strontium clinoptilolite showed a similar strontium selectivity as compared to the mineral (76).

G. CATION EXCHANGE IN NONAQUEOUS SOLVENTS

In order to determine exchange equilibria in nonaqueous solvents, the zeolite must be dehydrated. If the solvent molecule is too large to enter the channels of the dehydrated zeolite, cation exchange might not be expected to occur. In the potassium form of zeolite A, which does not adsorb ethanol, however, the $Na^+ \rightarrow K^+$ exchange can occur in etha-

CATION EXCHANGE IN NONAQUEOUS SOLVENTS 581

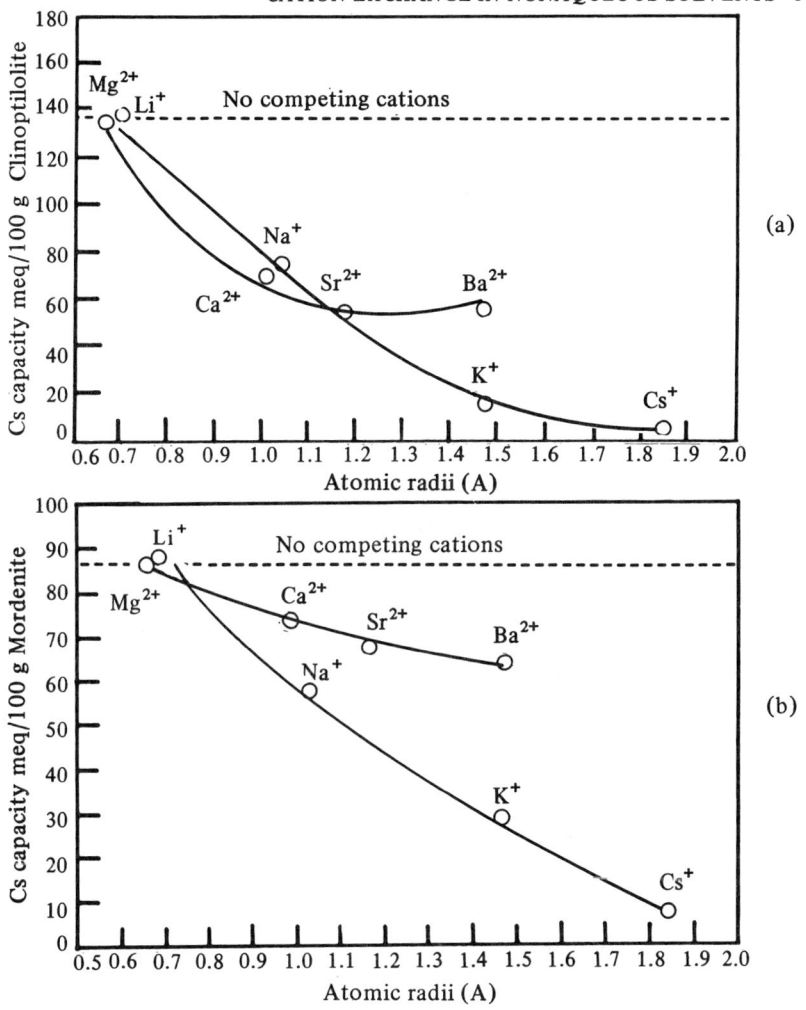

Figure 7.18 (a) Effect of competing cations on the cesium capacity of clinoptilolite (74).

Common influent solution	1.0 N competing cations
	0.01 N Cs$^+$, 1.74 x 10^{-8} N^{137} Cs$^+$
Common temperature:	25°C
Common influent pH:	5.0
Common flow rate:	294 ml/cm^2/hr
Common column:	50 g, 0.25 to 1.0 mm of sodium-exchanged clinoptilolite

(b) Effect of competing cations on the cesium capacity of mordenite

Common influent solution:	0.01 N competing cations
	0.01 N Cs$^+$, 1.74 x 10^{-8} N^{137} Cs$^+$
Common temperature:	25°C
Common influent pH:	7.0
Common flow rate:	473 ml/cm^2/hr
Common column	50 g, 0.25 to 1 mm sodium exchanged mordenite

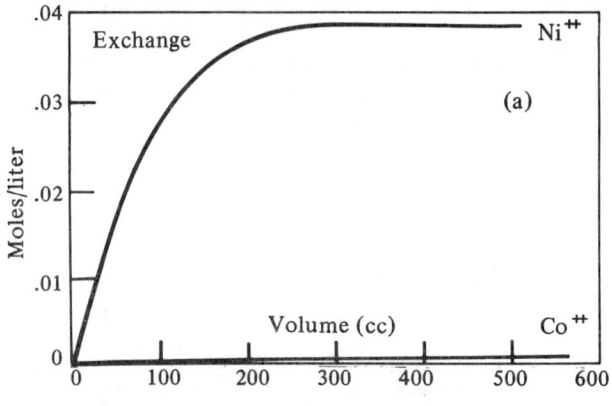

Figure 7.19 The separation of Co^{2+} and Ni^{2+} by zeolite A. Initial concentration 0.5M. (a) exchange of Co^{2+} and Ni^{2+} on a column 1.6 cm diameter by 92 cm long filled with pelletized zeolite A. Flow rate 1.4-1.5 cc/min; (b) elution of Co^{2+} and Ni^{2+} from the column with 5 M NaCl at the same flow rate (59).

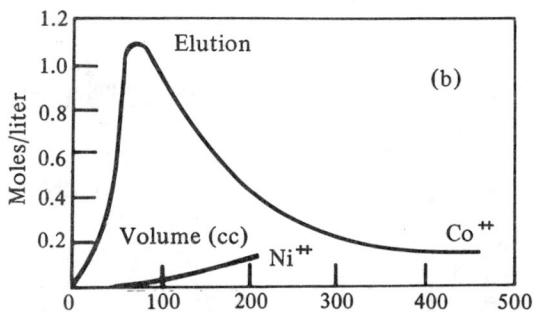

nol solution. The potassium exchanged form of zeolite A was observed to exchange with sodium from ethanol solution about as readily as it does in water. It was concluded that the ethanol solvent is adsorbed into the crystal as sodium exchange occurs (the sodium zeolite adsorbs ethanol) and proceeds inward through the zeolite crystal from the surface (59).

Some data for the exchange of univalent ions in zeolite A and zeolite X from ethanol and water are summarized in Table 7.20. In some instances, for example in ammonium exchange, the selectivity is reversed in ethanol solution.

Selectivity coefficients for the exchange of uni- and divalent ions in zeolite X from water and various alcohols are summarized in Table 7.21 (77). The potassium-sodium exchange reaction shows a reversal in selectivity from water to nonaqueous solvent. Otherwise, the selectivity decreases as expected with increasing reciprocal of the dielectric constant of the solvent.

Ion exchange rate data (see Table 7.22) show that exchange in these

Table 7.20 Ion Exchange in Ethanol (25°C, 0.2 N Total Concentration) [a]

Zeolite	Ion	A_s	A_z	α_B^A	Solvent
NaX	K⁺	0.54	0.46	0.72	Water
NaX	K⁺	0.36	0.64	3.2	Ethanol
	NH₄⁺	0.56	0.38	0.48	Water
	NH₄⁺	0.46	0.50	1.2	Ethanol
CaX	Na⁺	0.70	0.20	0.11	Water
	Na⁺	0.94	0.04	0.003	Ethanol
	K⁺	0.67	0.33	0.25	Water
	K⁺	0.92	0.08	0.075	Ethanol
NaA	NH₄⁺	0.58	0.38	0.44	Water
	NH₄⁺	0.45	0.50	1.22	Ethanol
	K⁺	0.54	0.46	0.72	Water
	K⁺	0.55	0.45	0.67	Ethanol
KA	Na⁺	0.38	0.40	1.1	Water
	Na⁺	0.64	0.23	0.17	Ethanol

[a] Unpublished results, Union Carbide Corp.

nonaqueous solvents is a slower process; also, the rate of exchange decreases with decreasing dielectric constant.

Zeolite A in Mixed Ion Exchange Media

Selectivity coefficients for the exchange of sodium by cesium in zeolite A and potassium by cesium and sodium in zeolite A were measured

Table 7.21 Exchange in Zeolite X of Various Ions in Alcohols [a] (77)

Alcohol or Solvent	Dielectric Constant ϵ	K_{Na}^M Exchanging Ion			
		K⁺	Li⁺	Ca²⁺	Ag⁺
Water	78.7	1.10	8.4×10^{-2}	0.44	45.9
Methanol	31.66	0.96	4.2×10^{-2}	4.5×10^{-3}	18.1
Ethanol	23.56	0.70	3.9×10^{-2}	1.1×10^{-3}	6.34
n-Propyl alcohol	19.8	0.50	1.5×10^{-2}		
Isopropyl alcohol	17.9	0.19	1.3×10^{-2}		
Isobutyl alcohol	17.1	0.19	1.1×10^{-2}		

[a] All salt solutions 0.05 M, 30 ± 0.02°C.
Zeolite X used in form of 1/16 in. pellets containing inert clay binder. Exchange capacity of the sodium zeolite as measured by Ag⁺ exchange is 4.77 meq/g of dry zeolite.

Table 7.22 Kinetic Data for Exchange of Various Ions in Zeolite X in Water, Methanol, and Ethanol (30°C) (77)

Ion	Solvent	$t_{1/2}$ (min)[a]
K$^+$	H$_2$O	18
K$^+$	CH$_3$OH	19
K$^+$	C$_2$H$_5$OH	48
Ag$^+$	H$_2$O	40
Ag$^+$	CH$_3$OH	120
Ag$^+$	C$_2$H$_5$OH	295
Ca^{2+}	H$_2$O	16

[a] $t_{1/2}$ = time required to attain 50% equilibrium.

in nonaqueous media (methanol, ethanol, and ethylene glycol) as well as mixed media. The exchanging ions were at radioactive tracer level concentrations. At high alcohol concentrations, the ion selectivity coefficient showed a sudden increase. This was attributed to the solvation of the zeolite by the alcohol at high alcohol concentrations (78).

The ion exchange of the alkali metal ions, lithium, sodium, etc., in zeolite A was studied in a mixed solvent system (water-methanol) as a function of the dielectric constant of the solution (80). Except for lithium, the addition of methanol to the solution increases the apparent equilibrium constant. With lithium the equilibrium constant decreases with increasing methanol concentration. This may be due to the high degree of hydration of the lithium ion. The corrected selectivity coefficient varied linearly with the cation fraction in the exchanger in accordance with the Kielland-Barrer relation. The standard free energies of exchange were all positive as they are in the case of the exchange of sodium by other alkali metals in water solution. It was concluded that the affinity of zeolite A for sodium is higher than for any other alkali metal ion whether in water or a mixed solvent system (80).

The sodium-lithium exchange in zeolite NaA was studied in water-dioxane and water-dimethyl sulfoxide mixtures (15, 79). Neither organic solvent can penetrate the zeolite pore system. The isotherms show an inflection which is attributed to two ion exchange processes involving two types of cation sites. Up to a level of exchange of Li$_z$ = 0.35 - 0.40, mobile lithium ions in the α-cage are involved, at higher levels, lithium ions in the type I sites in the single 6-rings, are exchanged. The relative selectivity for lithium decreases with increasing organic solvent concentration. At high concentrations of DMSO in the solution, the ion exchange capacity is reduced to 85% of the original capacity.

H. SALT OCCLUSION OR IMBIBITION

Mineral and synthetic aluminosilicates of the feldspathoid type (Chap. 2) are framework structures, and contain within the cavities of the anionic framework various species of inorganic compounds—NaCl in sodalite and $CaCO_3$ in scapolite. Synthetic types of feldspathoids, such as hydroxysodalite, which contains occluded NaOH, are common.

It is known that the occlusion of NaCl from aqueous solution may occur in zeolite X and that occlusion of $AgNO_3$ from the melt occurs in zeolite A (81). Zeolite A occludes 9 $AgNO_3$ molecules/unit cell. During occlusion, the zeolite is anhydrous, but the occluded $AgNO_3$ can be readily extracted with hot water. Interpretation of the x-ray powder data showed an ordered arrangement of silver and nitrate ion pairs which results in a super-structure with a doubled cubic unit cell dimension.

The occlusion of NaCl from aqueous solution in zeolite X is significant only at high salt concentrations (82). The imbibition or occlusion of NaCl is independent of temperature, or anion, except for an anion sieve effect. However, it was affected by the type of cation, the cattion density, and the zeolite framework structure. The zeolites X and Y differ in the amount of NaCl taken up. At 25°C, zeolite X occludes three times as much salt (5 N solution) as zeolite Y (0.5 g for zeolite X as compared to 0.17 g for zeolite Y). The occlusion is interpreted as a Donnan process.

In zeolite A, the kinetics of occlusion of NaCl and $CaCl_2$ show an energy barrier which is larger than that due to cation diffusion alone and results from the coupled migration of the anion. Figure 7.20 shows the occlusion isotherms for alkali chlorides in zeolite X. The order at low concentrations is KCl > NaCl > $CaCl_2$ > CsCl > LiCl. Significant occlusion does not occur except at high concentrations of salt in the aqueous solution. In zeolite A, an anion sieve effect toward SO_4^{2-} was observed. Saturated Na_2SO_4 solution in contact with zeolite A for a week at 190°C showed no occlusion of sulfate ion; the sulfate anion with a diameter of 5.5 Å is excluded. In the case of chlorides over the early stages of salt occlusion, the kinetics follow the $\sqrt{t}$ law with energies of activation of 25.8 kcal/mole for sodium chloride and 27.8 kcal/

586 ION EXCHANGE REACTIONS IN ZEOLITES

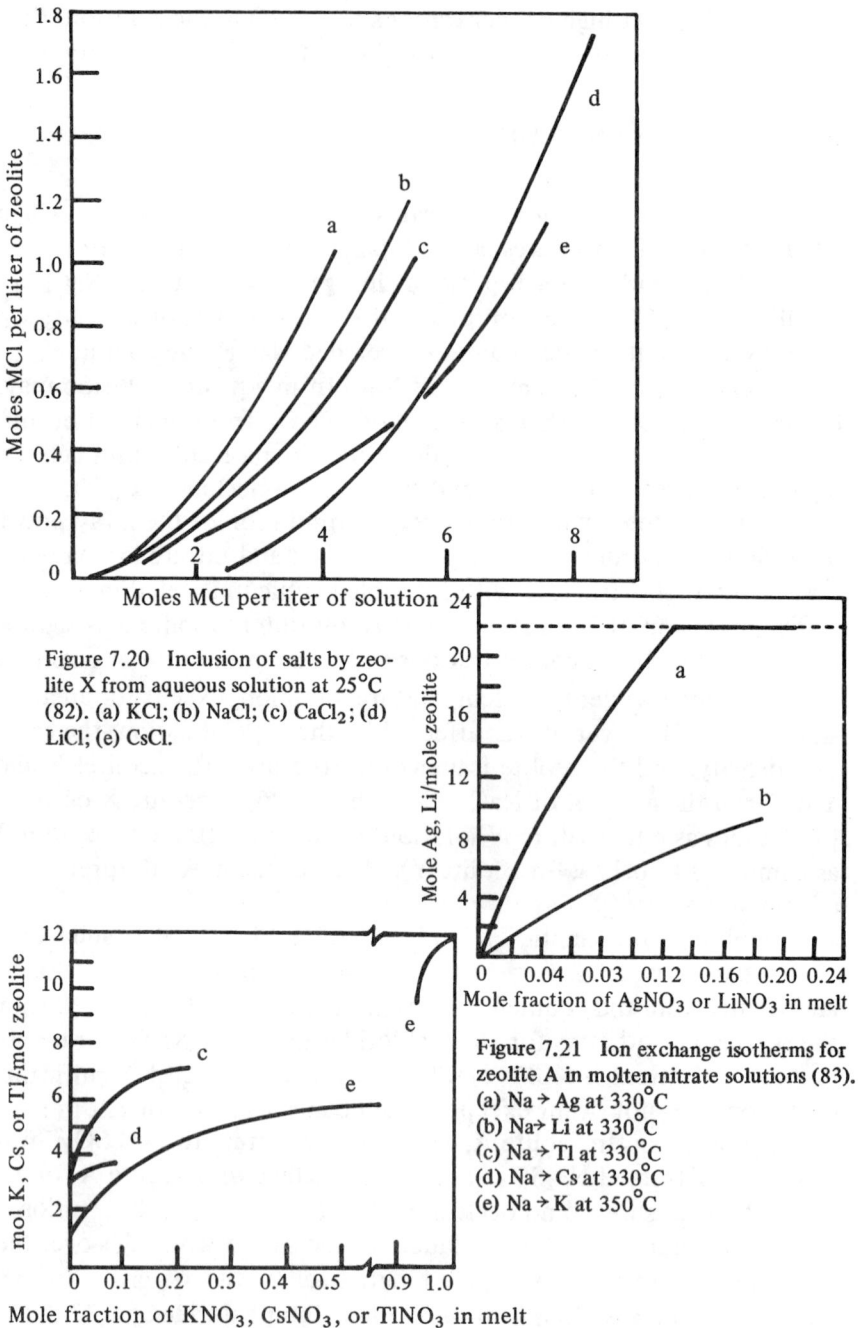

Figure 7.20 Inclusion of salts by zeolite X from aqueous solution at 25°C (82). (a) KCl; (b) NaCl; (c) CaCl$_2$; (d) LiCl; (e) CsCl.

Figure 7.21 Ion exchange isotherms for zeolite A in molten nitrate solutions (83).
(a) Na → Ag at 330°C
(b) Na → Li at 330°C
(c) Na → Tl at 330°C
(d) Na → Cs at 330°C
(e) Na → K at 350°C

mole for calcium chloride.

I. ION EXCHANGE IN FUSED SALTS

Zeolites with more open structures such as chabazite, zeolite A, etc., may occlude the salts within the larger cages after dehydration. The cation exchange of various uni- and divalent cations in nitrate melts has been studied with $LiNO_3$, $NaNO_3$, and KNO_3. Although related to a size-charge effect, the selectivity sequence varies with the salt. The size of the cation, the size of the solvent (salt) cation, and the size of the chabazite channels were all deemed important in determining the exchange behavior. A "two-way traffic" mechanism based on the additive effect of the diameters of the exchange and solvent cations has been postulated (84).

In chabazite, exchange of alkali metal ions from molten $NaNO_3$ follows the selectivity order of

$$Ba^{2+} > Cs^+ > Rb^+ > Na^+.$$

This sequence is like that found in dilute aqueous solutions. The distribution coefficient of the exchanging cation and the sequence of cation selectivity in chabazite varies with the salt used. The sequence of cation selectivity is related to the size of the cations and in terms of the "two-way traffic mechanism", it is related to the sum of the diameters of the exchanging cations and the cation of the solvent fused salt. The combined diameters of the solvent cation and the preferred exchange cation for each system were observed to lie in the range of 4.56 to 4.60 A. This is to be compared to the size of the main channels in chabazite which for the dehydrated zeolite is 3.1 to 4.4 A (84).

Zeolite A occludes 10 $NaNO_3$ per unit cell when brought into contact with molten $NaNO_3$ at 330°C (83). This results in total unit cell contents of $Na_{12}(AlO_2)_{12}(SiO_2)_{12} \cdot 10\,NaNO_3$. All of the 22 sodium ions are present within the zeolite structure, in the occluded $NaNO_3$, but they are not all completely exchangeable for other types of cations. Complete exchange occurs with silver ion in $NaNO_3$. Exchange of sodium by potassium, cesium, and thallium ions is incomplete; a plateau occurs in the exchange isotherms at 3 to 4 cesium ions, and 6 or 7 thallium and potassium ions per unit cell. There is no significant change in the total number of occluded nitrate ions with either cesium or thallium ions. (Fig. 7.21). The limited exchange is due to steric effects caused by the size of the larger cations. The radius of Cs^+ is 1.69 A and of Tl^+ is 1.44 A.

The exchange equilibria between zeolite A in the sodium form and lithium, potassium, cesium, silver and thallium ions show that when the large α-cage can occlude the nitrate, then all of the cations can be exchanged. In other cases, a limited exchange occurs. At trace concentrations in the zeolite A-nitrate system, the selectivity series is

$$Cs > Tl > Ag > K > Li > Na$$

The relative positions of the cesium, thallium, and potassium cations change as the degree of exchange increases.

For complete exchange from the sodium form of zeolite A to the silver exchanged form in molten $NaNO_3$, the standard free energy change, ΔG^0, was calculated to be -3.2 kcal/mole.

Uni-divalent exchanges in zeolite A (Ca^{2+} and Sr^{2+}) in molten sodium nitrate at 330°C have been studied for concentrations of $1 \times 10^{-3} - 10^{-4}$, mole fraction up to 0.04 - 0.06. The ion pair $CaNO_3^+$ exchanges for sodium after four calcium ions are exchanged per unit cell for 8 sodium ions. Exchange occurs with strontium by replacement of two sodium ions by one strontium ion. There is no evidence of exchange by a $SrNO_3^+$ pair (85).

J. ZEOLITE ION EXCHANGE APPLICATIONS

Treatment of Radioactive Wastes

Disposal of radioactive materials from the reprocessing of nuclear fuels is a serious problem. Studies at Battelle Northwest in Richland, Washington have shown that zeolites may be used to remove long-lived cesium and strontium isotopes. The isotopes are removed from the aqueous waste of the Purex process, converted to the anhydrous, cesium chloride or strontium fluoride, and encapsulated in metal cans for long-term storage. Clinoptilolite, Zeolon (mordenite), NaA, and AW-500 (see Chap. 9) have been used. Using Zeolon, kilocurie amounts of cesium-137 have been prepared in better than 98% purity (2, 86). Stability of the zeolite in the presence of high levels of radiation and resistance to chemical attack are required. Another method for storing these isotopes is based on selective removal by ion exchange, followed by drying and dehydration of the zeolite containing the isotope. The dehydrated zeolite containing the radioactive isotope is then sealed in containers for storage (87).

Ammonia Removal from Waste Water

Removal of ammonia, as ammonium ions, from municipal, agricultural, and industrial waste water is becoming increasingly necessary (88). Excess concentrations of ammonia in water are toxic to aquatic life and cause explosive algae growths leading to eutrophic conditions in lakes. Typical municipal waste water contains 30 mg of ammonia per liter. For reuse, the level must be below 0.5 mg/liter or less. The use of cation exchangers to remove ammonia has been studied extensively including amorphous, gel-type aluminosilicates and organic resins. Laboratory and pilot plant studies were conducted at Battelle Northwest using clinoptilolite which shows a selectivity for ammonium over sodium, calcium, and magnesium cations which are present in a typical waste water. Clinoptilolite is not selective for ammonium ion over potassium ion, however. Over 99% removal of ammonia was shown in laboratory column experiments. Regeneration of the zeolite column was accomplished by using a mixed $NaCl-CaCl_2$ solution in saturated $Ca(OH)_2$. Ammonia is then removed from the column by air stripping.

In two 500-gallon clinoptilolite columns in a mobile pilot plant, 97% removal of ammonia was obtained from a secondary effluent containing 16 mg/liter of nitrogen as ammonia.

The synthetic zeolite F is more effective than clinoptilolite in removing ammonium ion from waste water (89). Table 7.23 compares selectivities, as expressed by $\alpha_B^{NH_4}$ where B represents all the other competing ions. The solution contains 85 meq of NaCl, 50 meq of KCl, and 5 meq of $CaCl_2$ per liter. Zeolite F has a higher exchange capacity and greater selectivity for ammonium ion.

Table 7.23 NH_4 Selectivity Factors of Various Zeolites (89)

Zeolite	Exchange capacity (meq/g)	$\alpha_B^{NH_4}$
F (potassium form)	5.3	3.6
Y	4.0	0.95
X [b]	3.9	0.9
Clinoptilolite	1.2	1.7

[a] Selectivity factor determined at an NH_4 equivalent fraction in the zeolite, $[NH_4]_Z$ of 0.2
[b] Zeolite X in form of pellets with about 20% binder.

Table 7.24 Removal of Ammonium from Simulated Waste Water (89)

Zeolite	Nitrogen taken up (mg/g)
F, powder, K$^+$ form	3.9
X, pellets	1.2
Clinoptilolite (Hector, California)	1.9
Ion exchange resin (Amberlite IRC-84)	0.26

Laboratory experiments using a simulated waste water confirmed the ammonium selectivity shown by zeolite F. The solution contained ammonium (as nitrogen), 14 ppm; sodium, 58 ppm; potassium, 12 ppm; magnesium, 8 ppm; calcium, 34 ppm; at a pH of 7.45. Treatment of the solutions with zeolite F and other zeolites gave the results in Table 7.24. Zeolite F clearly showed a much higher ammonium exchange capacity.

REFERENCES

1. C. B. Amphlett, Inorganic Ion Exchangers, Elsevier, New York, 1964.
2. L. A. Bray, and H. T. Fullam, *Advan. Chem. Ser.*, **101**: 450 (1971).
3. F. Helfferich, Ion Exchange, McGraw-Hill, New York (1962) p. 185.
4. R. M. Barrer, and J. Klinowski, *Trans. Faraday Soc.*, **68**: 73 (1972).
5. R. M. Barrer, J. A. Davies, and L. V. C. Rees, *J. Inorg. Nucl. Chem.*, **30**: 3333 (1968).
6. R. M. Barrer, L. V. C. Rees, and M. Shamsuzzoha, *J. Inorg. Nucl. Chem.*, **28**: 629 (1966).
7. J. Kielland, *J. Soc. Chem. Ind.*, **54**: 232 (1935).
8. L. L. Ames, Jr., *Amer. Mineral.*, **49**: 1099 (1964).
9. R. M. Barrer and W. M. Meier, *Trans. Faraday Soc.*, **55**: 130 (1959).
10. E. F. Vansant and J. B. Uytterhoeven, *Trans. Faraday Soc.*, **67**: 2961 (1971).
11. R. M. Barrer, *J. Chem. Soc.*, 2342 (1950).
12. R. M. Barrer, L. V. C. Rees, and D. J. Ward, *Proc. Roy. Soc.*, **273A**: 180 (1963).
13. H. S. Sherry and H. F. Walton, *J. Phys. Chem.*, **71**: 1457 (1967).
14. D. W. Breck, W. G. Eversole, R. M. Milton, T. B. Reed, and T. L. Thomas, *J. Amer. Chem. Soc.*, **78**: 5963 (1956).
15. Z. Dizdar, *J. Inorg. Nucl. Chem.*, **34**: 1069 (1972).
16. H. S. Sherry, *J. Phys. Chem.*, **70**: 1332 (1966).
17. I. J. Gal, O. Jankovic, S. Malcic, P. Radovanov, and M. Todorovic, *Trans. Faraday Soc.*, **67**: 999 (1971).
18. M. V. Susic, V. M. Radak, N. A. Petranovic, and D. S. Veselinovic, *Trans. Faraday Soc.*, **62**: 3479 (1966).

19. H. S. Sherry, *J. Phys. Chem.*, **70**: 1158 (1966).
20. R. M. Barrer and W. M. Meier, *Trans. Faraday Soc.*, **54**: 1074 (1958).
21. N. F. Ermolenko, and L. P. Shirinskaya, *Zh. Fiz. Khim.*, **36**: 2432 (1962), *Translation.* 1317 (1962).
22. V. A. Federov, et. al., *Zh. Fiz. Khim.*, **38**(s): 1248 (1969); *Translation*, **38**: 679.
23. L. L. Ames, Jr., *Amer. Mineral.*, **49**: 127 (1964).
24. L. L. Ames, Jr., *Can. Mineral.*, **8**: 325 (1965).
25. B. K. G. Theng, E. Vansant, and J. B. Uytterhoeven, *Trans Faraday Soc.*, **64**: 3370 (1968).
26. H. S. Sherry, in Ion Exchange, J. A. Marinsky, Ed., Vol. 2, Marcel Dekker, N.Y., 1969, p. 89 - 133.
27. L. L. Ames, Jr., *J. Inorg. Nucl. Chem.*, **27**: 885 (1965).
28. H. S. Sherry, *J. Phys. Chem.*, **71**: 780 (1967).
29. H. S. Sherry, *J. Colloid Interfac. Sci.*, **28**: 288 (1968).
30. H. S. Sherry, *J. Phys. Chem.*, **72**: 4086 (1968).
31. R. M. Barrer and L. Hinds, *J. Chem. Soc.*, 1879 (1953).
32. R. M. Barrer and B. M. Munday, *J. Chem. Soc., A*, 2904 (1971).
33. R. M. Barrer and B. M. Munday, *J. Chem. Soc., A*, 2909 - 2916 (1971).
34. D. H. Olson and H. S. Sherry, *J. Phys. Chem.*, **72**: 4095 (1968).
35. P. J. Denny and R. Roy, Clays and Clay Minerals, Proc. Nat. Conf. Clays Clay Mineral, 567 (1963).
36. R. M. Barrer, R. Papadopoulos, and L. V. C. Rees, *J. Inorg. Nucl. Chem.*, **29**: 2047 (1967).
37. E. F. Vansant and J. B. Uytterhoeven, "Molecular Sieve Zeolites, *Advan. Chem. Ser., 101*, American Chemical Society, Washington, D. C., 1971, p. 426.
38. R. M. Barrer, J. A. Davies, and L. V. C. Rees, *J. Inorg. Nucl. Chem.*, **31**: 2599 (1969).
39. L. L. Ames, Jr., *Can. Mineral.*, **8**: 582 (1966).
40. L. L. Ames, Jr., *Can. Mineral.*, **8**: 572-81 (1966).
41. R. M. Barrer and D. C. Sammon, *J. Chem. Soc.*, 2838 (1955).
42. R. M. Barrer, J. A. Davies, and L. V. C. Rees, *J. Inorg. Nucl. Chem.*, **31**: 219 (1969).
43. R. M. Barrer and J. Klinowski, *J. Chem. Soc., Faraday Trans.*, **68**: 1956 (1972).
44. L. L. Ames, Jr., *Amer. Mineral.*, **49**: 127 (1964), *Amer. Mineral.*, **51**: 903 (1966).
45. D. G. Howery and H. C. Thomas, *J. Phys. Chem.*, **69**: 531 (1965).
46. G. R. Frysinger, *Nature*, **194**: 351 (1962).
47. H. S. Sherry, Ion Exch. Process Ind., Pap. Conf. (1969) p. 329.
48. R. M. Barrer and B. M. Munday, *J. Chem. Soc. A*, 2914 (1971).
49. F. Wolf, H. Fuertig, and H. Knoll, *Chem. Tech. (Leipzig)* **23**: 273 (1971).
50. G. Eisenman, *Biophys, J.*, **2**: 259 (1962).
51. R. M. Barrer and J. A. Davies, *J. Phys. Chem. Solids.*, **30**: 1921 (1969).
52. A. H. Keough and L. B. Sand, *J. Amer. Chem. Soc.*, **83**: 3536 (1961).
53. R. M. Barrer and M. B. Makki, *Can. Jour. Chem.*, **42**: 1481 (1964).
54. R. M. Barrer and D. L. Peterson, *Proc. Roy. Soc., Ser. A*, **280**: 466 (1964).
55. R. M. Barrer and B. Coughlan, Molecular Sieves, Society of Chemical Industry, London, 1968, p. 141.
56. J. Kibisz, *Mineral. Mag.*, **35**: 1071 (1966).
57. D. McConnell, *Science*, **141**: 171 (1963).
58. D. P. Roelofsen, E. R. J. Wils, H. Van Bekkum, *J. Inorg. Nucl. Chem.*, **34**: 1437 (1972).
59. D. W. Breck, *J. Chem. Ed.*, **41**: 678 (1964).
60. R. M. Barrer, R. Bartholomew, and L. V. C. Rees, *J. Phys. Chem. Solids*, **24**: 51-62 (1963).
61. R. M. Barrer, R. Bartholomew, and L. V. C. Rees, *J. Phys. Chem. Solids*, **24**: 309-317 (1963).

62. A. V. Rao and L. V. C. Rees, *Trans. Faraday Soc.*, **62**: 2505 (1966).
63. L. V. C. Rees and A. Rao, *Trans. Faraday Soc.*, **62**: 2103 (1966).
64. A. Dyer and J. M. Fawcett, *J. Inorg. Nucl. Chem.*, **28**: 615 (1966).
65. A. Dyer and R. B. Gettins, *J. Inorg. Nucl. Chem.*, **32**: 319 (1970).
66. L. M. Brown, H. S. Sherry, and F. J. Krambeck, *J. Phys. Chem.*, **75**: 3846 (1971).
67. H. S. Sherry, "Molecular Sieve Zeolites," *Adv. Chem. Ser.*, *101*, American Chemical Society, Washington, D. C., 1971, p. 350.
68. L. M. Brown and H. S. Sherry, *J. Phys. Chem.*, **75**: 3855 (1971).
69. A. Dyer, R. B. Gettins, and J. G. Brown, *J. Inorg. Chem.*, **32**: 2389 (1970).
70. E. Hoinkis, H. W. Levi, *Z. Naturforsch*, **A24**: 1784 (1969)..
71. A. Dyer and R. B. Gettins, Ion. Exch. Process Ind., Pap. Conf. (1969(p. 357.
72. A. Dyer and R. B. Gettins, *J. Inorg. Nucl. Chem.*, **32**. 2401 (1970).
73. R. M. Barrer, W. Buser, and W. F. Grutter, *Helv. Chim. Acta.*. **29**: 518 (1956).
74. L. L. Ames, Jr., *Amer. Mineral.*, **46**: 1120 (1961).
75. L. L. Ames, Jr., *Amer. Mineral.*, **45**: 689 (1960).
76. D. B. Hawkins, *Mat. Res. Bull.*, **2**: 1021 (1967).
77. P. C. Huang, A. Mizany, and J. L. Pauley, *J. Phys. Chem.*, **68**: 2575 (1966).
78. R. B. Barrett and J. A. Marinsky, *J. Phys. Chem.*, **75**: 85 (1971).
79. Z. Dizdar and P. Popovic, *J. Inorg. Nucl. Chem.*, **34**: 2633 (1972).
80. V. M. Radak and M. V. Susic, *J. Inorg. Nucl. Chem.*, **33**: 1927 (1971).
81. R. M. Barrer and W. M. Meier, *J. Chem. Soc.*, 299 (1958).
82. R. M. Barrer and A. J. Walker, *Trans. Faraday Soc.*, **60**: 171 (1964).
83. M. Liquornik and Y. Marcus, Israel Atomic Energy Comm., IA 810 (1963); *J. Phys. Chem.*, **72**: 2885 (1968).
84. C. M. Callahan, *J. Inorg. Nucl. Chem.*, **28**: 2743 (1966).
85. M. Liquornik and Y. Marcus, *J. Phys. Chem.*, **72**: 4704 (1968).
86. B. W. Mercer and L. L. Ames, *U.S.A.E.C. Report, HW 78461* (1963).
87. B. W. Mercer and W. C. Schmidt, *A.E.C. Accession No. 14466, Rept. No. RL-SA-58* (1965).
88. B. W. Mercer, L. L. Ames, and C. J. Tonhill, Preprints ACS Division of Water, Air and Waste Chemistry, 155th National Meeting, 1968. p. 97.
89. D. W. Breck, *U. S. Pat. 3,723,308* (1973).

Chapter Eight

ADSORPTION BY DEHYDRATED ZEOLITE CRYSTALS

Early observations of the adsorptive behavior of dehydrated zeolite crystals have been summarized in Chapter 1 and the origin of the term molecular sieve described. This term is wholly appropriate if one considers the exterior surface of the crystal as the "sieve"; only molecules of a specific size and/or shape may diffuse through the surface of the crystal into the intracrystalline voids and channels.

A representation of the surface of a zeolite crystal, in this case zeolite A, is illustrated in Fig. 8.1. Shown is a cubic, (100), face containing a regular network of 4.2 A apertures. This simplified illustration does not indicate surface defects which provide additional diffusion paths into the interior.

The basic principles of physical adsorption on solids are described in detail elsewhere (1-6); a detailed review is not considered appropriate in this book. In one respect, the adsorption of gases and vapors by zeolite solids is not as complex as adsorption phenomena on amorphous, less defined, solids. This is due to the regularity in the internal pore system which provides for the type I isotherm of the Brunauer-Emmett-Teller classification (2) or the Langmuir-type isotherm (Fig. 5.42 and 8.2a).

Void filling of the intracrystalline cavities and connecting channels of zeolites has been discussed in Chap. 5. It was concluded that although the pore-filling rule of Gurvitsch applies in many cases (1), there are exceptions which are caused by the interaction of the zeolite crystal and certain adsorbate molecules.

The usual concept of surface area in solids is not applicable to the microvoids of zeolite crystals since the total adsorption space is occupied by the zeolitic "fluid." The surface area of a zeolite as determined experimentally by physical adsorption, has no real physical

594 *ADSORPTION BY DEHYDRATED ZEOLITES CRYSTALS*

Figure 8.1 Representation of the surface of zeolite A crystals.
Upper left: Electron micrograph of zeolite A crystals, 2 microns on edge.
Bottom left: Cubic face of a single crystal of zeolite A magnified about 125,000 times.
Center: Surface is shown as it might appear if magnified 7 million times.
Right: Magnification of 60 million shows surface of crystal with uniform array of 4.2 A apertures leading into internal three-dimensional network of cavities.

meaning; it relates only to the adsorption of a given adsorbate under a set of specific conditions.

The term "monolayer equivalent" is sometimes used to express the adsorption capacity of zeolites; that is, all the molecules filling the intracrystalline voids are assumed to be spread out in a close-packed monolayer.

Many attempts have been made to calculate surface areas from the known crystal structure and compare these with adsorption data experimentally determined. Unfortunately the comparison has been poor in every case (7). However, the pore-filling rule described in Chap. 5 does result in calculated limiting adsorptive values which compare well with measured values.

The entire internal void volume of a zeolite represents a space where adsorption fields exist. In zeolites, the micropores are only a few molecular diameters in size and overlapping potential fields from opposite walls of the channels produce flat adsorption isotherms. This isotherm is characterized by a long horizontal section as p/p_0 approaches 1. The isotherm consequently intersects the line at $p/p_0 = 1$ at approximately a 90° angle.

In the crystalline zeolites, the adsorption isotherms do not exhibit hysteresis as do isotherms on many other noncrystalline, microporous solids. Adsorption and desorption are completely reversible so that the contour of the desorption isotherm follows that of adsorption.

In order to utilize the adsorption characteristics of zeolites in separation processes, commercial molecular sieve adsorbents are prepared as pelleted agglomerates containing a high percentage of the crystalline zeolite together with the necessary amount of an inert binder. The formation of the agglomerates produces macropores in the pellet (see Chap. 9) which may cause some capillary condensation at high adsorbate concentrations, that is, where the relative pressure, p/p_0, approaches unity.

In commercial molecular sieve adsorbents, the macrovoids contribute diffusion paths. However, the major part of the adsorption capacity is contained within the intracrystalline voids; the external area of the zeolite crystals is about 1% of the total equivalent surface area. As discussed in Chap. 2, the main channels by which diffusion occurs in zeolites are formed by uniform cavities connected by apertures or, in some cases, by one-dimensional tubelike structures. The walls of the channels which form the internal surface consist of a convoluted surface of oxygen ions. These channels contain a regular array of cations that bear a charge depending upon the inadequacy of the local coordination or screening by the oxygen ions of the framework.

These diffusion paths may be three-dimensional, two-dimensional, or one-dimensional paralleling a particular crystallographic direction. Needless to say, diffusion is less restricted in the zeolites with the three-dimensional network. There are several cases of zeolite structures which, theoretically, should provide for suitable three-dimensional or even two-dimensional diffusion paths if one neglects the effect of the cation. However, upon dehydration and removal of the water from the channels, the remaining cations are stranded at channel intersections or on channel walls where they seriously inhibit the diffusion of most molecules. Most zeolite structural analyses, as discussed in Chap. 2, were determined on the fully hydrated zeolite crystals. However, there is considerable information concerning the structures of certain important zeolites after dehydration.

Zeolites are high-capacity, selective adsorbents for two reasons:

1. They separate molecules based upon the size and configuration of the molecule relative to the size and geometry of the main apertures of the zeolite structure. (In these simplistic considerations the atoms of the molecule and the oxygen atoms of the zeolite framework are treated as hard spheres.)
2. Zeolites adsorb molecules, in particular those with a permanent dipole moment and which have other interaction effects, with a selectivity not found in other adsorbents.

A. EQUILIBRIUM ADSORPTION OF GASES AND VAPORS ON DEHYDRATED ZEOLITES

The amount of gas or vapor which is adsorbed by a dehydrated crystalline zeolite depends on the equilibrium pressure, p, on the temperature, t, the nature of the gas or vapor, and the nature of the micropores in the zeolite crystal. This is expressed by Eq. 8.1 (Fig. 8.2a).

$$x = f(p)_t \tag{8.1}$$

The nature of this function can be quite complex and generally is not predictable on totally theoretical grounds. At a given temperature, experimental measurement of the quantity of gas or vapor adsorbed, x, is made at equilibrium pressure, p. The plot of x as a function of p gives the experimental adsorption *isotherm*. Increasing the temperature of the gas-solid system at a constant pressure will decrease the quantity of x adsorbed. A plot of x versus t at constant pressure is the experimental adsorption *isobar* (Fig. 8.2b).

$$x = f(t)_p \tag{8.2}$$

A third method for representing the experimental data of adsorption is by the isostere, which is a plot of p as a function of t at constant x. (Fig. 8.2c).

$$p = f(t)_x \tag{8.3}$$

The process of adsorption involves a decrease in free energy and the

EQUILIBRIUM ADSORPTION OF GASES AND VAPORS 597

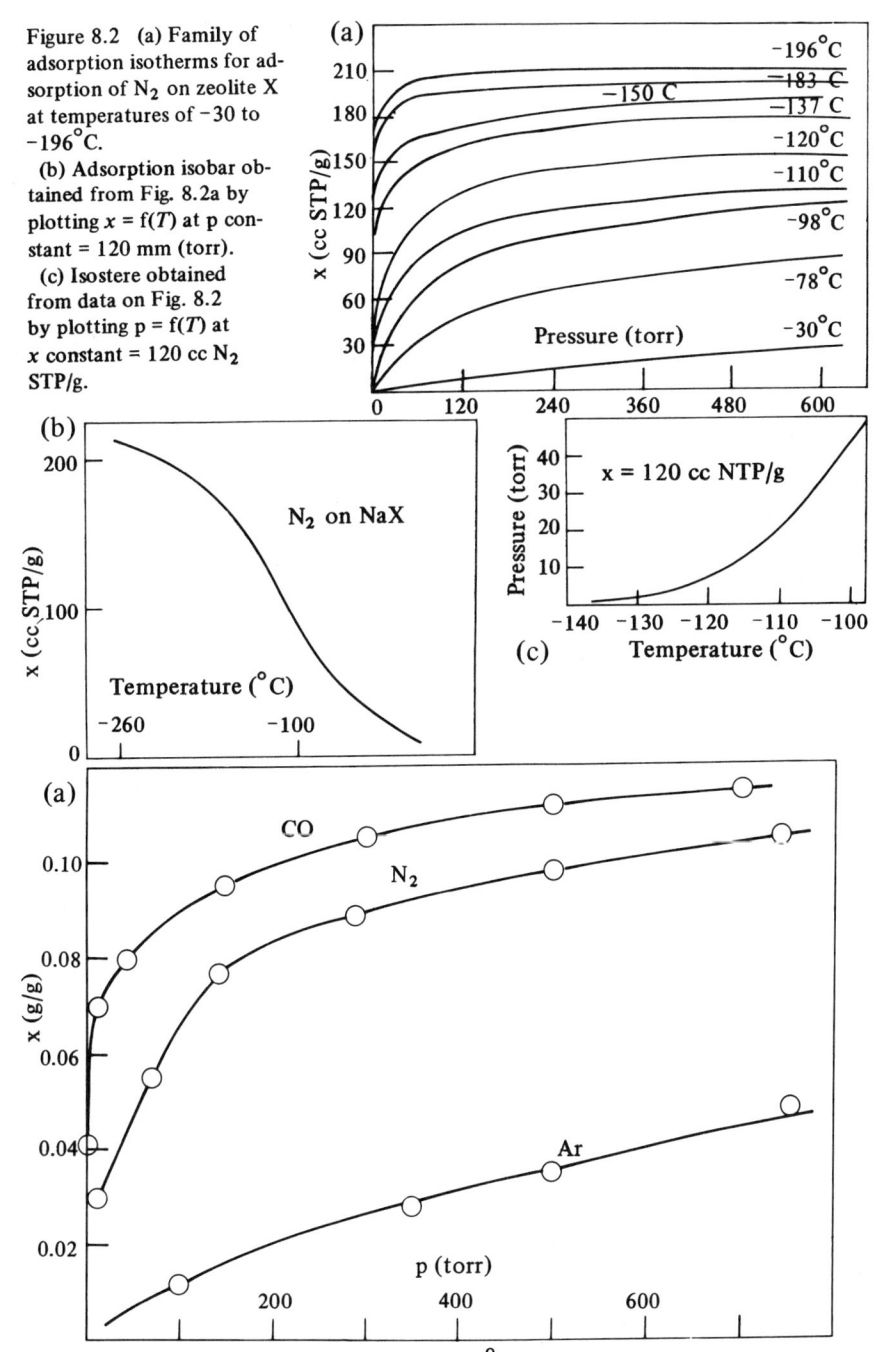

Figure 8.2 (a) Family of adsorption isotherms for adsorption of N_2 on zeolite X at temperatures of −30 to −196°C.

(b) Adsorption isobar obtained from Fig. 8.2a by plotting $x = f(T)$ at p constant = 120 mm (torr).

(c) Isostere obtained from data on Fig. 8.2 by plotting $p = f(T)$ at x constant = 120 cc N_2 STP/g.

Figure 8.3 (a) Adsorption of gases on NaA at 195°K (12).

change in enthalpy, ΔH, must also be negative since the change in entropy is negative; the adsorption process involves loss in degrees of freedom by the adsorbate molecules and the formation of a more ordered configuration in the intra-zeolitic state. Therefore the adsorption process is exothermic and heat is evolved. The heat of adsorption is derived from the isostere by use of the Clausius-Clapeyron equation (see Section D).

Considerable experimental data for the adsorption of gases and vapors on several zeolites are available. In this section, typical equilibrium data will be summarized and illustrated for several zeolites.

Synthetic zeolites A, X, Y, L, the minerals mordenite, erionite, and chabazite, and the related synthetic types, have received the most attention in fundamental investigations and the development of commercial processes. The adsorption data presented in this chapter are based upon the essentially pure, crystalline zeolite. Commercial molecular sieve adsorbents contain a binder which acts as an inert diluent (see Chap. 9). Consequently, the adsorption data for commercial materials are reduced by a dilution factor.

Equilibrium data are presented in tabular form with some isotherms for illustration. The tables are arranged by adsorbent. The adsorbates are of four types: (1) permanent gases, (2) polar, inorganic gases, (3) hydrocarbons, and (4) miscellaneous inorganic compounds.

Experimental Data for Adsorption Equilibria

The adsorption isotherm is generally determined by one of two methods. The first method, the volumetric method, determines the quantity of gas present in the system by measurement of the pressure, volume, and temperature. After exposing the activated adsorbent to a quantity of gas in the closed system, the quantity adsorbed is determined from the pressure, temperature, and volume when equilibrium is attained. The second method, the gravimetric method, measures the amount of gas or vapor adsorbed by weighing the sample in a closed system on a balance, generally of the quartz spring type. This balance was first used by McBain and is commonly referred to as the McBain adsorption balance (1). In the gravimetric method, a buoyancy correction must normally be applied which involves the determination of the volume occupied by the sample. These methods are described in various standard references (2, 8-10).

Abbreviated methods may also be used to determine adsorption capacities (11). In one method, the adsorbent samples are activated externally (to the system), weighed, and then transferred rapidly to a closed adsorption chamber.

The adsorption capacities—maximum quantity adsorbed—of various zeolites at different pressures and temperatures are summarized in Tables 8.1 - 8.13. These data are given in units of weight or volume depending upon the method used in measuring the isotherm. Adsorption capacities for some cation exchanged forms of various zeolites are given where a marked effect of the cation on the sieving characteristic is evident. The adsorbates include most of the common gases, both inorganic gases and vapors, and hydrocarbon vapors. The published literature dealing with the adsorption properties of zeolites is voluminous; consequently, the values listed in the tables are necessarily representative selections.

The effect of the type of cation on the sieving character of zeolite A is evident from the values given in Table 8.1 and Fig. 8.3 (12). The potassium form has a considerably restricted pore size while the calcium exchanged form has a larger pore size as shown by its ability to adsorb normal paraffin hydrocarbons. This "tailoring" effect due to cation exchange is discussed in Section C.

Zeolites X and Y have a three-dimensional channel system consisting of cavities separated by 12-membered oxygen rings with a structural diameter of about 7 A (see Chaps. 2 and 5) (13). These zeolites have nearly 50% of the volume of the crystal available for adsorption. As discussed in Chap. 5, this void volume is available to water but the space contained within the β-cages is not occupied by other gases and vapors. As shown by the data in Table 8.4 and Fig. 2.58, exchange of sodium by calcium reduces the effective pore size of zeolite X. (See also Tables 8.2 - 8.4, Fig. 8.4, and 8.5).

Unlike zeolites X and Y, zeolite L has a one-dimensional channel system. The structure of zeolite L provides for less void volume and the adsorptive capacity of zeolite L is less than that of zeolite Y; the pore size for adsorption is the same (Table 8.5, Fig. 8.6) (13 - 15).

Like zeolite A, chabazite exhibits a marked alteration in its sieving character by exchange with different cations (Table 8.6, Fig. 8.7). The sodium cation form shows diminished nitrogen adsorption and the po-

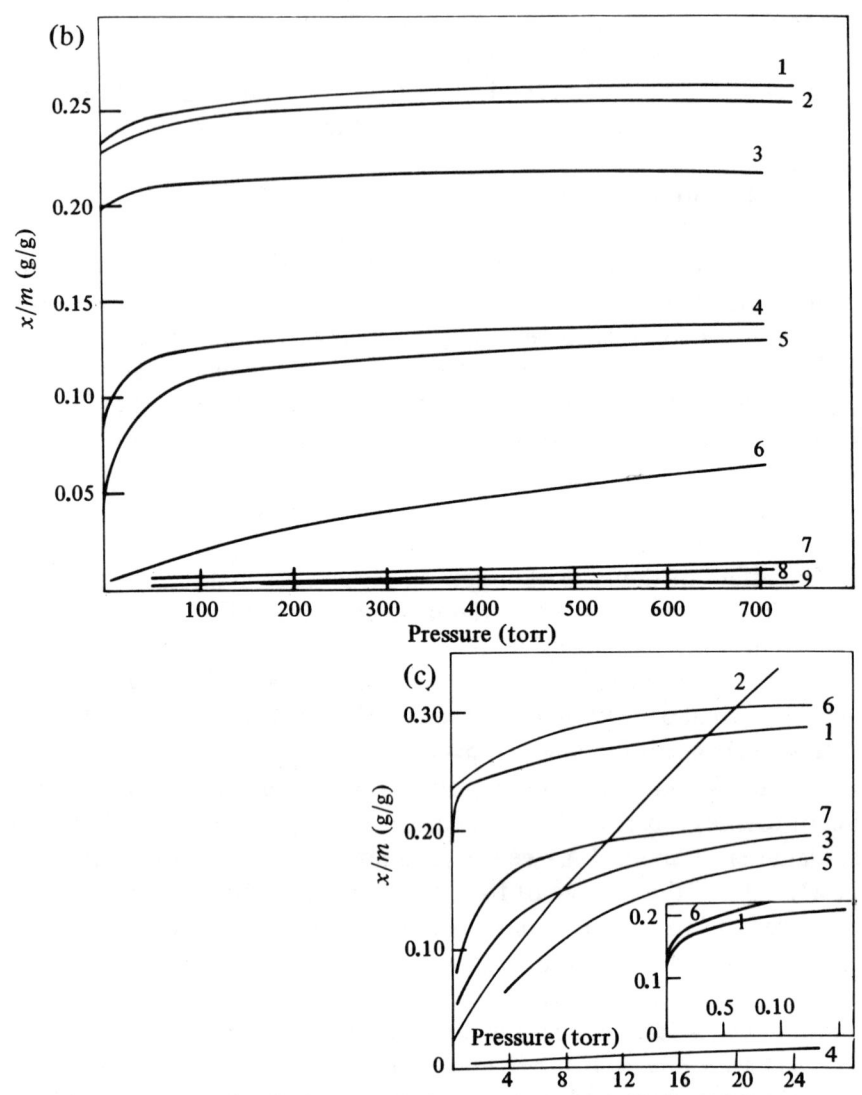

Figure 8.3 (b) Adsorption isotherms for zeolite A, chabazite, and silica gel. (1) O_2 on zeolite A, 90°K; (2) N_2 on calcium exchanged zeolite A, 77°K; (3) N_2 on chabazite, 77°K; (4) propylene on calcium exchanged A, 298°K; (5) propane on calcium exchanged A, 298°K; (6) propane on silica gel, 298°K; (7) propane on zeolite A, 298°K; (8) N_2 on zeolite A, 77°K; (9) isobutane on chabazite or calcium exchanged A at 298°K; where x/m = amount adsorbed (g) per gram of activated adsorbent.

(c) Adsorption of water on zeolite A, silica gel, and alumina (12). (1) zeolite A, 298°K; (2) silica gel, 298°K; (3) zeoliteA, 373°K; (4) silica gel, 373°K; (5) alumina, 298°K; (6) calcium exchanged A, 298°K; (7) calcium exchanged A, 373°K; insert shows adsorption at low pressures.

EQUILIBRIUM ADSORPTION OF GASES AND VAPORS 601

tassium exchanged form was found to adsorb no nitrogen. The adsorption of hydrocarbons is similarly affected (15 - 18).

Like zeolite L, mordenite and the synthetic type, Zeolon, also have a one-dimensional pore system and a modest void volume and adsorptive capacity (Table 8.7 and 8.8, Fig. 8.8) (21 - 24).

Zeolite omega has a one-dimensional channel system like zeolite L. After synthesis, zeolite omega contains large organic cations in the structure, consequently its adsorptive capacity is less than that provided by the framework structure. Removal of the large organic cations by calcination increases the adsorptive capacity (See Table 8.9).

Zeolite T is structurally related to offretite and erionite (see Chap. 2). Intergrowth of the erionite type structure with offretite restricts the effective pore size for adsorption (Tables 8.10 - 8.12).

Figure 8.4 (a) Isotherms for the adsorption of H_2O; (1) sodium X zeolite, 298°K; (2) faujasite, 298°K; (3) sodium X zeolite, 373°K; x/m = the amount adsorbed in g/g of activated zeolite.

(b) Adsorption of various gases at 195°K on sodium X zeolite.

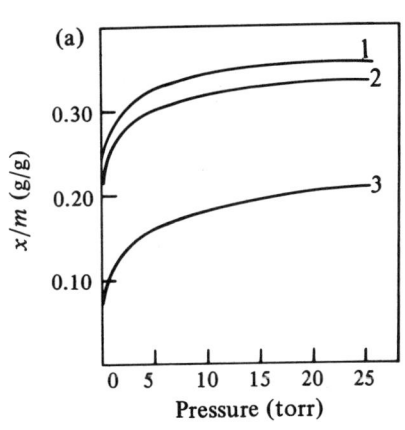

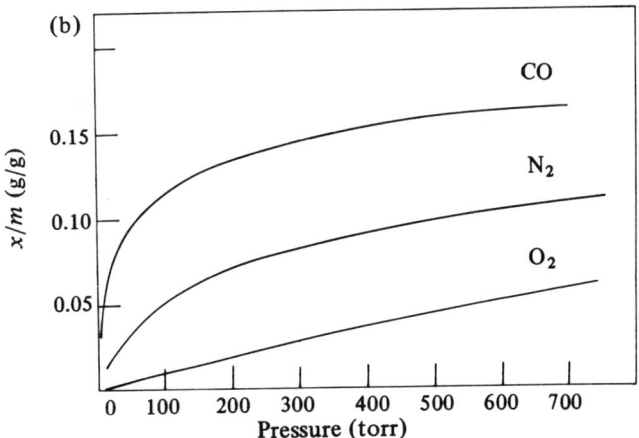

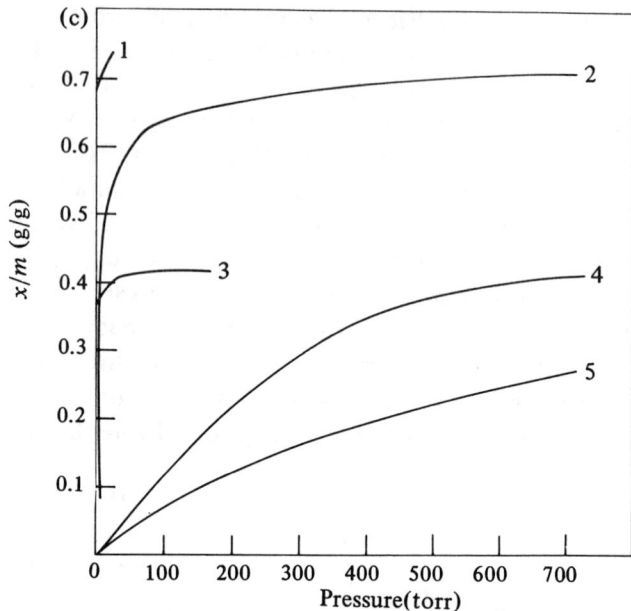

602 ADSORPTION BY DEHYDRATED ZEOLITE CRYSTALS

Figure 8.4 (c) Adsorption of rare gases on sodium X zeolite; (1) krypton, 90°K; (2) xenon, 195°K; (3) argon, 77°K; (4) krypton, 195°K; (5) xenon, 298°K.
(d) Adsorption of hydrocarbons on sodium X zeolite at 298°K; (1) cyclohexane; (2) benzene; (3) toluene; (4) neopentane; (5) methylcyclohexane; (6) n-pentane.

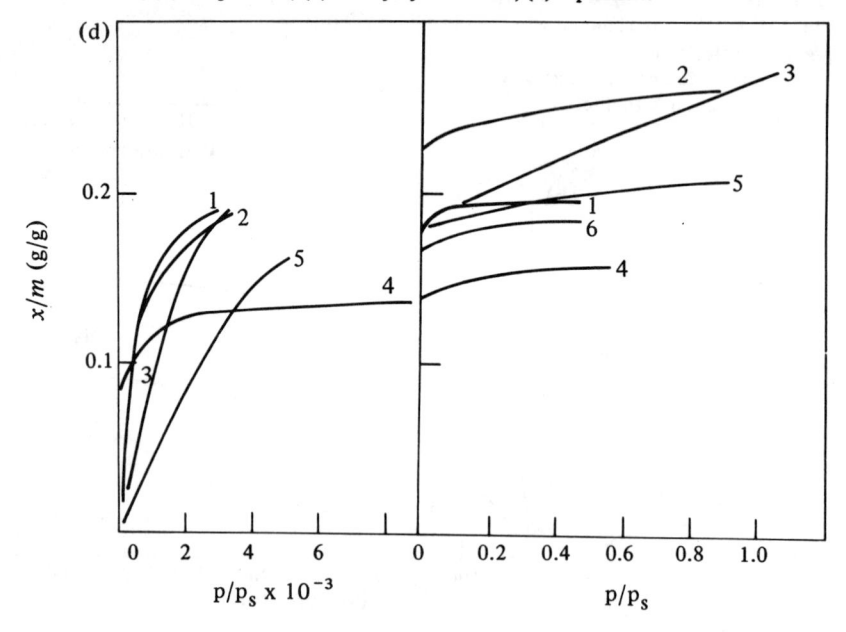

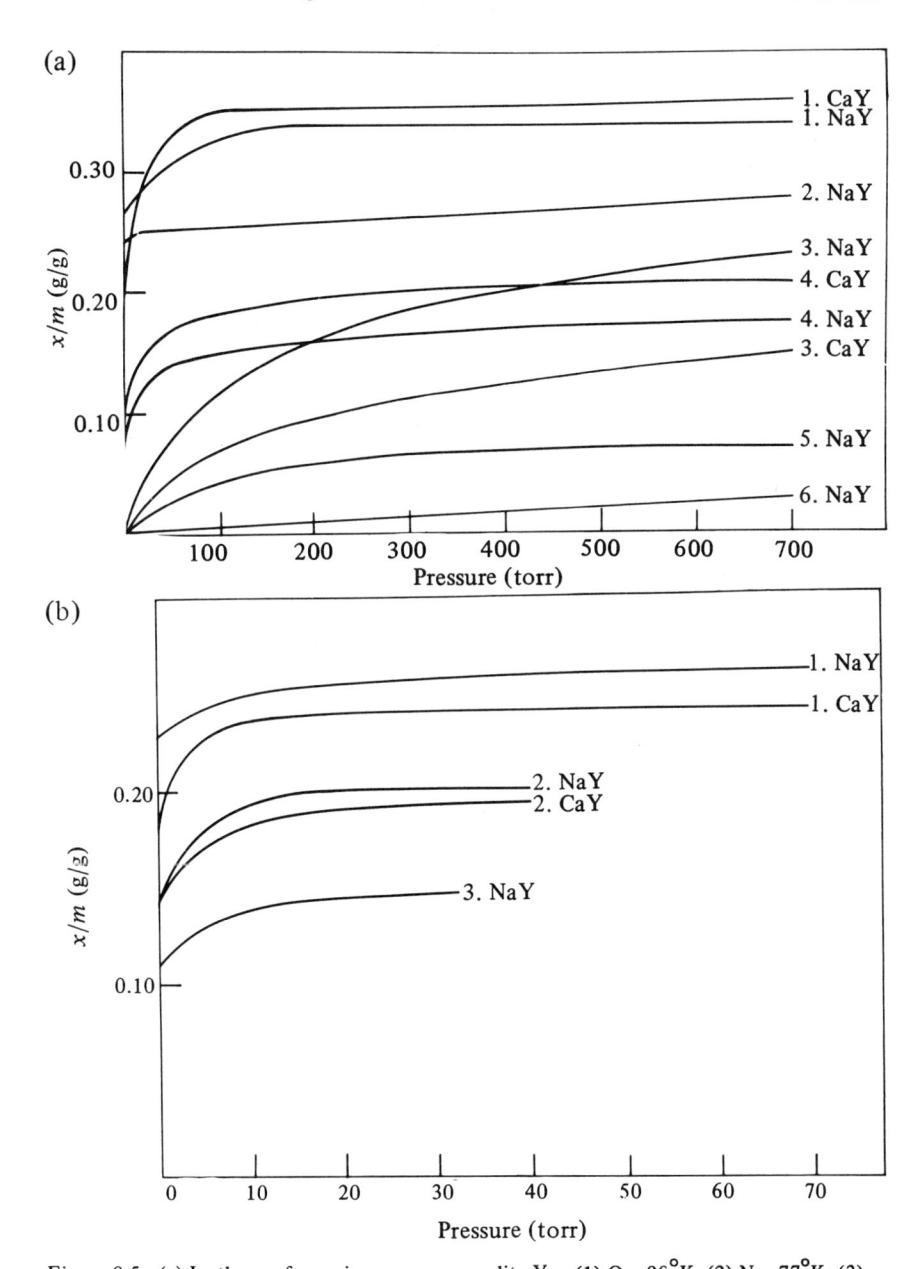

EQUILIBRIUM ADSORPTION OF GASES AND VAPORS 603

Figure 8.5 . (a) Isotherms for various gases on zeolite Y; (1) O_2, 90°K; (2) N_2, 77°K; (3) CO_2, 298°K; (4) NH_3, 298°K; (5) N_2, 195°K; (6) O_2, 195°K.
(b) Adsorption of hydrocarbons on zeolite Y: (1) benzene, 298°K; (2) n-heptane, 298°K; (3) n-heptane, 473°K.

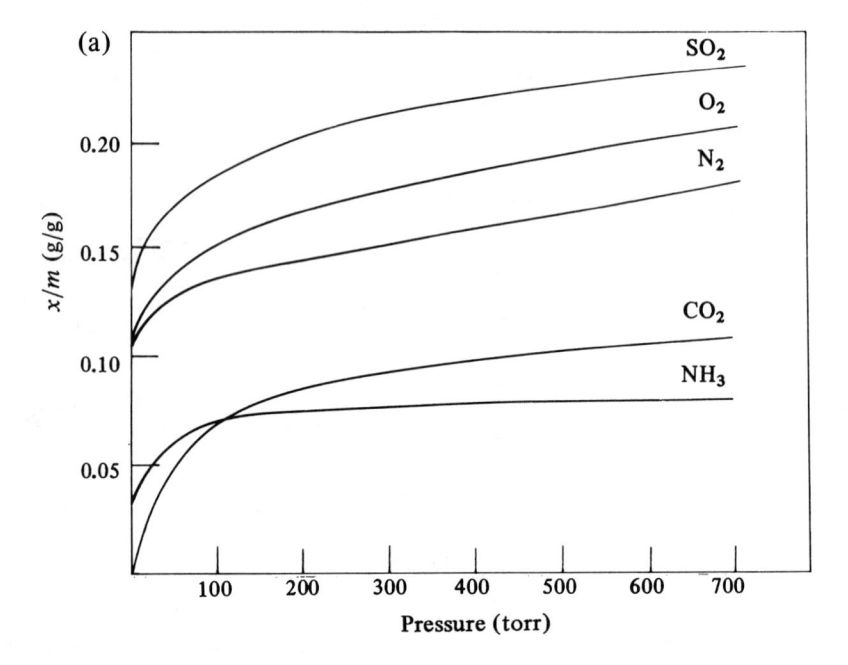

Figure 8.6 (a) Adsorption isotherms for various gases on zeolite KL: SO_2, CO_2, NH_3 at 298°K; O_2 at 90°K; N_2 at 77°K.
(b) Isotherms for hydrocarbons on zeolite KL at 298°K: (1) benzene; (2) neopentane; (3) propylene; (4) propane.

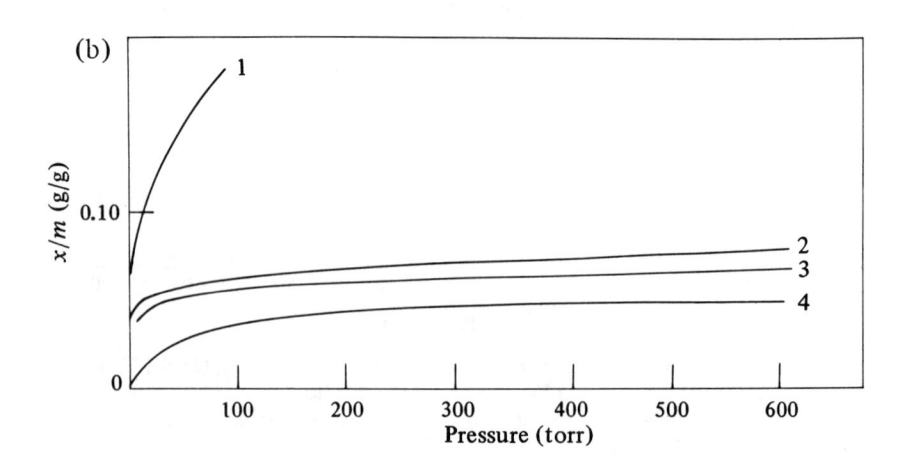

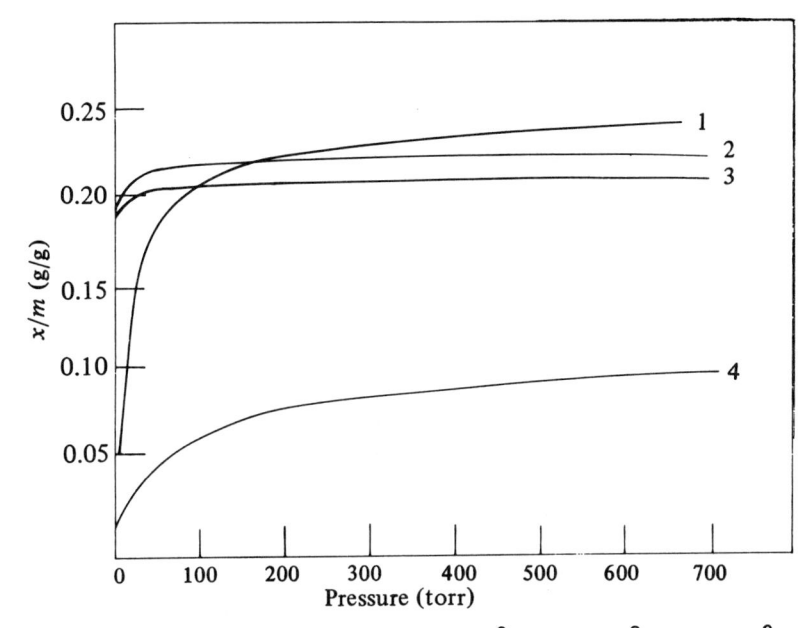

EQUILIBRIUM ADSORPTION OF GASES AND VAPORS 605

Figure 8.7 Isotherms for gases on chabazite: (1) CO_2, 298°K; (2) O_2, 90°K; (3) N_2, 77°K; (4) propylene, 298°K.

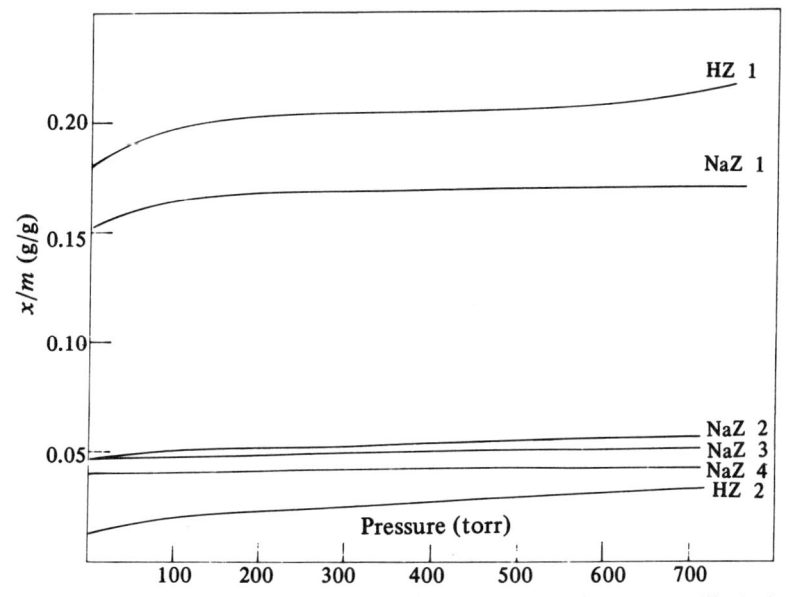

Figure 8.8 Isotherms for the adsorption of O_2 and hydrocarbon on LP mordenite (Zeolon). (1) oxygen at 90°K; (2) neopentane at 298°K; (3) n-butane at 298°K; and (4) isobutane at 298°K.

Table 8.4 Adsorption of Tertiary Amines by X and Y Zeolites (13)

	$(C_2F_5)_2NC_3F_7$	$(C_3H_7)_3N$	$(C_4H_9)_3N$	$(C_4F_9)_3N$ [a]
Critical dimension (Å)	7.7	8.1	8.1	10.2
Pressure (torr)	43	3	1	0.55
Temp (°C)	25	25	25	25
Zeolite X (Si/Al = 1.25)				
LiX (59)[b]	0.542	0.236	0.191	—
Zeolite X	0.521	0.229	0.227	0.031
CsX (51)	0.296	0.147	0.153	—
CaX (84)	0.487	0.018	0.012	0.016
BaX (93)	0.296	0.035	0.074	—
Zeolite Y (Si/Al = 2.4)				
Zeolite Y	—	—	0.228	0.018
CaY (85)	0.529	0.216	0.210	0.008
"Columbia" activated carbon	—	—	—	0.64

[a] Adsorption measured after four hours.
[b] Numbers in parentheses indicate mole % exchange of the Na form.

NOTE [Tables 8.1 - 8.13]:

The pressure, P, is expressed in millimeters of mercury (torr). For the adsorption of vapors below their critical temperatures, the ratio of P/P_0 where P_0 is the equilibrium vapor pressure, may be used. The quantity adsorbed, x/m, is given as the weight in grams per gram of dehydrated crystalline zeolite or as the volume in cc/g at standard conditions, STP. Usually, the zeolite was first outgassed at 350 - 400°C in vacuum.

NA = not adsorbed. All data in this table not otherwise referenced is from Union Carbide Corporation unpublished results.

Table 8.1a Zeolite: NaA

Adsorbate	t (°K)	P (torr)	x/m	P (torr)	x/m	P (torr)	x/m	P (torr)	x/m	Footnote	Figure	Ref.
Argon	77	100	<0.01							b		12
	195	100	5	300	15	700	30			a	8.3a	
	273	100	1	300	1.5	700	3.7			a		
Oxygen	90	0.2	0.11	1	0.17	700	0.26			b	8.3b	12
	195	40	0.003	150	0.01	700	0.044			b		
	195	100	6	300	18	700	34			a		
Nitrogen	77	700	<0.01									
	195	100	0.065	300	0.085	700	0.115			b	8.3b	12
	195	100	30	300	42	700	49			a	8.3a	80
H_2O	298	0.025	0.16	0.1	0.20	4	0.25	20	0.28	b	8.3c	12
	373	1	0.06	4	0.13	12	0.17	20	0.19	b	8.3c	12
NH_3	298	3	0.090	10	0.11	100	0.15	700	0.175	b		12
CO_2	198	10	0.25	700	0.30					b		
	298	2	0.070	10	0.12	100	0.165	700	0.20	b		12, 19, 31
	423	100	0.034	700	0.105					b		
CO	198	15	0.070	100	0.091	700	0.11			b	8:3a	12
	273	150	0.024	700	0.055					b		
SO_2	298	0.1	0.16	10	0.28	100	0.30	700	0.35	b		12

Table 8.1a Zeolite: NaA

Adsorbate	$t(°K)$	P (torr)	x/m	P (torr)	x/m	P (torr)	x/m	P (torr)	x/m	Footnote	Figure	Ref.
n-C$_4$H$_{10}$	298	NA										12
n-C$_6$H$_{14}$	298	NA										12
Isobutane	298	NA										12
Neopentane	298	NA										12
C$_2$H$_2$	298	10	0.072	50	0.094	100	0.10	700	0.11	b		12
	423	10	0.015	50	0.033	100	0.042	700	0.060	b		
C$_2$H$_4$	298	15	0.073	100	0.095	700	0.11			b		12
C$_3$H$_6$	298	10	0.11	100	0.135	700	0.14			b		12
Benzene	298	NA										
CH$_3$OH	298	0.01	0.066	0.1	0.13			100	0.19	b		12
H$_2$S	298	10	0.16	100	0.22	400	0.24			b		12
SF$_6$	298	NA										
CCl$_2$F$_2$	298	NA										
CH$_4$	198	150	0.023	700	0.058					b		80
	273	150	0.007	700	0.022					b		
C$_2$H$_6$	298	10	-0.010	100	0.048	700	0.080			b		12, 81

[a] x/m is in cc STP/g.
[b] x/m is in g/g.
Zeolite A does not adsorb molecules with a kinetic diameter greater than 3.6 A at 77°K and 4.0 A at 300°K.

Table 8.1b Zeolite: CaA

Adsorbate	t (°K)	P (torr)	x/m	P (torr)	x/m	P (torr)	x/m	P (torr)	x/m	Footnote	Figure	Source	Other Ref.
Argon	77	0.1	0.28	1	0.33	100	0.35			b			12, 32
	195	100	12	300	30	700	55			a			
	273	100	0.5	300	15	700	3.3			a			
Oxygen	77	0.1	0.16	0.2	0.26	100	0.31			b			12, 32
	195	100	0.035	300	0.065	700	0.095			b			
Nitrogen	77	3	0.21	10	0.22	700	0.24			b	8.3b		12.32
	195	100	0.07	300	0.085	700	0.11			b			
Krypton	273	11000	90	45000	120					a, c		37	32, 33
H_2O	298	0.025	0.19	0.1	0.22	4	0.27	20	0.30	b	8.3c		12
	373	1	0.11	4	0.17	12	0.19	20	0.20	b	8.3c		
NH_3	298	3	0.10	10	0.135	100	0.175	700	0.19	b			12
CO_2	298	1	0.066	10	0.13	100	0.20	700	0.24	b			12, 19
	373	15	0.060	100	0.12	700	0.20			b			
CO	198	15	0.077	100	0.11	700	0.15			b			12
	273	100	0.038	700	0.070					b			
SO_2	298	0.1	0.17	10	0.30	100	0.34	700	0.36	b			12
H_2S	298	10	0.21	100	0.28	100	0.30			b			12
SF_6	298	NA											
CCl_2F_2	See Footnote												12
I_2	393	90	0.86							b		86	
S	324	50	0.34							b		82	
CH_4	195	10	0.013							b			12
C_2H_6	298	25	0.025	100	0.053	700	0.085			b			12, 81
$n\text{-}C_4H_{10}$	298	2	0.07	10	0.10	700	0.13			b			12, 83

EQUILIBRIUM ADSORPTION OF GASES AND VAPORS 609

610 ADSORPTION BY DEHYDRATED ZEOLITE CRYSTALS

Table 8.1b Zeolite: CaA

Adsorbate	t(°K)	P (torr)	x/m	P (torr)	x/m	P (torr)	x/m	P (torr)	x/m	Footnote	Figure	Source	Ref.
n-C$_6$H$_{14}$	298	0.2	0.13	10	0.145	100	0.145			b			83, 84
n-heptane	278	0.003	0.085	0.01	0.11	0.1	0.14	10	0.175	b			84
	423		0	0.01	0.024	0.1	0.055	10	0.12	b			84
Isobutane	298	NA									8.3b		12
n-decane	457	100	0.14							b		83	
n-dodecane	488	300	0.15							b		83	
Neopentane	298	NA											
C$_2$H$_2$	298	5	0.055	50	0.080	100	0.088	700	0.11	b			46
C$_2$H$_4$	195	0.01	0.020	0.1	0.085	10	0.12			b			
Benzene	298	<0.006								b			12
CH$_3$OH	298	0.01	0.085	0.1	0.16	1	0.20	10	0.22	b			12

[a] x/m is in cc (STP)/g
[b] x/m is in g/g
[c] High pressure adsorption for Kr and Xe to 45 atm and over temperature range of 0 to 450°C from Ref. 37 absolute adsorption.

NA = not adsorbed

Zeolite CaA does not adsorb molecules with a kinetic diameter greater than 4.3 Å at 300°K and 4.4 Å at 420°K. The compound CF$_2$Cl$_2$, σ = 4.4 Å, is adsorbed at 420°K.

Table 8.2a Zeolite: NaX

Adsorbate	$t(°K)$	P (torr)	x/m	P (torr)	x/m	P (torr)	x/m	P (torr)	x/m	Footnote	Figure	Data Source	Other Ref.
Argon	78	1	0.33	10	0.40	100	0.42			b	8.4c		32, 85
	90	1	0.23	10	0.34	100	0.40	700	0.41	b			
	195	50	0.0043	100	0.0083	700	0.047			b			
Oxygen	90	20	0.30	400	0.31					b			32, 85
	195	60	0.005	100	0.083	700	0.054			b	8.4b		
Nitrogen	77	0.001	0.125	0.1	0.16	10	0.25			b			32, 85
	195	40	0.039	100	0.060	700	0.12	700	0.26	b	8.4b		
Krypton	195	10	0.01	50	0.05	100	0.10	600	0.41	b	8.4c		32, 35, 37, 39
H_2O	298	0.4	0.24	1	0.26	20	0.34			b	8.4a		
	373	0.5	0.095	1	0.11	10	0.18			b	8.4a		
He	4.2	3×10^{-4}	203	0.44	217	718	247			a, c		36	
NH_3	298	2	0.10	10	0.14	100	0.175	700	0.190	b			
	373	2	0.059	10	0.175	100	0.115	700	0.155	b			
CO_2	195	0.1	0.06	1	0.20	6	0.33	100	0.39	b			19, 31
	298	2	0.06	10	0.10	100	0.20	700	0.26	b			
CO	198	15	0.062	100	0.11	700	0.165			b	8.4b		
SO_2	298	0.05	0.14	1	0.31	10	0.39	700	0.44	b			
H_2S	298	10	0.224	100	0.294	400	0.34			b			
SF_6	298	10	0.04	100	0.23	300	0.33			b			34
CCl_2F_2	298	700	0.36							b			
Hg	596	50	0.002	200	0.010					b			38
I_2	393	90	1.11							b		38	86
P	324°C	60	0.44							b			82
S	320°C	60	0.41							b			82
CH_4	195	20	0.010	100	0.041	700	0.080			b			85
	273	700	0.020							b			

Table 8.2a Zeolite: NaX (con't)

Adsorbate	t(°K)	P (torr)	x/m	P (torr)	x/m	P (torr)	x/m	P (torr)	x/m	Footnote	Figure	Data Source	Other Ref.
C_2H_6	298	50	0.025	200	0.07	700	0.09			b, d			85, 87
C_3H_8	298	15	0.090	100	0.13	700	0.14			b			
$n\text{-}C_4H_{10}$	298	2	0.12	10	0.15	700	0.18			b			106
$n\text{-}C_6H_{14}$	298	0.2	0.17	10	0.19	100	0.20			b	8.4d		
n-heptane	298	0.01	0.175	1	0.21	40	0.21			b			
	423	0.01	0.015	0.1	0.076	1	0.11	30	0.15	b			
Isobutane	298	10	0.081	100	0.16	700	0.18			b			
n-pentane	298	1	0.16	100	0.19					b			85
Neopentane	298	0.1	0.082	1	0.11	50	0.14	700	0.156	b	8.4d		40
Isooctane	298	2	0.18	40	0.21					b			
C_2H_2	298	2	0.062	10	0.093	100	0.128	700	0.147	b			87
C_2H_4	298	10	0.050	100	0.085	700	0.105			b			88
Benzene	298	0.04	0.16	0.1	0.19	1	0.24	10	0.25	b	8.4d		
m-xylene	298	0.01	0.035	0.1	0.115	1	0.23	10	0.24	b			
Toluene													88
CH_3OH	298	0.01	0.074	0.1	0.15	1	0.20	10	0.23	b			

[a] x/m is in cc (STP)/g
[b] x/m is in g/g
[c] Zeolite NaX pellets, ~20% inert binder
NaX (Si/Al ≃ 1.25)

[d] Ref. 87 includes data for other cation forms.
[e] The zeolite NaX does not adsorb molecules with critical diameters greater than 8A at about 298°K.

Table 8.2b Zeolite: CaX

Adsorbate	$t(°K)$	P (torr)	x/m	P (torr)	x/m	P (torr)	x/m	P (torr)	x/m	Footnote	Data Source	Other Ref.
Kr	273	11000	90	45000	120					a, c	37	
Oxygen	90	20	0.32	700	0.38					b		
Nitrogen	77	30	0.26	700	0.27					b		
	238	20	0.016	100	0.031							
H_2O	300	3	0.31	20	0.36	700	0.056			b		
NH_3	298	2	0.10	10	0.14	100	0.185	700	0.220	b		
	373	2	0.065	10	0.080	100	0.12	700	0.180			
CO_2	298	20	0.125	100	0.20	700	0.29			b		19
	373	100	0.072	700	0.16							
CO	273	10	0.026	100	0.044	700	0.070			b		
SF_6	298	10	0.090	100	0.23	300	0.30			b		
CH_4	273	20	0.005	100	0.016	700	0.029			b		
Neopentane	298	1.5	0.07	50	0.138	700	0.153			b		

(Si/Al ≃ 1.25, Ca/Al$_2$ ≃ 0.85)

[a] x/m is in cc (STP)/g.
[b] x/m is in g/g.
[c] High pressure isotherms for Kr and Xe to 45 atm and over temperature range 0 to 450°C given in ref. 37. Absolute adsorption.

Zeolite CaX does not adsorb molecules with a critical kinetic diameter greater than 7.8 A at 300°K.
Data from UCC except where otherwise indicated.

614 ADSORPTION BY DEHYDRATED ZEOLITE CRYSTALS

Table 8.3a Zeolite: NaY

Adsorbate	t (°K)	P (torr)	x/m	P (torr)	x/m	P (torr)	x/m	P (torr)	x/m	Footnote	Figure	Data Source	Other Ref.
Argon													
Krypton	273	11000	105	45000	150					a, c		37	89
Oxygen	90	20	0.29	100	0.32	700	0.34			b	8.5a		
	195	100	0.005			700	0.033				8.5a		
Nitrogen	77	10	0.25	100	0.27	700	0.28			b	8.5a		89
	195	50	0.015	100	0.024	700	0.072				8.5a		
H$_2$O	298	0.1	0.080	1	0.25	20	0.35			b			
NH$_3$	298	2	0.080	10	0.11	100	0.15	700	0.18	b	8.5a		176
	373	2	0.040	10	0.060	100	0.095	700	0.135	b			176
CO$_2$	298	10	0.032	30	0.062	100	0.14	700	0.23	b	8.5a		19, 89
C$_2$H$_6$													89
n-pentane	298	1	0.145	100	0.180	400	0.180			b			
	373	1	0.035	10	0.095	100	0.140						
n-C$_6$H$_{14}$	298	0.3	0.175	10	0.190	100	0.190			b			
n-C$_7$H$_6$	298	2	0.16	10	0.20	40	0.21			b	8.5a		
Isooctane	298	2	0.17	40	0.20					b			
Neopentane	298	10	0.116	100	0.126	700	0.14			b			
C$_2$H$_4$													89
Benzene	298	0.1	0.17	1	0.23	10	0.25	70	0.26	b	8.5b		
(C$_4$F$_9$)$_3$N	323	0.5	0.028							b			

(Si/Al ≃ 2.4 - 2.5)

[a] x/m is in cc (STP)/g.
[b] x/m is in g/g.
[c] High pressure isotherms for Kr and Xe to 90 atm and 0 to 450°C. Absolute adsorption.

Table 8.3b Zeolite: CaY

Adsorbate	$t(°K)$	P (torr)	x/m	P (torr)	x/m	P (torr)	x/m	P (torr)	x/m	Figure	Other Ref.
Oxygen	90	1	0.085	10	0.25	700	0.37			8.5a	
	195	60	0.010	10	0.016	700	0.040				
Nitrogen	77	3	0.22	10	0.25	700	0.30				
	195	70	0.015	100	0.023	700	0.060				
H_2O	298	0.2	0.12	1	0.24	20	0.34	20	0.20		
	373	0.2	0.07	1	0.09	10	0.16				
NH_3	298	2	0.090	10	0.13	100	0.18	700	0.21	8.5a	
	373	2	0.062	10	0.080	100	0.12	700	0.17		
CO_2	298	10	0.026	30	0.042	100	0.070	700	0.15	8.5a	19
	373	30	0.026	100	0.044	700	0.095				
CO											91
n-pentane	298	1	0.135	10	0.160	100	0.175				
	373	3	0.055	10	0.082	100	0.130	400	0.15		
n-C_7H_{16}	298	2	0.16	10	0.20	40	0.20			8.5a	
Isooctane	298	2	0.16	40	0.19						
Benzene	298	0.1	0.16	1	0.22	10	0.23	70	0.25		

Degree of Ca exchange ≅ 80 mol %; Si/Al ≅ 2.4 - 2.5; all x/m values in g/g.

Table 8.5 Zeolite: (Na, K)L

Adsorbate	t(°K)	P (torr)	x/m	P (torr)	x/m	P (torr)	x/m	P (torr)	x/m	Figure	Other Ref.
Argon	88	0.1	0.034	1	0.08	10	0.125	150	0.195		14
Krypton	90	0.05	0.175	2	0.265	10	0.300				
Oxygen	90	10	0.124	100	0.149	500	0.183	750	0.214	8.6a	14
Nitrogen	88	0.1	0.054	1	0.10	100	0.135	700	0.18	8.6a	
H_2O	298	4	0.132	10	0.147	22	0.197				
NH_3	298	10	0.040	100	0.070	500	0.080	700	0.080	8.6a	48
CO_2	298	10	0.024	100	0.071	500	0.105	750	0.111	8.6a	48
SO_2	298	10	0.15	100	0.180	500	0.22	700	0.23	8.6a	
CH_4											14
C_2H_6											14
$n\text{-}C_4H_{10}$	298	10	0.065	100	0.075	300	0.085	700	0.090	8.6b	14
Isobutane	298	10	0.053	100	0.067	300	0.074	750	0.081		14
Neopentane	298	10	0.052	100	0.066	500	0.079	750	0.086	8.6b	14
Benzene	298	0.2	0.030	1	0.060	10	0.10	80	0.17	8.6b	14
$(C_3H_7)_3N$	323	2	0.119								
$(C_4H_9)_3N$	323	0.065	0.05								
$(C_4F_9)_3N$	323	0.5	0.05								

Composition of typical zeolite L is
$(Na, K)_2O \cdot Al_2O_3 \cdot 5\text{-}7\ SiO_2 \cdot 6\ H_2O$.
Zeolite L adsorbs molecules with a critical diameter up to 8.1 A. x/m is in g/g.
All data in Table 8.5 from UCC.

EQUILIBRIUM ADSORPTION OF GASES AND VAPORS 617

Table 8.6 Zeolite: Chabazite

Adsorbate	$t(^\circ K)$	P (torr)	x/m	P (torr)	x/m	P (torr)	x/m	P (torr)	x/m	Footnote	Figure	Source	Other Ref.
Argon	77	1	0.23	10	0.27	100	0.298			c			17, 18, 20
Krypton	423	7600	26	38000	50					b, d		42	
Xenon	423	7600	41	23000	50					b, d		42	
Oxygen	90	0.1	0.116	10	0.202	100	0.217	700	0.22	c	8.7		18, 41
Nitrogen	77	10	0.134	100	0.206	700	0.206			c	8.7	17, 1	17, 18
H_2O	.298	0.05	0.156	1	0.22	10	0.266			c			
NH_3													17
CO_2	298	1	0.037	20	0.14	100	0.218	700	0.24	c	8.7		19, 41
	195	10	0.26	100	0.29	700	0.31						
C_2H_4	298	50	0.07	200	0.08	700	0.085			c			
C_3H_8	298	NA											18
C_3H_6	298	2	0.010	50	0.050	200	0.073	700	0.095	c	8.7		
n-butene	298	700	<0.005							c			
n-C_4H_{10}	373	Adsorbed above 373°K								c			18
I_2	393	90	0.59							c		86	

[a] Chabazite mineral from Nova Scotia; composition $(Ca, Na_2)O \cdot Al_2O_3 \cdot 4.9 SiO_2$
[b] x/m is in cc(STP)/g.
[c] x/m is in g/g.
[d] High pressure adsorption isotherms for Kr and Xe to 50 atm and temperature range of 150 to 450°C. Absolute adsorption.

The mineral chabazite does not adsorb molecules with a kinetic diameter greater than 4.3 A at 400°K. The ellipsoidal apertures are 3.7 x 4.1 A in size.

Table 8.7 Zeolite: Mordenite

Adsorbate	$t(°K)$	P (torr)	x/m	P (torr)	x/m	P (torr)	x/m	Other Ref.
Argon	90		NA					
Oxygen	90	10	0.11	100	0.123	700	0.13	24
	195	700	0.063					
Nitrogen	77	100	NA					24
	195	700	0.061					
H_2O	298	1	0.095	5	0.136	20	0.15	28
CO_2	298	10	0.075	100	0.107	600	0.126	
CO	298	25	0.085	100	0.107	600	0.125	
CH_4	298	NA						
C_2H_4	298	700	0.034	slow				
C_2H_6	298	NA						
C_3H_8	298	NA						

Mordenite mineral from Nova Scotia. x/m is in g/g.
The typical mordenite mineral does not adsorb molecules with a critical kinetic diameter greater than 4.8 at 300°K.
All data from UCC.

Table 8.8 Mordenite (H-Zeolon)

Adsorbate	t (°K)	P (torr)	x/m	P (torr)	x/m	P (torr)	x/m	P (torr)	x/m	Footnote	Figure	Source	Other Ref.
Argon													21
Oxygen	90	10	0.182	100	0.198	700	0.212			b	8.8		92
Nitrogen	213	100	0.012	700	0.029					b			92
CO_2	298	100	0.035	700	0.070					b			19, 92
SO_2	273	50	0.083	200	0.18	700	0.24			f		92	
HCl	293	40	0.0835	120	0.104	200	0.117			e		43	
n-C_4H_{10}	298	10	0.045	100	0.053	500	0.058	700	0.060	b			
Isobutane	298	50	0.040	100	0.043	500	0.048	700	0.049	b			
Neopentane	298	50	0.044	100	0.046	500	0.052	700	0.054	f			
Zeolon—Na form													
Argon													23
Krypton	273	11000	60	45000	70					a, c		37	
Oxygen	90	10	0.154	100	0.165	700	0.171			b	8.8		23
Nitrogen	77	10	0.13	100	0.134	700	0.138			b			
	213	100	0.039	700	0.055					b			23
CO_2	298	100	0.10	700	0.12					b			19
SO_2	~300	10.7	0.17	34.0	0.175	83	0.187			b, d		47	

a x/m is in cc (STP)/g.
b x/m is in g/g.
c High pressure adsorption isotherms for Kr and Xe to 60 atm and 0 to 450°C. Absolute adsorption.
d SO_2 on three preparations of synthetic small-pore mordenite (Ref. 47).
e The H-Zeolon has a ratio of $SiO_2/Al_2O_3 = 13$.
f The H-Zeolon has a composition of $H_8Al_8Si_{40}O_{96}$. Data from UCC except where otherwise indicated.

Table 8.9 Zeolite: Na, TMA-Omega

Adsorbate	$t\ (°K)$	P (torr)	x/m	P (torr)	x/m	P (torr)	x/m	Other Reference
Oxygen	90	10	0.142	100	0.16	700	0.22	
Nitrogen	77	10	0.102	100	0.107	700	0.124	76
n-butane	298	10	0.030	100	0.035	700	0.041	
H$_2$O								76
n-butane	298	10	0.030	100	0.035	700	0.081	
CO$_2$								76
Isobutane	298	10	0.017	100	0.022	700	0.029	
Neopentane	298	10	0.025	100	0.033	700	0.043	
(C$_4$F$_9$)$_3$N	323	0.5	0.058					

Composition before calcination:
0.91 Na$_2$O·0.24 (TMA)$_2$O·Al$_2$O$_3$·7.1 SiO$_2$·5 H$_2$O.
Prior to outgassing at 400°C, the zeolite was calcined in air. All data in Table 8.9 from UCC.

Table 8.10 Zeolite: T

Adsorbate	$t\ (°K)$	$P\ (torr)$	x/m	$P\ (torr)$	x/m	$P\ (torr)$	$x/m\ (g/g)$	$P\ (torr)$	x/m	Data Source
Argon	77	0.1	0.080	10	0.135	100	0.16			49
Krypton	90	1	0.17	10	0.23	18	0.24			
Oxygen	77	0.02	0.040	3	0.125	50	0.15	100	0.16	
Xenon	195	3.8	0.20	slow rate						
Nitrogen	77	0.1	0.070	10	0.098	100	0.107	700	0.12	
H_2O	298	0.1	0.075	4	0.16	20	0.186			49
	373	16	0.11	22	0.12					
NH_3	298	0.4	0.015	10	0.047	100	0.066	700	0.098	
CO_2	298	20	0.048	100	0.12	300	0.14	700	0.15	
H_2S	298	1	0.049	50	0.085	100	0.043	150	0.10	
CCl_2F_2	298	NA								
$n\text{-}C_3H_8$	298	10	0.018	100	0.025	700	0.027			
$n\text{-}C_3H_6$	298	10	0.040	100	0.047	700	0.055			
$n\text{-}C_5H_{12}$	298	10	0.066	100	0.081	700	0.112			
Isobutane	298	NA								
Cyclohexane	298	NA								
Benzene	298	NA								

Typical composition of zeolite T:
0.3 Na_2O, 0.7 $K_2O \cdot Al_2O_3 \cdot 6\text{-}7\ SiO_2 \cdot 7\text{-}8\ H_2O$

Zeolite T does not adsorb molecules with a kinetic diameter greater than 4.3 A at 400°K.

622 ADSORPTION BY DEHYDRATED ZEOLITE CRYSTALS

Table 8.11 Zeolite: Erionite

Adsorbate	$t\,(°K)$	$P\,(torr)$	x/m	$P\,(torr)$	x/m	$P\,(torr)$	x/m	$P\,(torr)$	x/m	Footnote	Other Ref.
Argon	77	0.1	0.14	10	0.205	100	0.24			a	
Oxygen	77	0.020	0.11	3	0.18	50	0.207	100	0.232	a	
Xenon	195	70	0.27	100	0.32	700	0.35			c	
Nitrogen	77	0.1	0.12	10	0.145	100	0.16	700	0.194	a	27
	273	100	0.006	700	0.012	1400	0.05			b	
H_2O	298	0.1	0.10	1	0.13	20	0.22			a	28
NH_3	298	0.4	0.058	10	0.084	100	0.114	700	0.121	a	
CO_2	298	20	0.095	100	0.115	300	0.135	700	0.147	a	19
SO_2											47
CCl_2F_2	298	NA									
$n\text{-}C_3H_8$	298	10	0.050	100	0.063	700	0.070			a	
$n\text{-}C_4H_{10}$	298	10	0.076	100	0.090	700	0.105			a	
$n\text{-}C_6H_{14}$	371	25	0.018	350	0.026					c	27, 84
$n\text{-}C_5H_{12}$	298	10	0.093	100	0.108	400	0.145			b	27, 84
Isobutane	298	NA									
1-butene	298	10	0.083	100	0.096	700	0.12			a	
Neopentane	298	NA									a	
Benzene	298	NA									a	

x/m is in g/g of outgassed zeolite. All data from UCC.
[a] Erionite mineral from Durkee, Oregon.
[b] Sedimentary mineral from Pine Valley, Nevada.
[c] Sedimentary mineral from Rome, Orgeon. Reference includes data on K and Ca exchanged forms. Did not adsorb 2-methylpentane. Erionite does not adsorb molecules with a kinetic diameter greater than 4.3 Å at 400°K.

Table 8.12 Zeolite: O [Offretite] TMA, Na, K

Adsorbate	$t(°K)$	P (torr)	x/m	P (torr)	x/m	P (torr)	x/m
Oxygen	90	10	0.18	100	0.21	700	0.23
n-butane	298	10	0.10	100	0.115	700	0.122
Isobutane	298	10	0.058	100	0.067	700	0.080
Neopentane	298	10	0.055	100	0.058	700	0.069

Composition before calcination:
$(Na_2O)_{0.24}(K_2O)_{0.79}[(TMA)_2O]_{0.28}Al_2O_3 \cdot (SiO_2)_{7.08}H_2O_{5.7}$
Prior to outgassing at 400°C, the zeolite was calcined in air at 550°C.
x/m is in g/g. All data from UCC.

The zeolites reviewed in this section constitute the group of commercially important molecular sieve adsorbents. Many other zeolites (Table 2.4), have been studied. Table 8.13 lists the zeolites, minerals, and synthetic types for which some adsorption data are available, the adsorbates studied, and appropriate literature references.

Encapsulation and High Pressure Adsorption

Certain zeolites encapsulate gases at high pressures and elevated temperatures which are usually totally excluded due to the sieving effect. For example, zeolite KA will adsorb substantial quantities of gases such as methane, argon, and krypton at 350°C and 2000 - 4000 atm (Fig. 8.9). When quenched to room temperature, the gases remain trapped within the zeolite. In zeolite KA the average density of the encapsulated argon or krypton is about one-half the density of the liquid gases at their normal boiling point and about the same as the critical density. These zeolite gas encapsulates are stable for very long periods of time. To desorb the gas, either the structure is destroyed by chemical dissolution or the zeolite is heated to a higher temperature (50, 51).

The adsorption of neon, argon, and krypton, in heulandite and stilbite has been studied at pressures up to 10,000 psi and at elevated temperatures. Because the minerals are not stable toward activation at high temperatures, a lower temperature activation, 250°C, was used. Adsorption isotherms were determined at temperatures up to 300°C. After adsorption, the treatment of the crystals by surface rehydration caused more effective trapping of the argon and krypton.

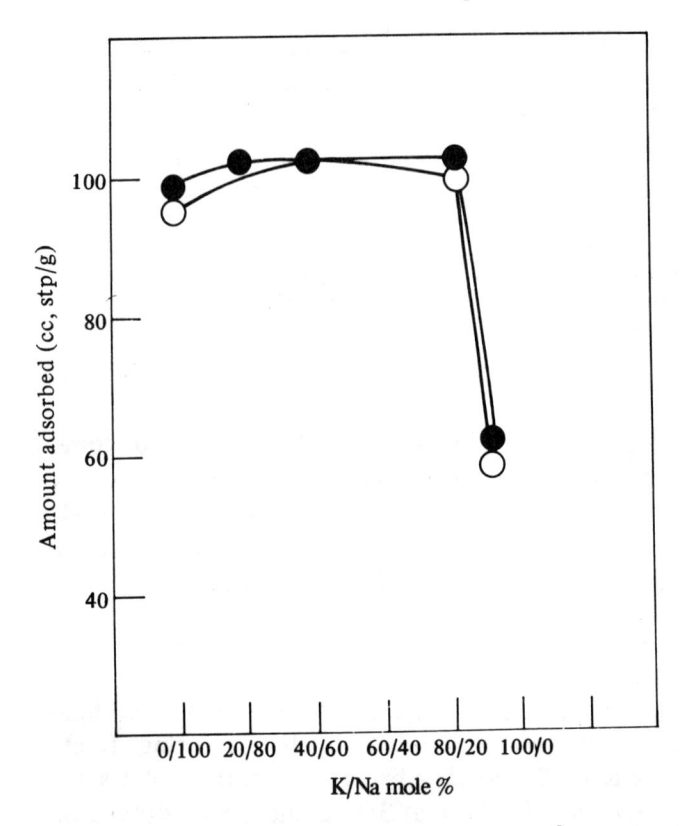

624 ADSORPTION BY DEHYDRATED ZEOLITE CRYSTALS

Figure 8.9 Encapsulation of krypton in zeolite A at 350°C and 62,500 psi. The particular cation composition, 40% potassium exchange, is optimum for the maximum storage capacity and stability over a long period of time. Argon and methane also encapsulate in the 40% potassium A (50).

The dark circles represent a sample shortly after encapsulation and the open circles are a sample measured 30 days after encapsulation.

Neon was not retained. The rate of diffusion of the gases out of the zeolite crystals was determined as a function of time and temperature and of surface rehydration (52). The encapsulation of argon and krypton in phillipsite and synthetic zeolite K-M* is short-lived (53).

The trapping or encapsulation of neon, argon, and krypton in synthetic sodalite and cancrinite hydrates at high pressures (700 - 1000 atm) and high temperatures (up to 440°C) has been reported. The structures of these materials (see Chap. 2) are based on the ABAB and ABC stacking of linked 6-rings. Argon and krypton are both adsorbed in the sodalite at high pressures and high temperatures. The

* Zeolite K-M is zeolite W (see Chap. 4, Table 4.26).

Table 8.13 Summary of Adsorption Equilibrium Data for Various Zeolites—Minerals and Synthetic

Zeolite	Adsorbate	Activation Temperature (°C)	T(°K)	P (torr)	x/m	P (torr)	x/m (g/g)	Footnote	Data Source	Other Ref.
Clinoptilolite (Hector, Calif.)	H_2O	350	298	1	0.10	20	0.16	a		93, 94
	CO_2	350	298	100	0.063	700	0.082	a, b		19, 93, 94
	O_2	350	90	700	0.019			a, b		
	N_2	350	77	700	0.014			a		
Faujasite (Kaiserstuhl)	H_2O	350	298	22	0.33			c	13	
	O_2	350	90	700	0.29			c		
	N_2	350	77	700	0.24			c		
	Ar	350	90	730	0.28			c		
	SF_6	350	298	710	0.23			c		
	$(C_2F_5)_3N$	350	298	40	0.24			c		
	$(C_4F_9)_3N$	350	298	0.07	0.009			c		
Ferrierite	O_2	350	90	700	0.105			d		
	C_2H_4	350	298	750	0.041			d		
	$n-C_4H_{10}$	350	298	NA				d		
	O_2	350	90	700	0.064			e		
	C_2H_4		298	700	0.025			e		
Gmelinite (Nova Scotia)	H_2O	350	298	4	0.23					
	NH_3	350	298	700	0.23					
	CO_2	350	298	700	0.14					
	SO_2	350	298	700	0.39					
	O_2	350	77	85	0.087					
	N_2	350	77	710	0.062					
	Ar	350	77	93	0.085					
	C_2H_4	350	298	760	0.037					24
	C_2H_6	350	298	700	0.030					24

Data from UCC except where otherwise indicated.
[a] Composition: 0.9 Na_2O · 0.1 K_2O · Al_2O_3 · 10 SiO_2 · 6.1 H_2O
[b] Some varieties of clinoptilolite adsorb up to 0.16 g/g of O_2 and, accordingly, more CO_2.
[c] Composition: 0.20 Na_2O · 0.41 CaO · 0.37 MgO · Al_2O_3 · 4.5 SiO_2 · 7 H_2O
[d] Ferrierite from Kamloops, B. C.
[e] Sedimentary ferrierite from Nevada.

Table 8.13 Summary of Adsorption Equilibrium Data for Various Zeolites—Minerals and Synthetics (con't)

Zeolite	Adsorbate	Activation Temperature (°C)	T(°K)	P (torr)	x/m (g/g)	Footnote	Data Source	Other Ref.
Heulandite (Iceland)	Ne	250	294	600 atm	15	f	52	
	NH_3	130	294	400	65	f		17
Laumontite	CO_2	350	298	700	<0.01		UCC	
	NH_3	200	298	700	15.2	f	226	
Levynite (Ireland)	CO_2	350	298	700	0.12			
	O_2, N_2	350	77	700	NA			
Paulingite (Washington)	H_2O	150	300	0.27, 22	0.14, 0.24			
	Ar	150	77	120	0.15			
	N_2	150	77	700	0.15			
	O_2	150	77	120	0.10			
	Kr	150	90	19	0.14			
	Xe	150	195	720	0.03			
Phillipsite (Pine Valley, Nev.)	Ar	400	673	816 atm	33	f	53	
Stellerite	H_2O	250	298	25	0.17			
	CO_2	250	298	707	0.097			
	O_2	250	77	NA				
Stilbite (Ireland)	CO_2	350	298	700	0.077			
	O_2	350	77	NA				
	Ar	180	453	800 atm	31	f	52	52
	Kr	200	473	300 atm	30	f	52	52
D-[Chabazite-type]	CO_2	250	298	700	0.16	g		
	O_2	350	90	700	0.21	g		
	N_2	250	77	700	0.15	g		
	Kr	250	90	16	0.32	g		
	C_3H_6	250	298	650	0.085	g		
	$n-C_5H_{12}$	250	298	400	0.10	g		

f x/m is in cc (STP)/g.
g Composition typically is 0.5 Na_2O · 0.5 K_2O · Al_2O_3 · 4.8 SiO_2 · 6.7 H_2O

Table 8.13 Summary of Adsorption Equilibrium Data for Various Zeolites—Minerals and Synthetic (con't)

Zeolite	Adsorbate	Activation Temperature (°C)	T(°K)	P (torr)	x/m (g/g)	Footnote	Data Source Ref.	Other Ref.
SrD-[Ferrierite-type]	Ar	340	77	700	62	f, h	44	
	Kr	340	90	$p = p_0$	18	f, h	44	
	O_2	340	77	80	66	f, h	44	
	CO_2	340	293	80	36	f, h	44	
W[K-M]	Ar	400	673	544 atm	22	f	53	
	NH_3	250	298	700	0.11			
	CO_2	250	298	700	0.057			
	N_2	250	77	700	NA			
S-[Gmelinite-type]	CO_2	350	298	700	0.14	i		
	N_2	350	77	700	0.06	i		
	NH_3	350	298	700	0.099	i		
	SO_2	350	298	700	0.21	i		
	n-C_5H_{12}	350	298	400	0.01			
F	CO_2	250	293	700	0.08	j		
	NH_3	250	298	700	0.07	j		
	SO_2	250	298	700	0.17	j		
	CH_3OH	250	298	120	0.11	j		
	O_2	250	77	130	~0.01	j		
ZK-5	H_2O	550	298	0.21	0.21	k	72	
	n-C_6H_{14}	550	298		0.14	k	72	
	Cyclo C_6H_{12}	550	298		0.01	k	72	

h Composition: 0.93 SrO · Al_2O_3 · 12.3 SiO_2 · 5.5 H_2O
The Li, Na, K, and Ca exchanged forms were prepared and isotherms measured for O_2, N_2, Ne, Ar, Kr, Xe, CH_4, and CO_2.
i Typical anhydrous composition of zeolite S is:
Na_2O · Al_2O_3 · 4.6 SiO_2.

j Typical composition: K_2O · Al_2O_3 · 2 SiO_2 · 2.9 H_2O Exchanged forms (Na$^+$, Ca^{2+}, Ba^{2+}) not stable at 250°C.
k Air calcined at 550°C before adsorption measurements.

quantity adsorbed at 1000 atm was about 45 cc/g. Although cancrinite has a one-dimensional structure of channels formed by 12-rings with a free diameter of 6.3 A, the location of cations in the channels produces a substantial restriction. The adsorption of these gases in cancrinite was very slow and the limiting capacity much less, about 14 cc/g. The equilibrium diameters of argon and krypton are about 3.8 and 3.9 A (see Section C.) Therefore, only one rare gas atom can occupy each cage in the hydrated sodalite or cancrinite. This allows a volume of 53 cc/g if the cages in sodalite do not contain NaOH, occluded during crystallization (54).

B. THE APPLICATION OF ISOTHERM EQUATIONS TO ZEOLITE ADSORPTION

In addition to standard isotherm equations (Henry's law, Langmuir equation, the standard BET, the Volmer isotherm, etc.) several attempts have been made to derive a standard adsorption equation which would apply to the adsorbed phase in zeolites under all conditions. Although several of the equations have some degree of broad applicability, no universal adsorption equation exists.

The Langmuir equation originally applied to the adsorption of molecules in a monolayer on an open surface; the classical interpretation of the type I isotherm is based upon this model. Since zeolites exhibit the type I isotherm, the Langmuir equation has been applied with some degree of success. Contrary to the usual interpretation of the type I isotherm, the break in the isotherm corresponds to the filling of the microvoids and not the completion of a monolayer. The Langmuir equation is given as follows:

$$\theta = \frac{x}{x_m} = \frac{Bp}{1 + Bp} \qquad (8.4)$$

The Langmuir model assumes that the adsorption potential is unifirm, that is, no energetic heterogeneity exists. This is not true in most cases of zeolite adsorption. In Eq. 8.4, the quantity adsorbed in the solid is given by x when the pressure is p. The coefficient B is a constant at constant temperature. The monolayer capacity or, in the case of zeolites, the point corresponding to void filling, is given by x_m. At very low pressures, where the term Bp may be neglected in comparison to unity, the equation simplifies into:

$$x = x_m B p \tag{8.5}$$

This means that adsorption is proportional to pressure (Henry's law) and the isotherm is a straight line passing through the origin. At high pressures, the value of Bp is very large relative to one and the equation reduces to $x = x_m$. The isotherm contour rises steeply and then becomes horizontal at high pressures.

The Langmuir equation may be rewritten in the form:

$$\frac{p}{x} = \frac{1}{x_m B} + \frac{p}{x_m} \tag{8.6}$$

The value of p/x plotted against p yields a straight line with a slope of $1/x_m$. Another method is to plot the Langmuir coefficient, given by B, against the adsorption expressed as θ. From 8.4,

$$B = \frac{\theta}{p(1-\theta)}$$

and a plot of $\log_{10} B$ versus θ should give a straight line since, if the Langmuir isotherm applies, the constant B does not vary with the level of coverage or adsorption.

The classical BET equation, although used for many years to evaluate surface areas and heats of adsorption on many adsorbent-adsorbate systems, is not applicable to zeolites. Basically, this equation was developed for multilayer adsorption; for a monolayer, such as in zeolites, it reduces to the Langmuir equation.

Reikert (55) has shown that for adsorption of CO_2 on H-Zeolon, the measured isotherm at $-80°C$ may be represented as the sum of two individual Langmuir isotherms, one for each of two different types of sites.

The Volmer equation (Eq. 8.7) assumes that the equilibrium gas phase behaves as a two-dimensional gas without any lateral interaction.

$$K_v = p \frac{(1-\theta)}{\theta} \exp[-\frac{\theta}{1-\theta}] \tag{8.7}$$

In order to test this equation, one plots the value of K_v against the coverage, θ, which should be a constant. This equation is not applicable in most cases, such as the adsorption of CO_2 on zeolite Y (19) and fluorine compounds on zeolite X (56).

In another model, the isotherm equation is based upon a virial equation by the application of solution thermodynamics. This includes a term, π, which is the pressure that would have to be applied to the dehydrated zeolite to reduce its chemical potential to that of the zeolite when adsorption had taken place. The equations are based on the osmotic ratio π/CRT (19, 57).

$$\frac{\pi}{CRT} = 1 + A_1 C + A_2 C^2 + \ldots A_n C^n \tag{8.8}$$

where C is the concentration of adsorbate in the zeolite.

The Polyani potential theory was one of the earliest empirical theories developed for treating adsorption (1-3). It was modified by Dubinin and Radushkevich to apply to microporous adsorption and void filling (58, 59) (See Chap. 5.). In the Polanyi theory the characteristic curve is the relationship between the work of adsorption, A, and the filled volume of the adsorption space, W. A is given by:

$$A = RT \ln \frac{p_s}{p} \tag{8.9}$$

and

$$W = a\,v^* \tag{8.10}$$

where a = the amount of adsorbate in millimoles/g at temperature T and equilibrium pressure p. The molar volume of the adsorbate is v^* and p_s is the saturation vapor pressure. This equation takes the form of:

$$W = W_0 \exp\left[-\frac{kA^2}{\beta^2}\right] \tag{8.11}$$

It has been applied to microporous adsorbents such as carbon and has also been found to apply in many instances to zeolites. The volume occupied by the adsorbed phase, W, at a given temperature, T, and pressure, p, is given by the equation:

$$W = W_0 \exp\left[-\frac{B}{\beta^2}(T \log_e \frac{p_s}{p})^2\right] = a v^* \tag{8.12}$$

W_0 is a constant and represents the volume of the adsorption space or pore volume. B is a constant which is independent of temperatures and is characteristic of the adsorbent porous structure. The constant β is an "affinity coefficient." At temperatures below the boiling point of the liquid, v^* can be taken as equal to the molar volume of the bulk liquid.

Experimentally, Eq. 8.12 is used in the following form (59, 99).

$$\log W = \log W_0 - \frac{B}{2.30\beta^2} (T \log p_s/p)^2 \qquad (8.13)$$

In order to evaluate W_0 and B, the equation is expressed as:

$$\log a = C\text{-}D \, (\log \frac{p_s}{p})^2 \qquad (8.14a)$$

When $\log a$ is plotted versus $(\log p_s/p)^2$ the curve should be linear, if the equation applies (Fig. 8.10).

$$C = \log \frac{W_0}{v^*} \quad \text{and} \quad D = \frac{0.434BT^2}{\beta^2} \qquad (8.14b)$$

A plot of A versus W gives a temperature-independent characteristic curve (Fig. 8.11).

Loughlin and Ruthven have applied the Dubinin equation to the adsorption of low molecular weight paraffin hydrocarbons on calcium exchanged zeolite A and have observed that the experimental data can be represented by a typical characteristic curve. A single characteristic curve was applied to all of the low molecular weight hydrocarbons including n-butane by using a reduced adsorption potential (60).

One of the earliest isotherm equations was attributed to Freundlich as given by the very simple expression (3):

$$v = kp^h \qquad (8.15)$$

and

$$\log v = h \log p + \text{constant} \qquad (8.16)$$

This equation has some limited applicability to zeolite adsorption (21).

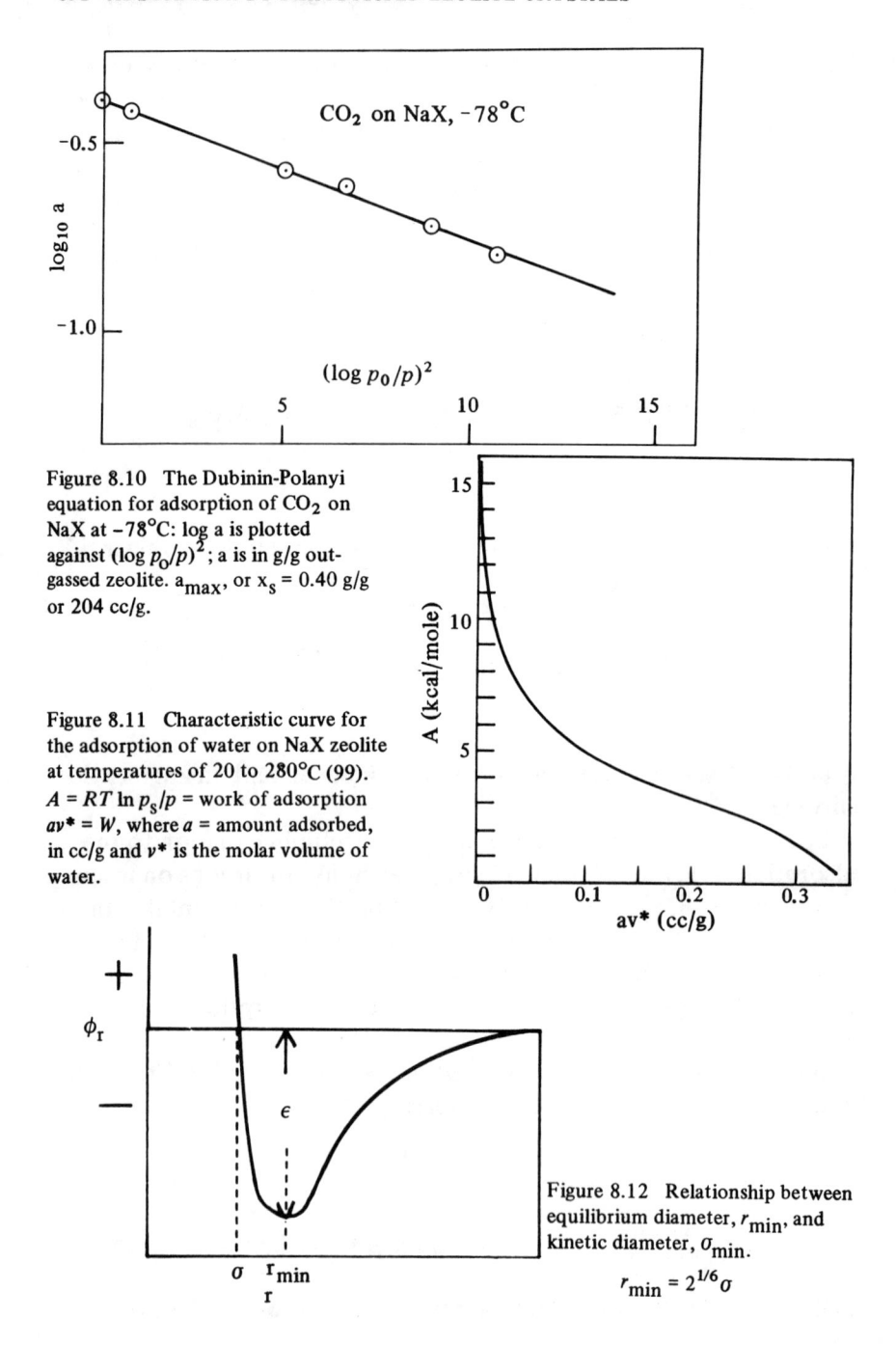

632 ADSORPTION BY DEHYDRATED ZEOLITE CRYSTALS

Figure 8.10 The Dubinin-Polanyi equation for adsorption of CO_2 on NaX at $-78°C$: log a is plotted against $(\log p_0/p)^2$; a is in g/g outgassed zeolite. a_{max}, or x_s = 0.40 g/g or 204 cc/g.

Figure 8.11 Characteristic curve for the adsorption of water on NaX zeolite at temperatures of 20 to 280°C (99). $A = RT \ln p_s/p$ = work of adsorption $av^* = W$, where a = amount adsorbed, in cc/g and v^* is the molar volume of water.

Figure 8.12 Relationship between equilibrium diameter, r_{min}, and kinetic diameter, σ_{min}.

$$r_{min} = 2^{1/6}\sigma$$

For example, in the adsorption of gases on zeolite A including nitrogen, oxygen, CO, CO_2, argon, ethane, and ethylene, linear or nearly linear plots were obtained by plotting the logarithm of the amount adsorbed versus the logarithm of the pressure. For O_2 and CO_2 the isotherms seem to be represented by two linear portions (61).

Based on the assumption that the adsorbed molecules form a mobile fluid which fills the adsorption space, a theoretical treatment was developed for the adsorption of noble gas molecules in zeolite A (62). Good agreement between experimental and calculated adsorption values over wide ranges of temperature and pressure were demonstrated for helium, neon, argon, and krypton adsorption. Whether or not this treatment is applicable to other adsorbates is not known.

In a more recent treatment, Loughlin and Ruthven have derived a new isotherm equation from the application of statistical thermodynamics (63). This was applied to the adsorption of low molecular weight hydrocarbons on zeolite A and calcium A including methane, ethane, propane, and n-butane. For low adsorption coverages, this equation enabled the calculation of Henry law constants in the region where experimental data are not available. This equation was found to be unsatisfactory for high adsorbate loadings or above a value of $\theta = 0.7$.

C. THE MOLECULAR SIEVE EFFECT

Molecular sieving by dehydrated zeolite crystals is caused by the size and/or shape differences between the crystal aperture (Chap. 2, Table 2.2) and the adsorbate molecule. Most information on crystal aperture dimensions was obtained by structural analysis of the hydrated crystals. As discussed in Chap. 2, the aperture size and shape may change during dehydration due to framework distortion or cation movement. In some instances, the apertures are circular (zeolite A or zeolite L), in other cases the aperture may take the form of an elipse (dehydrated chabazite or erionite). In the latter case, subtle differences in the adsorption of molecules occur based upon a shape factor.

In order to correlate the crystallographic aperture or pore size of a zeolite with the dimensional parameters of various adsorbate molecules, we must establish a scale of molecular dimensions. In early experiments, the molecular size was based upon the equilibrium diameter of the adsorbate molecule (12, 64). This was arrived at by calculation using the known molecular shape, bond distances, bond angles, and van der Waals radii (66). This approach was not very satisfactory since it was observed that certain molecules were freely adsorbed but were larger than the known aperture size of the zeolite crystal. In an improved treatment of this problem, Kington and MacLeod have utilized the collision or kinetic diameter (65).

The effective pore size of a zeolite molecular sieve can be determined from the sizes of molecules which are or are not adsorbed under a given set of conditions. For example, at temperatures of -183 to -196°C, zeolite A adsorbs oxygen freely while nitrogen is not adsorbed over long periods of time. At these temperatures, nitrogen diffusion is so slow that true equilibrium cannot be attained.

The Kinetic Diameter

For spherical and nonpolar molecules the potential energy of interaction, $\phi(r)$ is well described by the Lennard-Jones (6-12) potential

$$\phi(r) = 4\epsilon \left[\left(\frac{\sigma}{r}\right)^{12} - \left(\frac{\sigma}{r}\right)^{6} \right] \tag{8.17}$$

The parameters, σ and ϵ, are constants which are characteristic of the molecular species and are determined from second virial coefficients. At large separations the attractive component, $(\sigma/r)^6$, is dominant and describes the induced dipole-induced dipole interaction. At small separations the repulsive component is dominant (Fig. 8.12). When the potential equals 0 the diameter, r, is equal to σ. The *kinetic* or collision diameter is the intermolecular distance of closest approach for two molecules colliding with zero initial kinetic energy. The maximum energy of attraction, ϵ, occurs at r_{min}, where $r_{min} = 2^{1/6}\sigma$.

In assessing the apparent pore size of molecular sieve zeolites, the critical dimensions for spherical molecules are given by the value of r_{min}. For diatomic molecules, r_{min} is based upon the van der Waals length, and represents the molecule in all orientations. For long molecules, such as hydrocarbons, the dimension is the minimum cross-sectional diameter. It is preferred to use σ values, where available, for nonpolar spherical molecules and σ values obtained from minimum cross-sectional diameters for more complex molecules, such as *n*-paraffins.

For polar molecules, the Stockmayer potential function is widely used for describing the interaction between molecules for which dipole-quadrupole interactions are not important. Values of the parameter σ determined for certain polar molecules have been used. Kihara's data were used for spherocylindrical and ellipsoidal molecules (H_2 and N_2) (67).

Table 8.14 lists molecular dimensions calculated from Pauling along with r_{min} and σ, the minimum kinetic diameter. For complex molecules, the minimum equilibrium diameter of Pauling was used to compute σ. Adsorption phenomena indicate that the minimum equilibrium diameter should be used as r_{min} in computing σ. For example, the Lennard-Jones approach gives a value of σ = 4.05 for CO_2 and 3.64 for N_2. Under equivalent conditions, KA adsorbs CO_2 but not N_2. Therefore, the minimum equilibrium dimension of 3.7 A was used to compute a σ value of 3.3 A. The σ values for H_2O and NH_3 were obtained from data given by the Stockmayer potential.

Figure 8.13 compares σ, the minimum kinetic diameter, with the apparent pore diameter of various zeolites. Calcium A has a pore diameter of 4.2 A as determined by structural analysis. This compares well with the σ value of 4.3 for n-paraffins and 4.4 A for CF_2Cl_2 which is adsorbed at about 150°C. Thus, the apparent pore size of CaA varies from 4.2 - 4.4 A. NaA adsorbs C_2H_4 (slowly) and CH_4, σ = 3.9 and 3.8 A, respectively. At low temperatures, it does not adsorb N_2. The apparent pore diameter is 3.6 to 4.0 A, depending on temperature. The explanation for this variation has been based upon a process of activated diffusion. It has also been postulated that thermal vibration of the oxygen ions, and cations, in the zeolite lattice surrounding the apertures is responsible (70).

Examples of Molecular Sieving

KA, when completely exchanged, adsorbs some CO_2 and, at lower K exchanges, C_2H_2. A pore diameter of 3.3 A seems appropriate to these experimental results. Unusual results were obtained by Hamlen for the adsorption of H_2 on zeolite A as a function of potassium exchange (71). At 77°K, potassium A (20% exchanged) did not adsorb H_2. Sodium A adsorbed 160 cc STP/g at the same temperature. It thus appears that the apparent pore size of KA is temperature dependent and at 77°K is about 2.9 A, 0.3 - 0.4 A less than at room temperature.

636 ADSORPTION BY DEHYDRATED ZEOLITE CRYSTALS

Table 8.14 Table of Dimensions for Various Molecules (66)

	Pauling Length (Å)	Width (Å)	Lennard-Jones (6 - 12) r_{min}(Å)	σ(Å)[a]	Reference
He		~3	3.0	2.6	69
H_2	3.1	2.4	3.24	2.89	
Ne		3.2	3.08	2.75	68
Ar		3.84	3.84	3.40	68
O_2	3.9	2.8	4.02	3.46	68
N_2	4.1	3.0	4.09	3.64	67
Kr		3.96	3.96	3.60	68
Xe		4.36	4.45	3.96	68
NO	4.05	3.0	3.58	3.17	69
N_2O	4.2	3.7		3.3	
CO	4.2	3.7	4.25	3.76	68
CO_2	5.1	3.7		3.3	
Cl_2	5.6	3.6		3.2	
Br_2	6.2	3.9		3.5	
H_2O	3.9	3.15		2.65	68
NH_3	4.1	3.8		2.6	68
SO_2	5.28	4.0		3.6	
CH_4		4.2	4.25	3.8	68
C_2H_2	5.7	3.7	3.7	3.3	
C_2H_4	5.0	4.4		3.9	
C_3H_8	6.5	4.9		4.3	
n-C_4H_{10}		4.9		4.3	
HCl	4.29	3.6		3.2	
HBr	4.6	3.9		3.5	
H_2S	4.36	4.0		3.6	
Cyclopropane		4.75		4.23	
CS_2		4.0		3.6	
CF_2Cl_2		5.0		4.4	
CCl_4			6.65	5.9	69
Propylene			5.0	4.5	
Iso-C_4H_{10}		5.6		5.0	
Butene-1			5.1	4.5	
CF_4		4.9	5.28	4.7	68
SF_6		5.8	6.18	5.5	68
Neopentane		7.0		6.2	
$(C_4H_9)_3N$			9.1	8.1	
$(C_2F_5)_2NC_3F_7$		8.7		7.7	
$(C_4F_9)_3N$		11.5		10.2	
Benzene		6.6		5.85	
$(C_2H_5)_3N$			8.8	7.8	
Cyclohexane		6.7		6.0	

[a] Kinetic diameter, σ, calculated from the minimum equilibrium cross-sectional diameter.

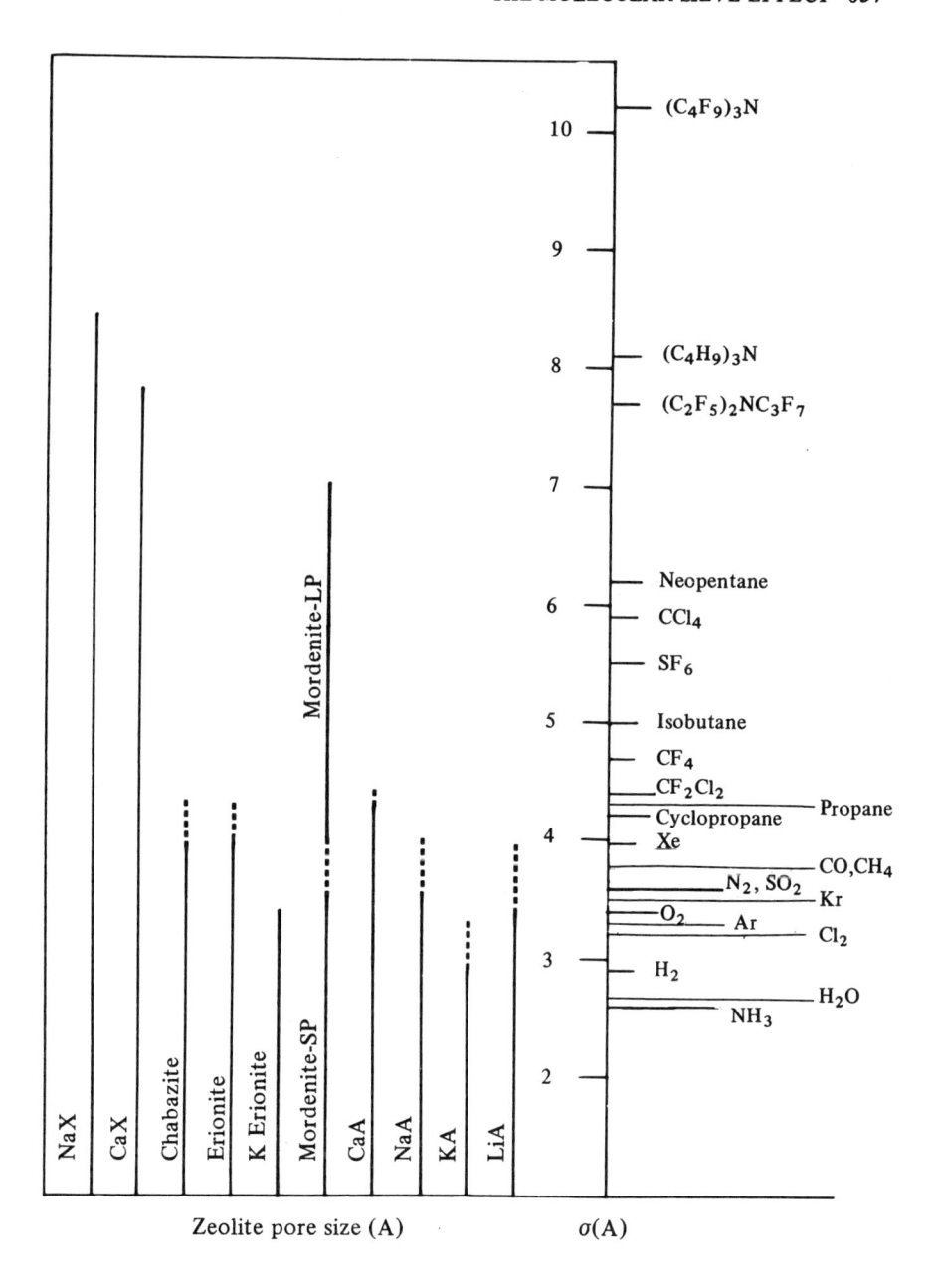

Figure 8.13 Chart showing a correlation between effective pore size of various zeolites in equilibrium adsorption over temperatures of 77° to 420°K (indicated by - - -), with the kinetic diameters of various molecules as determined from the L-J potential relation.

Distortion of the aluminosilicate lattice is also an important factor. When dehydrated, the aperture of calcium chabazite distorts from an essentially circular opening, 3.9 A in diameter, to an ellipsoidal shape, 3.7 x 4.2 A. During adsorption of polar molecules, it is known that the shape of the aperture changes *during* the adsorption process. With an ellipsoidal aperture measuring 3.6 x 4.8 A, erionite adsorbs xenon rapidly at − 78°C, however, chabazite adsorbs this molecule slowly. At ambient temperature, erionite adsorbs *n*-hexane but chabazite does not. Somewhat perplexing, however, is the observation that *n*-butane is *not* readily adsorbed by erionite at room temperature. It has been concluded that erionite has an apparent pore size slightly larger than that of chabazite. The minimum equilibrium diameter of *n*-paraffins may vary with length, i.e., the minimum cross-sectional diameter of *n*-hexane may be less than that of *n*-butane.

Zeolite NaX has an aperture 7.4 A in diameter. This compares well with a σ value of 8.1 A for $(C_4H_9)_3N$, adsorbed, and 10.2 A for $(C_4F_9)_3N$ which is not adsorbed. The adsorption behavior of CaX indicates an apparent pore size of 7.8 A. This slight difference is most likely due to distortion of the aluminosilicate framework (when dehydrated) rather than steric effects due to a "blocking" action by calcium ions.

Lithium A zeolite is another example of aperture contraction by cation effect. The unit cell dimension of LiA is about 12.0 A compared to 12.3 A for NaA, a 0.3 A overall contraction. The distance between centers of opposing oxygen atoms in the eight-ring in NaA is 7.0 A. The corresponding contraction of the ring due to lithium exchange amounts to 0.17 A and results in decreasing the *free* diameter of the aperture to 4.0 A. This may account for the experimental observation that O_2 is not adsorbed at −183°C by LiA. Some simple molecules are illustrated in Fig. 8.14.

The effect of temperature on molecular sieving is illustrated very dramatically by the behavior of oxygen, argon, and nitrogen on zeolite A at low temperatures. Figure 8.15 illustrates the adsorption isobars, that is, plots of the amount adsorbed versus temperature as determined from isotherms. Although the kinetic diameter of the nitrogen molecule is only 0.2 A larger than oxygen, this difference is apparently sufficient for exclusion, or an infinitely long rate of adsorption. Argon,

THE MOLECULAR SIEVE EFFECT 639

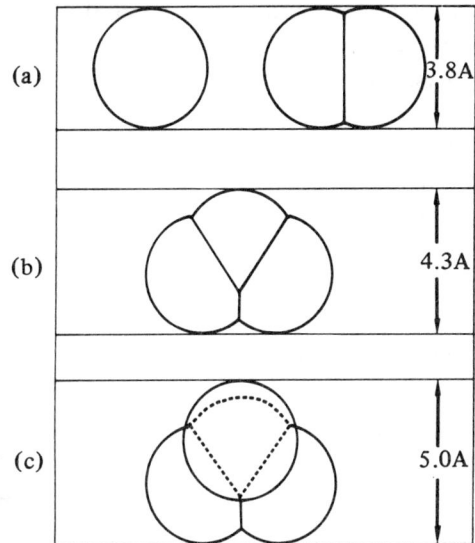

Figure 8.14 Illustration of the kinetic diameters of some simple molecules pertaining to zeolite adsorption: (a) methane, ethane; (b) propane; (c) isobutane.

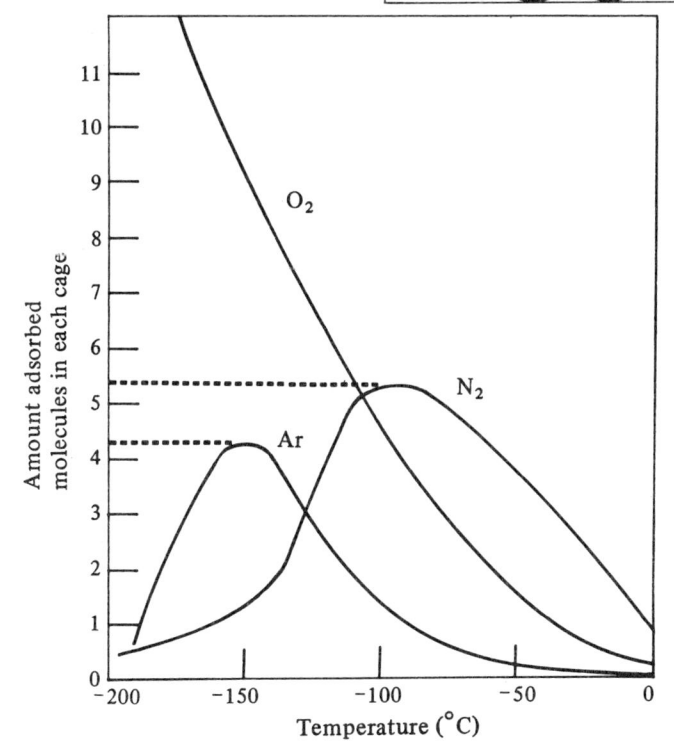

Figure 8.15 Gas adsorption isobars on zeolite NaA.

on the other hand, is intermediate in its behavior. Therefore, at the lower temperatures, nitrogen and argon diffuse into zeolite A with considerable difficulty and in any reasonable period of time do not attain adsorption equilibrium. One or both of two factors appear to be operative: (1) The process of activated diffusion that is a function of temperature or (2) the effect of increasing temperature on the vibration of the oxygen atoms in the zeolite crystal structure which surrounds the apertures. For example, a variation in vibrational amplitude of 0.1 to 0.2 A is to be expected over the temperature interval of 80° to 300°K. Consequently, with increasing temperature, the size of the aperture increases enough to permit the diffusion of nitrogen and argon (51).

At normal temperature, small, polar molecules such as NH_3 do not enter the sodalite, or β-cages of zeolites of A, X, or Y. Water, however, does occupy these smaller voids at ambient temperature. At elevated temperatures, it has been shown that ammonia does diffuse very slowly into the β-cages of zeolite X or Y (176). Below 373°K (100°C), the ammonia adsorption isotherm is reversible and equilibrium is attained rapidly. Above 473°K (200°C), activated adsorption of ammonia in the small cages takes place. At 300°C, equilibrium is reached in 5 hours. The quantity of ammonia adsorbed is equivalent to about 6 molecules per unit cell (less than one ammonia molecule/sodalite cage) or about 0.008 g of ammonia per g of outgassed zeolite Y at 300°C. The total ammonia adsorption is 20 molecules per unit cell or 0.027 g/g. In comparison, as shown in Table 8.3, at 700 torr and 100°C 0.14 g of ammonia are adsorbed per gram of zeolite (98 molecules per unit cell).

Another dramatic case of the true molecular sieve effect is exhibited by the adsorptive behavior of normal and branched-chain paraffin hydrocarbons on the calcium exchanged form of zeolite A. The replacement of four of the sodium ions in the type II sites of the zeolite A structure by two calcium ions permits the rapid diffusion of normal paraffin hydrocarbons into the zeolite's channels (Fig. 8.16). Normal paraffins may be separated from branch-chain paraffins on this zeolite. Consequently, the effective critical dimension of the molecule is the cross-sectional diameter. As given in Table 8.14, the kinetic cross-sectional diameter of a normal paraffin such as n-butane is 4.3 A which is very close to the known crystallographic diameter of the structure. The length of the paraffin hydrocarbon affects the rate of diffusion of the hydrocarbon into the zeolite structure.(See Section G.)

Important commercial separation processes involve the bulk separation of hydrocarbons. Other separations are based upon a partial molecular sieve action where, for example, both components of the binary system are adsorbed but at different rates and an activation energy for diffusion may be required.

In the case of the larger pore zeolites, such as zeolite X and mordenite, most molecules are rapidly adsorbed.

There are basically three methods for varying or tailoring the molecular sieve effect in zeolites. These variations of course are limited by the size of the zeolite apertures. These methods are discussed next.

Effect of Ion Exchange

As exhibited by the previous examples, changing the cation in a zeolite may effectively enlarge the pore openings by diminishing the cation population and/or a resiting of cations which are normally located near these openings. Variations in both a positive and negative direction may take place. In zeolite A divalent ion exchange opens the aperture to full diameter whereas exchange with a larger univalent ion diminishes the aperture size. Potassium ion exchange in zeolite A re-

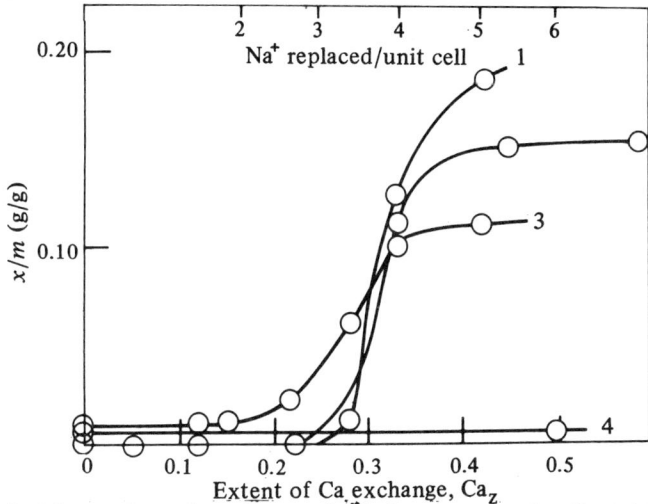

Figure 8.16 Effect of calcium exchange for sodium on the sieving properties of zeolite A (12): (1) nitrogen, 15 torr, $-196°C$; (2) n-heptane, 45 torr, $25°C$; (3) propane, 250 torr, $25°C$; (4) isobutane, 400 torr, $25°C$.

Table 8.15 Adsorption of 1,3,5-triethylbenzene (74)

Zeolite	x (cc/g)	Temp (°C)
NaX	0.178	20
	0.240	80
CaX	not adsorbed	20
	0.284	100

duces the effective adsorption pore size to the point where only small polar molecules are adsorbed (See Fig. 8.17).

This reduction does not occur gradually with increasing exchange but rather suddenly at a level of about 25% potassium exchange. Similarly, the increase in adsorption pore size by calcium exchange does not occur in a linear fashion but rather abruptly at a level of about 30% exchange (12). Similar effects are exhibited by the zeolite chabazite (73).

In contrast, zeolite X, when exchanged with calcium, exhibits a slight reduction in pore size. As was shown in Fig. 2.53, the adsorption of the tertiary amine is eliminated by calcium exchange.

The sieving effect by zeolite X was further confirmed by vapor phase adsorption of 1,3,5-triethylbenzene. NaX adsorbs this material at 20° while CaX does not. However, at higher temperatures of 100°C, extensive adsorption in CaX occurs (Table 8.15) (74).

The sieving behavior of synthetic and mineral mordenites is changed by the cation. As shown in Table 8.16, the sodium mineral has a smaller pore size than the synthetic large port variety. In both instances, calcium exchange reduces the pore size so that molecules of n-paraffins, kinetic diameter = 4.3 A, are not adsorbed. The reason for this is not clear since calcium exchange reduces the cation population in the channels. Ammonium exchange and conversion to the hydrogen form results in the adsorption of isobutane and benzene. Therefore, as in CaX, calcium positions must be such as to interfere with the intracrystalline diffusion of these larger molecules.

The type of cation exerts a pronounced effect on the sieving character of synthetic offretite-type zeolites (Chap. 4) (76). As synthesized, with a unit cell content of $K_2(TMA)_{1.9}[Al_{3.9}Si_{14.1}O_{36}] \cdot 6.9 H_2O$, the zeolite contains two potassium ions and two tetramethylammonium ions in each unit cell. The cation distribution was proposed to be one with potassium ions in the D6R units and ϵ-(cancrinite) cages with

THE MOLECULAR SIEVE EFFECT 643

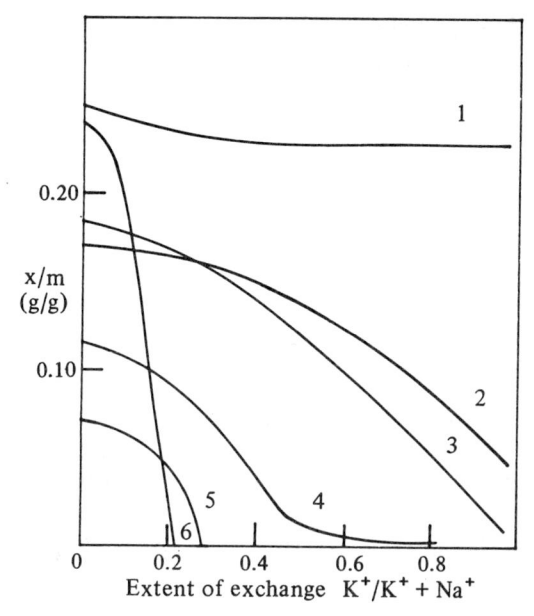

Figure 8.17 Effect of potassium exchange for sodium on the sieving properties of zeolite A (12): (1) water at 4.5 torr, 25°C; (2) methanol at 4 torr, 25°C; (3) carbon dioxide at 700 torr, 25°C; (4) ethylene at 700 torr, 25°C; (5) ethane at 700 torr, 25°C; (6) oxygen at 700 torr, −183°C.

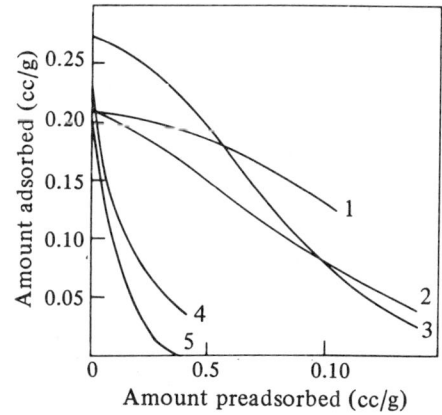

Figure 8.18 Effect of pre-adsorption on the adsorptive properties of zeolite A (12). (1) n-Butane on CaA with pre-adsorbed water; (2) n-butane on CaA with pre-adsorbed CH_3NH_2; (3) N_2 on CaA with pre-adsorbed CH_3NH_2; (4) O_2 on NaA with pre-adsorbed water; (5) O_2 on NaA with pre-adsorbed ammonia.

Amount pre-adsorbed is in units of cc of liquid of normal density per gram activated adsorbent. Amount adsorbed is in cc of liquid of normal density per gram of activated adsorbent.

Table 8.16 Effect of Cation on Adsorption in Mordenite (75)

		Amount Adsorbed (wt % at 25°C)					
		Synthetic Type[a]			Mineral[b]		
Adsorbate	p(torr)	Na	Ca	NH$_4$	Na	Ca	NH$_4$
n-butane	740	6.9	0.8	7.7	5.5	2.2	7.3
Isobutane	750	6.5	0.6	7.9	1.5	0.5	7.9
Benzene	68	6.8	1.1	7.8	2.2	0.3	7.2

[a]Synthesized "large-port" mordenite, Na form.
[b]Sedimentary mineral with some quartz and montmorillonite impurity-major cations are Na, Ca, Mg, and K.

tetramethylammonium ions in the gmelinite cages and main c-axis channels. Adsorption studies were conducted on the synthetic zeolite and modified forms prepared by removal of the tetramethylammonium ions (by oxidation in air at 650°C) and by exchange with ammonium ions followed by air oxidation at 550°C. This latter treatment resulted in a cation composition of one potassium and three hydrogens per unit cell. A third modification prepared by potassium exchange resulted in a composition of three potassium ions and one tetramethylammonium ion in each unit cell. Adsorption measurements gave the following results for each cation composition.

1. K$_2$(TMA)$_2$ —Adsorbs only H$_2$O and CO$_2$. The large TMA cations in both the gmelinite cages and large channels reduce the void volume.
2. K$_3$TMA—Adsorbs some n-hexane indicating replacement of the TMA ions in the large channels by potassium ions.
3. K$_2$H$_2$ —Adsorbs cyclohexane indicating that potassium ions in the large channels reduce the effective pore size.
4. KH$_3$ —Adsorbs cyclohexane and m-xylene (kinetic diameter 6.8 A). 1, 3, 5-trimethylbenzene (kinetic diameter = 7.6 A) was not adsorbed.

Decreasing the size of the cations by exchange with sodium and lithium increased the rate of n-hexane adsorption. The main c-axis channels in offretite have a free diameter of 6.3 A as determined from the crystal structure.

Preadsorption

A second method for altering the molecular sieving effect of a zeolite is by preadsorption of polar molecules. If small amounts of water or ammonia are preadsorbed on a dehydrated zeolite the adsorption of a second adsorbate such as oxygen is drastically reduced (see Fig. 8.18) (12, 77). It is assumed that the strong reaction between the zeolite cation and the dipole moment of ammonia or water produces a diffusion block by clustering of water or ammonia molecules about the cation in the channels. Similar results have been observed with chabazite and mordenite (78).

In the case of a large-pore zeolite such as zeolite X the effective pore size may be controlled by the formation of a stable inorganic complex. If copper is exchanged into the cation positions in the zeolite and subsequently treated with pyridine, a very stable pyridine-cation complex is formed. The adsorption of gases and vapors on this material are evidence of a considerable decrease in the pore size due to blocking by the cationic complex (Fig. 8.19) (79).

Pore Closure

A third method for varying the molecular sieve behavior of a zeolite is by the mechanism of hydrolytic pore closure (see Chap. 6). This technique reduces the effective pore size for critical molecules (95).

By steaming NaA at 550°C for 25 minutes the slow adsorption of the refrigerant CHF_2Cl is eliminated. Decomposition of the refrigerant CHF_2Cl does not occur since it is not adsorbed and NaA is used in refrigerant drying. More drastic steam treatment results in the elimination of all oxygen adsorption capacity at low temperature.

D. THE HEAT OF ADSORPTION

All adsorption processes involving physical adsorption are exothermic, that is, they evolve heat. For adsorption to take place the free energy change must be negative. From the thermodynamic relation

$$\Delta G = \Delta H - T\Delta S \tag{8.18}$$

the change in enthalpy, ΔH, is negative since the change in entropy, ΔS, is negative because the adsorbate molecules are in a more ordered configuration. There are three terms referring to the heat of adsorption. These are:

1. The isothermal integral heat of adsorption. This is the total heat

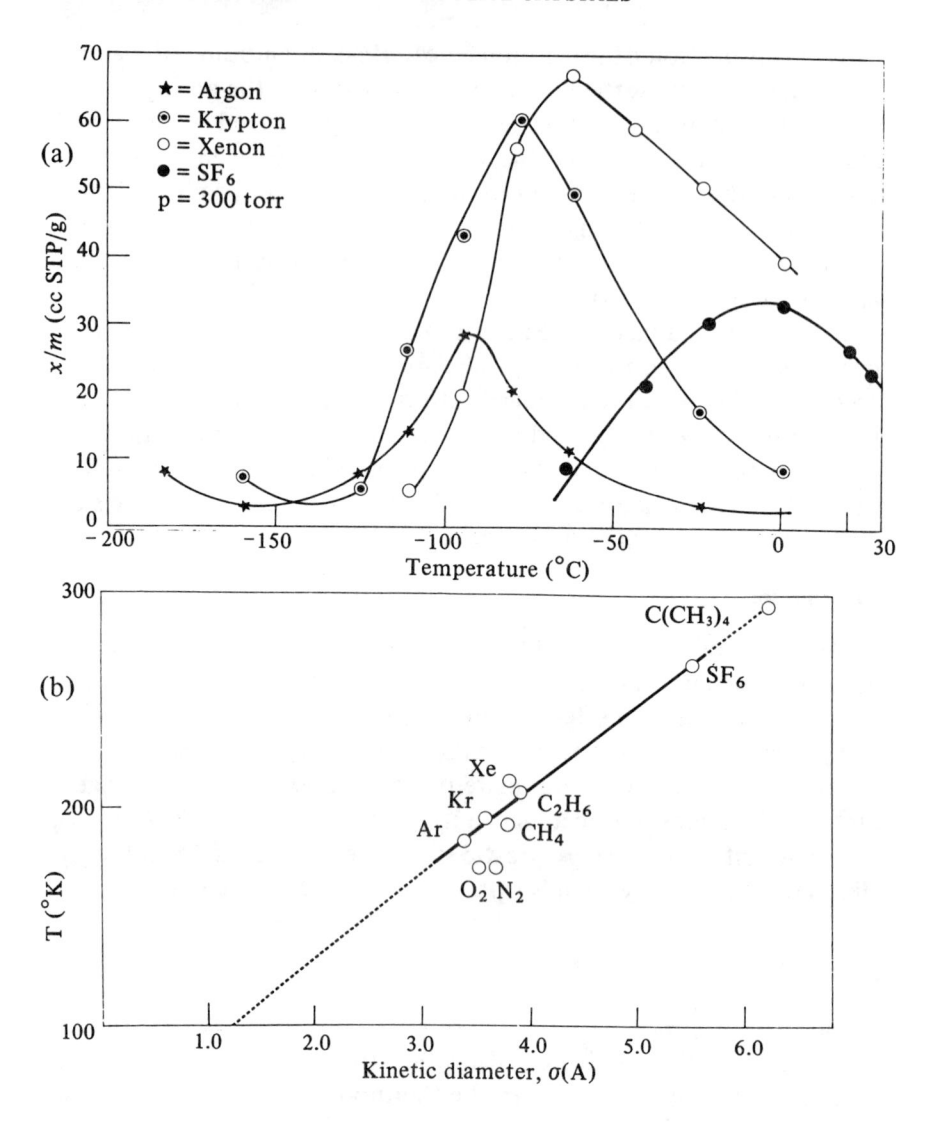

646 ADSORPTION BY DEHYDRATED ZEOLITE CRYSTALS

Figure 8.19 Effect of pyridine adsorption on the sieving property of copper-exchanged zeolite Y. The CuY was exchanged to about 80% and activated in vacuum at 350°C. Pyridine was admitted at room temperature and then reactivated at 250° in vacuum to give residual loading of 13 wt.%. This corresponds to 22 Cu^{2+}/unit cell and 21 pyridine/unit cell. (a) Apparent isobars for rare gases and SF_6; (b) Correlation of isobar maxima with kinetic diameters of molecules. The linear relation is expressed by the equation:

$T = m\sigma + b$

where T is in °K, m = 39°K/A, b = 52.5°K.

involved in the adsorption process from zero adsorbate loading to some final adsorbate loading at a constant temperature.

2. The differential heat of adsorption, $\overline{\Delta H_1}$. This is the change in integral heat of adsorption with a change in adsorbate loading. It may be defined by the following equation:

$$\overline{\Delta H_1} = (\overline{H_1} - \tilde{H}_g) \tag{8.19}$$

where $\tilde{H}_g$ is the molar enthalpy of the adsorbate gas and $\overline{H}_1$ is the partial molar enthalpy of the adsorbate. The differential heat of adsorption is dependent upon pressure, temperature, and adsorbate coverage or loading.

3. The isosteric heat of adsorption is derived from adsorption isosteres. It is obtained from a plot of the log of the pressure vs. the reciprocal of the absolute temperature, $1/T$, at a constant adsorbate loading by means of the Clausius-Clapyron equation (Fig. 8.20). From pairs of isotherms, q_{iso} may be calculated by:

$$q_{iso} = 4.58 \frac{T_1 T_2}{T_2 - T_1} \log \frac{p_2}{p_1} \tag{8.20}$$

The isosteric heat is related to the differential heat of adsorption by

Figure 8.20 Plot of $\log_{10} p$ vs $(1/T)$ for adsorption of N_2 on CaX at -98 to -137°C, with $x = 120$ cc/g. $Q_{iso} = 4700$ cal/mole.

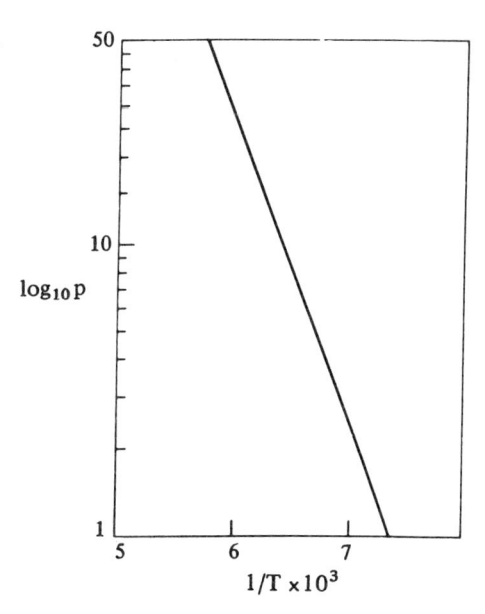

$$-q_{iso} = \overline{\Delta H_1} \tag{8.21}$$

It can be demonstrated that the isosteric and differential heats are identical. One can also obtain the integral heat of adsorption from isosteric heats. Isosteric adsorption heats which are obtained from a family of adsorption isosteres are plotted as a function of the adsorbate loading and by a subsequent graphical integration the isothermal integral heat of adsorption is evaluated.

The differential entropy of adsorption can be determined from the logarithmic plots since it is equal to the product of the absolute temperature and the isosteric heat of adsorption. The differential molar entropy of the adsorbate is given by

$$\overline{S}_a = \tilde{S}_g^o + R \ln(p^o/p) + \overline{\Delta H}/T \tag{8.22}$$

where $\tilde{S}_g^o$ is the molar entropy of the gas at standard pressure p^o, and temperature T, assuming a perfect gas. Knowing $\tilde{S}_g^o$ and the measured $\overline{\Delta H}$, one can calculate $\overline{S}_a$. The molar entropy varies with the amount adsorbed, as shown for NH_3 on NaX in Fig. 8.21.

Zeolite Adsorbate Interaction Energies

At low levels of adsorption, the initial adsorption heat, $\overline{\Delta H_1}$, has been related to several component interaction energies. These include dispersion and short range repulsion energies, polarization energy and additional components attributed to electrostatic interactions. These are caused by the interaction of the local electrostatic fields in the zeolite with molecules possessing permanent dipole moments or quadrupole quadrupole moments. The magnitude of these interactions has been estimated and related to $\overline{\Delta H_1}$.

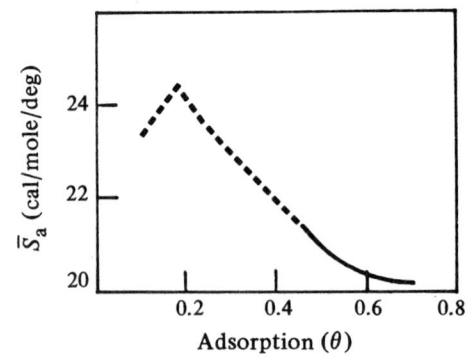

Figure 8.21 Plot of differential molar entropy of adsorption, $\overline{S}$, for NH_3 adsorbed on NaX zeolite at 100°C (100).

The dispersion and repulsion energy terms are found universally but the electrostatic terms depend on the characteristics of the adsorbate and adsorbent. At 0°K, $\overline{\Delta H_1}$, is given by

$$\overline{\Delta H_1} = \phi_D + \phi_R + \phi_P + \phi_{F-Q} + \phi_{F-\mu} \tag{8.23}$$

The dispersion and repulsion energies are given by

$$\phi_D = -\frac{A}{r^6} \tag{8.24}$$

$$\phi_R = \frac{B}{r^{12}} \tag{8.25}$$

where A and B are constants and r is the separation distance. Methods for estimating A and B have been given by Barrer and Gibbons (96, 97).

The polarization term is given by

$$\phi_P = -\frac{\alpha C^2}{2r^4} = \frac{\alpha F^2}{2} \tag{8.26}$$

where C is the charge, α is the polarizability and F the field. The electrostatic field-dipole interaction has been estimated by:

$$\phi_{F-\mu} = -\frac{C\mu}{r^2} \tag{8.27}$$

where μ is the dipole moment. The interaction energy for a single cation quadrupole pair is given by:

$$\phi_{F-Q} = \frac{CQ}{4r^3}(3\cos^2\theta - 1) \tag{8.28}$$

where C is the charge on the ion and θ is the angle between the axis of the quadrupole and the line between centers of the two species. The quantity θ_{F-Q} varies with the quadrupole orientation and is a maximum at $\theta = 0$.

$$\phi_{F-Q} = \frac{CQ}{2r^3} \tag{8.29}$$

Physical constants of various gases used in making these estimates are given in Table 8.17. The initial heats for adsorption of nitrogen and argon on zeolite X and a series of cation exchanged forms were analyzed in terms of dispersion, repulsion, polarization, and quadrupole interactions (98).

Table 8.17 Physical Constants of Some Adsorbate Gases

Gas	B.P. (°C)	Crit. Temp (°C)	Dipole (esu × 10¹⁸)	Quadrupole (Å³)	Polarizability (Å³) ∥ / ⊥	Ion. Pot. (volts)	Length (Å)	Width (Å)	Kinetic Diameter σ (Å)
Ar	−185.7	−122.4	—	—	1.6	15.7	1.92	1.92	3.4
Kr	−152.9	−62.6	—	—	2.1	13.9	1.98	1.98	3.6
Xe	−107.1	16.5	—	—	3.9	12.1	2.18	2.18	3.96
O₂	−183.0	−118.8	—	0.10	1.2 / 2.35	12.5	2.0	1.4	3.46
N₂	−195.8	−147.1	—	0.31	1.4 / 2.38	15.5	2.1	1.5	3.64
CH₄	−161.4	−82.5	—	—	2.6	14.5	2.0	2.0	3.8
CO	−192.0	−139	0.12	0.33	1.6 / 2.6	14.3	2.1	1.8	3.76
C₂H₄	−103.7	9.7	—	0.48	3.5	12.2	2.5	2.2	3.9
C₂H₆	−88.6	32.1	—	0.27	3.9	12.8	2.6	2.5	3.8
CO₂	−78.5	31.1	—	0.64	1.9 / 4.10	14.4	2.6	1.8	3.3
C₃H₆	−47.6	92.0	0.35	0.10	3.5	12.2	3.4	2.2	
C₃H₈	−42.3	96.8	—	—	5.0	12.8	3.3	2.5	4.3

Adsorption heats for nitrogen are larger and the zeolites show greater energetic heterogeneity toward nitrogen. For argon the main interaction energies are due to dispersion and polarization forces whereas for nitrogen the quadrupole interaction must be included and was estimated to be 2.5 and 1.5 kcal/mole in the lithium and sodium forms.

Heats of adsorption of ammonia on various cation forms of zeolite X (Li, Na, K, Rb and Cs) must include electrostatic components due to field-dipole interaction. Although calculated energies agreed with the observed sequence of initial adsorption heats, quantitative agreement was not obtained (100). In the case of CO_2 adsorption on type X zeolites, the quadrupole energy is the most important component of the CO_2 −zeolite interaction (96).

Experimentally it has been observed that the isoteric heat for a particular adsorbate varies with the type of exchange cation (Fig. 8.22a). A pronounced effect for the adsorption of nitrogen on zeolite X is shown in Fig. 8.22b. Isosteric heats of adsorption of nitrogen at 178°K are shown for the potassium, sodium, and lithium exchanged forms of zeolite X (97). The highest initial isosteric heat is shown by lithium X. Consequently, lithium X has the highest affinity for nitrogen at low loadings. In contrast, potassium X shows no change in isosteric heat with increasing adsorption and the surface is energetically homogeneous.

The variation in isosteric adsorption heats with loading is exemplified in Fig. 8.23 (65) for the adsorption of the gases argon, nitrogen, and carbon dioxide on chabazite. In the case of argon, the heat does not vary greatly with increasing adsorption. After a high initial heat the adsorption heat decreases for nitrogen and essentially levels out. For carbon dioxide two effects are shown: (1) a high initial adsorption heat characteristic of the quadrupole interaction followed by (2) a minimum and a maximum with increased loading. The increase in isosteric heat is due to interaction between the adsorbate molecules. The magnitude of the increase in isosteric heat in this instance is interpreted as due to a strongly directional interaction between four carbon dioxide molecules within one cage of the structure; some of the molecules are oriented in a T formation. Maxima in isosteric heat curves are shown by other adsorbate-zeolite combinations.

A particular example is the adsorption of argon on the zeolite

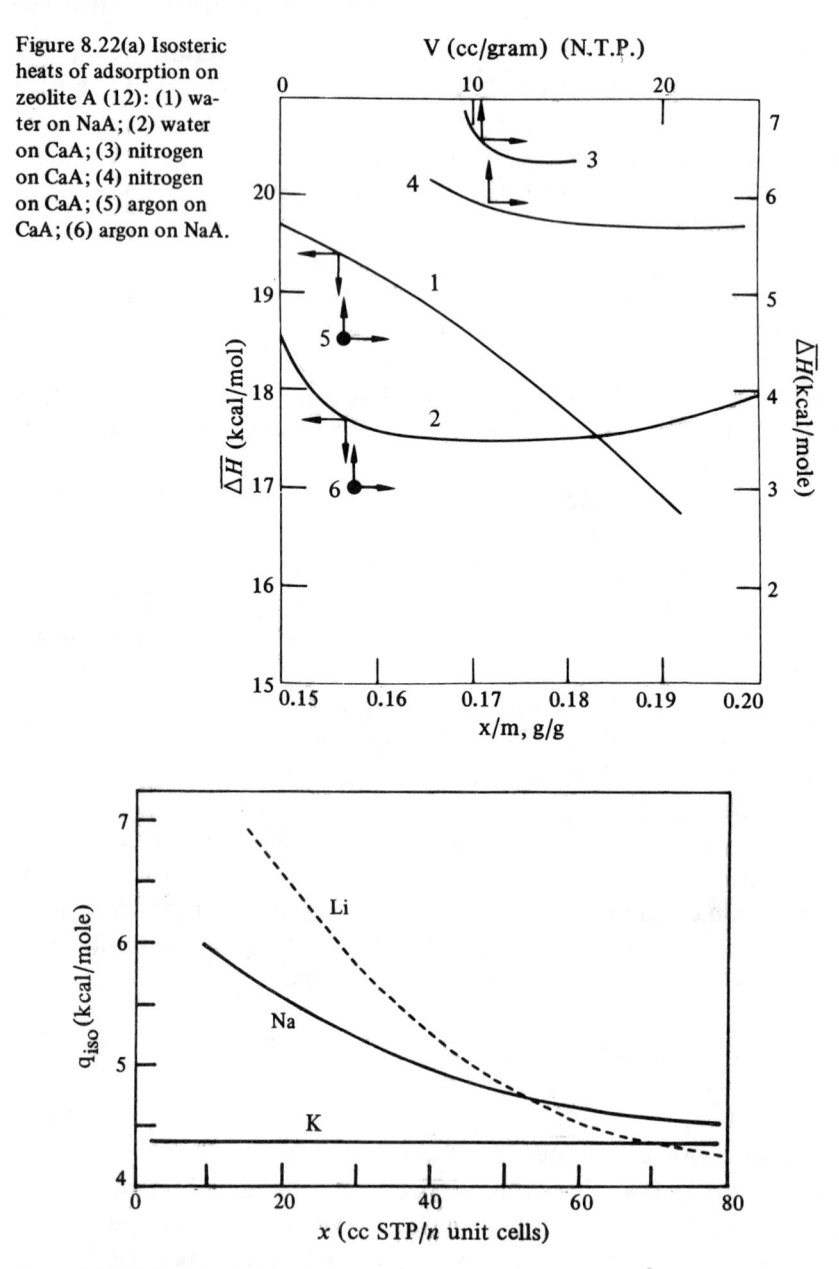

652 ADSORPTION BY DEHYDRATED ZEOLITE CRYSTALS

Figure 8.22(a) Isosteric heats of adsorption on zeolite A (12): (1) water on NaA; (2) water on CaA; (3) nitrogen on CaA; (4) nitrogen on CaA; (5) argon on CaA; (6) argon on NaA.

Figure 8.22(b) Isosteric heats of adsorption, q_{iso}, for nitrogen at 178°K on Li, Na and K exchanged X as a function of amount adsorbed; n is the number of unit cells per gram of hydrated sodium zeolite (97).

Figure 8.23 Heats of adsorption in chabazite (mineral) (65): (1) argon; (2) oxygen; (3) nitrogen; (4) carbon monoxide; (5) carbon dioxide.

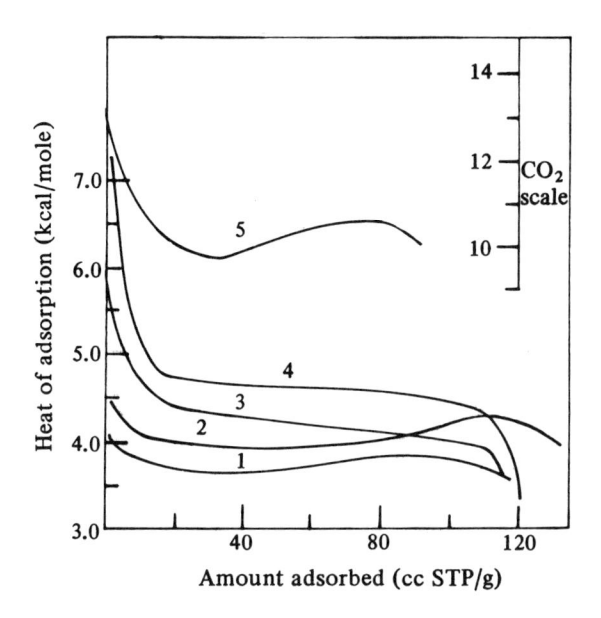

chabazite. The isosteric heat as a function of adsorption loading behaves in a similar manner, that is, the initial heat is high but drops off to a minimum then rises to a maximum as the pore volume becomes filled. In this instance, the maximum is the result of mutual interaction between the argon molecules contained in a rigid cage-type structure (20, 102).

Information on the magnitudes of the isosteric heats of adsorption and their variation with the level of adsorbate loading or coverage is useful (Tables 8.18 thru 8.23). In general, isosteric heats are given for two different cation-exchanged forms at a low level of adsorbate loading or essentially zero adsorbate loading (the initial isosteric heat) and additionally at a higher degree of adsorbate loading. This indicates the degree of homogeneity of the interior surface of the zeolite relative to the adsorbate concerned. The adsorbates are grouped in terms of the permanent gases, polar molecules such as water, ammonia, hydrocarbons, and miscellaneous inorganic materials such as iodine.

Detailed information concerning the entropies and free energies of adsorption is given in the References. However, the isosteric or differential heat, and accordingly the integral heat, are the more useful quantities to illustrate.

654 ADSORPTION BY DEHYDRATED ZEOLITE CRYSTALS

Table 8.18 Adsorption Heats: Zeolite A (Differential Heat, $-\overline{\Delta H}$ kcal/mol)

Adsorbate	NaA θ	$-\overline{\Delta H}$	θ	$-\overline{\Delta H}$	CaA θ	$-\overline{\Delta H}$	θ	$-\overline{\Delta H}$	Footnote	Data Source	Other References
H_2O	0.1	30	0.8	17.7	0.8	18			a	12, 133	125–128
CO_2	0.1	11	0.4	9.5	0.1	12.5	0.6	8	b	134	61, 140
NH_3					0.1	22	0.9	14	c	135	130
N_2	15	6.5			15	5.7			d	12	61, 131
Ar	4	3			4	4.5			d	12	61, 131
CH_4					0.1	5.2			e	136	61, 132
C_2H_6					0.1	6.6	0.8	9.5	f	136	61, 81, 132
C_3H_8					0.1	8.2	0.8	10.5	g	136	
n-C_4H_{10}					0.1	10.2	0.8	13	h	136	83
n-C_{12}					0.1	21			i	83	
n-C_{14}					0.8	28			i	83	
I_2					0.1	16	0.8	17	j	86	

[a] Reference 133 includes heats of immersion measured on the Li, K, Rb, Cs, and Mg exchanged zeolite.
[b] Temperature range was 30° - 190°C.
[c] Temperature was 23°C.
[d] θ is in cc STP/g, temperature was −78° to 0°C.
[e] Temperature, −90° to 0°C.
[f] Temperature, −40° to 72°C.
[g] Temperature, 0° to 125°C.
[h] Temperature, 50° to 225°C.
[i] Temperature, 317° to 390°C. For n-alkanes, $-\overline{\Delta H} = 5.44 + 1.55n$ where n = number of carbon atoms.
[j] Temperature, 120° - 300°C.

Table 8.19 Adsorption Heats for Zeolite X (Differential Heat, $-\overline{\Delta H}$ kcal/mole)

Adsorbate	θ	NaX $-\overline{\Delta H}$	θ	$-\overline{\Delta H}$	θ	CaX $-\overline{\Delta H}$	Footnote	Data Source	Other References
H$_2$O	0.1	22.7	0.8	16.9	0.4	14		133, 137	127, 129, 138, 139
NH$_3$	0.1	16	0.7	12	0.2	13	a	100	137, 138
CO$_2$	0.1	11	0.5	8.5	20	8	b	96, 140	134
O$_2$		3.3					c	101	
Ar	0.1	2.8	0.1	2.8			c	85, 97	
N$_2$	0	4.47			0.1	2.8	c	101	85
Kr	0	3.54					c	101	90
CO	0	5.63					c	101	
CH$_4$	0	4.21					c	101	
C$_2$H$_6$	40	7.3					d	85	81, 87, 90
C$_3$H$_8$	40	11.0					d	85	145
n-C$_4$H$_{10}$	40	14.5					d	85	
n-C$_5$H$_{12}$	40	17.2					d	85	144
n-C$_6$H$_{14}$	37	20.9					d	85	142, 143
Benzene	50	17.5	0	28			d, g	88, 141	142
Toluene	45	18.4					d, g	88	
Cyclopentane	49	18.5					d	88	
C$_2$H$_4$	0	9.2					d	87	
I$_2$	0.1	29	0.8	18			e	86	
P	→1	29.6					f	82	
S	→1	31.5					f	82	

[a] Temperature, 100°C; additional values in Ref. 100 for Li, K, Rb, Cs, Sr, and BaX.
[b] Temperature, 30° to 150°C. For CaX, θ is in cc STP/g.
[c] Temperature, 0° to 90°C, θ in cc STP/g.
[d] Temperature, 0° to about 60°C, θ in cc STP/g.
[e] Temperature 120° to 300°C.
[f] Temperature, 250° - 325°C.
[g] Pulse flow method on ion exchanged forms of zeolite X; see Ref. 30.

656 ADSORPTION BY DEHYDRATED ZEOLITE CRYSTALS

Table 8.20 Adsorption Heats for Zeolite Y (Differential Heat, $-\overline{\Delta H}$, kcal/mole)

Adsorbate	NaY θ	$-\overline{\Delta H}$	θ	$-\overline{\Delta H}$	CaY θ	$-\overline{\Delta H}$	θ	$-\overline{\Delta H}$	LaY θ	$-\overline{\Delta H}$	Footnote	Data Source	Other Ref.
H_2O	0.03	19.5	0.3	16	0.03	21	0.3	16	0.03	22	a	146	
NH_3	0.03	13.9			0.03	15.7			0.03	14.5	a, b	147	
Ar	~0	1.9			~0	2.6					c	89	
N_2	~0	4.24			~0	6.0					d	89, 101	
CO	~0	5.4			~0	9.3					c	89, 91	
Kr	~0	3.64									d	101	89
CO_2	~0	7.3			~0	11.0					c	89	
C_2H_4	~0	7.6			~0	9.4					c	89	
C_2H_6					~0	7.3					c	89	
n-Hexane	~0	11.0			~0	9.5			~0	9.0	e	143	
Propylene	~0	9.0			~0	13.3			~0	15.0	e	143	
Benzene	~0	17.0			~0	22.0			~0	26.0	e	143	
Cyclohexane	0.1	13	0.8	16	0.1	14	0.8	14			f	148	148

[a] Temperature 20°C. Microcalorimeter used. Si/Al$_2$ ratio of the Y zeolite = 5.1 and degree of Ca exchange = 84%, La^{3+} exchange = 70%.
[b] Degree of Ca exchange = 68%. Data also given for HY.
[c] CaY is 87% exchanged. Data also given for ZnY, SiO$_2$/Al$_2$O$_3$ = 4.9.
[d] Zeolite Y:SiO$_2$/Al$_2$O$_3$ = 5.27. Additional data for SiO$_2$/Al$_2$O$_3$ ratios of 2.36 to 4.13.
[e] Zeolite Y:SiO$_2$/Al$_2$O$_3$ = 4.2. Data for Mg, Sr, Ba, Cd, and NdY also given. Cation composition not given.
[f] Degree of Ca exchange = 90%. Calorimetric heats given.

Table 8.21 Adsorption Heats for Chabazite (Differential Heat, $-\overline{\Delta H}$, kcal/mole)

Adsorbate	Chabazite Mineral $-\overline{\Delta H}$	θ	$-\overline{\Delta H}$	H-Chabazite θ	$-\overline{\Delta H}$	θ	$-\overline{\Delta H}$	Footnote	Data Source	Other Ref.
H_2	~0	4		~0	2	80	1.5	a,b	16, 17	
Ar	~0	6		0	3.5	80	3.5	a,b	16, 17	
N_2	~0	4.1	80	3.8	0	3.8	80	4	a,b	16, 65	65
O_2	~0	6.1	80	4.6	0	5			a	16, 17	
NH_3	~0	4.5	80	4.0	0	3.8	80	3.4	a,b	16, 65	149
	3	28.6							a,b	17	
CO_2	0	8.2	80	6.5	0	9	80	8	a,b	16, 65	134
H_2O	0.1	30	0.8	18.9					c	133	
CH_4					0	4.5	80	5	a,b	15	
C_2H_6					0	7.0	80	6.2	a,b	15	
C_3H_8					0	8.8	50	9.4	a,b	15	
n-C_4H_{10}					0	10.7	40	9.0	a,b	15	
CO	0	7.1	80	4.5						65	

[a] Mineral (Nova Scotia) exchanged by NH_4^+ and treated in O_2 for two days at 633°K to give hydrogen chabazite $H_{3.35}Al_{3.35}Si_{8.65}O_{24}\cdot 9.26\ H_2O$. Data also given for Kr and Ne as well as isotherms.
[b] θ is in cc STP/g.
[c] Heat of immersion.

658 ADSORPTION BY DEHYDRATED ZEOLITE CRYSTALS

Table 8.23 Adsorption Heats for Mordenite-Zeolon (Differential Heat, $-\overline{\Delta H}$ kcal/mole)

Adsorbate	θ	Na $-\overline{\Delta H}$	θ	$-\overline{\Delta H}$	θ	H $-\overline{\Delta H}$	θ	$-\overline{\Delta H}$	Footnote	Data Source
N_2	0.1	6.5	0.8	4.2	0.1	5.2	0.8	3.0	a	21
O_2	0.1	4.6	0.8	3.5	0.1	4.3	0.8	3.2	a	21
Ar	0.1	4.5	—	—	0.1	4.3	0.8	2.5	a	21
H_2	0.3	2.0	0.8	1.6	0.1	2.0	0.5	1.5	a	21
Kr	—	—	0	5.8	0	5.8	—	—	a	21
Xe	—	—	—	—	0	8.4	—	—	a	93
CO_2	0	14.7	40	10.7	0	10.6	40	7.6	b	93
H_2O	100	15	—	—	100	12	—	—	c	93

[a] Composition of H-Zeolon given as $(H_3O)_{7-n}Na_nAl_7Si_{40}O_{92}(OH)_4 \cdot 15\ H_2O$ where $0.5 < n < 1$. The $(OH)_4$ group replaces one AlO_4 tetrahedron due to dealumination.

[b] θ is in cc STP/g.

[c] θ is in mg/g of outgassed adsorbent.

HEAT OF ADSORPTION 659

Table 8.22 Adsorption Heats for Zeolite L (Differential Heat, $-\overline{\Delta H}$, kcal/mole)

Adsorbate	Na, KL θ	Na, KL $-\overline{\Delta H}$	θ	HL $-\overline{\Delta H}$	θ	KL $-\overline{\Delta H}$	θ	KL $-\overline{\Delta H}$	Footnote	Data Source	Other Ref.
O_2	~0										
Ar	~0										
CH_4	~0	4.8	→1	—	~0	4.4	40	3.8	a, b	14, 15	150
C_2H_6	~0	7.2	→1	7.4	~0	6.9	30	6.3	a, b	14, 15	150
C_3H_8	~0	9.3	→1	9.8	~0	8.75	20	8.5	a, b	14, 15	150
n-C_4H_{10}	~0	11.3	→1	12.3	~0	10.7	15	9.5	a, b	14, 15	150
iso-C_4H_{10}	~0	11.7	→1	11.9					a	14, 15	150
neo-C_5H_{12}	~0	12.0	→1	13.3					a	14, 15	150
CO_2	1	10.5	5	11	~0	8.1		9	5	8.7	c, d	151
NH_3	1	16	7	10				13	7	10	c, d	151

[a] KL has compositions $K_2O \cdot 0.09\ Na_2O \cdot Al_2O_3 \cdot 6.2\ SiO_2 \cdot 5.0\ H_2O$. The HL was obtained from an 84% NH_4^+ exchanged zeolite L by treatment with O_2 at 633°K for 24 hrs. Isotherms also given in refs. 14 and 15.
[b] Initial $-\overline{\Delta H}$ follows relation $-\overline{\Delta H} = 2.5 + 2.0n$ for HL, where n = number of carbon atoms and $-\overline{\Delta H} = 2.6 + 2.2n$ for KL and $-\overline{\Delta H}$ is in kcal/mole.
[c] Compositions of zeolite L given as: Na, KL = $Na_{3.8}K_{5.2}Al_9Si_{27}O_{72}$. HL = $K_5H_{2.7}Al_{7.7}Si_{27}O_{69.4}$. KL = $K_9Al_9Si_{27}O_{72}$.
[d] θ in molecules per unit cell.

E. CHARACTER OF ADSORBED PHASE IN ZEOLITES

Techniques employed in characterizing the adsorbed molecules in zeolites include:

1. Spectroscopic methods—infrared specta (ir), raman spectra, reflectance spectra in the visible and ultraviolet regions, electron spin resonance (esr), and nuclear magnetic resonance (nmr).
2. X-ray crystal structure studies to determine the location of adsorbed molecules and their interaction with the internal zeolite surface atoms and cations.
3. Thermochemistry—Heats and entropies of adsorption as discussed in Section D.

Spectroscopic Methods

Results of infrared studies of adsorbed molecules were reviewed by Yates (103). These include carbon dioxide, carbon monoxide, ammonia, and various organic molecules. Infrared characterization of water and hydroxyl groups are discussed in Chaps. 5 and 6.

Carbon monoxide adsorbed on zeolites X and Y containing various cations exhibits two types of adsorption. One type is characteristic of the divalent forms and is characterized by a higher (than the gas phase) frequency of between 2217 and 2178 cm^{-1}. The origin of this band was attributed to polarization of the CO molecule by the electric field of the zeolite. The appearance of this high-frequency band was attributed to the location of divalent ions in surface sites II and III and it has been used to determine the presence of surface cations (104).

The infrared spectra of benzene and toluene adsorbed on a cobalt exchanged zeolite Y were used to interpret the nature of the zeolite-adsorbed molecule interaction. Changes in the spectra from that of the liquid were explained by the conclusion that the cations and surface oxygens of the zeolite interact with the π-orbital of the benzene ring which is oriented parallel to the internal surface (105).

The interaction of ethylene with various cation-exchanged forms of zeolite X has been studied by infrared spectroscopy (106). Heats of adsorption of ethylene measured calorimetrically on silver, cadmium, barium, calcium, and sodium X zeolite were related to their ir spectra. Heats of adsorption varied from 8.6 for NaX to 18.1 kcal/mole for

AgX. Ethylene could not be desorbed from AgX by evacuation at 200°C but it was readily desorbed from NaX by evacuation at room temperature. Ethylene is strongly adsorbed on CdX and AgX. It was concluded that, except in AgX, the adsorbed ethylene molecules were freely rotating. A bonding scheme to explain this difference was proposed and involves back donation of silver d electrons into the ethylene π^*-orbitals. In a further study, it was concluded that strongly held ethylene molecules interact with silver ions located on the walls (site III) of the supercages (4.4 molecules per cage) (107). Weakly held molecules (3.6 per cage) interact with silver ions in site II. Cation positions of silver ions in zeolite X are not known. However, in zeolite Y there is full occupancy of site II by silver ions.

Adsorption of nitrogen oxides has been intensively studied by ir spectroscopy and esr. The interest in nitrogen oxides is related to the use of zeolites in removing nitrogen oxides from gases such as the effluent from nitric acid plants. (See Section I.)

Nitrous oxide adsorbs on zeolite NaA. From infrared studies, it was concluded that the N_2O molecule moves in the α-cage at 40°C, but at low temperature, -50°C, the molecules remain near a sodium cation in site I (108).

When adsorbed on zeolites, nitric oxide was reported to disproportionate (109). The adsorption and reaction of NO on zeolite Y has been studied by infrared spectroscopy (110).

$$4NO = N_2O + N_2O_3$$

Studies at room temperature on CaY, NaY, hydrogen Y, and decationized Y identified N_2O, NO_2^+, N_2O_3, NO_3^-, a nitrito complex, and nitrite as adsorbed species. The final products were found to be N_2O and NO_2.

At low temperatures, three different adsorbed species were formed; N_2O_2, NO, and disproportionation reaction products. No reaction was found at -190°C. Adsorbed N_2O_2 was a key intermediate.

Nitric oxide was adsorbed on NaY, BaY, and ZnY zeolites and the esr spectra observed. The initial spectra were broad and poorly defined. On standing for several days the spectra became sharp and well defined. It was concluded that N_2O_3 is initially formed by the disproportionation reaction but ionizes due to the internal field of the zeolite (111).

$$N_2O_3 \rightarrow NO^+ + NO_2^-$$

Nitrogen dioxide in calcium X exhibits hindered motion. Adsorbed NO_2 molecules (about 11 per cavity) are essentially all dimerized. The single molecules, surrounded by several N_2O_4 molecules, are shielded from interacting with surface cations (112).

The polarization of paramagnetic species by the internal electrostatic fields of the zeolite has been observed mainly through esr techniques. From an esr study of ClO_2 and Cl_2^- adsorbed on NaX, it was concluded that the adsorbed species were associated with site II and site III cations. In comparison, a hydrogen Zeolon was found to have only one adsorption site. The internal fields in the Zeolon were observed to be smaller than those in NaX (113).

The interaction of various adsorbates with Cu^{2+} ions in a partially copper exchanged zeolite X was observed by esr spectroscopy. During dehydration, copper ions migrate to sites within the β-cages, site I' and II'. Strong interaction with adsorbed ammonia and pyridine, the latter at 473°K, was observed by copper ions in site II'. Since pyridine is too large to enter the β-cages, copper ions move towards the supercage. Heating the CuX in carbon monoxide at 623°K caused reduction of the Cu^{2+} to Cu^+. Reoxidation followed by carbon monoxide adsorption results in the formation of a Cu^{2+}-carbonyl complex (114, 115).

X-Ray Crystal Structure

The migration of Cu^{2+} ions from sites in the β-cages to sites in the supercages as the result of interaction with adsorbed pyridine or naphthalene was confirmed by an x-ray crystal structure study. Adsorbed ammonia, which can enter the β-cages, did not displace Cu^{2+} ions from site I'. Adsorption of pyridine on a zeolite Y of composition:

$$Cu_{16}Na_{24}Al_{56}Si_{136}O_{384}$$

resulted in the migration of 10 copper ions from the β-cages. It was further concluded that the copper ions are coordinated with pyridine molecules either through the nitrogen atoms or by π bonding (116).

The structure of zeolites containing various adsorbed molecules was reviewed by Smith (117). The structures include SO_2, Kr, Xe, Br_2, and I_2 adsorbed on zeolite A, and Cl_2 and Br_2 adsorbed on chabazite. In these studies, it was shown that most molecules occupy definite sites although complete interpretations were not possible. Although 12 concentrations of electron density (6 bromine molecules) were found in a study of bromine adsorbed in zeolite sodium A, there were two possible configurations of the bromine molecules (118). Chlorine adsorbed in chabazite gave similar results. The positions of cations in site I of zeolite A (4 Na and 4 Ca per unit cell) varied as the adsorbate was changed from xenon and krypton to polar molecules such as SO_2 and H_2O. The cations move toward the α-cage along the 3-fold axis. As discussed in Chap. 2, the cations in chabazite move from sites in the main cavities to sites within the double 6-rings as the water molecules are removed.

Acetylene molecules adsorbed on sodium A were found at two or more nonequivalent sites. These molecules are associated with sodium ions in site I, and three more are associated with the sodium ions in sites II and III. The principle interaction (ion-induced dipole) appears to be between the cation charge and the π-electron system of acetylene. Each sodium in site II is associated with an acetylene molecule (119).

Ammonia molecules adsorbed on sodium A were found in four sites. The four sites were occupied by 8, 4, 8, and 12 ammonia molecules (120). The crystal structure of sodium A containing adsorbed sulfur was determined by single crystal methods. In this case, the zeolite of composition $[Na_{11}Al_{11}Si_{13}O_{48}]$ contained only 11 cations per unit cell. Two parallel S_8 rings, each in a crown configuration, were found in each α-cage (121).

Nuclear Magnetic Resonance

A nuclear magnetic resonance study of sodium X and sodium Y zeolites with adsorbed hydrogen sulfide showed that the H_2S is more strongly held in sodium X. At low coverages the hydrogen sulfide appears to be dissociated into protons and HS⁻ ions (122).

Ethylene adsorbed on sodium Y and calcium exchanged Y was characterized by nmr. At low levels of adsorption, ∼3 cc/g, specific adsorption at the calcium ions produced a broad resonance. With increased adsorption, the line width decreased. At very high levels of a adsorption, polymerization of ethylene was evidenced by bands due to methyl groups and polyethylene (123).

F. SPECIFICITY OR ADSORPTION SELECTIVITY EFFECTS

The important types of interaction energies between adsorbate molecules in zeolites and the zeolite itself are shown in Eqs. 8.23-8.29. Physical properties of light gases which are utilized in estimating interaction energies are given in Table 8.17. The specificity shown by zeolite adsorbents toward an adsorbate molecule is determined by one or more of these different types of interaction energies. The dispersion and close-range repulsion energy terms are present in every case. The polarization energy term, however, is present only if the adsorbent is heteropolar, which is the case with all zeolites since the structures consist of positive and negative ions. The local electrostatic field in the zeolite is responsible for polarization of the adsorbate molecule. The ionic surface of the zeolite will also have an electric field which can interact with molecules which possess a quadrupole moment such as nitrogen or carbon monoxide.

Molecules which have permanent dipole moments such as water, and ammonia interact very strongly with the electrostatic field of the ionic zeolite structure. Initial heats of adsorption of water and ammonia in zeolites are in the range of 20 to 30 kilocalories per mole.

The dispersion energy of interaction may be very important in determining adsorption specificity in cases where the dipole and quadrupole interactions are absent. This energy of interaction increases with increased coordination number of the adsorbate molecule relative to the atoms of the adsorbent. In zeolites the adsorbate molecules may exist in closely fitting pockets within the framework of the zeolite oxygen atoms; and consequently, the dispersion energies may be quite high.

The dispersion energy may be as much as a factor of 8 greater depending upon the local environment and the fit of a molecule into a cavity or channel (98). Barrer has related the dispersion energy for Ar to the density of adsorbent atoms and the isosteric heats of adsorption (Table 8.24). He emphasizes that in porous zeolite crystals the density of adsorbent atoms immediately around an adsorbate molecule is low due to the presence of apertures. The total energy of interaction may be resolved into two types: (1) the type due to dispersion, repulsion and polarization and (2) the type due to the dipole and quadrupole interaction. Barrer has plotted the initial

Table 8.24 Some Heats of Adsorption and Intercalation of Argon (kcal/mole) (98)

		Isosteric heat	
Adsorbent	$N \times 10^{-22}$ (atoms of adsorbent/cc)	As $v/v_m \to 0$	As $v/v_m \to 0.5$ [or for v in cc(STP)/g for figs. in brackets]
β-quinol	2.27 C	4.8 [a]	4.8
	0.76 O		
	2.27 H		
Gas hydrate	2.66 H$_2$O	3.4 [b]	
		av. = 5.5	
		7.5 [b]	
Chabazite	3.0 O	6.0	3.0 [50]
(outgassed at 480°C)			
Chabazite	3.0 O	4.6	3.6 [40]
(outgassed at 300°C)			
H-mordenite	3.36 O	4.6	3.0
(outgassed at 350°C)			
NaX	2.47 O	2.7	2.7 [25]
Rutile	10.6 O	3.5	2.5
	5.3 Ti		
Carbon:			
(a) Spheron, heated to 2700°C	11.5 C as graphite	2.70	2.80
(b) Graphon		2.60	2.75
(c) Saran S84		3.50	2.63
(d) Saran S600H		3.90	3.63

[a] Calculated.
[b] Calculated using, respectively, London's, and Kirkwood and Muller's approximations for the dispersion energy constant. The former approximation tends to underestimate, and the latter to overestimate, the energy. Accordingly, the average value is given.

isosteric heat of adsorption for a series of nonpolar simple molecules against their molecular polarizability (Fig. 8.24). By interpolation, the dispersion, repulsion, and polarization contributions to the isosteric heats can be determined from the curves for other molecules which have a permanent electric moment, that is, a dipole moment or a quadrupole moment. Then, the dipole/quadrupole energy term may be derived by difference (Table 8.25).

Alteration of Selectivity

In zeolites, the specificity or selectivity shown toward particular

666 ADSORPTION BY DEHYDRATED ZEOLITE CRYSTALS

adsorbate molecules may be modified by methods which alter energy of interaction terms. There are basically three methods for altering adsorption selectivity. These are:

1. By introducing small amounts of a polar adsorbate (such as water) or preloading, which selectively locates on the most energetic sites and is adsorbed strongly enough that it may not be displaced by another less selective adsorbate molecule.
2. Cation exchange, in addition to altering the pore size as discussed in Sec. C, affects the heat of adsorption since the type, size, and location of the cation affects the local electric field as well as adsorbate polarization.
3. Decationization. Complete removal of the cations from the zeolite framework alters the local electric fields and field gradients and consequently reduces any interaction with a molecule with a permanent electric moment. Two methods have been used: the

Figure 8.24 Curve of initial adsorption heat, $-\Delta \overline{H}$, versus the polarizability of small adsorbed molecules (98). ● Ca-exchanged chabazite; ★ hydrogen Zeolon; ○ zeolite NaX; × graphitized carbon.

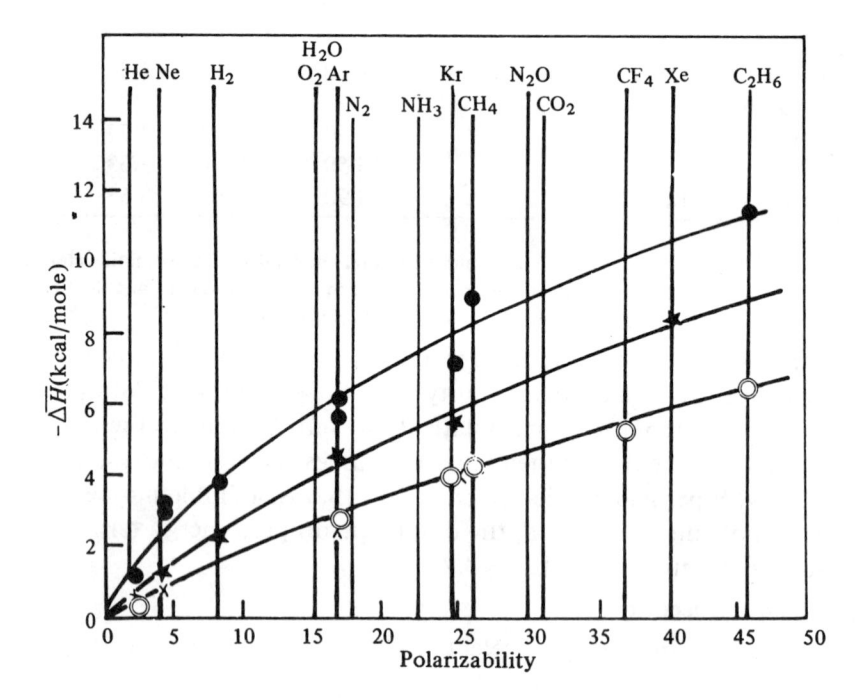

Table 8.25 Division of Components of Initial Heats, $\overline{\Delta H}$ (cal/mole) in Zeolitic Adsorbents (98)

Zeolite	Outgassed (°C)	Adsorbate	Total $-\overline{\Delta H}$	Dispersion + repulsion + polarization	Dipole + quadrupole energy
Chabazite	480	N_2	9,000	6,450	2,550
	450	N_2O	15,300	9,100	6,200
	480	NH_3	31,500	7,500	24,000
H-mordenite	350	N_2	6,200	4,500	1,700
		CO_2	11,100	6,750	4,350
Na-mordenite	350	N_2	7,000	4,500 [a]	2,500
		CO_2	15,700	6,750 [a]	8,950
NaX	350	N_2	6,500	3,100	3,400
		CO_2	12,200	4,200	8,000
		NH_3	18,000	3,750	14,250
		H_2O	~34,000	2,650	~31,350
NaY	350	CO_2	8,200	4,850 [b]	3,350

[a] Assuming dispersion + repulsion + polarization energy does not differ between Na- and H-mordenites.
[b] Assuming dispersion + repulsion + polarization energy does not differ between zeolites X and Y in the Na-forms.

thermal decomposition of an ammonium exchanged zeolite or treatment with an acidic proton including the removal of aluminum from tetrahedral sites (93). (See Chap. 6.) As the aluminum is removed the heat of adsorption is reduced for a molecule with a permanent electric moment such as CO_2, but is essentially not changed for a molecule with no permanent electric moment such as krypton (94).

The adsorption of carbon monoxide on calcium exchanged zeolite Y was shown by infrared spectroscopy to be cation specific, that is, the CO molecules are associated with the high electrostatic fields due to the divalent calcium ions (124). The specific adsorption is described by a Langmuir isotherm. It was further shown by Egerton and Stone that the Langmuir model gives a self-consistent description of the specific adsorption of CO on calcium exchanged zeolite Y (91). From stoichiometric considerations it was shown that the CO interacts with accessible calcium ions on a one-to-one basis and that every calcium ion in excess of the first 15 calcium ions enters a surface site which is accessible to carbon monoxide (such as site II). The first 15 calcium ions occupy site I which is unavailable to a carbon monoxide molecule. Thus, due to the differences in cation siting in certain zeolite structures such as zeolite Y, the mere fact of cation exchange may not necessarily affect adsorption specificity. In many cases the exchanged cation may be buried in the structure and will not evidence an effect on adsorption selectivity.

A number of methods are used to determine the variation in adsorption specificity with type of exchange cation. These variations are evidenced by the corresponding heats of adsorption such as the isosteric heats at low adsorption levels or the integral heats of adsorption (see Section D).

With the exception of ammonia, the polar molecule most studied in zeolites is water. The initial isosteric heats are very high and vary with the level of coverage. Several investigators have related the type of exchange cation in zeolite X with the specificity as determined by the heat of adsorption. In a study of Dzhigit and Kiselev the differential heats of adsorption, as determined calorimetrically, were determined to be due to the energy of interaction of a water molecule with the exchange cations as well as with negative oxygen ions of the zeolite framework (129). As shown in Fig. 8.25, the adsorption heat for water depends upon the cation radius and the number of water molecules in each large adsorption cavity. Due to their size the larger cations potassium, rubidium and cesium maximize the interactions between the water molecules, the cations, and the oxygen framework. At high water loadings the water molecules themselves interact by hydrogen bonding. The positions of the cations change with the increasing number of water molecules in the structure; at high water loadings the cations are fully hydrated and are displaced into the main cavities.

Effect of Cation Density on Selectivity

As the SiO_2/Al_2O_3 ratio of zeolite X and zeolite Y increases, the univalent cation density must decrease. With this decrease in cation density the adsorption selectivity for various gases also changes. Isosteric heats of adsorption at zero coverage or loading were measured utilizing a gas-solid chromatographic method (101). The sodium ion

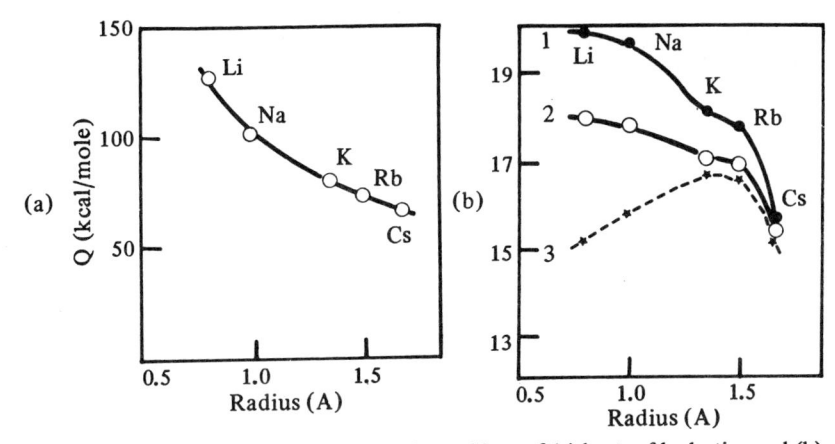

SPECIFICITY OR ADSORPTION SELECTIVITY EFFECTS 669

Figure 8.25 Dependence on the radius of alkali metal ions of (a) heats of hydration and (b) heats of adsorption of water for different numbers per large cavity of zeolite X: (1) 0.5; (2) 2; (3) 6 (129).

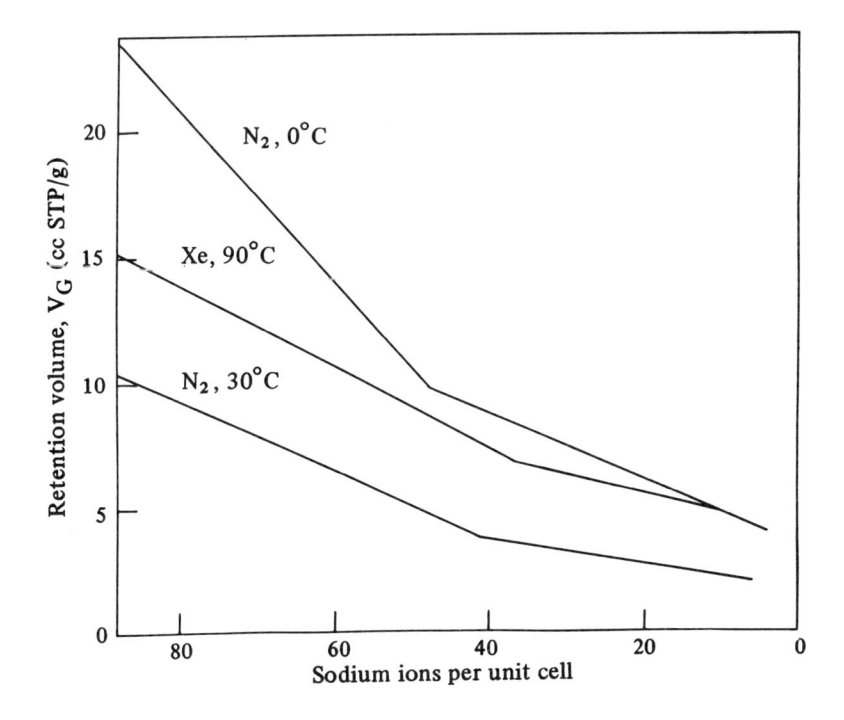

Figure 8.26 Retention volumes on NaX and NaY, and partially ammonium exchanged NaX and NaY (101).

content was varied by (1) variation in the silica/alumina ratio of the zeolites as synthesized and by (2) partial exchange with ammonium ion. Initial isosteric heats of adsorption were determined from the measured retention times of various gases including O_2, N_2, CO, CH_4 Xe, and Kr. The limiting isosteric heats did not vary with the sodium cation content of the zeolites studied. However, the retention volumes were found to be proportional to the number of sodium cations and decreased linearly with the cation content (Fig. 8.26). The results indicate that there are two types of cations on the internal surface in site II and site III which exhibit nearly equal adsorption energies; the number of S III cations decreases as the silica/alumina ratio increases.

From the measured isosteric heat, the contribution due to polarization interaction was estimated and, from this, the effective charge on the sodium cation was determined to be 0.66 esu (101).

The effect of the removal of cations from the zeolite clinoptilolite on adsorption selectivity was studied by treatment with acid solution, that is, exchange with hydronium, H_3O^+. The effect on adsorption selectivity of removal of tetrahedral aluminum ions through acid hydrolysis, and concomitant removal of framework charge, was studied. The change in the nature of the adsorption isotherms and heats of adsorption was determined using carbon dioxide which has a permanent electric moment, and krypton which is nonpolar (94).

On a volumetric basis, that is, in terms of a constant number of unit cells, the saturation adsorption capacities for carbon dioxide or krypton were invariant with the degree of decationization and aluminum removal. The isosteric heats of adsorption for carbon dioxide which were derived from the isotherms at various temperatures decreased with increasing removal of cation and framework charge (Fig. 8.27). In the ultimate case, the zeolite contained no cations or framework aluminum but was still crystalline as evidenced by x-ray diffraction studies. The surface is energetically homogeneous. On the other hand, the isosteric adsorption heats for the nonpolar gas, krypton, were essentially invariant with the degree of decationization. Although the initial clinoptilolite did not adsorb krypton, all of the acid treated zeolites adsorbed krypton readily.

The isotherm in every case was adequately represented by the Langmuir equation and a plot of the Langmuir constant, B (or K),

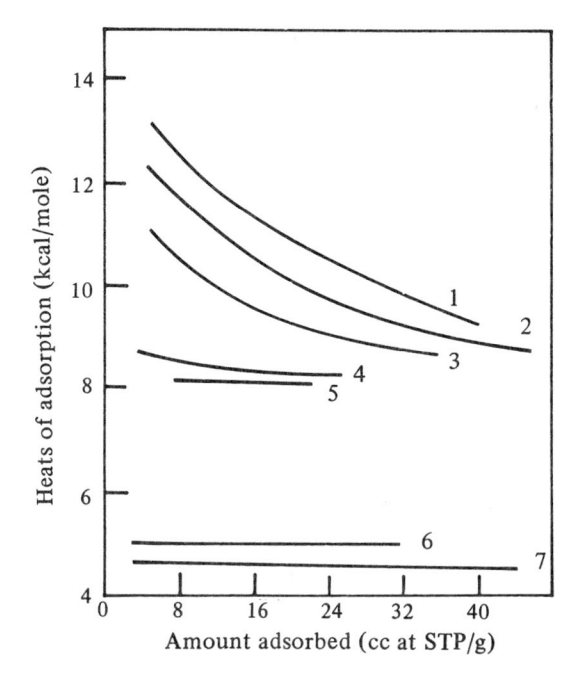

Figure 8.27 Isosteric heats of adsorption of CO_2 in clinoptilolite and HCl-treated clinoptilolite. (1) mineral, (2) 0.25 N, (3) 0.5 N, (4) 1 N, (5) 2 N acid; and for krypton, (6) 0.25 N, (7) 2 N acid. The SiO_2/Al_2O_3 ratio of the mineral 9.3:1; of the 0.25 N, 9.3:0.58; of the 0.5 N, 9.3:0.33; 1 N, 9.3:0.07; 2 N, 9.3:0.0 (94).

against the degree of coverage, θ, was constant. Only the completely decationized clinoptilolite exhibited an energetically homogeneous surface.

G. ADSORPTION KINETICS AND DIFFUSION

Excellent reviews of diffusion processes in crystalline zeolites have been presented by Barrer (156) and Walker et al. (154). Diffusion constants and activation energies have been determined for the adsorption of gases and medium chain length hydrocarbon molecules on mineral zeolites, synthetic zeolites and modifications of both. The modifications include the effect of cation replacement, decationization, and the effect of a pre-adsorbed polar compound such as ammonia. The results support the model of a diffusing molecule which encounters a periodic potential field within the zeolite. Various calculations of these potential fields have been made.

When the rate-controlling process in the adsorption of the gas on a zeolite is the activated diffusion of the adsorbate molecule through intracrystalline channels then the process may be described by the following equation:

$$Q_t - Q_0 = Bt^{1/2} + C \tag{8.30}$$

where Q_t = amount adsorbed at time t, Q_0 = amount adsorbed at $t = 0$, and B and C are constants.

This activated diffusion process is differentiated from competing processes by observing the increased rate of adsorption with an increase in temperature. In the case of the adsorption of nitrogen on zeolite NaA the activated diffusion process is observed over one temperature range and a different mechanism, which involves a negative temperature coefficient, predominates over another temperature range. This results in an isobar as illustrated in Fig. 8.15, which has a maximum; the position of this maximum depends on the time allowed for adsorption to occur and the particle size of the zeolite crystals. Based on Eq. 8.30 a plot of the amount adsorbed versus $t^{1/2}$ should yield a straight line except for a "foot," which may occur near the start of the curve and which results from the very rapid adsorption that occurs on the external surface of the zeolite crystal at the initiation of the experiment.

The diffusion characteristics of zeolite A are easily altered by exchange of the sodium ion by larger or smaller ions. The potassium form exhibits a smaller effective pore diameter whereas the calcium exchanged form exhibits the full structural diameter of the framework. Consequently, the diffusion of molecules in zeolite A has been extensively studied because one can alter these differences in a controlled manner. There are numerous publications concerned with diffusion and adsorption kinetics in the various ion forms of zeolite A involving adsorbates such as the permanent gases and hydrocarbon molecules. Illustrative data for the adsorption of argon and nitrogen on zeolite A powder are shown in Fig. 8.28 (153). The rate of the adsorption of both argon and nitrogen decreases rapidly with decreasing temperature. There is a linear portion of the curve which adheres to the $t^{1/2}$ diffusion law for the first several minutes. For an assembly of spherical or cubic particles and a small period of time, the dif-

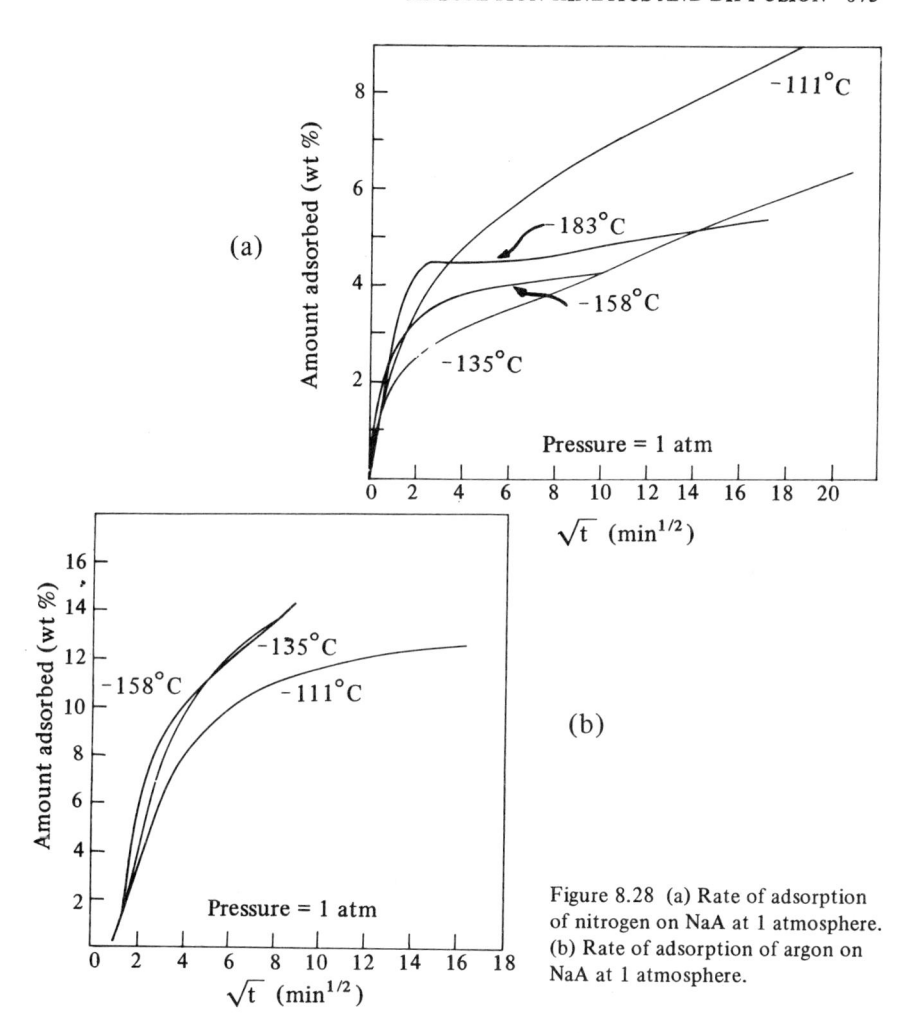

Figure 8.28 (a) Rate of adsorption of nitrogen on NaA at 1 atmosphere. (b) Rate of adsorption of argon on NaA at 1 atmosphere.

fusion equation at constant pressure is:

$$\frac{Q_t - Q_o}{Q_\infty - Q_o} = \frac{2A}{V}\left(\frac{Dt}{\pi}\right)^{1/2} = \frac{6}{r_o}\left(\frac{Dt}{\pi}\right)^{1/2} \quad (8.31)$$

where Q_∞ is the amount adsorbed at equilibrium, A is the external surface area of the particles in cm^2/g, V is the volume of the crystals in cc/g and D the diffusion coefficient. This applies for small amounts of adsorption and the application of Henry's law to the adsorption isotherms. Over large ranges of adsorbate concentration, D depends

674 ADSORPTION BY DEHYDRATED ZEOLITE CRYSTALS

on the concentration and the $t^{1/2}$ law is not applicable. The quantity $1/r_o$, where r_o is the radius of the spherical particles, can be replaced by $A/3V$. The total area A is determined by various methods, such as adsorption of N_2 or Kr on the external surface of the hydrated zeolite crystals. At constant volume the equation is:

$$\frac{Q_t - Q_o}{Q_\infty - Q_o} = \frac{2A}{V} \cdot \frac{1+K}{K} \left(\frac{Dt}{\pi}\right)^{1/2} \tag{8.32}$$

K is the ratio of the adsorbate in the gas phase to that in the crystals at equilibrium. Again, Henry's law must apply.

At very small times a "foot" is sometimes observed in the plot of $(Q_t - Q_o)/(Q_\infty - Q_o)$ versus $t^{1/2}$ for adsorptions at low temperatures. (see Fig. 8.29). This foot is attributed to surface adsorption on the external surface of the crystals. It may be used to measure the surface area of the crystalline powders by extrapolation of the $Q_t - Q_o$ plot to $t^{1/2} = 0$ (152).

The diffusion coefficient is determined from the slope of the $t^{1/2}$ plots by knowing A and V. The value of D so obtained is an average over the range of adsorption and varies with the quantity of adsorbate in the zeolite (Table 8.26).

For activated diffusion the variation in D with temperature is given by the following equation (157):

$$D = D_o e^{-E/RT} \tag{8.33}$$

where E is the activation energy for the diffusion process. The dif-

Table 8.26 Average Diffusion Coefficients in Chabazite (152)

Adsorbate	$t(°C)$	Q_o (ccSTP/g)	D (cm²/sec)	E_a (cal/mole)
C_3H_8	150	0.00	17.6 x 10⁻¹³	3100
		15.1	1.5 x 10⁻¹³	
	200	0.00	26.1 x 10⁻¹³	
		10.2	3.4 x 10⁻¹³	
$n\text{-}C_4H_{10}$	150	0.00	9.5 x 10⁻¹³	7300
		9.50	1.42 x 10⁻¹³	
	200	0.00	23.6 x 10⁻¹³	
		8.00	2.89 x 10⁻¹³	

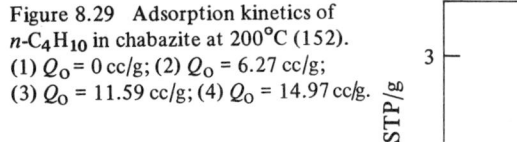

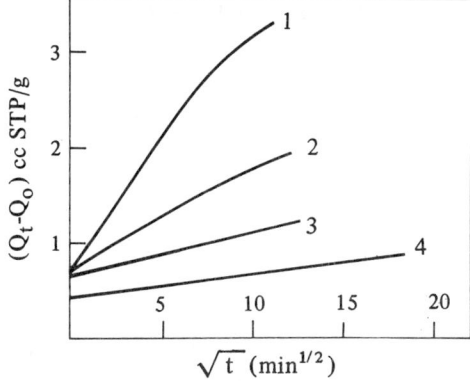

Figure 8.29 Adsorption kinetics of
n-C_4H_{10} in chabazite at 200°C (152).
(1) Q_0 = 0 cc/g; (2) Q_0 = 6.27 cc/g;
(3) Q_0 = 11.59 cc/g; (4) Q_0 = 14.97 cc/g.

fusion constants calculated for argon and nitrogen in zeolite NaA and plotted as a function of temperature are shown in Fig. 8.30. These yield activation energies of 2.7 kcal per mole for argon and 5.8 kcal per mole for nitrogen.

Typically it has been observed that at low temperatures the adsorption of oxygen from air is inhibited in NaA zeolite by the slowly adsorbing nitrogen which blocks the active sites and crystal openings. This mechanism, although not as severe, is probably similar to

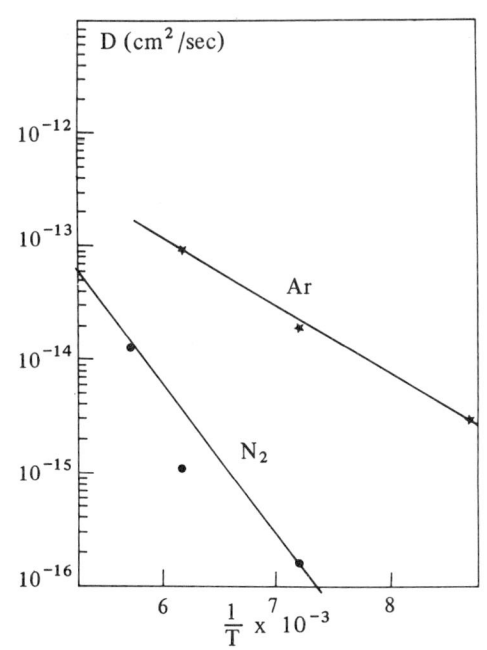

Figure 8.30 Diffusion constants for argon and nitrogen diffusion in zeolite NaA (153). V = 0.63 cc/g; A = 3.4 m²/g; average radius = 0.7/μm.

that caused by the preadsorption of ammonia. When exposed to air at −183°C, the adsorbed phase in the zeolite is greatly enriched in oxygen content to about 98 vol % (Table 8.27). However, the rate of adsorption of oxygen from air at these temperatures is so slow that any separation based on this scheme is not practical.

Using a differential experimental system, the diffusion of gases from potassium exchanged zeolite A at temperatures of 300°C and above was measured (154). At these temperatures, the adsorbate loading of these gases is very low. At low temperatures most of these gases are not adsorbed by potassium A. The magnitude of D_0 increases with the increasing kinetic diameter. However, nitrogen appears to be anomalous (Table 8.28).

Energies for interaction of the diffusing molecules with the main 8-ring apertures and the potassium ions were calculated. It was concluded that the position of the potassium ion adjacent to or near the 8-ring varied with the size of the diffusing molecule. The extent of the protrusion of the potassium ion in site II into the 8-ring aperture varied from 1.8 A to 2.5 A depending upon the kinetic diameter of the diffusing atom (xenon thru neon, respectively).

A cooperative effect between potassium ions in zeolite A and ethane molecules is indicated (158). The measured activation energy is much more than would be expected from extrapolation of results obtained on the lithium and sodium exchanged forms. The preadsorption of small amounts of water was found to increase the diffusion rate of ethane in LiA. This is attributed to occupancy by water molecules of adsorption sites near the lithium ions. The formation of this more uniform potential field results in a decrease in the activation energy for diffusion (158).

The diffusion of ethane in zeolite A in the form of crystalline powder and in the form of pellets containing 25% binder was compared by Kondis and Dranoff (159). Adsorption kinetics of ethane from a

Table 8.27 Adsorption of O_2 from Air by Zeolite A, Powder (153)

Temp.(°C)	Time (min)	Amount Adsorbed (wt%) O_2	N_2	O_2 in Adsorbate (vol %)
−183	60	17.0	0.2	99
−183	30	12.2	0.2	98
−158	30	8.9	0.9	90

Table 8.28 Diffusion Coefficients for Gases in KA (154)

Adsorbate Gas	Temp. Range(°C)	Kinetic Diameter(A)	D_0 (cm^2/sec)	E^a (kcal/mole)
Neon	50 - 200°	2.8	4.5×10^{-8}	7.0
Argon	300 - 500°	3.42	3.7×10^{-8}	12.6
Krypton	350 - 500°	3.61	60.3×10^{-8}	16.4
Xenon	400 - 600°	4.06	124×10^{-8}	19.2
Hydrogen	25 - 200°	2.92	4.5×10^{-6}	9.9
Carbon dioxide	200 - 600°	3.99	4.0×10^{-11}	4.3
Nitrogen	–	3.68	1.0×10^{-5}	16.2

[a] Based on $D = D_0 e^{-E/RT}$ and $r_0 = 2.0 \times 10^{-4}$ cm.

helium carrier gas on zeolite A (powder and pellets) were compared. Both adsorption rates and equilibrium data obtained on both forms of sample were the same. In the absence of a binder it was concluded that the rate of adsorption and desorption of ethane is controlled by the intracrystalline diffusion.

In another study, the rate of adsorption of ethane on pelletized zeolite A (with binder) was measured. Equilibrium isotherms agreed with those for the pure crystals when corrected for the binder content. Although adsorption-desorption is controlled by micropore diffusion it was found that the effective diffusivities of ethane into the 4A pellets were lower than for the pure crystal (160). The intracrystalline diffusivities were:

For A pellets: $Dc/r^2 = 0.293 \exp(-5230/RT)$

For A crystals: $Dc/r^2 = 3.22 \exp(-5660/RT)$

The activation energies are essentially the same.

The rate of adsorption of n-paraffin hydrocarbons through n-C$_8$ were measured on erionite and calcium A zeolites between 93° and 207°C (84). The erionite was a mineral sample and the calcium A was in a pure powder form but pressed into 1/8 inch diameter pellets which were subsequently reduced in particle size to a size range of of 420 to 840 microns in diameter. A gravimetric technique was used. Large differences in the adsorption rates between erionite and the calcium A were observed with the rates of adsorption in calcium A being considerably higher. This difference was attributed to the basic difference in the crystal structures. In zeolite A a normal para-

Figure 8.31 An *n*-butane molecule located in the "saddle" point in an 8-ring of erionite. The methylene group, A, is in the plane of the ring (155).

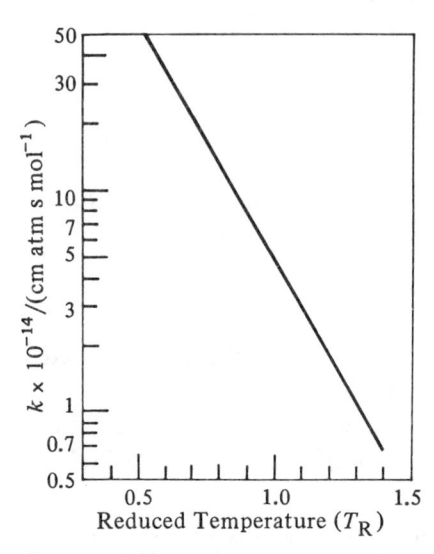

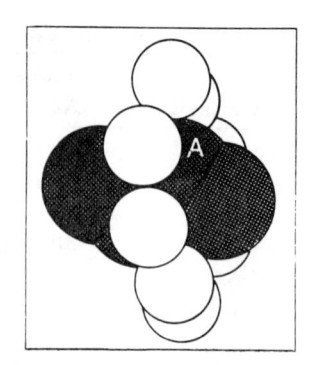

Figure 8.32 Temperature dependence of the drag coefficients for light paraffins in CaA zeolite (162). The relation for the hydrocarbons CH_4, C_2H_6, C_3H_8, and n-C_4H_{10} is represented by the equation: $k = 7.1 \times 10^{16} \exp(-4.97\, T_R)$ cm.

fin can diffuse through the channel system from cavity to cavity in a unidirectional manner whereas in erionite the hydrocarbon must diffuse through a zigzag channel in an irregular manner (see Chap. 2). Secondly, the elliptical aperture openings in erionite are more restricted than the essentially circular apertures in the calcium A zeolite.

A gross particle size effect in the case of the erionite was observed. Rates of adsorption on 1/8 inch pellets were compared to those obtained on 20 to 35 mesh particles. A five-fold difference in the apparent diffusion coefficient was observed. This difference was not as great as might be predicted by theory. It was again concluded that the main diffusion resistance is in the intracrystalline micropores rather than in the macropore system between individual crystallites (84).

Barrer and Peterson compared the role of the aperture size in three zeolites—chabazite, erionite, and calcium A—by calculating the energy of interaction between the main 8-ring apertures and a normal paraffin hydrocarbon chain (Fig. 8.31) (155). When a methylene group of the hydrocarbon chain moves through the plane of the ring, a "saddle" point in the interaction energy occurs. The maximum interaction potential at the saddle point was calculated for erionite, whereas zeolite A showed the lowest interaction potential energy. Some variation in this interaction potential may occur with movement of the oxygens in the rings from their equilibrium positions. Consequently, the relative diffusion rates may also change if the 8-ring is more easily deformable in one zeolite than another.

In one of the first studies of the kinetics of diffusion of hydrocarbons in zeolites, Barrer and Ibbitson found that the apparent activation energy for diffusion in chabazite increased with the chain length of the hydrocarbon (161). Additionally, the isosteric or differential heat of adsorption increases with increasing chain length. This same effect has been observed in zeolite A.

Adsorption kinetics for methane, ethane, propane, and normal butane on calcium A zeolite were studied gravimetrically by Ruthven and Loughlin (81, 162). The apparent activation energy and diffusivity are concentration dependent. More satisfactory correlations were obtained by introducing a drag coefficient. The drag coefficient is an empirical quantity which measures the total resistance of the zeolite structure to the flow of the adsorbate. Data for the kinetics of adsorption of all four hydrocarbons over a broad temperature range were correlated by the expression:

$$k = k_o \exp(-\beta T) \qquad (8.34)$$

where k is the drag coefficient. In Fig. 8.32, k is plotted as function of the reduced temperature, T_R ($T_R = T/T_c$).

In the calculation of diffusivities from experimental rate curves it has been common to use a mean equivalent spherical radius for the particle size. However, it has been shown that for correct interpretation, a detailed knowledge of the shape and size distribution of the zeolite crystals is necessary. Ruthven and Loughlin have compared experimental with calculated rates of adsorption curves based on the use of a spherical particle, and the use of a cubic particle. The basic diffusion equation fits the experimental data better if the particle size distribution and shape are considered (Fig. 8.33) (163).

In another kinetic study of the adsorption of normal butane on calcium A zeolite in the form of both crystals and pellets it was concluded that the binder behaves as a true inert diluent with no significant effect on the adsorption properties either in terms of equilibrium or adsorption kinetics (164). The equilibrium data were correlated well with a modified Polanyi equation (Fig. 8.34). If the concentra-

680 ADSORPTION BY DEHYDRATED ZEOLITE CRYSTALS

Figure 8.33 Observed and calculated curves for the adsorption of n-butane on crystals of the CaA zeolite (163). Desorption was at 60.4 torr and adsorption at 68 torr. $S = \mu/\sigma$ where μ, the mean deviation of the characteristic dimension is 1.8 microns and σ is the standard deviation.
$D = 7.7 \times 10^{-11} \text{cm}^2/\text{sec}$.

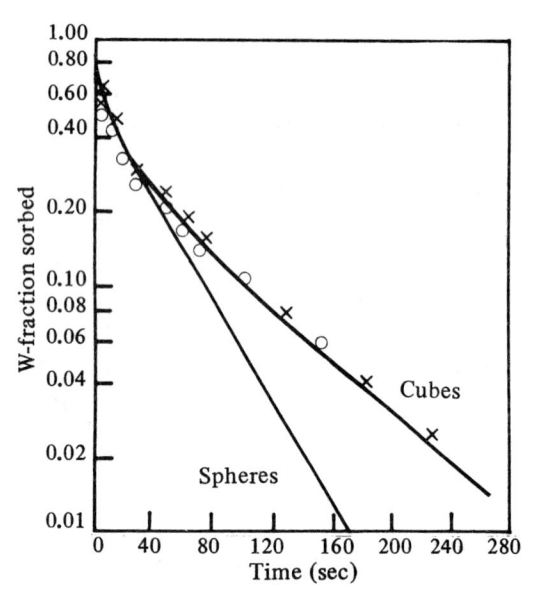

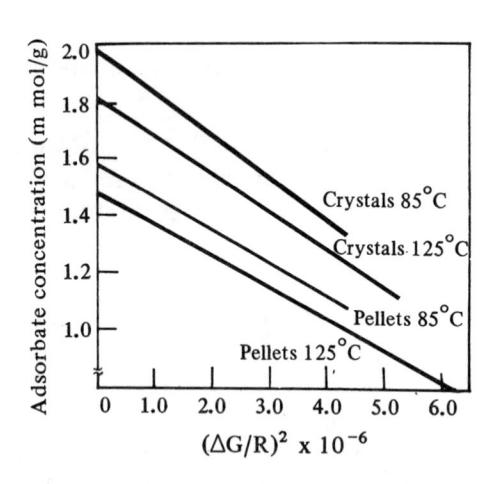

Figure 8.34 Isotherms for adsorption of n-butane on CaA zeolite, plotted according to the equation (164): $c/c_s = 1 - (\alpha' \Delta G)^2$ where c = adsorbate in the zeolite and c_s is the saturation loading. α' is a temperature-independent constant. $\Delta G = RT \ln (f/f_s)$ where f = the fugacity of the adsorbate vapor at equilibrium pressure p and f_s is the fugacity of the saturated adsorbate.

tion of the adsorbate—normal butane—in the pellets is corrected by a factor of 0.8 to allow for the presence of the pellet binder, then the calculated diffusivities on the crystals and pellets agree closely. Also, the empirical drag coefficient was determined on both pellets and crystals with no observed difference. Measured diffusion coefficients over the temperature range 0 to 225°C were within 10^{-9} to $10^{-10} sec^{-1}$.

For NaA zeolite, the diffusional resistance of pellets was observed to be six times greater than that of the crystals. This difference was attributed to partial structural breakdown of the zeolite crystal surface during the process of pelletization. Equilibrium properties are not affected (81).

The rate of adsorption of ethane from a helium carrier gas on commercial sodium A and calcium A pellets (with 20% clay binder) was determined at 25°C. The micropore and macropore diffusivities were calculated. In case of the sodium A pellet, the micropore diffusion is controlling but in the case of the calcium A pellet both micropore and macropore diffusion appear to be important. The activation energy for the calcium A-ethane diffusion is 4.45kcal/mole as compared to 6.40 kcal/mole observed on the sodium A pellets. This is consistent with the smaller pore size of sodium A (165, 166).

The counter diffusion characteristics of several liquid hydrocarbons in zeolite Y in three ion exchange forms were measured (167). Very little of this type of information is available although it may be important to an understanding of catalysis. The sodium, calcium, ammonium, and cerium exchanged forms of zeolite Y were studied. The diffusion measurements were made in the liquid phase with the zeolite presaturated with the desired hydrocarbon. The hydrocarbon saturated zeolite was placed in a stirred flask containing a known quantity of a second liquid hydrocarbon and samples taken as a function of time. The desorptive diffusion coefficients for 1-methylnaphthalene diffusing from the zeolite into cumene varied by two orders of magnitude over the various ion exchanged forms. The difference in diffusion coefficients was attributed to differences in the cation type, size, and population. For example, the rare earth cations primarily populate sites in the β-cages and would not be present on the intracrystalline surface to impede the molecular diffusion of a large hydrocarbon molecule. In contrast, there is a substantial population of sodium ions in surface sites II and interaction with a diffusing hydrocarbon molecule is expected. Initial activation energies calculated from diffusion coefficients obtained from the initial rates varied with the diffusing molecule and the cation form. The highest activation energies were found for sodium Y with the value for the diffusion of cumene into benzene being greater (17.4 kcal/mole) than that for the diffusion of 1-methyl naphthalene into cumene (13.8 kcal/mole). Counter-diffusion in zeolite Y occurs readily as compared to other types of zeolites such as mordenite.

The counter-diffusion behavior of aromatic hydrocarbons in the mordenite structure is in agreement with the pore structure which consists of nonintersecting arrays of parallel tubes that are somewhat larger than the diameter of the diffusing hydrocarbon molecules (168) The main channels (7.0 x 6.7 A) parallel the c-direction and connect in the b-direction via side pockets which have a diameter of 3.9 A and are too small to admit hydrocarbons of this size. In addition, these small pockets are occupied by sodium cations. Removal of the cations by hydrogen ion exchange should improve the diffusional characteristics of mordenite. The counter-diffusion of benzene and cumene requires movement of molecules in the tubular channel. Measured activation energies were much higher than those for diffusion of the hydrocarbons in liquid systems. They were also much greater than the activation energies for diffusion in other zeolites. The activation energy for the desorption of benzene was found to be much higher than the heat of adsorption on hydrogen Zeolon which is about 5.2 kcal/mole. In contrast the observed activation energy for counter-diffusion is 17.5 kcal/mole and is attributed to the blockage of cylindrical pores by a few hydrocarbon molecules which are strongly adsorbed on certain sites. The diffusion coefficient for cumene desorption into benzene decreased with the time of saturation before desorption. This decrease was caused by the slow formation of radical ions and diisopropylbenzene.

Adsorption equilibria and kinetics for the gaseous hydrocarbons CH_4 to $n\text{-}C_4H_{10}$ on a synthetic, sodium Zeolon, were studied (169, 170). Diffusion coefficients and activation energies are shown in Table 8.29 for a temperature of 25°C. The diffusion coefficients decrease with increasing molecular size. Using a method for measuring adsorption at small initial times, that is, a rapidly responding constant

Table 8.29 Diffusion of Hydrocarbons in Mordenite [a](170)

Compound	Temp (°C)	$P_{initial}$(torr)	F [b]	$D \times 10^{10}$ [c] (cm^2/sec)
CH_4	25	25 to 1.7	0.03 to 0.24	184 to 123
	100	22 to 12	0.01 to 0.02	294 to 304
n-C_4H_{10}	25	78 to 10	0.1 to 0.9	26.8 to 5.2
iso-C_4H_{10}	25	45 to 10	0.15 to 0.63	12.6 to 2.5
C_4F_{10}	25	176 to 9	0.01 to 0.14	11 to 2.7

[a] Sodium Zeolon: $Na_{6.72}H_{0.65}Al_{7.37}Si_{40.63}O_{96} \cdot x\ H_2O$; average crystal size, 21 x 21 x 33 μm.
[b] F = the fractional uptake of the adsorbate from the gas phase.
[c] D is calculated for a fraction adsorbed, $Q_t/Q_\infty = 0.5$.

volume system, the rates of adsorption of CH_4, n-C_4H_{10}, iso-C_4H_{10}, and n-C_4F_{10} were studied at 25 to 100°C. The diffusion coefficient for isobutane is the same as that for n-C_4F_{10} and about one-half of that of n-C_4H_{10}. The n-C_4F_{10} molecule is considerably larger, the kinetic diameter is 5.6 A as compared to 4.3 for n-C_4H_{10}, and would be expected to have a lower diffusion coefficient. The calculated activation energies vary with the initial pressure and the diffusivity.

The kinetics of the adsorption and desorption of hydrocarbons (ethane, n-butane, and n-pentane) as well as the interchange of carbon dioxide in ethane were measured on the synthetic zeolites T, Y, and "Zeolon" in the hydrogen form, the latter derived from the ammonium exchanged form (171). The time constant, τ, was used to compare the rate behavior (Table 8.30). The time constants are considerably longer for zeolite Y than for the hydrogen Zeolon.

Table 8.30 Diffusion of Hydrocarbons in Zeolites HY and H Zeolon (171)

	θ	Time Constants for Adsorption of C_2H_6 at $-80°C$ τ at $\Delta S/\Delta S_f =$		
		0.7	0.9	1
HY [a]	0 → 0.29	300	365	530
H-Zeolon [b]	0.22 → 0.54	12	43	150

[a] Composition of starting $NH_4Y = (NH_4)_{0.94}Na_{0.06}[(AlO_2)(SiO_2)_{2.86}]$
Hydrogen Y prepared by air calcination at 400°C.
[b] Composition of H-Zeolon = $H_{0.94}Na_{0.06}[(AlO_2)(SiO_2)_{6.4}]$
τ is given by $\tau = R^2 D$ in seconds.
R is characteristic of the crystal size.
$\Delta S/\Delta S_f$ is the degree of advancement of adsorbate.

684 ADSORPTION BY DEHYDRATED ZEOLITE CRYSTALS

Table 8.31 CO_2 and C_2H_6 Counter Diffusion in Zeolite HT at $-80°C$ (171)

Initial Adsorbate	Entering Gas	R^2/D_{AB} (sec)
C_2H_6, $\theta = 0.87 \to 0.72$	CO_2, $\theta = 0 \to 0.15$	6000
CO_2, $\theta = 0.85 \to 0.67$	C_2H_6, $\theta = 0 \to 0.72$	8000

For pure adsorbates R^2/D_{CO_2} = 500 sec ± 20% and $R^2/D_{C_2H_6}$ = 800 sec ± 30%.
Composition of starting $NH_4T = (NH_4)_{0.75}K_{0.25}[(AlO_2)(SiO_2)_{3.67}]$.
Hydrogen T prepared by air calcination at 400°C.

Consequently, the rate of diffusion of ethane into the hydrogen Zeolon is faster than into hydrogen Y which is counter to what one would expect.

The temperature dependence of the time constant was used to obtain activation energies. For n-C_5H_{12} in hydrogen T zeolite the activation energy was calculated to be 5.9 kcal/mole. This compares with an isosteric heat of adsorption of -16.6 kcal/mole. Thus, as has been generally observed, the activation energy for diffusion is smaller than the heat of adsorption. The counter-exchange of CO_2 and C_2H_6 in hydrogen T zeolite at $-83°C$ was studied. As shown in Table 8.31, the binary diffusion coefficients are considerably greater than the single diffusion coefficients (τ = 6000 to 8000 sec versus 500 to 800 sec).

The self diffusion of water has been measured in some zeolites but there is very little additional information on other adsorbents (156). Some results expressed as diffusion coefficients and activation energies are given in Table 8.32. The water molecule is small, about 2.8 Å,

Table 8.32 Self-Diffusion of Water in Zeolites (156)

Zeolite	$D_A*(cm^2 sec^{-1})$	$t(°C)$	D_0 $(cm^2 sec^{-1})$	E (kcal/mole)
Analcime	1.97×10^{-13}	46	1.52×10^{-1}	17.0 ± 0.3
Natrolite				15.0
Heulandite	2.07×10^{-8}	45	7.6×10^{-1}	11.0 ± 0.3
Chabazite	1.26×10^{-7}	45	1.2×10^{-1}	8.7 ± 0.3
Gmelinite	5.8×10^{-8}	45	2.0×10^{-2}	8.1 ± 0.3
NaX	2.11×10^{-5}	40		6.9
CaX	2.41×10^{-5}	40		6.8
CaY				5.6
Ice	1×10^{-10}	-2		13.5 ± 1.1
Liquid water	3.87×10^{-5}	45	5.6×10^{-2}	4.6

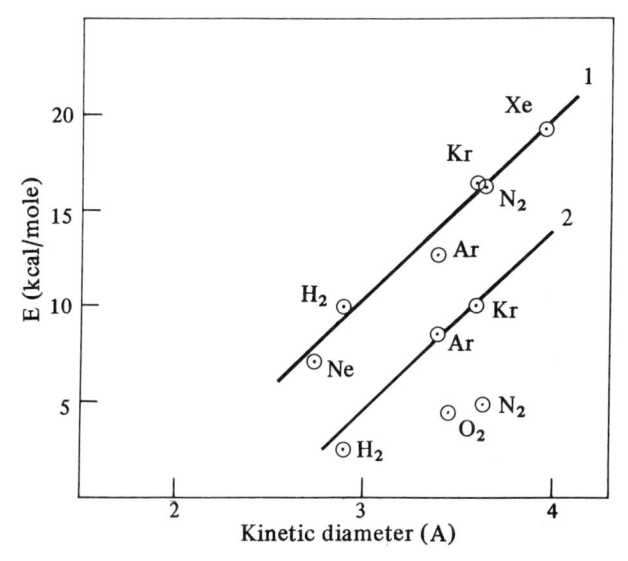

Figure 8.35 Energy of activation for the diffusion of gases in (1) KA and (2) potassium exchanged small-pore mordenite as a function of the molecular kinetic diameter (154, 156).

and consequently should not encounter a hindrance to diffusion based on the relative size of the channels in most zeolites. A few zeolites, such an analcime, however, have large energy barriers. The energy barriers indicated by the diffusion of water in zeolites are larger than those in liquid water. Water, like many other small polar molecules, when diffusing in a zeolite channel is subject to "sticking" on active sites and clustering around the channel-located cations. Sticking is due to the cation-dipole interaction (172).

The activation energy for self-diffusion of water in zeolites appears to vary inversely with the size of the intracrystalline channels which indicates that in the more open structures such as X and Y the zeolite water is liquid-like; in the tighter structures such as heulandite or analcime individual water molecules are coordinated with lattice oxygen and cations. In the latter case the diffusion mechanism must involve individual water molecules rather than clusters of water molecules. In general, the magnitude of the energy of activation is between that of water diffusion in liquid water and in ice. The more open zeolites behave like liquid water whereas the zeolites with tighter structures are more ice-like.

The activation energy for diffusion of a series of molecules of increasing diameter has been observed to increase in mordenite and in zeolite A. In Fig. 8.35 the activation energy for a series of gas molecules is plotted as a function of the molecular kinetic diameter. For both zeolites, the energy increases with increasing diameter. In zeolite A the relationship is quite good but in mordenite, it appears that nitrogen and oxygen are somewhat anomalous in their diffusion behavior. This may be related to the type of channel system and the cation positions. The cation-molecule cooperative effect which has been suggested to occur in zeolite A may not be operative in the unidimensional channels of mordenite.

The diffusivities of relatively large organic molecules from the liquid phase in NaY and HY zeolites are related to the molecular diameter and to the interaction of the adsorbate molecule with the zeolite (173). Diffusion coefficients for a series of hydrocarbons varying in size (equilibrium critical diameter) from 6.8 A to 9.4 A in NaY at 30°C are given in Table 8.33. The diffusion coefficients for 1, 3, 5-trimethylbenzene (mesitylene) and 2, 4, 6-trimethylaniline (mesidine) differ by a factor of 10. The sizes and shapes of these two compounds are essentially the same. The polar amino group reduces the diffusion rate of mesidine by interacting with the zeolite electrostatic field. The activation energies for diffusion were determined to be 9 kcal/mole for mesitylene and 17 kcal/mole for mesidine.

A summary of selected information on adsorption kinetics and diffusion of various gases and vapors in zeolites is presented in Table 8.34. Most of this information is concerned with zeolite A, mordenite (Zeolon), and chabazite.

Table 8.33 Diffusion coefficients in NaY [a] at 30°C (173)

Compound	Molecular Diameter(A) r_{min}	σ	Diffusion Coefficient $D_E \times 10^{13} (cm^2/sec)$ at M_t/M_D [b] $= 0.3$
1, 3, 5-trimethylbenzene	8.4	7.5	68
2, 4, 6-trimethylaniline	8.4	7.5	6.9
1, 3, 5-triethylbenzene	9.2	8.2	0.65
1, 3, 5-triisopropylbenzene	9.4	8.4	0.028
Cumene	6.8	5.9	>700

[a] NaY, $SiO_2/Al_2O_3 = 4.86$
[b] M_t/M_D = degree of adsorption

ADSORPTION KINETICS AND DIFFUSION 687

Table 8.34 Summary of Selected Adsorption Kinetics Data

Zeolite	Cation	Form	Adsorbate	Temp (°C)	E (kcal/mole)	D (cm^2/sec)	D_0 (cm^2/sec)	Remarks	Ref.
A	Li	PWD	C_2H_6	22 - 100	9.9		4.5×10^{-5}	Constant volume method	158
A	Na	PWD	C_2H_6	22 - 100	7.4		1.3×10^{-6}	Studied effect of presorbed H_2O	158
A	K	PWD	C_2H_6	22 - 100	9.0		2.2×10^{-6}		158
A	Mg	PWD	C_2H_6	22 - 100	7.0		1.1×10^{-6}		158
A	Ca	PWD	C_2H_6	22 - 100	7.5		1.4×10^{-6}		158
A	Na	PWD	N_2	−78	5.5			Effect of temperature change	180
A	Ca	PWD	C_3H_8	−78	3.0			Equilibrium data also	180
A	Ca	PWD	n-C_5 to n-C_8	93 to 207		$D/r^2 \approx 20 \times 10^{-4}$			84
A	K	Pellets	Ar, Kr	27 to	Ar 3.5, Kr 5.9			Chromatographic method	174
A	Ca	Pellets	Ar, Kr	160	4.07				174
A	Na	PWD	N_2	−79	4.07		2.3×10^{-9}	Equilibrium data included	80
A		PWD	CH_4	−79	7.42		5.8×10^{-8}		80
A	Na	PWD	C_2H_6	25 - 117	5.66				159
A	Na	Pellets	C_2H_6	25 - 117	5.23				160
A	Li, Na, K	PWD	C_2H_6	24, 78				Effect of presorbed H_2O	181
A	Ca	PWD, pellets	CH_4 to n-C_4H_{10}	−88 to 225				Concentration dependence	162
A	Ca	Pellet	CO_2	−25 to 25	~10^{-11}			Self diffusion and equilibrium data	182
A	K	PWD	Ne	50 to 200	7.0		4.5×10^{-8}	See Table 8.28	154
A	K	PWD	Ar	300 to 500	12.6		3.7×10^{-8}		154
A	K	PWD	Kr	350 to 500	16.4		60.3×10^{-8}		154
A	K	PWD	Xe	400 to 600	19.2		124×10^{-8}		154
A	K	PWD	H_2	25 to 200	9.9		4.5×10^{-6}		154
A	K	PWD	CO_2	200 to 600	4.3		4.0×10^{-11}		154
A	K	PWD	N_2		16.2		1.0×10^{-5}		154

688 ADSORPTION KINETICS AND DIFFUSION

Table 8.34 Summary of Selected Adsorption Kinetics Data *(continued)*

Zeolite	Cation	Form	Adsorbate	Temp (°C)	E (kcal/mole)	D (cm²/sec)	D_0 (cm²/sec)	Remarks	Ref.
A	Na	Pellets	C_2H_6	25 - 150	6.4	2.5-5.7×10^{-13}		Langmuir correlation	166
A	Ca	Pellets	C_2H_6	25 - 150	4.45	5 to 50×10^{-12}		Langmuir correlation	166
A	Ca	Pellets	C_2 to n-C_4	25 - 75		0.3-9×10^{-6}		Langmuir isotherms	177
			C_2H_2, C_3H_6						
A	Ca	PWD	n-alkanes	200 - 300	16.1	0.6-17×10^{-13}		Also gives isotherms ΔH	178
Chabazite		Mineral	C_3H_8	150 - 200	See Table 8.26			See Fig. 8.29	152
			n-C_4H_{10}	150 - 200					152
Chabazite			H_2, N_2, Ar	$-78, -183$	Vary with NH_3 content			Effect of preadsorbed NH_3	179
			C_2H_6	$-78, -183$					179
Chabazite		Mineral	CH_4	50		1.62-2.6×10^{-5}		Effect of particle size— low degassing temperature	175
Chabazite		Mineral	H_2O	4.5	6.7	4.8×10^{-6}	1.2×10^{-3}	See Table 8.32	156
Mordenite	Li	Mineral	Ar, Kr	-78	7.3 - 7.6			Effect of exchange on relative rates	152, 157
Mordenite	K		Ar, Kr, N_2, O_2	-78	4.4 - 10.0				152, 157
Mordenite	Na		Ar, Kr	20	9.3 - 11.0				152, 157
Mordenite	Ba		Ar, N_2	-78	6.6 - 9.8				152, 157
Mordenite	Ca		Ar, N_2, O_2	-78	8.1 - 11.5				152, 157
Zeolon	H	PWD	C_2H_6	-80				See Table 8.30	171
Zeolon	Na	PWD	CH_4 to n-C_4H_{10}	25 - 100				See Table 8.29	169, 170
X	Na	Pellets	Ar	154	2.9	180×10^{-3}		Chromatographic method	174
X	Na	Pellets	Kr	154	4.5	59×10^{-3}			174
X	Na	Pellets	SF_6	154	7.5	1×10^{-3}			174

H. ADSORPTION EQUILIBRIA FOR BINARY MIXTURES ON ZEOLITES

Until recently, the adsorption of gas mixtures by zeolites has received relatively little attention. The behavior of gas mixtures on zeolites was first reported by Barrer and Robbins in 1953 (183). The gas mixtures $H_2 + Ne$, $H_2 + N_2$ and $Ne + N_2$ were studied at $-183°C$ in mordenite and $Ar + O_2$ and $Ar + N_2$ were studied in the mineral chabazite at $-183°C$. Both components of the mixture must be capable of entering the intracrystalline channels of the zeolite. If, due to a sieving effect, one component of the mixture is excluded, a simple separation occurs. In some instances the pure component may inhibit or prevent the adsorption of the second component. For example, oxygen, as a single component, is readily adsorbed by zeolite A at $-183°C$ and nitrogen is excluded (12). In an $O_2 - N_2$ mixture at $-183°C$ nitrogen inhibits the adsorption of oxygen and little total adsorption occurs.

Experimentally the usual procedure is to circulate a mixture of the gas through a small bed of adsorbent until equilibrium is established, using an apparatus such as that of Basmadjian (184). A gas mixture of constant composition is passed over the adsorbent. When equilibrium is reached, the adsorbent-containing tube is isolated, the adsorbed mixture desorbed into a calibrated gas buret and analyzed. The data are generally presented in the form of a mixture isotherm together with the composition of the adsorbed phase.

To illustrate the behavior of a binary mixture, the adsorption of oxygen/nitrogen mixtures on zeolite X is shown (Figs. 8.36 and

Figure 8.36 This isothermal diagram for oxygen-nitrogen mixtures on NaX zeolite at $-78°C$ is a three-dimensional diagram which shows the amount adsorbed, the pressure, and the mixture composition at a constant temperature ($-78°C$). Sections through the isotherm diagram parallel to the X-Y plane are the mixture isotherms as shown in Fig. 8.37 (153).

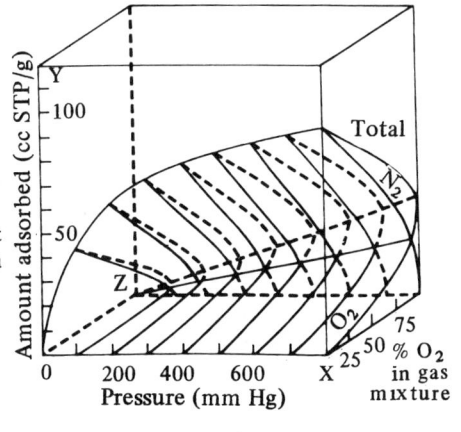

8.37) (153).

In a two-component mixture the relative selectivity or separation factor is derived from a plot of the equilibrium gas phase composition versus the composition of the adsorbed phase as in Fig. 8.38. If the composition of the adsorbed phase is the same as the gas phase, the adsorbent shows no selectivity for either component and the separation factor, α, as defined by the following equation, is unity.

$$\alpha = \frac{Y_a \cdot X_g}{Y_g \cdot X_a} \qquad (8.35)$$

where X_a, Y_a are the mole fractions of the two adsorbates, X, Y, in the adsorbed phase and X_g, Y_g in the gas phase.

There has been little progress made in predicting the behavior of gas mixtures from the adsorption isotherms of the pure components. One of the simplest applications involves the use of the Langmuir equation which has been applied in cases where the components of the mixture appear to conform to this adsorption equation:

$$\alpha = \frac{V_m^x \cdot b^x}{V_m^y \cdot b^y} \qquad (8.36)$$

where V_m^x and V_m^y are the monolayer capacities and b^x and b^y are the Langmuir constants (see Section B).

For this equation to apply the separation factor should be independent of the gas pressure and concentration. In the adsorption of oxygen-nitrogen mixtures on zeolite X, the composition of the adsorbed phase varies with pressure and the highest selectivity for nitrogen is found at low pressures and low nitrogen concentrations.

In a two-component system, at equilibrium, there are three degrees of freedom and a three-dimensional representation is needed, either isothermal (T constant) or isobaric (P constant) (Fig. 8.36). The application of the Langmuir equation is limited because it assumes that the surface is energetically homogeneous.

It is more suitable to interpret mixture adsorption on zeolites in terms of the relative importance of the energy terms involved in the interaction of individual molecules with the interior surface of the zeolite crystal. (This is discussed under selectivity effects in Section F.) The calculation of the absolute magnitude of the energy of adsorption in a zeolite is not practical. However, the total adsorption

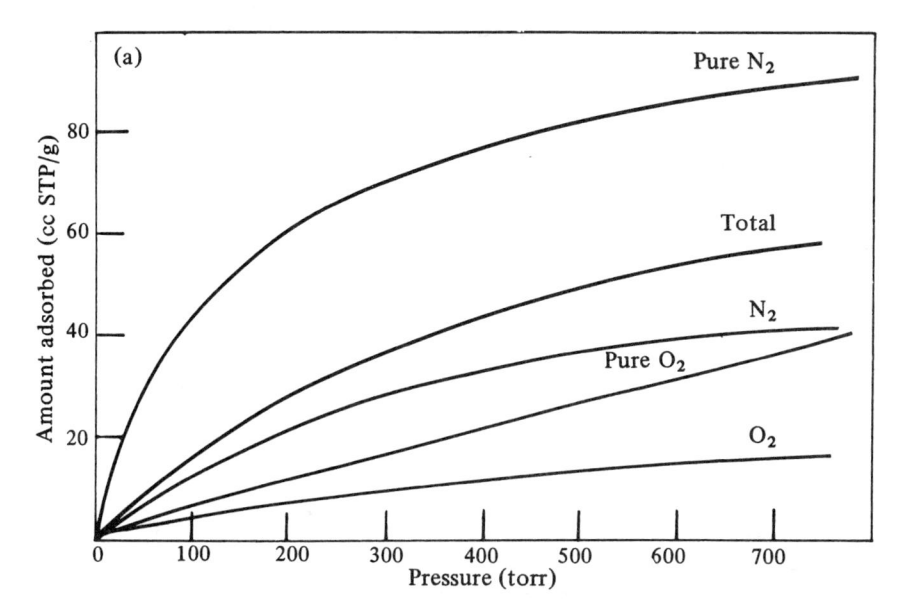

ADSORPTION EQUILIBRIA FOR BINARY MIXTURES ON ZEOLITES 691

Figure 8.37 Oxygen-nitrogen mixture isotherms on NaX zeolite at $-78°$; (a) 77.4 % O_2; (b) 50% O_2; (c) 21% O_2 (153).

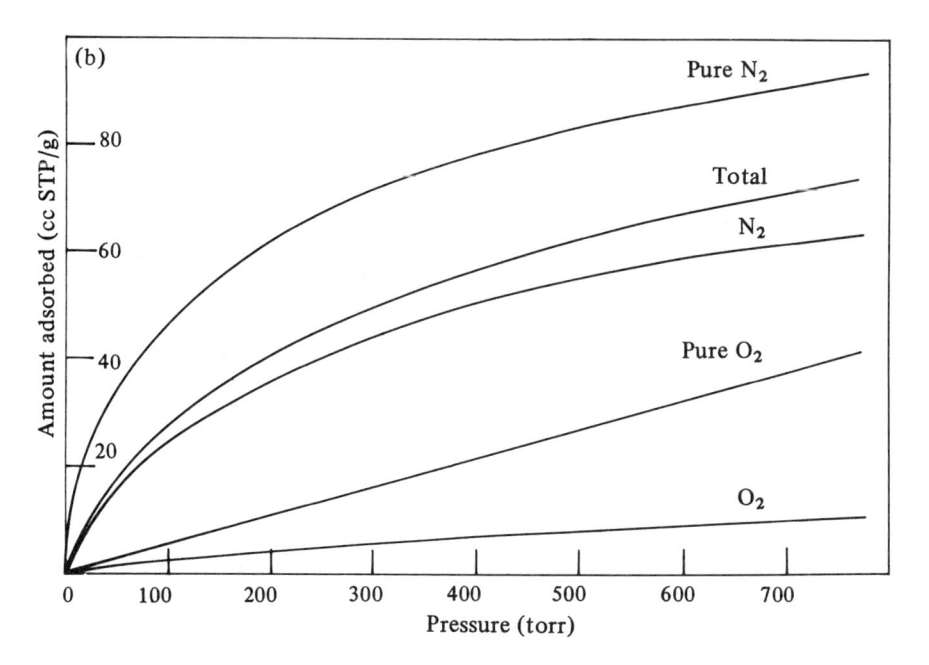

(c)

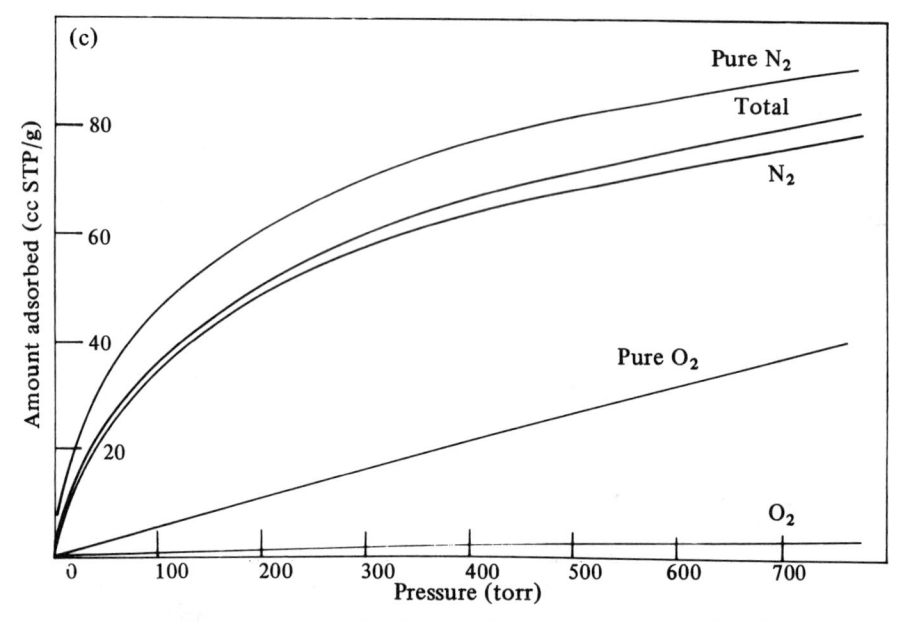

energy can be related to the interaction between the adsorbate molecules and the exchange cations present in the main cavity. For various cation exchanged forms of zeolite X, the relative numbers of oxygen and nitrogen molecules present in each unit cell is shown in Table 8.35. The number of nitrogen molecules decreases from LiX to CsX yet the number of oxygen molecules remains fairly constant. Since there are about 11 univalent cations in each cavity, these results indicate that there are many more cations than molecules in the adsorption cavities. If mutual interactions between adsorbate molecules are neglected, as well as the perturbing influence of adjacent cations, one may assume that the composition of the adsorbed gas at equilibrium depends on the relative heats of adsorption of the two different species. One can estimate the energy in a single cation-adsorbate interaction and then compare it with the composition of the adsorbate as a function of the cation. For a nonpolar molecule, the attraction potential consists mainly of induction and dispersion contributions. For nitrogen, an additional energy term due to the cation-quadrupole interaction must be considered. The interaction of the adsorbate quadrupole with the electrostatic field gradient contributes to the isosteric heat of adsorption and hence to the variation in the heat of adsorption, Q_{iso}, with the degree of filling of the

Table 8.35 Adsorption of a 25% Oxygen-75% Nitrogen Mixture on Zeolite X (153)

Cation	Exchange(%)	N₂ Adsorption cc/g	Molecules/ unit cell	O₂ Adsorption cc/g	Molecules/ unit cell
Li⁺	86	106	57.8	4.7	2.6
Na⁺	100	67	39.9	5.6	3.3
K⁺	80	56	36.0	5.3	3.4
Rb⁺	63	41	30.0	5.1	3.7
Cs⁺	50	21.8	17.3	4.6	3.6
Ag⁺	100	39.3	35.8	4.7	4.3

−78°C, 760 mm

microvoids.

A relationship between the separation or selectivity factor for the gas mixture and differences in the heats of adsorption of the constituent gases has been suggested. An equation of the following type has been applied to gas mixtures.

$$\alpha = K \exp \frac{\Delta q_{iso}}{RT} \qquad (8.37)$$

where Δq_{iso} = the difference in isosteric heats (184).

In Table 8.36 the maximum quadrupole cation interaction energies for nitrogen and the alkali metal cations used in this study are given. A plot of the interaction quadrupole energy versus the separation factor, α, is shown in Fig. 8.39. These data support the conclusion that the quadrupole moments of the molecules involved are very important in determining the behavior of zeolite adsorbents in gas mixtures.

The attraction between nitrogen and the zeolite is the basis for an adsorption process for the enrichment of air, i.e,. separation of oxygen from air, utilizing the preferred adsorption of nitrogen (187-189).

Pure component isotherms for the adsorption of H_2 and D_2 on calcium A zeolite and sodium X, both containing approximately 20% inert binder, showed that deuterium is adsorbed to a greater extent than hydrogen. The isosteric heat of adsorption for deuterium was greater than that of hydrogen at low adsorption loadings. The difference in the isosteric heat of adsorption for deuterium and hydrogen is considerably greater on zeolite adsorbents than on silica gel or charcoal (184).

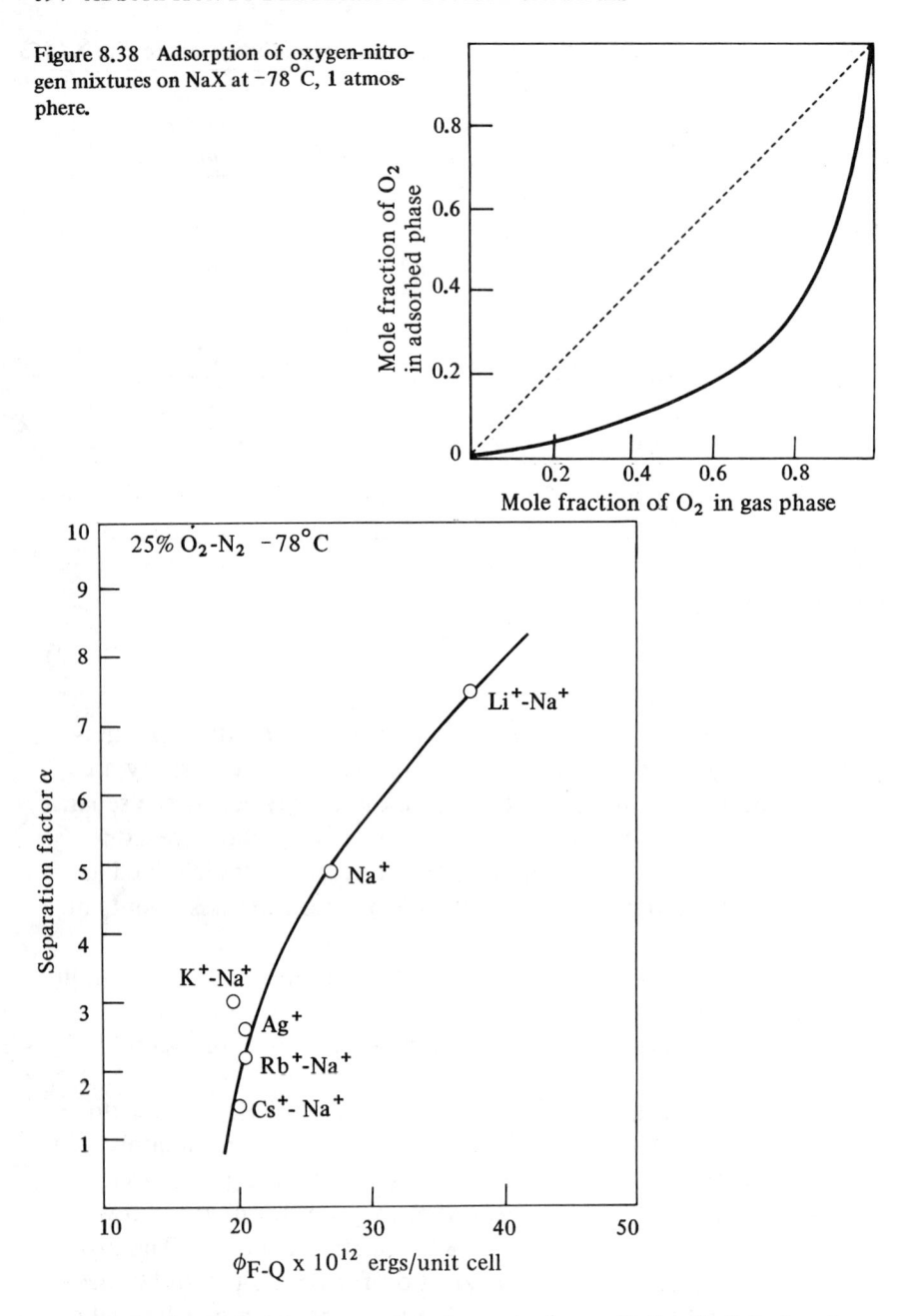

Figure 8.38 Adsorption of oxygen-nitrogen mixtures on NaX at $-78°C$, 1 atmosphere.

Figure 8.39 Correlation of cation quadrupole energies ϕ_{F-Q}, with the selectivity for nitrogen by univalent ion exchanged zeolite X. See also Table 8.36 (153).

Table 8.36 Selectivities of Alkali Metal-Exchanged Zeolite X (153)

Cation	r(A)[a]	Exchange (%)[b]	ϕ_{F-Q} (ergs)[c]	(ergs/unit cell)	Adsorption capacity (cc/g)[d]	Sep. Factor (α)
Li$^+$	0.60	86	45.6 × 10^{-14}	37.6 × 10^{-12}	111	7.5
Na$^+$	0.95	100	31.2	27.0	72.4	4.9
K$^+$	1.33	80	22.3	19.6	61.4	3.0
Rb$^+$	1.48	63	19.4	20.4	46.0	2.2
Cs$^+$	1.69	50	16.1	20.3	26.3	1.5
Ag$^+$	1.26	100	23.8	20.5	44.1	2.6

[a] The Pauling radius of the ion.
[b] The degree of cation exchange in zeolite X.
[c] The calculated cation-nitrogen quadrupole interaction energy.
[d] The total adsorption capacity for a 25% oxygen/nitrogen mixture at $-78°$ and 760 torr.

The binary adsorption equilibria in Fig. 8.40 show that the separation factors for D_2-H_2 adsorption on the zeolite are considerably greater than on carbon or silica gel. No sieving effect was observed for H_2 or D_2 on NaA or CaA, but adsorption on a potassium exchanged A was found to be negligible for both H_2 and D_2. These data confirm exclusion based on molecular size.

Many treatments of the correlation of binary with single component adsorption equilibria are based on the assumption that the single component isotherms conform to the Langmuir, Freundlich, or BET equations. In the case of H_2 - D_2 adsorption no conformation to these three equations was observed.

An empirical relation was used to correlate the adsorption of hydro-

Figure 8.40 Binary adsorption equilibria for hydrogen-deuterium mixtures on zeolites (184).

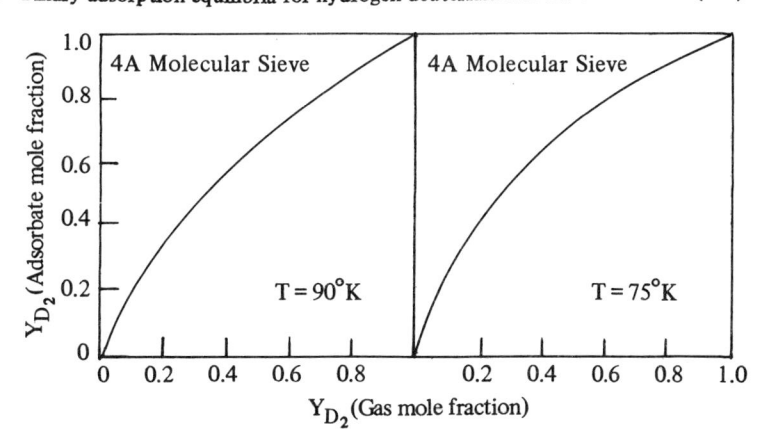

carbon mixtures with pure component isotherms on silica gel and carbon (10). The relation

$$N_1/N_1' + N_2/N_2' = 1 \qquad (8.38)$$

gave mixture adsorption values within 6% of experimental values where N_1 and N_2 are moles of each component adsorbed from the mixture and N_1' and N_2' the pure component capacities. A plot of N_1/N_1' versus N_2/N_2' gave a straight line. The relation was also applied to correlate the adsorption of O_2-N_2, $CO-N_2$, and $CO-O_2$ mixtures in CaA (Type 5A) and CaX (Type 10X) zeolites at low temperatures (−129°C) (178).

The two-component adsorption isotherms for carbon dioxide-nitrogen and sulfur dioxide-nitrogen mixtures were determined on hydrogen mordenite over the temperature range 0° to 100°C. The pure SO_2 adsorption isotherm exhibited a broad hysteresis loop whereas the other gases were totally reversible. In the SO_2-N_2 binary system the presence of nitrogen only slightly affects the adsorption of the very strongly adsorbed SO_2. Correlation of the two-component isotherms with single component isotherms was based upon a Freundlich equation for nitrogen and carbon dioxide. The SO_2 isotherm was fitted by the Langmuir equation.

The SO_2-CO_2 binary system is of interest since the relative selectivity for SO_2 over CO_2 increases with increasing temperature and reverses at a temperature of about 30°C. Below 30°C, CO_2 is the preferred species; above that temperature, SO_2 is the preferred species (92).

The adsorption of mixtures of ethane - n-butane and ethane-carbon dioxide on CaA (commercial type 5A) was interpreted by a solution theory (190). Ethane-normal butane mixtures showed a small negative deviation from Raoult's law. This treatment is called the Ideal Adsorbed Solution Theory (IAST), and is based on the assumption that the adsorbed mixture forms an ideal solution at a constant spreading pressure. A comparison of experimental data for the adsorption of n-butane - ethane and carbon dioxide-ethane mixtures with the adsorption equilibria derived by the ideal solution theory is shown in Figure 8.41. For n-butane - ethane agreement is good but for carbon dioxide-ethane the correlation is poor. Although such analyses may apply to molecules of the same type, such as saturated

ADSORPTION EQUILIBRIA FOR BINARY MIXTURES ON ZEOLITES

Figure 8.41 Adsorption of (a) ethane-n-butane mixtures; and (b) ethane-carbon dioxide mixtures on zeolite CaA (MS 5A) (63).

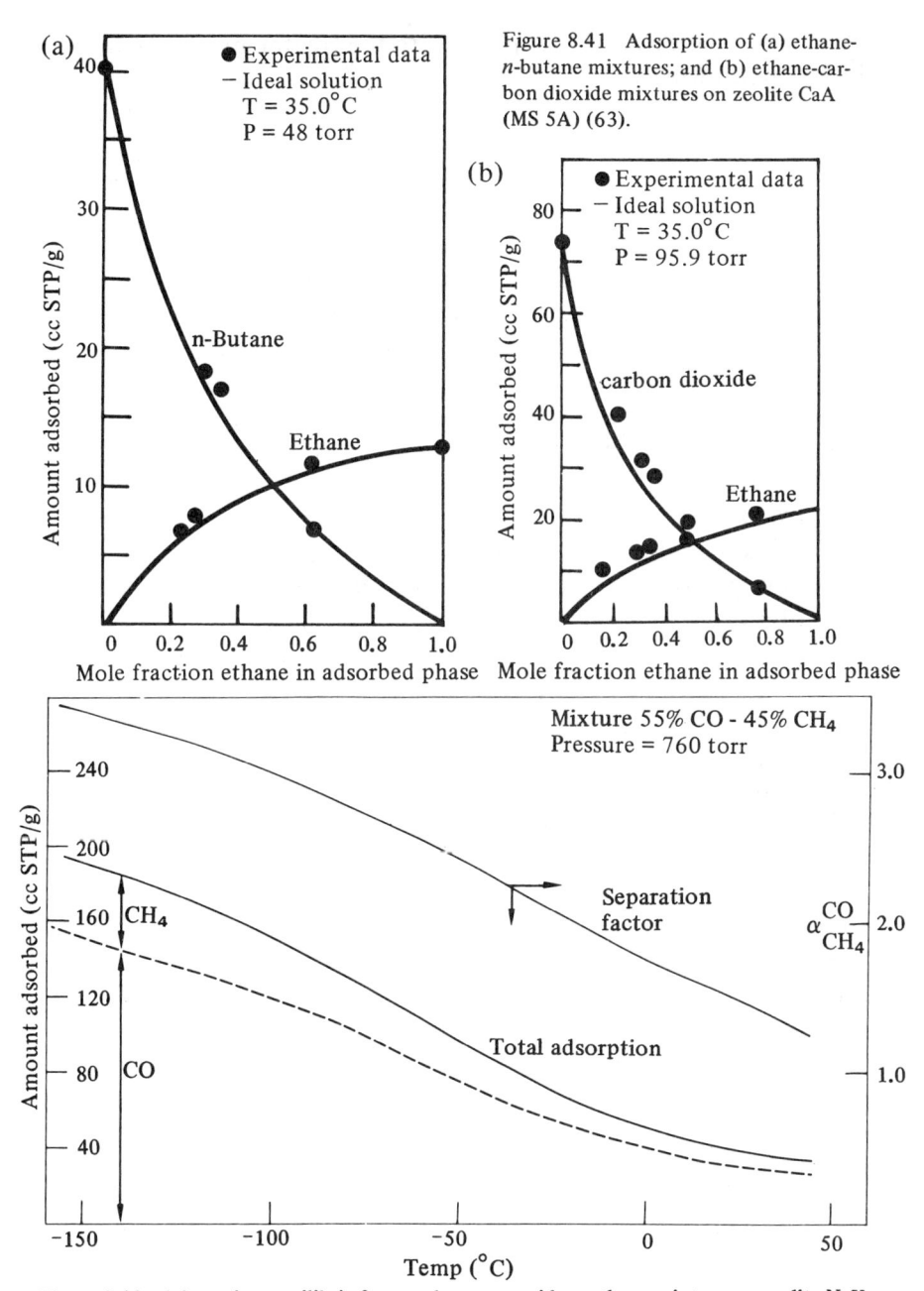

Figure 8.42 Adsorption equilibria for a carbon monoxide-methane mixture on zeolite NaX. Composition of gas phase 55% CO - 45% CH_4. Pressure = 760 torr.

hydrocarbons, they are not applicable to the treatment of mixtures which contain widely different molecular species.

As shown by Eq. 8.37, the selectivity factor for a component of a binary gas mixture is related to the difference in the heats of adsorption of the individual components. The heats of adsorption are related to the total interaction energy between adsorbate and adsorbent which is itself composed of several individual interaction energy terms (Section D). The importance of these various types of interaction energies is illustrated by the binary adsorption behavior of mixtures containing different types of molecules. In a mixture of methane and carbon monoxide, the dispersion and polarization attraction terms are greater for methane. However, the dipole and quadrupole interaction for carbon monoxide is larger and results in an overall selectivity for carbon monoxide over methane in the mixture (Fig. 8.42).

The adsorption of carbon monoxide–nitrogen mixtures on zeolite X at 30°C is shown in Fig. 8.43 (186). The dipole and quadrupole interaction of carbon monoxide is greater than the quadrupole interaction of nitrogen and consequently carbon monoxide is shown to be selectively adsorbed. The variation in selectivity with the exchange cation is illustrated as well as the decrease in selectivity with increasing pressure of the gas mixture. With increased pressure and simultaneous increased loading or coverage of the more active adsorption sites, the relative interaction of the field with the nitrogen quadrupole is diminished and the relative selectivity for carbon monoxide over nitrogen decreases.

In the adsorption of condensable hydrocarbon mixtures selectivity depends upon the unsaturated bonds. For example, in a mixture of n-hexane–benzene, the adsorption equilibria at 97°C show a strong selectivity for benzene over n-hexane on a potassium exchanged zeolite Y (191).

From measurements of the adsorption isotherms of benzene and cyclohexane on a zeolite Y (decationized by ammonium exchange and calcination at 500°C) over the temperature interval 20-200°C, it was found that benzene is much more strongly adsorbed above 150°C. At 150°C and 20 torr, the ratio of benzene to cyclohexane is 3.3, with a benzene saturation loading of 1.41 millimoles (0.11 g) per gram of adsorbent (192, 193).

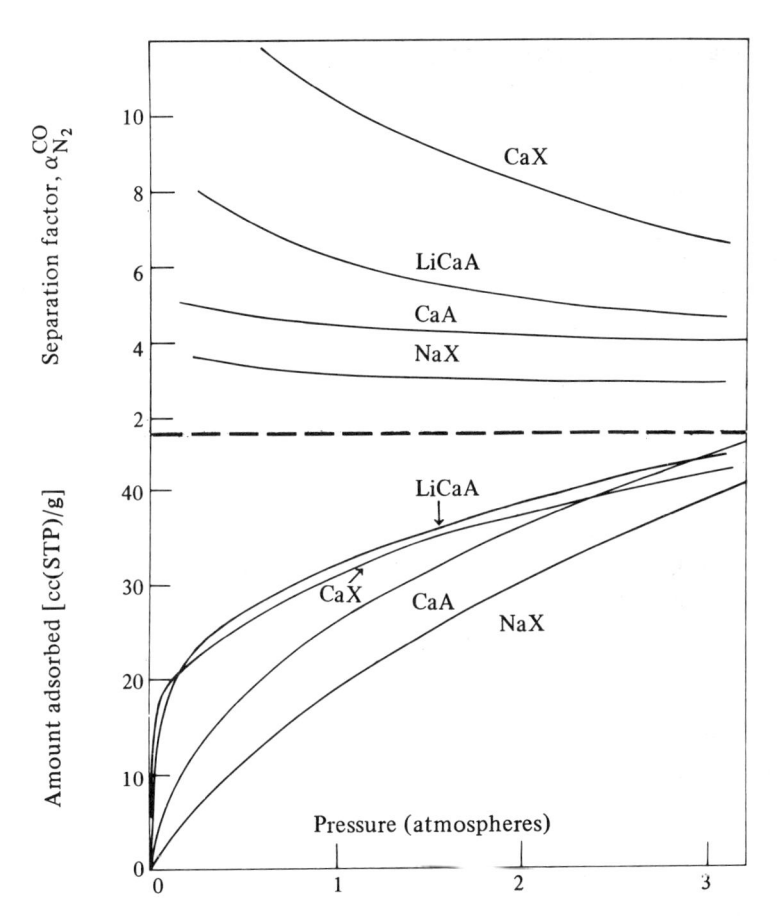

Figure 8.43 Adsorption of a nitrogen-carbon monoxide mixture on NaX, CaX, CaA, and LiCaA. Composition of gas phase - 50% N_2 - 50% CO, Temperature - 30°C (186).

I. ADSORPTION SEPARATION OF MIXTURES

The first separation of gas and liquid mixtures using crystalline zeolites was achieved by Barrer in 1945 (194). He demonstrated that by using three types of zeolite minerals the separation of gases, vapors, and liquids could be achieved. The crystalline zeolites then available did not include the large-pore zeolites such as zeolite X, zeolite Y, zeolite L, large port mordenite, etc. Consequently, his results were confined to the sieve separation of molecules which were no larger than the normal paraffin hydrocarbons. (See Chapter 1.)

Commercial molecular sieve adsorbents in use today in many types of separations are primarily those based on the A, X, and mordenite structures. Chabazite and erionite are utilized to some extent.

The separations, in principle, are based upon factors which we have discussed previously. These include the total sieving effect, a partial sieve effect, and relative adsorpiton selectivities. For obvious economic reasons, most adsorption separations require a regeneration of the adsorbent after it becomes saturated with the adsorbate (regenerative adsorption). From a practical point of view the adsorbent powder could be mixed into the fluid mixture and later filtered out; for most applications, however, separations are based on a fixed-bed of agglomerated zeolite particles. The microcrystalline zeolite is mixed with an inert clay binder and shaped into pellets (see Chap. 9). The particle size, shape and strength of these agglomerates determines the overall characteristics of the bed and the mechanical attrition resistance. Premature decrepitation of the zeolite pellets will produce a fine powder fraction which increases the pressure drop through the bed. This type of system is operated in a semicontinuous manner by using two or more beds in unison. While one bed is used for the adsorption part of a cycle, the other bed is regenerated and the adsorbed phase removed. The usual configuration is vertical and flow through the beds may be either up or down. Horizontal beds are also used.

A partial list of actual commercial separations which have been successfully achieved utilizing molecular sieves is given in Table 8.37 (195). Many hundreds of laboratory separations are reported in the literature. Zeolites have become a popular adsorbent for gas chromatographic separations and there are many references in the literature on analytical applications of this type (196-203).

Molecular Sieving of Water

All of the zeolites have a high affinity for water and other polar molecules and can in general be used for removing water from gases and liquids and for general drying. However, secondary reactions, such as polymerization of adsorbed olefins, etc., plus coadsorption of the hydrocarbon or olefin must be avoided.

The preferred zeolite for the dehydration of unsaturated hydrocarbon streams, such as cracked gas is potassium A (3A). The ef-

Table 8.37 Molecular Sieve Adsorption Processing (195) [a]

A. Petroleum Refining

Catalytic cracking	Unsaturate recovery
	Hydrogen recovery
Hydrocracking	*Recycle hydrogen drying*
	Recycle hydrogen purification
	Feedstock purification
	Hydrogen upgrading
Alkylation	*Feedstock drying*
	Feedstock desulfurization
	Normal/iso separation
Isomerization	*Feedstock drying*
	Feedstock desulfurization
	Normal/iso separation
Catalytic reforming	*Recycle hydrogen drying*
	Recycle hydrogen desulfurization
	Feedstock drying
	Regeneration gas drying
	Normal/iso separation
	Feedstock desulfurization
	Feedstock denitrogenation
	Hydrogen upgrading
Catalytic polymerization	Feedstock desulfurization
	Feedstock drying
Vapor recovery	*Adsorber oil drying*
	Refinery gas drying
	Ethylene recovery
	Propylene recovery
	Hydrogen recovery
	H_2S recovery
Heavy products finishing	*Transformer oil drying*
	Lube oil dewaxing
Light products finishing	*LPG drying*
	Butane drying
	Pentane drying
	Hexane drying
	Heptane drying
	Fuel oil drying
	Jet fuel drying
	Benzene drying
	Xylene drying
	Solvent drying
	LPG sweetening
	Butane sweetening
	Pentane desulfurization
	Propylene recovery
	Jet fuel desulfurization
	Jet fuel upgrading

Table 8.37 Molecular Sieve Adsorption Processing (195) [a] *(continued)*

	Diolefin removal
	Peroxide removal
	Hydrogen upgrading
B. Chemical and Petrochemical Industries	
Ammonia	*Feed desulfurization*
	Synthesis gas drying
	Synthesis gas purification
	NH_3 recovery from vent gas
	Bulk CO_2-removal
	Argon recovery from vent gas
	H_2 recovery from vent gas
Hydrogen	*Feed desulfurization*
	Hydrogen drying
	Hydrogen purification
	Hydrocarbon removal
	By-product hydrogen upgrading
	Recovery from dissociated NH_3
	Bulk CO_2 removal
Sulfuric acid	*SO_2 removal from plant vent gas*
Unsaturates	*Cracked gas drying*
	Ethylene drying
	Propylene drying
	Butene drying
	Butadiene drying
	Isoprene drying
	n-Butene from isobutylene
	CO_2 removal from ethylene
	Wulff furnace gas drying
	Acetylene drying
	Isoprene purification
	Propylene desulfurization
	Ethylene recovery from demethanizer overhead
	Ethylene recovery from ethylene oxide blow-off gas
	Acetylenes from butadiene
	Olefin recovery from refinery off-gas
	Ethylene recovery from coke oven gas
	Ethylene recovery from styrene off-gas
	Methyl acetylene from propadiene
Biodegradable detergents	*n-Paraffin recovery from kerosene*
	Kerosene drying
Aromatics	*Benzene drying*
	Toluene drying
	Xylene drying
	Cyclohexane drying
	Solvent drying
	Styrene drying
	n-Hexane from benzene

Table 8.37 Molecular Sieve Adsorption Processing (195) [a] *(continued)*

Other chemicals	*Butanol drying*
	Acetone drying
	CO_2 drying
	Carbon tetrachloride drying
	Propylene dichloride drying
	Ethylene dichloride drying
	Fluorocarbon refrigerant drying
	Phenol drying
	Acrylonitrile drying
	Pyridine drying
	Ethanol drying
	2-Ethylhexanol drying
	2-Ethylhexyl chloride drying
	n-Butyl chloride drying
	Amyl acetate drying
	Dimethylformamide drying
	Diethyl ether drying
	Isopropyl ether drying
	Isopropanol drying
	Tetrahydrofuran drying
	Methylene chloride drying
	Propylene oxide drying
	Vinyl chloride purification
	Trichloroethylene purification
	Solvent drying
	Nitrogen oxides from HNO_3 off-gas
	Chlorine drying
	Peppermint oil drying
	NH_3 from dissociated NH_3
	Cable oil drying

C. Atmospheric Gases

Oxygen, nitrogen, argon	*Drying of air feed*
	Carbon dioxide removal from air feed
	Nitrogen drying
	Argon drying
	Trace O_2 removal from argon
	Nitrogen oxides from air
	Nitrogen oxides from nitrogen
	Hydrocarbons from air feed
	Oxygen/nitrogen separation.
	Trace methane from liquid O_2
Inert gas	*Bulk carbon dioxide removal from flue gas*
	Sulfur dioxide removal

D. Natural Gas Industry

Production	*Natural gas drying*
	Field condensate drying
	Natural gas sweetening
	Condensate sweetening
	n-Paraffin recovery

704 ADSORPTION BY DEHYDRATED ZEOLITE CRYSTALS

Table 8.37 Molecular Sieve Adsorption Processing (195) [a] *(continued)*

Processing	
Processing	*Natural gas drying*
	Condensate drying
	Absorber oil drying
	LPG drying
	Butane drying
	Pentane plus drying
	Sour Canadian gas drying
	LPG sweetening
	Propane sweetening
	Butane sweetening
	Natural gas sweetening
	Ethane-propane mix purification
	Carbon dioxide removal from natural gas
	Normal/iso separation
	Natural gasoline sweetening
	Bulk H_2S recovery
	Normal paraffin recovery
	Ethane-plus recovery
	Aromatics recovery
	Bulk CO_2 removal
Transmission and storage	*Natural gas drying*
	Natural gas sweetening
	LNG plant feed drying
	LNG plant feed purification
	Helium plant feed drying
	Helium recovery

[a] Commercial applications are in italics.

fective adsorption pore size of 3 A results in the exclusion of all hydrocarbons including ethylene and other olefins which might tend to undergo secondary polymerization reactions if coadsorbed. The elimination of this coadsorption means that product consisting of ethylene, propylene, and butadiene is not lost during regeneration of the adsorbent bed by use of a purge gas. It is also excellent for drying polar liquids (such as ethanol) which are not coadsorbed (Fig. 8.44) (204, 205).

Another molecular sieve effect is utilized in the drying of chlorofluorocarbons used in refrigerants ("R-22", etc.) (206). In this instance zeolite A (4A) is used because the molecular size of the refrigerant molecules precludes adsorption. The "Refrigerant 12", CF_2Cl_2, has a kinetic diameter of 4.4 A and is not adsorbed by zeolite A. In refrigeration systems the adsorptive capacity of the zeolite is not affected by fluctuations in temperature or the presence of lubricating

Figure 8.44 Molecular sieve installation for cracked gas drying, Union Carbide Corporation, Taft, Louisiana.

oils which is always a possibility.

Because it excludes hydrocarbon molecules of the size of propane and larger, zeolite A (4 A) is used in the prepurification of natural gas. Prior to liquefaction, impurities in natural gas, such as water and carbon dioxide must be removed. Due to the high selectivity shown for water and carbon dioxide, zeolite A is used in this very large application. If left in the natural gas, the carbon dioxide and water freeze on the surfaces of the heat exchanger fins. This frost increases the heat transfer resistance and can even block off the flow of the gas (Fig. 8.45) (207-209).

706 ADSORPTION BY DEHYDRATED ZEOLITE CRYSTALS

The zeolite calcium A (5A) is uniquely suited to the separation and recovery of normal paraffin hydrocarbons from various hydrocarbon feedstocks such as natural gasoline and kerosene. The larger molecules such as the paraffin isomers and cyclic hydrocarbons are excluded. More than a half dozen plants are in operation to recover the normal paraffin from the feedstock. The recovered paraffin is utilized in the manufacture of biodegradable detergent (Fig. 8.46) (210-212).

Various commercial processes are in use for the separation of normal paraffin hydrocarbons that differ in the nature of the cycle. Initially the interest in this separation was prompted by a demand for upgrading the octane number of gasoline but later the requirement for normal paraffins as a raw material in detergents became more important. More recently, normal paraffins are required as a starting material for synthetic proteins. One type of process is based on

Figure 8.45 Installation for drying natural gas feed to helium plants. This dehydrator uses 1.25 million pounds of molecular sieve and removes 50,000 pounds of water per day from 500 million SCF per day of natural feed gas. It is located at Cities Service Helex, Inc., Hickok, Kansas.

ADSORPTION SEPARATION OF MIXTURES 707

Figure 8.46 Molecular sieve "IsoSiv" plant at Union Carbide Corporation, Texas City Texas. This plant uses calcium A (Type 5A) zeolite to recover high purity normal paraffins from kerosene. The normal hydrocarbons are used to manufacture biodegradable detergents.

vapor phase adsorption utilizing a pressure-swing desorption cycle. This separation is illustrated in a simple form in Fig. 8.47. A two-bed system is shown: one bed is on the adsorption part of the cycle while the other is being desorbed or regenerated and the product recovered. This process produces normal paraffin separations in the n-pentane to n-decane range with a 95-98% purity product and a 98% recovery. Ideally, the process should be operated so as to produce two products, normal paraffins for raw materials and nonadsorbed isoparaffins and aromatics to be used for octane improvement in gasoline. This separation is used for the removal of normal paraffin hydrocarbons in the higher molecular weight range, C_9 to C_{18}, for use in n-alkylbenzene production and the production of other chemical raw materials.

A molecular sieving effect may be utilized for the separation of large molecules by the zeolite sodium X (13X). Polynuclear aromatics may be separated. Mixtures of C_{16} to C_{20} hydrocarbons were passed

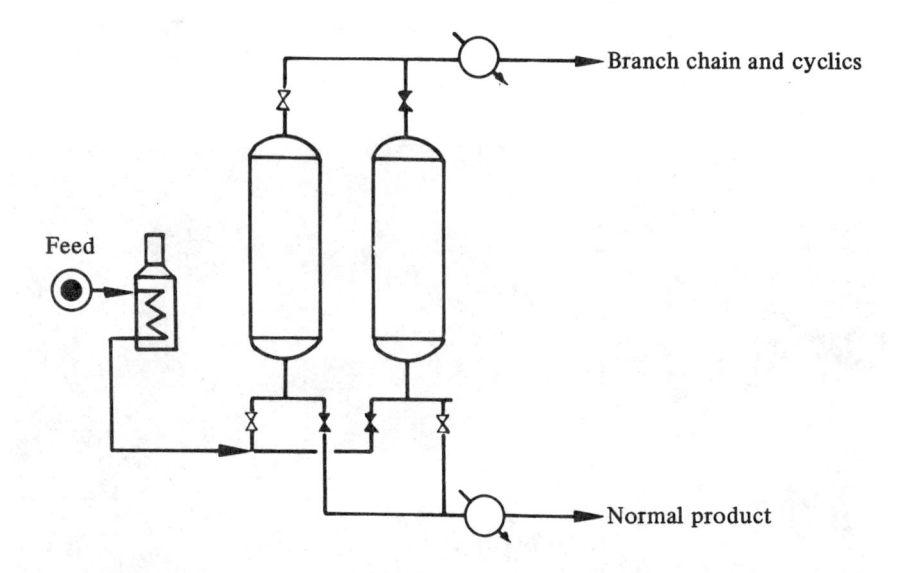

708 ADSORPTION BY DEHYDRATED ZEOLITE CRYSTALS

Figure 8.47 Pressure-swing process as applied to vapor phase, hydrocarbon separation. The left adsorbent bed is on the adsorption step. Normal paraffins are selectively adsorbed by the calcium A zeolite and isomers and cyclic hydrocarbons pass through the adsorber. The right adsorbent bed is on the desorption step; the normal hydrocarbons are desorbed and collected.

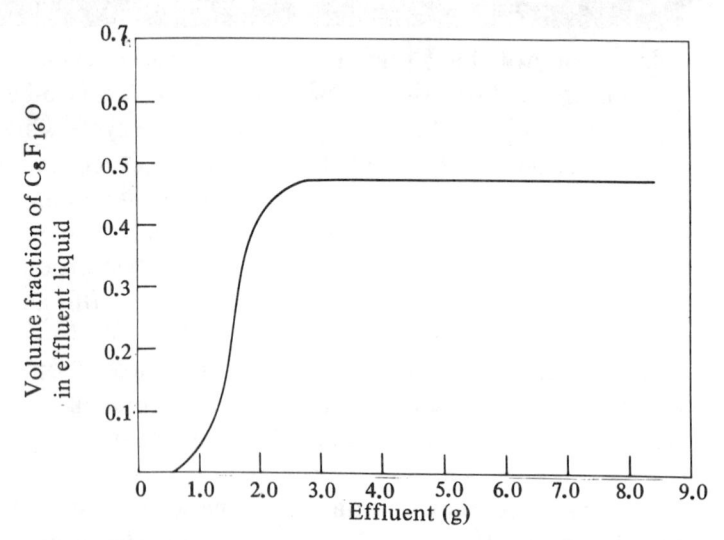

Figure 8.48 Separation of $C_8F_{16}O$ from $(C_4F_9)_3N$ by NaX zeolite in a column. Composition of feed is 50 vol % $C_8F_{16}O$. Loading of $C_8F_{16}O$ on the column at breakthrough was 30 wt.%.

Table 8.38 Percentage of Hydrocarbon Adsorbed on Molecular Sieve Adsorbents CaX and NaX (213)

Hydrocarbon	Formula	Approximate % Adsorbed On CaX	NaX
n-Decylbenzene	$C_{16}H_{26}$	100	100
1, 3, 5-Triethylbenzene	$C_{12}H_{18}$	6	100
6-Decyl- (1, 2, 3, 4-tetrahydronaphthalene)	$C_{20}H_{32}$	50	100
2-Butyl-1-hexylindan	$C_{19}H_{30}$	–	100
2-Butyl-5-hexylindan	$C_{19}H_{30}$	60	100
1, 2, 3, 4, 5, 6, 7, 8, 13, 14, 15, 16-dodecahydrochrysene	$C_{13}H_{24}$	5	100

through small columns of calcium X and sodium X. These mixtures, were analyzed at 25°C and the quantity adsorbed on each zeolite expressed as a percentage of the amount of the component in the inlet stream (Table 8.38) (213). All six of the complex hydrocarbons were completely adsorbed on the sodium zeolite but only n-decylbenzene was completely adsorbed on the calcium X. This confirms the pore size diminution effect for calcium X described earlier for fluorinated amines. In one experiment, an equivolume mixture of n-decylbenzene and dodecahydrochrysene was passed through a column of calcium X. The product, dodecahydrochrysene, contained no more than a trace of the n-decylbenzene which was retained on the zeolite column. The recovered adsorbate was 97% n-decylbenzene.

A sieve effect in a liquid system is shown by zeolite sodium X. When passed through a column of sodium X at room temperature, a mixture of the fluorocarbons $C_8F_{16}O$ (a cyclic ether with an estimated kinetic diameter of 6.7 A), and $(C_4F_9)_3N$, (σ = 10.2 A) were cleanly separated with a column loading of 30 wt% at breakthrough (Fig. 8.48).

Separation of Gases Based on Differences in Relative Selectivity

Many important commercial separations of gases and vapors are based upon the differences in relative adsorptive selectivity. An outstanding example of this type of separation is the production of oxygen-enriched

710 *ADSORPTION BY DEHYDRATED ZEOLITE CRYSTALS*

air by the selective adsorption of nitrogen on various zeolites including calcium A, calcium X, and various types of mordenite. As discussed earlier the selectivity for nitrogen, due to the quadrupole interaction, may be altered by varying the cation. Using this principle and a

"LINDOX" PSA OXYGEN GENERATING SYSTEM

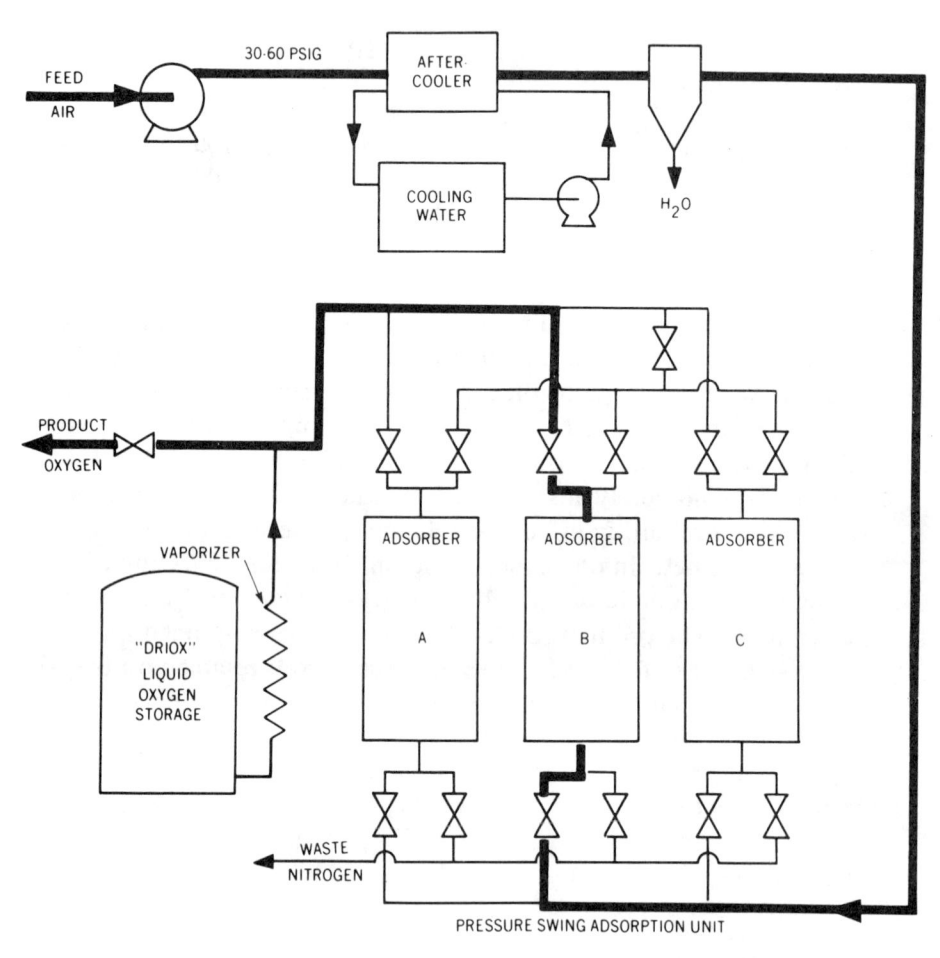

Figure 8.49 Flow diagram of the "LINDOX" PSA oxygen generating system. During the adsorption step, feed air flows through one of the adsorber vessels until the adsorbent is partially loaded. The feed is then switched to another adsorber and the first adsorber is regenerated in three steps: (1) depressurization to atmospheric pressure; (2) purging with product oxygen; and (3) the adsorber is repressurized to the adsorption pressure. In this diagram, feed air is passing through adsorber "B."

complex cycle based on a three-bed or four-bed adsorption system, 90% oxygen may be produced by operating between pressures of 55 psia to atmospheric. Recovery of the product oxygen is as much as 55% of the oxygen introduced in the air feed (189). This process is in use for the production of oxygen for secondary sewage treatment (Fig. 8.49 and 8.50).

Pressure-swing adsorption processes are used for the purification of hydrogen (214). A typical hydrogen raw material obtained from steam reforming contains carbon dioxide, water, methane, carbon monoxide, and nitrogen as impurities. These gas impurities are either polar, have a quadrupole moment, or a much higher molecular weight, than hydrogen. Impure hydrogen containing as low as 0.5% and up to 40% by volume of the impurities may be purified by this adsorption process.

The removal of water and carbon dioxide from air being fed to air separation plants is necessary to avoid plugging of heat exchangers by ice or solid carbon dioxide. Complete removal is needed and is readily accomplished with a molecular sieve zeolite such as NaX (215, 216).

The high adsorptive selectivity shown by zeolites for sulfur compounds such as hydrogen sulfide and mercaptans is utilized in the removal of sulfides from many streams such as hydrocarbons. The removal of sulfur compounds from reformer recycle hydrogen and isomerization feed is necessary in order to protect the catalysts which are sulfur sensitive. In natural gas processing, water, carbon dioxide and sulfur compounds are removed in a selective molecular sieve adsorption process. Residual levels are less than 0.5 ppm. Zeolite NaX is preferred in this application (Fig. 8.51) (217).

Zeolite CaX shows a selectivity effect for individual components of mixtures of aromatic hydrocarbons. In Table 8.39 the compounds in column A are reported to be more strongly adsorbed than the corresponding compound in column B. Separations were achieved in both the vapor phase and liquid phase at temperatures of 97 - 600°F and at pressures of 100 torr to 50 psia (218, 219).

Normally, zeolites such as zeolite X with a large pore size exhibit adsorption selectivity for the aromatic component of an aromatic-normal paraffin mixture. This selectivity, however, is reversed in the hydrogen form of an aluminum-deficient Zeolon. Removal of the

712 ADSORPTION BY DEHYDRATED ZEOLITE CRYSTALS

Figure 8.50 Pressure-swing adsorption plant for producing oxygen from air. The molecular sieve adsorbent removes water, carbon dioxide, and nitrogen. This plant is used in conjunction with the UNOX System for Waste Water Treatment. The main items are: (a) feed air compressor; (2) a PSA unit consisting of adsorber vessels pipe manifolded to segnencing valves; (3) cycle control system; (4) instrument air dryer.

ADSORPTION SEPARATION OF MIXTURES 713

Figure 8.51 Unit for removing mercaptans from natural gas. This unit was built by the El. Paso Natural Gas Company at Jal, New Mexico, and processes 200 MM SCFD of gas. Shallow beds of molecular sieve are located in the horizontal cylindrical adsorber vessels in order to minimize pressure drop.

Table 8.39 Adsorption of Aromatics on CaX (218, 219)

Strongly Adsorbed	Less Strongly Adsorbed
m-xylene	p-xylene
o-xylene	p-xylene
m-xylene	o-xylene
p-xylene	ethylbenzene
1, 2, 4-trimethylbenzene	1, 3, 5-trimethylbenzene
m-ethylmethylbenzene	p-ethylmethylbenzene
1, 2, 4-trimethylbenzene	o-ethyltoluene
1-methylnaphthalene	2-methylnaphthalene
o-diethylbenzene	p-diethylbenzene
p-diethylbenzene	m-diethylbenzene
2, 3-dimethylnaphthalene	2, 6-dimethylnaphthalene
2, 7-dimethylnaphthalene	2, 6-dimethylnaphthalene

cations removes the polarizing influence of the zeolite structure and in aromatic-paraffinic mixtures the paraffinic component is preferred. As shown in Table 8.40, hydrogen Zeolon prefers the aliphatic component, octane and heptane, in a mixture with an aromatic hydrocarbon. In contrast, a typical silica gel may prefer the aromatic component. The composition of the dealuminated mordenite is essentially SiO_2 (220).

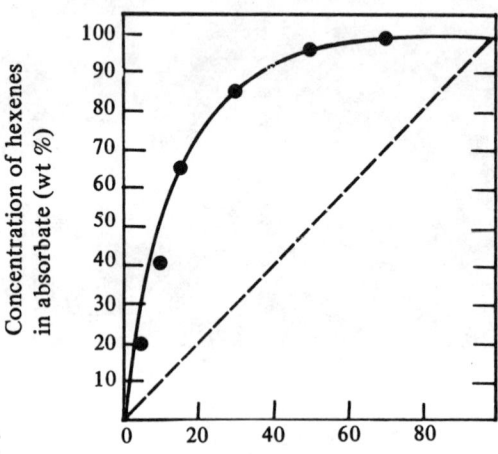

Figure 8.52 Adsorption equilibrium for mixture of n-hexenes and n-hexane on zeolite NaX (221).

Figure 8.53 Schematic illustration of the thermal swing and pressure swing cycles. In thermal swing, Δt, the differential loading, Δx, is given by (214): $\Delta x_t = x_{t_1} - x_{t_2}$ at p_2; in pressure swing: $\Delta x_p = x_{p_2} - x_{p_1}$ at t_1.

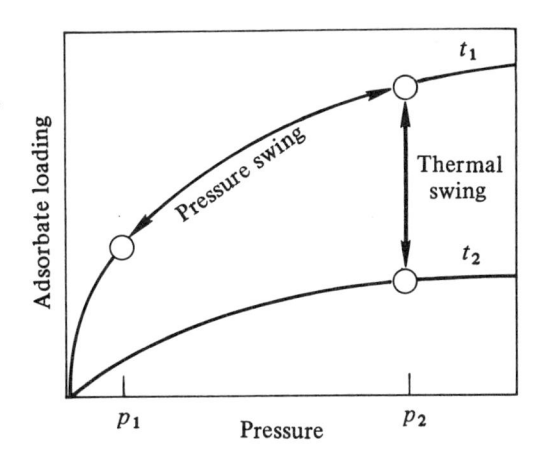

Zeolite sodium X will separate hexenes from mixtures containing normal hexane. The two-component equilibrium system is shown in Fig. 8.52. From mixtures containing about 60% hexenes, by vapor phase adsorption, yields of up to 98% of the hexenes were obtained using NaX as an adsorbent (221).

Adsorption Process Cycles

The basic adsorption process cycles which utilize fixed beds of pelletized molecular sieve zeolites are conveniently classified into four types (223). These types of process cycles differ primarily in the

Table 8.40 Separation of Hydrocarbon Mixtures at 200°F (220)

Mixture (mole %)		SiO_2 Gel Adsorption Capacity (mmole/g)	Preferred Component	H-Zeolon[a] Adsorption Capacity (mmole/g)	Preferred Component
Toluene	96.57	1.17	toluene	0.78	n-octane
n-Octane	3.43				
Benzene	95.28	1.28	n-heptane	0.95	n-heptane
n-Heptane	4.72				
Benzene	95.85			0.95	cyclohexane
Cyclohexane	4.15				

[a] The H-Zeolon was prepared from Na-Zeolon by NH_4NO_3 exchange and subsequently treated with HCl to remove aluminum. The resulting zeolite had a composition of $SiO_2/Al_2O_3 = 93$ with a pore volume of 0.37 cc/g.

desorption step of the process which is the most inefficient step. Generally, the adsorption part of any process cycle occurs rapidly and with a high degree of efficiency. The various process cycles are as follows:

1. Thermal Swing Cycles. As shown schematically in Fig. 8.53 thermal swing, as the name suggests, employs different temperature levels for adsorption (the lower temperature) and desorption (the higher temperature) and consequently operates between two isotherms. The quantity adsorbed in any one cycle is basically the difference in loading at the two different temperatures. The thermal swing cycle enables high adsorbent loadings but a cooling step is necessary to reduce the bed temperature. Heat is transferred to the bed by two methods: (1) direct heating, and (2) indirect heating. In direct heating, hot gases are passed through the adsorbent bed and the adsorbate is simultaneously stripped from the bed. In indirect heating the heat transfer fluid does not contact the solid adsorbent but heating coils are are imbedded and the fluid is passed through the heating coils.
2. Pressure Swing Cycles. In this type of cycle, as illustrated in Fig. 8.53, the process operates at essentially isothermal conditions between two pressures (189, 214, 222). The adsorption pressure is always higher than the desorption pressure and the difference in loading is that obtained between the two different pressures on the isotherm. Pressure swing cycles may operate between super-atmospheric pressures and atmospheric pressure; additional adsorbate may be recovered by vacuum desorption. A schematic flow diagram is shown in Fig. 8.47.

    The advantage of the pressure swing cycle is that no heating and cooling steps are involved; consequently the cycle time may be very short. A fast cycle reduces the size of the adsorbent bed. The main source of energy is gas compression and a high purity product may be produced directly. Relative to liquid phase processes, the hold-up of material in the macrovoids and interpellet space is considerably less.
3. Purge Gas Stripping Cycle. In this cycle a nonadsorbable purge gas is used for desorption. The purpose of the purge gas, like that of reduced pressure, is to reduce the partial pressure of the adsorbed components; desorption therefore occurs. The adsorb-

ate removed from the adsorbent bed is carried off by the purge gas. The purge gas may be either a condensable or a noncondensable fluid at normal conditions. Recycling of the purge gas requires that the adsorbate must be removed. Consequently a condensable purge gas is preferred since power requirements for circulation are reduced; a liquid pump may be used to circulate the fluid, the effluent stream may be condensed and the desorbed material separated from the purge fluid by distillation.

4. Displacement Cycle. In a displacement cycle an adsorbable fluid or vapor is used which displaces the adsorbate already loaded on the adsorbent bed. This fluid must be separated from the adsorbate and the nonadsorbed products. If a displacement fluid is used which is more strongly adsorbed than the adsorbate on the bed, it is adsorbed and will displace the adsorbate. Under these conditions no stripping action due to reduction in the partial pressure of the adsorbate occurs. If a less strongly adsorbed fluid is used, the desorption is due to a combination of displacement and the purge stripping process.

Use of a more strongly adsorbed displacement fluid adds the additional problem of removing it in the process cycle. The use of the displacement purging process is advantageous when the adsorbate to be recovered is sensitive to temperature.

Adsorption Engineering Concepts

In order to utilize zeolites in large scale separations, within recent years several concepts of an engineering nature have been developed. When adsorption occurs in a fixed bed of small diameter and the adsorbate is present as an impurity in low concentrations, the heat that is generated is small and is readily conducted out of the bed. In this case the process is essentially isothermal. In order to interpret dynamic adsorption data and in order to design adsorption systems a concept to handle this situation was developed and is referred to as the "Length of Unused Bed" (LUB) equilibrium section concept for fixed bed adsorption (223). The bed is considered as consisting of two sections: one the equilibrium section and the other the LUB section. The LUB is an additional quantity which is needed to compensate for the mass transfer zone during dynamic adsorption. This approach is best suited for separations where a stable mass transfer zone is formed.

In fixed beds of large diameters, large amounts of heat can be generated by adsorption. This heat is not easily conducted out of the bed. As a result, the heat is either carried out by the gas flowing through the bed or it remains in the bed. Therefore, the bed temperature may rise substantially and affect the adsorption step. The heat generated by adsorption within the transfer zone produces a temperature gradient along the length of the zone. For this situation a more complex procedure has been developed for interpreting and designing adsorption processes (224). A process using pentane as a displacement fluid has been used for the separation of propylene and propane. The process is both isothermal and isobaric. The net effect is that the mixture is split into two pentane-diluted streams, one containing propylene and the other propane. This has been termed vicinal exchange sorption, VES (225).

REFERENCES

1. J. W. McBain, The Sorption of Gases and Vapors by Solids, George Routledge and Sons, Ltd., London, 1932.
2. S. Brunauer, The Adsorption of Gases and Vapors, Princeton University Press, Princeton, New Jersey, 1945.
3. D. M. Young and A. D. Crowell, Physical Adsorption of Gases, T. Butterworth, London, 1962.
4. J. H. de Boer, The Dynamical Character of Adsorption, Oxford University Press, New York, 1953.
5. V..R. Deitz, "Bibliography of Solid Adsorbents, 1943 to 1953," National Bureau of Standards Circular 566, U. S. Government Printing Office, 1956.
6. V. R. Deitz, Bibliography of Solid Adsorbents, Lancaster Press, Lancaster, Pa., 1944.
7. M. M. Dubinin, *Adv. Colloid Interace Sci.,* **2**: 217 (1965).
8. C. Orr, Jr. and J. M. Dallavalle, Fine Particle Measurement, Size, Surface, and Pore Volume, Macmillan, New York, 1959.
9. R. A. Pierotti and H. E. Thomas, Surface and Colloid Science, Vol. 4, E. Matijevic, Ed., John Wiley & Sons, 1971, p. 93.
10. W. K. Lewis, E. R. Gilliland, B. Chertow, and W. P. Cadogan, *Ind. Eng. Chem.,* **42**: 1319, 1326 (1950).
11. George R. Landolt, *Anal. Chem.,* **43** (4): 613 (1971).
12. D. W. Breck, W. G. Eversole, R. M. Milton, T. B. Reed, and T. L. Thomas, *J. Amer. Chem. Soc.,* **78**: 5963 (1956).
13. D. W. Breck and E. M. Flanigen, Molecular Sieves, Society of Chemical Industry, London, 1968, p.47.
14. R. M. Barrer and J. A. Lee, *Surface Science,* **12** (2): 341 (1968).
15. R. M. Barrer and J. A. Davies, *Proc. Roy. Soc., Ser. A,* **1** (1971).
16. R. M. Barrer and J. A. Davies, *Proc. Roy. Soc., Ser. A,* 289 (1970).

17. R. M. Barrer, *Proc. Roy. Soc.*, **A167**: 392 (1938).
18. R. M. Barrer and D. A. Ibbitson, *Trans. Faraday Soc.*, **40**: 195 (1944).
19. R. M. Barrer and B. Coughlan, Molecular Sieves, Society of Chemical Industry, London, 1968, p. 233.
20. L. A. Garden, G. L. Kington, and W. Laing, *Trans. Faraday Soc.*, **51**: 1558 (1955).
21. R. M. Barrer and D. L. Peterson, *Proc. Roy. Soc.*, **A280**: 466 (1964).
22. P. E. Eberly, Jr., *J. Phys. Chem.*, **67**: 2404 (1963).
23. T. Takaishi and A. Yusa, *Trans. Faraday Soc.*, **67**: 3565 (1971).
24. R. M. Barrer, *Trans. Faraday Soc.*, **40**: 555 (1944).
25. E. M. Flanigen, *Netherlands Pat.*, 6,710,729 (1968).
26. D. W. Breck and N. A. Acara, *U. S. Pat.*, 2,950,952 (1960).
27. P. E. Eberly, Jr., *Amer. Mineral.*, **49**: 30 (1964).
28. D. L. Peterson and F. Helfferich, *J. Phys. Chem. Solids*, **26**: 835 (1965).
29. S. P. Zhadanov and B. G. Novikov, *Dokl. Akad. Nauk. SSR*, **166**: 1107 (1966).
30. P. E. Eberley, Jr., *J. Phys. Chem.*, **66**: 812 (1962).
31. A. Cointot, J. Cruchaudet, and M. H. Simonot-Grange, *Bull. Soc. Chim. Fr.*, 497, (1970).
32. G. Gnauck, E. Rosner, and E. Eichorst, *Chem. Tech. (Leipzig)*, **22**: 680 (1970).
33. O. P. Mahajan and P. L. Walker, Jr., *J. Colloid Interface Sci.*, **29**: 129 (1969).
34. D. Berg and W. M. Hickam, *J. Phys. Chem.*, **65**: 1911 (1961).
35. L. V. C. Rees and C. J. Williams, *Trans. Faraday Soc.*, **60**: 1973 (1964).
36. J. G. Daunt and C. Z. Rosen, *J. Low Temp. Phys.*, **3**: 89 (1970).
37. R. M. Barrer, R. Papadopoulos, and J. D. F. Ramsay, *Proc. Roy. Soc., Ser. A*, 331 (1972).
38. R. M. Barrer and J. L. Whiteman, *J. Chem. Soc.*, **A**: 19 (1967).
39. B. G. Aristov, V. Bosacek and A. V. Kiselev, *Trans. Faraday Soc.*, **63**: 2057 (1967).
40. O. M. Dzhigit, A. V. Kiselev, and L. G. Ryabuikhina, *Zh, Fiz, Khim.*, **44**: 1790 (1970).
41. E. Rabinowitch and W. C. Wood, *Trans. Faraday. Soc.*, **32**: 947 (1936).
42. R. M. Barrer and R. Papadopoulos, *Proc. Roy. Soc., Ser. A*, 315, (1972).
43. E. B. Krasnyi, L. M. Iozefson, and L. I. Piguzova, *Zh. Prikl. Khim. (Leningrad)*, **43**: 2449 (1970).
44. R. M. Barrer and J. A. Lee, *J. Colloid Interface Sci.*, **30**: 111 (1969).
45. W. E. Addison, J. Plummer, and A. Walton, *J. Chem. Soc.*, 4728 (1962).
46. R. I. Derrah, K. F. Loughlin, and D. M. Ruthven, *J. Chem. Soc., Faraday Trans.*, **68**: 1947 (1972).
47. J. C. Gupta, Y. H. Ma, and L. B. Sand, *Amer. Inst. Chem. Eng. Symp. Ser.*, **117**: 51 (1971).
48. S. S. Khvoshchev, S. P. Zhadanov, and M. A. Shubaeva, *Izv. Akad. Nauk SSR, Ser. Khim.*, 1004 (1972).
49. D. W. Breck and N. A. Acara, *U. S. Pat.*, 2,950,952 (1960).
50. W. J. Sesny, L. H. Shaffer, *U. S. Pat.*, 3,316,691 (1967).
51. D. W. Breck, *J. Chem. Educ.*, **41**: 678 (1964).
52. R. M. Barrer and D. E. W. Vaughan, *Surface Sci.*, **14**: 77 (1969).
53. R. M. Barrer and D. E. W. Vaughan, *Trans. Faraday Soc.*, **67**: 2129 (1971).
54. R. M. Barrer and D. E. W. Vaughan, *J. Phys. Chem. Solids*, **32**: 731 (1971).
55. L. Riekert, *Adv. Catal. Relat. Subj.*, **21**: 281 (1970).
56. R. M. Barrer and P. J. Reucroft, *Proc. Roy. Soc.*, **258A**: 431 (1960), *Proc. Roy. Soc.*, **258A**: 449 (1960).
57. A. V. Kiselev, "Molecular Sieve Zeolites," *Adv. Chem. Ser. 102*, American Chemical Society, Washington, D. C., 1971, p. 37.
58. M. M. Dubinin and L. V. Radushkevich, *Proc. Acad. Sci. USSR*, **55**: 327 (1947).
59. M. M. Dubinin, *Chem. Phys. Carbon*, **2**: 51 (1966).

60. K. F. Loughlin and D. N. Ruthven, *J. Colloid. Interface Sci.*, **39**: 331 (1972).
61. R. J. Harper, G. R. Stifel, and R. B. Anderson, *Can. J. Chem.*, **47**: 4661 (1969).
62. A. K. Lee and D. Basmadjian, *Can J. Chem. Eng.*, **48**: 682 (1970).
63. K. F. Loughlin and D. M. Ruthven, *J. Phys. Chem. Solids*, **32**: 2451 (1971).
64. R. M. Barrer, *Quart. Rev.*, **3**: 293 (1949).
65. G. L. Kington and A. C. Macleod, *Trans. Faraday Soc.*, **55**: 1799 (1959).
66. L. Pauling, Nature of the Chemical Bond, 3rd Ed., Cornell University Press, Ithaca, New York (1960).
67. T. Kihara, *J. Phys. Soc., Japan*, **6**:289 (1951).
68. J. O. Hirschfelder, C. F. Curtiss, and R. B. Bird, Molecular Theory of Gases and Liquids, John Wiley & Sons, New York, 1954, pp. 215, 1110.
69. H. A. Stuart, Die Struktur des Freien Moleküls, Vol. 1, Springer, Berlin, 1952, p. 88.
70. D. W. Breck and J. V. Smith, *Sci. Amer.*, **200**: 85 (1959).
71. R. P. Hamlen, Union Carbide Corporation, *Unpublished Results*.
72. G. T. Kerr, *Inorg. Chem.*, **5**: 1539 (1966).
73. R. M. Barrer and D. W. Riley, *Trans. Faraday Soc.*, **46**: 853 (1950).
74. M. M. Dubinin, K. M. Nikolaev, N. S. Polyakov, and N. I. Seregina, *Izv. Akad. Nauk SSR, Ser. Khim.*, 1871 (1971).
75. Y. Nishimura and H. Takahashi, *Kolloid. Z. Polym.*, **245**: 415 (1971).
76. R. Aiello, R. M. Barrer, J. Arthur, and I. S. Kerr, *Trans. Faraday Soc.*, **66**: 1610 (1970).
77. L. V. C. Rees and T. Berry, Molecular Sieves, Society of Chemical Industry, London, 1968, p. 149.
78. R. M. Barrer and L. V. C. Rees, *Trans. Faraday Soc.*, **50**: 852 (1954).
79. D. W. Breck and R. J. Clark, Union Carbide Corporation, *Unpublished Results*.
80. H. W. Habgood, *Can. J. Chem.*, **36**: 1384 (1958).
81. K. F. Loughlin and D. M. Ruthven, *Chem. Eng. Sci.*, **27**: 1401 (1972).
82. R. M. Barrer and J. L. Whiteman, *J. Chem. Soc., Ser. A*, 13 (1967).
83. W. Schirmer, G. Friedrich, A. Grossman, and H. Stark, Molecular Sieves, Society of Chemical Industry, London, 1968, p. 252.
84. P. E. Eberly, Jr., *Ind. Eng. Chem., Prod. Res. Develop.*, **8**: 140 (1969).
85. R. M. Barrer and J. W. Sutherland, *Proc. Roy. Soc.*, **A237**: 439 (1956).
86. R. M. Barrer and S. Wasilewski, *Trans. Faraday Soc.*, **57**: 1140 (1961), *Trans. Faraday Soc.*, **57**: 1153 (1961).
87. A. G. Bezus and A. V. Kiselev, *Zh. Fiz. Khim.*, **40**: 1773 (1966).
88. R. M. Barrer, F. W. Bultitude, and J. W. Sutherland, *Trans. Faraday Soc.*, **53**: 1111 (1957).
89. T. A. Egerton and F. S. Stone, *J. Colloid Interface Sci.*, **38**: 195 (1972).
90. B. G. Aristov, V. Bosacek, and A. V. Kiselev, *Trans. Faraday Soc.*, **63**: 2957 (1967).
91. T. A. Egerton and F. S. Stone, *Trans. Faraday Soc.*, **66**: 2364 (1970).
92. J. I. Joubert and I. Zwiebel, Molecular Sieve Zeolites, Vol. 102, American Chemical Society, Washington, D. C., 1971, p. 209.
93. R. M. Barrer and E. V. T. Murphy, *J. Chem. Soc.*, **A**: 2506 (1970).
94. R. M. Barrer and B. Coughlan, Molecular Sieves, Society of Chemical Industry, London 1968, p. 141.
95. R. A. Jones, *U. S. Pat.*, 3,224,167 (1965).
96. R. M. Barrer and R. M. Gibbons, *Trans. Faraday Soc.*, **61**: 948 (1965).
97. R. M. Barrer and W. I. Stuart, *Proc. Roy. Soc.*, **A249**: 464 (1959).
98. R. M. Barrer, *J. Colloid Interface Sci.*, **21**: 415 (1966).
99. M. M. Dubinin, *J. Colloid Interface Sci.*, **23**: 487 (1967).
100. R. M. Barrer and R. M. Gibbons, *Trans. Faraday Soc.*, **59**: 2569 (1963).
101. R. J. Neddenriep, *J. Colloid. Interface Sci.*, **28**: 293 (1968).

102. L. A. Garden, G. L. Kington, and W. Laing, *Proc. Roy. Soc.*, **234A**: 35 (1956).
103. D. J. C. Yates, Molecular Sieves, Pap. Conf. 1967 (pub. 1968), Soc. Chem. Ind., 334.
104. C. L. Angell and P. C. Schaffer, *J. Phys. Chem.*, **70**: 1413 (1966).
105. C. L. Angell and M. V. Howell, *J. Colloid. and Interface Sci.*, **28**: 279 (1968).
106. J. L. Carter, D. J. C. Yates, P. J. Lucchesi, J. J. Elliott, and V. Kevorkian, *J. Phys. Chem.*, **70**: 1126 (1966).
107. D. J. C. Yates, *J. Phys. Chem.*, **70**: 3693 (1966).
108. E. Cohen de Lara,and J. Vincent-Geisse, *J. Phys. Chem.*, **76**: 1972).
109. W. E. Addison and R. M. Barrer, *J. Chem. Soc.*, **757** (1955).
110. C. C. Chao and J. H. Lunsford, *J. Am. Chem. Soc.*, **93**: 71 (1971), *J. Am. Chem. Soc.*, **93**: 6794 (1971).
111. P. H. Kasai and R. J. Bishop, Jr., *J. Am. Chem. Soc.*, **94**: 5560 (1972).
112. T. M. Pietzak and D. E. Wood, *J. Chem. Phys.*, **53**: 2754 (1970).
113. J. A. R. Coope, C. L. Gardner, C. A. McDowell, and A. I. Perlman, *Molecular Physics*, **21**: 1043 (1971).
114. I. R. Leith and H. F. Leach, *Proc. Roy. Soc. Lond. A.*, **330**: 242 (1972).
115. L. Naccache and Y. Ben Taarit, *Chem. Physics Letters*, **11**: 11 (1971).
116. P. Gallezot, Y. Ben Taarit, and B. I. Melik, *J. Catalysis*, **26**: 295 (1972).
117. J. V. Smith, Molecular Sieves, Society of Chemical Industry, London, 1968, p. 28.
118. W. M. Meier and D. P. Shoemaker, *Z. Kristallogr.*, **123**: 357 (1966).
119. A. A. Amaro and K. Seff, *J. Chem. Soc.. Chem. Comm.* 1201 (1972).
120. R. Y. Yanagida and K. Seff, *J. Phys. Chem.*, **76**: 2597 (1972).
121. K. Seff, *J. Phys. Chem.*, **76**: 2601 (1972).
122. H. Lechert and H. J. Hennig, *Z. Phys. Chem.*, **76**: 319 (1971).
123. T. A. Egerton and R. D. Green, *Trans. Faraday Soc.*, **67**: 2699 (1971).
124. J. A. Rabo, C. L. Angell, P. H. Kasai, and V. Schomaker, *Disc. Faraday Soc.*, **41**: 328 (1966).
125. A. V. Kiselev and A. A. Lopatkin, Molecular Sieves, Society of Chemical Industry, London, 1968, p. 252.
125. O. M. Dzhigit and A. V. Kiselev, *Trans. Faraday Soc.*, **67**: 458 (1971).
126. M. M. Dubinin, A. A. Isirikyan, A. I. Sarakhov, and V. V. Serpinskii, *Izv. Akad. Nauk SSR, Ser. Khim.*, 2355 (1969).
127. N. N. Avgul, *Zh, Fiz. Khim.*, **42**: 768 (1968).
128. B. Morris, *J. Colloid Interface Sci.*, **28**: 149 (1968)..
129. O. M. Dzhigit and A. V. Kiselev, *Trans. Faraday Soc.*, **67**: 458 (1971).
130. H. Stach, T. Peinze, K. Fiedler, and W. Schirmer, *Z. Chem.*, **10**: 229 (1970).
131. N. Dupont-Pavlovsky, and J. Bastick, *Bull. Soc. Chim. Fr.*, **1**: 24 (1970).
132. V. L. Keibal, *Zh. Fiz. Khim.*, **41**: 1203 (1967).
133. R. M. Barrer and P. J. Cram, "Molecular Sieve Zeolites," *Adv. Chem. Ser. 102*, American Chemical Society, Washington, D. C., 1971, p. 105.
134. R. M. Barrer, and B. Coughlan, Molecular Sieves, Society of Chemical Industry, London, 1968, p. 241.
135. K. H. Sichhart, P. Kolsch, and W. Schirmer, "Molecular Sieve Zeolites, *Adv. Chem. Ser. 102*, American Chemical Society, Washington, D. C., 1971, p. 132.
136. D. M. Ruthven, and K. F. Loughlin, *Trans. Faraday Soc.*, **68**: 696 (1972).
137. R. M. Barrer, and C. G. Bratt, *Phys. Chem. Solids*, **12**: 130 (1960), *J. Phys. Chem. Solids*, **12**: 146 (1960), *J. Phys. Chem. Solids*, **12**: 154 (1960)
138. N. N. Avgul, and A. V. Kiselev, *Russ. J. Phys. Chem.*, **42**: 96 (1968).
139. O. M. Dzhigit, A. V. Kiselev, K. N. Mikos, and G. G. Muttik, *Russ. J. Phys. Chem.*, **38**: 973 (1964).

140. N. N. Avgul, B. G. Aristov, A. V. Kiselev, and L. Ya. Kurdyukova, *Zh. Fiz. Khim.*, **42**: 2678 (1968).
141. A. T. Khudiev, *Izv. Akad. Nauk SSR, Ser. Khim.*, 717 (1968).
142. N. N. Avgul, and A. V. Kiselev, *Kolloid Zhur.*, **25**: 129 (1963).
143. V. I. Bogmolov, M. Kh. Minachev, N. V. Mirzabekova, and I. Ya. Isakov, *Izv. Akad. Nauk SSR, Ser. Khim.* 41 (1968).
144. O. M. Dzhigit, S. P. Zhdanov, A. V. Kiselev, T. A. Mel'nikova, K. N. Mikos, and G. G. Muttik, *Zh. Fiz. Khim.*, **41**: 1431 (1967).
145. O. M. Dzhigit, K. Karpinskii, A. V. Kiselev, T. A. Mel'nikova, K. N. Mikos, and G. G. Muttik, *Zh. Fiz. Khim.*, **42**: 198 (1968).
146. B. V. Romanovskii, K. V. Topchieva, L. V. Stolyarova, and A. M. Alekseev, *Kinet. Katal.*, **11**: 1525 (1970).
147. B. V. Romanovskii, K. V. Topchieva, L. V. Stolyarova, and A. M. Alekseev, *Kinet. Katal.*, **12**: 1003 (1971).
148. A. L. Kliyachko-Gurvich, A. T. Khudiev, I. Ya. Isakov, and A. M. Rubinshtein, *Izv. Akad. Nauk SSR, Ser. Khim.*, 1355 (1967).
149. L. A. Garden, and G. L. Kington, *Trans. Faraday Soc.*, **52**: 1397 (1956).
150. R. M. Barrer, J. A. Lee, *Surface Sci.*, **12**: 354 (1968).
150. R. M. Barrer, and J. A. Lee, *Surface Sci.*, **12**: 354 (1968).
151. S. S. Khvoshchev, S. P. Zhdanov, and M. A. Shubaeva, *Dokl. Akad. Nauk SSR*, **196**: 1391 (1971), *Izv. Akad. Nauk SSR,Ser Khim.*, 1004 (1972).
152. R. M. Barrer, and D. W. Brook, *Trans. Faraday Soc.*, **49**: 1049 (1953).
153. D. W. McKee, and R. P. Hamlen, Union Carbide Corporation *Unpublished Results*.
154. P. L. Walker, Jr., L. G. Austin, and S. P. Nandi, *Chem. Phys. Carbon*, **2**: 257 (1966).
155. R. M. Barrer, and D. L. Peterson, *J. Phys. Chem.*, **68**: 3427 (1964).
156. R. M. Barrer, "Molecular Sieve Zeolites," *Adv. Chem. Ser. 102*, American Chemical Society, Washington D. C., 1971, p. 1.
157. R. M. Barrer, *Trans. Faraday Soc.*, **45**: 358 (1949).
158. W. W. Brandt, and W. Rudloff, *J. Phys. Chem. Solids*, **26**: 741 (1965).
159. E. F. Kondis, and J. S. Dranoff, "Molecular Sieve Zeolites," *Adv. Chem. Ser. 102*, lical American Chemical Society, Washington, D. C., 1971, p. 171.
160. E. F. Kondis, and J. S. Dranoff, *Ind. Eng. Chem., Process Des. Develop.*, **10**: 108 (1971).
161. R. M. Barrer, and D. A. Ibbitson, *Trans. Faraday Soc.*, **40**: 206 (1944).
162. D. M. Ruthven, and K. F. Loughlin, *Trans. Faraday Soc.*, **67**: 1661 (1971).
163. D. M. Ruthven, and K. F. Loughlin, *Chem. Eng. Sci.*, **26**: 577 (1971).
164. D. M. Ruthven, and K. F. Loughlin, *Chem. Eng. Sci.*, **26**: 1145 (1971).
165. C. R. Antonson, and J. S. Dranoff, *Chem. Eng. Prog. Symp. Ser.*, **74**: 61 (1967).
166. C. R. Antonson, and J. S. Dranoff, *Chem. Eng. Prog. Symp. Ser.*, **65**: 27 (1969).
167. C. N. Satterfield, and J. R. Katzer, "Molecular Sieve Zeolites," *Adv. Chem. Ser. 102*, American Chemical Society, Washington, D. C., 1971, p. 193.
168. C. N. Satterfield, J. R. Katzer, and W. R. Vieth, *Ind. Eng. Chem., Fundam.*, **10**: 478 (1971).
169. C. N. Satterfield, and A. J. Frabetti, Jr., *J. Amer. Instit. Chem. Eng.*, **13**: 731 (1967).
170. C. N. Satterfield, and W. G. Margetts, *J. Amer. Instit. Chem. Eng.*, **17**: 295 (1971).
171. L. Riekert, *J. Amer. Instit. Chem. Eng.*, **17**: 446 (1971).
172. R. M. Barrer, and B. E. F. Fender, *J. Phys. Chem. Solids*, **21**: 1 (1961)
173. C. N. Satterfield, and C. S. Cheng, *Amer. Instit. Chem. Eng. Symp. Ser.*, **117**: 43 (1971).
174. P. E. Eberly, Jr., *Ind. Eng. Chem., Fundam.*, **8**: 25 (1969).
175. W. W. Brandt, and W. Rudloff, *J. Phys. Chem.*, **71**: 3948 (1967).
176. S. S. Khvoshchev and S. P. Zhdanov, *Dokl. Akad. Nauk SSR*, **200**: 1156 (1971).

177. G. R. Youngquist, J. L. Allen, and J. Eisenberg, *Ind. Eng. Chem., Prod. Res. Develop.*, **10**: 308 (1971).
178. W. Schirmer, G. Fiedrich, A. Grossman, and H. Stach, Molecular Sieves, Society of Chemical Industry, London, 1968, p. 276.
179. R. M. Barrer, and L. V. C. Rees, *Trans. Faraday Soc.*, **50**: 989 (1954).
180. J. D. Eagan, B. Kindl, R. B. Anderson, "Molecular Sieve Zeolites," *Adv. Chem. Ser. 102*, American Chemical Society, Washington, D. C., 1971, p. 164.
181. W. Rudloff, and W. W. Brandt, *J. Phys. Chem.*, **71**: 3689 (1967).
182. R. W. H. Sargent, and C. J. Whitford, "Molecular Sieve Zeolites," *Adv. Chem. Ser. 102*, American Chemical Society, Washington, D. C., 1971, p. 155.
183. R. M. Barrer, and A. B. Robbins, *Trans. Faraday Soc.*, **49**: 807 (1953), *Trans. Faraday Soc.*, **49**: 929 (1953).
184. I. D. Basmadjian, *Can. J. Chem.*, **38**: 141 (1960).
185. G. S. Petryaeva, N. V. Keltsev, and E. A. Timofeeva, *Izv. Akad. Nauk SSR, Ser. Khim.*, 1860 (1967).
186. R. J. Neddenriep, Union Carbide Corporation, *Unpublished Results*.
187. D. W. McKee, *U. S. Pat.*, 3,140,933 (1964).
188. D. W. McKee, *U. S. Pat.*, 3,140,932 (1964).
189. L. B. Batta, *U. S. Pat.*, 3,564,816 (1968), *U. S. Pat.*, 3,636,679 (1972).
190. A. J. Glessner, and A. L. Myers, *Chem. Eng. Progr., Symp. Ser.*, **65**: 73 (1969).
191. P. E. Eberly, C. N. Kimberlin, and L. E. Baker, *J. Appl. Chem.*, **17**: 44 (1967).
192. K. I. Slovetskaya, T. R. Brueva, and A. M. Rubinshtein, *Izv. Akad. Nauk SSR, Ser. Khim.*, 37 (1968).
193. K. I. Slovetskaya, T. R. Brueva, and A. M. Rubinshtein, *Izv. Akad. Nauk SSR, Ser. Khim.*, 249 (1968).
194. R. M. Barrer, *J. Soc. Chem. Ind.*, **64**: 130 (1945).
195. J. J. Collins, *Chem. Eng. Prog.*, **64**: 66 (1968).
196. T. G. Andronikashvili, G. V. Tsitsishvili, and Sh. D. Sabelashvili, *J. Chromatogr.*, **58**: 47 (1971).
197. V. Bosacek, *Coll. Czechoslov. Chem. Communic.*, **29**: 1797 (1964).
198. G. Eppert, Molecular Sieves Society of Chemical Industry, London, 1968, p. 182.
199. A. V. Kiselev, Yu. L. Chernen'kova, and Ya. I. Yashin, *Gazov, Khromatogr.*, 38 (1966).
200. H. A. Szymanski, *J. Gas Chromatogr.*, **2**: 154 (1964).
201. G. V. Tsitsishvili, T. G. Andronikashvili et al., *Dokl Chem.*, **194**: 774 (1970).
202. G. V. Tsitsishvili, and T. G. Andronikashvili, "Molecular Sieve Zeolites," *Adv. Chem. Ser. 102*, American Chemical Society, Washington, D. C., 1971, p. 217.
203. J. A. J. Walder, *Nature*, **209**: 197 (1966).
204. G. E. Hales, *Chem. Eng. Progr.*, **67**: 49 (1971).
205. J. E. Pierce, and D. L. Steinhagen, *Hydrocarbon Process. Petrol. Refiner.*, **45**: 170 (1966).
206. R. L. Mays, *ASH REA J.*, **4**: 73 (1962).
207. J. P. Fris, and A. Kessock, *Hydrocarbon Process. Petr. Ref.*, **44**: 123 (1965).
208. G. H. Weyermuller, J. D. Harlan, and D. Roberts, *Chem. Process.*, 37 (1966).
209. E. L. Clark, *Oil Gas J.*, **57**: 120 (1959).
210. G. J. Griesmer, W. F. Avery, and M. N. Y. Lee, *Hydrocarbon Process. Petrol. Refiner.*, **44**: 147 (1965).
211. M. J. Sterba, *Hydrocarbon Process Petrol. Refiner.*, **44**: 151 (1965).
212. L. J. LaPlante, *Oil Gas J.*, **68**: 55 (1970).
213. B. J. Mair, and M. Shamaienger, *Anal. Chem.*, **30**: 276 (1958).
214. H. A. Stewart, and J. L. Heck, *Chem. Eng. Progr.*, **65**: 78 (1969).
215. R. E. Latimer, 59th Annual Meeting Amer. Inst. Chem. Engrs. (1966).
216. D. A. Webber, *Chem. Eng. (London).*, **1**: 18 (1972).

217. S. A. Conviser, *Oil Gas. J.*, **63**: 130 (1965).
218. P. E. Eberly, Jr., and W. F. Arey, *U. S. Pat.*, 3,126,425 (1964).
219. R. N. Fleck, and C. G. Wright, *U. S. Pat.*, 3,114,782 (1963).
220. P. E. Eberly, *Ind. Eng. Chem., Prod. Res. Develop.*, **10**: 433 (1971).
221. G. S. Petryaeva, E. A. Timofeeva, and N. I. Shuikin, *Dokl. Adad. Nauk SSR*, **172**: 361 (1967).
222. D. Domine, and L. Hay, Molecular Sieves, Society of Chemical Industry, London, 1968, p. 204.
223. J. J. Collins, *Chem. Eng. Progr., Symp. Ser.*, **63**: 31 (1967).
224. F. W. Leavitt, *Chem. Eng. Progr.*, **58**: 54 (1962).
225. D. L. Peterson, F. G. Helfferich, and R. K. Griep, Molecular Sieves, Society of Chemical Industry, London, 1968, p. 217.
226. J. Sameshima, and H. Hemmi, *Bull. Chem. Soc. Japan*, **9**: 27 (1934).

Chapter Nine

MANUFACTURE AND PROPERTIES OF COMMERCIAL MOLECULAR SIEVE ADSORBENTS

The structural chemistry and related physical and chemical properties of many species of synthetic zeolites and zeolite minerals have been discussed and reviewed. There are more than 30 known structural types and at least an equal number of unknown structural types of zeolites. It is significant that relatively few types of zeolites are actually utilized in the many commercial applications. The majority of the known types of zeolites are laboratory curiosities with as yet unknown applications.

This chapter reviews the commercially important zeolite products and relates these to the basic zeolite species as discussed in previous chapters. Part of the substantial development that led to the use of zeolites in important chemical processes and separation processes is discussed. The commercialization of zeolites as molecular sieves has stimulated extensive scientific interest leading to the discovery of new chemical and structural properties.

A. MANUFACTURING PROCESSES

Outside of the patent literature there are few publications which are concerned with the processes of manufacturing molecular sieve zeolites on a commercial scale (1). Processes for the manufacture of commercial molecular sieve products may be classified into three groups.
1. The preparation of molecular sieve zeolites as high purity crystalline powders or as preformed pellets from reactive aluminosilicate gels or hydrogels.
2. The conversion of clay minerals—kaolin, in particular—into zeo-

lites either in the form of high purity powders or as binderless high purity preformed pellets.
3. Processes based on the use of other naturally occurring raw materials.

The hydrogel and clay conversion processes may also be used to manufacture products which contain the zeolite as a major or a minor component in a gel matrix, a clay matrix, or a clay-derived matrix. These three processes are summarized in Table 9.1.

Hydrogel Processes

The first commercial process for preparing zeolites on a large scale was based on the results of original laboratory syntheses using amorphous hydrogels. (See Chap. 4.) The typical starting materials included sodium silicate in aqueous solution, sodium aluminate solution, and sodium hydroxide. In Table 9.2, the hydrogel processes are further classified into three categories.

1. Processes which use homogeneous gels, that is, hydrogels prepared from solutions of soluble reactants.
2. Processes based on the use of heterogeneous hydrogels which

Table 9.1 Molecular Sieve Zeolite Preparation Processes

Process	Reactants	Products
Hydrogel	Reactive oxides Soluble silicates Soluble aluminates Caustic	High purity powders Gel preform Zeolite in gel matrix
Clay conversion	Raw kaolin Metakaolin Calcined kaolin Soluble silicate Caustic Sodium chloride	Low to high purity powder Binderless, high purity preform Zeolite in clay-derived matrix
Other	Natural SiO_2 Acid-treated clay Amorphous minerals Volcanic glass Caustic $Al_2O_3 \cdot 3 H_2O$	Low to high purity powder Zeolite on ceramic support Binderless preforms

Table 9.2 Hydrogel Processes

Process	Reactants	Product	Reference
Homogeneous gel	Sodium silicate Sodium aluminate Caustic $Al_2O_3 \cdot 3 H_2O$	Type A—powder Type X—powder Type Y—powder	2-6 7, 8 9
Heterogeneous gel	$Al_2O_3 \cdot H_2O$ Sodium aluminate Silica sol Amorphous solid silica Sodium silicate Caustic	Zeolon—mordenite powder Type X—powder Type Y—powder	18 7 9-17, 72
Gel preform	Sodium aluminate Amorphous silica Sodium silicate	Type A—preformed spheres Type Y—preformed spheres Mordenite preform Type Y—in a gel matrix	19-23 24-26 27 28-30

are prepared from reactive alumina or silica in a solid form, for example, solid amorphous silica powder.

3. Gel-preform processes in which the reactive aluminosilicate gel is first formed into a pellet. This is reacted with sodium aluminate solution and caustic solution to crystallize the zeolite *in situ* within the pellet, leaving essentially no binder or crystallizing the zeolite as a component in an unconverted amorphous matrix. This binderless amorphous matrix is desirable as a base for many catalysts such as cracking catalysts.

The literature source for these processes is based on available patent literature, specifically in issued United States patents. This does not imply that all of these processes are currently practiced. Table 9.2 correlates technical information given in various patents in terms of the process and reactant with the particular type of crystalline zeolite product that is prepared. A photo of the Union Carbide molecular sieve plant in Mobile, Alabama is shown in Fig. 9.1.

Figure 9.1 Union Carbide Corporation Molecular Sieve Plant in Mobile, Alabama. (a) Aerial view of the whole plant and (b) hopper cars used to transport zeolite molecular sieves in the form of a filter cake.

A process flowsheet for the manufacture of type 4A, type 13X, and type Y, in high purity, crystalline zeolite powders from reactant hydrogels is shown in Fig. 9.2.

The process flowsheet illustrates the type of equipment and the raw materials used. A typical material balance, and chemical compositions, in terms of oxide mole ratios for the reactant hydrogel and the zeolite product is given. The raw materials are metered into the make-up tanks in the proper ratios. The crystallization step is conducted in a separate crystallizer after an intermediate aging step, where required, is conducted at ambient temperature. As discussed in Chap. 4, the aging step may be required for the synthesis of certain zeolites in high purity.

Although the process appears to be quite simple in terms of equipment and experimental conditions, due to the meta-stability of zeolite species formed from typical reactant systems, problems arise when large scale synthesis is attempted. These problems are related to (a) mixing large masses of materials homogeneously and (b) transfer of heat which may cause the nucleation and growth of undesirable zeolite phases. Generally, the crystallization temperature is near the boiling point of water; in some instances, such as in the synthesis of mordenite-type zeolites, higher temperatures are required.

Techniques have been developed in the laboratory for following the crystallization process in the plant. An important control technique is the use of x-ray powder diffraction patterns of samples taken at regular intervals. Other quality controls may include special types of adsorption measurements as well as chemical analysis. From an evaluation of a combination of these control data, the quality or purity of the manufactured zeolite can be ascertained. After the digestion period, the slurry of crystals in the mother liquor is filtered in a rotary filter as illustrated in Fig. 9.2. Occluded liquor in the filter cake is high in caustic, or some other alkali hydroxide, and may contain some excess silica. This is removed from the filter cake by additional water washing in the rotary filter. In some instances, the excess silica in the mother liquor may be recycled. Washing generally is carried out until the pH of the slurry is about 9.

In order to utilize zeolite crystals commercially, they usually are bonded into agglomerates or pellets by one of various techniques.

730 COMMERCIAL MOLECULAR SIEVE ADSORBENTS

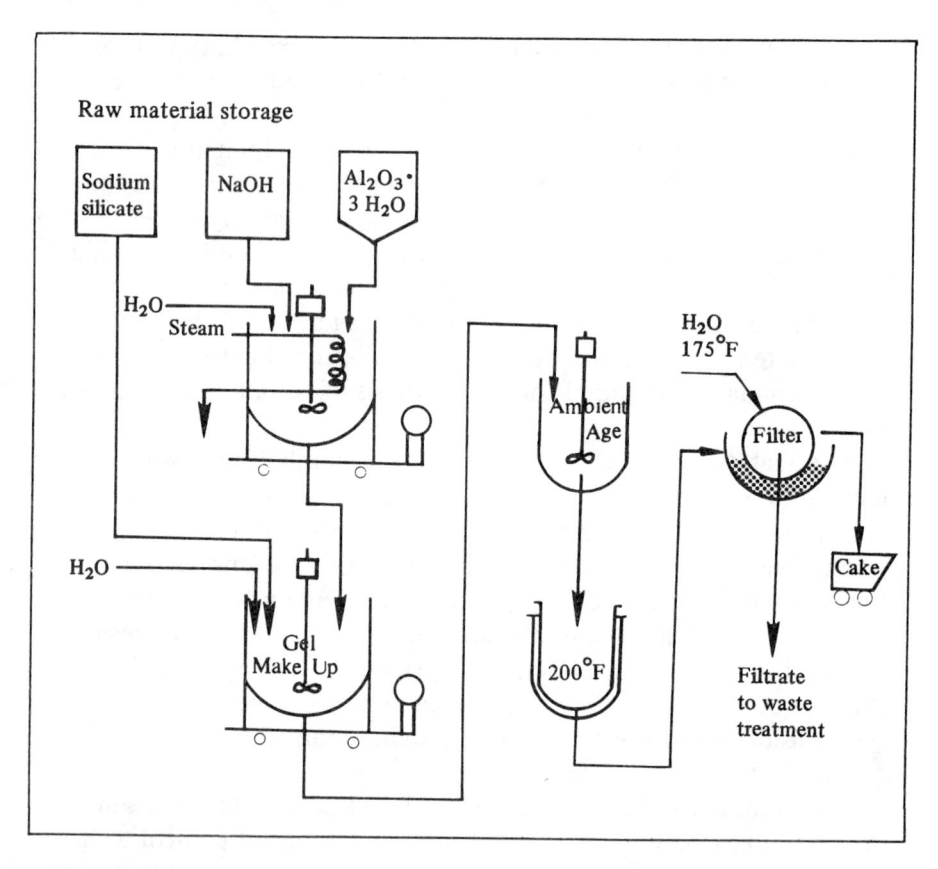

Figure 9.2 Hydrogel Process

Raw Materials	Weight (To Produce 1000 lbs, Dry Basis)		
	4A	13X	Y
Sodium Silicate [a]	1350	2000	–
SiO_2 powder [b]	–	–	1450
Alumina Trihydrate [c]	575	500	340
Caustic, 50% NaOH	870	1600	1400
Water	3135	7687	5300
Gel composition (moles)			
Na_2O	2.04	4.09	4.0
Al_2O_3	1.0	1.00	1.0
SiO_2	1.75	3.0	10.6
H_2O	70	176	161

[a] 9.4% Na_2O, 28.4% SiO_2.
[b] 95% SiO_2.
[c] 65% Al_2O_3, 35% H_2O.

For the manufacture of commercial zeolite type Y from a heterogeneous hydrogel, the equipment is basically the same. Provision is made for preparing a slurry of solid amorphous silica prior to introduction into the gel make-up tank where the other reactants are mixed. In this case the ambient aging step is necessary to prepare a high purity powder product. Other zeolites which are prepared from heterogeneous hydrogels include zeolite X and the mordenite-type Zeolon materials.

Clay Conversion Processes (Table 9.3)

The major clay-type starting material used in the manufacture of molecular sieves zeolite is kaolin. The kaolin is usually dehydroxylated to form metakaolin. The dehydroxylation of kaolin is generally accomplished by air calcination. The structural and chemical changes of kaolin have been extensively studied and were discussed in Chap. 4, Section F. Depending upon the calcination temperature, a product is obtained which may be more or less suitable for conversion to a zeolite. In general, two types of calcined kaolin have been used—metakaolin, calcined at 550°C and kaolin calcined at 925°C (31).

Table 9.3 Clay Conversion Processes

Process	Reactants	Product	Reference
1. Slurry—high purity powder	Caustic Metakaolin Sodium silicate Allophane (36)	Type A Type X Type Y	32 - 37 34, 37 37
2. *In situ* crystallization of preform to yield high purity, binderless pellet	Metakaolin Caustic Sodium silicate Diatomaceous earth	Type A Type X Type Y Zeolon-mordenite	38 - 43, 78 41, 42 41, 42 44
3. Partial, *in situ* conversion of preformed particle to yield zeolite in clay-derived matrix	Caustic Metakaolin Calcined kaolin Sodium silicate Raw kaolin	Type X Type Y	45 - 48 45, 46, 48-59

At 550 to 600°C kaolin undergoes dehydration (endothermic dehydroxylation) to metakaolin according to the following reaction:

$$2\,Al_2Si_2O_5(OH)_4 \rightarrow 2\,Al_2Si_2O_7 + 4\,H_2O \qquad (9.1)$$

The theoretical weight loss for this change is 13.95%. The product aluminosilicate has the correct stoichiometry in terms of the ratio of SiO_2 to Al_2O_3 for zeolite A. For the conversion to zeolites with higher silica contents, additional SiO_2 must be added to the reactant mixture. Metakaolin is amorphous to x-rays, but has a residual type of ordered structure. At 800°C the density is only 1.52% less than that of the original kaolin because the interlayer spacing decreases from 7.15 A to 6.3 A.

As the calcination temperature is raised the unstable reactive metakaolin transforms to a defect aluminum-silicon spinel structure (also referred to as a gamma alumina phase) at 925°C according to the following reaction:

$$2\,Al_2Si_2O_7 \rightarrow Si_3Al_4O_{12} + SiO_2 \qquad (9.2)$$

This spinel phase is cubic and persists over the temperature range of 925-1075°C. The transformation in Eq. 9.2 produces an additional mole of SiO_2 in a very reactive form. At still higher temperatures, 1050-1100°C, this spinel phase transforms to a mullite of uncertain composition with the additional elimination of SiO_2 which appears as cristobalite according to the following equation:

$$3\,Si_3Al_4O_{12} \rightarrow 2\,Si_2Al_6O_{13} + 5\,SiO_2 \qquad (9.3)$$
$$\text{spinel} \qquad\qquad \text{mullite } + \text{ cristobalite}$$

At higher temperatures, the mullite and cristobalite continue to form. The overall reaction of kaolin to mullite at this elevated temperature is given by the following equation:

$$6\,Al_2Si_2O_5(OH)_4 \rightarrow 2\,Si_2Al_6O_{13} + 8\,SiO_2 + 12\,H_2O \qquad (9.4)$$

Therefore, 1000 grams of kaolin, formula weight 222, will produce 360 grams of SiO_2.

Processes for making zeolites from kaolin are summarized in Table 9.3. A process flowsheet for the manufacture of zeolite A in the form of a high purity powder is illustrated in Fig. 9.3. This process is based upon the use of kaolin clay that is calcined at 550°C. In making

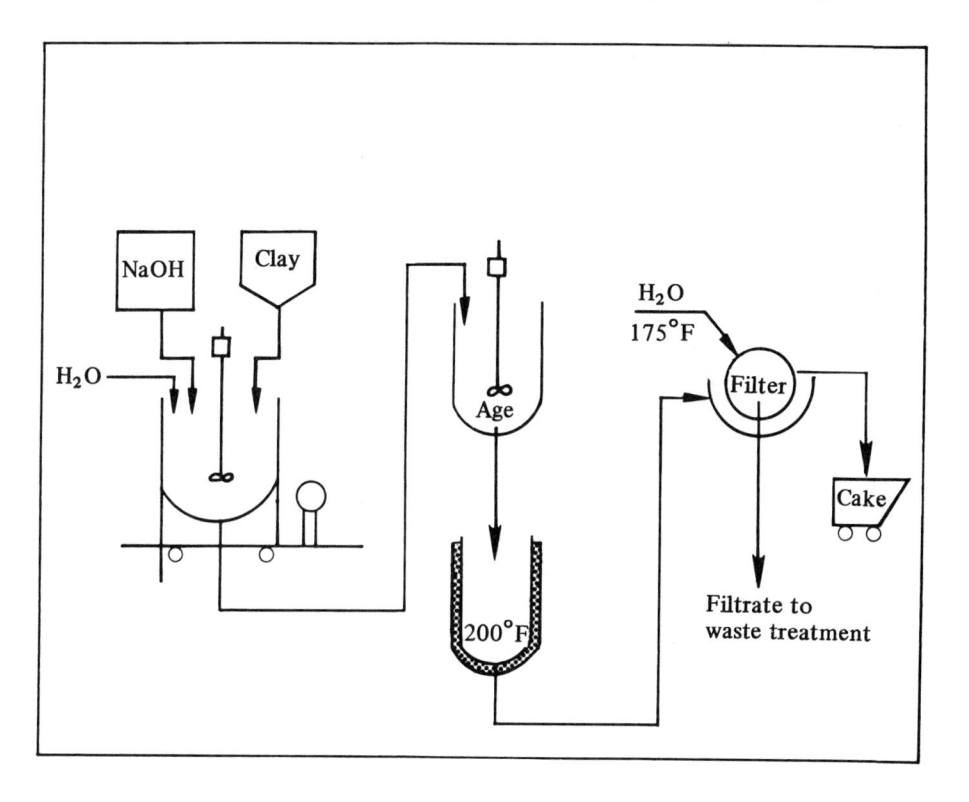

Fig. 9.3 Clay Process for Producing Type 4A

Raw Materials	Weight (To Produce 1000 lb. of 4A)
Calcined kaolin [a]	800
Caustic, 50% NaOH	1100
Water	3755
Reactant Composition (moles)	
Na_2O	2.0
Al_2O_3	1.0
SiO_2	2.0
H_2O	70

[a] 44.8% Al_2O_3, 52.8% SiO_2

zeolite A, the only necessary reactants are caustic, calcined kaolin clay and water which are mixed in the gel make-up tank. The ambient aging step is also used prior to the crystallization step.

Calcined kaolin is the starting material for making several zeolites such as types A, X, and Y (37). Complete conversion of the clay to zeolite powder in high purity is achieved. In order to crystallize zeolite type X or type Y, additional silica is needed to increase the SiO_2/Al_2O_3 ratio of the reactant mixture. In the preparation of zeolite Y a larger quantity of additional silica is generally added as sodium silicate. Addition of inorganic salts enhances formation of zeolite Y from kaolin (42). For example, sodium chloride may be added to the reactant mixture in the ratio of 2 moles of NaCl per mole of alumina.

Several methods are concerned with the preparation of zeolites as essentially binderless preformed pellets starting with kaolin as a raw material. The kaolin is shaped in the desired form of the finished product as an extruded pellet or a sphere, and is converted *in situ* in the pellet by treatment with suitable alkali hydroxide solutions. One manufacturing process is based upon the use of kaolin as a starting material for the preparation of preformed pellets of zeolite A, zeolite X, and zeolite Y. For preformed pellets of zeolite Y, it is advantageous to use added sodium chloride to enhance crystallization (42). Figure 9.4 illustrates a variation of the basic preform process. Crystalline zeolite powder in the form of a filter cake is mixed with the calcined clay, blended extruded into pellets, and dried as shown in Fig. 9.8. The raw kaolin component is then calcined to metakaolin. The metakaolin component in the calcined pellets is then crystallized to zeolite A to form a preformed pellet which contains essentially pure zeolite A. The preformed pellets may be converted by ion exchange to other forms such as moleuclar sieve Type 5A (Fig. 9.4). Table 9.3 includes processes which are based on the partial conversion of a preformed particle to form zeolite within a clay-derived matrix.

In one example, kaolin is used to prepare a product containing about 75% of zeolite Type Y (45). The reaction mixture composition is given in Table 9.4. A commercially available kaolin pre-calcined to form an incipient mullite phase (commercially known as Satintone No. 1) and a metakaolin are used. The components and composition of the reactant mixture are shown in Table 9.4. The product contained 75% zeolite of composition $Na_2O \cdot Al_2O_3 \cdot 3.43\ SiO_2$. Based on the overall stoichiometry, this yield can be obtained only if all of the silica contained in the reactant mixture is incorporated in the zeolite. This would then leave about 145 g of Al_2O_3 in the slurry.

An *in situ* crystallization of zeolite Y has been described (59). In this process a zeolite "precursor" is prepared from typical reactants

Figure 9.4 Clay Preform Conversion Process

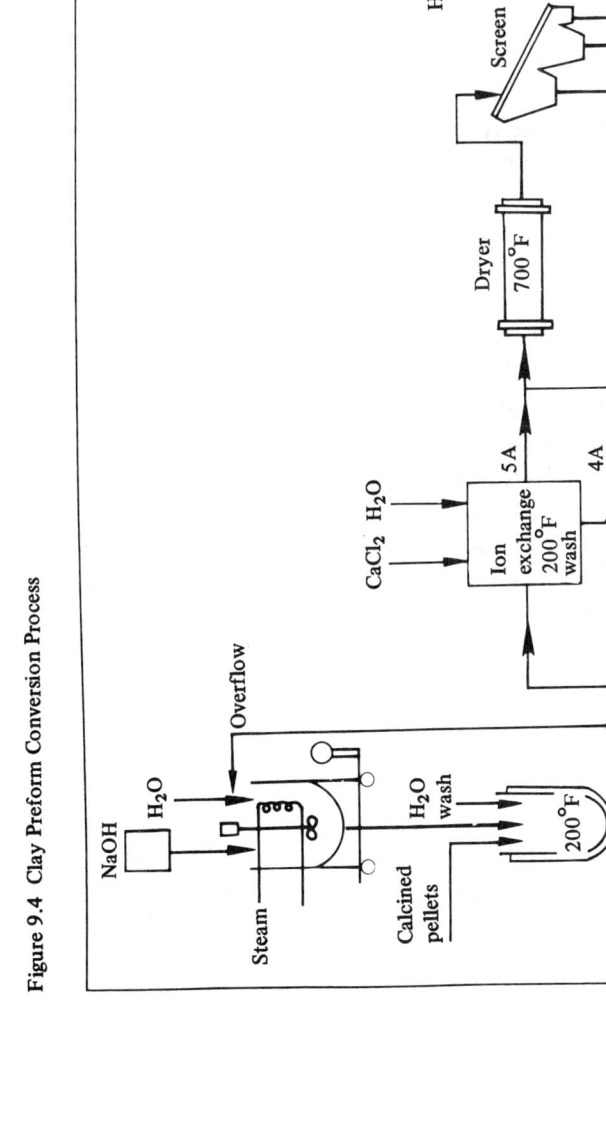

Table 9.4 Zeolite Y from Kaolin (45)

Ingredient	Na$_2$O	Al$_2$O$_3$	SiO$_2$	H$_2$O
Satintone No. 1, 999 g		4.36	8.7	
Satintone No. 2, 333 g		1.45	2.9	
20% NaOH, 1200 ml	3.67			65
Total moles	3.67	4.81	11.6	65
Moles/Al$_2$O$_3$	0.765	1.00	2.42	14.3

such as alumina trihydrate and amorphous silica. After aging this composition at room temperature it is further mixed with kaolin clay and the mixture extruded into pellets. The pellets are than treated at about 100°C for 24 hours and zeolite crystallizes within the clay matrix. The zeolite content is low, typically about 25%. These processes are directed at the manufacture of a zeolite binder composite, but by utilizing reactants that form the zeolite in the presence of an unreative matrix material such as a crystalline kaolin clay.

Another process is described which is said to result in the formation of zeolite aggregates directly in a particle size range of 50 to 200 microns and suitable for fluidized bed applications (48). A slurry prepared from a calcined kaolin clay and a sodium silicate solution is first spray dried to form particles in the range of 50 to 200 microns. The spray-dried particles are first aged in sodium hydroxide solution, at ambient temperature, followed by a high temperature crystallization in the neighborhood of 80 - 120°C. The zeolite forms *in situ* within the particle which, for the most part, appears to retain its physical integrity. The process is reported to be useful in the preparation of zeolite X and zeolite Y.

A process which is aimed at the preparation of a cracking catalyst containing a zeolite X or zeolite Y component in an ammonium or other polyvalent metal ion-exchanged form is described (47). This process is based upon the partial conversion of clay to zeolite by hydrothermal treatment with sodium hydroxide. A product containing 85% zeolite X or Y is prepared from uncalcined kaolin clay. The clay, however, is first treated with a concentrated sodium hydroxide and sodium silicate solution in a furnace at a temperature of approximately 300°C. The resulting solid is ground and further treated hydrothermally in water at about 100°C.

In many instances, it is desired to have the zeolite in a final product, as in a cracking catalyst, contained in a matrix. This may be achieved by several methods. The zeolite component is first manufactured in an essentially pure form and then blended or mixed with a suitable matrix material such as a clay or silica-alumina amorphous gel. Other methods are based upon crystallizing the zeolite from suitable precursor materials such as clays, amorphous silica and various alumina sources *in situ* in the parent material. By achieving partial conversion, the end product contains the proper amount of zeolite component formed in a matrix, such as calcined kaolin clay.

Various processes have been reported in the patent literature for achieving this type of commercial product. These products are largely directed toward use in cracking catalysts where, in general, the zeolite component is present to the extent of 10 to 20 wt%. Consequently, although these processes are unsuitable for preparing the zeolite in sufficiently pure form for many applications, they may be applicable for preparing the zeolite as a minor component.

In some cases, kaolin is calcined at rather high temperatures prior to conversion (55). The kaolin is calcined at 985° to 1010°C so that the high temperature DTA exotherm is essentially not present and there is no evidence of crystallization to mullite. Kaolin calcined in this way is mixed with sodium hydroxide and water in the weight ratios of 5 kaolin, 1 NaOH, 5 H_2O. This produces a thick mass described as a batter. This mass is permitted to age statically with no agitation, at a temperature of about 90°C, and partial conversion to a zeolite of the faujasite type results. The SiO_2/Al_2O_3 ratio of the zeolite component is reported to be approximately 5. The amount of zeolite component was determined by x-ray powder diffraction and was reported to be 17% or higher in some instances. The determination of the SiO_2/Al_2O_3 ratio is best made by x-ray diffraction.

In order to increase the silica-alumina ratio of the zeolite in the product, additional silica is added in the form of a suitable sodium silicate or preferably, an amorphous solid silica (56,57). Kaolin calcined at 1000°C is slurried in aqueous sodium hydroxide and mixed with a reactive silica. Crystallization at 100°C resulted in zeolite Y with a SiO_2/Al_2O_3 ratio of 5.0, compared to 4.3 without use of the added silica. In one example, the reaction mixture composition of the calcined kaolin-sodium hydroxide slurry was 0.6 Na$_2$O·Al$_2$O$_3$ · 2SiO$_2$ 14.3 H$_2$O. It is obvious that this composition was far removed from the stoichiometry of a typical zeolite Y—Na$_2$O·Al$_2$O$_3$· 3-6 SiO$_2$· H$_2$O. However, a product was obtained which was formed as a hard cake and contained about 42% of a NaY with a SiO$_2$/Al$_2$O$_3$ ratio of about 5.1. This means that the unconverted product is sodium aluminate and that essentially all of the silica in the kaolin is contained in the zeolite component. Conversion into a suitable cracking catalyst is achieved by grinding the hard cake obtained and adding a sodium silicate as a binder. The amount of binder is said to be about 19% of the initial slurry. Thus, the silica added approximates 16% of the final cracking catalyst. Conversion to the cracking catalyst is achieved by ion exchange treatment of the finely ground material with ammonium nitrate solution.

Some Other Processes

Other processes for the manufacture of molecular sieve zeolites are listed in Table 9.5. In general, these are based upon heterogeneous reactant mixtures, some utilizing natural sources of silica and alumina. In some instances, an amorphous, mineral aluminosilicate, allophane, has been used as a reactant. This is the principal clay-type constituent in soils of volcanic ash origin. Allophane has a chemical composition which may range from Al$_2$O$_3$· 2 SiO$_2$· 3 H$_2$O to Al$_2$O$_3$· SiO$_2$· 2 H$_2$O. The proposed structural scheme for the first composition consists of an SiO$_4$ tetrahedral chain with an AlO$_6$ octahedral chain sharing corners. By adding another AlO$_6$ chain the composition changes to one with Si/Al equal to 1/2 (60).

The manufacture of large port mordenite, commercially referred to as Zeolon, is based on the hydrothermal treatment (175°C for 24 hours) of a reaction mixture which is prepared from sodium aluminate and various types of amorphous silica materials such as diatomaceous earth or pumice—a naturally occurring volcanic glass. The pressure utilized in this process is the saturated vapor pressure of water at the temperature concerned. The hydrogen form, H-Zeolon, is prepared from the sodium form by treatment with hydrochloric acid at room temperature (65).

A modified process may be used to manufacture the silica-rich zeolites such as zeolite Y from kaolin. Typically, kaolin with a SiO$_2$/Al$_2$O$_3$

Table 9.5 Some Miscellaneous Molecular Sieve Processes

Process	Reactants	Product	Reference
Heterogeneous gel	Calcined, acid-washed- diatomite, sodium aluminate	Type Y, powder	61
Heterogeneous gel-seeded kaolin conversion	Metakaolin + hydrogel	Type Y, powder	62
Heterogeneous gel	Calcined acid leached meta-kaolin + caustic	Type Y, powder	63, 64
Heterogeneous gel mix	Natural silicas, e.g., diatomite, volcanic glass	Zeolon-mordenite	65
Heterogeneous gel mixed with raw clay	$Al_2O_3 \cdot 3 H_2O$ Amorphous silica caustic raw kaolin	Type Y in a clay matrix	59
Hydrogel impregnated into pores of a ceramic	Sodium silicate sodium aluminate caustic $\alpha - Al_2O_3 \cdot 3 H_2O$	Zeolite on a ceramic support Type A Type X	66, 67

ratio of 2, is unsuited for the preparation of a zeolite Y with a higher ratio of, for example, 5. Partial conversion may occur but complete conversion to a pure product is impossible. It is first necessary to eliminate some of the alumina or to add additional silica, in the form of an amorphous silica, to obtain a correct stoichiometry for the reactant mixture.

Calcined kaolin may first be treated with a mineral acid in order to provide a material having a higher silica content useful for the manufacture of the more siliceous zeolites (63). A process is described for the manufacture of zeolite Y which utilizes a calcined kaolin clay that has been treated with hydrochloric acid. Treatment of the activated or calcined clay with the hydrochloric acid removes alumina and, consequently, results in a reactive material that is high in silica. Ratios of SiO_2/Al_2O_3 up to 20/1 are obtained. This reactive material can then be utilized in a typical manufacturing process at 100°C as the source of silica for making zeolite Y.

In one process metakaolin is seeded with a small amount of an amorphous zeolite precursor slurry with a particle size of less than 0.1 micron. The amorphous aluminosilicate "seeds" were first prepared by aging a sodium silicate-sodium aluminate mixture at ambient temperature for 16 hours. The metakaolin-sodium silicate slurry was added to another sodium hydroxide-sodium silicate solution which was seeded with the amorphous zeolite seeding mixture. Typically, crystallization is completed at 100°C in 18 to 36 hours. The formation of zeolite Y with a ratio of 4.9 to 5.9 is reported. However, what happens in the absence of the amorphous seeding mixture is not disclosed (62).

Diatomite silica may be used to make zeolite Y. The diatomite is first washed with hydrochloric acid and calcined at 430°C for 2 hours. This is used together with the usual sodium hydroxide-sodium aluminate combination. A typical reactant composition which results in the formation of zeolite Y is reported to be about $3.6\ Na_2O \cdot Al_2O_3 \cdot 9\ SiO_2 \cdot 97\ H_2O$ (61).

The manufacture of the mordenite-type zeolite from amorphous aluminosilicate precursors may be preferred. The precursor is prepared from sodium silicate solution ($SiO_2/Na_2O = 3.2$) and aluminum sulfate. Although the precursor precipitate is washed to remove the sodium sulfate which is formed, it is reported that such washing is unnecessary. Crystallization of the mixture to mordenite is accomplished hydrothermally at 300°C (87 bars) in 15 hours. The ratio of SiO_2/Al_2O_3 in the initial reaction mixture varied from 9 to 13. The direct preparation of agglomerated, that is, pelletized mordenite, is also reported; this was accomplished by extruding the amorphous aluminosilicate precursor in a paste followed by drying and calcining at 600°C. Subsequently, the extruded particles were hydrothermally treated at 300°C for 15 hours (27).

The Processing of Zeolite Minerals into Commercial Products

The most extensive application of zeolite minerals in commercial products seems to have been made in Japan. The two zeolite minerals occurring in usable deposits in Japan are mordenite and clinoptilolite. Zeolite production in Japan is reported to be 5,000 to 6,000 tons per month (68). Zeolite minerals are used in Japan in the following appli-

MANUFACTURING PROCESSES 741

cations: (a) in papermaking; (b) in air separation adsorption processes, (c) as desiccants; (d) in fertilizers; (e) as soil conditioners—this application is based upon the ability of the zeolite to ion exchange with soil nutrients; (f) in agricultural applications—these include the use as deodorants in the raising of pigs and chickens; (g) in the treatment of radioactive wastes; (h) as carriers for toxic materials; and (i) as adsorbents for toxic gases.

The distribution of zeolite deposits in Japan was discussed in Chap. 3. The Zieglite Chemical Company operates a plant at Itagaya. The main material mined is a potassium-rich clinoptilolite which exists in a thick white tuff approximately 250 meters thick (Fig. 9.5). This is one of the larger deposits and is estimated to contain about 500,000 tons. Most of the material is mined from shafts and is taken to a plant where it is crushed, dried, powdered, and then meshed into three grades based on particle size (Figs. 9.6, 9.7)

Figure 9.5 View of zeolite deposit and processing plant at Itagaya, Japan. (Photograph courtesy of Professor H. Minato.)

Figure 9.6 Zeolite plant at Itagaya, Japan. (Photograph courtesy of Professor H. Minato.)

B. THE PELLETIZATION OF SYNTHETIC ZEOLITE POWDERS

High purity zeolite crystals to be used in adsorption processing, must be formed into agglomerates having high physical strength and attrition resistance. Methods for forming the crystalline powders into agglomerates include the addition of an inorganic binder, generally a clay, to the high purity zeolite powder in wet mixture (77). The blended clay zeolite mixture is extruded into cylindrical type pellets or formed into beads which are subsequently calcined in order to convert the clay to an amorphous binder of considerable mechanical strength (Fig. 9.8). As binders, clays of the kaolin type are generally used. In some instances, the use of metal powders has been reported (69).

The zeolite may be formed into binderless particles by hot pressing techniques. When subjected to high pressures at an elevated temperature, a mass of the zeolite crystals may self-bond into a 100% zeolite pellet. This method, although utilized for various laboratory applications, has not achieved extensive industrial use.

A third method might be to form the zeolite in single crystal form in particle size ranges suitable for actual use. Although single crystals up to 100 microns in size have been reported, these processes have not yet achieved commercial significance. Single 100-micron crystals of zeolite may be suitable for fluidized-bed adsorption processes.

PELLETIZATION OF SYNTHETIC ZEOLITE POWDERS 743

Figure 9.7 (a) Roller mill for crushing zeolite minerals; (b) Rotary kiln for drying zeolite. (Photographs courtesy of Professor H. Minato.)

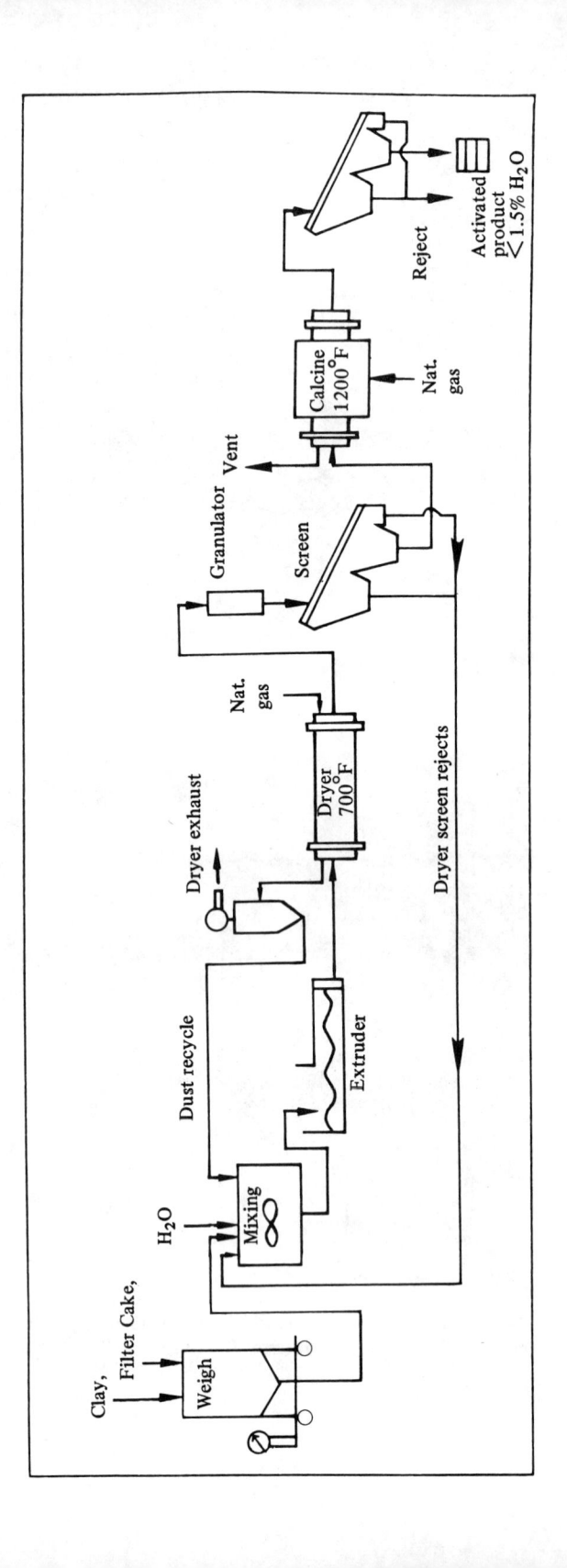

Figure 9.8 Pelletization Process

In many applications using pelleted molecular sieves, the physical resistance of the pellets to mechanical attrition is important since it will determine the economic life of the material in typical adsorption processes.

The preparation of zeolite-binder agglomerates as spheres or cylindrical pellets which have high mechanical attrition resistance would not be difficult. However, in order to utilize the zeolite in a process of adsorption or catalysis, the diffusion characteristics must not be unduly interfered with. Consequently, the binder component in the pellet must be such that a macroporosity is maintained which does not increase the diffusion resistance. The problem, therefore, is one of optimizing the binder-zeolite combination so as to achieve a pellet of maximum density (to produce a high adsorption capacity) with maximum mechanical attrition resistance and minimum diffusion resistance. Some applications are based upon using the equilibrium adsorption capacity of the zeolite component with essentially infinite time being allowed for adsorption (for example, as a cartridge in a refrigeration system for removing trace amounts of water). In adsorption processes involving the use of fixed-beds of molecular sieves, suitable macroporosity must be maintained in the pelleted conglomerate.

Binderless pellets of zeolite A are manufactured by first preparing a zeolite A-silicic acid granule composite which contains 15% silica on a dry basis followed by treating these granules by cold aging in sodium aluminate solution for about one day. This is followed by a digestion treatment at 80°C in aqueous NaOH for several hours (70).

In another method, the zeolite is first synthesized from an aqueous gel system prepared from caustic, alumina trihydrate and colloidal silica (71). The zeolite crystals are recovered by filtration, washed with water and converted to the ammonium form by ion exchange and then mixed with kaolin clay which is activated in steam at 700 - 750°C for several hours. These final activated pellets contain 10% zeolite. In this case, steam activation probably produces a stabilized form of zeolite Y. (See Chap. 6.)

A slurry of zeolite contained in the mother liquor from which it was crystallized is dispersed in an acidified sodium silicate solution so that the final pH is about 4-4.5 (73). This produces a silicate gel which after aging for 30 minutes is added to a mixture of aluminum sulfate and ammonium hydroxide. A silica-alumina matrix is formed surrounding the zeolite crystals. The final zeolite content in this composite is about 10% by weight, in an amorphous silica matrix comprising 92.5% silica and 7.5% alumina. Additional ion exchange treatment of the zeolite can be performed within this matrix.

Zeolite Y has been formed into a composite containing 10 wt% zeolite in an amorphous silica matrix. This is accomplished by spray drying a slurry of the zeolite in a silica hydrogel which is prepared by acidifying sodium silicate to a pH of 4.3. Prior to the spray drying the pH is adjusted to 3.7 and aged for 6 hours (74).

One process results in a composite primarily designed for use as a catalyst containing zeolite Y in an amorphous silica matrix. After crystallization the zeolite in the mother liquor is slurried directly with an acidified sodium silicate solution so that the final pH of the mix is 4.0-4.5. After spray drying the composite particles are converted to an ammonium form by ion exchange for catalytic use (75).

Zeolite-clay composites may be prepared as spheres (76). The spherical agglomerates are formed by dry blending the clay and the zeolite powder and then adding controlled amounts of water in a rotating drum. The spheroidal bodies form during tumbling. The surface of these spheres may be further hardened by soaking in a soluble silicate solution and drying the coating on the surface. After impregnation, the coated spherical pellet is calcined to harden the surface silicate coating.

C. PROPERTIES OF COMMERCIAL MOLECULAR SIEVES

Commercially available molecular sieve products and related materials are listed in Table 9.6. These products are listed according to the basic zeolite structure types which were discussed in Chap. 2. The major cation species, the pore size, the particle sizes, typical bulk density values, the crushing strength of the pellet, and the water adsorption capacity are shown. In most cases the water content is less than 1.5 to 2.5 wt%; certain products, however, are sold as fully hydrated crystalline powders. The water capacity given in the last column is the total water adsorption capacity on an activated basis. In those products where the binder is present to the extent of about 20 wt% of the activated composite, the water capacity is equal to about 80% of the water adsorption capacity of the pure crystalline zeolite powder.

Adsorption properties and other properties of commercial products are affected by the binder. Generally, however, the binder acts as a

Table 9.6 Commercial Molecular Sieves

Product	Structure Type	Major Cation	Pore Size (Å)	Particle Size	Bulk Density (lb/cu ft)	Crush Strength (lb)	Water Content (wt %)	Water Capacity (wt %)	Ref.
Type 3A Linde	A	K^+	3	1/16" pellet	40	6.4	<1.5	20	F21A (80)
Type 4A Linde	A	Na^+	4	1/8" pellet	40	14.5	<1.5	20	
				Powder <10 µm			<2.5	23	
				1/16" pellet	41	10.4	<1.5	22	F22 (80)
				0.0575 0 0.0775 in					
				1/8" pellet	41	21	<1.5	22	
				0.115 - 0.135 in					
				14 x 30 mesh	40		<1.5	22	
				0.0232 - 0.0555 in					
				8 x 12 beads	42	6.9	<1.5	22	
				0.0661 - 0.0937 in					
				4 x 8 beads	40	18	<1.5	22	
				0.0937 - 0.187 in					
				Powder <10 µm			<2.5	28.5	
Type 4A -30 Linde	A	Na^+	4	1/16" pellet	50	20		25	F3206 (80)
				0.0575 - 0.0775 in					
Type 4A -XH-2 Linde	A	Na^+	4	8 x 12 mesh bead	0.082 g/cc	10			F2309 (80)
				4 x 8 mesh bead	0.77 g/cc	19			
Type 5A Linde	A	Ca^{2+}	5	1/16" pellet	45	5.8	<1.5	21.5	F2184B (80)
				0.0575 - 0.0775 in					
				1/8" pellet	45	12.6	<1.5	21.5	
				0.115 - 0.135 in					
				Powder <10 µm			<2.5	28	

Table 9.6 Commercial Molecular Sieves (continued)

Product	Structure Type	Major Cation	Pore Size (Å)	Particle Size	Bulk Density (lb/cu ft)	Crush Strength (lb)	Water Content (wt %)	Water Capacity (wt %)	Ref.
Type 10X Linde	X	Ca	8	1/16" pellet 0.0575 - 0.0775 in 1/8" pellet 0.115 - 0.135 in Powder $<$ 10 μm	40 [c] 40 [c]	9 22	$<$1.5 $<$1.5 $<$2.5	31.6 [b] 31.6 [b] 35 [b]	3097 (80)
Type 13X Linde	X	Na	10	1/16" pellet 0.0575 - 0.0775 in 1/8" pellet Powder $<$ 10 μm	38 38	12 25	$<$1.5 $<$1.5 $<$2.5	28.5 [a] 28.5 34	F23 (80)
Type AW-300, Linde	mordenite	Na	4	1/8, 1/16" pellet	55.4			10	F2308 (80)
Type AW-500, Linde	chabazite	Na, Ca	5	1/8, 1/16" pellet	45.4			11	
SK40 Linde	Y	Na		Powder 1 - 2 μm Extrudate 1/8" x 1/4"	38	30			F-10 (80)
SK 41 Linde	Y	NH_4^+		Powder 1 - 2 μm					F-4 (80)
SK 45 Linde	L	K^+		Powder 5 - 10 μm			33		F2988 (80)
SK 120 Linde	Y	Rare earth, 0.5% Pd		1/8" extrudate	38 - 45	15			F12 (80)
SK 500 Linde	Y	Rare earth		1/8 - 1/4" extrudate	38 - 45	15			F-15 (80)
Zeolon-100, Norton	mordenite	Na	7	Powder 5 - 12 μm	40			14	Z-50 (81)
		H	8 - 9	Powder 5 - 12 μm	42			15	

PROPERTIES OF COMMERCIAL MOLECULAR SIEVES   749

Table 9.6 Commercial Molecular Sieves *(continued)*

Product	Structure Type	Major Cation	Pore Size (Å)	Particle Size	Bulk Density (lb/cu ft)	Crush Strength (lb)	Water Content (wt %)	Water Capacity (wt %)	Ref.
Zeolon-200, Norton	mordenite	H	8 - 9	1/8" extrudate 1/16" extrudate	38			12	Z-50 (81)
Zeolon-300, Norton	mordenite	mixed cations	3 - 4	1/8" extrudate 1/16" extrudate	48			5.5	Z-50 (81)
Zeolon-500, Norton	chabazite erionite	mixed cations	4 - 5	1/8" extrudate	47			13.5	Z-50 (81)
Zeolon-900, Norton	mordenite	Na	7	1/8" extrudate 1/16" extrudate self-bonded 20 - 50 mesh	40			10	Z-50 (81)
		H	8 - 9	1/8" extrudate 1/16" extrudate 20 - 50 mesh	45			11	Z-50 (81)
Type 3A Davison	A	K	3	8 - 12 mesh, spheres 4 - 9 mesh, spheres Powder, 3 - 5 μm	46 46 32	7 11 –	1.5 1.5 1.5	21.0 21.0 23.0	82 82 82
Type 4A Davison	A	Na	4	8 - 12 mesh, spheres 4 - 8 mesh, spheres 14 - 30 mesh, granules 10 - 16 mesh, spheres Powder, 3 - 5 μm	44 44 – – 30	7 11 – – –	1.5 1.5 1.5 1.5 1.5	23.0 23.0 23.0 23.0 28.5	82 82 82 82 82
Type 5A Davison	A	Ca	5	8 - 12 mesh, spheres 4 - 8 mesh, spheres 20 - 50 mesh, spheres Powder, 3 - 5 μm	44 44 44 30	7 11 – –	1.5 1.5 1.5 1.5	21.7 21.7 21.7 28.0	82 82 82 82
Type 13X Davison	X	Na	10	8 - 12 mesh, spheres 4 - 8 mesh, spheres Powder, 3 - 5 μm	43 43 30	6 10 –	1.5 1.5 1.5	29.5 29.5 36.0	82 82 82

[a] lbs H$_2$O/100 lbs activated adsorbent at 25°C, 17.5 torr. [b] lbs H$_2$O/100 lbs activated adsorbent at 25°C, 25 torr. [c] Settled.

diluent. The binder affects adsorption by contributing secondary porosity in the form of macropores.

Although detailed information on the distribution of voids in composite pellets of molecular sieves is not available in every case, some typical data are shown in Tables 9.7 and 9.8. In Table 9.7 the distribution of voids in a single, typical molecular sieve pellet is given in terms of two components, the void volume present in the crystals themselves and the void volume present between the crystals. This, together with the volume fraction occupied by the zeolite crystal framework and the solid binder total 100%.

When these pellets are packed in a bed, the distribution of voids as typically found is shown in Table 9.8. Again, the breakdown is in terms of the two types of voids present, those in the individual pellet together with the voids present between the pellets in the packed bed. For example, in the case of the 1/16-inch pellets, the voids total nearly 74 vol% of the packed bed.

Typical physical properties of the molecular sieve type 5A in two sizes of pellets, 1/16 and 1/8 inch, are shown in Table 9.9. Typical distribution curves of a lot of molecular sieve type 5A pellets are shown in Fig. 9.9 for two sizes of pellets. The distributions are shown in terms of the length of the pellets. In either case, the average length to diameter ratio is about 1.6 to 1.7. The heat capacity of the molecular sieve pellets is important in considering adsorption processes and a typical curve showing the heat capacity of 5 A pellets versus temperature is given in Fig. 9.10. The heat capacities were calculated from enthalpy values obtained over a temperature range of 32 to 572°F (79). The sample was first fully activated for 12 hours at 662°F at a pressure of 0 to 1 torr.

Table 9.7 Volume Distribution in Single Molecular Sieve Pellets

	1/16-inch Pellet (%)	1/8-inch Pellet (%)
Intracrystal voids	28.1	28.8
Intercrystal voids	33.4	31.5
Solid portion of crystals	29.2	30.1
Solid portion of binder	9.3	9.6
	100.0	100.0

PROPERTIES OF COMMERCIAL MOLECULAR SIEVES 751

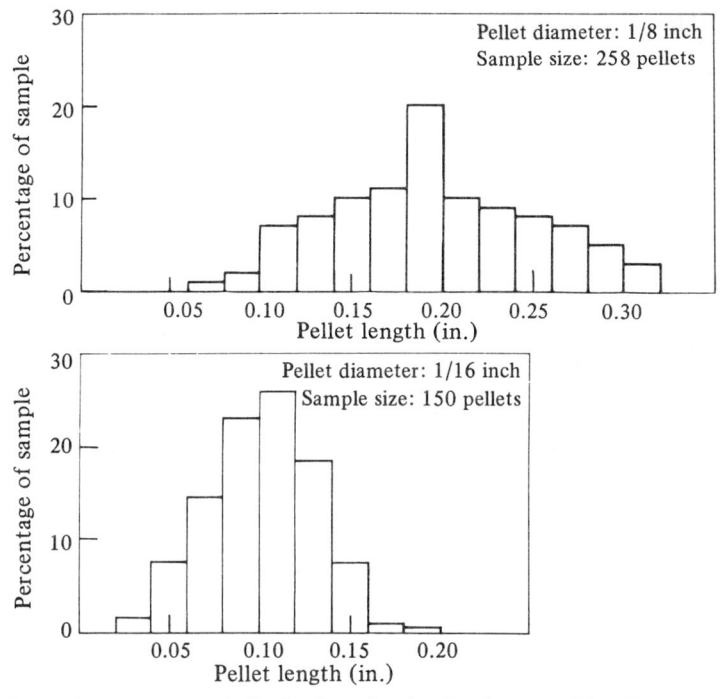

Figure 9.9 Typical length distribution of molecular sieve type 5A pellets.

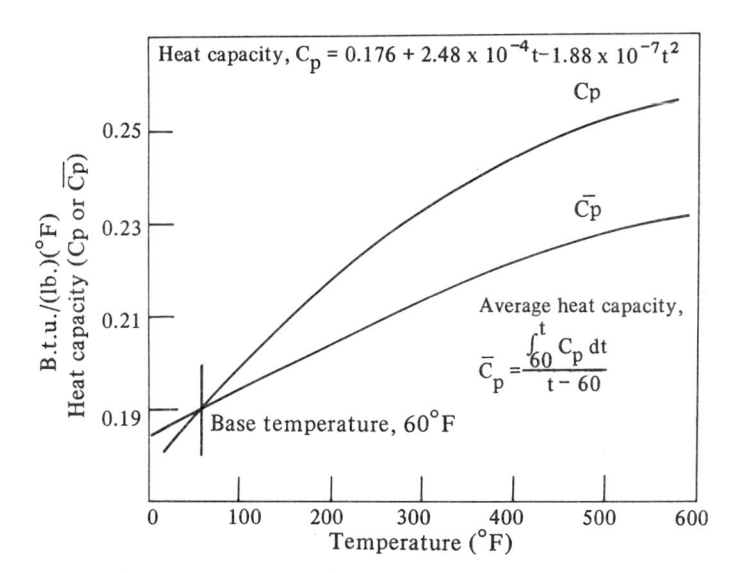

Figure 9.10 Heat capacity of molecular sieve type 5A pellets.

Heat capacity, $C_p = 0.176 + 2.48 \times 10^{-4} t - 1.88 \times 10^{-7} t^2$

Average heat capacity,
$$\bar{C}_p = \frac{\int_{60}^{t} C_p \, dt}{t - 60}$$

Base temperature, 60°F

Table 9.8 Volume Distribution in Beds Packed with Molecular Sieve Pellets [a]

	Bed Packed with 1/16-inch Pellets (%)	Bed Packed with 1/8-inch Pellets (%)
Intracrystal voids	19.1	19.6
Intercrystal voids	22.7	21.4
Interpellet voids	32.0	32.0
Solid portion of crystals	19.9	20.5
Solid portion of binder	6.3	6.5
	100.0	100.0

[a] Based on a bed packed with an external void fraction, ϵ, equal to 0.32 (interpellet voids).

Pellet densities of activated molecular sieve pellets were determined using a mercury displacement technique. The external void fraction in a packed bed is determined directly from the bulk density and particle density. The external void fraction, ϵ, is given by:

$$\epsilon = 1 - \frac{\text{Bulk density}}{\text{Particle density}}$$

The external void portion, determined by packing the pellets in columns of up to 8 inches in diameter and 5 feet in length, is generally in the range of 0.32 to 0.36. Initial packing is not as dense as that obtained after the bed has settled or been in operation for some time. Therefore, the void fraction in a settled bed will be less. The initial value for the void fraction in a large packed bed is probably about 0.38 but, after settling, generally falls to a value of 0.32.

Table 9.9 Typical Physical Properties of Molecular Sieve Type 5A Pellets

Physical Property	Units	Pellets 1/16-inch	Pellets 1/8-inch
Pellet density	lb/cu ft	70.0	72.0
Bulk density, newly packed	lb/cu ft	43.4	44.6
Bulk density, settled bed	lb/cu ft	47.5	48.8
External void fraction, newly packed	—	0.38	0.38
External void fraction, settled bed	—	0.32	0.32
Average pellet length	ft	0.00900	0.0173
Pellet length distribution	—	See Fig. 9.9.	
Average pellet diameter	ft	0.00522	0.0107
Average ratio of length to diameter	—	1.72	1.63
External surface area of pellets	sq ft/cu ft	655	370
Effective particle diameter (based on Ergun's expression)	ft	0.00606	0.0122
Effective particle diameter (based on sphere with equal external surface area)	ft	0.00777	0.0155
Heat capacity	Btu/(lb)(°F)	See Fig. 9.10.	

REFERENCES

1. R. A. Labine, *Chem. Eng.*, 104 (1959).
2. R. M. Milton, *U. S. Pat.*, 2,882,243 (1959).
3. E. E. Sensel, *U. S. Pat.*, 2,841,471 (1958).
4. J. H. Estes, *U. S. Pat.*, 2,847,280 (1958).
5. H. Weber, *U. S. Pat.*, 3,058,805 (1962).
6. M. Michel and D. Paper, *U. S. Pat.*, 3,433,588 (1969).
7. R. M. Milton, *U. S. Pat.*, 2,882,244 (1959).
8. G. T. Kerr, *U. S. Pat.*, 3,321,272 (1967).
9. D. W. Breck, *U. S. Pat.*, 3,130,007 (1964).
10. C. V. McDaniel and P. K. Maher, *U. S. Pat,*, 3,264,059 (1966).
11. H. E. Robson, *U. S. Pat.*, 3,343,913 (1967).
12. C. V. McDaniel, P. K. Maher, and W. E. Waxter, *U. S. Pat.*, 3,374,058 (1968).
13. J. Ciric and L. J. Reid, Jr., *U. S. Pat.*, 3,433,589 (1959).
14. D. Domine and J. Quobex, *U. S. Pat.*, 3,481,699 (1969).
15. S. G. Hindin and J. C. Dettling, *U. S. Pat.*, 3,484,194 (1969).
16. E. E. Jenkins, *U. S. Pat.*, 3,492,090 (1970).
17. S. G. Hindin and J. C. Dettling, *U. S. Pat.*, 3,510,258 (1970).
18. L. J. Reid, Jr. *U. S. Pat.*, 3,334,964 (1967).
19. F. J. Dzierzanowski and W. L. Haden, Jr., *U. S. Pat.*, 3,094,383 (1963).
20. E. Michalko, *U. S. Pat.*, 3,348,911 (1967).
21. E. Michalko, *U. S. Pat.*, 3,356,451 (1967).
22. E. Michalko, *U. S. Pat.*, 3,359,068 (1967).
23. E. Michalko, *U. S. Pat.*, 3,386,802 (1968).
24. E. Michalko, *U. S. Pat.*, 3,428,574 (1969).
25. C. J. Plank and E. J. Rosinski, *U. S. Pat.*, 3,459,501 (1969).
26. K. D. Vesely, *U. S. Pat.*, 3,492,089 (1970).
27. D. Domine and J. Quobex, *U. S. Pat.*, 3,574,539 (1971).
28. R. C. Hansford, *U. S. Pat.*, 3,227,660 (1966).
29. E. M. Gladrow and W. J. Mattox, *U. S. Pat.* 3,329,627 (1967).
30. E. M. Gladrow and W. J. Mattox, *U. S. Pat.*, 3,329,628 (1967).
31. G. W. Brindley and M. Nakahira, *J. Amer. Ceram. Soc.*, **42**: 311, 314, 319 (1959).
32. E. E. Sensel, *U. S. Pat.*, 3,009,776 (1961).
33. P. A. Howell, *U. S. Pat.*, 3,114,603 (1963).
34. P. K. Maher, *U. S. Pat.*, 3,185,544 (1965).
35. P. K. Maher and E. J. Nealon, *U. S. Pat.*, 3,205,037 (1965).
36. T. Veda, K. Sato, and Y. Nakamura, *U. S. Pat.*, 3,535,075 (1970).
37. P. A. Howell, N. A. Acara, and M. K. Towne, Jr., *Brit. Pat.*, 980,891 (1965).
38. W. L. Haden, Jr. and F. J. Dzierzanowski, *U. S. Pat.*, 2,992,068 (1961).
39. W. L. Haden, Jr. and F. J. Dzierzanowski, *U. S. Pat.*, 3,065,054 (1962).
40. W. L. Haden, Jr. and F. J. Dzierzanowski, *U. S. Pat.*, 3,100,684 (1963).
41. L. L. Taggart and G. L. Ribaud, *U. S. Pat.*, 3,119,659 (1964).
42. P. A. Howell and N. A. Acara, *U. S. Pat.*, 3,119,660 (1964).
43. E. Eichorn and L. G. Garrison, *U. S. Pat.*, 3,370,917 (1968).
44. O. J. Whittemore, Jr., *U. S. Pat.*, 3,445,184 (1969).
45. W. L. Haden, Jr. and F. J. Dzierzanowski, *U. S. Pat.*, 3,338,672 (1967).
46. W. L. Haden, Jr. and F. J. Dzierzanowski, *U. S. Pat.*, 3,367,886 (1968).
47. C. J. Plank and E. J. Rosinski, *U. S. Pat.*, 3,431,218 (1969).

48. C. V. McDaniel and P. K. Maher, *U. S. Pat.*, 3,472,617 (1969).
49. W. L. Haden, Jr. and F. J. Dzierzanowski, *U. S. Pat.*, 3,367,847 (1968).
50. W. L. Haden, Jr. and F. J. Dzierzanowski, *U. S. Pat.*, 3,391,994 (1968).
51. W. L. Haden, Jr. and F. J. Dzierzanoswki, *U. S. Pat.*, 3,433,587 (1969).
52. J. S. Lapides and J. E. McEvoy, *U. S. Pat.*, 3,458,454 (1969).
53. W. L. Haden, Jr., and F. J. Dzierzanowski, *U. S. Pat.*, 3,503,900 (1970).
54. W. L. Haden, Jr. and F. J. Dzierzanowski, *U. S. Pat.*, 3,506,594 (1970).
55. W. H. Flank, *U. S. Pat.*, 3,515,511 (1970).
56. W. H. Flank, *U. S. Pat.*, 3,515,680 (1970).
57. W. H. Flank, J. E. McEvoy, and G. A. Mills, *U. S. Pat.*, 3,515,681 (1970).
58. W. H. Flank, J. E. McEvoy, and G. A. Mills, *U. S. Pat.*, 3,515,682 (1970).
59. J. E. McEvoy, *U. S. Pat.*, 3,450,645 (1969).
60. K. Wada, *Amer. Mineral*, **52**: 690 (1970).
61. D. A. Young, *U. S. Pat.*, 3,341,284 (1967).
62. C. V. McDaniel and H. C. Duecker, *U. S. Pat.*, 3,574,538 (1971).
63. P. A. Howell, *U. S. Pat.*, 3,340,958 (1968).
64. P. K. Maher, *U. S. Pat.*, 3,393,045 (1968).
65. L. B. Sand, *U. S. Pat.*, 3,436,174 (1969).
66. A. B. Schwartz, *U. S. Pat.*, 3,244,643 (1966).
67. E. L. Cole and E. C. Knowles, *U. S. Pat.*, 3,468,815 (1969).
68. H. Minato, *Koatsu Gasu*, **5**: 536 (1968), "Molecular Sieve Zeolites," *Adv. Chem. Ser. 101*, American Chemical Society, Washington, D. C., 1971, p. 311.
69. D. W. Breck, *U. S. Pat.*, 3,181,231 (1965).
70. G. Heinze, *U. S. Pat.*, 3,356,450 (1967).
71. E. B. Cornelius and J. E. McEvoy, *U. S. Pat.*, 3,382,188 (1968).
72. W. W. Weber, *U. S. Pat.*, 3,594,121 (1971).
73. K. D. Vesely and E. Michalko, *U. S. Pat.*, 3,472,792 (1969).
74. E. Michalko, *U. S. Pat.*, 3,499,846 (1970).
75. E. Michalko, *U. S. Pat.*, 3,503,874 (1970).
76. W. J. Mitchell and W. F. Moore, *U. S. Pat.*, 2,973,327 (1961).
77. G. L. Ribaud, *U. S. Pat.*, 3,219,590 (1965).
78. W. Drost, *U. S. Pat.*, 3,394,989 (1968).
79. H. T. Spengler and W. S. Tamplin, *Anal. Chem.*, **24**: 941 (1952).
80. Union Carbide Corporation, Linde Division Technical Bulletin designations.
81. Norton Company Technical Bulletin.
82. W. R. Grace and Company Technical Bulletin.

APPENDIX

GENERAL REFERENCES AND CONFERENCES

M. M. Dubinin and T. G. Plachenov, Eds.
Zeolites, Their Synthesis, Properties, and Applications
Second All-Union Conference on the Synthesis, Study, and Applications of Adsorbents, USSR Academy of Sciences, held in 1964, Proceedings published in 1965, English translation published by the International Information Institute in 1967.

R. M. Barrer, Chairman
Molecular Sieves
Papers read at the Conference held in April, 1967. Proceedings published by the Society of Chemical Industry, London, 1968.

L. B. Sand and E. M. Flanigen, Chairpeople
Molecular Sieve Zeolites, *Advances in Chemistry Series, Vol. 101, 102*
Second International Conference cosponsored by the Division of Colloid and Surface Chemistry, Division of Petroleum Chemistry, and Division of Physical Chemistry of the American Chemical Society and Worcester Polytechnic Institute, Massachusetts, in September 1970. Proceedings published by the American Chemical Society, Washington, D.C. in 1971.

W. M. Meier and J. B. Uytterhoeven, Editors
Molecular Sieves, *Advances in Chemistry Series, Vol. 121*
The Third International Conference on Molecular Sieves cosponsored by the Eidgenossische Technische Hochschule and the Swiss Chemical Society in Zurich, Switzerland in September, 1973. Proceedings published by the American Chemical Society, Washington, D.C. in 1973.

R. L. Mays and T. L. Thomas
Separations With Molecular Sieves
in Physical Methods in Chemical Analysis, W. G. Berl, Ed., Academic Press New York, 1961, pp. 45 - 97.

R. M. Barrer
Some Aspects of Molecular Sieve Science and Technology
Chem. Ind., 1203 - 1213(1968).

R. M. Milton
Commercial Development of Molecular Sieve Technology
Molecular Sieves, Society of Chemical Industry, London, 1968, pp. 199 - 203.

D. W. Breck
Recent Advances in Zeolite Science
Molecular Sieve Zeolites, *Advan. Chem. Ser.*, Vol. 101, American Chemical Society, Washington, D.C., 1971, pp. 1 - 17.

STRUCTURE

K. F. Fischer and W. M. Meier
Kristallchemie der Zeolithe
Fortschr. Miner. *42*, 50-86 (1965)

W. M. Meier
Zeolite Structures
Molecular Sieves, Society of Chemical Industry, London, 1968, pp. 10-27.

W. M. Meier and D. H. Olson
Zeolite Frameworks
Molecular Sieve Zeolites, *Advan. Chem. Ser.*, Vol. 101, 155-70 (1971).

W. M. Meier and D. H. Olson
Atlas of Zeolite Structure Types
Intl. Zeolite Assoc., 1978, dist. by Polycrystal Book Service,
P. O. Box 11567, Pittsburgh, Pa. 15238.

J. V. Smith
Faujasite-type Structures. Aluminosilicate framework. Positions of cations and molecules. Nomenclature.
Molecular Sieve Zeolites, *Advan. Chem. Ser.*, Vol. 101, 171-200 (1971).

MINERALS

R. L. Hay
Zeolites and Zeolitic Reactions in Sedimentary Rocks
GSA Special Paper No. 85, 1966.

R. A. Sheppard
Zeolites in Sedimentary Deposits of the United States
Molecular Sieve Zeolites, *Advan. Chem. Ser.*, Vol. 101, 279-316 (1971).

SYNTHESIS

E. E. Senderov and N. I. Khitarov,
Zeolites, their Synthesis and Conditions of Formation in Nature, Nauka, Moscow (1970)

R. M. Barrer,
Mineral Synthesis by the Hydrothermal Technique
Chemistry in Britain *2*, 380-394 (1966).

R. M. Barrer,
Some Researches on Silicates
Mineral Syntheses and Metamorphoses
Trans. Brit. Ceram. Soc., 56, 155-184 (1957).

ION-EXCHANGE

R. M. Barrer
Crystalline Ion-Exchangers
Proc. Chem. Soc., (1958), 99-112

R. M. Barrer,
Some Features of Ion Exchange in Crystals
Molecular Sieves, Society of Chemical Industry, London, 1962, pp. 1258-66.

M. S. Sherry
Cation Exchange in Zeolites
Molecular Sieve Zeolites, *Advan. Chem. Ser.*, Vol. 101, American Chemical Society, Washington, D.C., 1971, pp. 350-379.

CATALYSIS

R. S. Mays and P. E. Pickert
Molecular Sieve Catalysts; Their Properties and Applications
Molecular Sieves, Society of Chemical Industry, London, 1968, pp. 112-116.

J. Turkevich
Zeolites as Catalysts
Catalysis Reviews, 1, 259-371(1967).

P. B. Venuto and P. S. Landis
Organic Catalysis over Crystalline Aluminosilicates
Molecular Sieve Zeolites, *Advan. in Catalysis*, Vol. 18, American Chemical Society, Washington, D.C., 1968, pp. 259-371.

P. B. Venuto
Some Perspectives on Zeolite Catalysis
Molecular Sieve Zeolites, *Advan. Chem. Ser.*, Vol. 102, American Chemical Society, Washington, D. C., 1971, pp. 260-283.

D. M. Nace
Catalytic Cracking Over Crystalline Aluminosilicates. I. Instantaneous Rate Measurements for Hexadecane Cracking. II. Application of Microreactor Technique to Investigation of Structural Effects of Hydrocarbon Reactions
Ind. Eng. Chem., Prod. Res. Develop., 8:24-38 (1969).

D. M. Nace, S. E. Voltz, and V. W. Weekman, Jr.
Application of a Kinetic Model for Catalytic Cracking
Ind. Eng. Chem. Process Des. Develop., 10:530-538 (1971).

J. A. Rabo and M. L. Poutsma
Structural Aspects of Catalysis with Zeolites: Cracking of Cumene and Hexane
Molecular Sieve Zeolites, *Advan. Chem. Ser.*, Vol. 102, American Chemical Society, Washington, D. C., 1971, pp. 284-314.

J. Turkevich, F. Nozaki, and D. Stamires
Nature of Active Centers and Mechanism of Heterogeneous Catalysis
Proc. Internat. Congr. Catalysis, 3rd, Amsterdam, 1:586-595(1964).

C. L. Thomas and D. S. Barmby
The Chemistry of Catalytic Cracking with Molecular Sieve Catalysts
J. Catal., 12:341-346 (1968).

S. E. Tung and E. McIninch
Zeolitic Aluminosilicates I. Surface Ionic Diffusion, Dynamic Field, and Catalytic Activity with Hexane on CaY. II. Surface Oxide Diffusion, Dynamic (Time Variant) Lewis Acids, and Catalytic Activity with Hexane on Decationized Y.
J. Catal., 10:166-182 (1968).

P. B. Venuto
Organic Molecules and Zeolite Crystal at the Interface
Chem. Tech., 215-224 (1971).

INDEX

A

Adsorbed phase:
  nature of, 660-663
    by ir, 660-662, 667
    by nmr, 412-414, 663
    by spectroscopic methods, 660-662
    by x-ray crystal structure, 662-663
Adsorbents:
  nonzeolites, 3, 6-10
  types of, 3-4
  zeolite (see individual zeolites), 3
Adsorption:
  diffusion during, 671
  diffusion of water, 684
  effect of ion exchange on, 641-644
  encapsulation, 623-628
  engineering concepts, 717-718
  entropy of, 648
  equilibria for binary mixtures, 689-699
  equilibria for single components, 596-627
  heat of, 645-659
    interaction energies, 648-650, 664, 667†
    argon in various adsorbents, 665†
  high pressure, 623-628
  isotherm, 596
  isobar, 596, 639*
  isostere, 596
  kinetic diameter, 632*, 634-635, 636†, 637†
  kinetics, 671-689
  measurement of, 598-599
  pore closure, effect on, 645
  selectivity effects, 596, 664-671
    alteration of, 665-668
    cation density effect on, 668
    polarizability on, 666*
    preadsorption on, 643*, 644-645, 646*
  separation of mixtures, 699-718
    hydrocarbons, 640, 706-707
    process applications, 701-704†
    process cycles in, 708*, 710*, 715-717
    relative selectivity in, 709-715
    separation factor in, 609, 693, 694*
    with water, 700
  sieving effects in, 599, 633-645, 700
Adsorption isotherms:
  (see individual zeolites)
  determination of, 598-599
  equations of, 628-633

Air separation, 709-711
Alkylammonium zeolites, 304-312
Alumina, adsorption, 600
Aluminosilicates:
  synthetic amorphous ion exchangers, 11-13
Aluminum, solution species, 514†, 515*
Ammonia removal:
  from waste water by ion exchange, 589-590
Ammonium zeolites:
  Deammoniation, 474
  DTA, 479, 481, 482*
Analcime:
  adsorption by, 18
  early synthesis results, 251
  infrared spectra, 422*
  ion exchange by potassium, 68
  structural data, 135†
  structure, 66*, 67-68
  synthesis of, 251-252, 260, 267, 278
  x-ray and physical property data, 209†
Aperture sizes, 64-67

B

Benitoite, structure, 34*
Beryl, structure, 34*
Bikitaite:
  structural data, 136†
  structure, 128, 131*
  synthetic type, synthesis of 259

---

* Denotes a figure; † denotes a table.

761

762 INDEX

x-ray and physical
property data, 210†
Binary mixtures:
adsorption equilibria,
689-699
Brewsterite
structural data, 137†
structure, 130-132,
132*
x-ray and physical
property data, 211†

C

Cancrinite:
hydroxy, synthesis of,
272, 319
structure, 66*
Carbons, 8-10
Catalysts:
cracking, 2
Cation coordination
with oxygen, 29-31,
30†
Cation hydrolysis, 460-
473
Cation reduction, 519-
523
Cation sieve effects:
(see ion sieve effects)
Chabazite:
adsorption by, 18,
109
adsorption data, 15*,
617†
adsorption isotherms,
600*, 605*
diffusion in, 674†
early synthesis results,
252-253
heats of adsorption,
652†, 653*, 657†,
667†
infrared spectra, 422*
ion exchange in,
554-558, 573, 555†
557*
pore volume, 432†
rates of adsorption
in, 675*, 688†
structural data, 138†
structure, 79*, 107-
108

cation positions,
109
hydrated, 108
dehydrated, 108-
109
composition, 110
substitutes, synthesis,
256
x-ray & physical prop-
erty data, 212†
surface reactions,
408
Channel structure in
zeolites, 59-62
Chrysotile structure,
37*
Classification:
early classifications,
15-16
mineral zeolites, 19-
22, 20†
synthetic zeolites,
22-26
Clinoptilolite:
adsorption data, 625
625*
adsorption heats in,
670-671†
ion exchange in, 558,
559†, 560*, 581*
mineral processing,
740-741
structural data, 139†
structure, 129
synthetic type,
synthesis, 260, 303†
x-ray data, 355
x-ray and physical
property data, 213†
Coals:
adsorption on, 6
Color, 390-392
Commercial processes:
(see manufacturing
processes)
Commercial products:
binder effects, 595
macropores, 595
properties of, 746-752
Composition:
effect on properties,
390*

uniformity of, 389-
390
Conferences on Mole-
cular Sieve Zeolites, 1
Cracking catalysts, 2
Crystal growth 8, 344-
346
Crystallization:
crystal growth during,
342*, 343*, 340-
344
kinetics and mechan-
ism, 333-344

D

Dachiardite:
structural data, 140†
structure, 124-127,
126*
x-ray and physical
property data, 214†
Decationization, 475,
476*, 507
Defect structures, 507-
523
Dehydration:
of zeolites, summary,
449-451†
of group 1 zeolites,
447-454
of group 2 zeolites,
459
of group 3 zeolites,
454-455
of group 4 zeolites,
455-457
of group 5 zeolites,
457
of group 6 zeolites,
457-458
of group 7 zeolites,
458-459
of group 8 zeolites,
459
Dehydroxylation: 462-
472, 478-482
Stoichiometry, 478-
482
Density, 388
framework, 62-63
Dielectric properties,
392-397

Differential thermal
analysis (DTA), 443-
447
Diffusion:
of cations, 400*, 401,
404*
in adsorption, 595, 667-
684
in ion exchange,
Dubinin-Polanyi relation,
436, 630-632*

E

Edingtonite:
structural data, 141†
structure, 119-122,
120*, 121*
x-ray and physical
property data, 215†
Effective pore size in
zeolites, 637*
Electrical conductivity:
activation energy,
400*, 401*
by surface reaction on
crystals, 408-410
effect of adsorbed
ammonia on, 405
effect of cation on,
399-401, 406*
effect of adsorbed
water, 401-404
effect of nonpolar
molecules on, 405-
406
effect of temperature,
399*
mechanisms of, 406-
408
of crystal - salt solu-
tion interface, 408-
410
Encapsulation:
adsorption, 623-624*,
628
Epistilbite
structural data, 142†
structure, 126*, 127-
128
synthetic type, 295

x-ray and physical
property data, 216†
Erionite:
adsorption data, 622†
ion exchange in, 563-
564†
rates of adsorption in,
678
structural data, 143†
structure, 77-79, 78*,
79*
synthetic type, syn-
thesis of, 309-310
x-ray and physical
property data, 217†

F

Faujasite:
adsorption, 625†
early attempted syn-
thesis, 254
structural data, 145†
structure, cation
positions, 98†
x-ray and physical
property data, 218†
Faujasite-type structures:
92
cation positions by
esr, 105-106
composition, 93
electrostatic poten-
tial, 106-107, 111†
framework, 85*, 93-
95
hydrated cation
positions, 96*, 95-
97
dehydrated cation
positions, 96*,
97-103
lattice constant, 94*
related physical
properties, 103-
105
Feldspar structure, 39*
40-41
Feldspathoid structure,
41-44, 41†
Ferrierite:

INDEX 763

adsorption, 625†
structural data, 146†
structure, 125*, 126*,
127
synthetic type syn-
thesis, 273, 275†,
296, 298
x-ray data, 358†
x-ray and physical
property data, 219†
Framework density,
62-63, 436-438

G

Garronite:
structural data, 148†
structure, 73-74
synthetic type, 295
x-ray and physical
property data, 220†
Gases, properties of,
658†
Gismondine:
structural data, 149†
structure, 70-72, 71*
synthetic type, syn-
thesis of, 311
x-ray and physical
property data, 221†,
206*
Glasses, 7
Gmelinite:
adsorption data, 625†
structural data, 150†
structure, 79*, 110-
111, 112*
x-ray and physical
property data, 222†
Gonnardite:
structural data, 151†
structure, 119
x-ray and physical
property data, 223†
Gurvitsch rule, 426

H

Halloysite in synthesis,
320
Hardness, 388
Harmotome:

structural data, 153†
structure, 69-70, 71*
x-ray and physical
  property data, 224†,
  206*
Heat of adsorption,
  645-659
Heat capacity of commercial products, 751
Hemimorphite, structure, 33*
Herschelite:
  structure, 110
  x-ray and physical
    property data, 225†
Heulandite:
  adsorption data, 626†
  infrared spectra, 423*
  structural data, 154†
  structure, 128-129,
    131*
  synthetic type, 295
  x-ray data, 359
  x-ray and physical
    property data, 226†
High pressure, effect on
  zeolites, 497-498
Hydrocarbons:
  adsorption isotherms,
    600*, 602*, 605*
Hydroxyl processes,
  726-731
Hydrogen exchange,
  569-571
Hydrolysis of cations,
  460-468
Hydrolysis of framework aluminum, 509
Hydrothermal transformation, 483-490,
  487†, 489†
Hydroxyl groups:
  surface, 460
  structural, 460-483
  in univalent cation
    zeolites, 461
  in divalent cation zeolites, 461-462
  dehydroxylation of,
    462-465*
  in trivalent cation

zeolites, 465-468
dehydroxylation of,
  468, 470*
structural studies,
  468-470
spectral studies,
  471
interaction with
  bases, 471
other observations,
  473
from ammonium zeolites, 474
location, 475
dehydroxylation,
  478-482
from alkylammonium
  zeolite, 482
infrared observations,
  485-486†
Hydroxysodalite:
  (see sodalite hydrate)

I

Infrared spectra:
  of adsorbed water,
    424-425
  of phosphate zeolites,
    327
  of structural hydroxyl groups, 460, 462,
    464*, 465, 469,
    471-472, 475,
    478*, 481*, 485-
    486†
  of zeolites, 415-424,
    416-417†
Interaction energies:
  and heat of adsorption, 648-650, 664,
    667†
Intercalation compounds:
  alkali, graphite, 10
Ion diffusion:
  and conductivity,
    400*, 401, 404*
  in aqueous ion exchange, 571-578
  in nonaqueous solvents, 578-579
Ion exchange:

alkylammonium ions,
  552†
applications of, 588-
  590
by soil minerals, 11
comparison of zeolites X and Y, 553-
  554
effect on adsorption,
  641-644
effect on zeolite color,
  390-391
equilibrium constant
  in, 533-536
exchange capacity in,
  536-537†
hydrogen exchange,
  569-571
hysteresis in, 531
in chabazite, 554-558,
  573*, 555†, 557*
in clinoptilolite, 558,
  559†, 560, 581*
in erionite, 563-564†
in fused salts, 587-
  588
in mixed solvent, 583-
  585
in mordenite, 563-
  565, 574†, 575†,
  581*
in nonaqueous solvents, 578-579,
  580-585
in phillipsite, 561,
  562†
in stilbite, 563, 564†
in Zeolite A, 537-540,
  541†, 539*
complex ion effects,
  540
in mixed solvents,
  583-585
in Zeolite P, 561-562†
in Zeolite T, 558-561†
in Zeolite X, 537†, 5
  540-553, 543†, 544
  544*, 546*, 571*,
  576†
in nonaqueous solvents, 578-579,

583-584
in Zeolite Y, 537†, 540-553, 550†, 545†, 546*
in Zeolite Z, 562, 563†, 565*
in Zeolite ZK-4, 576-577
ion diffusion in, 571-579
isotherm, types, 532*
Kielland coefficient in, 584, 533-534
kinetics, 571-579
occlusion of salts in, 585-586*
of ammonia in waste water, 589-590
of radioactive ions, 529, 588
selectivity coefficient, 532-534, 567*
separation factor, 531
theory of, 530-536
thermodynamics of, 565-569
uni-univalent, 535-536
Ion sieve effects, 579-580, 581*, 582*
Isostructural zeolite minerals, 26†
Isotherm adsorption: 596
Dubinin-Polanyi relation, 436,
shape, 15
type I, 7, 14, 15*, 425-426*
Isotherm equations, 628-633

J

Jadeite:
structure, 35*
formation from zeolite Y, 497

K

Kaolin:
reaction with alkali hydroxides, 314
structure, 37*
thermal transitions, 314-315, 732
zeolite synthesis from, 313-320
Kehoeite: 68-69
structure, 68
x-ray and physical property data, 227†
Kielland coefficient:
in ion exchange, 533-534, 584
Kinetic diameter, 632*, 634-635, 636†, 637*
Kinetics:
of ion exchange, 571-579
Kinetics of adsorption, 671-689

L

Langmuir equation, 628
Laumontite:
adsorption data, 626†
infrared spectra 423*
structure, 75, 76*
structural data, 158†
x-ray and physical property data, 228†
Levynite: 80*, 81
adsorption data, 626
structural data, 159†
structure, 80*, 81
x-ray and physical property data, 229†
Luminescence, 392

M

Magnesium in zeolites, 298
Manufacturing processes: 725-746
clay conversion type, 731-738
forming methods, 742-746
hydrogel type, 726-731
mineral zeolites, 740-741

Mesolite:
structural data, 161†
structure, 117-118
x-ray and physical property data, 230†
Metakaolin
(see kaolin)
Metallic dispersions in zeolites, 519-523
Metastability, 248
Mineral zeolites:
early classification, 20†
glossary of terms for, 207-208
identification of, 205
igneous, 189-191
list of, 188†
occurrence, types of, 187-188
origin of, 200-205
chemical composition in, 203
pressure and temperature in, 199†, 199-202
by sedimentation, 204
silica concentration in, 203
from volcanic glass, 204
physical properties of, 205-207
processing, 740-741
sedimentary, 19, 192-200
sedimentary deposits, 193†
Molecular sieve behavior:
of alkali graphite intercalation compounds, 10†
of carbons, 8-9
of coals, 6†
of glasses, 7
of oxides, 6
of silica gels, 7†
Molecular sieve effect, 599, 633-645, 700
Molecular sieve types, 4-10

Molecular sieve zeolites,
  categories, 16†
Montmorillonite, structure, 38*
Mordenite:
  adsorption data, 618-619†
  adsorption isotherms, 605*
  adsorption of gas mixtures in, 696
  commercial product properties, 748-749†
  diffusion in, 682-686, 683†, 685*
  early synthesis work, 253-254
  heats of adsorption, 659†, 667†
  infrared spectra, 423*
  ion exchange in, 563-565, 574†, 575†, 581*
  large-port, synthesis of, 264, 262†, 363†
  manufacturing processes, 738-740
  pore volume, 433
  rates of adsorption in, 682-686, 688†
  small-port, synthesis of, 260, 261-265, 280, 295
  structural data, 162†, 163†
  structure, 120*, 122-124, 124*
  synthetic types, x-ray data, 363, 364
  x-ray and physical property data, 231†
Morphology of crystals, 379-384, 379-383*
Mossbauer spectroscopy, 472

N

Natrolite:
  infrared spectra, 422*
  structural data, 164†
  structure, 115*, 117-118, 119*
  synthetic type, synthesis of, 266-267
  x-ray and physical property data, 232†
Nepheline:
  structure, 43
nmr data, 477
  on hydrated zeolites, 412-414
  studies of hydrogen in zeolites, 479
Nomenclature:
  mineral zeolites, 21
  synthetic zeolites, 22-24

O

Offretite, 79-81
  structural data, 166†
  structure, 79, 80*
  synthetic, adsorption data, 623†
  synthetic type, synthesis of, 311-312
  x-ray and physical property data, 233†
Optical properties, 390-392

P

Particle size, 384-386, 384*, 388*
  commercial products, 747-749, 751*
Paulingite, 74*, 75, 76*
  adsorption data, 626†
  structural data, 169†
  structure, 74*, 75, 76*
  x-ray and physical properties, 234†
Pelletization, 742-746
Permutites, 11-13
  "crystalline," synthesis of, 257
Phillipsite:
  adsorption data, 626†
  ion exchange in, 561, 562†
  structural data, 170†
  structure, 69-70

synthetic type, synthesis of, 260
  x-ray and physical properties, 235†, 206*
Phosphate zeolites, 322-328
Phosphorus-containing zeolites, 322-328
Pore closure, 490-492
Pore filling, 425
Pore size in zeolites, 637†
Pore size distribution:
  nonzeolite, 3, 4*
  zeolite, 3, 4*
Pore volume, 425-438
  correlation, 434-438, 435
Porous glasses, 7
Porous oxides, 6-7
Preadsorption:
  effect on adsorption, 643*, 644-645, 646*
Preformed shapes, 315
Proton exchange, 569-571

R

Radiation effects, 523
Radioactive ions:
  ion exchange of, 529, 588
Reaction diagrams, 249
Reactions of zeolites:
  categories of, 441
  cation hydrolysis, 460-473
  deammoniation, 474
  dehydration, 442-459
    thermal analyses of, 443-447, 481-482*
  dehydroxylation, 478-482
  in solution,
    with chelating agents, 505-507
    with strong acids, 502-504
    with strong bases, 504-505
  radiation effects on,

INDEX 767

522-523
transformation,
at high pressures,
497-498
with volatile inorganic
compounds, 498-
with bases, 470
with volatile inorganic
compounds, 498-502
with water vapor, 490-
493

S

Salt-containing zeolites:
synthesis of, 331-333
Salt occlusion, 585-586
Scapolite structure, 44
Scolecite:
structural data, 171†
structure, 117-118
synthetic type, 297
x-ray and physical
property data, 236†
Secondary building units,
46*
Selectivity coefficient of
ion exchange, 532-535,
567*
Selectivity effects:
alteration of, 665-668
cation density effect,
668
in adsorption, 664-671
Self-diffusion:
(see adsorption)
(see ion diffusion)
Sieving effects, in adsorption, 633-645, 700
Silica gels, 7†, 600*
Silica solubility, 338,
339†
Silica structures, 39, 42*
Silicate structures, 29-44
chain structures, 35
framework structures,
38-44
island structures, 32
isolated group structures, 33
polyhedra in, 32†
sheet structures, 35-38
structural classes, 31

structure models, 31
Simplexity principle, 247
Sodalite structure, 42-43*
Sodalite hydrate:
(see zeolite HS)
synthesis of, 269-272,
275†, 277, 319
Stability to water vapor,
492-493
Stabilization, 507-519
mechanism, 515-518
of structure,
by adsorbed or occluded molecules
and salts, 410
Stellerite:
adsorption data, 626†
structure, 130
Stilbite:
adsorption data, 626†
infrared spectra, 423*
ion exchange in, 563,
564†
structural data, 172†
structure, 129, 131*
x-ray and physical
property data, 237†
Structural formula, 5
Structures:
unknown, 52†, 132
of zeolites by infrared
spectroscopy, 415-
424
Superstable zeolites, 507-
519
Surface area of zeolites,
594
Surface hydroxyl groups,
460
Synthesis:
complexing agents in,
328-331
crystal growth in, 342-
343*, 344-346
early work in, 251-257
free energy relations,
248, 279
from alkali aluminosilicate gels, 257-
294
from alkaline earth

aluminosilicate gels,
294-304
from alkylammonium
compounds, 304-312
from barium aluminosilicate gels, 300-301,
303†
from calcium aluminosilicate gels, 294-298,
296†
from cesium aluminosilicate gels, 293
from clays, 313-320
using sodium hydroxide, 313-319,
316†
using alkali and alkaline earth hydroxides, 324†,
319-320, 317†
from lithium aluminosilicate gels, 258-260
from magnesium aluminosilcate gels, 298
from organic bases,
304-312
from potassium aluminosilicate gels,
280-287
from rubidium aluminosilicate gels, 293
from aluminosilicate
gels, 259-278
from sodium aluminosilicate gels, 260-280
at low temperatures,
267-276
based on colloidal
silica, 277-280
from sodium-calcium
aluminosilicate gels,
301-302
from sodium-lithium
aluminosilicate gels,
280-281
from sodium-potassium
aluminosilicate gels,
287-293
from strontium aluminosilicate gels, 298-
300, 299†

768 INDEX

general conditions for, 249-250
kinetics and mechanism of, 333-344
metastability in, 248
methods of, 245-250
low temperature, 246
nucleation, 339, 344
permutites, 11-13
of salt-containing zeolites, 331-333
rate of crystallization, 279, 333-334, 335†
reaction diagrams, 249, 267-268
starting materials, 268†
simplexity principle in, 247, 279
using chromium, 322
using gallium, 321, 329
using germanium, 321
using phosphorous, 322-331
using titanium, 322
using zirconium, 322
Synthetic zeolites:
with other framework atoms, 320-331
physical properties of, 344, 345-349†
summary list, 346-352
x-ray data, 347, 353-374

T

Tetramethylammonium zeolites, 304
Thermal expansion, 386-388
Thermal stability, 493-496*, 495*
Thermochemical data, 410-412*, 414*
Thermodynamics:
of adsorption, 565-569
of ion exchange processes, 565-569
Thomsonite:
infrared spectra, 422*

structural data, 174†
structure, 119, 120*
x-ray and physical properties, 238†
TMA offretite:
(see zeolite O)
structural data, 165†
Truncated cuboctahedron, 46, 84*
Truncated octahedron, 46, 83*

U

Ultrastable zeolites, 507-519

V

Virial equation, 630
Viseite:
structure, 68
x-ray and physical property data, 239†
Void space in zeolites, 62-63
Volmer equation, 629
Volume distribution in molecular sieve pellet, 750†, 742†

W

Wairakite:
structure, 68
synthetic-type, 295
x-ray and physical properties, 240†
Waste water treatment, 589-590
Water:
adsorption isotherms, 600*, 601*
in hydrated zeolites, 412-415

Y

Yugawaralite:
structural data, 178†
structure, 75-77, 78*
x-ray and physical property data, 241†

Z

Zeolite A:
adsorption data, 607†, 610†, 624*
adsorption isobars, 639*
adsorption isotherms, 594*, 597*, 600*
adsorption selectivity, 695
commercial product properties, 743†, 748†, 749†, 747-752
composition, 86-87
conversion to P, 276
dielectric properties, 392-395
diffusion in, 672-673*, 675*, 677*, 685*
electron micrographs of, 342*-343*, 592*
encapsulation of gases in, 624
heats of adsorption, 652*, 654†
ion exchange in, 537-540, 541†, 576†, 582*
complex ion effects, 540
in mixed solvents, 583-585
ion diffusion, 576†, 576, 577
manufacturing processes, 741, 726-730, 733, 735
pore volume, 427-428†
rates of adsorption in, 672-673*, 676-681, 687†, 688†
silica-rich, synthesis of, 273
stability to water vapor, 490-492
structural data, 133†
alpha cages, 84, 86*
cation positions, 87-90
composition, 86-87
structure, 83-90
cation forms, 87†
dehydrated, 88-90

electrostatic potential, 90
framework, 83-85, 83*, 85*
hydrated, 87-88
ordering, 267-272
synthesis of, 278
from metakaolin, 313, 316†
void space, 89
x-ray data, 353†
Zeolite N-A:
structural data, 134†
structure, 90
synthesis of, 304-308, 305†
x-ray data, 354†
Zeolite P-A:
synthesis, 323-328
x-ray data, 354†
Zeolite α:
synthesis of, 309
x-ray data, 354†
Zeolite Li-A:
synthesis of, 258, 259†, 317†, 319
x-ray data, 355†
Zeolite B
(see Zeolite P)
Zeolite Na-B
synthesis of, 261-264
Zeolite Beta:
synthesis of, 309
Zeolite P-B:
synthesis of, 323-324
x-ray data, 355†
Zeolite P-C:
synthesis of, 323-328
x-ray data, 355†
Zeolite D:
adsorption data, 626†
synthesis, 287-293
x-ray data, 356†
Zeolite Ca-D:
synthesis, 294-296
x-ray data, 356†
Zeolite Na-D:
synthesis of, 261-264
x-ray data, 356†
Zeolite Rb-D:
synthesis, 294, 317†

Zeolite Sr-D:
adsorption data, 627†
synthesis, 298-300
x-ray data, 356†
Zeolite E:
synthesis of, 287-288
x-ray data, 356†
Zeolite Ca-E:
synthesis of, 294-296
x-ray data, 357†
Zeolite K-E:
synthesis of, 282†, 284
x-ray data, 357†
Zeolite TMA-E:
x-ray data, 357†
Zeolite F:
adsorption data, 627†
structural data, 144†
synthesis of, 282, 284
x-ray data, 357†
Zeolite K-F:
synthesis from kaolin, 319-320, 317†
(see Zeolite Z)
Zeolite Sr-F:
synthesis of, 298-300
x-ray data, 357†
Zeolite G:
synthesis of, 282, 285, 286
from kaolin, 317†, 319-320
x-ray data, 358†
Zeolite Ba-G:
structural data, 147†
structure, 116
synthesis, 300-301, 317†, 319
x-ray data, 359
Zeolite P-G:
synthesis, 323-328
x-ray data, 358†
Zeolite Sr-G:
synthesis of, 298-300
x-ray data, 359†
Zeolite H:
structural data, 152†
synthesis, 282, 285
x-ray data, 359
Zeolite Li-H:
synthesis of, 258, 259†,

317†
x-ray data, 356†
Zeolite HS, 82
(see sodalite hydrate)
structural data, 155†
structure, 82
synthesis, 269, 279
x-ray data, 360
Zeolite Ca-I:
synthesis of, 294-296
x-ray data, 360
Zeolite Sr-I:
synthesis of, 298-300
x-ray data, 360
Zeolite J:
synthesis of, 283†
x-ray data, 360
Zeolite Ba-J:
synthesis of, 300-301,
x-ray data, 360
Zeolite Ca-J:
synthesis of, 294-296
x-ray data
Zeolite Ba-K:
synthesis of, 300, 301
x-ray data, 361†
Zeolite L:
adsorption data, 616†
commercial product properties, 748
heats of adsorption, 658†
isotherms, 604*
pore volume, 431-432†
structural data, 156†
structure, 113-115
framework, 115*
cation positions, 114*, 116
synthesis of, 283†, 287, 289-293, 316†,
x-ray data, 361
Zeolite Ca-L:
synthesis of, 294-296
x-ray data, 361
Zeolite P-L, 365
Zeolite Losod, 82
structural data, 160†
structure, 82
synthesis of, 272, 312
x-ray data, 362

Zeolite M:
  synthesis of, 282
  x-ray data, 362
Zeolite Ba-M:
  systhesis of, 300-301
  x-ray data, 363
Zeolite K-M:
  (see zeolite W)
Zeolite Sr-M:
  synthesis of, 299-300
  x-ray data, 363
Zeolite N:
  structural data, 163†
  synthesis of, 311
  x-ray data, 364
Zeolite Ba-N:
  synthesis of, 317†, 319
  x-ray data, 364
Zeolite O:
  adsorption data, 623†
  structural data, 165†
  synthesis of, 311-312
  x-ray data, 365
Zeolite Omega:
  adsorption data, 620†
  structural data, 167†
  structure, 80*, 81-82
  synthesis of, 310
  x-ray data, 364
Zeolite P:
  cation exchanged
    forms, unit cell data,
    73†
  composition, 276
  ion exchange, 561-562†
  structural data, 168†
  structure, 72-73, 74*
  synthesis of, 266-267,
    270, 273-276, 289-
    293, 309
    from kaolin, 316†
  x-ray data, 355, 365
Zeolite N-P:
  synthesis of, 304-308
Zeolite P-(Cl):
  synthesis of, 332-333
  x-ray data, 365
Zeolite Q:
  synthesis of, 283, 287
  x-ray data, 366
Zeolite Q-(Br):

synthesis of, 332-333
x-ray data, 366
Zeolite Ca-Q:
  synthesis of, 283, 287
  x-ray data, 366
Zeolite Sr-Q:
  synthesis of, 298-300
  x-ray data, 366
Zeolite structure:
  classification, 45-47,
    48-50†
  theoretical structures,
    47-58
  from archimedean
    polyhedra, 55-58,
    58†
  from eight-rings, 53-
    54
  from epsilon-cages,
    58, 60*
  from five-rings, 54,
    55*
  from four-rings, 53-
    54
  from polyhedra, 55-
    58
  from six-rings, 54-
    55, 56*, 57†
  (see individual zeolite)
Zeolite R:
  synthesis of, 267, 274†
  x-ray data, 366
Zeolite P-R:
  synthesis of, 323-328
  x-ray data, 367
Zeolite Sr-R:
  synthesis of, 298-300
  x-ray data, 367
Zeolite S:
  adsorption data, 627†
  synthesis of, 274†,
    316†
  x-ray data, 367
Zeolite T:
  adsorption data, 621†
  diffusion in, 683-684
  ion exchange in, 558-
    561†
  rates of adsorption in,
    683-684
  structural data, 173†

structure, 81
synthesis of, 287-292
x-ray data, 367
Zeolite Ba-T:
  synthesis of, 317†, 319
  x-ray data, 368
Zeolite W:
  adsorption, 624
  structural data, 175†
  synthesis of, 283-285,
    318-319, 317†
  x-ray data, 368
Zeolite P-W:
  synthesis of, 323-328
  x-ray data, 367
Zeolite X:
  adsorption data, 606†,
    612†, 611-613†
  adsorption isotherms,
    597*, 601*, 602*,
    632*, 691*
  adsorption of mixtures
    on, 689-699, 714
  characteristic curve
    for water on, 632*
  commercial product
    properties, 748†,
  dielectric properties,
    395
  heats of adsorption,
    652*, 655†, 667†,
  ion exchange in, 537†,
    540-553, 543†, 544†,
    546*, 571*, 576†
  in nonaqueous sol-
    vents, 582-584,
    583†, 584†
  manufacturing pro-
    cesses, 727-740
  pore volume, 429-431
  rates of adsorption in,
    688†
  stability to water va-
    por, 492-493, 496*
  structural data, 176†
  structure, 85*, 92, 95*
  cation positions by
    esr, 105-106
  composition, 93
  dehydrated cation
    positions, 97-103,

97*
  electrostatic potential and field, 106-107, 111†
  framework, 93-95
  hydrated cation positions, 95-97, 97*
  lattice constant, 94*, 104*
  related physical properties, 103-105
  synthesis of, 267-273, 274, 277-278, 289
  from metakaolin, 316†, 315
  x-ray data, 369
Zeolite N-X:
  synthesis of, 304-308, 305†
  x-ray data, 369†
Zeolite Y:
  adsorption data, 606†, 615-615†
  adsorption isotherms, 603*
  commercial product properties, 748†
  diffusion in, 681-686, 683†, 686†
  heats of adsorption, 656†, 667†
  ion exchange in, 537†, 540-553, 545*, 546*, 550†
  manufacturing processes, 726-730, 727-740
  stability to water vapor, 492-493, 496*
  structural data, 177†
  structure, 85*, 92
  composition, 93
  cation positions by esr, 105-106
  dehydrated cation positions, 97-103, 98†
  electrostatic potential and field, 106-107, 111†
  framework, 93-95
  hydrated cation positions, 95-97, 98†
  lattice constant, 94*, 104*
  related physical properties, 103-105
  synthesis of, 277-279, 329
  from calcined kaolins, 316†, 318
  x-ray data, 369
Zeolite N-Y:
  synthesis of, 304-308
  x-ray data, 369
Zeolite Z:
  ion exchange in, 562-563†, 565*
  synthesis of, 282, 284, 320, 332
  x-ray data, 370
Zeolite Z-21:
  synthesis of, 272-273
  x-ray data, 370
Zeolite Zh, 269
  (see HS and sodalite hydrate)
Zeolite ZK-4:
  ion exchange in, 576-577
  structural data, 179†
  structure, 90-92
  synthesis of, 308
  x-ray data, 370
Zeolite ZK-5:
  adsorption data, 627†
  structural data, 180
  structure, 111-113, 112*
  synthesis of, 309-310
  x-ray data, 370
Zeolite ZK-19:
  synthesis of, 288, 293, 328
  x-ray data, 370
Zeolite ZK-20:
  synthesis of, 311
  x-ray data, 320
Zeolite ZK-21:
  synthesis of, 331
  x-ray data, 371
Zeolite ZK-22:
  synthesis of, 330†, 331
Zeolite ZSM-2:
  synthesis of, 259*, 260
  x-ray data, 372
Zeolite ZSM-3:
  synthesis of, 280-281
  x-ray data, 372
Zeolite ZSM-4:
  synthesis of, 312
  x-ray data, 373
Zeolite ZSM-5:
  synthesis of, 312
  x-ray data, 373
Zeolite ZSM-8:
  synthesis of, 312
  x-ray data, 373
Zeolite ZSM-10:
  synthesis of, 312
  x-ray data, 374
Zeolon:
  (see mordenite)
Zunyite structure, 59*